Excursions in
Modern Mathematics

Excursions in Modern Mathematics

Sixth Edition

Peter Tannenbaum

California State University—Fresno

PEARSON
Prentice
Hall

Upper Saddle River, NJ 07458

Library of Congress Cataloging-in-Publication Data
Tannenbaum, Peter (date)
   Excursions in modern mathematics.—6th ed. / Peter Tannenbaum.
     p. cm.
   Includes bibliographical references and index.
   ISBN 0-13-187363-6
   1. Mathematics.   I. Title.

   QA36.T35   2007
   510—dc22   2006044809

Editor in Chief: *Sally Yagan*
Project Manager: *Michael Bell*
Production Editor: *Barbara Mack*
Assistant Managing Editor: *Bayani Mendoza de Leon*
Senior Managing Editor: *Linda Mihatov Behrens*
Executive Managing Editor: *Kathleen Schiaparelli*
Manufacturing Buyer: *Maura Zaldivar*
Manufacturing Manager: *Alexis Heydt-Long*
Director of Marketing: *Patrice Jones*
Marketing Manager: *Wayne Parkins*
Marketing Assistant: *Jennifer de Leeuwerk*
Editorial Assistant/Print Supplements Editor: *Jennifer Urban*
Art Director: *Heather Scott*
Interior Designers: *Susan Anderson-Smith, Tamara Newnam*
Cover Designer: *Tamara Newnam*
Art Editor: *Thomas Benfatti*
Creative Director: *Juan R. López*
Director of Creative Services: *Paul Belfanti*
Director, Image Resource Center: *Melinda Patelli*
Manager, Rights and Permissions: *Zina Arabia*
Manager, Visual Research: *Beth Brenzel*
Manager, Cover Visual Research & Permissions: *Karen Sanatar*
Image Permission Coordinator: *Angelique Sharps*
Photo Researcher: *Julie Tesser*
Cover Photo: *Gavriel Jecan/CORBIS*
Art Studio: *Scientific Illustrators*
Compositor: *Preparé, Inc.*

Printed in the United States of America

10 9 8 7 6 5 4 3 2 1

ISBN 0-13-187363-6

Pearson Education LTD., *London*
Pearson Education Australia PTY, Limited, *Sydney*
Pearson Education Singapore, Pte. Ltd
Pearson Education North Asia Ltd, *Hong Kong*
Pearson Education Canada, Ltd., *Toronto*
Pearson Educación de Mexico, S.A. de C.V.
Pearson Education—Japan, *Tokyo*
Pearson Education Malaysia, Pte. Ltd

To the members of the board of Last Tango

Contents

part 2 Management Science

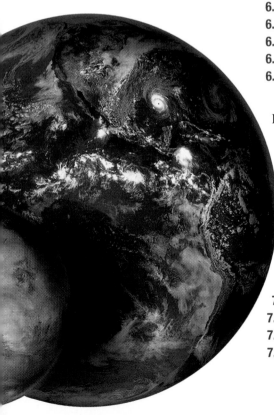

part 3 Growth and Symmetry

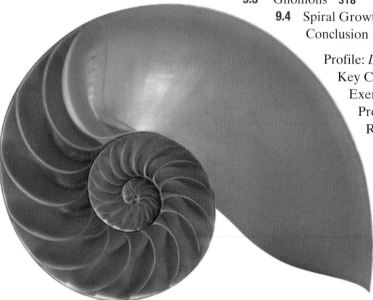

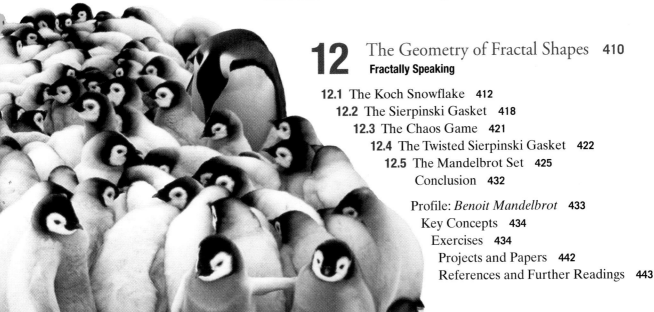

part 4   Statistics

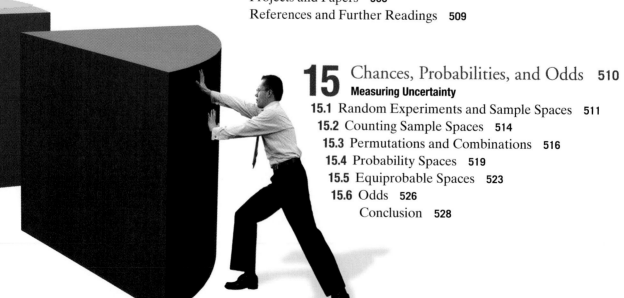

16 Normal Distributions   540
Everything Is Back to Normal (Almost)

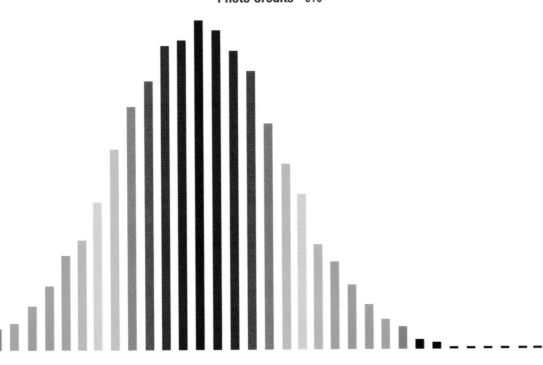

Preface

Excursions in Modern Mathematics is, as I hope the title might suggest, a collection of "trips" into that vast and alien frontier that many people perceive mathematics to be. While the formal purpose of this book is quite conventional—it is intended to serve as a textbook for a college-level liberal arts mathematics course—its philosophy and contents are not. My goal is to show the open-minded reader that mathematics is a lively, interesting, useful, and surprisingly rich human activity.

The "excursions" in this book represent a collection of topics chosen to meet the following simple criteria.

- **Applicability.** No need to ask here that great existential question of college mathematics: "What is this stuff good for?" The connection between the mathematics presented in these excursions and down-to-earth, concrete real-life problems is transparent and immediate.

- **Accessibility.** As a general rule, the excursions in this book do not presume a background beyond standard high school mathematics—by and large, Intermediate Algebra and a little Euclidean Geometry are appropriate and sufficient prerequisites. (In the few instances in which more advanced concepts are unavoidable, an effort has been made to provide enough background to make the material self-contained.) A word of caution—this does not mean that the excursions in this book are easy! In mathematics, as in many other walks of life, simple and straightforward is not synonymous with easy and superficial.

- **Modernity.** Unlike much of traditional mathematics, which is often hundreds of years old, most of the mathematics in this book has been discovered within the last 100 years, and in some cases within the last couple of decades. Modern mathematical discoveries do not have to be the exclusive province of professional mathematicians.

- **Aesthetics.** The notion that there is such a thing as beauty in mathematics is surprising to most casual observers. There is an important aesthetic component in mathematics and, just as in art and music (which mathematics very much resembles), it often surfaces in the simplest ideas. A fundamental objective of this book is to develop an appreciation for the aesthetic elements of mathematics.

Outline

The excursions in the book are divided into four independent parts. Each of these parts consists of four chapters plus a "Mini-Excursion." (Mini-Excursions are a new feature added to this edition. More details on these Mini-Excursions are given in the "What's New in the Sixth Edition" section on page xvii.)

- **Part 1 The Mathematics of Social Choice.** This part deals with mathematical applications in social science. How are elections decided? (Chapter 1); How can power be measured? (Chapter 2); How can competing claims on property be resolved in a fair and equitable way? (Chapter 3); How are seats apportioned in a legislative body? (Chapter 4); and, What is the current method for apportioning seats in the United States House of Representatives? (Mini-Excursion 1).

■ **Part 2 Management Science.** This part deals with methods for solving problems involving the organization and management of complex activities—that is, activities involving either a large number of steps and/or a large number of variables. Typically, these problems revolve around questions of efficiency: some limited or precious resource (time, money, raw materials) must be managed in such a way that waste is minimized. Among the many applications discussed in this part are: What is a good delivery route for a mail carrier? (Chapter 5); How can a salesman organize a multi-city sales trip and save money? (Chapter 6); What is the best way to design a communication network? (Chapter 7); How do we create a schedule for constructing a home in the least amount of time? (Chapter 8); and, How do we color a map using the fewest number of colors? (Mini-Excursion 2).

■ **Part 3 Growth and Symmetry.** This part deals with nontraditional notions of growth and shape. What are the mechanisms that govern spiral growth in natural shapes? (Chapter 9); How can mathematics be used to predict the rise and fall of populations? (Chapter 10); How can we analyze the symmetries of an object or a pattern? (Chapter 11); What type of geometric shape is a head of cauliflower, our circulatory system, or a mountain range? (Chapter 12); and How does time affect the value of money? (Mini-Excursion 3).

■ **Part 4 Statistics.** In one way or another, statistics affects all of our lives. Government policy, insurance rates, our health, our diet, and public opinion are all governed by statistical data. This part deals with some of the most basic aspects of statistics. How should data be collected? (Chapter 13): How can large amounts of data be summarized so that the data makes some sense? (Chapter 14); How can we measure the uncertainty produced by random phenomena? (Chapter 15); How can we use statistical laws to predict long-term patterns in random events? (Chapter 16); and How do we separate foolish risks from risks that are worth taking? (Mini-Excursion 4).

Exercises and Projects

An important goal for this book is that it be flexible enough to appeal to a wide range of readers in a variety of settings. The exercises, in particular, have been designed to convey the depth of the subject matter by addressing a broad spectrum of levels of difficulty—from the routine drill to the ultimate challenge. For convenience (but with some trepidation) the exercises are classified into three levels of difficulty:

■ **Walking.** These exercises are meant to test a basic understanding of the main concepts, and they are intended to be within the capabilities of students at all levels.

■ **Jogging.** These are exercises that can no longer be considered as routine—either because they use basic concepts at a higher level of complexity, or they require slightly higher-order critical thinking skills, or both.

■ **Running.** This is an umbrella category for problems that range from slightly unusual or slightly above average in difficulty to problems that can be a real challenge to even the most talented of students.

A **Projects and Papers** section following the exercise sets at the end of each chapter offers some potential topics and ideas for further exploration and enrichment. In most cases, the projects are well suited for group work, be it a handful of students or an entire small class.

What's New in the Sixth Edition

While much has changed in this book since the first edition, I have made a conscious effort to resist changing its overall structure. Four parts consisting of four chapters each has a nice symmetry to it, and users have made it clear that they like the flexibility that this structure offers. At the same time, many users have expressed a desire to have additional topics covered. In an effort to balance these competing interests, a new feature makes its debut in this edition—the "Mini-Excursion."

If one thinks of a chapter in this book as a snorkeling trip along a colorful coral reef, then a Mini-Excursion is a free dive into somewhat deeper waters along a particularly interesting section of the reef. In this edition I have introduced four Mini-Excursions, one for each of the four parts:

- Mini-Excursion 1 (Apportionment Today: *The Huntington-Hill Method*) discusses the current method used to apportion the United States House of Representatives, and enriches (both mathematically and politically) the material introduced in Chapter 4.

- Mini-Excursion 2 (A Touch of Color: *Graph Coloring*) discusses the dual concepts of graph coloring and chromatic number, and some of the applications of these concepts. This is an enrichment topic tangentially related to some of the ideas introduced in Chapters 5 and 6.

- Mini-Excursion 3 (The Time Value of Money: *Annuities*) builds on the exponential growth model and its applications to financial mathematics (introduced in Chapter 10) to explore the concepts of fixed and deferred annuities.

- Mini-Excursion 4 (The Cost of Risk-Taking: *Expected Value*) explores the idea that risks can be quantified and managed. This Mini-Excursion is an enrichment topic for the material introduced in Chapter 15.

The Mini-Excursions are available as printable PDF files at the book's Web site *www.prenhall.com/tannenbaum,* as well as in hard copy form in the Student Resource Guide.

As with previous editions, numerous changes were made throughout the text to improve the flow and readability and make the material more user-friendly. Some of the more noticeable and significant changes are:

- The introduction of marginal notes for suggestions, references, and comments.

- The end of each chapter opener now includes a brief narrative of the chapter's organization.

- Approximately 250 new exercises.

- Many old examples have been updated or expanded. Many new examples have been added.

- Chapter 7 has undergone a significant but not extreme makeover. The chapter now uses the concept of a *network* as the underlying paradigm for the study of connectivity and trees.

Ancillary Materials Available with This Edition

- **Companion Web Site (*www.prenhall.com/tannenbaum*)**   Features a syllabus manager, online quizzes, Internet projects, graphing calculator help, spreadsheet basics and Excel projects, PowerPoint Presentation downloads, Java applets, the Mini-Excursions, and dozens of additional resource links.

- **Student Resource Guide**   (0-13-187382-2) In addition to the worked out solutions to odd-numbered exercises from the text, this guide contains "selected hints" that point the reader in one of many directions leading to a solution. Mini-Excursions in full color and keys to student success, including lists of skills that will help prepare for chapter exams, are also included.

- **Instructor's Resource Manual**   (0-13-187381-4) Contains solutions to all the exercises in the text. Also contains an Instructor's Guide that includes Learning Outcomes, Skill Objectives, Ideas for the Classroom, and Worksheets.

- **Test Item File**   (0-13-187380-6) PDF version of TestGen. Available for download from the Instructor's Resource Center.

- **TestGen**   (0-13-187379-2) Test-generating software that creates randomized tests and offers an onscreen LAN-based testing environment, complete with Instructor Gradebook.

- **Course Compass Access Card**   (0-13-242929-2)

- **Blackboard Access Card**   (0-13-224927-8)

- **WebCT Access Card**   (0-13-156424-2) All course management cartridges contain course compatible content, including the Student Resource Guide, technology help, quizzes, an e-book, and lecture content.

- **Tutor Center Access Card**   (0-13-064604-0) Tutors provide one-on-one tutoring for any problem with an answer at the back of the book. Students access the Tutor Center via phone, fax, or e-mail.

Acknowledgments

In developing this new edition, I consulted extensively with Dale Buske of St. Cloud State University. Dale read the manuscript several times and made numerous comments and suggestions. In addition, Dale contributed greatly to the development of the new mini-excursions, as well as to the expanded exercise sets. Dale's contributions to this edition went above and beyond the ordinary and I owe Dale a special debt of gratitude.

I was also fortunate to have an excellent group of reviewers who reviewed the 5th edition and made many excellent suggestions as to how it might be improved:

Lowell Abrams, *George Washington University*
Guanghwa Andy Chang, *Youngstown State University*
Kimberly A. Conti, *SUNY College at Fredonia*
Abdi Hajikandi, *Buffalo State College, State University of New York*
Cynthia Harris, *Triton Community College*
Peter D. Johnson, *Auburn University*
Stephen Kenton, *Eastern Connecticut State University*
Thomas Lada, *North Carolina State University*

Many people contributed to previous editions of this book. Special thanks go to Robert Arnold, my original co-author, to Dale Buske (again), and to the following reviewers: Teri Anderson, Carmen Artino, Donald Beaton, Terry L. Cleveland, Leslie Cobar, Crista Lynn Coles, Ronald Czochor, Nancy Eaton, Lily Eidswick, Kathryn E. Fink, Stephen I. Gendler, Marc Goldstein, Josephine Guglielmi, William S. Hamilton, Harold Jacobs, Tom Kiley, Jean Krichbaum, Kim L. Luna, Mike Martin, Thomas O'Bryan, Daniel E. Otero, Philip J. Owens, Matthew Pickard, Lana Rhoads, David E. Rush, Kathleen C. Salter, Theresa M. Sandifer, Paul Schembari, Marguerite V. Smith, William W. Smith, David Stacy, Zoran Sunik, John Watson, Sarah N. Ziesler.

Many tasks go into overseeing the production of a book from beginning to end, and it is the responsibility of the production editor to see that everything falls into place. Barbara Mack, my production editor, succeeded admirably in smoothly bringing this project to a timely completion. Barbara is, by far, the best production editor I have ever worked with.

Heather Scott was responsible for the design of this edition. Thanks to her talent we managed a rare feat—a book that is better looking, has more content, and is 100 pages shorter than its predecessor! And then, there is Sally Yagan, editor extraordinaire. There is an editor behind every book, but few that can match her vision, "can-do" attitude, and leadership.

A Final Word

This book grew out of the conviction that a liberal arts mathematics course should teach students more than just a collection of facts and procedures. The ultimate purpose of this book is to instill in the reader an overall appreciation of mathematics as a discipline and an exposure to the subtlety and variety of its many facets: problems, ideas, methods, and solutions. Last, but not least, I have tried to show that mathematics can be interesting and fun.

Peter Tannenbaum

part 1

The Mathematics
of Social Choice

The Mathematics of Voting

The Paradoxes of Democracy

It's not the voting that's democracy; it's the counting.

Tom Stoppard

Vote! In a democracy, the rights and duties of citizenship are captured in that simple one-word mantra. And, by and large, as citizens of a democracy we should and do vote. We vote in presidential elections, gubernatorial elections, local elections, school bonds, stadium bonds, American Idol selection, and initiatives large and small. The paradox is that the more opportunities we have to vote, the less we seem to appreciate and understand the meaning of voting. Why should we vote? Does our vote really count? How does it count?

We can best answer these questions once we understand the full story behind an election. The reason we have elections is that we don't all think alike. Since we cannot all have things our way, we vote. But *voting* is only the first half of the story, the one we are most familiar with.

As playwright Tom Stoppard suggests, it's the second half of the story—the *counting*—that is at the very heart of the democratic process. How does it work, this process of sifting through the many voices of individual voters to find the collective voice of the group? And even more importantly, how well does it work? Is the process always fair? Answering these questions and explaining a few of the many intricacies and subtleties of **voting theory** are the purpose of this chapter.

But wait just a second! Voting theory? Why do we need a fancy theory to figure out how to count the votes? It all sounds pretty simple: We have an election; we count the ballots. Based on that count, we decide the outcome of the election in a consistent and fair manner. Surely, there must be a reasonable way to accomplish all of this! Surprisingly, there isn't.

In the late 1940s, the mathematical economist Kenneth Arrow discovered a remarkable fact: For elections involving three or more candidates, there is no consistently fair democratic method for choosing a winner. In fact, Arrow demonstrated that *a method for determining election results that is democratic and always fair is a mathematical impossibility*. This, the most famous fact in voting theory, is known as **Arrow's impossibility theorem**. In 1972, Kenneth Arrow was awarded the Nobel Prize in Economics for his pioneering work in what is now known as social-choice theory.

This chapter is organized as follows: We will start with a general discussion of elections and ballots in Section 1.1. This provides the backdrop for the remaining sections, which are the heart of the chapter. In Sections 1.2 through 1.5 we will explore four of the most commonly used **voting methods**—how they work, what their implications are, and how they measure up in terms of a few basic tests of fairness called **fairness criteria**. In Section 1.6 we generalize these methods to elections where we want to produce not just a winner but a complete or partial ranking of the candidates. The entire excursion will be sprinkled with plenty of examples depicting real-life situations or metaphors thereof. At the end of the chapter, with some gained insight, we will come back to Arrow's impossibility theorem.

For more details, see the biographical profile of Arrow at the end of this chapter.

1.1 Preference Ballots and Preference Schedules

We kick off our discussion of voting theory with a simple but important example, which we will revisit several times throughout the chapter. (You may want to think of this example as a mathematical parable—its importance being not in the story itself, but in what lies hidden behind it. As you will soon see, there is a lot more than first meets the eye behind this simple example.)

> ### EXAMPLE 1.1   The Math Club Election

The Math Appreciation Society (MAS) is a student organization dedicated to an unsung but worthy cause—that of fostering the enjoyment and appreciation of mathematics among college students. The Tasmania State University chapter of MAS is holding its annual election for president. There are four candidates running for president: Alisha, Boris, Carmen, and Dave (A, B, C, and D for short). Each of the 37 members of the club votes by means of a ballot indicating his or her first, second, third, and fourth choice. The 37 ballots submitted are shown in Fig. 1-1. Once the ballots are in, it's decision time. Who should be the *winner* of the election? Why?

Ballot	Ballot	Ballot	Ballot	Ballot	Ballot	Ballot	Ballot	Ballot	Ballot	Ballot	Ballot	Ballot	Ballot	Ballot	Ballot	Ballot	Ballot
1st A	1st B	1st A	1st C	1st B	1st B	1st A	1st C	1st B	1st C	1st A	1st C	1st A	1st A	1st C	1st A	1st C	1st D
2nd B	2nd D	2nd B	2nd B	2nd D	2nd B	2nd B	2nd B	2nd D	2nd B	2nd C	2nd B	2nd B	2nd B	2nd B	2nd B	2nd B	2nd C
3rd C	3rd C	3rd C	3rd D	3rd C	3rd D	3rd C	3rd C	3rd D	3rd C	3rd D	3rd B	3rd C	3rd C	3rd D	3rd C	3rd D	3rd B
4th D	4th A	4th D	4th A	4th A	4th A	4th D	4th A	4th A	4th D	4th A	4th A	4th D	4th D	4th A	4th D	4th A	4th A

Ballot	Ballot	Ballot	Ballot	Ballot	Ballot	Ballot	Ballot	Ballot	Ballot	Ballot	Ballot	Ballot	Ballot	Ballot	Ballot	Ballot	Ballot	Ballot
1st C	1st A	1st D	1st D	1st C	1st C	1st D	1st A	1st D	1st C	1st A	1st D	1st C	1st A	1st D	1st B	1st A	1st C	1st A	1st A	1st D	1st A
2nd B	2nd B	2nd C	2nd C	2nd B	2nd B	2nd C	2nd B	2nd C	2nd B	2nd B	2nd C	2nd D	2nd B	2nd D	2nd B	2nd B	2nd B	2nd C	2nd B
3rd D	3rd C	3rd B	3rd B	3rd D	3rd D	3rd B	3rd C	3rd B	3rd D	3rd C	3rd B	3rd C	3rd C	3rd B	3rd C	3rd C	3rd B	3rd C	3rd C
4th A	4th D	4th A	4th A	4th A	4th A	4th A	4th D	4th A	4th A	4th D	4th A	4th A	4th D	4th A	4th A	4th D	4th D	4th A	4th D

FIGURE 1-1 The 37 anonymous ballots for the Math Club election

Before we try to answer these two deceptively simple questions, let's step back and take a brief look at elections in general. The essential ingredients of every election are a set of *voters* (in our example the members of the Tasmania State chapter of MAS), and a set of *candidates* or *choices* (in our example A, B, C, D). Typically, the word *candidate* is used when electing people, and the word *choice* is associated with nonhuman alternatives (cities, colleges, pizza toppings, etc.), but in this chapter we will use the two words interchangeably.

Individual voters are asked to express their opinions through *ballots*, which can come in many forms. In the Math Club election, the ballots asked the voters to rank each of the candidates in order of preference, and ties were not allowed. A ballot in which the voters are asked to rank the candidates in order of preference is called a **preference ballot**. A ballot in which ties are not allowed is called a **linear ballot**.

In this chapter we will illustrate all of our examples using linear preference ballots as the preferred (no pun intended) format for voting. While it is true that the preference ballot is not the most typical way we cast our vote (in most elections for public office, for example, we are asked to choose the top candidate), it is also true that it is one of the best ways to vote since it allows us to express our opinion on the relative merit of *all* the candidates.

A quick look at Fig. 1-1 is all it takes to see that there are many repeats among the ballots submitted—there are, after all, only a few different ways that the four candidates can be ranked. Thus, a logical way to organize the ballots is to group together identical ballots (Fig. 1-2), and this leads in a rather obvious way to Table 1-1, called the **preference schedule** for the election.

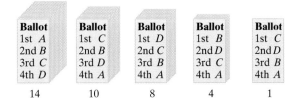

FIGURE 1-2 The 37 Math Club election ballots organized into piles

TABLE 1-1 Preference Schedule for the Math Club Election					
Number of voters	14	10	8	4	1
1st choice	A	C	D	B	C
2nd choice	B	B	C	D	D
3rd choice	C	D	B	C	B
4th choice	D	A	A	A	A

If we think of an election as a process that takes an *input* (the ballots) and produces an *output* (the winner or winners of the election), then the preference schedule is the input, neatly packaged and organized and ready for us to do the counting. From now on, all of our election examples will be given in the form of a preference schedule.

Transitivity and Elimination of Candidates

There are two important facts that we need to keep in mind when we work with preference ballots. The first is that a voter's preferences are **transitive**, a fancy way of saying that a voter who prefers candidate A over candidate B and prefers candidate B over candidate C automatically prefers candidate A over candidate C. A useful consequence of this observation is this: If we need to know which candidate a voter would vote for if it came down to a choice between just X and Y, all we have to do is look at where X and Y are placed on that voter's ballot—whichever is higher would be the one getting the vote. We will use this observation throughout the chapter.

Ballot
1st C
2nd B
3rd D
4th A

FIGURE 1-3

The other important fact is that the relative preferences of a voter are not affected by the elimination of one or more of the candidates. Take, for example, the ballot shown in Fig. 1-3 and pretend that candidate B drops out of the race right before the ballots are submitted. How would this voter now rank the remaining three candidates? As Fig. 1-4 shows, the relative positions of the remaining candidates are unaffected: C remains the first choice, D moves up to the second choice, and A moves up to the third choice.

Ballot
1st C
~~2nd B~~
3rd D
4th A
➡️
Ballot
1st C
2nd D
3rd A

FIGURE 1-4

Let's now return to the business of deciding the outcome of elections in general and the Math Club election in particular.

1.2 The Plurality Method

Perhaps the best known and most commonly used method for finding a winner in an election is the **plurality method**. With the plurality method, all we care about is first-place votes—the candidate with the *most* first-place votes (called the **plurality candidate**) wins. Thus, in an election conducted under the plurality method we don't really need each voter to rank the candidates—the only information that we need is the voter's first choice.

▶ EXAMPLE 1.2 The Math Club Election (Plurality Method)

If the Math Club election (see Table 1-1) is decided under the plurality method, then

  A gets 14 first-place votes,
  B gets 4 first-place votes,
  C gets 11 first-place votes,
  D gets 8 first-place votes,

and the results of the election are clear—the winner is *A* (Alisha).

The vast majority of elections for political office in the United States are decided using the plurality method. With political elections the allure of the plurality method lies in its simplicity (voters have little patience for complicated procedures) and in the fact that plurality is a natural extension of the principle of **majority rule**: *In a democratic election between two candidates, the candidate with a majority (more than half) of the votes should be the winner.* (Not surprisingly, the candidate with a majority of first-place votes is called the **majority candidate**.)

With two candidates, a plurality candidate is also a majority candidate and everything works out well. The problem with the majority rule is that when there are *three* or more candidates there is no guarantee that there is going to be a majority candidate. In the Math Club election, for example, to get a majority would require at least 19 first-place votes (out of 37). Alisha, with 14 first-place votes, had a *plurality* (more than any other candidate) but was far from being a majority candidate. In closely contested elections with many candidates, the percentage of the vote needed to win under plurality can be ridiculously low.

The Majority Criterion

One of the fundamental principles of a democratic election is the notion that if there is a majority candidate, then that candidate should be the winner of the election. This notion is the basis for our first basic fairness criterion known as the *majority criterion*.

> **The Majority Criterion**
>
> If candidate *X* has a *majority* of the first-place votes, then candidate *X* should be the *winner* of the election.

The preceding criterion is *satisfied* by the plurality method: A majority candidate is guaranteed to be the winner of the election by the simple fact that a

In presidential elections this can be a serious problem as well: In 1992 Bill Clinton was elected President with only 43% of the popular vote. See also Exercises 15 and 16.

majority always implies a plurality (if you have more than half of the votes you automatically have more than anyone else).

As we will soon see, there are widely used voting methods that can produce *violations* of the majority criterion. Specifically, a violation of the majority criterion occurs in an election where there is a majority candidate but that candidate does not win the election. If this can happen under some voting method, then we say that the voting method itself *violates* the majority criterion. Be careful how you interpret this—it only means that violations *can* happen, not that they *must* happen! Thus, there is a subtle but important difference between an election violating a criterion and a voting method violating a criterion. The former is an individual instance, the latter is a general property of the voting method itself, and an undesirable one at that! (Hold this thought, because it will come up again!)

The plurality method satisfies the majority criterion—that's good! Other than that, the method has very little going for it and is generally considered a very poor method for choosing the winner of an election when there are more than two candidates.

The Condorcet Criterion

The principal weakness of the plurality method is that it fails to take into consideration a voter's other preferences beyond first choice, and in so doing can lead to some very bad election results. To underscore the point, consider the following example.

> **EXAMPLE 1.3** The Marching Band Goes Bowling

Tasmania State University has a superb marching band. They are so good that this coming bowl season they have invitations to perform at five different bowl games: The Rose Bowl (R), the Hula Bowl (H), the Fiesta Bowl (F), the Orange Bowl (O), and the Sugar Bowl (S). An election is held among the 100 members of the band to decide in which of the five bowl games they will perform. A preference schedule giving the results of the election is shown in Table 1-2.

TABLE 1-2 Preference Schedule for the Band Election			
Number of voters	49	48	3
1st choice	R	H	F
2nd choice	H	S	H
3rd choice	F	O	S
4th choice	O	F	O
5th choice	S	R	R

Under the plurality method the winner of the election is the Rose Bowl (R), with 49 first-place votes. It's hard not to notice that this is a rather bad outcome, as there are 51 voters that have the Rose Bowl as their last choice. By contrast, the Hula Bowl (H) has 48 first-place votes and 52 second-place votes. Simple common sense tells us that the Hula Bowl is a far better choice to represent the wishes of the entire band. In fact, we can make the following persuasive argument in favor of the Hula Bowl: If we compare the Hula Bowl to any other bowl on a *head-to-head* basis, the Hula Bowl is always the preferred choice. Take, for example, a comparison between the Hula Bowl and the Rose Bowl. There are 51 votes for the Hula Bowl (48 from the second column plus the 3 votes in the last column) versus 49 votes for the Rose Bowl. Likewise, a comparison between the Hula Bowl and the Fiesta Bowl would result in 97 votes for the Hula Bowl (first and second columns) and 3 votes for the Fiesta Bowl. And when the Hula Bowl is compared to either the Orange Bowl or the Sugar Bowl, it gets all 100 votes. Thus, no matter which bowl we compare the Hula Bowl with, there is always a majority of the band that prefers the Hula Bowl. ◀◀

Marie Jean Antoine Nicolas Caritat, Marquis de Condorcet (1743–1794). Condorcet was a French aristocrat, mathematician, philosopher, economist, and social scientist, and one of the great rationalist thinkers of the eighteenth century.

A candidate preferred by a majority of the voters over every other candidate when the candidates are compared in head-to-head comparisons is called a **Condorcet candidate** (named after French mathematician and philosopher Le Marquis de Condorcet). Not every election has a Condorcet candidate, but if there is one, it is a good sign that this candidate represents the voice of the voters better than any other candidate. In Example 1.3 the Hula Bowl is a Condorcet candidate—it is not unreasonable to expect that it should be the winner of the election.

In 1785, Condorcet introduced a basic criterion for fairness in an election—if there is a Condorcet candidate, then that candidate should be the winner of the election. This important fairness criterion is now known as the *Condorcet criterion*.

The Condorcet Criterion

If candidate X is preferred by the voters over each of the other candidates in a head-to-head comparison, then candidate X should be the winner of the election.

We can now see the problem with Example 1.3 in a new light—the example illustrates a violation of the Condorcet criterion. The fact that this can happen using the plurality method means that *the plurality method violates the Condorcet criterion*.

Once again, an important point to remember: Just because violations of the Condorcet criterion *are possible* under the plurality method does not imply that they *must happen*. The marching band example was extreme in the sense that there was one very polarizing choice (the Rose Bowl) which was either loved or hated by the voters. One of the major flaws of the plurality method is that it sometimes leads us to pick such a choice even when there is a much more reasonable choice—one preferred by a majority of the voters over each of the others.

Insincere Voting

We will return to the concept of head-to-head comparisons between the candidates shortly. In the meantime, we conclude our discussion of the plurality method by pointing out another one of its major flaws: the ease with which election results can be manipulated by a voter or a block of voters through insincere voting.

The idea behind **insincere voting** (also known as **strategic voting**) is simple: If we know that the candidate we really want doesn't have a chance of winning, then rather than "wasting our vote" on our favorite candidate we can cast it for a lesser choice that has a better chance of winning the election. In closely contested elections, a few insincere voters can completely change the outcome of an election.

To illustrate how easy it is to manipulate the results of an election decided under the plurality method, let's take a second look at Example 1.3, the marching band example.

> **EXAMPLE 1.4** The Marching Band Election Gets Manipulated

We know that given the preference schedule shown in Table 1-2 (shown again as Table 1-3A) under the plurality method the winner of the marching band election is the Rose Bowl, a result sure to disappoint most of the band members. It so happens that three of the band members (the Dorsey triplets from Tempe, Arizona) realize that there is no chance that their first choice, the Fiesta Bowl, can win this election, so rather than waste their votes they decide to make a strategic move—they cast their votes for the Hula Bowl by switching the first and second choices in their ballots (Table 1-3B).

TABLE 1-3A The Real Preferences			
Number of voters	**49**	**48**	**3 Dorseys**
1st choice	R	H	F
2nd choice	H	S	H
3rd choice	F	O	S
4th choice	O	F	O
5th choice	S	R	R

TABLE 1-3B The Actual Votes			
Number of voters	**49**	**48**	**3 Dorseys**
1st choice	R	H	H
2nd choice	H	S	F
3rd choice	F	O	S
4th choice	O	F	O
5th choice	S	R	R

The impact of this simple switch on the outcome of the election is major—the new preference schedule gives 51 votes, and thus the win, to the Hula Bowl. On the plane ride to Hawaii, the Dorsey triplets are smiling smugly, knowing that their simple manipulation changed the outcome of the election.   ◀◀

While all voting methods are manipulable and can be affected by insincere voting (see Project D in Projects and Papers), the plurality method is the one that can be most easily manipulated, and insincere voting is quite common in real-world elections. For Americans, the most significant cases of insincere voting occur in close presidential or gubernatorial races between the two major party candidates and a third candidate ("the spoiler") who has little or no chance of winning. These are exactly the circumstances describing the 2000 and 2004 presidential elections—in both cases the race was very tight and in both cases the third-party candidate, Ralph Nader, lost many votes from insincere voters that liked him best but did not want to "waste their vote." Likewise, in the 2004 Washington state race for governor, Republican Dino Rossi won the original election by a few hundred votes. After two separate recounts, Democrat Christine Gregoire was declared the winner by a mere 129 votes. No one knows how this election would have gone had all voters voted sincerely, but it is clear that insincere voting might have had a major impact in the outcome of this election.

Insincere voting not only hurts small parties and fringe candidates, it has unintended and often negative consequences on the political system itself. The history of American political elections is littered with examples of independent candidates and small parties that never get a fair voice or a fair level of funding (it takes 5% of the vote to qualify for federal funds for the next election) because of the "let's not waste our vote" philosophy of insincere voters. The ultimate consequence of the plurality method is an entrenched two-party system that often gives the voters little real choice.

The idea that the plurality method invariably leads to a two-party system is known as *Duverger's law* after the French sociologist Maurice Duverger.

1.3 The Borda Count Method

Jean-Charles de Borda (1733–1799) was a brilliant scientist (he made contributions on such diverse subjects as mathematics, astronomy, and voting theory) as well as a captain in the French navy who fought against the British in the American War of Independence.

A commonly used method for determining the winner of an election is the **Borda count method**, named after the Frenchman Jean-Charles de Borda. In this method each place on a ballot is assigned points. In an election with N candidates we give 1 point for *last* place, 2 points for *second from last* place, and so on. At the top of the ballot, a *first-place* vote is worth N points. The points are tallied for each candidate separately, and the candidate with the highest total is the winner. We will call such a candidate the *Borda winner*.

While not common, two or more candidates could tie with the highest point count, and in these cases we must either break the tie using some predetermined tie-breaking procedure or allow the tie to stand. To simplify the presentation, unless otherwise noted we will assume in this chapter that ties are allowed to stand. (A more detailed discussion of ties and tie-breaking procedures can be found in Appendix 1.)

> ### EXAMPLE 1.5 The Math Club Election (Borda Count)

Let's use the Borda count method to choose the winner of the Math Appreciation Society election first introduced in Example 1.1. Table 1-4 shows the point values under each column based on first place worth 4 points, second place worth 3 points, third place worth 2 points, and fourth place worth 1 point.

TABLE 1-4 Borda Points for the Math Club Election

Number of voters	14	10	8	4	1
1st choice: 4 points	A: 56 pts	C: 40 pts	D: 32 pts	B: 16 pts	C: 4 pts
2nd choice: 3 points	B: 42 pts	B: 30 pts	C: 24 pts	D: 12 pts	D: 3 pts
3rd choice: 2 points	C: 28 pts	D: 20 pts	B: 16 pts	C: 8 pts	B: 2 pts
4th choice: 1 point	D: 14 pts	A: 10 pts	A: 8 pts	A: 4 pts	A: 1 pt

When we tally the points,

A gets $56 + 10 + 8 + 4 + 1 = 79$ points;
B gets $42 + 30 + 16 + 16 + 2 = 106$ points;
C gets $28 + 40 + 24 + 8 + 4 = 104$ points;
D gets $14 + 20 + 32 + 12 + 3 = 81$ points.

The Borda winner of this election is Boris! (Wasn't Alisha the winner of this election under the plurality method?) ◀◀

What's Wrong with the Borda Count Method?

See Exercise 68 for details.

In contrast to the plurality method, the Borda count method takes into account all the information provided in the voters' preference ballots, and the Borda winner is the candidate with the best average ranking, the best compromise candidate if you will. On its face, the Borda count method seems like an excellent way to take full consideration of the voters' preferences, so indeed, what's wrong with it?

The next example illustrates some of the problems with the Borda count method.

> ## EXAMPLE 1.6 A School Principal Selection Goes Awry

The last principal at Washington Elementary School has just retired and the School Board must hire a new principal. The four finalists for the job are Mrs. Amaro, Mr. Burr, Mr. Castro, and Mrs. Dunbar (*A*, *B*, *C*, and *D*, respectively). After interviewing the four finalists, each of the 11 school board members gets to rank the candidates by means of a preference ballot, and the Borda winner gets the job. Table 1-5 shows the preference schedule for this election.

TABLE 1-5 Preference Schedule for Example 1.6

Number of voters	6	2	3
1st choice	*A*	*B*	*C*
2nd choice	*B*	*C*	*D*
3rd choice	*C*	*D*	*B*
4th choice	*D*	*A*	*A*

A simple count tells us that Mr. Burr is the Borda winner with 32 points. (You may want to check this out on your own—it will just take a minute or two.) According to the school bylaws, he gets the principal's job. But what about Mrs. Amaro? She is the majority candidate, with 6 out of the 11 first-place votes, as well as the Condorcet candidate, since with 6 first-place votes she beats each of the other three candidates in a pairwise comparison. Needless to say, Mrs. Amaro feels very strongly that the only reasonable outcome of this election is that she should be getting the principal's job. After consulting an attorney, who encourages her to sue, she consults a voting theory expert, who advises her that she has no case—Mr. Burr won the election fair and square. ≪

The sad tale of the unlucky Mrs. Amaro in Example 1.6 is really a story about the Borda count method and how the method violates two basic criteria of fairness—*the majority criterion and the Condorcet criterion*. It is tempting to conclude from this that the Borda count method is a terrible voting method. It is not! In fact, in spite of its flaws, experts in voting theory consider the Borda count method one of the best, if not the very best, method for deciding elections with many candidates. Two things can be said in defense of the Borda count method: (1) although violations of the majority criterion can happen, they do not happen very often, and when there are many candidates such violations are rare; and (2) violations of the Condorcet criterion automatically follow violations of the majority criterion, since a majority candidate is automatically a Condorcet candidate. (If there is no violation of the majority criterion, then a violation of the Condorcet criterion is still possible, but it is also rare.)

In real life, the Borda count method (or some variation of it) is widely used in a variety of settings, from individual sports awards (Heisman trophy winner, NBA Rookie of the Year, NFL MVP, etc.) to college football polls, to music industry awards, to the hiring of school principals, university presidents, and corporate executives.

See, for example, Exercises 69, 70, and 71.

1.4 The Plurality-with-Elimination Method (Instant Runoff Voting)

In most municipal and local elections, a candidate needs a majority of the first-place votes to get elected. When there are three or more candidates running, it is often the case that no candidate gets a majority. Typically, the candidate (or candidates) with the fewest first-place votes is eliminated, and a runoff election is held. Since runoff elections are expensive to both the candidates and the municipality, this is an inefficient and cumbersome method for choosing a mayor or a county supervisor.

A much more efficient way to implement the same process without needing separate runoff elections is to use preference ballots, since a preference ballot tells us not only which candidate the voter wants to win but also which candidate the voter would choose in a runoff between any pair of candidates. The idea is simple but powerful: From the original preference schedule for the election we can eliminate the candidates with the fewest first-place votes one at a time until one of them gets a majority. This method has become increasingly popular in the last few years and is nowadays fashionably known as **instant runoff voting** (IRV). Other names had been used in the past and in other countries for the same method, including **plurality with elimination** and the *Hare method*. For the sake of clarity, we will call it the *plurality-with-elimination method*—it is the most descriptive of the three names.

A formal description of the plurality-with-elimination method goes like this:

Plurality-with-Elimination Method

- **Round 1.** Count the first-place votes for each candidate, just as you would in the plurality method. If a candidate has a majority of first-place votes, that candidate is the winner. Otherwise, eliminate the candidate (or candidates if there is a tie) with the *fewest* first-place votes.

- **Round 2.** Cross out the name(s) of the candidates eliminated from the preference schedule and recount the first-place votes. (Remember that when a candidate is eliminated from the preference schedule, in each column the candidates below it move up a spot.) If a candidate has a majority of first-place votes, declare that candidate the winner. Otherwise, eliminate the candidate with the fewest first-place votes.

- **Rounds 3, 4, etc.** Repeat the process, each time eliminating one or more candidates until there is a candidate with a majority of first-place votes. That candidate is the winner of the election.

> **EXAMPLE 1.7** The Math Club Election (Plurality with Elimination)

Let's apply the plurality-with-elimination method to the Math Club election. For the reader's convenience Table 1-6 shows the preference schedule again—it is exactly the same as Table 1-1.

TABLE 1-6 Preference Schedule for the Math Club Election					
Number of voters	**14**	**10**	**8**	**4**	**1**
1st choice	A	C	D	B	C
2nd choice	B	B	C	D	D
3rd choice	C	D	B	C	B
4th choice	D	A	A	A	A

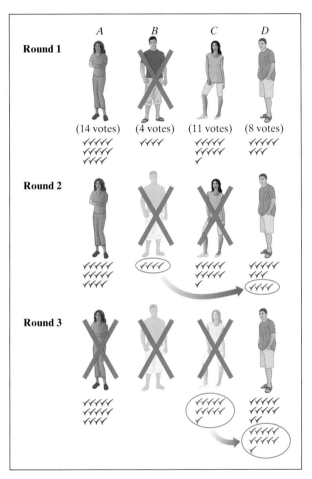

FIGURE 1-5 Boris is eliminated first, then Carmen, and then Alisha. The last one standing is Dave.

Round 1.

Candidate	A	B	C	D
Number of first-place votes	14	4	11	8

Since B has the fewest first-place votes, he is eliminated first.

Round 2. Once B is eliminated, the four votes that originally went to B in round 1 will now go to D, the next-best candidate in the opinion of these four voters (see Fig. 1-5). The new tally is shown below:

Candidate	A	B	C	D
Number of first-place votes	14		11	12

In this round C has the fewest first-place votes and is eliminated.

Round 3. The 11 votes that went to C in round 2 now go to D (just check the relative positions of D and A in the second and fifth columns of Table 1-5). This gives the following:

Candidate	A	B	C	D
Number of first-place votes	14			23

We now have a winner, and lo and behold, it's neither Alisha nor Boris. The winner of the election, with 23 first-place votes, is Dave!   ≪

Applying the Plurality-with-Elimination Method

The next two examples are intended primarily to illustrate a few subtleties that can come up when applying the plurality-with-elimination method.

> **EXAMPLE 1.8** Electing the Mayor of Kingsburg

Five candidates (A, B, C, D, and E) are running for mayor in the small town of Kingsburg. Being at the cutting edge of electoral reform, Kingsburg chooses its mayor using instant runoff voting, in other words, plurality with elimination. Table 1-7 shows the preference schedule for the election.

TABLE 1-7 Preference Schedule for the Kingsburg Mayoral Election

Number of voters	93	44	10	30	42	81
1st choice	A	B	C	C	D	E
2nd choice	B	D	A	E	C	D
3rd choice	C	E	E	B	E	C
4th choice	D	C	B	A	A	B
5th choice	E	A	D	D	B	A

Since the number of voters is $93 + 44 + 10 + 30 + 42 + 81 = 300$, a candidate needs 151 or more votes to win. We now eliminate candidates one by one until someone gets 151 or more votes.

Round 1.

Candidate	A	B	C	D	E
Number of first-place votes	93	44	40	42	81

Here C has the fewest number of first-place votes and is eliminated first. Of the 40 votes originally cast for C, 10 go to A (look at the third column of Table 1-7) and 30 go to E (from the fourth column of Table 1-7).

Round 2.

Candidate	A	B	C	D	E
Number of first-place votes	103	44		42	111

In this round D has the fewest first-place votes and is eliminated. The 42 votes originally cast for D would next go to C (look at the fifth column of Table 1-7), but C is out of the picture at this point, so we dip further down into the column. Thus, the 42 votes go to E, the next top candidate in that column.

Round 3.

Candidate	A	B	C	D	E
Number of first-place votes	103	44			153

At this point we are finished, since E has a majority of the votes and is therefore the winner.   ≪

EXAMPLE 1.9 The School Principal Election Revisited

We are going to take another look at the controversial election for the principal's job at Washington Elementary School. Due to a change in the school's bylaws, the principal is now going to be chosen using plurality with elimination. The preference schedule is shown again in Table 1-8.

TABLE 1-8 Preference Schedule for the Principal Election

Number of voters	6	2	3
1st choice	A	B	C
2nd choice	B	C	D
3rd choice	C	D	B
4th choice	D	A	A

Well, in this case there are no rounds to go through—candidate A is the majority candidate and thus automatically the winner of the election under plurality with elimination.

The simple but important moral of Example 1.9 is that *the plurality-with-elimination method satisfies the majority criterion.*

What's Wrong with the Plurality-with-Elimination Method?

The main problem with the plurality-with-elimination method is quite subtle and is illustrated by the next example.

EXAMPLE 1.10 A Mess of Olympic Proportions

Three cities, Athens (A), Barcelona (B), and Calgary (C), are competing to be the host city for the 2020 Summer Olympic Games. The final decision is made by a secret vote of the 29 members of the Executive Council of the International Olympic Committee, and the winner is to be chosen using the plurality-with-elimination method.

Straw poll: *An unofficial vote or poll indicating the trend of opinion on a candidate or issue.*

American Heritage Dictionary

Two days before the actual election is to be held, a straw poll is conducted by the Executive Council just to see how things stand. The results of the straw poll are shown in Table 1-9.

TABLE 1-9 Results of the Straw Poll

Number of voters	7	8	10	4
1st choice	A	B	C	A
2nd choice	B	C	A	C
3rd choice	C	A	B	B

Based on the results of the straw poll Calgary is going to win the election. (In the first round Athens has 11 votes, Barcelona has 8, and Calgary has 10. Barcelona is eliminated and in the second round Barcelona's 8 votes go to Calgary.)

Although the results of the straw poll are supposed to be secret, the word gets out that it looks like Calgary is going to host the next Summer Olympics. Since everybody loves a winner, the four delegates represented by the last column of Table 1-9 decide as a block to switch their votes and vote for Calgary first and Athens second. Calgary is going to win, so there is no harm in that, is there? Well, let's see ...

The results of the official vote are shown in Table 1-10. The only changes between the straw poll in Table 1-9 and the official vote are the four votes that were switched in favor of Calgary. (To get Table 1-10, switch A and C in the last column of Table 1-9 and then combine columns 3 and 4 into a single column.)

TABLE 1-10 Official Election Results

Number of voters	7	8	14
1st choice	A	B	C
2nd choice	B	C	A
3rd choice	C	A	B

When we apply the plurality-with-elimination method to Table 1-10, Athens gets eliminated in the first round, and the 7 votes originally going to Athens go to Barcelona in the second round. Barcelona, with 15 votes in the second round, gets to host the next Summer Olympics! ◀◀

How could this happen? How could Calgary lose an election it was winning in the straw poll just because it got additional first-place votes in the official election? While you will never convince the conspiracy theorists in Calgary that the election was not rigged, double-checking the figures makes it clear that everything is on the up and up—Calgary is simply the victim of a quirk in the plurality-with-elimination method: the possibility that you can actually do worse by doing better! Example 1.10 illustrates what in voting theory is known as a *violation of the monotonicity criterion*.

The Monotonicity Criterion

If candidate X is a winner of an election and, in a reelection, the only changes in the ballots are changes that favor X (and only X), then X should remain a winner of the election.

Based on Example 1.10, we know that the plurality-with-elimination method *violates the monotonicity criterion*. It is not hard to come up with examples showing that the plurality-with-elimination method also *violates the Condorcet criterion*.

See Exercises 33 and 60.

In spite of its flaws, the plurality-with-elimination method is used in many real-world elections. While Example 1.10 was just a simple dramatization, it was realistic to the extent that the International Olympic Committee does use the plurality-with-elimination method to choose the Olympic host cities. These high-stakes elections take place every four years and are typically surrounded by much drama and intrigue. In addition, plurality with elimination (under the name *instant runoff voting*) is becoming an increasingly popular method in municipal elections for mayor, city council, district attorney, etc. In 2002, the voters in San Francisco approved the use of instant runoff voting in all San Francisco municipal elections, and in 2005 a similar proposition was passed by the voters of Burlington, Vermont. Many other municipalities across the United States (for example, Berkeley, California and Ferndale, Michigan) are in the process of implementing instant runoff voting in their local elections. Instant runoff voting is also used in Australia to elect members of the House of Representatives.

The most current information on the use of instant runoff voting can be found at www.fairvote.org, the Web site of the Center for Voting and Democracy.

1.5 The Method of Pairwise Comparisons (Copeland's Method)

So far, all three voting methods we have discussed violate the Condorcet criterion, but this is not an insurmountable problem. Our next method called the **method of pairwise comparisons** (sometimes also called **Copeland's method**) is the classic example of a voting method that satisfies the Condorcet criterion.

The method of pairwise comparisons is like a round-robin tournament in which every candidate is matched head-to-head against every other candidate. Each of these head-to-head matches is called a **pairwise comparison**. In a pairwise comparison between X and Y every vote is assigned to either X or Y, *the vote going to whichever of the two candidates is listed higher on the ballot*. The winner is the one with the most votes; if the two candidates split the votes equally, the pairwise comparison ends in a tie. The winner of the pairwise comparison gets 1 point and the loser gets none; in case of a tie each candidate gets 1/2 point. The winner of the election is the candidate with the most points after all the pairwise comparisons are tabulated. (Overall point ties are common under this method, and as with other methods, the tie is broken using a predetermined tie-breaking procedure or the tie can stand if multiple winners are allowed.)

Once again, we will illustrate the method of pairwise comparisons using the Math Club election.

> **EXAMPLE 1.11** The Math Club Election (Pairwise Comparisons)

Let's start with a pairwise comparison between A and B. For convenience we will think of A as the *red* candidate and B as the *blue* candidate—the other candidates do not come into the picture at all. (Don't read too much into the colors—they have no political significance and their only purpose is to help us visualize the comparison.) The first column of Table 1-11 represents 14 red voters (they rank A above B); the last four columns of Table 1-11 represent 23 blue voters (they rank B above A). Consequently, the winner is the blue candidate B. We summarize this result as follows:

A versus *B*: 14 votes to 23 votes (*B* wins). *B* gets 1 point.

TABLE 1-11 Pairwise Comparison Between A and B

Number of voters	14	10	8	4	1
1st choice	A	C	D	B	C
2nd choice	B	B	C	D	D
3rd choice	C	D	B	C	B
4th choice	D	A	A	A	A

Let's next look at the pairwise comparison between *C* and *D*. We'll call *C* the *red* candidate and *D* the *blue* candidate. The first, second, and last columns of Table 1-12 represent red votes (they rank *C* above *D*); the third and fourth columns represent blue voters (they rank *D* above *C*). Here red beats blue 25 to 12.

C versus *D*: 25 votes to 12 votes (*C* wins). *C* gets 1 point.

TABLE 1-12 Pairwise Comparison Between C and D

Number of voters	14	10	8	4	1
1st choice	A	C	D	B	C
2nd choice	B	B	C	D	D
3rd choice	C	D	B	C	B
4th choice	D	A	A	A	A

If we continue in this manner, comparing in all possible ways two candidates at a time, we end up with the following scoreboard (the devil is in the details, so go get 'em!):

A versus *B*: 14 votes to 23 votes (*B* wins). *B* gets 1 point.

A versus *C*: 14 votes to 23 votes (*C* wins). *C* gets 1 point.

A versus *D*: 14 votes to 23 votes (*D* wins). *D* gets 1 point.

B versus *C*: 18 votes to 19 votes (*C* wins). *C* gets 1 point.

B versus *D*: 28 votes to 9 votes (*B* wins). *B* gets 1 point.

C versus *D*: 25 votes to 12 votes (*C* wins). *C* gets 1 point.

The final tally produces 0 points for *A*, 2 points for *B*, 3 points for *C*, and 1 point for *D*. Can it really be true? Yes! The winner of the election under the method of pairwise comparisons is Carmen! ◀◀

The method of pairwise comparisons satisfies all three of the fairness criteria discussed so far in the chapter. A majority candidate automatically wins every pairwise comparison and thus will win the election under the method of pairwise comparisons, so the method *satisfies the majority criterion*. A Condorcet candidate wins every pairwise comparison by definition, so the method clearly *satisfies the Condorcet criterion*. It can also be shown that the method of pairwise comparisons *satisfies the monotonicity criterion*.

See Exercise 65.

So, What's Wrong with the Method of Pairwise Comparisons?

So far, the method of pairwise comparisons looks like the best method we have, at least in the sense that it satisfies the three fairness criteria we have studied. Unfortunately, the method does violate a fourth fairness criterion. Our next example illustrates the problem.

> **EXAMPLE 1.12** The NFL Draft

As the newest expansion team in the NFL, the Los Angeles LAXers will be getting the number-one pick in the upcoming draft. After narrowing the list of candidates to five players (Allen, Byers, Castillo, Dixon, and Evans), the coaches and team executives meet to discuss the candidates and eventually choose the team's first pick on the draft. By team rules, the choice must be made using the method of pairwise comparisons.

Table 1-13 shows the preference schedule after the 22 voters (coaches, scouts and team executives) turn in their preference ballots.

TABLE 1-13 LAXer's Draft Choice Election

Number of voters	2	6	4	1	1	4	4
1st choice	A	B	B	C	C	D	E
2nd choice	D	A	A	B	D	A	C
3rd choice	C	C	D	A	A	E	D
4th choice	B	D	E	D	B	C	B
5th choice	E	E	C	E	E	B	A

There are 10 possible pairwise comparisons to look at, and the results of the comparisons are as follows (this is a good place to take a break from your reading and verify these calculations on your own!):

A versus B: 7 votes to 15 votes. B gets 1 point.
A versus C: 16 votes to 6 votes. A gets 1 point.
A versus D: 13 votes to 9 votes. A gets 1 point.
A versus E: 18 votes to 4 votes. A gets 1 point.
B versus C: 10 votes to 12 votes. C gets 1 point.
B versus D: 11 votes to 11 votes. B gets 1/2 point, D gets 1/2 point.
B versus E: 14 votes to 8 votes. B gets 1 point.
C versus D: 12 votes to 10 votes. C gets 1 point.
C versus E: 10 votes to 12 votes. E gets 1 point.
D versus E: 18 votes to 4 votes. D gets 1 point.

The final tally produces 3 points for A, $2\frac{1}{2}$ points for B, 2 points for C, $1\frac{1}{2}$ points for D, and 1 point for E. It looks as if Allen (A) is the lucky young man who will make millions of dollars playing for the Los Angeles LAXers.

The interesting twist to the story surfaces when it is discovered right before the actual draft that Castillo had accepted a scholarship to go to medical school and will not be playing professional football. Since Castillo was not the top choice, the fact that he is choosing med school over pro football should not affect the fact that Allen is the number-one draft choice. Does it?

Suppose we eliminate Castillo from the original preference schedule (we can do this by eliminating *C* from Table 1-13 and moving the players below *C* up one slot). Table 1-14 shows the results of the revised election when Castillo is not a candidate.

TABLE 1-14   LAXer's Revised Draft Choice Election

Number of voters	2	6	4	1	1	4	4
1st choice	*A*	*B*	*B*	*B*	*D*	*D*	*E*
2nd choice	*D*	*A*	*A*	*A*	*A*	*A*	*D*
3rd choice	*B*	*D*	*D*	*D*	*B*	*E*	*B*
4th choice	*E*	*E*	*E*	*E*	*E*	*B*	*A*

Now we have only four players and six pairwise comparisons to consider. The results are as follows:

A versus *B*: 7 votes to 15 votes. *B* gets 1 point.

A versus *D*: 13 votes to 9 votes. *A* gets 1 point.

A versus *E*: 18 votes to 4 votes. *A* gets 1 point.

B versus *D*: 11 votes to 11 votes. *B* gets 1/2 point, *D* gets 1/2 point.

B versus *E*: 14 votes to 8 votes. *B* gets 1 point.

D versus *E*: 18 votes to 4 votes. *D* gets 1 point.

In this new scenario *A* has 2 points, *B* has $2\frac{1}{2}$ points, *D* has $1\frac{1}{2}$ points, and *E* has 0 points, and the winner is Byers. In other words, when Castillo is not in the running, then the number-one pick is Byers, not Allen. Needless to say, the coaching staff is quite perplexed. Do they draft Byers or Allen? How can the presence or absence of Castillo in the candidate pool be relevant to this decision?   ◀◀

The strange happenings in Example 1.12 help illustrate an important fact: The method of pairwise comparisons violates a fairness criterion known as *the independence-of-irrelevant-alternatives criterion*.

The Independence-of-Irrelevant-Alternatives Criterion (IIA)

If candidate *X* is a winner of an election and in a recount one of the non-winning candidates is removed from the ballots, then *X* should still be a winner of the election.

An alternative way to interpret the independence-of-irrelevant-alternatives criterion is to switch the order of the two elections (think of the recount as the original election and the original election as the recount). Under this interpretation, the IIA says that *if candidate X is a winner of an election and in a reelection another candidate that has no chance of winning (an "irrelevant alternative") enters the race, then X should still be the winner of the election.*

The method of pairwise comparisons can sometimes be quite indecisive—it is not unusual to have multiple ties for first place. Our next example shows an extreme case of this.

▶ EXAMPLE 1.13 The Hockey Team Is Out to Lunch

The Icelandia State University varsity hockey team is on a road trip. An important decision needs to be made: Where to go out to lunch? In the past, this has led to some heated arguments, so this time they decide to hold an election. The choices boil down to three restaurants: Andreychuk's (*A*), Bure's (*B*), and Chelios's (*C*). The decision is to be made using the method of pairwise comparisons. Table 1-15 shows the results of the voting by the 15 players on the squad.

TABLE 1-15 Restaurant Election Results				
Number of voters	4	2	6	3
1st choice	*A*	*B*	*C*	*B*
2nd choice	*B*	*C*	*A*	*A*
3rd choice	*C*	*A*	*B*	*C*

Here is the surprising part: *A* beats *B* 10 votes to 5, *B* beats *C* 9 votes to 6, and *C* beats *A* 8 votes to 7. This results in a three-way tie for first place. What now? Since the coaches insist that the team must eat together, it becomes necessary to break the tie. If plurality is used to break the tie, then the winner is *C*; if the Borda count is used to break the tie, then the winner is *A*; and if plurality with elimination is used to break the tie, the winner is *B*. Clearly, we are no better off now than when we started! ◀◀

How Many Pairwise Comparisons?

One practical difficulty with the method of pairwise comparisons is that as the number of candidates grows, the number of comparisons grows even faster: With 5 candidates we have a total of 10 pairwise comparisons, with 10 candidates we have a total of 45 pairwise comparisons, and (heaven forbid) with 100 candidates we have a total of 4950 pairwise comparisons. Computing all of these comparisons by hand would not be fun.

In this section we will establish the exact connection between the number of candidates and the number of pairwise comparisons. In so doing we will develop a simple but elegant mathematical formula that has many other applications. We will start with a seemingly unrelated example.

> ### EXAMPLE 1.14 Long Sums Made Short

We would like to find the sum of the first 99 positive integers, namely, $1 + 2 + 3 + \cdots + 98 + 99$. We could certainly do it by adding all these numbers, but that would be tedious and wouldn't teach us anything. Instead, we will call the sum S, and write S twice, once forward and once backward, as follows:

$$1 +\ \ 2 +\ \ 3 + \cdots + 98 + 99 = S$$
$$99 + 98 + 97 + \cdots +\ \ 2 +\ \ 1 = S$$

You may have noticed that when we add the numbers column by column, each column totals 100. Thus, adding the 99 columns together totals $99 \times 100 = 9900$. At the same time, this total equals $2S$, so we can conclude that $2S = 9900$ and therefore $S = 4950$. ⋘

The argument used in Example 1.14 can be easily generalized to any sum of the form $S = 1 + 2 + 3 + \cdots + L$, where L is a positive integer (think of L as short for "last" integer). When we write the sum twice, once forward and once backward, we get L columns each of which totals $(L + 1)$. This implies that $2S = L(L + 1)$, and $S = \dfrac{L(L + 1)}{2}$. This give us an extremely nice and useful formula for the sum of all integers from 1 to L:

$$1 + 2 + 3 + \cdots + L = \frac{L(L + 1)}{2}$$

See Exercises 53 and 54.

With a little care the preceding formula can also be used to find the sum of any set of consecutive integers, even when the starting term is bigger than 1.

> ### EXAMPLE 1.15 Counting Pairwise Comparisons

Consider an election with 10 candidates: A, B, C, D, E, F, G, H, I, and J. Let's count the pairwise comparisons:

- Compare A with each of the other 9 candidates (B through J). This gives 9 pairwise comparisons. We are now done with A.
- Compare B with each of the remaining 8 candidates (C through J). This gives 8 pairwise comparisons. We are now done with B.
- Compare C with each of the remaining 7 candidates (D through J). This gives 7 pairwise comparisons. We are now done with C.
- We keep going this way. At the end, when we get to I, we have one comparison left, between I and J.

This gives a total of $1 + 2 + 3 + 4 + 5 + 6 + 7 + 8 + 9 = 45$ pairwise comparisons. ⋘

Generalizing the argument used in Example 1.15 is straightforward: The number of pairwise comparisons with N candidates is given by the sum $1 + 2 + 3 + 4 + \cdots + (N - 1)$. [Here L, which represents the "last" number in the sum, is $(N - 1)$.] Using the formula for the sum of the first L integers and substituting $(N - 1)$ for L gives us the following:

The Number of Pairwise Comparisons

In an election with N candidates the total number of pairwise comparisons between candidates is $\dfrac{(N-1)N}{2}$.

> ### EXAMPLE 1.16 Scheduling a Round-Robin Tournament

In a round-robin tournament, every player (team) plays every other player (team) once. Imagine you are in charge of scheduling a round-robin Ping-Pong tournament with 32 players. The tournament organizers agree to pay you $1 per match for running the tournament—you want to know how much you are going to make.

Since each match is like a pairwise comparison, we can think of the players as candidates and use the formula for pairwise comparisons. We can conclude that there will be a total of $\dfrac{31 \times 32}{2} = 496$ matches played. Not a bad gig for a weekend job! «

1.6 Rankings

Quite often it is important not only to know who wins the election but also to know who comes in second, third, and so on. Let's consider once again the Math Club election. Suppose now that instead of electing just the president we need to elect a board of directors consisting of a president, a vice president, and a treasurer. The club's bylaws state that rather than having separate elections for each office, the winner of the election gets to be the president, the second-place candidate gets to be the vice president, and the third-place candidate gets to be the treasurer. In a situation like this, we need a voting method that gives us not just a winner but also a second place, a third place, and so on—in other words, a **ranking** of the candidates.

Extended Ranking Methods

Each of the four voting methods we discussed earlier in this chapter has a natural extension that can be used to produce a ranking of the candidates.

> ### EXAMPLE 1.17 The Math Club Election (Extended Plurality)

For the reader's convenience, the preference schedule is shown again in Table 1-16.

TABLE 1-16 Preference Schedule for the Math Club Election					
Number of voters	14	10	8	4	1
1st choice	A	C	D	B	C
2nd choice	B	B	C	D	D
3rd choice	C	D	B	C	B
4th choice	D	A	A	A	A

The number of first-place votes is

A: 14 first-place votes,
B: 4 first-place votes,
C: 11 first-place votes,
D: 8 first-place votes.

We know that under the plurality method Alisha is the winner. Who should be second? Clearly Carmen, with the second most first-place votes (11). Likewise, Dave is third (8 votes) and Boris (4 votes) last. The ranking of all the candidates under the *extended* plurality method is shown in Table 1-17.

TABLE 1-17 Ranking Under Extended Plurality

Office	Place	Candidate	First-place votes
President	1st	Alisha	14
Vice president	2nd	Carmen	11
Treasurer	3rd	Dave	8
	4th	Boris	4

 EXAMPLE 1.18 The Math Club Election (Extended Borda Count)

Ranking the candidates using the extended Borda count method is equally simple. Recall that the point totals under the Borda count (Example 1.5) were

A: 79 Borda points,
B: 106 Borda points,
C: 104 Borda points,
D: 81 Borda points.

The resulting ranking, based on the *extended* Borda count method, is given in Table 1-18.

TABLE 1-18 Ranking Under Extended Borda Count

Office	Place	Candidate	Borda points
President	1st	Boris	106
Vice president	2nd	Carmen	104
Treasurer	3rd	Dave	81
	4th	Alisha	79

 EXAMPLE 1.19 The Math Club Election
(Extended Plurality with Elimination)

Table 1-19 shows the ranking of the candidates under the *extended* plurality-with-elimination method. The idea is that we rank the candidates in reverse order of elimination: Since Boris was eliminated in the first round (see Example 1.7), we rank

In elections where a candidate gets a majority of votes before all the rounds of elimination are completed, we continue the process of elimination to rank the remaining candidates.

Boris last. Carmen was eliminated in the second round, so she gets third place; Alisha was eliminated in the third round, so she gets second place, and Dave takes first place.

TABLE 1-19 Ranking Under Extended Plurality with Elimination

Office	Place	Candidate	Eliminated in
President	1st	Dave	
Vice president	2nd	Alisha	3rd round
Treasurer	3rd	Carmen	2nd round
	4th	Boris	1st round

 EXAMPLE 1.20 The Math Club Election
(Extended Pairwise Comparisons)

TABLE 1-20 Ranking Under Extended Pairwise Comparisons

Office	Place	Candidate	Points
President	1st	Carmen	3
Vice president	2nd	Boris	2
Treasurer	3rd	Dave	1
	4th	Alisha	0

We rank the candidates under the *extended* method of pairwise comparisons according to the total number of points in their comparisons with the other candidates (recall that a win counts as 1 point, a loss as 0 points and a tie as $\frac{1}{2}$ point). Table 1-20 shows the results for the Math Club election.

A summary of the results of the Math Club election using the different extended ranking methods is shown in Table 1-21. The most striking thing about Table 1-21 is the wide discrepancy of results. While it is somewhat frustrating to see this much equivocation, it is important to keep things in context: This is the exception rather than the rule. One purpose of the Math Club example is to illustrate how crazy things can get in some elections, but in most real-life elections there tends to be much more consistency among the various methods.

TABLE 1-21 The Math Club Election: A Tale of Four Methods

Method	Ranking			
	1st	2nd	3rd	4th
Extended plurality	A	C	D	B
Extended Borda count	B	C	D	A
Extended plurality with elimination	D	A	C	B
Extended pairwise comparisons	C	B	D	A

Recursive Ranking Methods

We will now discuss a different, somewhat more involved strategy for ranking the candidates, which we will call the *recursive* approach. (A word to the wise—recursive processes will show up again several times later in this book and are extremely useful and important.) The idea behind a recursive process is similar to that of a feedback loop: At each step of the process the output of the process determines the input to the next step of the process. In the case of an election the process is to find the winner and remove the winner's name from the preference schedule, thus creating a new preference schedule. We then start the process all over again.

The **recursive ranking** approach follows a consistent strategy, but the details depend on the voting method we want to use. Let's say we are going to use the recursive version of some generic voting method X to rank the candidates in an election. We first use method X to find the winner of the election. So far, so good. We then remove the name of the winner on the preference schedule and obtain a new preference schedule with one less candidate on it. We apply method X once again to find the "winner" based on this new preference schedule, and this candidate is ranked second. (A similar process takes place in some Olympic events where after the gold medal winner is determined the other finalists continue to compete to determine the silver and bronze medalists.) We can continue this process to rank as many of the candidates as we need to.

We will illustrate recursive ranking with a couple of examples, both based on the Math Club election.

▶ EXAMPLE 1.21 The Math Club Election (Recursive Plurality)

Here is how we rank the four candidates in the Math Club election using the *recursive* plurality method.

Step 1. (Choose the winner and remove.) Table 1-22 shows the original preference schedule. We know the winner is A with 14 first-place votes. Table 1-23 shows the preference schedule when A is removed. This will be the input to the next step.

Step 2. (Choose second place and remove.) The winner of the "election" shown in Table 1-23 is B with 18 votes. Thus, *second place goes to B.* When we remove B from Table 1-23 we get the preference schedule shown in Table 1-24. This will be the input to the next step.

Step 3. (Choose third and fourth places.) The last step is to find the results of the two-candidate election shown in Table 1-24. Here, C wins with 25 votes. Thus, *third place* goes to C and *last place* goes to D.

The final ranking of the candidates under the *recursive* plurality method is shown in Table 1-25. It is worth noting how different this ranking is from the ranking shown in Table 1-17 based on the *extended*

TABLE 1-22 Step 1 Input

Number of voters	14	10	8	4	1
1st choice	A	C	D	B	C
2nd choice	B	B	C	D	D
3rd choice	C	D	B	C	B
4th choice	D	A	A	A	A

TABLE 1-23 Step 2 Input

Number of voters	14	10	8	4	1
1st choice	B	C	D	B	C
2nd choice	C	B	C	D	D
3rd choice	D	D	B	C	B

TABLE 1-24 Step 3 Input

Number of voters	14	10	8	4	1
1st choice	C	C	D	D	C
2nd choice	D	D	C	C	D

TABLE 1-25 Ranking Under Recursive Plurality

Office	Place	Candidate
President	1st	Alisha
Vice president	2nd	Boris
Treasurer	3rd	Carmen
	4th	Dave

Ranking the Math Club election using recursive Borda count and recursive pairwise comparisons is left as exercises for the reader—see Exercises 45 and 46.

plurality method. In fact, except for first place (which will always be the same), all the other positions turned out to be different.

For our final example, we will illustrate how to rank the candidates in the Math Club election using the *recursive* plurality-with-elimination method.

> ### EXAMPLE 1.22 The Math Club Election (Recursive Plurality with Elimination)

We have to be careful to distinguish the two different types of "elimination" that take place in this method. Under the plurality-with-elimination method, candidates are *eliminated* in rounds until there is a winner left. Under the recursive approach, the winner at each step of the recursion is *removed* from the preference schedule so that we can move on to the next step. Thus, each step consists of several rounds of elimination and, at the end, the removal of the winner. Here is how it works:

TABLE 1-26 Step 1 Input

Number of voters	14	10	8	4	1
1st choice	A	C	D	B	C
2nd choice	B	B	C	D	D
3rd choice	C	D	B	C	B
4th choice	D	A	A	A	A

TABLE 1-27 Step 2 Input

Number of voters	14	10	8	4	1
1st choice	A	C	C	B	C
2nd choice	B	B	B	C	B
3rd choice	C	A	A	A	A

TABLE 1-28 Step 3 Input

Number of voters	14	10	8	4	1
1st choice	A	B	B	B	B
2nd choice	B	A	A	A	A

Step 1. (Choose the winner and remove.) Table 1-26 shows the original preference schedule. After several rounds of elimination, the winner is *D*. (We did this in Example 1.7.) Table 1-27 shows the preference schedule when *D* is removed. This is the input to Step 2.

Step 2. (Choose second place and remove.) For the election given in Table 1-27 no rounds of elimination are needed, since *C* has 19 votes (a majority). Thus, second place goes to *C*. We now remove *C* and get the preference schedule shown in Table 1-28. This is the input to the last step.

Step 3. (Choose third and fourth places.) The last step is to find the results of the two-candidate election shown in Table 1-28. Clearly, *B* wins with 23 votes. Thus, *third place* goes to *B* and *last place* goes to *A*.

The final ranking of the candidates under the *recursive* plurality-with-elimination method is shown in Table 1-29. Once again, if we compare this ranking with the ranking under the *extended* plurality-with-elimination method (Table 1-19), the differences in how second, third, and fourth places are ranked are striking.

TABLE 1-29 Ranking Under Recursive Plurality with Elimination

Office	Place	Candidate
President	1st	Dave
Vice president	2nd	Carmen
Treasurer	3rd	Boris
	4th	Alisha

It is clear from the last two examples that other than first place, recursive ranking methods and extended ranking methods can produce very different results. Which one produces better rankings? As with everything else in election theory, there is no simple answer to this question. It is true, however, that in real-life elections, extended ranking methods are almost always used. Recursive ranking methods, while mathematically interesting, are of little practical use. But don't fret—what we learned about recursion will come in handy in Chapters 9, 10, and 12.

Conclusion: Elections, Fairness, and Arrow's Impossibility Theorem

There are several broad themes running through this chapter. One of them is that elections are more than just for choosing our president or our governor— "small" elections of various kinds play a pervasive and important role in all of our lives, from getting a job to deciding where to go to dinner. The second is that there are many different formal methods that can be used to decide the outcome of an election. In this chapter we focused on four basic methods— mostly because they are simple and commonly used—but there are many others, some quite exotic (see, for example, the discussion of the Academy Awards in Appendix 2). Third, we saw that the results of an election can change if we change the voting method. The Math Club example dramatically illustrated this point—each of the four voting methods produced a different winner. Since there were four candidates, we can say that each of them won the election (just pick the "right" voting method).

Given an abundance of voting methods, how do we determine which are "good" voting methods and which are not so good? In a democratic society, the most important quality we seek in a voting method is that of *fairness*. This leads us to the last, but probably most important, theme of this chapter—the notion of *fairness criteria*. Fairness criteria set the basic standards that a fair election should satisfy. We discussed four of them in this chapter—let's review what they were:

- **The Majority Criterion:** *A majority candidate should always win the election.* (After all, it seems clearly unfair when a candidate with a majority of the first-place votes does not win.)

- **The Condorcet Criterion:** *A Condorcet candidate should always win the election.* (When the candidates are compared two at a time, the Condorcet candidate beats each of the other candidates. How could it be fair to give the win to a different candidate?)

- **The Monotonicity Criterion:** *Suppose candidate X is a winner of the election, but for one reason or another, there is a new election. If the only changes in the ballots are changes in favor of candidate X (and only X), then X should win the new election.* (Think of the election as a card game: If at some point in the game X has the winning hand and then X's hand improves while all the other players' hands stay the same or get worse, then X should still have the winning hand. By any reasonable standard, if X were to lose under these circumstances it would not be a fair game.)

- **The Independence-of-Irrelevant-Alternatives Criterion (IIA).** *Suppose candidate X is a winner of the election but for one reason or another, there is a new election. If the only changes are that one of the other candidates withdraws or is disqualified, then X should win the new election.* (Clearly, it would be unfair to penalize the winner of an election because of the mere fact that one of the losers scratches out of the race.) The flip side of this criterion is that a winner of the election should not be penalized by the introduction of irrelevant new candidates who have no chance of winning.

For additional fairness criteria, see Exercises 74–78.

Other fairness criteria beyond the ones just listed have been proposed, but these four are sufficient to establish a minimum standard of fairness—*a fair voting method should clearly satisfy all four of the above.* Surprisingly, none of the four voting methods we discussed in this chapter meets this minimum standard—far from it! Plurality violates the Condorcet criterion and the IIA criterion; Borda count violates the majority criterion, the Condorcet criterion, and the IIA criterion; plurality with elimination violates the Condorcet criterion, the monotonicity criterion, and the IIA criterion; pairwise comparison violates the IIA criterion.

An obviously relevant question then is, What voting method satisfies the minimum standards of fairness set forth by the preceding four criteria? For democratic elections involving three or more candidates, the answer is that *there is no such voting method.* At first glance, this fact seems a little surprising. Finding a voting method that satisfies a few simple fairness criteria does not seem like an impossible task, and given the obvious importance of having fair elections in a democracy, how is it possible that no one has come up with such a voting method? Up until the late 1940s this was one of the most challenging questions in social-choice theory. Finally, in 1949, the mathematical economist Kenneth Arrow demonstrated that this is indeed an impossible task. **Arrow's impossibility theorem** essentially says (his exact formulation of the theorem is slightly different from the one given here) that *it is mathematically impossible for a democratic voting method to satisfy all of the fairness criteria.* Ironically, making decisions in a consistently fair way is inherently impossible in a democratic society.

The search of the great minds of recorded history for the perfect democracy, it turns out, is the search for a chimera, a logical self-contradiction.

Paul Samuelson

Profile Kenneth J. Arrow (1921–)

Kenneth Arrow is one of the best known and most versatile mathematical economists of the twentieth century. Born and raised in New York City, Arrow completed his undergraduate studies at City College of New York, where at the age of 19 he received a Bachelor's degree in Social Science and Mathematics. He continued his graduate studies at Columbia University, where he received a Master of Arts (M.A.) in Mathematics in 1941. At this time he became interested in mathematical economics, and by 1942 he completed all the necessary course work for a Ph.D. in Economics at Columbia. His graduate studies at Columbia were interrupted by World War II, and between 1942 and 1946 he was assigned to serve as a weather officer in the U.S. Army Air Corps. As a result of his weather work for the Army, Arrow was able to publish his first mathematical research paper—*On the Optimal Use of Winds for Flight Planning.* Arrow's versatility and ability to see significant mathematical ideas in seemingly nonmathematical subjects became one of the hallmarks of his work.

In 1946 Arrow returned to Columbia to write his doctoral thesis, which he completed in 1949. (By then he was already working as a research associate and Assistant Professor of Economics at the University of Chicago.) Arrow's doctoral thesis, entitled *Social Choice and Individual Values*, became a landmark work in mathematical economics and would eventually lead to his being awarded the 1972 Nobel Prize in Economics. Using the mathematical principles of game theory, Arrow proved his famous *impossibility theorem*, which in essence says that it is impossible for a democratic society to take the individual opinions of its voters (the "individual values") and always make a collective decision (the "social choice") that is fair. As one might imagine for a doctoral thesis and Nobel Prize–quality work, the details are quite technical, but the mathematics itself was not difficult. Arrow's genius was to find a way to redefine the idea of "fairness" in a way that lends itself to mathematical treatment.

In 1949 Arrow was appointed Assistant Professor of Economics and Statistics at Stanford University, eventually becoming Professor of Economics, Statistics, and Operations Research. Between 1968 and 1979 he was a Professor of Economics at Harvard University, but in 1979 he returned to Stanford, where he is the Joan Kenney Professor of Economics Emeritus and Professor of Operations Research. Even in his eighty's, Professor Arrow remains active and is pursuing new areas research. (His latest research focuses on environmental economics, and he is the principal investigator in a multidisciplinary project—*The Research Initiative on the Environment, the Economy and Sustainable Welfare*—which aims to study the role of national and international economic policies on the environment and the climate system.)

Key Concepts

Arrow's impossibility theorem, **3**
Borda count method, **10**
Condorcet candidate, **8**
Condorcet criterion, **8**
extended ranking methods, **23**
independence-of-irrelevant-
    alternatives criterion, **20**

insincere voting, **8**
linear ballot, **4**
majority candidate, **6**
majority criterion, **6**
method of pairwise comparisons, **17**
monotonicity criterion, **16**
plurality candidate, **6**

plurality method, **6**
plurality-with-elimination
    method, **12**
preference ballot, **4**
preference schedule, **5**
ranking, **23**
recursive ranking methods, **25**

Exercises

WALKING

A. Ballots and Preference Schedules

1. The management of the XYZ Corporation has decided to treat their office staff to dinner. The choice of restaurants is The Atrium (*A*), Blair's Kitchen (*B*), The Country Cookery (*C*), and Dino's Steak House (*D*). Each of the 12 staff members is asked to submit a preference ballot listing his or her first, second, third, and fourth choices among these restaurants. The resulting preference ballots are as follows:

Ballot	**Ballot**	**Ballot**	**Ballot**	**Ballot**	**Ballot**	**Ballot**	**Ballot**	**Ballot**	**Ballot**	**Ballot**	**Ballot**
1st *A*	1st *C*	1st *B*	1st *C*	1st *C*	1st *C*	1st *A*	1st *C*	1st *A*	1st *A*	1st *C*	1st *A*
2nd *B*	2nd *B*	2nd *B*	2nd *B*	2nd *B*	2nd *B*	2nd *B*	2nd *B*	2nd *B*	2nd *B*	2nd *B*	2nd *B*
3rd *C*	3rd *D*	3rd *C*	3rd *A*	3rd *A*	3rd *D*	3rd *C*	3rd *A*	3rd *C*	3rd *C*	3rd *D*	3rd *C*
4th *D*	4th *A*	4th *A*	4th *D*	4th *D*	4th *A*	4th *D*	4th *D*	4th *D*	4th *D*	4th *A*	4th *D*

(a) How many first-place votes are needed for a majority?

(b) Which restaurant has the most first-place votes? Is it a majority or a plurality?

(c) Write out the preference schedule for this election.

2. The Latin Club is holding an election to choose its president. There are three candidates, Arsenio, Beatrice, and Carlos (A, B, and C for short). Following are the votes of the 11 members of the club that voted.

Voter	Sue	Bill	Tom	Pat	Tina	Mary	Alan	Chris	Paul	Kate	Ron
1st choice	C	A	C	A	B	C	A	A	C	B	A
2nd choice	A	C	B	B	C	B	C	C	B	C	B
3rd choice	B	B	A	C	A	A	B	B	A	A	C

(a) How many first-place votes are needed for a majority?

(b) Which candidate has the most first-place votes? Is it a majority or a plurality?

(c) Write out the preference schedule for this election.

3. An election is held to choose the Chair of the Mathematics Department at Tasmania State University. The candidates are Professors Argand, Brandt, Chavez, Dietz, and Epstein (A, B, C, D, and E). The preference schedule for the election is as follows:

Number of voters	5	3	5	3	2	3
1st choice	A	A	C	D	D	B
2nd choice	B	D	E	C	C	E
3rd choice	C	B	D	B	B	A
4th choice	D	C	A	E	A	C
5th choice	E	E	B	A	E	D

(a) How many people voted in this election?

(b) How many first-place votes are needed for a majority?

(c) Which candidate had the most first-place votes?

(d) Which candidate had the most last-place votes?

4. A math class is asked by the instructor to vote among four possible times for the final exam—A (December 15, 8:00 A.M.), B (December 20, 9:00 P.M.), C (December 21, 7:00 A.M.), and D (December 23, 11:00 A.M.). The following is the class preference schedule.

Number of voters	3	4	9	9	2	5	8	3	12
1st choice	A	A	A	B	B	B	C	C	D
2nd choice	B	B	C	C	A	C	D	A	C
3rd choice	C	D	B	D	C	A	B	D	A
4th choice	D	C	D	A	D	D	A	B	B

(a) How many students in the class voted?

(b) How many first-place votes are needed for a majority?

(c) Which alternative(s) had the least first-place votes?

(d) Which alternative(s) had the least last-place votes?

5. This exercise refers to the election for Mathematics Department Chair discussed in Exercise 3. Suppose that the election rules are that when there is a candidate with a majority of the first-place votes, he/she is the winner. Otherwise, all candidates with 20% or less of the first-place votes are eliminated and the ballots are recounted.

(a) Which candidates are eliminated in this election?

(b) Find the preference schedule for the recount.

(c) Which candidate is the majority winner after the recount?

6. The student body at Eureka High School is having an election for Homecoming Queen. The candidates are Alicia, Brandy, Cleo, and Dionne (A, B, C, D, and E). The preference schedule for the election is as follows:

Number of voters	153	102	55	202	108	20	110	160	175	155
1st choice	A	A	A	B	B	B	C	C	D	D
2nd choice	C	B	D	D	C	C	A	B	A	B
3rd choice	B	D	C	A	D	A	D	A	C	C
4th choice	D	C	B	C	A	D	B	D	B	A

Suppose that the election rules are that when there is a candidate with a majority of the first-place votes, she is the winner. Otherwise, all candidates with 25% or less of the first-place votes are eliminated and the ballots are recounted.

(a) Which candidates are eliminated in this election?

(b) Find the preference schedule for the recount.

(c) Which candidate is the majority winner after the recount?

7. The Demublican Party is holding its annual convention. The 1500 voting delegates are choosing among three possible party platforms: L (a liberal platform), C (a conservative platform), and M (a moderate platform). Seventeen percent of the delegates prefer L to M and M to C. Thirty-two percent of the delegates like C the most and L the least. The rest of the delegates like M the most and C the least. Write out the preference schedule for this election.

8. The Epicurean Society is holding its annual election for President. The three candidates are A, B, and C. Twenty percent of the voters like A the most and B the least; forty percent of the voters like B the most and A the least. Of the remaining voters there are 225 that prefer C to B and B to A, and 675 voters that prefer C to A and A to B. Write out the preference schedule for this election.

In this chapter we used preference ballots that list ranks (1st choice, 2nd choice, etc.) and asked the voter to put the name of a candidate or choice next to each rank. Exercises 9 and 10 refer to an alternative format for preference ballots in which the names of candidates appear in some order and the voter is asked to put a rank (1, 2, 3, etc.) next to each name.

9. Rewrite the following preference schedule in the conventional format used in the book.

Number of voters	47	36	24	13	5
A	3	1	2	4	3
B	1	2	1	2	5
C	4	4	5	3	1
D	5	3	3	5	4
E	2	5	4	1	2

10. Rewrite the following conventional preference schedule in the alternative format. Assume the candidates are listed in alphabetical order on the ballots.

Number of voters	47	36	24	13	5
1st choice	A	B	D	C	B
2nd choice	C	A	B	A	D
3rd choice	B	D	C	E	E
4th choice	E	C	E	B	A
5th choice	D	E	A	D	C

B. Plurality Method

11. This exercise refers to the election for Homecoming Queen at Eureka High School discussed in Exercise 6.

(a) Find the winner(s) of the election under the plurality method.

(b) Suppose that in case of a tie, the winner is decided by choosing the candidate with the fewest last-place votes. In this case, which candidate would win the election?

12. This exercise refers to the final exam election discussed in Exercise 4.

(a) Which alternative would win under the plurality method?

(b) Suppose that in case of a tie the alternative with the fewest last-place votes wins. In this case, when would the final exam be given?

13. An election with 4 candidates (A, B, C, D) and 150 voters is to be decided using the plurality method. After 120 ballots have been recorded, A has 26 votes, B has 18 votes, C has 42 votes, and D has 34 votes.

(a) What is the smallest number of the remaining 30 votes that A must receive to guarantee a win for A? Explain.

(b) What is the smallest number of the remaining 30 votes that C must receive to guarantee a win for C? Explain.

14. An election with 4 candidates (A, B, C, D) and 150 voters is to be decided using the plurality method. After 120 ballots have been recorded, A has 26 votes, B has 18 votes, C has 42 votes, and D has 34 votes.

(a) What is the smallest number of the remaining 30 votes that B must receive to guarantee a win for B? Explain.

(b) What is the smallest number of the remaining 30 votes that D must receive to guarantee a win for D? Explain.

15. Consider an election with 721 voters.

(a) What is the smallest number of votes needed to be a majority candidate?

(b) If there are 5 candidates, what is the smallest number of votes that a plurality candidate could have?

(c) If there are 10 candidates, what is the smallest number of votes that a plurality candidate could have?

16. Consider an election with 1025 voters.

(a) What is the smallest number of votes needed to be a majority candidate?

(b) If there are 4 candidates, what is the smallest number of votes that a plurality candidate could have?

(c) If there are 8 candidates, what is the smallest number of votes that a plurality candidate could have?

C. Borda Count Method

17. This exercise refers to the following preference schedule. (This is the election for Mathematics Department chair discussed in Exercise 3. The candidates are Professors Argand, Brandt, Chavez, Dietz, and Epstein.)

Number of voters	5	3	5	3	2	3
1st choice	A	A	C	D	D	B
2nd choice	B	D	E	C	C	E
3rd choice	C	B	D	B	B	A
4th choice	D	C	A	E	A	C
5th choice	E	E	B	A	E	D

(a) Find the winner of the election under the Borda count method.

(b) Suppose that before the votes are counted, Professor Epstein withdraws from the race. Find the preference schedule for a new election held without Professor Epstein as a candidate, and then find the winner under the Borda count method.

(c) Explain why this election shows that the Borda count method violates the *independence-of-irrelevant-alternatives* criterion.

18. This exercise refers to the following preference schedule. (This is the election for Homecoming Queen discussed in Exercises 6 and 11. The candidates are Alicia, Brandy, Cleo, and Dionne.)

Number of voters	153	102	55	202	108	20	110	160	175	155
1st choice	A	A	A	B	B	B	C	C	D	D
2nd choice	C	B	D	D	C	C	A	B	A	B
3rd choice	B	D	C	A	D	A	D	A	C	C
4th choice	D	C	B	C	A	D	B	D	B	A

(a) Find the winner of the election under the Borda count method. (You will probably want to use a calculator to do the arithmetic.)

(b) Suppose that before the votes are counted, Dionne is found to be ineligible because of her grades. It is decided that Dionne's name will be removed from the original preference schedule. Find the preference schedule when Dionne's name is removed, and then find the winner of this new election under the Borda count method.

(c) Explain why this election shows that the Borda count method violates the *independence-of-irrelevant-alternatives* criterion.

19. The editorial board of *Gourmet* magazine is having an election to choose the "Restaurant of the Year." The candidates are Andre's, Borrelli, Casablanca, Dante,

and Escargot. The preference schedule for the election is given in the following table.

Number of voters	8	7	6	2	1
1st choice	A	D	D	C	E
2nd choice	B	B	B	A	A
3rd choice	C	A	E	B	D
4th choice	D	C	C	D	B
5th choice	E	E	A	E	C

(a) Find the winner under the Borda count method.

(b) Explain why this election shows that the Borda count method violates the *majority* criterion.

(c) Explain why this election shows that the Borda count method violates the *Condorcet* criterion.

20. The members of the Tasmania State University soccer team are having an election to choose the captain of the team from among the four seniors—Anderson, Bergman, Chou, and Delgado. The preference schedule for the election is given in the following table.

Number of voters	4	1	9	8	5
1st choice	A	B	C	A	C
2nd choice	B	A	D	D	D
3rd choice	D	D	A	B	B
4th choice	C	C	B	C	A

(a) Find the winner under the Borda count method.

(b) Explain why this election shows that the Borda count method violates the *majority* criterion.

(c) Explain why this election shows that the Borda count method violates the *Condorcet* criterion.

21. An election is held among four candidates (A, B, C, D). Each column in the following preference schedule shows the percentage of voters voting that way.

Percentage of voters	40%	25%	20%	15%
1st choice	A	C	B	B
2nd choice	D	B	D	A
3rd choice	B	D	A	D
4th choice	C	A	C	C

(a) Assuming there are 200 voters, find the number of voters for each column in the preference schedule and then find the winner of the election under the Borda count method.

(b) Assuming the number of voters is given by $20N$, where N is a positive integer, find the number of voters for each column in the preference schedule (in terms of N) and then find the winner of the election under the Borda count method.

(c) Does your answer in part (b) depend on N? Explain.

22. An election is held among four candidates (A, B, C, D). Each column in the following preference schedule shows the percentage of voters voting that way.

Percentage of voters	48%	24%	16%	12%
1st choice	A	C	B	B
2nd choice	D	B	D	A
3rd choice	B	D	A	D
4th choice	C	A	C	C

(a) Assuming there are 75 voters, find the number of voters for each column in the preference schedule and then find the winner of the election under the Borda count method.

(b) Assuming the number of voters is given by $25N$, where N is a positive integer, find the number of voters for each column in the preference schedule (in terms of N) and then find the winner of the election under the Borda count method.

(c) Does your answer in part (b) depend on N? Explain.

23. An election with four candidates and 50 voters is to be determined using the Borda count method.

(a) What is the maximum number of points a candidate can receive?

(b) What is the minimum number of points a candidate can receive?

24. An election with three candidates and 100 voters is to be determined using the Borda count method.

(a) What is the maximum number of points a candidate can receive?

(b) What is the minimum number of points a candidate can receive?

25. An election is to be decided using the Borda count method. There are four candidates (A, B, C, D) in this election.

(a) How many points are given out by one ballot?

(b) If there are 110 voters in the election, what is the total number of points given out to the candidates?

(c) If candidate A gets 320 points, candidate B gets 290 points, and candidate C gets 180 points, how many points did candidate D get?

26. An election is to be decided using the Borda count method. There are five candidates (A, B, C, D, E) and 40 voters. If candidate A gets 139 points, candidate B gets 121 points, candidate C gets 80 points, and candidate D gets 113 points, who is the winner of the election?

(**Hint:** *If you have trouble with this exercise, try Exercise 25 first.*)

D. Plurality-with-Elimination Method

27. This exercise refers to the following preference schedule. (This is the Mathematics Department Chair election discussed in Exercises 3, 11, and 17.)

Number of voters	5	3	5	3	2	3
1st choice	A	A	C	D	D	B
2nd choice	B	D	E	C	C	E
3rd choice	C	B	D	B	B	A
4th choice	D	C	A	E	A	C
5th choice	E	E	B	A	E	D

(a) Find the winner of the election under the plurality-with-elimination method.

(b) Suppose that before the votes are counted, C withdraws from the race. Find the preference schedule when C is removed and then find the winner of this new election under the plurality-with-elimination method.

(c) The results of (a) and (b) show that the plurality-with-elimination method violates one of the fairness criteria discussed in the chapter. Which one?

28. Find the winner of the election under the plurality-with-elimination method. (This is the Homecoming Queen election discussed in Exercises 6 and 18.)

Number of voters	153	102	55	202	108	20	110	160	175	155
1st choice	A	A	A	B	B	B	C	C	D	D
2nd choice	C	B	D	D	C	C	A	B	A	B
3rd choice	B	D	C	A	D	A	D	A	C	C
4th choice	D	C	B	C	A	D	B	D	B	A

29. This exercise refers to the following preference schedule. (This is the "Restaurant of the Year" election discussed in Exercise 19.)

Number of voters	8	7	6	2	1
1st choice	A	D	D	C	E
2nd choice	B	B	B	A	A
3rd choice	C	A	E	B	D
4th choice	D	C	C	D	B
5th choice	E	E	A	E	C

(a) Find the winner of the election under the plurality-with-elimination method.

(b) Explain why the winner in (a) can be determined in the first round.

(c) Based on your observations in (b), explain why the plurality-with-elimination method satisfies the *majority* criterion.

30. This exercise refers to the following preference schedule. (This is the Tasmania State University soccer captain election discussed in Exercise 20.)

Number of voters	4	1	9	8	5
1st choice	A	B	C	A	C
2nd choice	B	A	D	D	D
3rd choice	D	D	A	B	B
4th choice	C	C	B	C	A

(a) Find the winner of the election under the plurality-with-elimination method.

(b) Explain why the winner in (a) can be determined in the first round.

31. Find the winner of the election under the plurality-with-elimination method. (This is the same election as in Exercise 21.)

Percentage of voters	40%	25%	20%	15%
1st choice	A	C	B	B
2nd choice	D	B	D	A
3rd choice	B	D	A	D
4th choice	C	A	C	C

32. Find the winner of the election under the plurality-with-elimination method. (This is the same election as in Exercise 22.)

Percentage of voters	48%	24%	16%	12%
1st choice	A	C	B	B
2nd choice	D	B	D	A
3rd choice	B	D	A	D
4th choice	C	A	C	C

33. The Professional Surfing Association executive committee is having an election to choose the location of the next QEST (Quicksilver Extreme Surfing Tournament). The choices are A (Año Nuevo, California), B (Banzai Pipeline, Hawaii), C (Cloudbreak, Fiji), and D (Dungeons, South Africa). The preference schedule for the election is given in the following table.

Number of voters	10	6	5	4	2
1st choice	A	B	B	C	D
2nd choice	C	D	C	A	C
3rd choice	B	C	A	D	B
4th choice	D	A	D	B	A

(a) Find the winner of the election under the plurality-with-elimination method.

(b) There is a Condorcet candidate in this election. Find it.

(c) The results of (a) and (b) show that the plurality-with-elimination method violates one of the fairness criteria discussed in the chapter. Which one?

34. The 17 members of a small Iowa precinct are preparing for the Iowa caucuses held in January. The candidates running for president this year are Anderson, Buford, Clinton, and Dorfman. The results of a straw poll taken prior to the official vote are given in the following preference schedule.

Number of voters	4	5	6	2
1st choice	A	B	C	A
2nd choice	D	C	A	C
3rd choice	B	A	D	D
4th choice	C	D	B	B

(a) Find the winner of the straw poll under the plurality-with-elimination method.

(b) In the official vote, everyone votes the same as in the straw poll except for the two voters in the last column of the table—they switch their votes and move Clinton ahead of Anderson in their ballots. Find the winner of the official vote under the plurality-with-elimination method.

(c) The results of (a) and (b) show that the plurality-with-elimination method violates one of the fairness criteria discussed in the chapter. Which one?

E. Pairwise Comparisons Method

35. The 26 members of the Tasmania State University chess team are holding an election to choose a senior captain. The five candidates are Alberto, Bill, Chad, Dora, and Eunice. The results of the election are shown in the following preference schedule.

Number of voters	8	6	5	5	2
1st choice	C	A	E	D	D
2nd choice	B	E	C	C	A
3rd choice	A	D	D	A	E
4th choice	D	B	B	E	B
5th choice	E	C	A	B	C

(a) Find the winner of the election under the method of pairwise comparisons.

(b) Just after the election, it is discovered that Alberto has failed Math 101 and is therefore ineligible to be captain of the chess team. Find the preference schedule for a recount without Alberto, and then find the winner of this recount under the method of pairwise comparisons.

(c) The results of (a) and (b) show that the method of pairwise comparisons violates one of the fairness criteria discussed in the chapter. Which one?

36. Find the winner of the election under the method of pairwise comparisons. (This is the final exam election discussed in Exercises 4 and 12.)

Number of voters	3	4	9	9	2	5	8	3	12
1st choice	A	A	A	B	B	B	C	C	D
2nd choice	B	B	C	C	A	C	D	A	C
3rd choice	C	D	B	D	C	A	B	D	A
4th choice	D	C	D	A	D	D	A	B	B

37. Find the winner of the election under the method of pairwise comparisons. (This is the Math Department Chair election discussed in Exercises 3, 17, and 27.)

Number of voters	5	3	5	3	2	3
1st choice	A	A	C	D	D	B
2nd choice	B	D	E	C	C	E
3rd choice	C	B	D	B	B	A
4th choice	D	C	A	E	A	C
5th choice	E	E	B	A	E	D

38. Find the winner of the election under the method of pairwise comparisons. (This is the "Restaurant of the Year" election discussed in Exercises 19 and 29.)

Number of voters	8	7	6	2	1
1st choice	A	D	D	C	E
2nd choice	B	B	B	A	A
3rd choice	C	A	E	B	D
4th choice	D	C	C	D	B
5th choice	E	E	A	E	C

39. An election with five candidates A, B, C, D, and E is held under the method of pairwise comparisons. Partial results of the pairwise comparisons are as follows: A wins two pairwise comparisons, B wins two and ties one, C wins one, and D wins one and ties one. Find the winner of the election.

40. An election with six candidates A, B, C, D, E, and F is held under the method of pairwise comparisons. Partial results of the pairwise comparisons are as follows: A wins three pairwise comparisons, B and C both win two, D and E both win two and tie one.

(a) Find the winner of the election.

(b) Give the result of the pairwise comparison between D and E.

F. Ranking Methods

41. For the election given by the following preference schedule,

Number of voters	4	1	9	8	5
1st choice	A	B	C	A	D
2nd choice	C	A	D	B	C
3rd choice	B	D	A	D	B
4th choice	D	C	B	C	A

(a) rank the candidates using the extended plurality method.

(b) rank the candidates using the extended Borda count method.

(c) rank the candidates using the extended plurality-with-elimination method.

(d) rank the candidates using the extended pairwise comparisons method.

42. For the election given by the following preference schedule (this is the Professional Surfing Association election discussed in Exercise 33),

Number of voters	10	6	5	4	2
1st choice	A	B	B	C	D
2nd choice	C	D	C	A	C
3rd choice	B	C	A	D	B
4th choice	D	A	D	B	A

(a) rank the candidates using the extended plurality method.

(b) rank the candidates using the extended Borda count method.

(c) rank the candidates using the extended plurality-with-elimination method.

(d) rank the candidates using the extended pairwise comparisons method.

43. For the election given by the following preference schedule (this is the election discussed in Exercise 21),

Percentage of voters	40%	25%	20%	15%
1st choice	A	C	B	B
2nd choice	D	B	D	A
3rd choice	B	D	A	D
4th choice	C	A	C	C

(a) rank the candidates using the extended plurality method.

(b) rank the candidates using the extended Borda count method.

(c) rank the candidates using the extended plurality-with-elimination method.

(d) rank the candidates using the extended pairwise comparisons method.

44. For the election given by the following preference schedule (this is the election discussed in Exercise 22),

Percentage of voters	48%	24%	16%	12%
1st choice	A	C	B	B
2nd choice	D	B	D	A
3rd choice	B	D	A	D
4th choice	C	A	C	C

(a) rank the candidates using the extended plurality method.

(b) rank the candidates using the extended Borda count method.

(c) rank the candidates using the extended plurality-with-elimination method.

(d) rank the candidates using the extended pairwise comparisons method.

45. Use the recursive Borda count method to rank the candidates in the Math Club election (this is the election introduced in Example 1.1).

46. Use the recursive pairwise comparisons method to rank the candidates in the Math Club election.

47. For the election given by the following preference schedule (this is the election discussed in Exercise 41),

Number of voters	4	1	9	8	5
1st choice	A	B	C	A	D
2nd choice	C	A	D	B	C
3rd choice	B	D	A	D	B
4th choice	D	C	B	C	A

(a) rank the candidates using the recursive plurality method.

(b) rank the candidates using the recursive Borda count method.

(c) rank the candidates using the recursive plurality-with-elimination method.

(d) rank the candidates using the recursive pairwise comparisons method.

48. For the election given by the following preference schedule (this is the election discussed in Exercise 42),

Number of voters	10	6	5	4	2
1st choice	A	B	B	C	D
2nd choice	C	D	C	A	C
3rd choice	B	C	A	D	B
4th choice	D	A	D	B	A

(a) rank the candidates using the recursive plurality method.

(b) rank the candidates using the recursive Borda count method.

(c) rank the candidates using the recursive plurality-with-elimination method.

(d) rank the candidates using the recursive pairwise comparisons method.

49. For the election given by the following preference schedule (this is the election discussed in Exercise 43),

Percentage of voters	40%	25%	20%	15%
1st choice	A	C	B	B
2nd choice	D	B	D	A
3rd choice	B	D	A	D
4th choice	C	A	C	C

(a) rank the candidates using the recursive plurality method.

(b) rank the candidates using the recursive Borda count method.

(c) rank the candidates using the recursive plurality-with-elimination method.

(d) rank the candidates using the recursive pairwise comparisons method.

50. For the election given by the following preference schedule (this is the election discussed in Exercise 44),

Percentage of voters	48%	24%	16%	12%
1st choice	A	C	B	B
2nd choice	D	B	D	A
3rd choice	B	D	A	D
4th choice	C	A	C	C

(a) rank the candidates using the recursive plurality method.

(b) rank the candidates using the recursive Borda count method.

(c) rank the candidates using the recursive plurality-with-elimination method.

(d) rank the candidates using the recursive pairwise comparisons method.

G. Miscellaneous

Find the sums in Exercises 51–54.

51. $1 + 2 + 3 + \cdots + 498 + 499 + 500$

52. $1 + 2 + 3 + \cdots + 3218 + 3219 + 3220$

53. $501 + 502 + 503 + \cdots + 3218 + 3219 + 3220$
(***Hint:*** *Do Exercises 51 and 52 first.*)

54. $1801 + 1802 + 1803 + \cdots + 8843 + 8844 + 8845$
(***Hint:*** *Do Exercise 53 first.*)

55. In an election with 15 candidates,

(a) how many pairwise comparisons are there?

(b) if it takes one minute to calculate a pairwise comparison, approximately how long would it take to calculate the results of the election using the method of pairwise comparisons?

56. Suppose that 21 players sign up for a round-robin Ping-Pong tournament.

(a) How many matches must be scheduled for the entire tournament?

(b) If six matches can be scheduled per hour and the tournament hall is available for 12 hours each day, for how many days would the tournament hall have to be reserved?

57. In an election with three candidates, what is the maximum number of columns possible in the preference schedule?

58. In an election with four candidates,

(a) what is the maximum number of columns possible in the preference schedule?

(b) how many different ways are there to choose the first and second choices?

59. Consider the election given by the following preference schedule.

Number of voters	7	4	2
1st choice	A	B	D
2nd choice	B	D	A
3rd choice	C	C	C
4th choice	D	A	B

(a) Find the Condorcet candidate in this election.

(b) Find the winner of this election under the Borda count method.

(c) Suppose that C drops out of the race. Find the winner under the Borda count method when C is removed from the preference schedule.

(d) The results of (a), (b), and (c) show that the Borda count method violates several of the fairness criteria discussed in the chapter. Which ones? Explain.

60. Consider the election given by the following preference schedule. (This is the Professional Surfing Association election discussed in Exercises 33 and 42.)

Number of voters	10	6	5	4	2
1st choice	A	B	B	C	D
2nd choice	C	D	C	A	C
3rd choice	B	C	A	D	B
4th choice	D	A	D	B	A

(a) Find the Condorcet candidate in this election.

(b) Find the winner of this election under the plurality-with-elimination method.

(c) Suppose that D drops out of the race. Find the winner under the plurality-with-elimination method when D is removed from the preference schedule.

(d) The results of (a), (b), and (c) show that the plurality-with-elimination method violates several of the fairness criteria discussed in the chapter. Which ones? Explain.

JOGGING

61. Two-candidate elections. Explain why when there are only two candidates, the four voting methods we discussed in this chapter give the same winner and the winner is determined by straight majority. (Assume there are no ties.)

62. Give an example of an election with four candidates (A, B, C, and D) satisfying the following: (i) No candidate has a majority of the first-place votes; (ii) C is a Condorcet candidate but has no first-place votes; (iii) B is the winner under the Borda count method; (iv) A is the winner under the plurality method.

63. Explain why the plurality method satisfies the monotonicity criterion.

64. Explain why the Borda count method satisfies the monotonicity criterion.

65. An election is held using the method of pairwise comparisons. Suppose that X is the winner of the election under the method of pairwise comparisons, but due to an election irregularity, there is a reelection. In the reelection, the only changes are changes that favor X and only X (specifically, you can interpret this to mean that the only changes involve voters that move X up in their ballot without altering the relative order of any of the other candidates). Explain why candidate X must be the winner of the reelection (i.e., the method of pairwise comparisons satisfies the monotonicity criterion).

66. Equivalent Borda count (Variation 1). The following simple variation of the Borda count method described in the chapter is sometimes used: A first place is worth $N - 1$ points, second place is worth $N - 2$ points, ..., last place is worth 0 points (where N is the number of candidates). The candidate with the most points is the winner.

(a) Suppose that a candidate gets p points using the Borda count as originally described in the chapter and q points under this variation. Explain why if k is the number of voters, then $p = q + k$.

(b) Explain why this variation is equivalent to the original Borda count described in the chapter (i.e., it produces exactly the same election results).

67. Equivalent Borda count (Variation 2). Another commonly used variation of the Borda count method described in the chapter is the following: A first place is worth 1 point, second place is worth 2 points, ..., last place is worth N points (where N is the number of candidates). The candidate with the fewest points is the winner, second fewest points is second, and so on.

(a) Suppose that a candidate gets p points using the Borda count as originally described in the chapter and r points under this variation. Explain why if k is the number of voters, then $p + r = k(N + 1)$.

(b) Explain why this variation is equivalent to the original Borda count described in the chapter (i.e., it produces exactly the same election results).

68. The average ranking. The average ranking of a candidate is obtained by taking the place of the candidate on each of the ballots, adding these numbers, and dividing by the number of ballots. Explain why the candidate with the best average ranking is the Borda winner.

69. The 2004 college football AP Poll. The following table shows the top three college football teams in the Associated Press poll at the end of the regular 2004 college football season (prior to the January bowl games). This poll is based on the votes of 65 sportswriters, each one of whom ranks the top 25 teams. (The remaining 22 teams are not shown because they are irrelevant to this exercise.) A team gets 25 points for each first-place vote, 24 points for each second-place vote, 23 points for each third-place vote, and so on.

Team	Points	Number of first-place votes
1. USC	1610	52
2. Oklahoma	1540	7
3. Auburn	1530	6

(a) Based on the information given in the table, it is possible to conclude that all 65 writers had USC, Oklahoma, and Auburn in some order as their top three choices. Explain why this is true.

(b) Find the number of second- and third-place votes for each of the three teams.

70. The women's college basketball All-American selection. Each year, a group of Associated Press sportswriters selects a women's NCAA basketball All-American team using a variation of the Borda count method. Each voter gets to select five players for first-team, five players for second-team, and five players for third-team All-American. In the 2003–2004 selection, the top three candidates and the number of votes they received on the 47 ballots submitted were as follows.

Player	Total Points	Votes
Alana Beard	235	47 (1st team)
Diana Taurasi	231	45 (1st team); 2 (2nd team)
Nicole Ohlde	207	36 (1st team); 8 (2nd team); 3 (3rd team)

(a) Based on the information given in the table, determine how many points are awarded for each first-team, second-team, and third-team vote in this election. (You may assume that these numbers are different positive integers.)

(b) The 2003–2004 All-American team also included Nicole Powell of Stanford. Nicole received 183 total points and 25 first-team votes. Assuming that Nicole was on each of the 47 ballots cast, how many second-team votes did she receive? How many third-team votes did she receive?

71. The 2003–2004 NBA Rookie of the Year vote. Each year, a panel of broadcasters and sportswriters selects an NBA rookie of the year using a variation of the Borda count method. The following table shows the results of the balloting for the 2003–2004 season.

Player	1st place	2nd place	3rd place	Total points
LeBron James	78	39	1	508
Carmelo Anthony	40	76	2	430
Dwayne Wade	0	3	108	117

Based on the information above, determine how many points are awarded for each first-place, second-place, and third-place vote in this election. (You may assume that these numbers are different positive integers.)

72. Plurality with a runoff. This is a simple variation of the plurality-with-elimination method. Here, if a candidate has a majority of the first-place votes, then that candidate wins the election; otherwise we eliminate *all* candidates except the two with the most first-place votes. The winner is chosen between these two by recounting the votes in the usual way.

(a) Use the Math Club election (Example 1.1) to show that plurality with a runoff can produce a different outcome than plurality with elimination.

(b) Give an example that shows that plurality with a runoff violates the monotonicity criterion.

(c) Give an example that shows that plurality with a runoff violates the Condorcet criterion.

73. The Coombs method. This method is just like the plurality-with-elimination method except that in each round we eliminate the candidate with the *largest number of last-place votes* (instead of the one with the fewest first-place votes).

(a) Find the winner of the Math Club election (Example 1.1) using the Coombs method.

(b) Give an example showing that the Coombs method violates the Condorcet criterion.

(c) Give an example showing that the Coombs method violates the monotonicity criterion.

RUNNING

74. Consider the following fairness criterion: *If a majority of the voters have candidate X ranked last, then candidate X should not be a winner of the election.*

(a) Give an example to show that the plurality method violates this criterion.

(b) Give an example to show that the plurality-with-elimination method violates this criterion.

(c) Explain why the method of pairwise comparisons satisfies this criterion.

(d) Explain why the Borda count method satisfies this criterion.

75. Suppose the following was proposed as a fairness criterion: *If a majority of the voters prefer candidate X to candidate Y, then the results of the election should have X ranked above Y.* Give an example to show that all four of the extended voting methods discussed in the chapter can violate this criterion.

(*Hint: Consider an example with no Condorcet candidate.*)

76. The Pareto criterion. The following fairness criterion was proposed by the Italian economist Vilfredo Pareto (1848–1923): *If every voter prefers candidate X to candidate Y, then X should be ranked above Y.*

(a) Explain why the extended Borda count method satisfies the Pareto criterion.

(b) Explain why the extended pairwise comparisons method satisfies the Pareto criterion.

77. Explain why all four of the recursive ranking methods (recursive plurality, recursive plurality with elimination, recursive Borda count, and recursive pairwise comparisons) satisfy the Pareto criterion as defined in Exercise 76.

78. The Condorcet loser criterion. *If there is a candidate that loses in a one-to-one comparison to each of the other candidates, then that candidate should not be the winner of the election.* (This fairness criterion is a sort of mirror image of the regular Condorcet criterion.)

(a) Give an example that shows that the plurality method violates the Condorcet loser criterion.

(b) Give an example that shows that the plurality-with-elimination method violates the Condorcet loser criterion.

(c) Explain why the Borda count method satisfies the Condorcet loser criterion.

79. Consider a variation of the Borda count method in which a first-place vote in an election with N candidates is worth F points (where $F > N$) and all other places in the ballot are the same as in the ordinary Borda count: $N - 1$ points for second place, $N - 2$ points for third place, ..., 1 point for last place. By choosing F large enough, we can make this variation of the Borda count method satisfy the majority criterion. Find the smallest value of F (expressed in terms if N) for which this happens.

80. Suppose that the candidates in an election are ranked using the extended pairwise comparisons method, and suppose that there is no Condorcet candidate. Explain why if the number of voters is odd then there must be a tie somewhere in the rankings. Explain why the result does not have to be true with an even number of voters.

Projects and Papers

A. Ballots, Ballots, Ballots!

In this chapter we discussed elections in which the voters cast their votes by means of linear preference ballots. There are many other types of ballots used in real-life elections, ranging from the simple (winner only) to the exotic (each voter has a fixed number of points to divide among the candidates any way he or she sees fit). In this project you are to research other types of ballots; how, where, and when they are used; and what are the arguments for and against their use.

B. Sequential Voting

Sequential voting is a voting method used by legislative bodies and committees to choose one among a list of alternatives. In sequential voting the alternatives are presented in some order A, B, C, D, and so on. The voters then choose between A and B, the winner is then matched against C, the winner of that vote against D, and so on. In sequential voting, the *agenda* (the order in which the options are presented) can have a critical impact on the outcome of the vote. Write a research paper on sequential voting, its history, and its use, paying particular attention to the issue of the agenda and its impact on the outcome. Illustrate your points with examples that *you* have made up.

C. Instant Runoff Voting

Imagine you are a political activist in your community. The city council is having hearings to decide if the *method of instant runoff voting* (plurality with elimination) should be adopted in your city. Stake out a position for or against instant runoff voting, and prepare a brief to present to the city council that justifies that position. To make an effective case, your argument should include mathematical, economic, political, and social considerations. (Remember that your city council members are not as well versed as you are on the mathematical aspects of elections. Part of your job is to educate them.)

D. Manipulability of an Election

A voter is said to *manipulate* the results of an election if he or she is able to change the outcome of an election by voting insincerely. (We touched briefly on *insincere voting* in our discussion of the plurality method, but this is a big subject.) In this project you should research the issue of manipulability of elections and its political and social costs. Your research should include the mathematical aspects of manipulability, culminating with a description and interpretation of the *Gibbard-Satterthwaite manipulability theorem* concerning the impossibility of voting methods that are immune to manipulation. (A suggested start for this project is reference 20 at the end of the chapter.)

E. The 2000 Presidential Election and the Florida Vote

The unusual circumstances surrounding the 2000 presidential election and the Florida vote are a low point in American electoral history. Write an analysis paper on the 2000 Florida vote, paying particular attention to what went wrong and how a similar situation can be prevented in the future. You should touch on technology issues (outdated and inaccurate vote tallying methods and equipment, poorly designed ballots, etc.), political issues (the two-party system, the Electoral College, etc.), and, as much as possible, on issues related to concepts from this chapter (can presidential elections be improved by changing to preference ballots, using a different voting method, etc.).

F. Short Story

Write a fictional short story using an election as the backdrop. Weave into the dramatic structure of the story elements and themes from this chapter (fairness, manipulation, monotonicity, and independence of irrelevant alternatives all lend themselves to good drama). Be creative and have fun.

Appendix 1 Breaking Ties

By and large, most of the examples given in the chapter were carefully chosen to avoid tied winners, but of course in the real world ties are bound to occur.

In this appendix we will discuss very briefly the problem of how to break ties when necessary. Tie-breaking methods can raise some fairly complex issues, and our purpose here is not to study such methods in great detail but rather to make the reader aware of the problem and give some inkling as to possible ways to deal with it.

For starters, consider the election with preference schedule shown in Table A-1. If we look at this preference schedule carefully, we can see that there is complete *symmetry* in the positions of the three candidates. Essentially, this means that we could interchange the names of the candidates and the preference schedule would not change. Given the complete symmetry of the preference schedule, it is clear that no rational voting method could choose one candidate as the winner over the other two. In this situation, a tie is inevitable regardless of the voting method used. We call this kind of tie an **essential tie**. We cannot break essential ties using a rational tie-breaking procedure and must instead rely on some sort of outside intervention such as chance (draw straws, roll the dice, play rock-paper-scissors, etc.), a third party (the judge, mom, etc.), or even some outside factor (experience, age, etc.).

TABLE A-1 A Three-Way Essential Tie

Number of voters	7	7	7
1st choice	A	B	C
2nd choice	B	C	A
3rd choice	C	A	B

Most ties are not essential ties, and we can often break them in more rational ways: either by implementing some tie-breaking rule or by using a different voting method to break the tie. To illustrate some of these ideas, let's consider as an example the election with preference schedule shown in Table A-2.

TABLE A-2 A Tie That Could Be Broken

Number of voters	5	3	5	3	2	4
1st choice	A	A	C	D	D	B
2nd choice	B	B	E	C	C	E
3rd choice	C	D	D	B	B	A
4th choice	D	C	A	E	A	C
5th choice	E	E	B	A	E	D

If we decide this election using the method of pairwise comparisons, we have the following:

A versus B: 13 votes to 9 votes. A gets 1 point.

A versus C: 12 votes to 10 votes. A gets 1 point.

A versus D: 12 votes to 10 votes. A gets 1 point.

A versus E: 10 votes to 12 votes. E gets 1 point.

B versus C: 12 votes to 10 votes. B gets 1 point.

B versus D: 12 votes to 10 votes. B gets 1 point.

B versus E: 17 votes to 5 votes. B gets 1 point.

C versus D: 14 votes to 8 votes. C gets 1 point.

C versus E: 18 votes to 4 votes. C gets 1 point.

D versus E: 13 votes to 9 votes. D gets 1 point.

In this election A and B, with three wins each, tie for first place. How could we break this tie? Here is just a sampler of the many possible ways:

- Use the results of the pairwise comparison between the winners. In the preceding example, since A beats B 13 votes to 9, the tie would be broken in favor of A.

- Use the total point differentials. For example, since A beats B 13 to 9, the point differential for A is +4, and since A lost to E 10 to 12, the point differential for A is −2. Computing the total point differentials for A gives $4 + 2 + 2 - 2 = 6$. Likewise, the total point differential for B is $2 + 2 + 12 - 4 = 12$. In this case the point differentials favor B, so B would be declared the winner.

- Use first-place votes. In the example, A has 8 and B has 4. With this method, the winner would be A.

- Use Borda count points to choose between the two winners. Here

  A has $5 \times 8 + 3 \times 4 + 2 \times 7 + 1 \times 3 = 69$ points;

  B has $5 \times 4 + 4 \times 8 + 3 \times 5 + 1 \times 5 = 72$ points;

and the tie would be broken in favor of B.

In Groveland, Florida, the 2004 election for city council between G. P. Sloan and Richard Flynn ended in a tie. By mutual agreement, the tie was broken by a coin toss. Sloan called heads and lost.

By now we should not be at all surprised that different tie-breaking methods produce different winners and that there is no single *right* method for breaking ties. In retrospect, flipping a coin might not be such a bad idea! (In fact, it's happened before.)

Appendix 2 A Sampler of Elections in the Real World

Olympic Games. When a city is chosen to host the Olympic Games it's a big deal. In addition to prestige, the award can have enormous social, economic, and political impact for the winning city, and it goes without saying that the choice always generates a fair amount of controversy. The long, drawn-out competition (cities start their planning years in advance) culminates with an election very much like some of the ones described in this chapter. (See Example 1.10.) The voters in this election are the members of the International Olympic Committee, and the *voting method* used to select the winner is the *plurality-with-elimination* method with a minor twist: Instead of indicating their preferences all at once, the voters let their preferences be known one round at a time.

On July 6, 2005, the members of the International Olympic Committee met in Singapore to vote on the selection of the host city for the 2012 Summer Olympics. The finalists were London, Paris, Madrid, New York, and Moscow. Because of the prestige of the bidding cities, the competition was particularly fierce, and the stakes were even higher than usual. At the end, London was the winner. Here is how the voting went, round by round.

- **Round 1.**

	London	Paris	Madrid	New York	Moscow	Abstain
Votes	22	21	20	19	15	7

Moscow eliminated in round 1.

- **Round 2.**

	London	Paris	Madrid	New York	Abstain
Votes	27	25	32	16	4

New York eliminated in round 2.

- **Round 3.**

	London	Paris	Madrid	Abstain
Votes	39	33	31	1

Madrid eliminated in round 3.

- **Round 4.**

	London	Paris
Votes	54	50

Paris eliminated in round 4. London gets the 2012 Olympics.

The Academy Awards. The Academy of Motion Picture Arts and Sciences gives its annual Academy Awards ("Oscars") for various achievements in connection with motion pictures (best picture, best director, best actress, etc.). Eligible members of the Academy elect a winner in each category. The election process varies slightly from award to award and is quite complicated. For the sake of brevity we will describe the election process for best picture. (The process is almost identical for each of the major awards.) The election takes place in two stages: (1) the nomination stage, in which the five top pictures are nominated, and (2) the final balloting for the winner.

We describe the second stage first because it is so simple: Once the five top pictures are nominated, each eligible member of the Academy votes for one candidate, and the winner is chosen using the plurality method. Because the number of voters is large (somewhere in the neighborhood of 5000), ties are not likely to occur, but if they do, they are not broken. Thus, it is possible for two candidates to share an award, as in 1968 when Katherine Hepburn and Barbra Streisand shared the award for Best Actress.

The process for selecting the five nominations is considerably more complicated and is based on a voting method called **single transferable voting**. Each eligible member of the Academy submits a preference ballot with the names of his or her top five choices ranked from first to fifth. Based on the total number of valid ballots submitted, the minimum number of votes needed to get a nomination (called the **quota**) is established, and any picture with enough first-place votes to make the quota is automatically nominated.

The quota is always chosen to be a number over one-sixth (16.66%) but not more than one-fifth (20%) of the total number of valid ballots cast. (Setting the quota this way ensures that it is impossible for six or more pictures to get automatic nominations.) While in theory it is possible for five pictures to make the quota right off the bat and get an automatic nomination (in which case the nomination process is over), this has never happened in practice. In fact, what usually happens is that there are no pictures that make the quota automatically. Then the picture with the fewest first-place votes (say X) is eliminated, and on all the ballots that originally had X as the first choice, X's name is crossed off the top and all the other pictures are moved up one spot. The ballots are then counted again. If there are still no pictures that make the quota, the process of elimination is repeated. Eventually, there will be one or more pictures that make the quota and are nominated.

The moment that one or more pictures are nominated, there is a new twist: Nominated pictures "give back" to the other pictures still in the running (not nominated but not eliminated either) their "surplus" votes. This process of giving back votes (called a **transfer**) is best illustrated with an imaginary example. Suppose that the quota is 800 (a nice, round number) and at some point a picture (say Z) gets 1000 first-place votes, enough to get itself nominated. Now there are $1000 - 800 = 200$ surplus votes for Z, votes that Z doesn't really need. For this reason the 200 surplus votes are taken away from Z and divided fairly among the second-place candidates on the 1000 ballots cast for Z. The way this is done may seem a little bizarre, but it makes perfectly good sense. Since there are 200 surplus votes to be divided into 1000 equal shares, each second-place vote on the 1000 ballots cast for Z is worth $\frac{200}{1000} = \frac{1}{5}$ of a vote. While one-fifth of a vote may not seem like much, enough of these fractional votes can make a difference and help some other picture or pictures make the quota. If that's the case, then once again the surplus or surpluses are transferred back to the remaining pictures following the procedure described previously; otherwise, the process of elimination is started up again. Eventually, after several possible cycles of eliminations and transfers, five pictures get enough votes to make the quota and be nominated, and the process is over.

The method of single transferable voting is not unique to the Academy Awards and is used in both Australia and Ireland to elect members of the Senate.

Corporate Boards of Directors. In most corporations and professional societies, the members of the Board of Directors are elected by a method called **approval voting**. In approval voting, a voter does not cast a preferential ballot but rather votes for as many candidates as he or she wants. Each of these votes is simply a yes vote for the candidate, and it means that the voter approves of that candidate. The candidate with the most approval votes wins the election. In the last few years a case has been made suggesting that for political elections, approval voting is an improvement over the more traditional voting methods.

Table A-3 shows an example of a small hypothetical election based on approval voting. The candidates are A, B, and C, and the voters are labeled 1 through 8.

TABLE A-3 An Election Based on Approval Voting								
	Voters							
	#1	#2	#3	#4	#5	#6	#7	#8
A	Yes		Yes	Yes	Yes		Yes	Yes
B			Yes		Yes	Yes		
C	Yes				Yes	Yes	Yes	

The results of this election are as follows: winner, *A* (6 approval votes); second place, *C* (4 approval votes); last place, *B* (3 approval votes). Note that a voter can cast anywhere from no approval votes at all (voter #2) to approval votes for all the candidates (voter #5). It is somewhat ironic that the impact of their votes is exactly the same.

References and Further Readings

1. Arrow, Kenneth J., *Social Choice and Individual Values*. New York: John Wiley & Sons, Inc., 1963.

2. Baker, Keith M., *Condorcet: From Natural Philosophy to Social Mathematics.* Chicago: University of Chicago Press, 1975.

3. Brams, Steven J., and Peter C. Fishburn, *Approval Voting*. Boston: Birkhäuser, 1982.

4. Dummett, Michael A., *Voting Procedures*. New York: Oxford University Press, 1984.

5. Farquharson, Robin, *Theory of Voting*. New Haven, CT: Yale University Press, 1969.

6. Fishburn, Peter C., and Steven J. Brams, "Paradoxes of Preferential Voting," *Mathematics Magazine*, 56 (1983), 207–214.

7. Gardner, Martin, "Mathematical Games (From Counting Votes to Making Votes Count: The Mathematics of Elections)," *Scientific American*, 243 (October 1980), 16–26.

8. Hill, Steven, *Fixing Elections: The Failure of America's Winner Take All Politics*. New York: Routledge, 2003.

9. Kelly, Jerry S., *Arrow Impossibility Theorems*. New York: Academic Press, 1978.

10. Merrill, Samuel, *Making Multicandidate Elections More Democratic*. Princeton, NJ: Princeton University Press, 1988.

11. Niemi, Richard G., and William H. Riker, "The Choice of Voting Systems," *Scientific American*, 234 (June 1976), 21–27.

12. Nurmi, Hannu, *Comparing Voting Systems*. Dordretch, Holland: D. Reidel, 1987.

13. Saari, Donald G., *Basic Geometry of Voting*. New York: Springer-Verlag, 1995.

14. Saari, Donald G., *Chaotic Elections: A Mathematician Looks at Voting.* Providence, RI: American Mathematical Society, 2001.

15. Saari, Donald G., *Decisions and Elections: Explaining the Unexpected*. Cambridge, U.K.: Cambridge University Press, 2001.

16. Saari, Donald G., "Suppose You Want to Vote Strategically," *Math Horizons*, (Nov. 2000), 5–10.

17. Saari, Donald G., and F. Valognes, "Geometry, Voting, and Paradoxes," *Mathematics Magazine*, 71 (Oct. 1998), 243–259.

18. Straffin, Philip D., Jr., *Topics in the Theory of Voting*, UMAP Expository Monograph. Boston: Birkhäuser, 1980.

19. Taylor, Alan, *Mathematics and Politics: Strategy, Voting, Power and Proof*. New York: Springer-Verlag, 1995.

20. Taylor, Alan, "The Manipulability of Voting Systems," *The American Mathematical Monthly,* 109 (April 2002), 321–337.

2

Weighted Voting Systems

The Power Game

In a democracy we take many things for granted, not the least of which is the idea that we are all equal. When it comes to voting rights, the democratic ideal of equality translates into the principle of *one person–one vote*. But is the principle of *one person–one vote* always justified? Must it apply when the *voters* are something other than individuals, such as organizations, states, and even countries?

In a diverse society, it is in the very nature of things that voters—be they individuals or institutions—are not equal, and sometimes it is actually desirable to recognize their differences by giving them different amounts of say over the outcome of an election. What we are talking about here is the exact opposite of the principle of *one voter–one vote*, a principle best described as *one voter–x votes*, and more formally known as **weighted voting**.

One of the best known and most controversial examples of weighted voting is the Electoral College, that uniquely American institution bequeathed to the nation by the Founding Fathers through Article II, Section 1 of the U.S. Constitution. In the Electoral College, each of the 50 states controls a number of votes equal to its number of Representatives plus Senators in Congress (in addition, the District of Columbia controls 3 votes). At one end of the spectrum is California, an electoral heavyweight with 55 electoral votes; at the other end of the spectrum are the little guys (Alaska, Montana, North Dakota, etc.) with a meager 3 electoral votes each. The other states fall somewhere in between. (See the appendix at the end of this chapter for full details.) The 2000 and 2004 presidential elections brought to the surface, in a very dramatic way, the vagaries and complexities of the Electoral College system, and, in particular, the pivotal role that a single state (Florida in 2000, Ohio in 2004) can have in the final outcome of the election.

Even in democratic societies, weighted voting is quite common—we can find examples of weighted voting in shareholder elections, in business partnerships, in international legislative bodies, such as the United Nations Security Council and the European Union Council of Ministers, and, of course, in American presidential elections. In this chapter we will look at the mathematics behind weighted voting. We will start by introducing and illustrating the concept of a *weighted voting system*, a simple mathematical formalism that will allow us to describe most weighted voting situations (Section 2.1). In the remaining sections we will focus on two different mathematical measures of a voter's power in a weighted voting system. Sections 2.2 and 2.3 discuss what is known as the Banzhaf power index of a voter; Sections 2.4 and 2.5 discuss what is known as the Shapley-Shubik power index of a voter. In so doing we will be introduced to several basic but important mathematical ideas—*coalitions*, *sequential coalitions*, and *factorials*. Enjoy the ride!

Each State shall appoint a Number of Electors equal to the whole Number of Senators and Representatives to which the state may be entitled in the Congress.

Article II, Section 1

2.1 Weighted Voting Systems

We start this section by introducing some terminology. We will use the term **weighted voting system** to describe any formal voting arrangement in which voters are not necessarily equal in terms of the number of votes they control. To keep things simple, we will only consider *yes-no* votes, generally known as **motions** (a vote between two candidates or alternatives can be rephrased as a *yes-no* vote and thus is a motion).

Every weighted voting system is characterized by three elements:

- **The players.** The voters in a weighted voting system are called *players*. (From now on we will stick to the usual convention of using "voters" when we are dealing with a *one person–one vote* situation as in Chapter 1, and "players" in the case of a weighted voting system.) Note that in a weighted voting system the players may be individuals, but they may also be institutions, agencies, corporations, and even nations. We will let N denote the number of players and $P_1, P_2, \ldots, P_N$ the names of the players. (Think of P_1 as short for "player 1," P_2 as short for "player 2," etc.—it is a little less personal but a lot more convenient than using Archie, Betty, Charlie, etc.)

- **The weights.** The hallmark of a weighted voting system is that each player controls a certain number of votes, called the *weight* of the player. We will assume that the weights are all positive integers, and we will let $w_1, w_2, \ldots, w_N$ represent the weights of $P_1, P_2, \ldots, P_N$, respectively. (A weighted voting system is only interesting when the weights are not all the same, but in principle we do not preclude the possibility that the weights are all equal. Painted with this broad brush, weighted voting includes ordinary voting as a special case—just set all the weights equal to 1.)

- **The quota.** The *quota* is the minimum number of votes needed to pass a motion. We will use q to denote the quota. It is important to realize that the quota q can be something other than a simple majority of the votes. There are many voting situations in which the requirement to pass a motion is set higher than just a majority. In the U.S. Senate, for example, it takes a simple majority to pass an ordinary law, but it takes a minimum of 60 votes to stop a filibuster, and it takes a minimum of two-thirds of the votes to override a presidential veto. In other organizations the rules may stipulate that three-fourths (75%) of the votes are needed, or four-fifths (80%), or even unanimity (100%). In fact, any integer q is a possible choice for the quota as long as (i) it is more than 50% of the total number of votes, and (ii) it is not more than 100% of the votes.

Notation and Examples

Every weighted voting system can be described by the generic form $[q: w_1, w_2, \ldots, w_N]$. In this notation, we use square brackets to indicate we are dealing with a weighted voting system. Inside the square brackets we always list the quota first, followed by a colon and then the respective weights of the individual players. *It is customary to list the weights in numerical order, starting with the highest, and we will adhere to this convention throughout the chapter.*

We will now look at a few examples to illustrate some basic concepts in weighted voting.

EXAMPLE 2.1 Weighted Voting in a Partnership

Four partners—which we will call P_1, P_2, P_3, and P_4—decide to start a new business venture. In order to raise the $200,000 needed as startup money, they issue 20 shares worth $10,000 each. Suppose that P_1 owns 8 shares, P_2 owns 7 shares, P_3 owns 3 shares, and P_4 owns 2 shares, and that each share equals one vote in the partnership. Imagine now that they set up the rules of the partnership so that two-thirds of the partner's votes are needed to pass any motion. Using our new, simplified notation, this partnership can be described as the weighted voting system [14: 8, 7, 3, 2]. (Note that the quota is 14 because 14 is the first integer larger than two-thirds of 20.) ◀◀

 EXAMPLE 2.2 The Quota Can't Be Too Small . . .

Imagine the same partnership discussed in Example 2.1, with the only difference being that the quota is changed to 10 votes. We might be tempted to think that this arrangement can be described by [10: 8, 7, 3, 2], but this is not a viable weighted voting system. The problem here is that the quota is not big enough to allow for any type of decision making. Imagine that an important decision needs to be made and P_1 and P_4 vote yes and P_2 and P_3 vote no. Now we have a stalemate, since both the Yes's and the No's have enough votes to win. This is a mathematical version of anarchy. ◀◀

Example 2.2 illustrates why in a weighted voting system *the quota must be more than half of the total number of votes*. This can be expressed mathematically by the inequality

$$q > (w_1 + w_2 + \cdots + w_N)/2.$$

 EXAMPLE 2.3 . . . or Too Large

Once again, let's look at the partnership introduced in Example 2.1, but this time with a quota $q = 21$. Given that there are only 20 votes to go around, even if every partner were to vote Yes, a motion is doomed to fail. This is the mathematical version of gridlock. ◀◀

Example 2.3 illustrates why in a weighted voting system *the quota cannot be more than the total number of votes*. This can be expressed mathematically by the inequality

$$q \le (w_1 + w_2 + \cdots + w_N).$$

EXAMPLE 2.4 One Partner–One Vote?

Let's consider the partnership introduced in Example 2.1 one final time. This time the quota is set to be $q = 19$. Here we can describe the partnership as the weighted voting system [19: 8, 7, 3, 2]. What's interesting about this weighted voting system is that the *only way a motion can pass is by the unanimous support of all the players*. (Note that P_1, P_2, and P_3 together can only muster 18 votes—they still need P_4's votes to pass a motion.) In a practical sense, this weighted voting system is no different from a weighted voting system where each partner has 1 vote and it takes the unanimous agreement of the four partners to pass a motion (i.e., $[4: 1, 1, 1, 1]$).

The surprising conclusion of Example 2.4 is that the weighted voting system $[19: 8, 7, 3, 2]$ describes a one person–one vote situation in disguise. This seems like a contradiction only if we think of *one person–one vote* as implying that all players have an *equal number of votes rather than an equal say in the outcome of the election*. Apparently, these two things are not the same! As Example 2.4 makes abundantly clear, just looking at the number of votes a player owns can be very deceptive.

Power; More Terminology; More Examples

Let's look at a few more examples of weighted voting systems and start to informally focus on the notion of power.

 EXAMPLE 2.5 The Making of a Dictator

Consider the weighted voting system $[11: 12, 5, 4]$. Here one of the players (P_1) owns enough votes to carry a motion singlehandedly. In this situation, P_1 is in complete control—if P_1 is for the motion, the motion will pass; if P_1 is against it, the motion will fail. Clearly, in terms of the power to influence decisions, P_1 has *all* of it. Not surprisingly, we will say that P_1 is a *dictator*. ◀◀

In general, a player is a **dictator** if *the player's weight is bigger than or equal to the quota*. Notice that there can only be one dictator (having two players with weights bigger than or equal to the quota would contradict the requirement that the quota be bigger than half of the total number of votes). This observation, coupled with the assumption that we list the players by descending weights, implies that *if there is a dictator, it must be P_1*.

When P_1 is a dictator, all the other players, regardless of their weights, have absolutely no say on the outcome of the voting—there is never a time when any of their votes are needed. A player that never has a say in the outcome of the voting is a player that has no power and is called a **dummy**. When there is a dictator, all the other players are dummies, but there can be dummies even when there is no dictator. This is illustrated in our next example.

 EXAMPLE 2.6 The Curse of the Dummy

Four college students (P_1, P_2, P_3, and P_4) decide to go into business together. P_1, P_2, and P_3 each invest \$10,000 in the business, and each gets 10 votes in the partnership. P_4 is a little short on cash, so he invests only \$9000 and thus gets 9 votes. Suppose the quota is set at $q = 30$ (don't ask why). Under these assumptions the partnership can be described as the weighted voting system $[30: 10, 10, 10, 9]$. Everything seems right, right?

Wrong! In this weighted voting system, P_4 *turns out to be a dummy*! Why? Notice that a motion can only pass if P_1, P_2, and P_3 are for it, and then it makes no difference whether P_4 is for or against it. Thus, there is never going to be a time when P_4's 9 votes are going to make a difference in the outcome of the voting. Poor P_4—in spite of having almost as many votes as the other partners, he has no power to affect any partnership decisions! ◀◀

▶ EXAMPLE 2.7 The Power of the Veto

In the weighted voting system [12: 9, 5, 4, 2], P_1 plays an interesting role—while not having enough votes to be a dictator, he has the power to obstruct by *preventing any motion from passing*. This happens because without P_1's votes, a motion cannot pass—even if all the remaining players were to vote in favor of the motion. In a situation like this we will say that P_1 has *veto power*. ◀◀

In general, a player P that is not a dictator is said to have **veto power** if a *motion cannot pass unless P votes in favor of the motion*. (This, of course, does not mean that if P votes in favor of the motion the motion must pass!)

John F. Banzhaf III is Professor of Public Interest Law at George Washington University, as well as founder and executive director of the national antismoking group Action on Smoking and Health.

2.2 The Banzhaf Power Index

From the preceding set of examples we can already draw an important lesson—in weighted voting, a player's weight does not always tell the full story of how much power the player holds. Sometimes a player with lots of votes can have little or no power (see Example 2.6), and conversely, a player with just a couple of votes can have a lot of power (see Example 2.4). And yet, we have not even formally defined what having "power" means. We will do so in this section, introducing a mathematical interpretation of power first suggested by the American lawyer John Banzhaf III in 1965.

To get us started, let's look at a simple example.

▶ EXAMPLE 2.8 The Weirdness of Parliamentary Politics

The Parliament of Icelandia has 200 members, divided among three political parties—the Red Party (P_1), the Blue Party (P_2), and the Green Party (P_3). The Red Party has 99 seats in Parliament, the Blue Party has 98, and the Green Party has only 3. Decisions are made by majority vote, which in this case requires 101 out of the total 200 votes. Since in Icelandia members of Parliament always vote along party lines (voting against the party line is extremely rare in parliamentary governments), we can think of Icelandia's Parliament as the weighted voting system [101: 99, 98, 3].

By just looking at the numbers of votes, we would guess that most of the power is shared between the Red and Blue parties, with the Green Party having very little power, if any. On closer inspection, however, a strikingly different story emerges when we look at the different ways that a motion might pass.

Let's consider all the possible *winning* combinations (remember that we are assuming that voting is strictly along party lines). There are only four:

(i)   Reds and Blues vote Yes; Greens vote No. The motion passes 197 to 3.

(ii) Reds and Greens vote Yes; Blues vote No. The motion passes 102 to 98.

(iii) Blues and Greens vote Yes; Reds vote No. The motion passes (just barely) 101 to 99.

(iv) Reds, Blues, and Greens all vote Yes. The motion passes 200 to 0.

The surprising fact that emerges from the preceding list is that for a motion to pass it must have the support of two of the parties, and it makes no difference which two. Based on this observation we conclude that in the Icelandia Parliament the little Green Party holds as much power as either of the other two parties! ◀◀

Building on Example 2.8, we now introduce some additional notation and terminology.

- **Coalitions.** We will use the term **coalition** to describe any set of players that might join forces and vote the same way. In principle, we can have a coalition with as few as *one* player and as many as *all* players. The coalition consisting of all the players is called the **grand coalition**. Since coalitions are just sets of players, the most convenient way to describe coalitions mathematically is to use *set* notation. For example, the coalition consisting of players P_1, P_2, and P_3 can be written as the set $\{P_1, P_2, P_3\}$ (or $\{P_3, P_1, P_2\}$, or $\{P_2, P_1, P_3\}$, etc.—the order in which the members of a coalition are listed is irrelevant).

- **Winning coalitions.** Some coalitions have enough votes to win and some don't. Quite naturally, we call the former **winning coalitions** and the latter **losing coalitions**. A single-player coalition can be a winning coalition only when that player is a dictator, so under the assumption that there are no dictators in our weighted voting systems (dictators are boring) a winning coalition must have at least two players. At the other end of the spectrum, the grand coalition is always a winning coalition, since it controls all the votes. In some weighted voting systems (see Example 2.4) the grand coalition is the only winning coalition.

- **Critical players.** In a winning coalition, a player is said to be a **critical player** for the coalition if the coalition must have that player's votes to win. In other words, when we subtract a critical player's weight from the total weight of the coalition, the total of the remaining votes drops below the quota. (In more formal terminology, *P is a critical player for a winning coalition if and only if* $W - w < q$, where W is the total weight of the coalition and w is the weight of P.) Sometimes a winning coalition has no critical players (the coalition has enough votes that no single player's desertion can keep it from winning), sometimes a winning coalition has several critical players, and when the coalition has just enough votes to make the quota, then every player is critical.

▶ EXAMPLE 2.8(B) Example 2.8 Revisited

We can think of Example 2.8 in a purely mathematical context and leave the details of Icelandian politics out of the picture. We have a weighted voting system that can be described by [101: 99, 98, 3]. In this weighted voting system we have four winning coalitions, listed in the first column of Table 2-1. In the two-player coalitions, both players are critical players (without both players the coalition wouldn't win); in the grand coalition no player is critical, since any two out of three would have enough votes to win. Each of the three players is critical for two coalitions. End of story.

TABLE 2-1 Winning Coalitions in [101: 99, 98, 3]		
Coalition	**Votes**	**Critical players**
$\{P_1, P_2\}$	197	P_1 and P_2
$\{P_1, P_3\}$	102	P_1 and P_3
$\{P_2, P_3\}$	101	P_2 and P_3
$\{P_1, P_2, P_3\}$	200	None

When John Banzhaf introduced his mathematical interpretation of power in 1965, he had a key insight—*a player's power should be measured by how often the player is a critical player*. Thus, the key to getting a measure of a player's power is to count the number of winning coalitions in which that player is critical. From Table 2-1 we can clearly see that in [101: 99, 98, 3] each player is critical twice. Since there are three players, each critical twice, we can say that each player holds two out of six, or one-third of the power.

The preceding ideas lead us to the final definitions of this section.

■ **The Banzhaf power index.** For the sake of simplicity, let's start with P_1— exactly the same idea applies to each of the other players. To measure the power (according to Banzhaf) of P_1, we first count the total number of winning coalitions in which P_1 is a critical player. Let's call this number B_1. While useful, B_1 does not tell us how much power P_1 has in relation to the other players. A better way to tell the story is to think of power as a pie, and to describe P_1's power as a fraction or percentage of that pie. To do this we find how many times each of the other players is critical $(B_2, B_3, \ldots, B_N)$, and the total number of times *all* players are critical $(T = B_1 + B_2 + \cdots + B_N)$. The ratio B_1/T (number of times P_1 is a critical player over the total number of times *all* players are critical) is called the **Banzhaf power index** of P_1. This number measures the size of P_1's "slice" of the "Banzhaf power pie" and can be expressed either as a fraction or decimal between 0 and 1 or, equivalently, as a percent between 0 and 100%. For convenience, we will use the symbol β_1 (read "beta-one") to denote the Banzhaf power index of P_1.

■ **The Banzhaf power distribution.** Just like P_1, each of the other players in the weighted voting system has a Banzhaf power index, which we can find in a similar way. The complete list of power indexes $\beta_1, \beta_2, \ldots, \beta_N$ is called the **Banzhaf power distribution** of the weighted voting system. The sum of all the β's is 1 (or 100% if they are written as percentages).

Following is a summary of the steps needed to compute the Banzhaf power distribution of a weighted voting system with N players.

Computing a Banzhaf Power Distribution

- **Step 1.** Make a list of all possible *winning* coalitions.
- **Step 2.** Within each winning coalition determine which are the *critical* players. (To determine if a given player is critical or not in a given winning coalition, we subtract the player's weight from the total number of votes in the coalition—if the difference drops below the quota q, then that player is critical. Otherwise, that player is not critical.)
- **Step 3.** Count the number of times that P_1 is critical. Call this number B_1. Repeat for each of the other players to find $B_2, B_3, \ldots, B_N$.
- **Step 4.** Find the total number of times *all* players are critical. This total is given by $T = B_1 + B_2 + \cdots + B_N$.
- **Step 5.** Find the ratio $\beta_1 = B_1/T$. This gives the *Banzhaf power index* of P_1. Repeat for each of the other players to find $\beta_2, \beta_3, \ldots, \beta_N$. The complete list of β's gives the *Banzhaf power distribution* of the weighted voting system.

In the next set of examples we will illustrate how to implement the preceding sequence of steps.

EXAMPLE 2.9 Banzhaf Power in [4: 3, 2, 1]

Let's find the Banzhaf power distribution of the weighted voting system $[4:3,2,1]$ using Steps 1 through 5.

Step 1. There are three winning coalitions in this weighted voting system. They are $\{P_1, P_2\}$ with 5 votes, $\{P_1, P_3\}$ with 4 votes, and the grand coalition $\{P_1, P_2, P_3\}$ with 6 votes.

Step 2. The critical players in each winning coalition are shown in the following table.

Winning coalition	t	Critical players
$\{P_1, P_2\}$	5	P_1 and P_2
$\{P_1, P_3\}$	4	P_1 and P_3
$\{P_1, P_2, P_3\}$	6	P_1 only

Step 3. $B_1 = 3$ (P_1 is critical in three coalitions); $B_2 = 1$ and $B_3 = 1$ (P_2 and P_3 are critical once each).

Step 4. $T = 3 + 1 + 1 = 5$.

Step 5. $\beta_1 = B_1/T = 3/5$; $\beta_2 = B_2/T = 1/5$; $\beta_3 = B_3/T = 1/5$. (If we want to express the β's in terms of percents, then $\beta_1 = 60\%$, $\beta_2 = 20\%$, $\beta_3 = 20\%$.) $\ll$

Of all the steps we must carry out in the process of computing Banzhaf power, by far the most demanding is Step 1. When we have only three players, as in Example 2.9, we can list the winning coalitions on the fly—there simply aren't that many—but as the number of players increases, the number of possible winning coalitions grows rapidly, and it becomes necessary to adopt some form of strategy to come up with a list of *all* the winning coalitions. This is important, because if we miss a single one, we are in all likelihood going to get the wrong Banzhaf power distribution. One conservative strategy is to make a list of *all* possible coalitions and then cross out the losing ones. The next example illustrates how we would use this approach.

▶ EXAMPLE 2.10 Banzhaf Power and the NBA Draft

When NBA teams prepare for the annual draft of college players, the decision on which college basketball players to draft may involve many people, including the management, the coaches, and the scouting staff. Typically, not all of these people have an equal voice in the process—the head coach's opinion is worth more than that of an assistant coach, and the general manager's opinion more than that of a scout. In some cases, this arrangement is formalized in the form of a weighted voting system. Let's use a fictitious team—the Akron Flyers—for the purposes of illustration.

In the Akron Flyers draft system, the head coach (P_1) has 4 votes, the general manager (P_2) has 3 votes, the director of scouting operations (P_3) has 2 votes, and the team psychiatrist (P_4) has 1 vote. Of the 10 votes cast, a simple majority of 6 votes is required for a yes vote on a player to be drafted. In essence, the Akron Flyers operate as the weighted voting system $[6: 4, 3, 2, 1]$.

We will now find the Banzhaf power distribution of this weighted voting system using Steps 1 through 5.

Step 1. Table 2-2 starts with the complete list of *all* possible coalitions (first column) and the number of votes in each (second column). From this information we can immediately determine which are the winning coalitions. (Notice that the list of coalitions is organized systematically—one-player coalitions first, two-player coalitions next, and so on. Also notice that within each coalition the players are listed in numerical order from left to right. Both of these are good bookkeeping strategies, and you are encouraged to use them when you do your own work.)

TABLE 2-2 Coalitions in [6: 4, 3, 2, 1]

Coalition	Number of votes	Coalition	Number of votes
$\{P_1\}$	4	$\{P_2, P_3\}$	5
$\{P_2\}$	3	$\{P_2, P_4\}$	4
$\{P_3\}$	2	$\{P_3, P_4\}$	3
$\{P_4\}$	1	$\{\underline{P_1}, P_2, P_3\}^*$	9
$\{\underline{P_1}, \underline{P_2}\}^*$	7	$\{\underline{P_1}, \underline{P_2}, P_4\}^*$	8
$\{\underline{P_1}, \underline{P_3}\}^*$	6	$\{\underline{P_1}, \underline{P_3}, P_4\}^*$	7
$\{P_1, P_4\}$	5	$\{\underline{P_2}, \underline{P_3}, \underline{P_4}\}^*$	6
		$\{P_1, P_2, P_3, P_4\}^*$	10

Note: * indicates winning coalitions.

Step 2. Now we disregard the losing coalitions and go to work on the winning coalitions only. For each winning coalition, we determine which players are critical. This is indicated in Table 2-2 by underlining the critical players. (Don't take someone else's word for it—please check that these are all the critical players!)

Step 3. We now count carefully how many times each player is underlined in Table 2-2. The counts are $B_1 = 5$, $B_2 = 3$, $B_3 = 3$, and $B_4 = 1$.

Step 4. $T = 5 + 3 + 3 + 1 = 12$.

Step 5. $\beta_1 = 5/12 = 41\frac{2}{3}\%$; $\beta_2 = 3/12 = 25\%$; $\beta_3 = 3/12 = 25\%$, and $\beta_4 = 1/12 = 8\frac{1}{3}\%$.

An interesting and unexpected result of these calculations is that the team's general manager (P_2) and the director of scouting operations (P_3) have the same Banzhaf power index—not exactly the arrangement originally intended. ◀◀

How Many Coalitions?

Attention! Brief mathematical detour begins here!

Before we go on to the next example, let's consider the following mathematical question: For a given number of players, how many different coalitions are possible? Here, our identification of coalitions with sets will come in particularly handy. Except for the empty subset { }, we know that every other subset of the set of players can be identified with a different coalition. This means that we can count the total number of coalitions by counting the number of subsets and subtracting one. So how many subsets does a set have?

A careful look at Table 2-3 shows us that each time we add a new element we are doubling the number of subsets—the same subsets we had before we added the element plus an equal number consisting of each of these subsets but with the new element thrown in.

TABLE 2-3 The Subsets of a Set

Set	$\{P_1, P_2\}$	$\{P_1, P_2, P_3\}$		$\{P_1, P_2, P_3, P_4\}$		$\{P_1, P_2, P_3, P_4, P_5\}$
Number of subsets	4	8		16		32
Subsets	{ }	{ }	$\{P_3\}$	{ }	$\{P_4\}$	The 16 subsets from the previous column along with each of these with P_5 thrown in.
	$\{P_1\}$	$\{P_1\}$	$\{P_1, P_3\}$	$\{P_1\}$	$\{P_1, P_4\}$	
	$\{P_2\}$	$\{P_2\}$	$\{P_2, P_3\}$	$\{P_2\}$	$\{P_2, P_4\}$	
	$\{P_1, P_2\}$	$\{P_1, P_2\}$	$\{P_1, P_2, P_3\}$	$\{P_1, P_2\}$	$\{P_1, P_2, P_4\}$	
				$\{P_3\}$	$\{P_3, P_4\}$	
				$\{P_1, P_3\}$	$\{P_1, P_3, P_4\}$	
				$\{P_2, P_3\}$	$\{P_2, P_3, P_4\}$	
				$\{P_1, P_2, P_3\}$	$\{P_1, P_2, P_3, P_4\}$	

Since each time we add a new player we are doubling the number of subsets, we will find it convenient to think in terms of powers of 2. Table 2-4 summarizes what we have learned.

TABLE 2-4 The Number of Coalitions

Players	Number of subsets	Number of coalitions
P_1, P_2	$4 = 2^2$	$2^2 - 1 = 3$
P_1, P_2, P_3	$8 = 2^3$	$2^3 - 1 = 7$
P_1, P_2, P_3, P_4	$16 = 2^4$	$2^4 - 1 = 15$
P_1, P_2, P_3, P_4, P_5	$32 = 2^5$	$2^5 - 1 = 31$
$\vdots$	$\vdots$	$\vdots$
$P_1, P_2, \ldots, P_N$	2^N	$2^N - 1$

End of detour.

We will now return to the problem of computing Banzhaf power distributions. Given what we now know about the rapid growth of the number of coalitions, the strategy used in Example 2.10 (list all possible coalitions and then eliminate the losing ones) can become a little tedious (to say the least) when we have more than a handful of players. Sometimes we can save ourselves a lot of work by figuring out directly which are the winning coalitions.

EXAMPLE 2.11 Cutting Loose the Losing Coalitions

The disciplinary committee at George Washington High School has five members: the principal (P_1), the vice principal (P_2), and three teachers $(P_3, P_4, \text{and } P_5)$. When voting on a specific disciplinary action, the principal has three votes, the vice principal has two votes, and each of the teachers has one vote. A total of five votes are needed for any disciplinary action. Formally speaking, the disciplinary committee is the weighted voting system $[5: 3, 2, 1, 1, 1]$.

There are $2^5 - 1 = 31$ possible coalitions. We can save a fair amount of effort by skipping the losing coalitions and listing only the winning coalitions. A little organization will help: We will go through the winning coalitions systematically according to the number of players in the coalition. Obviously, there are no one-player winning coalitions—we would have to have a dictator for that to happen. There is only one two-player winning coalition, namely $\{P_1, P_2\}$. The only three-player winning coalitions are those that include the principal P_1. All four-player coalitions are winning coalitions, and so is the grand coalition.

Steps 1 and 2. Table 2-5 shows *all* the winning coalitions, with the critical players underlined. (You should double-check and make sure that these are the right critical players in each coalition—it's good practice!)

TABLE 2-5 Winning Coalitions in $[5: 3, 2, 1, 1, 1]$ and Critical Players

Two players	Three players	Four players	Five players
$\{\underline{P_1}, \underline{P_2}\}$	$\{\underline{P_1}, \underline{P_2}, P_3\}$	$\{\underline{P_1}, P_2, P_3, P_4\}$	$\{P_1, P_2, P_3, P_4, P_5\}$
	$\{\underline{P_1}, \underline{P_2}, P_4\}$	$\{\underline{P_1}, P_2, P_3, P_5\}$	
	$\{\underline{P_1}, \underline{P_2}, P_5\}$	$\{\underline{P_1}, P_2, P_4, P_5\}$	
	$\{\underline{P_1}, \underline{P_3}, \underline{P_4}\}$	$\{\underline{P_1}, P_3, P_4, P_5\}$	
	$\{\underline{P_1}, \underline{P_3}, \underline{P_5}\}$	$\{\underline{P_2}, \underline{P_3}, \underline{P_4}, \underline{P_5}\}$	
	$\{\underline{P_1}, \underline{P_4}, \underline{P_5}\}$		

Step 3. The critical player counts are $B_1 = 11$, $B_2 = 5$, $B_3 = 3$, $B_4 = 3$, and $B_5 = 3$.

Step 4. $T - 25$.

Step 5. $\beta_1 = 11/25 = 44\%$; $\beta_2 = 5/25 = 20\%$; $\beta_3 = \beta_4 = \beta_5 = 3/25 = 12\%$.

▶ EXAMPLE 2.12 Weighted Voting Without Weights

The Tasmania State University Promotion and Tenure committee consists of five members: the dean (D) and four other faculty members of equal standing (F_1, F_2, F_3, and F_4). (For convenience we are using a slightly different notation for the players.) In this committee faculty members vote first, and motions are carried by simple majority. The dean *only* votes to break a 2-2 tie. Is this a weighted voting system? If so, what is the Banzhaf power distribution?

The answer to the first question is Yes, and the answer to the second question is that we can apply the same steps we used before even though we don't have specific weights for the players.

Step 1. The possible winning coalitions fall into two groups: (i) three or more faculty members vote Yes. In this case the dean does not vote. These winning coalitions are listed in the first column of Table 2-6. (ii) Only two faculty members vote Yes, but the dean breaks the tie with a Yes vote. These winning coalitions are listed in the second column of Table 2-6.

TABLE 2-6 Winning Coalitions and Critical Players	
$\{\underline{F_1}, \underline{F_2}, \underline{F_3}\}$	$\{\underline{D}, \underline{F_1}, \underline{F_2}\}$
$\{\underline{F_1}, \underline{F_2}, \underline{F_4}\}$	$\{\underline{D}, \underline{F_1}, \underline{F_3}\}$
$\{\underline{F_1}, \underline{F_3}, \underline{F_4}\}$	$\{\underline{D}, \underline{F_1}, \underline{F_4}\}$
$\{\underline{F_2}, \underline{F_3}, \underline{F_4}\}$	$\{\underline{D}, \underline{F_2}, \underline{F_3}\}$
$\{F_1, F_2, F_3, F_4\}$	$\{\underline{D}, \underline{F_2}, \underline{F_4}\}$
	$\{\underline{D}, \underline{F_3}, \underline{F_4}\}$

Step 2. In the winning coalitions consisting of just three faculty members, all faculty members are critical. In the coalition consisting of all four faculty members, no faculty member is critical. In the coalitions consisting of two faculty members plus the dean, all players, including the dean, are critical. Table 2-6 shows the details, with the critical players underlined.

Step 3. The dean is critical six times (in each of the six coalitions in the second column). Each of the faculty members is critical three times in the first column and three times in the second column. Thus, all players are critical an equal number of times.

Steps 4 and 5. No counts are really necessary here. Since all five players are critical an equal number of times, they all have the same Banzhaf power index: $\beta_1 = \beta_2 = \beta_3 = \beta_4 = \beta_5 = 20\%$.

The surprising conclusion of Example 2.12 is that although the rules appear to set up a special role for the dean (the role of tie-breaker), in practice the dean is no different than any of the faculty members. A larger scale version of Example 2.12 occurs in the U.S. Senate, where the Vice President of the United States can only vote to break a tie. An analysis similar to the one used in Example 2.12 shows that as a member of the Senate the vice president has the same Banzhaf power index as an ordinary senator.

2.3 Applications of Banzhaf Power

The Nassau County Board of Supervisors John Banzhaf first introduced the concept of Banzhaf power in 1965. In an article entitled *Weighted Voting Doesn't Work* (see reference 2) Banzhaf used mathematics to argue that the weighted voting system used by the Board of Supervisors of Nassau County, New York, violated the "equal protection" guarantees of the Fourteenth Amendment.

In the 1960s, Nassau County was divided into six uneven districts, four with high populations and two rural districts with low populations. Table 2-7 shows the names of the districts and their weights based on their respective populations. The quota was a simple majority of 58 (out of 115) votes.

When we look at these weights carefully, as Banzhaf did, we discover something remarkable—all the power in the County Board was concentrated in the hands of the largest three districts (Hempstead #1, Hempstead #2, and Oyster Bay), with Glen Cove, Long Beach and North Hempstead unwittingly playing the role of *dummies*! North Hempstead a dummy? How is it possible? Here is how the numbers happen to work out: (i) Any two of the top three districts can form a winning coalition, and (ii) there can be no winning coalition without two out of the top three districts. The implications of this are that none of the last three districts can ever be critical.

TABLE 2-7 Nassau County Board (1964)	
District	**Weight**
Hempstead #1	31
Hempstead #2	31
Oyster Bay	28
North Hempstead	21
Long Beach	2
Glen Cove	2

Banzhaf's analysis of power in the Nassau County Board set the stage for a long series of legal challenges to the use of weighted voting in county boards, culminating with a federal court decision in 1993 abolishing weighted voting in New York State (see Project D). In 1996, after a protracted fight, the Nassau County Board of Supervisors became a 19-member legislature, each member having just one vote and representing districts of roughly equal population.

The United Nations Security Council The main body responsible for maintaining the international peace and security of nations is the United Nations Security Council. The Security Council is a classic example of a weighted voting system. It currently consists of fifteen voting nations—five of them are the *permanent* members (Britain, China, France, Russia, and the United States); the other ten nations are *nonpermanent* members appointed for a two-year period on a rotating basis. To pass a motion in the Security Council requires a Yes vote from each of the permanent members (in effect giving each permanent member *veto power*) plus additional Yes votes from at least four of the ten nonpermanent members. Thus, a winning coalition must include all five of the permanent members plus four or more nonpermanent members.

All in all, there are 848 possible winning coalitions in the Security Council—too many to list one by one. However, it's not too difficult to figure out the critical player story: In the winning coalitions with nine players (five permanent members plus four nonpermanent members) *every* member of the coalition is a critical player; in all the other winning coalitions (ten or more players) only the permanent members are critical players. Using a few simple (but not elementary) calculations, we can find that of the 848 winning coalitions, there are 210 with nine members and the rest have ten or more members. Carefully piecing together this information leads to the following surprising conclusion: The Banzhaf power index of each permanent member is 848/5080 (roughly 16.7%), while the Banzhaf power index of each nonpermanent member is 84/5080 (roughly 1.65%). Notice the discrepancy in power between the permanent and nonpermanent members: A permanent member has more than ten times as much power as a nonpermanent member. One has to wonder if this was the original intent of the United Nations charter or perhaps a miscalculation based on a less than clear understanding of the mathematics of weighted voting.

The European Union (EU) The EU is a political and economic confederation of 25 member nations, a sort of United States of Europe. The main legislative body of the EU is the Council of Ministers, and this council operates as a weighted voting system where different member nations are assigned weights based on a formula that takes into account their populations as well as other historical and economic factors. The first column of Table 2-8 lists the 25 member nations of the EU, and the second column shows each nation's weight in the Council of Ministers. The total number of votes in the Council of Ministers is 321, and the quota is $q = 232$ (roughly 72.3% of the votes). The "Relative weight" column shows the weights expressed as a percent of the 321 total votes. For example, France has 29 out of 321 votes, for a relative weight of 9.03%. The last column in Table 2-8 shows the Banzhaf power index of each member nation.

For more details see Exercise 67.

The calculations were done using a computer program available at this book's Web site *http://prenhall.com/tannenbaum*. (From the "Jump to" menu, jump to Chapter 2 and then click on Applets.)

TABLE 2-8 The European Union Council of Ministers (2005)

Country	Weight	Relative weight	Banzhaf power index
France, Germany, Italy, United Kingdom	29	9.03%	8.57%
Spain, Poland	27	8.41%	8.13%
Netherlands	13	4.05%	4.23%
Belgium, Greece, Czech Republic, Hungary, Portugal	12	3.74%	3.91%
Austria, Sweden	10	3.12%	3.27%
Denmark, Ireland, Lithuania, Slovakia, Finland	7	2.18%	2.31%
Cyprus, Estonia, Latvia, Luxembourg, Slovenia	4	1.25%	1.32%
Malta	3	0.93%	0.99%

One way to determine how well a weighted voting system accomplishes its intended purpose is by matching relative weights with power indexes. Comparing the "Relative weight" and "Banzhaf power index" columns of Table 2-8, we see that there is a very close match between the respective entries in the two columns, an indication that as a weighted voting system, the European Union Council of Ministers—unlike the Nassau County Board of Supervisors—works pretty much the way it was meant to work.

2.4 The Shapley-Shubik Power Index

Lloyd Shapley is Professor Emeritus of Mathematics and Economics at UCLA (see the biographical profile of Shapley at the end of this chapter). Martin Shubik is Professor of Economics at Yale University.

In this section we will discuss a different approach to measuring power, first proposed by American economists Lloyd Shapley and Martin Shubik in 1954. The key difference between the Shapley-Shubik measure of power and the Banzhaf measure of power has to do with how coalitions are formed. In the Shapley-Shubik method, the assumption is that coalitions are formed sequentially: Players join the coalition and cast their votes in an orderly sequence (there is a first player, then comes a second player, then a third, and so on). If the order in which the players join the coalition is taken into consideration, we call the coalition a *sequential coalition*.

Let's illustrate the difference with a simple example.

> ### EXAMPLE 2.13 Three-Player Sequential Coalitions

When we think of the *coalition* involving three players P_1, P_2, and P_3, we think in terms of a set. For convenience we write the set as $\{P_1, P_2, P_3\}$, but we can also write in other ways such as $\{P_3, P_2, P_1\}$, $\{P_2, P_3, P_1\}$, etc. In a coalition the only thing that matters is who are the members—the order in which we list them is irrelevant.

In contrast, with three players P_1, P_2, and P_3 we can form six different sequential coalitions: $\langle P_1, P_2, P_3 \rangle$ (this means that P_1 is the first player, then P_2 joined in, and last came P_3); $\langle P_1, P_3, P_2 \rangle$; $\langle P_2, P_1, P_3 \rangle$; $\langle P_2, P_3, P_1 \rangle$; $\langle P_3, P_1, P_2 \rangle$; $\langle P_3, P_2, P_1 \rangle$. The six sequential coalitions of three players are graphically illustrated in Fig. 2-1. [Note the change in notation. From now on the notation $\langle\ \rangle$ will indicate that we are dealing with a *sequential coalition*, and the order in which the players are listed does matter!]

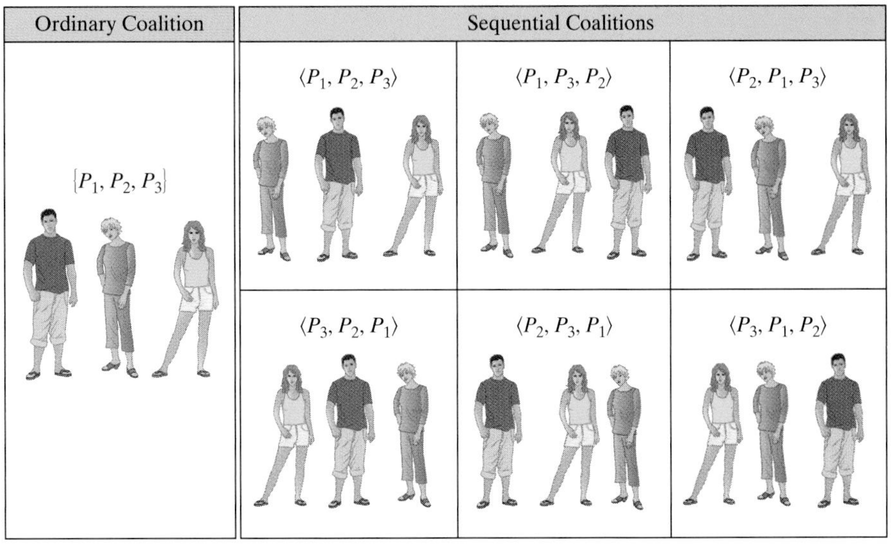

Ordinary Coalition	Sequential Coalitions		
$\{P_1, P_2, P_3\}$	$\langle P_1, P_2, P_3 \rangle$	$\langle P_1, P_3, P_2 \rangle$	$\langle P_2, P_1, P_3 \rangle$
	$\langle P_3, P_2, P_1 \rangle$	$\langle P_2, P_3, P_1 \rangle$	$\langle P_3, P_1, P_2 \rangle$

FIGURE 2-1 Three players form six different sequential coalitions.

The Shapley-Shubik approach to measuring power is parallel to the Banzhaf approach, but it is based on the concept of the *pivotal player* in a sequential coalition.

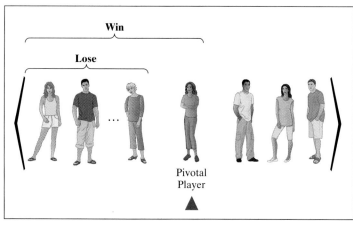

FIGURE 2-2 The pivotal player tips the scales.

- **Pivotal player.** In every sequential coalition there is a player that contributes the votes that turn what was a losing coalition into a winning coalition—we call such a player the **pivotal player** of the sequential coalition. Every sequential coalition has *one and only one* pivotal player. To find the pivotal player we just add the players' weights from left to right, one at a time, until the tally is bigger or equal to the quota q. The player whose votes tip the scales is the pivotal player. This idea is illustrated in Fig. 2-2.

- **The Shapley-Shubik power index.** For a given player P, the **Shapley-Shubik power index** of P is obtained by counting the number of times P is a *pivotal* player and dividing this number by the total number of times *all* players are pivotal (just like what we did with the Banzhaf power index but now using pivotal players instead of critical players and sequential coalitions instead of ordinary coalitions). As in the case of the Banzhaf power index, the Shapley-Shubik power index of a player can be thought of as a measure of the size of that player's "slice" of the "power pie" and can be expressed either as a fraction between 0 and 1 or as a percent between 0 and 100%. We will use σ_1 ("sigma-one") to denote the Shapley-Shubik power index of P_1, σ_2 to denote the Shapley-Shubik power index of P_2, and so on.

- **The Shapley-Shubik power distribution.** The complete listing of the Shapley-Shubik power indexes of all the players is called the **Shapley-Shubik power distribution** of the weighted voting system. It tells the complete story of how the (Shapley-Shubik) power pie is divided among the players.

Following is a summary of the steps needed to compute the Shapley-Shubik power distribution of a weighted voting system with N players.

Computing a Shapley-Shubik Power Distribution

- **Step 1.** Make a list of all possible sequential coalitions of the N players. Let T be the number of such coalitions. (We will have a lot more to say about T soon!)
- **Step 2.** In each sequential coalition determine *the* pivotal player.
- **Step 3.** Count the total number of times that P_1 is pivotal. Call this number SS_1. Repeat for each of the other players to find $SS_2, SS_3, \ldots, SS_N$.
- **Step 4.** Find the ratio $\sigma_1 = SS_1/T$. This gives the *Shapley-Shubik power index* of P_1. Repeat for each of the other players to find $\sigma_2, \sigma_3, \ldots, \sigma_N$. The complete list of σ's gives the *Shapley-Shubik power distribution* of the weighted voting system.

This is the beginning of a brief but extremely important mathematical detour. At the end of this detour we will be able to answer the question, How many sequential coalitions are possible?

The Multiplication Rule and Factorials

We will start by introducing one of the most useful rules of basic mathematics, stunning in its simplicity:

> **The Multiplication Rule**
>
> If there are m different ways to do X, and n different ways to do Y, then X and Y together can be done in $m \times n$ different ways.

EXAMPLE 2.14 Cones and Flavors

An ice cream shop offers 2 different choices of cones and 3 different flavors of ice cream (this is a boutique ice cream shop!). Using the multiplication rule, we can conclude that there are $2 \times 3 = 6$ different choices for a *single* (cone and one scoop of ice cream) order. The choices are illustrated in Fig. 2-3.

Figure 2-3 also clarifies the logic behind the multiplication rule: If we had m different types of cones and n different flavors of ice cream, the corresponding arrangement would show the cone/flavor combinations laid out in m rows and n columns, illustrating the $m \times n$ possibilities. ◀◀

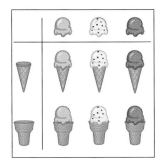

FIGURE 2-3 The scoop on the multiplication rule.

EXAMPLE 2.15 Cones and Flavors Plus Toppings

A bigger ice cream shop offers 5 different choices of cones, 31 different flavors of ice cream (this is *not* a boutique ice cream shop!), and 8 different choices of topping. The question is, If you are going to choose a cone and a single scoop of ice cream but then add a topping for good measure, how many orders are possible? The number of choices is too large to list individually, but we can find it by using the multiplication rule twice: First, there are $5 \times 31 = 155$ different cone/flavor combinations, and each of these can be combined with one of the 8 toppings into a grand total of $155 \times 8 = 1240$ different cone/flavor/topping combinations. The implications—as well as the calories—are staggering! ◀◀

We will now use the multiplication rule to count sequential coalitions. Specifically, we want to know, how many sequential coalitions with N players are possible?

EXAMPLE 2.16 Counting Sequential Coalitions

Let's start with a warm-up—how do we count the number of sequential coalitions with four players? Using the multiplication rule, we can argue as follows: We can choose any one of the *four* players to go first, then choose any one of the remaining *three* players to go second, then choose any one of the remaining *two* players to go third, and finally the *one* player left goes last. Using the multiplication rule, we get a total of $4 \times 3 \times 2 \times 1 = 24$ sequential coalitions with four players.

If we have 5 players, following up on our previous argument, we can count on a total of $5 \times 4 \times 3 \times 2 \times 1 = 120$ sequential coalitions, and with N players the number of sequential coalitions is $N \times (N - 1) \times \cdots \times 3 \times 2 \times 1$. The number $N \times (N - 1) \times \cdots \times 3 \times 2 \times 1$ is called the **factorial** of N and is written in the shorthand form $N!$.

This is the end of our mathematical detour. We will return to factorials in Chapter 6, but for the time being, if you have a business or scientific calculator, you might want to find the factorial key and familiarize yourself with its use. This is also a good time to give Exercises 35–40 a try!

The Number of Sequential Coalitions

The number of sequential coalitions with N players is $N! = N \times (N - 1) \times \cdots \times 3 \times 2 \times 1$.

Back to the Shapley-Shubik Power Index

We will illustrate the procedure for computing Shapley-Shubik power by revisiting a couple of examples done earlier in this chapter.

> ### EXAMPLE 2.17 Shapley-Shubik Power in [4: 3, 2, 1]

In this example we will compute the Shapley-Shubik power distribution of the weighted voting system $[4: 3, 2, 1]$. (We computed the Banzhaf power distribution of this weighted voting system in Example 2.9, so at the end we will be able to compare the results and see if there is a difference.)

Steps 1 and 2. Table 2-9 shows the six sequential coalitions of three players, with the pivotal player in each sequential coalition underlined. The second column shows the weights as they are being added from left to right until the tally reaches or exceeds the quota $q = 4$.

TABLE 2-9 Sequential Coalitions for [4: 3. 2. 1] (Pivotal Players Underlined)

Sequential coalition	Weight tallies (from left to right)
$\langle P_1, \underline{P_2}, P_3 \rangle$	$3 + \mathbf{2} = \mathbf{5}$
$\langle P_1, \underline{P_3}, P_2 \rangle$	$3 + \mathbf{1} = \mathbf{4}$
$\langle P_2, \underline{P_1}, P_3 \rangle$	$2 + \mathbf{3} = \mathbf{5}$
$\langle P_2, P_3, \underline{P_1} \rangle$	$2 + 1 + \mathbf{3} = \mathbf{6}$
$\langle P_3, \underline{P_1}, P_2 \rangle$	$1 + \mathbf{3} = \mathbf{4}$
$\langle P_3, P_2, \underline{P_1} \rangle$	$1 + 2 + \mathbf{3} = \mathbf{6}$

Step 3. A count of how many times each player is pivotal yields $SS_1 = 4$, $SS_2 = 1$, and $SS_3 = 1$.

Step 4. The Shapley-Shubik power distribution is given by $\sigma_1 = \frac{4}{6} = 66\frac{2}{3}\%$, $\sigma_2 = \frac{1}{6} = 16\frac{2}{3}\%$, and $\sigma_3 = \frac{1}{6} = 16\frac{2}{3}\%$.

When we compare this power distribution with the Banzhaf power distribution found in Example 2.9 ($\beta_1 = \frac{3}{5} = 60\%$, $\beta_2 = \frac{1}{5} = 20\%$, and $\beta_3 = \frac{1}{5} = 20\%$), we see that the two interpretations of power are indeed different. «

> ## EXAMPLE 2.18 Shapley-Shubik Power and the NBA Draft

We will now revisit Example 2.10, the NBA draft example. The weighted voting system in this example is [6: 4, 3, 2, 1], and we will now find its Shapley-Shubik power distribution.

> **Steps 1 and 2.** Table 2-10 shows the 24 sequential coalitions of P_1, P_2, P_3, and P_4. In each sequential coalition the pivotal player is underlined. (Notice there is a method to the madness of Table 2-9—each column corresponds to the sequential coalitions with a given first player. You may want to use the same or a similar pattern when you do the exercises on Shapley-Shubik power distributions.)

TABLE 2-10 Sequential Coalitions for [6: 4. 3. 2. 1] (Pivotal Players Underlined)

$\langle P_1, \underline{P_2}, P_3, P_4 \rangle$	$\langle P_2, \underline{P_1}, P_3, P_4 \rangle$	$\langle P_3, \underline{P_1}, P_2, P_4 \rangle$	$\langle P_4, P_1, \underline{P_2}, P_3 \rangle$
$\langle P_1, \underline{P_2}, P_4, P_3 \rangle$	$\langle P_2, \underline{P_1}, P_4, P_3 \rangle$	$\langle P_3, \underline{P_1}, P_4, P_2 \rangle$	$\langle P_4, P_1, \underline{P_3}, P_2 \rangle$
$\langle P_1, \underline{P_3}, P_2, P_4 \rangle$	$\langle P_2, P_3, \underline{P_1}, P_4 \rangle$	$\langle P_3, P_2, \underline{P_1}, P_4 \rangle$	$\langle P_4, P_2, \underline{P_1}, P_3 \rangle$
$\langle P_1, \underline{P_3}, P_4, P_2 \rangle$	$\langle P_2, P_3, \underline{P_4}, P_1 \rangle$	$\langle P_3, P_2, \underline{P_4}, P_1 \rangle$	$\langle P_4, P_2, \underline{P_3}, P_1 \rangle$
$\langle P_1, P_4, \underline{P_2}, P_3 \rangle$	$\langle P_2, P_4, \underline{P_1}, P_3 \rangle$	$\langle P_3, P_4, \underline{P_1}, P_2 \rangle$	$\langle P_4, P_3, \underline{P_1}, P_2 \rangle$
$\langle P_1, P_4, \underline{P_3}, P_2 \rangle$	$\langle P_2, P_4, \underline{P_3}, P_1 \rangle$	$\langle P_3, P_4, \underline{P_2}, P_1 \rangle$	$\langle P_4, P_3, \underline{P_2}, P_1 \rangle$

> **Step 3.** The pivotal player counts are $SS_1 = 10$, $SS_2 = 6$, $SS_3 = 6$, and $SS_4 = 2$.
>
> **Step 4.** The Shapley-Shubik power distribution of the weighted voting system is given by $\sigma_1 = \frac{10}{24} = 41\frac{2}{3}\%$, $\sigma_2 = \frac{6}{24} = 25\%$, $\sigma_3 = \frac{6}{24} = 25\%$, and $\sigma_4 = \frac{2}{24} = 8\frac{1}{3}\%$.

If you compare this result with the Banzhaf power distribution obtained in Example 2.10, you will notice that here the two power distributions are the same. If nothing else, this shows that it is not impossible for the Banzhaf and Shapley-Shubik power distributions to agree. In general, however, for randomly chosen real-life situations, it is very unlikely that the Banzhaf and Shapley-Shubik methods will give the same answer.

> ## EXAMPLE 2.19 Shapley-Shubik in the City

In some cities, the city council operates under what is known as the "strong-mayor" system. Under this system, the city council can pass a motion under simple majority, but the mayor has the power to veto the decision. The mayor's veto can then be overruled by a "super-majority" of the council members. As an illustration of the strong-mayor system we will consider the city of Cleansburg. In Cleansburg the city council has four members plus a strong mayor that has a vote as well as the power to veto motions supported by a simple majority of the council members. On the other hand, the mayor cannot veto motions supported by all four council members. Thus, a motion can pass if the mayor plus two or more council members support it, or, alternatively, if the mayor is against it but the four council members support it.

Common sense tells us that under these rules, the four council members have the same amount of power but the mayor has more. We will compute the Shapley-Shubik power distribution of this weighted voting system to figure out exactly how much more.

With 5 players, this weighted voting system has $5! = 120$ sequential coalitions to consider. Obviously, we would prefer not to have to write them all down, and, in fact, we can skip that step and just use our imagination. Specifically, we will imagine a table in which the 120 sequential coalitions are listed in 5 columns. The first column has all the sequential coalitions where the mayor (let's call him P_1) appears as the first player in the coalition, the second column has all the sequential coalitions where P_1 is in second place, the third column has all sequential coalitions with P_1 in third place, and so on. Now close your eyes and try to visualize this table in your mind's eye for a moment or two. Most likely, you visualized a table that had the same number of sequential coalitions in each of the columns. This is a key observation, as it allows us to conclude that there are $120 \div 5 = 24$ sequential coalitions in each column.

See Exercise 68 for an important generalization of these observations.

We are now in a position to find the pivotal player count of the mayor (P_1). We first observe that the mayor is the pivotal player only when he is the third or fourth player in a sequential coalition, as illustrated in Figs. 2-4 (a) and (b) respectively. This means that in our imaginary table, P_1 is pivotal in every sequential coalition in columns 3 and 4, and no others. So there we have it—the pivotal count for the mayor is $24 + 24 = 48$, and his Shapley-Shubik power index is $48/120 = 2/5 = 40\%$.

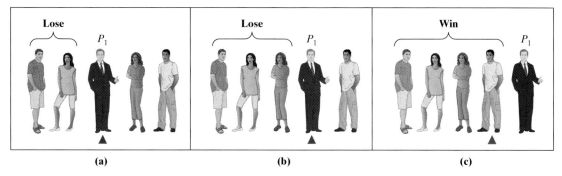

FIGURE 2-4

Since the Shapley-Shubik power index of the mayor equals 40% and the four council members must share the remaining 60% of the power equally, it follows that each council member has a Shapley-Shubik power index of 15%. In conclusion, the Shapley-Shubik power distribution of the Cleansburg city council is $\sigma_1 = 40\%$, $\sigma_2 = \sigma_3 = \sigma_4 = \sigma_5 = 15\%$.

For the purposes of comparison, the reader is encouraged to calculate the Banzhaf power distribution of the Cleansburg city council–see Exercise 71.

In this example it's not so much the conclusion but how we got it that is worth remembering.

2.5 Applications of Shapley-Shubik Power

The Electoral College   Calculating the Shapley-Shubik power index of the states in the Electoral College is no easy task. There are 51! sequential coalitions, a number so large (67 digits long) we don't even have a name for it. Individually checking all possible sequential coalitions is out of the question, even for the world's fastest computer. There are, however, some sophisticated mathematical shortcuts, which, when coupled with the right kind of software, allow the calculations to be done by an ordinary computer in a matter of seconds (see reference 16 for details).

The appendix at the end of this chapter shows both the Banzhaf and the Shapley-Shubik power indexes for each of the 50 states and the District of Columbia. Comparing the Banzhaf and the Shapley-Shubik power indexes shows that there is a very small difference between the two. This example shows that in some situations the Banzhaf and Shapley-Shubik power indexes give essentially the same answer. The next example illustrates a very different situation.

The United Nations Security Council As mentioned in Section 2.3, the United Nations Security Council consists of fifteen member nations—five are permanent members and ten are nonpermanent members appointed on a rotating basis. For a motion to pass it must have a Yes vote from each of the five permanent members plus at least four of the ten nonpermanent members. It can be shown that this arrangement is equivalent to giving the permanent members 7 votes each, the nonpermanent members 1 vote each, and making the quota equal to 39 votes.

We will sketch a rough outline of how the Shapley-Shubik power distribution of the Security Council can be calculated. The details, while not terribly difficult, go beyond the scope of this book.

1. There are 15! sequential coalitions of 15 players (roughly about *1.3 trillion*).

2. A nonpermanent member can be pivotal only if it is the ninth player in the coalition, preceded by all five of the permanent members and three nonpermanent members. (There are approximately *2.44 billion* sequential coalitions of this type.)

3. From (1) and (2) we can conclude that the Shapley-Shubik power index of a nonpermanent member is approximately 0.19% (2.44 billion/1.3 trillion $\approx$ 0.0019 = 0.19%). For the purposes of comparison it is worth noting that there is a big difference between this Shapley-Shubik power index and the corresponding Banzhaf power index of 1.65% obtained in Section 2.3.

4. The ten nonpermanent members (each with a Shapley-Shubik power index of 0.19%) have together 1.9% of the power pie, leaving the remaining 98.2% to be divided equally among the five permanent members. Thus, the Shapley-Shubik power index of each permanent member is approximately 98.2/5 = 19.64%.

This analysis shows the enormous difference between the Shapley-Shubik power of the permanent and nonpermanent members of the Security Council—permanent members have roughly 100 times the Shapley-Shubik power of nonpermanent members!

The European Union We introduced the European Union Council of Ministers in Section 2.3 and observed that the Banzhaf power index of each member nation matches very closely the percentage of votes that it holds. What about the Shapley-Shubik power index? Once again, there isn't much future in trying to calculate the power indexes directly—there is a total of 25! sequential

If you are curious as to exactly how many years, try Exercise 35(d).

coalitions, and even for a supercomputer it would take years to list them all. On the other hand, through some mathematical shortcuts, the same calculations can be done in a few seconds by even an ordinary computer. (An applet that can be used to carry out the calculations is available at this book's Web site *http://prenhall.com/tannenbaum*. From the "Jump to" menu, jump to Chapter 2 and then click on Applets.) The results are shown in the last column of Table 2-11. Notice, once again, how nicely the power indexes match the relative weights, confirming our earlier observation that the EU Council is an excellent working weighted voting system.

TABLE 2-11 The European Union Council of Ministers (2005)

Country	Weight	Relative weight	SS power index
France, Germany, Italy, United Kingdom	29	9.03%	9.29%
Spain, Poland	27	8.41%	8.61%
Netherlands	13	4.05%	3.98%
Belgium, Greece, Czech Republic, Hungary, Portugal	12	3.74%	3.65%
Austria, Sweden	10	3.12%	3.02%
Denmark, Ireland, Lithuania, Slovakia, Finland	7	2.18%	2.10%
Cyprus, Estonia, Latvia, Luxembourg, Slovenia	4	1.25%	1.19%
Malta	3	0.93%	0.89%

Conclusion

In any society, no matter how democratic, some individuals and groups have more power than others. This is simply a consequence of the fact that individuals and groups are not all equal. Diversity is the inherent reason why the concept of power exists.

Power itself comes in many different forms. We often hear clichés such as "In strength lies power" or "Money is power" (and the newer cyber version, "Information is power"). In this chapter we discussed the notion of power as it applies to formal voting situations called *weighted voting systems* and saw how mathematical methods allow us to measure the power of an individual or group by means of a *power index*. In particular, we looked at two different kinds of power indexes: the *Banzhaf power index* and the *Shapley-Shubik power index*. These indexes provide two different ways to measure power, and while they occasionally agree, they often differ significantly. Of the two, which one gives the more accurate measure of a player's power?

Unfortunately, there is no simple answer. Both of them are useful, and in some sense the choice is subjective. Perhaps the best way to evaluate them is to think of them as being based on a slightly different set of assumptions. The idea behind the Banzhaf interpretation of power is that players are free to come and go, negotiating their allegiance for power (somewhat like professional athletes since the advent of free agency). Underlying the Shapley-Shubik interpretation of power is the assumption that when a player joins a coalition, he or she is making a commitment to stay. In the latter case a player's power is generated by his or her ability to be in the right place at the right time.

In practice, the choice of which method to use for measuring power is based on which of the assumptions better fits the specifics of the situation. Contrary to what we've often come to expect, mathematics does not give us the answer, just the tools that might help us make an informed decision.

Profile Lloyd S. Shapley (1923–   )

Lloyd Shapley is one of the giants of modern *game theory*— which, in a nutshell, is the mathematical study of *games*. (The word *game* is used here as a metaphor for any situation involving competition and cooperation among individuals or groups each trying to further their own separate goals. This broad interpretation encompass- es not only real games like poker and Monopoly, but also economic, political, and military "games," thus making game theory one of the most important branches of modern applied mathematics.)

Lloyd Shapley was born in Cambridge, Massachusetts to a prominent scientific family. His father, Harlow Shapley, was a renowned astronomer at Harvard University and director of the Harvard Observatory. Shapley entered Harvard in 1942, but he interrupted his studies to join the Army during World War II. From 1943 to 1945 he served in the Army Air Corps in China, where he was awarded a Bronze Star for breaking the Japanese weather code. In 1945 he returned to Harvard, receiving a Bachelor of Arts in Mathematics in 1948.

Upon graduation from Harvard, Shapley joined the Rand Corporation, a famous think tank in Santa Monica, California, where much of the pioneering research in game theory was being conducted. After a year at Rand, and with a reputation as a budding young star in game theory in tow, Shapley went to Princeton University as a graduate student in mathematics. At Princeton he joined a circle that included some of the most brilliant mathematical minds of his time, including John von Neumann, the father of game theory, and fellow graduate student John Nash, the mathematical genius whose life is portrayed in the Academy Award–winning movie *A Beautiful Mind* (with Russell Crowe playing John Nash).

As a graduate student, Shapley had a reputation as a fierce competitor in everything he took on, including card games and board games, some of his own invention. One of the games Shapley invented with a couple of other graduate students (one of which was Nash and the other was Martin Shubik, who at the time was an economics graduate student as well as Shapley's roommate) was a fiendish board game called *So Long Sucker,* in which players formed coalitions to gang up and double-cross each other. Some of the experiences and observations that Shapley and Shubik (who became life-long friends) derived from their graduate student game-playing days eventually led to their development of the Shapley-Shubik index for measuring power in a weighted voting system (a weighted voting system, after all, is just a simple *game* in which players form coalitions in order to pass or block a specific action by the group).

After receiving his Ph.D. in Mathematics from Princeton in 1953, Shapley returned to the Rand Corporation, where he worked as a research mathematician until 1981. During this time he made many major contributions to game theory, including the invention of *convex* and *stochastic* games, the *Shapley value* of a game, and the *Shapley-Shubik power index*. The citation when he was awarded the Von Neumann Prize in Game Theory in 1981 partly read, "His individual work and his joint research with Martin Shubik has helped build bridges between game theory, economics, political science, and practice."

In 1981 Shapley joined the Mathematics department at UCLA, where he is currently Professor Emeritus of Mathematics and Economics.

Key Concepts

Exercises

WALKING

A. Weighted Voting Systems

1. Consider the weighted voting system $[13: 7, 4, 3, 3, 2, 1]$. Find

   (a) the total number of players

   (b) the total number of votes

   (c) the weight of P_2

   (d) the minimum percentage of the votes needed to pass a motion (rounded to the next whole percent)

2. Consider the weighted voting system $[62: 10, 10, 10, 10, 8, 5, 5, 5, 5, 4, 4, 3, 3, 3, 2]$. (This weighted voting system describes the voting in the European Union Council of Ministers prior to 2004.) Find

   (a) the total number of players

   (b) the total number of votes

   (c) the weight of P_6

   (d) the minimum percentage of the votes needed to pass a motion (rounded to the next whole percent)

3. Consider the weighted voting system $[q: 10, 6, 5, 4, 2]$.

   (a) What is the smallest value that the quota q can take?

   (b) What is the largest value that the quota q can take?

   (c) What is the value of the quota if *at least* two-thirds of the votes are required to pass a motion?

   (d) What is the value of the quota if *more* than two-thirds of the votes are required to pass a motion?

4. Consider the weighted voting system $[q: 6, 4, 3, 3, 2, 2]$.

   (a) What is the smallest value that the quota q can take?

   (b) What is the largest value that the quota q can take?

   (c) What is the value of the quota if *at least* three-fourths of the votes are required to pass a motion?

   (d) What is the value of the quota if *more* than three-fourths of the votes are required to pass a motion?

5. A committee has four members (P_1, P_2, P_3, and P_4). In this committee P_1 has twice as many votes as P_2; P_2 has twice as many votes as P_3; P_3 has twice as many votes as P_4. Describe the committee as a weighted voting system when the requirements to pass a motion are

   (a) at least two-thirds of the votes

   (b) more than two-thirds of the votes

   (c) at least 80% of the votes

   (d) more than 80% of the votes

6. A committee has six members (P_1, P_2, P_3, P_4, P_5, and P_6). In this committee P_1 has twice as many votes as P_2; P_2 and P_3 each has twice as many as P_4; P_4 has twice as many votes as P_5; P_5 and P_6 have the same number of votes. Describe the committee as a weighted voting system when the requirements to pass a motion are

   (a) a simple majority of the votes

   (b) at least three-fourths of the votes

   (c) more than three-fourths of the votes

   (d) at least two-thirds of the votes

   (e) more than two-thirds of the votes

7. In each of the following weighted voting systems, determine which players, if any, (i) are dictators; (ii) have veto power; (iii) are dummies.

   (a) $[6: 4, 2, 1]$

   (b) $[6: 7, 3, 1]$

   (c) $[6: 5, 5, 1]$

8. In each of the following weighted voting systems, determine which players, if any, (i) are dictators; (ii) have veto power; (iii) are dummies.

   (a) $[95: 95, 80, 10, 2]$

   (b) $[95: 65, 35, 30, 25]$

   (c) $[48: 32, 16, 8, 4, 2, 1]$

9. In each of the following weighted voting systems, determine which players, if any, (i) are dictators; (ii) have veto power; (iii) are dummies.

(a) $[19: 9, 7, 5, 3, 1]$

(b) $[15: 16, 8, 4, 1]$

(c) $[17: 13, 5, 2, 1]$

(d) $[25: 12, 8, 4, 2]$

10. In each of the following weighted voting systems, determine which players, if any, (i) are dictators; (ii) have veto power; (iii) are dummies.

(a) $[27: 12, 10, 4, 2]$

(b) $[22: 10, 8, 7, 2, 1]$

(c) $[21: 23, 10, 5, 2]$

(d) $[15: 11, 5, 2, 1]$

B. Banzhaf Power

11. Consider the weighted voting system $[10: 6, 5, 4, 2]$.

(a) What is the weight of the coalition formed by P_1 and P_3?

(b) Write down all winning coalitions.

(c) Which players are critical in the coalition $\{P_1, P_2, P_3\}$?

(d) Find the Banzhaf power distribution of this weighted voting system.

12. Consider the weighted voting system $[5: 3, 2, 1, 1]$.

(a) What is the weight of the coalition formed by P_1 and P_3?

(b) Which players are critical in the coalition $\{P_1, P_2, P_3\}$?

(c) Which players are critical in the coalition $\{P_1, P_3, P_4\}$?

(d) Write down all winning coalitions.

(e) Find the Banzhaf power distribution of this weighted voting system.

13. (a) Find the Banzhaf power distribution of the weighted voting system $[6: 5, 2, 1]$.

(b) Find the Banzhaf power distribution of the weighted voting system $[3: 2, 1, 1]$. Compare your answers in (a) and (b).

14. (a) Find the Banzhaf power distribution of the weighted voting system $[7: 5, 2, 1]$.

(b) Find the Banzhaf power distribution of the weighted voting system $[5: 3, 2, 1]$. Compare your answers in (a) and (b).

15. (a) Find the Banzhaf power distribution of the weighted voting system $[10: 5, 4, 3, 2, 1]$.

(b) Find the Banzhaf power distribution of the weighted voting system $[11: 5, 4, 3, 2, 1]$.

(***Hint:*** *Note that the only change from (a) is in the quota, and use this fact to your advantage.*)

16. (a) Find the Banzhaf power distribution of the weighted voting system $[9: 5, 5, 4, 2, 1]$.

(b) Find the Banzhaf power distribution of the weighted voting system $[9: 5, 5, 3, 2, 1]$.

17. Consider the weighted voting system $[q: 8, 4, 2, 1]$. Find the Banzhaf power distribution of this weighted voting system when

(a) $q = 8$

(b) $q = 9$

(c) $q = 10$

(d) $q = 12$

(e) $q = 14$

18. Consider the weighted voting system $[q: 5, 3, 1]$. Find the Banzhaf power distribution of this weighted voting system when

(a) $q = 5$

(b) $q = 6$

(c) $q = 7$

(d) $q = 8$

(e) $q = 9$

19. A business firm is owned by 4 partners, A, B, C, and D. When making decisions, each partner has one vote and the majority rules, except in the case of a 2-2 tie. Then, the coalition that contains D (the partner with the least seniority) loses. What is the Banzhaf power distribution in this partnership?

20. Consider the weighted voting system $[8: 5, 3, 1, 1, 1]$.

(a) Make a list of all winning coalitions.

(***Hint:*** *There aren't too many!*)

(b) Using (a), find the Banzhaf power distribution of this weighted voting system.

(c) Suppose that P_1, with 5 votes, sells one of her votes to P_2, resulting in the weighted voting system $[8: 4, 4, 1, 1, 1]$. Find the Banzhaf power distribution of this system.

(d) Compare the power index of P_1 in (b) and (c). Describe the paradox that occurred.

Exercises 21 and 22 refer to the Nassau County (N.Y.) Board of Supervisors, as discussed in this chapter.

21. The Nassau County Board of Supervisors (1960s version). In the 1960s, the Nassau County Board of Supervisors operated as the weighted voting system $[58: 31, 31, 28, 21, 2, 2]$. Assume the players are P_1 through P_6.

(a) List the winning coalitions involving *only* P_1, P_2, or P_3.

(b) List the winning coalitions in which P_4 is a member.

(c) Use the results in (b) to explain why P_4 is a dummy.

(d) Use the results in (a), (b), and (c) to find the Banzhaf power distribution of the weighted voting system.

22. The Nassau County Board of Supervisors (1990s version). In the early 1990s, after a series of court decisions, the Nassau County Board of Supervisors was changed to operate as the weighted voting system $[65: 30, 28, 22, 15, 7, 6]$.

(a) List all the *three-player* winning coalitions and the critical players in each.

(b) List all the *four-player* winning coalitions and the critical players in each.

(**Hint:** *There are 11 four-player winning coalitions.*)

(c) List all the *five-player* winning coalitions and the critical players in each.

(d) Use the results in (a), (b), and (c) to find the Banzhaf power distribution of the weighted voting system.

C. Shapley-Shubik Power

23. Consider the weighted voting system [16: 9, 8, 7].

(a) Write down all the sequential coalitions, and in each sequential coalition underline the pivotal player.

(b) Find the Shapley-Shubik power distribution of this weighted voting system.

24. Consider the weighted voting system [8: 7, 6, 2].

(a) Write down all the sequential coalitions, and in each sequential coalition underline the pivotal player.

(b) Find the Shapley-Shubik power distribution of this weighted voting system.

25. Find the Shapley-Shubik power distribution of the weighted voting system [5: 3, 2, 1, 1].

26. Find the Shapley-Shubik power distribution of the weighted voting system [60: 32, 31, 28, 21].

27. Find the Shapley-Shubik power distribution of each of the following weighted voting systems.

(a) [8: 8, 5, 1]

(b) [8: 7, 5, 2]

(c) [8: 7, 6, 1]

(d) [8: 6, 5, 1]

(e) [8: 6, 5, 3]

28. Find the Shapley-Shubik power distribution of each of the following weighted voting systems.

(a) [6: 4, 3, 2, 1]

(b) [7: 4, 3, 2, 1]

(c) [8: 4, 3, 2, 1]

(d) [9: 4, 3, 2, 1]

(e) [10: 4, 3, 2, 1]

29. Consider the weighted voting system [q: 5, 3, 1]. Find the Shapley-Shubik power distribution of this weighted voting system when

(a) $q = 5$

(b) $q = 6$

(c) $q = 7$

(d) $q = 8$

(e) $q = 9$

30. Consider the weighted voting system [q: 4, 3, 2]. Find the Shapley-Shubik power distribution of this weighted voting system when

(a) $q = 5$

(b) $q = 6$

(c) $q = 7$

(d) $q = 8$

(e) $q = 9$

31. Find the Shapley-Shubik power distribution of each of the following weighted voting systems.

(a) [51: 40, 30, 20, 10]

(b) [59: 40, 30, 20, 10]

[**Hint:** *Compare this situation with the one in (a).*]

(c) [60: 40, 30, 20, 10]

32. Find the Shapley-Shubik power distribution of each of the following weighted voting systems.

(a) [41: 40, 10, 10, 10]

(b) [49: 40, 10, 10, 10]

[**Hint:** *Compare this situation with the one in (a).*]

(c) [60: 40, 10, 10, 10]

33. A business firm is owned by four partners, A, B, C, and D. When making decisions, each partner has one vote and the majority rules. In case of a 2-2 tie, the tie is broken by going against D (i.e., if D votes yes, the decision is no, and vice versa). Find the Shapley-Shubik power distribution in this partnership.

34. A business firm is owned by four partners, A, B, C, and D. When making decisions, each partner has one vote and the majority rules. In case of a 2-2 tie, the tie is broken in favor of A (the senior partner). Find the Shapley-Shubik power distribution in this partnership.

D. Miscellaneous

For Exercises 35 and 36 you should use a calculator with a factorial key (typically it's a key labeled either x! or n!). All scientific calculators and most business calculators have such a key.

35. Use a calculator to compute each of the following.

(a) 13!

(b) 18!

(c) 25!

(d) Suppose that you have a supercomputer that can list 1 million sequential coalitions per second. Estimate (in years) how long it would take the computer to list all sequential coalitions of 25 players.

36. Use a calculator to compute each of the following.

(a) 12!

(b) 15!

(c) 20!

(d) Suppose that you have a supercomputer that can list 10,000 sequential coalitions per second. Estimate (in years) how long it would take the computer to list all sequential coalitions of 20 players.

The purpose of Exercises 37 through 40 is for you to learn how to manipulate factorials numerically. If you use a calculator to answer these questions, you are defeating the purpose of the exercise. Please answer Exercises 37 through 40 without using a calculator.

37. (a) Given that $10! = 3,628,800$, find $9!$.

  (b) Find $11!/10!$.

  (c) Find $11!/9!$.

  (d) Find $9!/6!$.

  (e) Find $101!/99!$.

38. (a) Given that $20! = 2,432,902,008,176,640,000$, find $19!$.

  (b) Find $20!/19!$.

  (c) Find $201!/199!$.

  (d) Find $11!/8!$.

39. Find the value of each of the following. Give the answer in decimal form.

  (a) $(9! + 11!)/10!$

  (b) $(101! + 99!)/100!$

40. Find the value of each of the following. Give the answer in decimal form.

  (a) $(19! + 21!)/20!$

  (***Hint:*** $1/20 = 0.05$)

  (b) $(201! + 199!)/200!$

41. Consider a weighted voting system with six players (P_1 through P_6).

  (a) Find the total number of coalitions in this weighted voting system.

  (b) How many coalitions in this weighted voting system do not include P_6?

  (***Hint:*** Think of all the possible coalitions of the remaining players.)

  (c) How many coalitions in this weighted voting system do not include P_5 or P_6?

  (d) How many coalitions in this weighted voting system include both P_5 and P_6?

  [***Hint:*** Use your answers of (a) and (c).]

42. Consider a weighted voting system with seven players (P_1 through P_7).

  (a) Find the total number of coalitions in this weighted voting system.

  (b) How many coalitions in this weighted voting system do not include P_7?

  (***Hint:*** Think of all the possible coalitions of the remaining players.)

(c) How many coalitions in this weighted voting system do not include P_7 or P_6?

(d) How many coalitions in this weighted voting system include both P_7 and P_6?

  [***Hint:*** Use your answers of (a) and (c).]

43. Consider a weighted voting system with six players (P_1 through P_6).

  (a) Find the number of sequential coalitions in this weighted voting system.

  (b) How many sequential coalitions in this weighted voting system have P_1 as the last player?

  (***Hint:*** Think of all the possible sequential coalitions of the remaining players.)

  (c) How many sequential coalitions in this weighted voting system do not have P_1 as the last player?

  [***Hint:*** Use your answers of (a) and (b).]

44. Consider a weighted voting system with seven players (P_1 through P_7).

  (a) Find the number of sequential coalitions in this weighted voting system.

  (b) How many sequential coalitions in this weighted voting system have P_7 as the first player?

  (***Hint:*** Think of all the possible sequential coalitions of the remaining players.)

  (c) How many sequential coalitions in this weighted voting system do not have P_7 as the first player?

  [***Hint:*** Use your answers of (a) and (b).]

45. Consider the weighted voting system $[q: 6, 5, 4, 4, 3, 2]$.

  (a) Find all the possible values of q for which no player has veto power. Explain.

  (b) Find all the possible values of q for which P_1 is the only player with veto power.

  (c) Find all the possible values of q for which every player has veto power.

  (d) Find all values of q such that the only winning coalition is the grand coalition.

  (e) What is the connection between the answers to (c) and (d)? Explain.

46. Consider the weighted voting system $[q: 10, 8, 7, 5, 4, 4, 3]$.

  (a) Find all the possible values of q for which no player has veto power. Explain.

  (b) Find all the possible values of q for which P_1 is the only player with veto power.

  (c) Find all the possible values of q for which every player has veto power.

  (d) Find all values of q such that the only winning coalition is the grand coalition.

  (e) What is the connection between the answers to (c) and (d)? Explain.

JOGGING

47. In a weighted voting system with five players (P_1 through P_5) the winning coalitions are as follows: $\{P_1, P_2, P_3\}$, $\{P_1, P_2, P_4\}$, $\{P_1, P_2, P_3, P_4\}$, $\{P_1, P_2, P_3, P_5\}$, $\{P_1, P_2, P_4, P_5\}$, and $\{P_1, P_2, P_3, P_4, P_5\}$. Find the Banzhaf power distribution of this weighted voting system.

48. Consider the weighted voting system $[24: 6, 5, 4, 4, 3, 2]$. (Note that here the quota equals 100% of the votes.)

    (a) Write down the winning coalition(s).

    (b) Find the Banzhaf power index of each player in this weighted voting system.

    (c) Explain why the following is true: If the grand coalition is the only winning coalition, the Banzhaf power index of each player is $1/N$, where N denotes the number of players.

49. Consider the weighted voting system $[24: 6, 5, 4, 4, 3, 2]$.

    (a) How many different sequential coalitions are there?

    (b) In how many sequential coalitions is P_6 pivotal?

    [**Hint:** See Exercise 43(b).]

    (c) What is the Shapley-Shubik power index of P_6?

    (d) What is the Shapley-Shubik power index of each of the other players? Explain.

    (e) A generalization of the results in (d) is that in any weighted voting system where the quota equals 100% of the votes, the Shapley-Shubik power index of each player is $1/N$, where N denotes the number of players. Explain why this is true.

    [**Hint:** Generalize your observations in (a) through (d).]

50. The disciplinary board at Tasmania State University is composed of five members, two of which must be faculty and three of which must be students. To pass a motion requires at least three votes, and at least one of the votes must be from a faculty member.

    (a) Find the Banzhaf power distribution of the disciplinary board.

    (b) Describe the disciplinary board as a weighted voting system $[q: f, f, s, s, s]$.

51. A professional basketball team has four coaches, a head coach (H), and three assistant coaches (A_1, A_2, A_3). Player personnel decisions require at least three yes votes, one of which must be H's.

    (a) If we use $[q: h, a, a, a]$ to describe this weighted voting system, find q, h, and a.

    (b) Find the Shapley-Shubik power distribution of the weighted voting system.

52. **Veto power.** We defined a player P to have *veto power* if the coalition formed by all the other players is a losing coalition. Explain why each of the following is true.

    (a) If P has veto power, then P is a member of every winning coalition.

    (b) If P is a critical member of every winning coalition, then P has veto power.

53. **Dummies.** We defined a *dummy* as a player that is never a critical player. Explain why each of the following is true.

    (a) If P is a dummy, then any winning coalition that contains P would also be a winning coalition without P.

    (b) If P is a dummy, then P cannot be a member of every winning coalition.

    (c) If P is a dummy, then P cannot be a pivotal member in any sequential coalition.

    (d) If P is never a pivotal member in a sequential coalition, then P must be a dummy.

54. (a) Consider the weighted voting system $[22: 10, 10, 10, 10, 1]$. Are there any dummies? Explain your answer.

    (b) Without doing any work [but using your answer for (a)], find the Banzhaf and Shapley-Shubik power distributions of this weighted voting system.

    (c) Consider the weighted voting system $[q: 10, 10, 10, 10, 1]$. Find all the possible values of q for which P_5 is not a dummy.

    (d) Consider the weighted voting system $[34: 10, 10, 10, 10, w]$. Find all positive integers w which make P_5 a dummy.

55. Consider the weighted voting system $[q: 8, 4, 1]$.

    (a) What are the possible values of q?

    (b) Which values of q result in a dictator? (Who? Why?)

    (c) Which values of q result in exactly one player with veto power? (Who? Why?)

    (d) Which values of q result in more than one player with veto power? (Who? Why?)

    (e) Which values of q result in one or more dummies? (Who? Why?)

56. Consider the weighted voting system $[9: w, 5, 2, 1]$.

    (a) What are the possible values of w?

    (b) Which values of w result in a dictator? (Who? Why?)

    (c) Which values of w result in a player with veto power? (Who? Why?)

    (d) Which values of w result in one or more dummies? (Who? Why?)

57. (a) Verify that the weighted voting systems $[12: 7, 4, 3, 2]$ and $[24: 14, 8, 6, 4]$ result in exactly the same Banzhaf power distribution. (If you need to make calculations, do them for both systems side by side and look for patterns.)

    (b) Based on your work in (a), explain why the two proportional weighted voting systems $[q: w_1, w_2, \ldots, w_N]$ and $[cq: cw_1, cw_2, \ldots, cw_N]$ always have the same Banzhaf power distribution.

58. (a) Verify that the weighted voting systems [12: 7, 4, 3, 2] and [24: 14, 8, 6, 4] result in exactly the same Shapley-Shubik power distribution. (If you need to make calculations, do them for both systems side by side and look for patterns.)

(b) Based on your work in (a), explain why the two proportional weighted voting systems $[q: w_1, w_2, \ldots, w_N]$ and $[cq: cw_1, cw_2, \ldots, cw_N]$ always have the same Shapley-Shubik power distribution.

59. Consider the weighted voting system [14: 10, 1, 1, 1, 1, 1].

(a) When P_1 is the last player in a sequential coalition, then P_1 is pivotal. In how many ways can P_1 be the last player in a sequential coalition?

(**Hint:** See Exercise 43.)

(b) When P_1 is the fifth player in a sequential coalition, then P_1 is pivotal. In how many ways can P_1 be the fifth player in a sequential coalition?

(c) Without listing any sequential coalitions, determine the Shapley-Shubik power index of P_1.

(d) Without listing any sequential coalitions, determine the Shapley-Shubik power distribution of this weighted voting system.

60. Consider the generic weighted voting system $[q: w_1, w_2, \ldots, w_N]$. (Assume $w_1 \geq w_2 \geq \cdots \geq w_N$.)

(a) Find all the possible values of q for which no player has veto power.

(**Hint:** See Exercises 45 and 46.)

(b) Find all the possible values of q for which every player has veto power.

(**Hint:** See Exercises 45 and 46.)

61. The weighted voting system [6: 4, 2, 2, 2, 1] represents a partnership among five people (P_1, P_2, P_3, P_4, and you!). You are the last player (the one with 1 vote), which in this case makes you a dummy! Not wanting to remain a dummy, you offer to buy one vote. Each of the other four partners is willing to sell you one of their votes, and they are all asking the same price. Which partner should you buy from in order to get as much power for your buck as possible? Use the Banzhaf power index for your calculations. Explain your answer.

62. The weighted voting system [27: 10, 8, 6, 4, 2] represents a partnership among five people (P_1, P_2, P_3, P_4, and P_5). You are P_5, the one with two votes. You want to increase your power in the partnership and are prepared to buy one share (1 share = 1 vote) from any of the other partners. P_1, P_2, and P_3 are each willing to sell cheap ($1000 for one share), but P_4 is not being quite as cooperative—she wants $5000 for one share. Given that you still want to buy one share, whom should you buy it from? Use the Banzhaf power index for your calculations. Explain your answer.

63. The weighted voting system [18: 10, 8, 6, 4, 2] represents a partnership among five people (P_1, P_2, P_3, P_4, and P_5). You are P_5, the one with two votes. You want to increase your power in the partnership and are prepared to buy shares (1 share = 1 vote) from any of the other partners.

(a) Suppose that each partner is willing to sell one share and they are all asking the same price. Assuming that you decide to buy only one share, which partner should you buy from? Use the Banzhaf power index for your calculations.

(b) Suppose that each partner is willing to sell two shares and they are all asking the same price. Assuming that you decide to buy two shares from a single partner, which partner should you buy from? Use the Banzhaf power index for your calculations.

(c) If you have the money and the cost per share is fixed, should you buy one share or two shares (from a single person)? Explain.

64. Sometimes in a weighted voting system, two or more players decide to merge—that is to say, to combine their votes and always vote the same way. (Notice that a merger is different from a coalition—coalitions are temporary, whereas mergers are permanent.) For example, if in the weighted voting system [7: 5, 3, 1] P_2 and P_3 were to merge, the weighted voting system would then become [7: 5, 4]. In this exercise, we explore the effects of mergers on a player's power.

(a) Consider the weighted voting system [4: 3, 2, 1]. In Example 9 we saw that P_2 and P_3 each have a Banzhaf power index of 1/5. Suppose that P_2 and P_3 merge and become a single player P^*. What is the Banzhaf power index of P^*?

(b) Consider the weighted voting system [5: 3, 2, 1]. Find first the Banzhaf power indexes of players P_2 and P_3 and then the Banzhaf power index of P^* (the merger of P_2 and P_3). Compare.

(c) Rework the problem in (b) for the weighted voting system [6: 3, 2, 1].

(d) What are your conclusions from (a), (b), and (c)?

65. Decisive voting systems. A weighted voting system is called **decisive** if for every losing coalition, the coalition consisting of the remaining players (called the *complement*) must be a winning coalition.

(a) Show that the weighted voting system [5: 4, 3, 2] is decisive.

(b) Show that the weighted voting system [3: 2, 1, 1, 1] is decisive.

(c) Explain why any weighted voting system with a dictator is decisive.

(d) Find the number of winning coalitions in a decisive voting system with N players.

66. **Equivalent voting systems.** Two weighted voting systems are **equivalent** if they have the same number of players and exactly the same winning coalitions.

    (a) Show that the weighted voting systems [8: 5, 3, 2] and [2: 1, 1, 0] are equivalent.

    (b) Show that the weighted voting systems [7: 4, 3, 2, 1] and [5: 3, 2, 1, 1] are equivalent.

    (c) Show that the weighted voting system discussed in Example 2.12 is equivalent to [3: 1, 1, 1, 1, 1].

    (d) Explain why equivalent weighted voting systems must have the same Banzhaf power distribution.

    (e) Explain why equivalent weighted voting systems must have the same Shapley-Shubik power distribution.

67. **The United Nations Security Council.** The U.N. Security Council consists of 15 member countries—5 permanent members and 10 nonpermanent members. A motion can pass only if it has the vote of *all* five of the permanent members plus at least four of the nonpermanent members.

    (a) Describe the critical players in a winning coalition.

    (***Hint:*** *Consider separately nine-member winning coalitions and winning coalitions with ten or more members.*)

    (b) Use your answer in (a), together with the fact that there are 210 nine-member coalitions and 638 coalitions with ten or more members, to explain why the total number of times all players are critical is 5080.

    (c) Using the results of (a) and (b), show that the Banzhaf power index of a permanent member is given by the ratio 848/5080.

    (d) Using the results of (a), (b), and (c), show that the Banzhaf power index of a nonpermanent member is given by the ratio 84/5080.

    (e) Explain why the U.N. Security Council is equivalent to a weighted voting system in which each nonpermanent member has 1 vote, each permanent member has 7 votes, and the quota is 39 votes.

RUNNING

68. Consider a weighted voting system with N players, P_1 through P_N.

    (a) Choose a random player P. In how many sequential coalitions is P in the first place? Second place? Last place?

    (b) Choose randomly two players. Call them P and Q. In how many sequential coalitions is P in the first place and Q in the second place? How about P in the last place and Q in the first place? How about P in place j and Q in place k, where j and k are any two distinct numbers between 1 and N?

    (c) Choose randomly three players. Call them P, Q, and R. Choose randomly three different numbers between 1 and N. Call them j, k, and m. In how many sequential coalitions is P in the jth place, Q in the kth place, and R in the mth place?

    (d) Generalize the results of (a), (b), and (c).

69. **Minimal voting systems.** A weighted voting system is called **minimal** if there is no equivalent weighted voting system with a smaller quota or with a smaller total number of votes. (For the definition of equivalent weighted voting systems, see Exercise 66.)

    (a) Show that the weighted voting system [3: 2, 1, 1] is minimal.

    (b) Show that the weighted voting system [4: 2, 2, 1] is not minimal and find an equivalent weighted voting system that is minimal.

    (c) Show that the weighted voting system [8: 5, 3, 1] is not minimal and find an equivalent weighted voting system that is minimal.

    (d) Given a weighted voting system with N players and a dictator, describe the minimal voting system equivalent to it.

70. **The Nassau County Board of Supervisors revisited.** Between 1994 and 1995, the Nassau County Board of Supervisors operated as the weighted voting system [65: 30, 28, 22, 15, 7, 6] (see Exercise 22). Show that the weighted voting system [15: 7, 6, 5, 4, 2, 1] is

    (a) equivalent to [65: 30, 28, 22, 15, 7, 6].

    (b) minimal (see Exercise 69).

    (***Hint:*** *First show that all six weights must be positive and all must be different. Then examine possible quotas that would give the correct results for the three coalitions* $\{P_1, P_2, P_5\}$, $\{P_1, P_2, P_6\}$, *and* $\{P_2, P_3, P_4\}$. *Conclude that the players' weights cannot be 6, 5, 4, 3, 2, 1. Use the same coalitions to conclude that* $w_3 + w_4 > w_1 + w_6$, *and finally that if* $w_1 = 7$, *then* $w_4 = 4$.)

71. **The Cleansburg City Council.** Find the Banzhaf power distribution in the Cleansburg City Council. (See Example 2.19 for details.)

72. **The Fresno City Council.** In Fresno, California, the city council consists of seven members (the mayor and six other council members). A motion can be passed by the mayor and at least three other council members, or by at least five of the six ordinary council members.

    (a) Describe the Fresno City Council as a weighted voting system.

    (b) Find the Shapley-Shubik power distribution for the Fresno City Council.
    (***Hint:*** *See Example 2.19 for some useful ideas.*)

73. Suppose that in a weighted voting system there is a player A who hates another player P so much that he will always vote the opposite way of P, regardless of the issue. We will call A the **antagonist** of P.

(a) Suppose that in the weighted voting system $[8; 5, 4, 3, 2]$, P is the player with two votes and his antagonist A is the player with five votes. What are the possible coalitions under these circumstances? What is the Banzhaf power distribution under these circumstances?

(b) Suppose that in a generic weighted voting system with N players there is a player P who has an antagonist A. How many coalitions are there under these circumstances?

(c) Give examples of weighted voting systems where a player A can

(i) increase his Banzhaf power index by becoming an antagonist of another player

(ii) decrease his Banzhaf power index by becoming an antagonist of another player

(d) Suppose that the antagonist A has more votes than his enemy P. What is a strategy that P can use to gain power at the expense of A?

74. (a) Give an example of a weighted voting system with four players and such that the Shapley-Shubik power index of P_1 is $3/4$.

(b) Show that in any weighted voting system with four players, a player cannot have a Shapley-Shubik power index of more than $3/4$ unless he or she is a dictator.

(c) Show that in any weighted voting system with N players, a player cannot have a Shapley-Shubik power index of more than $(N - 1)/N$ unless he or she is a dictator.

(d) Give an example of a weighted voting system with N players and such that P_1 has a Shapley-Shubik power index of $(N - 1)/N$.

75. (a) Give an example of a weighted voting system with three players and such that the Shapley-Shubik power index of P_3 is $1/6$.

(b) Explain why in any weighted voting system with three players, a player cannot have a Shapley-Shubik power index of less than $1/6$ unless he or she is a dummy.

(c) Give an example of a weighted voting system with four players and such that the Shapley-Shubik power index of P_4 is $1/12$.

(d) Explain why in any weighted voting system with four players, a player cannot have a Shapley-Shubik power index of less than $1/12$ unless he or she is a dummy.

76. (a) Give an example of a weighted voting system with N players having a player with veto power who has a Shapley-Shubik power index of $1/N$.

(b) Explain why in any weighted voting system with N players, a player with veto power must have a Shapley-Shubik power index of at least $1/N$.

77. (a) Give an example of a weighted voting system with N players having a player with veto power who has a Banzhaf power index of $1/N$.

(b) Explain why in any weighted voting system with N players, a player with veto power must have a Banzhaf power index of at least $1/N$.

78. An alternative way to compute Banzhaf power. As we know, the first step in computing the Banzhaf power index of player P is to compute B, the total number of times P is a critical player. The following formula gives an alternative way to compute B: $B = 2 \cdot W_P - W$ where W is the number of winning coalitions and W_P is the number of winning coalitions containing P. Explain why the preceding formula is true.

Projects and Papers

A. The Johnston Power Index

The Banzhaf and Shapley-Shubik power indexes are not the only two mathematical methods for measuring power. The Johnston power index is a subtle but rarely used variation of the Banzhaf power index in which the power of a player is based not only on how often he or she is critical in a coalition, but also on the number of other players in the coalition.

Specifically, being a critical player in a coalition of 2 players contributes $1/2$ toward your power score; being critical in a coalition of 3 players contributes $1/3$ toward your power score; and being critical in a coalition of 10 contributes only $1/10$ toward your power score.

A player's *Johnston power score* is obtained by adding all such fractions over all coalitions in which the player is critical. The player's *Johnston power index* is his or her

Johnston power score divided by the sum of all players' power scores. (A more detailed description of how to compute the Johnston power index is given in reference 17.)

Prepare a presentation on the Johnston power index. Include a mathematical description of the procedure for computing Johnston power, give examples, and compare the results with the ones obtained using the Banzhaf method. Include your own personal analysis on the merits of the Johnston method compared with the Banzhaf method.

B. The Past, Present, and Future of the Electoral College

Starting with the Constitutional Convention of 1776 and ending with the Bush-Gore presidential election of 2000, give a historical and political analysis of the Electoral College. You should address some or all of the following issues: How did the Electoral College get started? Why did some of the Founding Fathers want it? How did it evolve? What has been its impact over the years in affecting presidential elections? (Pay particular attention to the 2000 presidential election.) What does the future hold for the Electoral College? What are the prospects that it will be reformed or eliminated?

C. Mathematical Arguments in Favor of the Electoral College

As a method for electing the president, the Electoral College is widely criticized as being undemocratic. At the same time,

different arguments have been made over the years to support the case that the Electoral College is not nearly as bad as it seems. Massachusetts Institute of Technology physicist Alan Natapoff has recently used mathematical ideas (many of which are connected to the material in this chapter) to make the claim that the Electoral College is a better system than a direct presidential election. Summarize and analyze Natapoff's mathematical arguments in support of the Electoral College. (Natapoff's arguments are nicely described in reference 6.)

D. Banzhaf Power and the Law

John Banzhaf was a lawyer, and he made his original arguments on behalf of his mathematical method to measure power in court cases, most of which involved the Nassau County Board of Supervisors in New York State. (See the discussion in Section 2.3.) Among the more significant court cases were *Graham v. Board of Supervisors* (1966); *Bechtle v. Board of Supervisors* (1981); *League of Women Voters v. Board of Supervisors* (1983); and *Jackson v. Board of Supervisors* (1991). Other important legal cases based on the Banzhaf method for measuring power but not involving Nassau County were *Ianucci v. Board of Supervisors of Washington County* and *Morris v. Board of Estimate* (U.S. Supreme Court, 1989). Choose one or two of these cases, read their background, arguments, and the court's decision, and write a brief for each. This is a good project for pre law and political science majors, but it might require access to a good law library.

Appendix 1 Power in the Electoral College (2001–2010)

| State | Electoral votes | | Power index (%) | |
	Number	Percent	Shapley-Shubik	Banzhaf
Alabama	9	1.673	1.639	1.640
Alaska	3	0.558	0.540	0.546
Arizona	10	1.859	1.824	1.823
Arkansas	6	1.115	1.086	1.092
California	55	10.223	11.036	11.402
Colorado	9	1.673	1.639	1.640

Connecticut	7	1.301	1.270	1.274
Delaware	3	0.558	0.540	0.546
District of Columbia	3	0.558	0.540	0.546
Florida	27	5.019	5.087	5.012
Georgia	15	2.788	2.761	2.744
Hawaii	4	0.743	0.722	0.728
Idaho	4	0.743	0.722	0.728
Illinois	21	3.903	3.910	3.865
Indiana	11	2.045	2.010	2.007
Iowa	7	1.301	1.270	1.274
Kansas	6	1.115	1.086	1.092
Kentucky	8	1.487	1.454	1.457
Louisiana	9	1.673	1.639	1.640
Maine	4	0.743	0.722	0.728
Maryland	10	1.859	1.824	1.823
Massachusetts	12	2.230	2.197	2.190
Michigan	17	3.160	3.141	3.116
Minnesota	10	1.859	1.824	1.823
Mississippi	6	1.115	1.086	1.092
Missouri	11	2.045	2.010	2.007
Montana	3	0.558	0.540	0.546
Nebraska	5	0.929	0.904	0.910
Nevada	5	0.929	0.904	0.910
New Hampshire	4	0.743	0.722	0.728
New Jersey	15	2.788	2.761	2.744
New Mexico	5	0.929	0.904	0.910
New York	31	5.762	5.888	5.795
North Carolina	15	2.788	2.761	2.744
North Dakota	3	0.558	0.540	0.546
Ohio	20	3.717	3.717	3.677
Oklahoma	7	1.301	1.270	1.274

State	Electoral votes		Power index (%)	
	Number	**Percent**	**Shapley-Shubik**	**Banzhaf**
Oregon	7	1.301	1.270	1.274
Pennsylvania	21	3.903	3.910	3.865
Rhode Island	4	0.743	0.722	0.728
South Carolina	8	1.487	1.454	1.457
South Dakota	3	0.558	0.540	0.546
Tennessee	11	2.045	2.010	2.007
Texas	34	6.320	6.499	6.393
Utah	5	0.929	0.904	0.910
Vermont	3	0.558	0.540	0.546
Virginia	13	2.416	2.384	2.375
Washington	11	2.045	2.010	2.007
West Virginia	5	0.929	0.904	0.910
Wisconsin	10	1.859	1.824	1.823
Wyoming	3	0.558	0.540	0.546

References and Further Readings

1. Banzhaf, John F., III, "One Man, 3.312 Votes: A Mathematical Analysis of the Electoral College," *Villanova Law Review*, 13 (1968), 304–332.

2. Banzhaf, John F., III, "Weighted Voting Doesn't Work," *Rutgers Law Review*, 19 (1965), 317–343.

3. Brams, Steven J., *Game Theory and Politics*. New York: Free Press, 1975, chap. 5.

4. Felsenthal, Dan, and Moshe Machover, *The Measurement of Voting Power: Theory and Practice, Problems and Paradoxes*. Cheltenham, England: Edward Elgar, 1998.

5. Grofman, B., "Fair Apportionment and the Banzhaf Index," *American Mathematical Monthly*, 88 (1981), 1–5.

6. Hively, Will, "Math Against Tyranny," *Discover*, November 1996, 74–85.

7. Imrie, Robert W., "The Impact of the Weighted Vote on Representation in Municipal Governing Bodies of New York State," *Annals of the New York Academy of Sciences*, 219 (November 1973), 192–199.

8. Lambert, John P., "Voting Games, Power Indices and Presidential Elections," *UMAP Journal*, 3 (1988), 213–267.

9. Merrill, Samuel, "Approximations to the Banzhaf Index of Voting Power," *American Mathematical Monthly*, 89 (1982), 108–110.

10. Meyerson, Michael I., *Political Numeracy: Mathematical Perspectives on Our Chaotic Constitution*. New York: W. W. Norton, 2002, chap. 2.

11. Riker, William H., and Peter G. Ordeshook, *An Introduction to Positive Political Theory*. Englewood Cliffs, NJ: Prentice-Hall, Inc., 1973, chap. 6.

12. Shapley, Lloyd, and Martin Shubik, "A Method for Evaluating the Distribution of Power in a Committee System," *American Political Science Review*, 48 (1954), 787–792.

13. Sickels, Robert J., "The Power Index and the Electoral College: A Challenge to Banzhaf's Analysis," *Villanova Law Review*, 14 (1968), 92–96.

14. Straffin, Philip D., Jr., "The Power of Voting Blocs: An Example," *Mathematics Magazine*, 50 (1977), 22–24.

15. Straffin, Philip D., Jr., *Topics in the Theory of Voting, UMAP Expository Monograph*. Boston: Birkhäuser, 1980, chap. 1.

16. Tannenbaum, Peter, "Power in Weighted Voting Systems," *The Mathematica Journal*, 7 (1997), 58–63.

17. Taylor, Alan, *Mathematics and Politics: Strategy, Voting, Power and Proof*. New York: Springer-Verlag, 1995, chaps. 4 and 9.

3

Fair Division

The Mathematics of Sharing

> Grief can take care of itself, but to get the full value of a joy you must have someone to divide it with.
>
> Mark Twain

We start to learn about sharing at a very young age—sharing toys, sharing treats, sharing attention. As we get older, we learn about more abstract forms of sharing such as sharing duties, responsibilities, and even blame. This business of dividing things—be they "good" things (food, toys, love) or "bad" things (chores, responsibility, guilt)—is among the most social of all human interactions. Even social animals share, although not always in agreeable ways (if we are to believe Aesop's fable on the opposing page). Of course, our long history of war and conquest bears witness to the fact that we humans often share just as badly, possibly worse. But we can also do it better, quite well in fact, when we set our minds to it. Dividing things fairly using reason and logic, instead of bullying our way to a solution, is one of the great achievements of social science, and, once again, we can trace the roots of this achievement to simple mathematics.

We take our first mathematical stab at *fair division* somewhere around the third or fourth grade. A typical problem goes like this: There are 20 pieces of candy to be divided among four equally deserving children. What is a fair solution? We know, of course, the standard answer: Give each child five pieces. The problem with this answer is that it may not produce a fair division after all. What if the pieces of candy are not all the same? Say the 20 pieces are made up of a wide variety of goodies: M&M's®, Milk Duds®, Crunch® bars, Reese's® cups, and so on, with some pieces clearly more desirable than others, and with each child having a different set of preferences. Can we take these diverse opinions into account and still divide the pieces fairly? And by the way, what does *fairly* mean in this situation? These are some of the many thought-provoking questions we will discuss in this chapter.

Why are these questions important, you may wonder? After all, it's only candy and kids. Not exactly. Just like the booty in Aesop's fable, the candy is a metaphor—we could just as well be dividing family heirlooms, land, works of art, and other items of real value, as is often the case in an inheritance or a divorce proceeding. And, on an even grander scale, questions of fair division may involve issues of global significance like dividing an entire country (consider the division of the former Yugoslavia in the 1990s), or dividing the rights to mine the ocean's resources (consider the 1982 Convention of the Law of the Sea), or even dividing the global responsibilities for reducing greenhouse gas emissions (consider the 1997 Kyoto treaty).

The basic issue in all fair-division problems can be stated in reasonably simple terms: How can something that must be shared by a set of competing parties be divided among them in a way that ensures that each party receives a fair share? It could be argued that a good general answer to this question would go a long way in solving most of the problems of humankind, but unfortunately, good general answers are not possible. Under the right set of circumstances, however, it is possible to solve special types of fair-division problems using basic mathematical ideas. We will explore a few of these ideas in this chapter.

The Lion, the Fox, and the Ass
One of Aesop's Fables

The Lion, the Fox and the Ass entered into an agreement to assist each other in the hunt. Having secured a large booty, the Lion on their return from the forest asked the Ass to allot its due portion to each of the three partners in the treaty.

The Ass carefully divided the spoil into three equal shares and modestly requested the two others to make the first choice.

The Lion, bursting out into a great rage, devoured the Ass. Then he requested the Fox to do him the favor to make the division. The Fox accumulated all that they had killed into one large heap and left to himself the smallest possible morsel.

The Lion said, "Who has taught you, my very excellent fellow, the art of division? You are perfect to a fraction," to which the Fox replied, "I learned it from the Ass, by witnessing his fate."

The chapter is organized in a manner very similar to that of Chapters 1 and 2. We will start with an introduction to the key concepts and definitions (Section 3.1), followed by a description and analysis of several specific *fair-division methods* that have been developed over the years (Sections 3.2 through 3.7). In the conclusion, we recap the main ideas of the chapter and take a brief look at the "big picture"—what is fair division all about and why should we really care.

3.1   Fair-Division Games

In this section we will introduce some of the basic concepts and terminology of fair division. Much as we did in Chapter 2 when we studied weighted voting systems, we will think of fair-division problems in terms of games—with players, goals, rules, and strategies.

The underlying elements of every fair-division game are as follows:

Obligations, responsibilities, and chores are items with a negative value—here the name of the game is to get the smallest possible share. The division of items of negative value is discussed in Exercises 78 and 79.

- **The goods (or "booty").** This is the informal name we will give to the item or items being divided. Typically, these items are tangible physical objects with a positive value, such as candy, cake, pizza, jewelry, art, land, and so on. In more exotic situations the items being divided may be intangible things such as rights (water rights, drilling rights, broadcast licenses, etc.), or obligations (chores, taxes, environmental cleanup, etc.). Regardless of the nature of the items being divided, we will use the symbol S throughout this chapter to denote the booty.

- **The players.** In every fair-division game there is a set of parties entitled to share the booty. They are the *players* in the game. Most of the time the players in a fair-division game are individuals, but it is worth noting that some of the most significant applications of fair division occur when the players are *institutions* (ethnic groups, political parties, states, and even nations).

- **The value systems.** The fundamental assumption we will make is that each player has an internalized *value system* that gives it the ability to quantify the value of the booty or any of its parts. Specifically, this means that each player can look at the set S or any subset of S and assign to it a value—either in absolute terms ("to me, that's worth $147.50"), or in relative terms ("to me, that piece is worth 30% of the total value of S").

Like most games, fair-division games are predicated on certain assumptions about the players. For the purposes of our discussion, we will make the following four assumptions:

- **Rationality:** Each of the players is a thinking, rational agent seeking to maximize his or her share of the booty S. We will further assume that in the pursuit of this goal, a player's moves are based on reason alone (we are taking emotion, psychology, mind games, and all other nonrational elements out of the picture).

- **Cooperation:** The players are willing participants and accept the rules of the game as binding. The rules are such that after a *finite* number of moves by the players the game terminates with a division of S. (There are no outsiders such as judges or referees involved in these games—just the players and the rules.)

- **Privacy:** Players have *no* useful information on the other players' value systems, and thus, of what kinds of moves they are going to make in the game. (This assumption does not always hold in real life, especially if the players are siblings or friends.)

- **Symmetry:** Players have *equal* rights in sharing the set S. A consequence of this assumption is that at a minimum, each player is entitled to a *proportional* share of S—when there are two players, each is entitled to at least one-half of S, with three players each is entitled to at least one-third of S, and so on.

Fair-division methods can also be used in situations where different players are entitled by right (or might) to different-sized shares of the booty. For more details on asymmetric fair division, see Project B.

Fair Shares and Fair-Division Methods

Given the booty S, and players $P_1, P_2, P_3, \ldots, P_N$, each with his or her own value system, the ultimate goal of the game is to end up with a **fair division** of S, that is, to divide S into N shares and *assign shares to players in such a way that each player gets a fair share*. This leads us to what is perhaps the most important definition of this section.

> **Fair Share**
>
> Suppose that s denotes a share of the booty S and P is one of players in a fair division game with N players. We will say that s is a **fair share to player P** if s is worth *at least* $1/N$th of the total value of S *in the opinion of P*. (Such a share is often called a *proportional fair share*, but for simplicity we will refer to it just as a *fair share*.)

For example, suppose that we have four players, and S has been divided into four shares s_1, s_2, s_3, and s_4. Paul (one of the four players) values s_1 at 12% of S, s_2 at 27% of S, s_3 at 18% of S, and s_4 at 43% of S. In this situation, s_2 *and* s_4 are both fair shares to Paul, since the threshold for a fair share when dividing among four players is 25% of S. Note that a share can be a fair share without being the most valuable share—clearly Paul would be most happy if he got s_4, but he may very well have to settle for s_2.

A fair division in which each player gets a share that he or she considers to be the best share is called an envy-free fair division. For more on envy-free fair division, see Project A.

In the coming sections of this chapter, we will discuss several different **fair-division methods** (also known as fair-division *protocols* or fair-division *schemes*). Essentially, we can think of a *fair-division method* as the *set of rules* that define how the game is to be played. Thus, in a fair-division game we must consider not only the booty S and the players $P_1, P_2, P_3, \ldots, P_N$ (each with his or her own opinions about how S should be divided), but also a specific method by which we plan to accomplish the fair division.

There are many different fair-division methods known, but in this chapter we will only discuss a few of the classic ones. Depending on the nature of the set S, a fair-division game can be classified as one of three types: *continuous, discrete,* or *mixed*, and the fair-division methods used depend on which of these types we are facing.

In a **continuous** fair-division game the set S is divisible in infinitely many ways, and shares can be increased or decreased by arbitrarily small amounts. Typical examples of continuous fair-division games involve the division of land, a cake, a pizza, and so on.

A fair-division game is **discrete** when the set S is made up of objects that are indivisible like paintings, houses, cars, boats, jewelry, and so on. (What about pieces of candy? One might argue that with a sharp enough knife a piece of candy could be chopped up into smaller and smaller pieces, but that's messy and impractical. As a practical matter let's agree that candy is indivisible and therefore dividing candy is a discrete fair-division game.)

A *mixed* fair-division game is one in which some of the components are continuous and some are discrete. Dividing an estate consisting of jewelry, a house, and a parcel of land is a mixed fair-division game.

Fair-division methods are classified according to the nature of the problem involved. Thus, there are *discrete fair-division* methods (used when the set S is made up of indivisible, discrete objects), and there are *continuous fair-division* methods (used when the set S is an infinitely divisible, continuous set). Mixed fair-division games can usually be solved by dividing the continuous and discrete parts separately, so we will not study them in this chapter.

3.2 Two Players: The Divider-Chooser Method

The **divider-chooser method** is undoubtedly the best known of all continuous fair-division methods. This method can be used anytime the fair-division game involves two players and a *continuous* set S (a cake, a pizza, a plot of land, etc.). Most of us have unwittingly used it at some time or another, and informally it is best known as the *you cut—I choose method*. As this name suggests, one player, called the *divider*, divides the *cake* (a metaphor for any continuous set S) into two pieces, and the second player, called the *chooser*, picks the piece he or she wants, leaving the other piece to the divider. When played properly, this method guarantees that both players will get a share they believe to be worth *at least one-half* of the total. The divider can guarantee this by dividing the cake into two halves of equal value, and the chooser is guaranteed a fair share by choosing the piece he or she likes best.

> **EXAMPLE 3.1 Damian and Cleo Divide a Cheesecake**

On their first date, Damian and Cleo go to the county fair. They buy jointly a $2 raffle ticket, and as luck would have it, they win the chocolate-strawberry cheesecake shown in Fig. 3-1(a).

To Damian, chocolate and strawberry are equal in value—he has no preference for one over the other. Thus, in Damian's eyes the value of the cake is distributed evenly between the chocolate and strawberry parts [Fig. 3-1(b)]. On the

FIGURE 3-1 A chocolate-strawberry cake. The values are in the eyes of the beholder.

(a) (b) (c)

other hand, Cleo hates chocolate (she is allergic to it and gets sick if she eats any). Thus, in Cleo's eyes the value of the cake is concentrated entirely in the strawberry half; the chocolate half has *zero value* [Fig. 3-1(c)]. Since this is their first date, we can assume neither one of them knows anything about the other's likes and dislikes.

Let's now see how Damian and Cleo might divide this cake using the divider-chooser method. Damian volunteers to go first and be the divider. His cut is shown in Fig. 3-2(a). Note that this is a perfectly rational cut based on Damian's value system—each piece is worth one-half of the total value of the cake. It is now Cleo's turn to choose, and her choice is obvious—she will pick the piece having the largest strawberry part [Fig. 3-2(b)]. Notice that while Damian gets a piece that to him is worth 50% of the cake, Cleo ends up with a much sweeter deal—a piece that to her is worth about 67% of the cake.

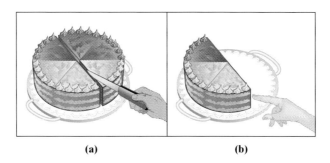

FIGURE 3-2 (a) Damian cuts (b) Cleo picks.

Hugo Steinhaus (1887–1972). For more on Steinhaus see the biographical profile at the end of the chapter.

Example 3.1 illustrates why, given a choice, it is always *better to be the chooser than the divider*—the divider is guaranteed a piece worth exactly one-half of the total, but the chooser has a chance to get a piece worth more than one-half. Since a fair-division method should treat all players equally, both players should have an equal chance of being the chooser. This is best done by means of a coin toss, with the winner of the coin toss getting the privilege of making the choice to be divider or chooser.

The idea behind the divider-chooser method is simple, and ancient. But what do we do when there are three players? Four? *N*? Surprisingly, going from two to three or more players is not simple, and generalizations of the divider-chooser method did not come about until the 1940s, primarily as a result of the pioneering work of one man, the Polish mathematician Hugo Steinhaus.

In the next few sections we will discuss three methods that can be used to extend the divider-chooser method to *continuous* fair-division games involving *three or more* players.

3.3 The Lone-Divider Method

The first important breakthrough in the mathematics of fair division came in 1943, when Steinhaus came up with a clever way to extend some of the ideas in the divider-chooser method to the case of a continuous fair division game with *three* players, one of which plays the role of the *divider* and the other two play the role of *choosers*. Steinhaus's approach was subsequently generalized to any number of players N (one divider, $N - 1$ choosers) by Princeton mathematician Harold Kuhn. In either case, we will refer to this method as the **lone-divider method**.

We start this section with a description of Steinhaus's *lone-divider method* for the case of $N = 3$ players.

The Lone-Divider Method for Three Players

■ **Preliminaries.** One of the three players will be the divider; the other two players will be choosers. Since it is better to be a chooser than a divider, the decision of who is what is made by a random draw (rolling dice, drawing cards from a deck, etc.). We'll call the divider D and the choosers C_1 and C_2.

■ **Step 1. (Division).** The divider D divides the cake into three pieces (s_1, s_2, and s_3). D will get one of these pieces, but at this point does not know which one. (Not knowing which of the pieces will be his share is critical—it forces D to divide the cake in three shares that to him have equal value.)

See Exercise 70.

■ **Step 2. (Bidding).** C_1 declares (usually by writing on a slip of paper) which of the three pieces are fair shares to her. Independently, C_2 does the same. These are the choosers' *bid lists*. A chooser's bid list should include *every piece that he or she values to be a fair share* (i.e., worth one-third or more of the cake)—it may be tempting to bid only for the best share, but this is a strategy that can easily backfire. To preserve the privacy requirement, it is important that the bids be made independently, without the choosers being privy to each other's bid lists.

■ **Step 3. (Distribution).** Who gets which piece? The answer, of course, depends on the bid lists. For convenience, we will separate the pieces into two groups: *chosen* pieces that appear in either one or both of the choosers' bid lists (let's call them C-pieces), and *unwanted* pieces that did not appear in either bid list (let's call them U-pieces). Expressed in terms of the choosers' value systems, a U-piece is a piece valued at less than $33\frac{1}{3}\%$ of the cake by *both* choosers, and a C-piece is a piece valued at $33\frac{1}{3}\%$ or more by *at least* one of the choosers. Since each bid list must have at least one piece, there is always at least one C-piece, usually more. There are two separate cases to consider:

Case 1. When there are two or more C-pieces, there is always a way to give each chooser a different piece from her bid list. (The details will be covered in Examples 3.2 and 3.3.) Once each chooser gets her piece, the divider gets the last remaining piece. At this point every player has received a fair share, and our goal of fair division has been met. (Sometimes we might end up in a situation where C_1 likes C_2's piece better than her own, and vice versa. In this case it is perfectly reasonable to add a final, informal step and let them swap pieces—this would make each of them happier than they already were, and who could be against that?)

Case 2. There is only one C-piece, which means that both choosers are bidding on the same piece. The solution here requires a little more creativity. First, we take care of the divider D—to whom all pieces are equal in value— by giving him one of the pieces that neither chooser wants, as shown in Fig. 3-3(a). (If the two choosers can agree on the least desirable piece, then so much the better; if they can't agree, the choice of which piece to give the divider can be made randomly.) After D gets his piece, there are two pieces left, the "good" piece that both choosers want (the C-piece), and the "not so good" piece that neither chooser wants (the remaining U-piece). We now recombine these two pieces into a single piece [the B-piece in Fig. 3-3(b)] and *use the divider-chooser method* to divide the piece between the two choosers as

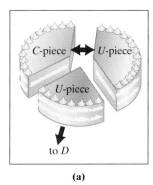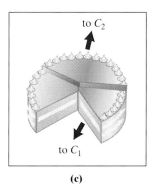

(a) **(b)** **(c)**

FIGURE 3-3 Case 2 in the lone-divider method (three players). (a) One of the U-pieces goes to the divider D. (b) The C-piece and the remaining U-piece are recombined into the B-piece. (c) The B-piece is divided in two shares using the divider-chooser method.

shown in Fig. 3-3(c) (C_1 cuts the B-piece in two; C_2 chooses the share she likes best). We now claim that both choosers end up with a fair share of the original cake. Why? The key observation is that in the eyes of both choosers the B-piece is worth *more than two-thirds of the value of the original cake* [think of the B-piece as the original cake (100%) minus a U-piece (worth less than $33\frac{1}{3}\%$)], so that when we divide *it* fairly into *two* shares, each party is guaranteed *more than one-third* of the original cake. We will come back to this point in Example 3.4.

We will now illustrate the details of Steinhaus's lone-divider method for three players with several examples. In all of these examples, we will assume that the divider has already divided the cake into three pieces s_1, s_2, and s_3. In each example, the values that each of the three players assigns to the pieces, expressed as percentages of the total value of the cake, are shown in the form of a table. The reader should remember, however, that this information is never available in full to the players—an individual player only knows the percentages on his or her row.

> EXAMPLE 3.2 Lone-Divider with Three Players (Case 1, Version 1)

Dale, Cindy, and Cher are dividing a cake using Steinhaus's lone-divider method. They draw cards from a well-shuffled deck of cards and Dale draws the low card (bad luck!) and has to be the divider.

Step 1. (Division). Dale divides the cake into three pieces s_1, s_2, and s_3. Table 3-1 shows the values of the three pieces in the eyes of each of the players.

TABLE 3-1			
	s_1	s_2	s_3
Dale	$33\frac{1}{3}\%$	$33\frac{1}{3}\%$	$33\frac{1}{3}\%$
Cindy	35%	10%	55%
Cher	40%	25%	35%

Step 2. (Bidding). From Table 3-1 we can assume that Cindy's bid list is $\{s_1, s_3\}$, and Cher's bid list is also $\{s_1, s_3\}$.

Step 3. (**Distribution**). The C-pieces are s_1 and s_3. There are two possible distributions. One distribution would be, Cindy gets s_1, Cher gets s_3, and Dale gets s_2. An even better distribution (the *optimal* distribution) would be, Cindy gets s_3, Cher gets s_1, and Dale gets s_2. In the case of the first distribution, both Cindy and Cher would benefit by swapping pieces, and there is no rational reason why they would not do so. Thus, using the rationality assumption, we can conclude that in either case the final result will be the same: Cindy gets s_3, Cher gets s_1, and Dale gets s_2. ◀◀

> **EXAMPLE 3.3** Lone-Divider with Three Players (Case 1, Version 2)

We'll use the same setup as in Example 3.2—Dale is the divisor, Cindy and Cher are the choosers.

Step 1. (**Division**). Dale divides the cake into three pieces s_1, s_2, and s_3. Table 3-2 shows the values of the three pieces in the eyes of each of the players.

TABLE 3-2

	s_1	s_2	s_3
Dale	$33\frac{1}{3}\%$	$33\frac{1}{3}\%$	$33\frac{1}{3}\%$
Cindy	30%	40%	30%
Cher	60%	15%	25%

Step 2. (**Bidding**). Here Cindy's bid list is $\{s_2\}$ only, and Cher's bid list is $\{s_1\}$ only.

Step 3. (**Distribution**). This is the simplest of all situations, as there is only one possible distribution of the pieces: Cindy gets s_2, Cher gets s_1, and Dale gets s_3. ◀◀

> **EXAMPLE 3.4** Lone-Divider with Three Players (Case 2)

The gang of Examples 3.2 and 3.3 is back at it again.

Step 1. (**Division**). Dale divides the cake into three pieces s_1, s_2, and s_3. Table 3-3 shows the values of the three pieces in the eyes of each of the players.

TABLE 3-3

	s_1	s_2	s_3
Dale	$33\frac{1}{3}\%$	$33\frac{1}{3}\%$	$33\frac{1}{3}\%$
Cindy	20%	30%	50%
Cher	10%	20%	70%

Step 2. (**Bidding**). Here Cindy's and Cher's bid lists consist of just $\{s_3\}$.

Step 3. (**Distribution**). The only C-piece is s_3. Cindy and Cher talk it over, and, without giving away any other information, agree that of the two U-pieces, s_1

is the least desirable, so they all agree that Dale gets s_1. (Dale doesn't care which of the three pieces he gets, so he has no rational objection.) The remaining pieces (s_2 and s_3) are then recombined to form the B-piece, to be divided between Cindy and Cher using the divider-chooser method (one of them divides the B-piece into two shares, the other one chooses the share she likes best). Regardless of how this plays out, both of them will get a very healthy share of the cake: Cindy will end up with a piece worth at least 40% of the original cake (the B-piece is worth 80% of the original cake to Cindy) and Cher will end up with a piece worth at least 45% of the original cake (the B-piece is worth 90% of the original cake to Cher). ⏪

The Lone-Divider Method for More Than Three Players

In 1967 Harold Kuhn, a mathematician at Princeton University, was able to extend Steinhaus's lone-divider method to any number of players $N > 3$. The first two steps of Kuhn's method are a straightforward generalization of Steinhaus's lone-divider method for three players, but the distribution step requires some fairly sophisticated mathematical ideas and is rather difficult to describe in full generality, so we will only give an outline here and illustrate the details with a couple of examples for $N = 4$ players. (For a complete description of Kuhn's lone-divider method, see reference 12.)

- **Preliminaries.** One of the players is chosen to be the divider D, and the remaining $N - 1$ players are going to be all choosers. As always, it's better to be a chooser than a divider, so the decision should be made by a random draw.

- **Step 1. (Division).** The divider D divides the set S into N shares $s_1, s_2, s_3, \ldots, s_N$. D is guaranteed of getting one of these shares, but doesn't know which one.

- **Step 2. (Bidding).** Each of the $N - 1$ choosers independently submits a bid list consisting of *every share that he or she considers to be a fair share* (i.e., worth $1/N$th or more of S).

- **Step 3. (Distribution).** The bid lists are opened. Much as we did with three players, we will have to consider two separate cases, depending on how these bid lists turn out.

  Case 1. If there is a way to assign a different share to each of the $N - 1$ choosers, then that should be done. (Needless to say, the share assigned to a chooser should be from his or her bid list.) The divider, to whom all shares are presumed to be of equal value, gets the last unassigned share. At the end, players may choose to swap pieces if they want.

  Case 2. There is a *standoff*—in other words, there are two choosers both bidding for just one share, or three choosers bidding for just two shares, or K choosers bidding for less than K shares. This is a much more complicated case, and what follows is a rough sketch of what to do. To resolve a standoff, we first set aside the shares involved in the standoff from the remaining shares. Likewise, the players involved in the standoff are temporarily separated from the rest. Each of the remaining players (including the divider) can be assigned a fair share from among the remaining shares and sent packing. All the shares left are recombined into a new booty S to be divided among the players involved in the standoff, and the process starts all over again.

The following two examples will illustrate some of the ideas behind the lone-divider method in the case of $N = 4$ players. The first example is one without a standoff; the second example involves a standoff.

> ## EXAMPLE 3.5 Lone-Divider with Four Players (Case 1)

We have one divider, Demi, and three choosers, Chan, Chloe, and Chris.

Step 1. (Division). Demi divides the cake into four shares s_1, s_2, s_3, and s_4. Table 3-4 shows how each of the players values each of the four shares. Remember that the information on each row of Table 3-4 is private and only known to that player.

TABLE 3-4

	s_1	s_2	s_3	s_4
Demi	25%	25%	25%	25%
Chan	30%	20%	35%	15%
Chloe	20%	20%	40%	20%
Chris	25%	20%	20%	35%

Step 2. (Bidding). Chan's bid list is $\{s_1, s_3\}$; Chloe's bid list is $\{s_3\}$ only; Chris's bid list is $\{s_1, s_4\}$.

Step 3. (Distribution). The bid lists are opened. It is clear that for starters Chloe must get s_3—there is no other option. This forces the rest of the distribution: s_1 must then go to Chan, and s_4 goes to Chris. Finally, we give the last remaining piece s_2 to Demi.

This distribution results in a fair division of the cake, although it is not entirely "envy free"—Chan wishes he had Chloe's piece (35% is better than 30%) but Chloe is not about to trade pieces with him, so he is stuck with s_1. (From a strictly rational point of view, Chan has no reason to gripe—he did not get the best piece, but got a piece worth 30% of the total, much more than he had the right to expect.) ◀◀

> ## EXAMPLE 3.6 Lone-Divider with Four Players (Case 2)

Once again, we will let Demi be the divisor, and Chan, Chloe, and Chris be the three choosers (same players, different game).

Step 1. (Division). Demi divides the cake into four shares s_1, s_2, s_3, and s_4. Table 3-5 shows how each of the players values each of the four shares.

TABLE 3-5

	s_1	s_2	s_3	s_4
Demi	25%	25%	25%	25%
Chan	20%	20%	20%	40%
Chloe	15%	35%	30%	20%
Chris	22%	23%	20%	35%

Step 2. (Bidding). Chan's bid list is $\{s_4\}$; Chloe's bid list is $\{s_2, s_3\}$; Chris's bid list is $\{s_4\}$.

Step 3. (Distribution). The bid lists are opened, and the players can see that there is a standoff brewing in the horizon—Chan and Chris are both bidding for s_4. The first step is to set s_4 aside and assign Chloe and Demi a fair share from s_1, s_2, and s_3. Chloe could be given either s_2 or s_3. (She would rather have s_2, of course, but it's not for her to decide.) A coin toss is used to determine which one. Let's say Chloe ends up with s_3 (bad luck!). Demi could be now given either s_1 or s_2. Another coin toss, and Demi ends up with s_1. The final move is . . . you guessed it!—recombine s_2 and s_4 into a single piece to be divided between Chan and Chris using the divider-chooser method. Since $(s_2 + s_4)$ is worth 60% to Chan and 58% to Chris (you can check it out in Table 3-5), regardless of how this final division plays out they are both guaranteed a final share worth more than 25% of the cake. When all is said and done, every player ends up with a fair share. ◀◀

3.4 The Lone-Chooser Method

A completely different approach for extending the divider-chooser method was proposed in 1964 by A. M. Fink, a mathematician at Iowa State University. In this method, one player plays the role of chooser, all the other players start out playing the role of divisors. For this reason, the method is known as the **lone-chooser method**. Once again, we will start with a description of the method for the case of three players.

The Lone-Chooser Method for Three Players

- **Preliminaries.** We have one chooser and two dividers. Let's call the chooser C and the dividers D_1 and D_2. As usual, we decide who is what by a random draw.

- **Step 1. (Division).** D_1 and D_2 divide S [Fig. 3-4(a)] between themselves into *two* fair shares. To do this, they use the divider-chooser method. Let's say that D_1 ends up with s_1 and D_2 ends up with s_2 [Fig. 3-4(b)].

- **Step 2. (Subdivision).** Each divider divides his or her share into three subshares. Thus, D_1 divides s_1 into three subshares, which we will call s_{1a}, s_{1b}, and

 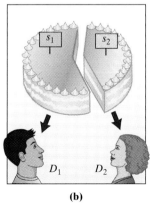 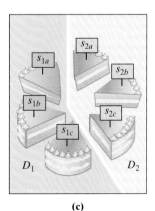 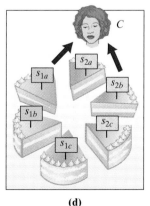

| (a) | (b) | (c) | (d) |

FIGURE 3-4 (a) The original cake, (b) first division, (c) second division, and (d) selection

s_{1c}. Likewise, D_2 divides s_2 into three subshares, which we will call s_{2a}, s_{2b}, and s_{2c} [Fig. 3-4(c)].

- **Step 3. (Selection).** The chooser C now selects one of D_1's three subshares and one of D_2's three subshares (whichever she likes best). These two subshares make up C's final share. D_1 then keeps the remaining two subshares from s_1, and D_2 keeps the remaining two subshares from s_2 [Fig. 3-4(d)].

Why is this a fair division of S? D_1 ends up with two-thirds of s_1, and s_1 is worth at least one-half of S, so s_1 is worth at least one-third of S—a fair share. The same argument applies to D_2. What about C's share? We don't know what s_1 and s_2 are each worth to C, but it really doesn't matter. Let's say for the sake of argument that in C's eyes s_1 was worth only 30% of S. This automatically implies that s_2 was worth 70%. Now C got a subshare from s_1 worth at least one-third of 30% (10%), and another subshare from s_2 worth at least one-third of 70% $\left(23\frac{1}{3}\%\right)$. Between the two, C got a fair share. The argument works no matter how C values s_1 and s_2.

The following example illustrates in detail how the lone-chooser method works with $N = 3$ players.

> ## EXAMPLE 3.7 Lone-Chooser with 3 Players

David, Dinah, and Cher are dividing the orange-pineapple cake shown in Fig. 3-5(a) using the lone-chooser method. The cake is valued by each of them at $27. Here are their individual value systems (not known by the other players): David likes pineapple and orange the same. [To him, value equals size, so in his eyes the cake looks like Fig. 3-5(b).] Dinah likes orange but hates pineapple. [To her the entire value of the cake is concentrated in the orange half, so in her eyes the cake looks like Fig.3-5(c).] Cher likes pineapple twice as much as she likes orange. [In her eyes the cake looks like Fig. 3-5(d).]

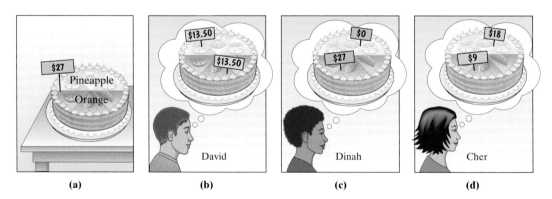

FIGURE 3-5 (a) The original cake, (b) David's view of the cake, (c) Dinah's view of the cake, (d) Cher's view of the cake

After a random selection, Cher gets to be the chooser and thus gets to sit out Steps 1 and 2.

Step 1. (Division). David and Dinah start by dividing the cake between themselves using the divider-chooser method. After a coin flip, David cuts the cake in two pieces, as shown in Fig. 3-6. Since Dinah doesn't like pineapple, she will take the share with the most orange.

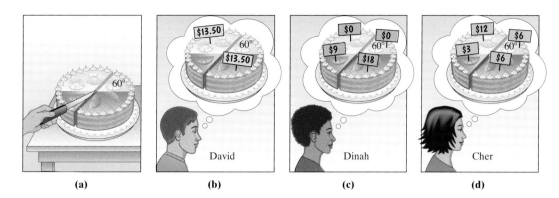

FIGURE 3-6
Division.

(a) (b) (c) (d)

Step 2. (Subdivision). David divides his share into three subshares that in his opinion are of equal value. Notice that the subshares [Fig. 3-7(a)] are all the same size. Dinah also divides her share into three smaller subshares

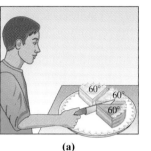

that in her opinion are of equal value [Fig. 3-7(b)]. (Remember that Dinah hates pineapple. Thus, she has made her cuts in such a way as to have one-third of the orange in each of the subshares.)

FIGURE 3-7 Subdivision.

(a) (b)

Step 3. (Selection). It's now Cher's turn to choose one subshare from David's three and one subshare from Dinah's three. Figure 3-8 shows the

values of the subshares in Cher's eyes. It's clear what her choices will be: She will choose one of the two pineapple wedges from David's subshares and the big orange-pineapple wedge from Dinah's subshares [Fig. 3-9(c)].

FIGURE 3-8 Selection.

(a) (b)

FIGURE 3-9 Final Division.

(a) (b) (c)

The final division of the cake is shown in Fig. 3-9. Since each player receives a share worth at least $9, the division is a fair division of the original cake. «

The Lone-Chooser Method for *N* Players

In the general case of *N* players, the lone chooser method involves one chooser *C* and $N - 1$ dividers $D_1, D_2, \ldots, D_{N-1}$. As always, it is preferable to be a chooser than a divider, so the chooser is determined by a random draw. The method is based on an inductive strategy—if you can do it for three players then you can do it for four players, if you can do it for four then you can do it for five, and so on. Thus, when we get to *N* players, we can assume that we can use the lone-chooser method with $N - 1$ players.

- **Step 1. (Division).** $D_1, D_2, \ldots, D_{N-1}$ divide fairly the set *S* among themselves, as if *C* didn't exist. This is a fair division among $N - 1$ players, so each one gets a share they consider worth at least $1/(N - 1)$th of *S*.
- **Step 2. (Subdivision).** Each divider subdivides his or her share into *N* subshares.
- **Step 3. (Selection).** The chooser *C* finally gets to play. *C* selects one subshare from each divider—one subshare from D_1, one from D_2, and so on. At the end, *C* ends up with $N - 1$ subshares, which make up *C*'s final share, and each divider gets to keep the remaining $N - 1$ subshares in his or her subdivision.

When properly played, the lone-chooser method guarantees that everyone, dividers and chooser alike, ends up with a fair share (see Exercise 81).

3.5 The Last-Diminisher Method

The last-diminisher method was proposed by the Polish mathematicians Stefan Banach and Bronislaw Knaster in the 1940s. The basic idea behind this method is that throughout the game, the set *S* is divided into two pieces—a piece currently "owned" by one of the players (we will call that piece the *C*-piece and the player claiming it the "claimant") and the rest of *S*, "owned" jointly by all the other players. We will call this latter piece the *R*-piece and the remaining players the "nonclaimants." The interesting part of the game is that the entire arrangement is temporary—as the game progresses, each nonclaimant has a chance to become the current claimant (and bump the old claimant back into the nonclaimant group) by making changes to the *C*-piece and, consequently, to the *R*-piece. Thus, the claimant, the nonclaimants, the *C*-piece, and the *R*-piece all keep changing throughout the game.

Here are the specific details of how the last-diminisher method works:

- **Preliminaries.** Before the game starts the players are randomly assigned an order of play (like in a game of Monopoly this can be done by rolling a pair of dice). We will assume P_1 plays first, P_2 second, $\ldots$, P_N last, and the players will play in this order throughout the game. The game is played in rounds, and at the end of each round there is one fewer player and a smaller *S* to be divided.
- **Round 1.**
  - P_1 kicks the game off by "cutting" for herself a $1/N$th share of *S* (i.e., a share whose value equals $1/N$th of the value of *S*). This will be the current *C*-piece, and P_1 is its claimant. P_1 does not know whether or not she will end up with this share, so she must be careful that her claim is neither too small (in case she does) nor too large (in case someone else does).
  - P_2 comes next and has a choice: *pass* (remain a nonclaimant), or *diminish* the *C*-piece into a share that is a $1/N$th share of *S*. Obviously, P_2 can be a diminisher only if he thinks that the value of the current *C*-piece is *more*

than $1/N$th *the value of S*. If P_2 *diminishes*, several changes take place: P_2 becomes the new claimant; P_1 is bumped back into the nonclaimant group; the diminished C-piece becomes the new current C-piece; and the "trimmed" piece is added to the old R-piece to form a new, larger R-piece (there is a lot going on here and the best way to visualize it is by taking a close look at Fig. 3-10).

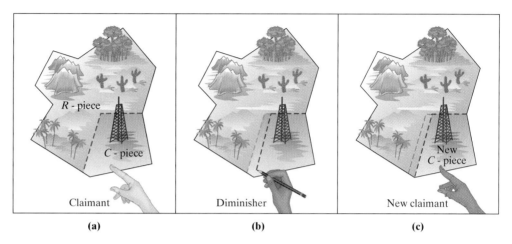

Claimant	Diminisher	New claimant
(a)	**(b)**	**(c)**

FIGURE 3-10 (a) *C*-piece claimed by claimant. (b) Diminisher "trims" *C*-piece. (c) Diminisher becomes new claimant.

- P_3 comes next and has exactly the same opportunity as P_2: *Pass* or *diminish* the current C-piece. If P_3 passes, then there are no changes, and we move on to the next player. If P_3 diminishes (only because in her value system the current C-piece is worth more than $1/N$th of S), she does so by trimming the C-piece to a $1/N$th share of S. The rest of the routine is always the same: The trimmed piece is added to the R-piece, and the previous claimant (P_1 or P_2) is bumped back into the nonclaimant group.

- The round continues this way, each player in turn having an opportunity to pass or diminish. The last player P_N can also pass or diminish, but if he chooses to diminish he has a certain advantage over the previous players—knowing that there is no player behind him that could further diminish his claim. In this situation the logical move for P_N is to become a claimant by trimming the tiniest possible sliver from the current C-piece—a sliver so small that for all practical purposes it has zero value. (Remember that a player's goal is to maximize the size of his or her share.) We will describe this move as "trimming by 0%," although in practice there has to be something trimmed, however small it may be. At the end of Round 1, the current claimant, or *last diminisher*, gets to keep the C-piece (it's her fair share) and is out of the game. The remaining players (the nonclaimants) move on to the next round, to divide the R-piece among themselves. At this point everyone is happy—the last diminisher got his or her claimed piece and the nonclaimants are happy to move to the next round, where they will have a chance to divide the R-piece among themselves.

- **Round 2.** The R-piece becomes the new S and a new version of the game is played with the new S and the $N - 1$ remaining players [this means that the new standard for a fair share is a value of $1/(N - 1)$th or more of the new S]. At the end of this round, the last diminisher gets to keep the current C-piece and is out of the game.

> ■ **Rounds 3, 4, etc.** Repeat the process, each time with one fewer player and a smaller S, until there are just two players left. At this point, divide the remaining piece between the final two players using the *divider-chooser method.*

▶ EXAMPLE 3.8 The Castaways

A new reality TV show called *The Castaways* is making its debut this season. In the show, five contestants (let's call them P_1, P_2, P_3, P_4, and P_5) are dropped off on a deserted tropical island in the middle of nowhere and left there for a year to manage on their own. After one year, the player that has succeeded and prospered the most wins the million-dollar prize. (The producers are counting on the quarreling, double-crossing, and backbiting among the players to make for great reality TV!) In the first episode of the show, the players are instructed to divide up the island among themselves any way they see fit. Instead of quarreling and double-dealing, as the producers were hoping for, these five players choose to divide the island using the lone-divider method. This being reality TV, pictures speak louder than words, and the whole episode unfolds in Figs. 3-11 through 3-15.

Players: P_1 P_2 P_3 P_4 P_5

Move 1 (by P_1) CLAIM	**Current Status** Claimant: P_1 Nonclaimants: P_2, P_3, P_4, P_5

Comments: P_1 considers C to be worth 20% and R to be worth 80% of the total value of the island.

Move 2 (by P_2) PASS	**Current Status** Claimant: P_1 Nonclaimants: P_3, P_4, P_5, P_2

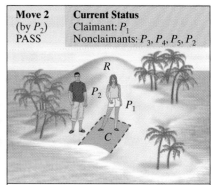

Comments: P_2 passes (he considers C to be worth *less* than or equal to 20% of the total value of the island).

Move 3 (by P_3) DIMINISH	**Current Status** Claimant: P_3 Nonclaimants: P_4, P_5, P_2, P_1

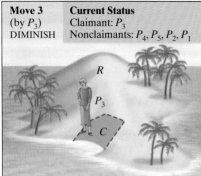

Comments: P_3 considers P_1's claim to be worth *more* than 20% of the total. P_3 diminishes it to a new C worth exactly 20% of the total. P_1 becomes a nonclaimant in contention for a fair share of the new R.

Move 4 (by P_4) DIMINISH	**Current Status** Claimant: P_4 Nonclaimants: P_5, P_2, P_1, P_3

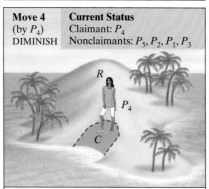

Comments: P_4 considers C to be worth *more* than 20% of the total. P_4 diminishes it to a new C worth 20% of the total. P_3 becomes a nonclaimant in contention for a fair share of the new R.

Move 5 (by P_5) PASS	**Current Status** Claimant: P_4 Nonclaimants: P_2, P_1, P_3, P_5

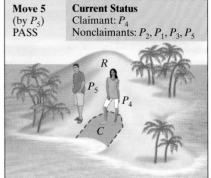

Comments: P_5 considers C to be worth *less* than 20% of the total value of the island and passes. All players have now had a chance to diminish or pass. Round 1 is over, with C going to the last diminisher (P_4).

FIGURE 3-11 Example 3.8, Round 1.

The rest of
the island
(new S)

P_4

Players:

P_1 P_2 P_3 P_5

Move 1	**Current Status**
(by P_1)	Claimant: P_1
CLAIM	Nonclaimants: P_2, P_3, P_5

R

C P_1

P_4

Comments: P_1 considers C to be worth 25% and R to be worth 75% of the value of the new S.

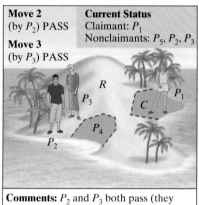

Move 2	**Current Status**
(by P_2) PASS	Claimant: P_1
Move 3	Nonclaimants: P_5, P_2, P_3
(by P_3) PASS	

R

P_3 C P_1

P_4

P_2

Comments: P_2 and P_3 both pass (they consider C to be worth *less* than 25% of the total value of the new S).

Move 4	**Current Status**
(by P_5)	Claimant: P_5
DIMINISH	Nonclaimants: P_2, P_3, P_1

R P_5

C

P_4

Comments: P_5 considers C to be worth *more* than 25% of the value of the new S. P_5 trims C by 0%. P_1 becomes a nonclaimant and the claim goes to P_5.

FIGURE 3-12 Example 3.8, Round 2 (four players left).

The rest of
the island
(new S) P_5

P_4

Players:

P_1 P_2 P_3

Move 1	**Current Status**
(by P_1)	Claimant: P_1
CLAIM	Nonclaimants: P_2, P_3

P_1

C

R P_5

P_4

Comments: P_1 considers C to be worth $33\frac{1}{3}$% and R to be worth $66\frac{2}{3}$% of the total value of the new S.

FIGURE 3-13 Example 3.8, Round 3 (three players left).

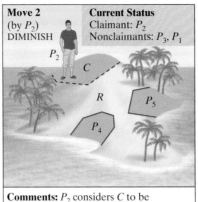

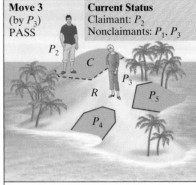

Move 2 (by P_2) DIMINISH — **Current Status** Claimant: P_2 Nonclaimants: P_3, P_1

Move 3 (by P_3) PASS — **Current Status** Claimant: P_2 Nonclaimants: P_1, P_3

FIGURE 3-13 Example 3.8, Round 3 (*continued*).

Comments: P_2 considers C to be worth *more* than $33\frac{1}{3}\%$ of the value of S. P_2 diminishes it to a new C worth exactly $33\frac{1}{3}\%$ of the value of S. P_1 goes back to being a nonclaimant.

Comments: P_3 passes (she considers C to be worth *less* than $33\frac{1}{3}\%$ of the value of S). The claim C goes to P_2.

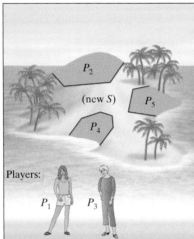

Players:

FIGURE 3-14 Example 3.8, last round (divider-chooser method).

Comments: P_1 and P_3 divide S using the divider-chooser method. Since P_1 goes first, P_1 is the divider and P_3 is the chooser.

FIGURE 3-15 The final division of the island.

In the next two sections we will discuss *discrete* fair-division methods—methods for dividing a booty S consisting of indivisible objects such as art, jewels, candy, and so on. As a general rule of thumb, discrete fair division is harder to achieve than continuous fair division because there is a lot less flexibility in the division process, and discrete fair divisions that are truly fair are only possible under a limited set of conditions. Thus, it is important to keep in mind that while both of the methods we will discuss in the next two sections have limitations, they still are the best methods we have available. Moreover, when they work, both methods work remarkably well and produce surprisingly good fair divisions.

3.6 The Method of Sealed Bids

The **method of sealed bids** was originally proposed by Hugo Steinhaus and Bronislaw Knaster around 1948. The best way to illustrate how this method works is by means of an example.

> ## EXAMPLE 3.9 Settling Grandma's Estate

In her last will and testament, Grandma plays a little joke on her four grandchildren (Art, Betty, Carla, and Dave) by leaving just three valuable items—a house, a Rolls Royce, and a Picasso painting—with the stipulation that the items must remain with the grandchildren (not sold to outsiders) and must be divided fairly in equal shares among them. How can we possibly resolve this conundrum? The method of sealed bids will give an ingenious and elegant solution.

Step 1. (Bidding). Each of the players makes a bid (in dollars) for each of the items in the estate, giving his or her honest assessment of the actual value of each item. To satisfy the privacy assumption, it is important that the bids are done independently, and no player should be privy to another player's bids before making his or her own. The easiest way to accomplish this is for each player to submit his or her bid in a sealed envelope. When all the bids are in, they are opened. Table 3-6 shows each player's bid on each item in the estate.

TABLE 3-6 The Players' Bids				
	Art	**Betty**	**Carla**	**Dave**
House	220,000	250,000	211,000	198,000
Rolls Royce	40,000	30,000	47,000	52,000
Picasso	280,000	240,000	234,000	190,000

Not to worry—it all works out at the end!

Step 2. (Allocation). Each item will go to the highest bidder for that item. (If there is a tie, the tie can be broken with a coin flip.) In this example, the house will go to Betty, the Rolls Royce will go to Dave, and the Picasso painting will go to Art. Notice that Carla gets nothing. In some situations, one player can end up with all the items (if he or she is the highest bidder on all of them) and the rest of the players with none.

Step 3. (First Settlement). It's now time to settle things up. Depending on what items (if any) a player gets in Step 2, he or she will owe money to or be owed money by the estate. To determine how much a player owes or is owed, we first calculate each player's *fair-dollar share* of the estate. A player's *fair-dollar share* is found by adding that player's bids and dividing the total by the number of players. For example, Art's bids on the three items add up to $540,000. Since there are four equal heirs, Art realizes he is only entitled to one-fourth of that—his fair-dollar share is therefore $135,000. The last row of Table 3-7 shows the fair dollar share of each player.

TABLE 3-7				
	Art	**Betty**	**Carla**	**Dave**
Home	220,000	250,000	211,000	198,000
Rolls Royce	40,000	30,000	47,000	52,000
Picasso	280,000	240,000	234,000	190,000
Total	540,000	520,000	492,000	440,000
Fair-dollar share	135,000	130,000	123,000	110,000

The fair-dollar shares are the baseline for the settlements—if the total value of the items that the player gets in Step 2 is more than his or her fair-dollar share, the player *pays* the estate the difference. If the total value of the items that the player gets is *less* than his or her fair-dollar share, the player *gets* the difference in cash. Here are the details of how the settlement works out for each of our four players.

Art: As we have seen, Art's fair-dollar share is $135,000. At the same time, Art is getting a Picasso painting worth $280,000, so Art must pay the estate the difference of $145,000 ($280,000 − $135,000). The combination of getting the $280,000 Picasso painting but paying $145,000 for it in cash results in Art getting a fair share of the estate.

Betty: Betty's fair-dollar share is $130,000. Since she is getting the house, which she values at $250,000, she must pay the estate the difference of $120,000. By virtue of getting the $250,000 house for $120,000, Betty also ends up with a fair share of the estate.

Carla: Carla's fair-dollar share is $123,000. Since she is getting no items from the estate, she receives her full $123,000 in cash. Clearly, she is getting her fair share of the estate.

Dave: Dave's fair-dollar share is $110,000. Dave is getting the Rolls, which he values at $52,000, so he has an additional $58,000 coming to him in cash. The Rolls plus the cash constitute Dave's fair share of the estate.

At this point each of the four heirs has received a fair share, and we might consider our job done, but this is not the end of the story—there is more to come (good news mostly!). If we add Art and Betty's payments to the estate and subtract the payments made by the estate to Carla and Dave, we discover that there is a surplus of $84,000! ($145,000 and $120,000 came in from Art and Betty; $123,000 and $58,000 went out to Carla and Dave.)

Step 4. (Division of the Surplus). The surplus is common money that belongs to the estate, and thus to be divided equally among the players. In our example each player's share of the $84,000 surplus is $21,000.

Step 5. (Final Settlement). The final settlement is obtained by adding the surplus money to the first settlement obtained in Step 3.

Art: Gets the Picasso painting and pays the estate $124,000—the original $145,000 he had to pay in Step 3 minus the $21,000 he gets from his share of the surplus. (Everything done up to this point could be done on paper, but now finally, real money needs to change hands!)

Betty: Gets the house and has to pay the estate only $99,000 ($120,000 − $21,000).

Carla: Gets $144,000 in cash ($123,000 + $21,000).

Dave: Gets the Rolls Royce plus $79,000($58,000 + $21,000).

The method of sealed bids works so well because it tweaks a basic idea in economics. In most ordinary transactions there is a buyer and a seller, and the buyer knows the other party is the seller and vice versa. In a sense, this puts both parties at a disadvantage. In the method of sealed bids, each player is simultaneously a buyer and a seller, without actually knowing which one until all the bids are opened. This keeps the players honest and, in the long run, works out to everyone's advantage.

The method of sealed bids works well as long as the following two important conditions are satisfied.

- Each player must have enough money to play the game. If a player is going to make honest bids on the items, he or she must be prepared to pay for some or all of them, which means that he or she may have to pay the estate certain sums of money. A player that does not have the cash to cover bids is at a definite disadvantage in playing the game.

- Each player must accept money (if it is a sufficiently large amount) as a substitute for any item. This means that no player can consider any of the items priceless.

The method of sealed bids takes a particularly simple form in the case of two players and one item. Consider the following example:

▷ EXAMPLE 3.10 Splitting Up the House

Al and Betty are getting a divorce. The only joint property of any value is their house. Rather than hiring attorneys and going to court to figure out how to split up the house, they agree to give the method of sealed bids a try.

Al's bid on the house is $340,000; Betty's bid is $364,000. Their fair-dollar shares of the "estate" are $170,000 and $182,000, respectively. Since Betty is the highest bidder, she gets to keep the house and must pay Al cash for his share. The computation of how much cash Betty pays Al can be done in two steps: In the first settlement, Betty owes the estate $182,000. Of this money, $170,000 pays for Al's fair share, leaving a surplus of $12,000 to be split equally between them. The bottom line is that Betty ends up paying $176,000 to Al for his share of the house, and both come out $6000 ahead.

The method of sealed bids can provide an excellent solution not only to settlements of property in a divorce but also to the equally difficult and often contentious issue of splitting up a partnership. The catch is that in these kinds of splits we can rarely count on the rationality assumption to hold. A divorce or partnership split devoid of emotion, spite, or hard feelings is a rare thing indeed!

3.7 The Method of Markers

The *method of markers* is a discrete fair-division method proposed in 1975 by William F. Lucas, a mathematician at the Claremont Graduate School. The method has the great virtue that it does not require the players to put up any of their own money. On the other hand, unlike the method of sealed bids, this method cannot be used effectively unless (i) there are many more items to be divided than there are players in the game, and (ii) the items are reasonably close in value.

In this method, we start with the items lined up in an *array* (a fixed sequence which cannot be changed). For convenience, think of the array as a string of objects. Each player independently bids for segments of consecutive items in the array by "cutting" the string. If there are N players, then each player must cut the string into N segments, each of which represents an acceptable share of the entire set of items. Notice that to cut a string into N sections, we need $N - 1$ cuts. In practice, one way to make the "cuts" is to lay markers in the places where the cuts are made. Thus, each player can make his or her bids by placing $N - 1$ markers so that they divide the array into N segments. To insure privacy, no player should see the markers of another player before laying down his or her own.

What the method of markers essentially accomplishes is to guarantee that each player ends up with one of his or her bid segments (a section between two consecutive markers). The easiest way to explain how the method works is with an example.

FIGURE 3-16 The Halloween leftovers.

> ### EXAMPLE 3.11 Dividing the Halloween Leftovers

Alice, Bianca, Carla, and Dana want to divide the Halloween leftovers shown in Fig. 3-16 among themselves. There are 20 pieces, but having each randomly choose 5 pieces is not likely to work well—the pieces are too varied for that. Their teacher, Mrs. Jones, offers to divide the candy for them, but the children reply that they just learned about a cool fair division game they want to try, and they can do it themselves, thank you.

As a preliminary step, the 20 pieces are arranged in an array (Fig. 3-17). For convenience, we will label the pieces of candy 1 through 20. The order in which the pieces are lined up should be random (the easiest way to do this is to dump the pieces into a paper bag, shake the bag, and take the pieces out of the bag one at a time).

FIGURE 3-17 The items lined up in an array.

Step 1. (Bidding). Each child writes down independently on a piece of paper exactly where she wants her three markers. (Three markers divide the array into four sections.) The bids are opened, and the results are shown in Fig. 3-18. The A-labels indicate the position of Alice's markers (A_1 denotes her first marker, A_2 her second marker, and A_3 her third and last marker).

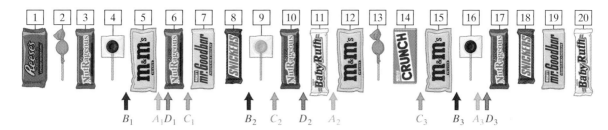

FIGURE 3-18 The bids.

Alice's bid means that she is willing to accept one of the following as a fair share of the candy: (i) pieces 1 through 5 (first segment), (ii) pieces 6 through 11 (second segment), (iii) pieces 12 through 16 (third segment), or (iv) pieces 17 through 20 (last segment). Bianca's bid is shown by the B-markers and indicates how she would break up the array into four segments that are fair shares; ditto for Carla's bid (shown by the C-markers), and Dana's bid (shown by the D-markers).

Step 2. (Allocations). Here is how to allocate a fair share to each child: Scan the array from left to right until the first *first marker* comes up. Here the first first marker is Bianca's (B_1). This means that Bianca will be the first player to get fair share—the first segment in her bid (pieces 1 through 4, Fig. 3-19). Bianca is done now, and her markers can be removed since they are no

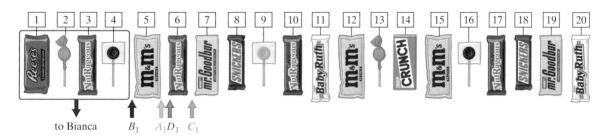

FIGURE 3-19 B_1 is the first 1-marker. Bianca gets her first segment..

longer needed. Continue scanning from left to right looking for the first *second marker*. Here the first second marker is Carla's (C_2). This means that Carla will be the second player taken care of: Carla gets the second segment in her bid (pieces 7 through 9, Fig. 3-20). Carla's remaining markers can now be removed. Continue scanning from left to right looking for the first *third marker*. Here there is a tie between Alice's A_3 and Dana's D_3. As usual, a coin toss is used to break the tie and Alice will be the third player to get a

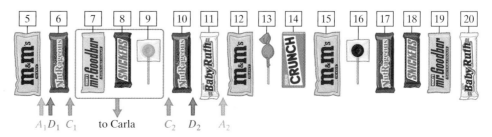

FIGURE 3-20 C_2 is the first 2-marker (among A's, C's, and D's). Carla gets her second segment.

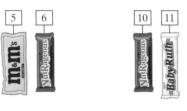

FIGURE 3-21 A_3 and D_3 are tied as the first 3-marker. After a coin toss, Alice gets her third segment.

share—she will get the third segment in her bid (pieces 12 through 16, Fig. 3-21). Dana is the last player and gets the last segment in her bid (pieces 17 through 20, Fig. 3-22). Now each player has gotten one of her chosen segments. The amazing part is that there is *leftover candy*!

FIGURE 3-22 Dana is last, gets her last segment.

Step 3. (Dividing Leftovers). Usually, there are just a few pieces of candy left over, not enough to play the game all over again. The simplest thing to do is randomly draw lots and let the players go in order picking one piece at a time until there is no more candy left. Here the leftover pieces are 5, 6, 10, and 11 (Fig. 3-23). The players now draw lots; Carla gets to choose first and takes piece 11. Dana chooses next and takes piece 5. Bianca and Alice receive pieces 6 and 10, respectively.

FIGURE 3-23 The leftovers (to be given randomly to the players one at a time) are a bonus.

The ideas behind Example 3.11 can be easily generalized to any number of players. We now give the general description of the **method of markers** with N players and M discrete items.

- **Preliminaries.** The items are arranged randomly into an array. For convenience, label the items 1 through M, going from left to right.
- **Step 1. (Bidding).** Each player independently divides the array into N segments (segments 1, 2, ..., N) by placing $N - 1$ markers along the array. These segments are assumed to represent the fair shares of the array in the opinion of that player.
- **Step 2. (Allocations).** Scan the array from left to right until the first *first marker* is located. The player owning that marker (let's call him P_1) goes first, and gets the first segment in his bid. (In case of a tie, break the tie randomly.) P_1's markers are removed, and we continue scanning from left to right, looking for the first *second marker*. The player owning that marker (let's call her P_2) goes second and gets the second segment in her bid. Continue this process, assigning to each player in turn one of the segments in her bid. The last player gets the last segment in her bid.

■ **Step 3. (Dividing Leftovers).** The leftover items can be divided among the players by some form of lottery, and, in the rare case that there are many more leftover items than players, the method of markers could be used again.

In spite of its simple elegance, the method of markers can be used only under some fairly restrictive conditions. In particular, the method assumes that every player is able to divide the array of items into segments in such a way that each of the segments has approximately equal value. This is usually possible when the items are of small and homogeneous value, but almost impossible to accomplish when there is a combination of expensive and inexpensive items (good luck trying to divide 19 candy bars and an iPod using the method of markers!).

Conclusion

The Judgment of Solomon, by Nicolas Poussin (1649).

Problems of fair division are as old as humankind. One of the best-known and best-loved biblical stories is built around one such problem: Two women, both claiming to be mothers of the same baby, make their case to King Solomon. As a solution, King Solomon proposes to cut the baby in two and give each woman a share. (Basically, King Solomon was proposing a continuous solution to a discrete fair-division problem!) This solution is totally unacceptable to the true mother, who would rather see the baby go to the other woman than be cut into pieces. The final settlement, of course, is that the baby is returned to its rightful mother.

The problem of dividing an object or set of objects among the members of a group is a practical problem that comes up regularly in our daily lives. When the object is a pizza, a cake, or a bunch of candy, we don't always pay a great deal of attention to the issue of fairness, but when the object is an estate, land, jewelry, or some other valuable asset, dividing things fairly becomes a critical issue.

In this chapter we looked at fair division from a mathematical perspective. Within mathematics, problems of fair division are considered part of *game theory* (a branch of mathematics that provides the tools to analyze "games" of competition or cooperation between individuals or groups), and we approached fair-division problems within this context. We found that, when certain conditions are satisfied (especially assumptions about the behavior and rationality of the players), mathematics can provide solutions that guarantee that each player will always receive a fair share.

Over the years, mathematicians have developed different *methods* for solving fair-division problems. Each of these methods works under some restricted set of circumstances (there is no single method that will work in every possible situation), so it is important not only to know how the method works but to understand under what circumstances it can and cannot be used.

In our analysis we classified fair-division problems into *continuous* and *discrete* and discussed different fair-division methods for each type of problem. In the case of a continuous fair-division problem with two players, the *divider-chooser method* is always the method of choice. For continuous fair-division problems with three or more players, there are many possible methods that can be considered, and we discussed three of these: the *lone-divider method*, the *lone-chooser method*, and the *last-diminisher method*. (There are several other lesser-known methods that we did not discuss.)

Discrete fair-division problems are inherently harder than continuous problems, and the number of methods available to tackle them is very limited. The best-known discrete fair-division method is the *method of sealed bids*, and this method has many useful applications in real life. An alternative is the *method of markers*, which works well but only for very limited types of objects.

This chapter gave an overview of what is an interesting and surprising application of mathematics to one of the fundamental questions of social science—how to get humans to share in a reasonable and fair way. Mathematics can provide only some answers, and these answers work under a fairly limited set of circumstances, but when they do work, they can work remarkably well. Remember this next time you must divide an inheritance, a piece of real estate, or even some of the chores around the house. It may serve you well.

Profile Hugo Steinhaus (1887–1972)

Dividing things, be it the spoils of war or the fruits of peace, is an issue as old as man. Throughout most of human history, fairness in dividing things was seen as a problem far removed from the mathematical arena—the province of kings, priests, judges, and politicians. It wasn't until the 1940s that the revolutionary notion that mathematical methods could be used to tackle successfully many types of fair-division problems came about. This breakthrough can be traced to the inspiration and creativity of one man—the great Polish mathematician Hugo Steinhaus.

Hugo Steinhaus was born in Jaslo, Poland. After completing his secondary education in his homeland, Steinhaus went to Germany to study mathematics at the University of Göttingen, which in the early 1900s was the most prestigious center of mathematical research in the world. At Göttingen, Steinhaus studied under many famous mathematicians, including David Hilbert, his doctoral supervisor and arguably the most famous mathematician of his time. Steinhaus was awarded a doctorate in mathematics (with distinction) in 1911.

After completing four years of military service in World War I, Steinhaus returned to Poland, where he began a distinguished academic career as a professor of mathematics, first in Krakow and then in Lvov, where he founded the famous Lvov school of mathematics. Steinhaus was a prolific and versatile mathematician who made important contributions to many different branches of mathematics, including functional analysis, trigonometric series, and probability theory. But Steinhaus's interests went beyond conventional mathematical research—he loved to discuss and expound on simple but interesting real-life applications of mathematics and to foster an appreciation for the power and beauty of mathematics among young people. In 1937, Steinhaus wrote a classic book called *Mathematical Snapshots*, a collection of vignettes intended to appeal, in Steinhaus's own words, to the *scientist in the child and the child in the scientist*. After many editions, *Mathematical Snapshots* is still in print and widely quoted.

Sometime in the late 1930s, Steinhaus made one of his more famous mathematical discoveries, a theorem informally known as the *Ham Sandwich Theorem*. Essentially, the theorem can be paraphrased as follows: If you have a sandwich consisting of *three* ingredients (say bread, ham, and cheese), then there is a way to slice the sandwich (using a single straight cut) into two parts each of which has exactly half of the volume of bread, half of the ham, and half of the cheese—you can share the sandwich with your friend and each of you gets exactly half of each of the three ingredients. (You can do this with even the funkiest of sandwiches, as long as you stick to just three ingredients. Unfortunately, if you want to divide *n* ingredients equally, you need to live in the *n*th dimension!)

The Ham Sandwich Theorem is a beautiful theoretical result but has little practical value—it says that the desired cut is possible in theory, but it doesn't give even the slightest hint as to how to do it. It promises a fair division but it cannot deliver it. The limitations of the Ham Sandwich Theorem forced Steinhaus to think about a different, truly practical approach to the problem of dividing things *equally*, and this led to the foundations of the theory of fair division as we know it today. By the late 1940s, Steinhaus, together with Stefan Banach and Bronislaw Knaster, two of his students and collaborators, had developed many of the most important fair-division methods we know today, including the *last-diminisher method*, the *method of sealed bids*, and a continuous version of the *method of markers*.

Hugo Steinhaus died in 1972, at the age of 85. His mathematical legacy includes over 170 articles, 5 books, and the invaluable insight that mathematics can help humankind solve some of its disputes.

Key Concepts

continuous fair-division game, **87**
discrete fair-division game, **87**
divider-chooser method, **88**
fair division, **87**

fair-division method, **87**
fair share, **87**
last-diminisher method, **98**
lone-chooser method, **95**

lone-divider method, **89**
method of markers, **106**
method of sealed bids, **103**

Exercises

WALKING

A. Shares, Fair Shares, and Fair Divisions

Exercises 1 and 2 refer to the following figure.

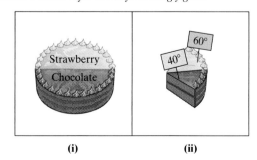

(i)   (ii)

1. Alex buys the chocolate-strawberry mousse cake shown in (i) for $12. Alex values chocolate three times as much as he values strawberry.

(a) What is the value of the chocolate half of the cake to Alex?

(b) What is the value of the strawberry half of the cake to Alex?

(c) A piece of the cake is cut as shown in (ii). What is the value of the piece to Alex?

2. Jody buys the chocolate-strawberry mousse cake shown in (i) for $13.50. Jody values strawberry four times as much as she values chocolate.

(a) What is the value of the chocolate half of the cake to Jody?

(b) What is the value of the strawberry half of the cake to Jody?

(c) A piece of the cake is cut as shown in (ii). What is the value of the piece to Jody?

Exercises 3 and 4 refer to the following figure.

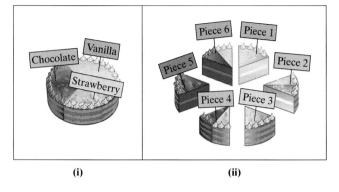

(i)   (ii)

3. Kala buys the chocolate-strawberry-vanilla cake shown in (i) for $12. Kala values strawberry twice as much as vanilla and values chocolate three times as much as vanilla.

   (a) What is the value of the chocolate part of the cake to Kala?

   (b) What is the value of the strawberry part of the cake to Kala?

   (c) What is the value of the vanilla part of the cake to Kala?

   (d) If the cake is cut into the six equal sized pieces shown in (ii), find the value to Kala of each of the six pieces.

4. Malia buys the chocolate-strawberry-vanilla cake shown in (i) for $11.20. Malia values strawberry twice as much as chocolate and values chocolate twice as much as vanilla.

   (a) What is the value of the chocolate part of the cake to Malia?

   (b) What is the value of the strawberry part of the cake to Malia?

   (c) What is the value of the vanilla part of the cake to Malia?

   (d) If the cake is cut into the six equal sized pieces shown in (ii), find the value to Malia of each of the six pieces.

5. Three players (Ana, Ben, and Cara) must divide a cake among themselves. Suppose the cake is divided into three slices (s_1, s_2, and s_3). The values of the entire cake and of each of the three slices in the eyes of each of the players are shown in the following table.

	Whole cake	s_1	s_2	s_3
Ana	$12.00	$3.00	$5.00	$4.00
Ben	$15.00	$4.00	$4.50	$6.50
Cara	$13.50	$4.50	$4.50	$4.50

(a) Indicate which of the three slices are fair shares to Ana.

(b) Indicate which of the three slices are fair shares to Ben.

(c) Indicate which of the three slices are fair shares to Cara.

(d) Describe a fair division of the cake.

6. Three players (Alex, Betty, and Cindy) must divide a cake among themselves. Suppose the cake is divided into three slices (s_1, s_2, and s_3). The following table shows the percentage of the value of the entire cake that each slice represents to each player.

	s_1	s_2	s_3
Alex	30%	40%	30%
Betty	35%	25%	40%
Cindy	$33\frac{1}{3}$%	50%	$16\frac{2}{3}$%

(a) Indicate which of the three slices are fair shares to Alex.

(b) Indicate which of the three slices are fair shares to Betty.

(c) Indicate which of the three slices are fair shares to Cindy.

(d) Describe a fair division of the cake.

7. Four partners (Adams, Benson, Cagle, and Duncan) jointly own a piece of land. The land is subdivided into four parcels s_1, s_2, s_3, and s_4. The following table shows the relative value of the parcels (as a percentage of the value of the land) in the eyes of each partner.

	s_1	s_2	s_3	s_4
Adams	30%	24%	20%	26%
Benson	35%	25%	20%	20%
Cagle	25%	15%	40%	20%
Duncan	20%	20%	20%	40%

(a) Indicate which of the four parcels are fair shares to Adams.

(b) Indicate which of the four parcels are fair shares to Benson.

(c) Indicate which of the four parcels are fair shares to Cagle.

(d) Indicate which of the four parcels are fair shares to Duncan.

(e) Assuming that the four parcels cannot be changed or further subdivided, describe a fair division of the land.

8. Abe, Betty, Cory, and Dana are dividing a cake among themselves. The cake is divided into four slices s_1, s_2, s_3, and s_4. The values of the cake and of each of the four slices in the eyes of each of the players are shown in the following table.

	Whole cake	s_1	s_2	s_3	s_4
Abe	$15.00	$3.00	$5.00	$5.00	$2.00
Betty	$18.00	$4.50	$4.50	$4.50	$4.50
Cory	$12.00	$4.00	$3.50	$1.50	$3.00
Dana	$10.00	$2.75	$2.40	$2.45	$2.40

(a) Indicate which of the four slices are fair shares to Abe.

(b) Indicate which of the four slices are fair shares to Betty.

(c) Indicate which of the four slices are fair shares to Cory.

(d) Indicate which of the four slices are fair shares to Dana.

(e) Using s_1, s_2, s_3, and s_4 as the four shares, describe a fair division of the cake.

9. Abe, Betty, Cory, and Dana are dividing a cake among themselves. The cake is divided into four slices s_1, s_2, s_3, and s_4.

(a) To Abe, the value of s_1 is $3.00, the value of s_4 is $3.50, s_2 and s_3 have equal value, and the value of the entire cake is $15.00. Indicate which of the four slices are fair shares to Abe.

(b) To Betty, all slices have equal value. Indicate which of the four slices are fair shares to Betty.

(c) To Cory, s_1, s_2, and s_4 have equal value, and the value of s_3 is three times the value of s_1. Indicate which of the four slices are fair shares to Cory.

(d) To Dana, s_2, s_3, and s_4 have equal value, the value of the entire cake is $18.00, and the value of s_1 is $2.00 more than the value of s_2. Indicate which of the four slices are fair shares to Dana.

(e) Using s_1, s_2, s_3, and s_4 as the four shares, describe a fair division of the cake.

10. Four partners (Adams, Benson, Cagle, and Duncan) jointly own a piece of land with a market value of $400,000. The land is subdivided into four parcels s_1, s_2, s_3, and s_4.

(a) To Adams, the value of s_1 is $40,000 more than the value of s_2, s_2, and s_3 have equal value, and the value of s_4 is $20,000 more than the value of s_1. Indicate which of the four parcels are fair shares to Adams.

(b) To Benson, the value of s_1 is $40,000 more than the value of s_2, the value of s_4 is $8000 more than the value of s_3, and together s_4 and s_3 have a combined value equal to 40% of the value of the land. Indicate which of the four parcels are fair shares to Benson.

(c) To Cagle, the value of s_1 is $40,000 more than the value of s_2 and $20,000 more than the value of s_4, and the value of s_3 is twice the value of s_4. Indicate which of the four slices are fair shares to Cagle.

(d) To Duncan, the value of s_1 is $4000 more than the value of s_2; s_2 and s_3 have equal value; and together, s_1, s_2, and s_3 have a combined value of $280,000. Indicate which of the four slices are fair shares to Duncan.

(e) Assuming that the four parcels cannot be changed or further subdivided, describe a fair division of the land.

B. The Divider-Chooser Method

11. Jared and Karla are planning to divide the half meatball, half vegetarian submarine sandwich shown in the figure using the divider-chooser method. Jared likes meatball subs three times as much as vegetarian subs. Karla is a strict vegetarian and does not eat meat at all. Jared and Karla have just met and know nothing about each other's likes and dislikes.

Vegetarian   Meatball

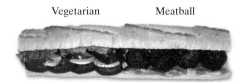

(a) If Jared is the divider, describe how he would cut the sandwich (assume he makes only one cut perpendicular to the sandwich's length).

(b) If Karla is the divider, describe how she would cut the sandwich (assume she can make only one cut perpendicular to the sandwich's length).

(c) Suppose the value of the sandwich is $8.00 to both Jared and Karla. If Jared is the divider and Karla is the chooser, give the value of Jared's fair share (in Jared's eyes) and the value of Karla's fair share (in Karla's eyes).

(d) Suppose the value of the sandwich is $8.00 to both Jared and Karla. If Karla is the divider and Jared is the chooser, give the value of Jared's fair share (in Jared's eyes), and the value of Karla's fair share (in Karla's eyes).

12. Martha and Nick are planning to divide the giant sub sandwich shown in the figure using the divider-chooser method. Martha likes ham subs twice as much as she likes turkey subs and likes turkey and roast beef subs the same. Nick likes roast beef subs twice as much as he likes ham subs and likes ham and turkey subs the same.

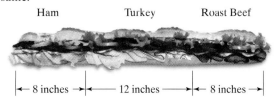

Ham   Turkey   Roast Beef

← 8 inches → ← 12 inches → ← 8 inches →

(a) If Martha is the divider, describe how she would cut the sandwich (assume she makes only one cut perpendicular to the sandwich's length).

(b) If Nick is the divider, describe how he would cut the sandwich (assume he can make only one cut perpendicular to the sandwich's length).

(c) Suppose the value of the sandwich is $9.00 to both Nick and Martha. If Martha is the divider and Nick is the chooser, give the value of Martha's fair share (in Martha's eyes) and the value of Nick's fair share (in Nick's eyes).

(d) Suppose the value of the sandwich is $13.50 to both Nick and Martha. If Nick is the divider and Martha is the chooser, give the value of Martha's fair share (in Martha's eyes) and the value of Nick's fair share (in Nick's eyes).

13. David and Paul are planning to divide the pizza shown in the figure using the divider-chooser method. David likes pepperoni, sausage, and mushrooms equally well, but hates anchovies. Paul likes anchovies, mushrooms, and pepperoni equally well, but hates sausage. Neither one knows anything about the other one's likes and dislikes (they are new friends).

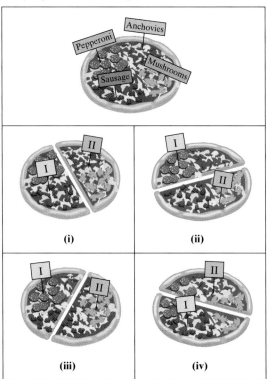

(a) If David is the divider, which of the cuts in (i) through (iv) is consistent with David's value system?

(b) For each of the cuts consistent with David's value system, which of the two pieces is Paul's best choice?

14. Raul and Karli are planning to divide a chocolate-strawberry mousse cake in the figure using the divider-chooser method. Raul values chocolate three times as much as he values strawberry. Karli values chocolate twice as much as she values strawberry.

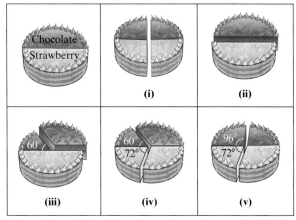

(a) If Raul is the divider, which of the cuts shown in figures (i) through (v) are consistent with Raul's value system?

(b) For each of the cuts consistent with Raul's value system, indicate which of the pieces is Karli's best choice.

15. This exercise is a continuation of Exercise 13.

(a) Suppose Paul is the divider. Draw three different cuts that are consistent with his value system.

(b) For each of the cuts in (a), indicate which of the pieces is David's best choice.

16. This exercise is a continuation of Exercise 14.

(a) Suppose Karli is the divider. Draw three different cuts that are consistent with her value system.

(b) For each of the cuts in (a), indicate which of the pieces is Raul's best choice.

Exercises 17 and 18 refer to the following figure.

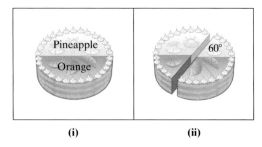

17. Jamie and Mo want to divide the orange-pineapple cake shown in the figure using the divider-chooser method. Jamie values orange four times as much as he values pineapple. Mo is the divider and cuts the cake as shown in figure (ii).

(a) What percent of the value of the cake is the pineapple half in Mo's eyes?

(b) What percent of the value of the cake is each piece in (ii) in Jamie's eyes?

(c) Describe the final fair division of the cake.

18. Susan and Veronica want to divide the orange-pineapple cake shown in the figure using the divider-chooser method. Susan values orange four times as much as she values pineapple. Veronica is the divider and cuts the cake as shown in figure (ii).

(a) What percent of the value of the cake is the pineapple half in Veronica's eyes?

(b) What percent of the value of the cake is each piece in (ii) in Susan's eyes?

(c) Describe the final fair division of the cake.

C. The Lone-Divider Method

19. Three partners (Divine, Chase, and Chandra) are dividing a plot of land among themselves using the lone-divider method. Using a map, the divider (Divine) divides the property into three parcels $s_1, s_2,$ and s_3. When the choosers' bid lists are opened, Chase's bid list is $\{s_2, s_3\}$ and Chandra's bid list is $\{s_1, s_3\}$.

(a) Describe the fair division where Divine's fair share is s_1.

(b) Describe the fair division where Divine's fair share is s_2.

(c) Describe the fair division where Divine's fair share is s_3.

20. Three partners (Divine, Chase, and Chandra) are dividing a plot of land among themselves using the lone-divider method. Using a map, the divider (Divine) divides the property into three parcels $s_1, s_2,$ and s_3. When the choosers' bid lists are opened, Chase's bid list is $\{s_1, s_2, s_3\}$ and Chandra's bid list is $\{s_1\}$.

(a) Describe the fair division where Divine's fair share is s_2.

(b) Describe the fair division where Divine's fair share is s_3.

21. Four partners (DiPalma, Childs, Choate, and Chou) are dividing a plot of land among themselves using the lone-divider method. Using a map, DiPalma divides the land into four parcels $s_1, s_2, s_3,$ and s_4. When the choosers' bid lists are opened, Childs's bid list is $\{s_2, s_3\}$, Choate's bid list is $\{s_3, s_4\}$, and Chou's bid list is $\{s_1, s_4\}$.

(a) Describe the fair division where DiPalma's fair share is s_1.

(b) Describe the fair division where DiPalma's fair share is s_2.

(c) Describe the fair division where DiPalma's fair share is s_3.

(d) Describe the fair division where DiPalma's fair share is s_4.

22. Four partners (DiPalma, Childs, Choate, and Chou) are dividing a plot of land among themselves using the lone-divider method. Using a map, DiPalma divides the land into four parcels $s_1, s_2, s_3,$ and s_4. When the choosers' bid lists are opened, Childs's bid list is $\{s_2, s_3\}$, Choate's bid list is $\{s_3, s_4\}$, and Chou's bid list is $\{s_4\}$.

(a) Describe a fair division of the land.

(b) Explain why your answer in (a) is the only possible fair division of the land using the four given parcels.

23. Desi, Cher, Cheech, and Chong are dividing a cake among themselves using the lone-divider method. Desi cuts the cake into four slices $s_1, s_2, s_3,$ and s_4. When the choosers' bid lists are opened, Cher's bid list is $\{s_2, s_3\}$, Cheech's bid list is $\{s_1, s_3\}$, and Chong's bid list is $\{s_1, s_2\}$.

(a) Describe the fair division where Cher's fair share is s_2.

(b) Describe the fair division where Chong's fair share is s_2.

(c) Explain why Desi's fair share has to be s_4.

24. Desi, Cher, Cheech, and Chong are dividing a cake among themselves using the lone-divider method. Desi cuts the cake into four slices $s_1, s_2, s_3,$ and s_4. When the choosers' bid lists are opened, Cher's bid list is $\{s_1, s_2\}$, Cheech's bid list is $\{s_1, s_2\}$, and Chong's bid list is $\{s_2\}$. Describe how to proceed to obtain a fair division of the cake.

25. Five players $(D, C_1, C_2, C_3,$ and $C_4)$ are dividing a cake among themselves using the lone-divider method. D cuts the cake into five slices s_1, s_2, s_3, s_4, s_5. When the choosers' bid lists are opened, C_1's bid list is $\{s_2, s_4\}$, C_2's bid list is $\{s_2, s_4\}$, C_3's bid list is $\{s_2, s_3, s_4\}$, and C_4's bid list is $\{s_2, s_3, s_5\}$.

(a) Describe the fair division where C_1's fair share is s_2.

(b) Describe the fair division where C_1's fair share is s_4.

(c) Explain why the fair divisions given in parts (a) and (b) are the only possible fair divisions of the cake.

26. Five players $(D, C_1, C_2, C_3,$ and $C_4)$ are dividing a cake among themselves using the lone-divider method. D cuts the cake into five slices s_1, s_2, s_3, s_4, s_5. When the choosers' bid lists are opened, C_1's bid list is $\{s_2, s_5\}$, C_2's bid list is $\{s_1, s_2\}$, C_3's bid list is $\{s_1, s_4, s_5\}$, and C_4's bid list is $\{s_1, s_5\}$.

(a) Describe the fair division where C_4's fair share is s_1.

(b) Describe the fair division where C_2's fair share is s_1.

(c) Explain why C_3's fair share has to be s_4.

27. Six players (D, C_1, C_2, C_3, C_4, and C_5) are dividing a cake among themselves using the lone-divider method. D cuts the cake into six slices s_1, s_2, s_3, s_4, s_5, and s_6. When the choosers' bid lists are opened, C_1's bid list is $\{s_2, s_3, s_5\}, C_2$'s bid list is $\{s_1, s_5, s_6\}, C_3$'s bid list is $\{s_3, s_5, s_6\}, C_4$'s bid list is $\{s_2, s_3\}$, and C_5's bid list is $\{s_3\}$.

  (a) Describe a fair division of the cake.

  (b) Explain why the answer in part (a) is the only possible fair division of the cake.

28. Six players (D, C_1, C_2, C_3, C_4, and C_5) are dividing a cake among themselves using the lone-divider method. D cuts the cake into six slices s_1, s_2, s_3, s_4, s_5, and s_6. When the choosers' bid lists are opened, C_1's bid list is $\{s_1\}, C_2$'s bid list is $\{s_2, s_3\}, C_3$'s bid list is $\{s_4, s_5\}, C_4$'s bid list is $\{s_4, s_5\}$, and C_5's bid list is $\{s_1\}$. Describe how to proceed to obtain a fair division of the cake.

29. Four partners (Egan, Fine, Gong, and Hart) are dividing a piece of land valued at \$480,000 among themselves using the lone-divider method. Using a map, the divider divides the property into four parcels s_1, s_2, s_3, and s_4. The following table shows the value of the four parcels in the eyes of each partner, but some of the entries in the table are missing.

	s_1	s_2	s_3	s_4
Egan	\$80,000	\$85,000		\$195,000
Fine		\$100,000	\$135,000	\$120,000
Gong	\$120,000		\$120,000	
Hart	\$95,000	\$100,000		\$110,000

  (a) Who was the divider? Explain.

  (b) Describe the chooser's respective bid lists.

  (c) Describe a fair division of the property.

  (d) Explain why the answer in part (c) is the only possible fair division of the property.

30. Four partners (Egan, Fine, Gong, and Hart) are dividing a piece of land among themselves using the lone-divider method. Using a map, the divider divides the property into four parcels s_1, s_2, s_3, and s_4. The following table shows the relative value of the parcels (as a percentage of the total value of the land) in the eyes of each partner, but some of the entries in the table are missing.

	s_1	s_2	s_3	s_4
Egan		30%	21%	22%
Fine	35%	20%		20%
Gong		26%	28%	21%
Hart	25%	25%		

  (a) Who was the divider? Explain.

  (b) Describe the chooser's respective bid lists.

  (c) Describe a fair division of the property where Gong's fair share is s_2.

D. The Lone-Chooser Method

Exercises 31 through 34 refer to the following fair-division game: Angela, Boris, and Carlos are dividing a vanilla-strawberry cake using the lone-chooser method. The dollar value of the vanilla and strawberry parts in the eyes of each player is shown in the following figure.

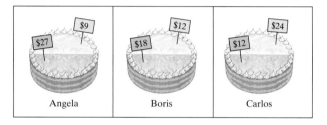

Angela	Boris	Carlos

31. Suppose that Angela and Boris are the dividers and Carlos is the chooser. In the first division, the cake is cut vertically through the center. Angela gets the left half, and Boris gets the right half, as shown in the figure.

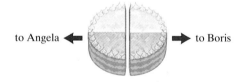

  to Angela ← → to Boris

  (a) Describe the second division that Angela might make of the left half of the cake.

  (b) Describe the second division that Boris might make of the right half of the cake.

  (c) Based on the second divisions you gave in (a) and (b), describe a possible final fair division of the cake.

  (d) For the final fair division you described in (c), find the dollar value of each share in the eyes of the player receiving it.

32. Suppose that Carlos and Angela are the dividers and Boris is the chooser. In the first division, the cake is cut vertically through the center. Angela gets the left half, and Carlos gets the right half, as shown in the figure.

  to Angela ← → to Carlos

  (a) Describe the second division that Carlos might make of the right half of the cake.

(b) Describe the second division that Angela might make of the left half of the cake.

(c) Based on the second divisions you gave in (a) and (b), describe a possible final fair division of the cake.

(d) For the final fair division you described in (c), find the dollar value of each share in the eyes of the player receiving it.

33. Suppose that Angela and Boris are the dividers and Carlos is the chooser. In the first division Angela cuts the cake in two shares as shown in the figure, and Boris chooses the share he likes best.

(a) Describe the second division that Boris might make on his share of the cake.

(b) Describe the second division that Angela might make on her share of the cake.

(c) Based on the second divisions you gave in (a) and (b), describe a possible final fair division of the cake.

(d) For the final fair division you described in (c), find the dollar value of each share in the eyes of the player receiving it.

34. Suppose that Carlos and Angela are the dividers and Boris is the chooser. In the first division, Carlos cuts the cake as shown in the figure, and Angela chooses the share she likes best.

(a) Describe the second division that Angela might make on her share of the cake.

(b) Describe the second division that Carlos might make on his share of the cake.

(c) Based on the second divisions you gave in (a) and (b), describe a possible final fair division of the cake.

(d) For the final fair division you described in (c), find the dollar value of each share in the eyes of the player receiving it.

Exercises 35 through 38 refer to the following fair-division game: Arthur, Brian, and Carl are dividing the cake shown in the following figure using the lone-chooser method. The players value the different parts of the cake as follows: (i) Arthur likes chocolate and orange equally well, but hates strawberry and vanilla; (ii) Brian likes chocolate and strawberry equally well, but hates orange and vanilla; (iii) Carl likes chocolate and vanilla equally well, but hates orange and strawberry.

35. Suppose that Arthur and Brian are the dividers, and Carl is the chooser. In the first division Arthur makes the first cut as shown in the figure, and Brian chooses the share he likes best.

(a) Describe the second division that Brian might make on his share of the cake.

(b) Describe the second division that Arthur might make on his share of the cake.

(c) Based on the second divisions you gave in (a) and (b), describe a possible final fair division of the cake.

(d) For the final fair division you described in (c), find the value of each share (as a percentage of the total value of the cake) in the eyes of the player receiving it.

36. Suppose that Carl and Arthur are the dividers, and Brian is the chooser. In the first division Carl makes the first cut as shown in the figure, and Arthur chooses the share he likes best.

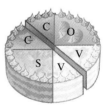

(a) Describe the second division that Arthur might make on his share of the cake.

(b) Describe the second division that Carl might make on his share of the cake.

(c) Based on the second divisions you gave in (a) and (b), describe a possible final fair division of the cake.

(d) For the final fair division you described in (c), find the value of each share (as a percentage of the total value of the cake) in the eyes of the player receiving it.

37. Suppose that Brian and Carl are the dividers, and Arthur is the chooser. In the first division Brian makes the first cut as shown in the figure, and Carl chooses the share he likes best.

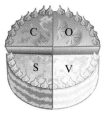

(a) Describe the second division that Carl might make on his share of the cake.

(b) Describe the second division that Brian might make on his share of the cake.

(c) Based on the second divisions you gave in (a) and (b), describe a possible final fair division of the cake.

(d) For the final fair division you described in (c), find the value of each share (as a percentage of the total value of the cake) in the eyes of the player receiving it.

38. Suppose that Arthur and Carl are the dividers, and Brian is the chooser. In the first division Arthur makes the first cut as shown in the figure, and Carl chooses the share he likes best.

(a) Describe the second division that Carl might make on his share of the cake.

(b) Describe the second division that Arthur might make on his share of the cake.

(c) Based on the second divisions you gave in (a) and (b), describe a possible final fair division of the cake.

(d) For the final fair division you described in (c), find the value of each share (as a percentage of the total value of the cake) in the eyes of the player receiving it.

Exercises 39 and 40 refer to the following fair-division game: Jared, Karla, and Lori are dividing a sub sandwich (half meatball, half vegetarian) using the lone-chooser method. Jared likes the vegetarian and meatball parts equally well, Karla is a strict vegetarian and does not eat meat at all, and Lori likes the meatball part twice as much as she likes the vegetarian part. Assume that all the cuts are perpendicular to the length of the sandwich, as shown in the figure.

Vegetarian   Meatball

39. Suppose that Karla and Jared are the dividers, and Lori is the chooser. In the first division Karla makes the first cut, and Jared chooses the share he likes best.

(a) Describe the first division (exactly where Karla cuts the sandwich and who ends up with which piece).

(b) Describe the second division that Jared makes on his share of the sandwich.

(c) Describe the second division that Karla makes on her share of the sandwich.

(d) Describe the final fair division of the sandwich, and give the value of each share as a percentage of the total value of the sandwich in the eyes of the player receiving it.

40. Suppose that Lori and Karla are the dividers, and Jared is the chooser. In the first division Lori makes the first cut, and Karla chooses the share she likes best.

(a) Describe the first division (exactly where Lori cuts the sandwich and who ends up with which piece).

(b) Describe the second division that Karla makes on her share of the sandwich.

(c) Describe the second division that Lori makes on her share of the sandwich.

(d) Describe the final fair division of the sandwich, and give the value of each share as a percentage of the total value of the sandwich in the eyes of the player receiving it.

E. The Last-Diminisher Method

41. A cake valued at \$30 is divided among five players ($P_1, P_2, P_3, P_4,$ and P_5) using the last-diminisher method. The players play in a fixed order, with P_1 first, P_2 second, and so on. In round 1, P_1 makes the first cut and makes a claim on a C-piece. For each of the remaining players, the value of the *current* C-piece at the time it is their turn to play is given in the following table.

	P_2	P_3	P_4	P_5
Value of the current C-piece	\$7.00	\$4.50	\$6.50	\$4.00

(a) Which player gets his or her share at the end of round 1?

(b) What is the value of the share to the player receiving it?

(c) Which player makes the first cut to start round 2?

42. A cake valued at \$30 is divided among four players ($P_1, P_2, P_3,$ and P_4) using the last-diminisher method. The players play in a fixed order, with P_1 first, P_2 second, and so on. In round 1, P_1 makes the first cut and makes a claim on a C-piece. For each of the remaining players, the value of the *current C*-piece at the time it is their turn to play is given in the following table.

	P_2	P_3	P_4
Value of the current *C*-piece	$9.00	$8.50	$6.50

(a) Which player gets his or her share at the end of round 1?

(b) What is the value of the share to the player receiving it?

(c) Which player makes the first cut to start round 2?

43. A cake valued at $30 is divided among five players (P_1, P_2, P_3, P_4, and P_5) using the last-diminisher method. The players play in a fixed order, with P_1 first, P_2 second, and so on. In round 1, P_1 makes the first cut and makes a claim on a *C*-piece. For each of the remaining players, the value of the *current C*-piece at the time it is their turn to play is given in the following table.

	P_2	P_3	P_4	P_5
Value of the current *C*-piece	$5.00	$5.50	$8.50	$7.00

(a) Which player gets his or her share at the end of round 1?

(b) What is the value of the share to the player receiving it?

(c) Which player makes the first cut to start round 2?

44. A cake valued at $30 is divided among four players (P_1, P_2, P_3, and P_4) using the last-diminisher method. The players play in a fixed order, with P_1 first, P_2 second, and so on. In round 1, P_1 makes the first cut and makes a claim on a *C*-piece. For each of the remaining players, the value of the *current C*-piece at the time it is their turn to play is given in the following table.

	P_2	P_3	P_4
Value of the current *C*-piece	$6.50	$8.50	$8.00

(a) Which player gets his or her share at the end of round 1?

(b) What is the value of the share to the player receiving it?

(c) Which player makes the first cut to start round 2?

45. A cake is divided among 12 players ($P_1, P_2, P_3, \ldots, P_{12}$) using the last-diminisher method. The players play in a fixed order, with P_1 first, P_2 second, and so on. In round 1, after P_1 makes the first cut, P_3, P_7, and P_9 are the only diminishers. In round 2, after the first claim is made, P_5

is the only diminisher. In round 3, after the first claim is made, all the other players pass.

(a) Which player gets his or her share at the end of round 1?

(b) Which player makes the first cut to start round 2?

(c) Which player gets his or her share at the end of round 2?

(d) Which player gets his or her share at the end of round 3?

(e) Which player makes the first cut to start round 4?

46. A cake is divided among six players ($P_1, P_2, P_3, P_4, P_5, P_6$) using the last-diminisher method. The players play in a fixed order, with P_1 first, P_2 second, and so on. In round 1, after P_1 makes the first cut, P_2, P_5, and P_6 are the only diminishers. In round 2, after the first claim is made, there are no diminishers. In round 3, after the first claim is made, each successive player is a diminisher.

(a) Which player gets his or her share at the end of round 1?

(b) Which player makes the first cut to start round 2?

(c) Which player gets his or her share at the end of round 2?

(d) Which player gets his or her share at the end of round 3?

(e) Which player makes the first cut to start round 4?

47. Arthur, Brian, Carl, and Damian are dividing the tutti-frutti cake shown in the figure using the last-diminisher method.

The players value systems are as follows: (i) Arthur likes chocolate and orange equally well, but hates strawberry and vanilla; (ii) Brian likes chocolate and strawberry equally well, but hates orange and vanilla; (iii) Carl likes chocolate and vanilla equally well, but hates orange and strawberry; (iv) Damian likes all flavors the same. The order of play will be alphabetical. Arthur begins play by cutting an all chocolate wedge. (Assume all claims are pie wedges cut from the center of the cake out and that the first claim in each round is a piece of only one flavor.)

(a) Describe the size of Arthur's claim (in degrees).

(b) Which players are diminishers in round 1?

(c) Describe the share received by the last diminisher in round 1.

(d) Which player gets his share at the end of round 2?

(e) Which player gets his share at the end of round 3?

48. Angela, Boris, Carlos, and Dale are dividing the vanilla-strawberry cake shown in the figure using the last-diminisher method. The players value systems are as follows: (i) Angela likes strawberry twice as much as she likes vanilla; (ii) Boris loves vanilla and hates strawberry; (iii) Carlos likes strawberry and vanilla equally well; (iv) Dale likes strawberry three times as much as he likes vanilla. The order of play is Dale first, Carlos second, Boris third, and Angela last. Assume the following: (i) All claims are pie wedges cut from the center of the cake out; (ii) in round 1, the first claim is made up entirely of vanilla; and (iii) in round 2, the first claim is made up entirely of strawberry.

(a) Describe the *C*-piece claimed by Dale.

(b) Which players are diminishers in round 1?

(c) Describe the share received by the last diminisher in round 1.

(d) Which players are diminishers in round 2?

(e) Describe the share received by the last diminisher in round 2.

Exercises 49 and 50 refer to the following fair-division game: Jared, Karla, and Lori are dividing the half meatball, half vegetarian sub sandwich shown in the figure using the last-diminisher method. Jared likes the vegetarian part and the meatball part equally. Karla is a vegetarian and will not eat any of the meatball part. Lori likes the vegetarian part three times as much as the meatball part. Assume all players make their cuts perpendicular to the length of the sandwich.

Vegetarian   Meatball

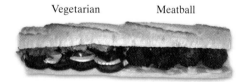

49. Suppose that the order of play is Lori first, Karla second, and Jared last. Lori starts by cutting a *C*-piece on the vegetarian part of the sandwich.

(a) What fraction of the vegetarian half is the *C*-piece cut by Lori?

(b) Which player is the last diminisher in round 1?

(c) Describe Karla's fair share of the sandwich.

(d) Describe Lori's fair share of the sandwich.

(e) Describe Jared's fair share of the sandwich.

50. Suppose that the order of play is Jared first, Lori second, and Karla last. Jared starts by cutting a *C*-piece on the vegetarian part of the sandwich.

(a) Which player is the last diminisher in round 1?

(b) Describe Karla's fair share of the sandwich.

(c) Describe Lori's fair share of the sandwich.

(d) Describe Jared's fair share of the sandwich.

F. The Method of Sealed Bids

51. Ana, Belle, and Chloe are dividing four pieces of furniture using the method of sealed bids. Their bids on each of the items are given in the following table.

	Ana	Belle	Chloe
Dresser	$150	$300	$275
Desk	$180	$150	$165
Vanity	$170	$200	$260
Tapestry	$400	$250	$500

(a) Describe the first settlement of this fair division and compute the surplus.

(b) Describe the final settlement of this fair-division problem.

52. Robert and Peter are dividing an estate consisting of a cabin and a classic car using the method of sealed bids. Robert bids $29,200 on the car and $60,900 on the cabin. Peter bids $33,200 on the car and $65,300 on the cabin.

(a) Describe the first settlement of this fair division and compute the surplus.

(b) Describe the final settlement of this fair-division problem.

53. Bob, Ann, and Jane are equal partners in a small flower shop. They can't get along anymore, but they don't want to sell the flower shop to an outsider so they decide to settle using the method of sealed bids. Bob bids $240,000 for the business, Ann bids $210,000, and Jane bids $225,000. Describe the final settlement of this fair-division problem.

54. Andre, Bea, and Chad are dividing an estate consisting of a house, a small farm, and a painting, using the method of sealed bids. Their bids on each of the items are given in the following table.

	Andre	Bea	Chad
House	$150,000	$146,000	$175,000
Farm	$430,000	$425,000	$428,000
Painting	$50,000	$59,000	$57,000

(a) Describe the first settlement of this fair division and compute the surplus.

(b) Describe the final settlement of this fair-division problem.

55. Five heirs (A, B, C, D, and E) are dividing an estate consisting of six items using the method of sealed bids. The heirs' bids on each of the items are given in the following table.

	A	B	C	D	E
Item 1	$352	$295	$395	$368	$324
Item 2	$98	$102	$98	$95	$105
Item 3	$460	$449	$510	$501	$476
Item 4	$852	$825	$832	$817	$843
Item 5	$513	$501	$505	$505	$491
Item 6	$725	$738	$750	$744	$761

(a) Describe the first settlement of this fair division and compute the surplus.

(b) Describe the final settlement of this fair-division problem.

56. Alan, Bly, and Claire are dividing five items using the method of sealed bids. Their bids on each of the items are given in the following table.

	Alan	Bly	Claire
Item 1	$14,000	$12,000	$22,000
Item 2	$24,000	$15,000	$33,000
Item 3	$16,000	$18,000	$14,000
Item 4	$16,000	$16,000	$18,000
Item 5	$18,000	$24,000	$20,000

(a) Describe the first settlement of this fair division and compute the surplus.

(b) Describe the final settlement of this fair-division problem.

57. After breaking off their engagement, Angelina and Brad agree to divide their common assets (a plasma TV, a digital camera, a laptop computer, and an MP3 player) using the method of sealed bids. Their bids on the items are shown in the following table, but Angelina's bid on the laptop computer is missing.

	Angelina	Brad
Plasma TV	$2100	$2200
Camera	$500	$580
Laptop	?	$1600
MP3 player	$300	$260

In the final settlement Angelina got the laptop computer, the MP3 player, and $355 in cash. Determine Angelina's bid on the laptop.

58. Alan, Bly, and Claire are dividing three items using the method of sealed bids. Their bids on each of the items are given in the following table, but Claire's bid on item 2 is missing.

	Alan	Bly	Claire
Item 1	$3400	$4000	$3600
Item 2	$500	$580	?
Item 3	$1800	$1600	$1500

In the final settlement Claire got item 2 and $1480 in cash. Determine Claire's bid on item 2.

G. The Method of Markers

59. Three children (A, B, and C) are dividing the array of 13 candy pieces shown in the accompanying figure using the method of markers. The players' bids are indicated in the figure.

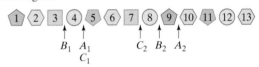

(a) Describe the allocation of candy to each child.

(b) Which items are left over?

60. Three children (A, B, and C) are dividing the array of 13 candy pieces shown in the following figure using the method of markers. The players' bids are indicated in the figure.

(a) Describe the allocation of candy to each child.

(b) Which items are left over?

61. Three children (A, B, and C) are dividing the array of 12 candy pieces shown in the following figure using the method of markers. The players' bids are indicated in the figure.

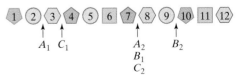

(a) Describe the allocation of candy to each child.

(b) Which items are left over?

62. Three children (A, B, and C) are dividing the array of 12 candy pieces shown in the following figure using the method of markers. The players' bids are indicated in the figure.

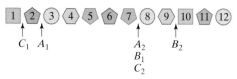

(a) Describe the allocation of candy to each child.

(b) Which items are left over?

63. Five players (A, B, C, D, and E) are dividing the array of 20 items shown in the following figure using the method of markers. The players' bids are indicated in the figure.

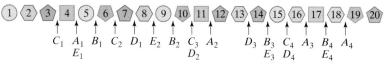

(a) Describe the allocation of items to each player.

(b) Which items are left over?

64. Four players (A, B, C, and D) are dividing the array of 15 items shown in the following figure using the method of markers. The players' bids are indicated in the figure.

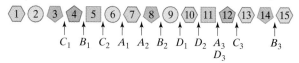

(a) Describe the allocation of items to each player.

(b) Which items are left over?

65. Quintin, Ramon, Stephone, and Tim are dividing a collection of 18 classic superhero comic books using the method of markers. The comic books are randomly lined up in the array shown in the following figure. (The W's are Wonder Woman comic books, the S's are Spiderman comic books, the G's are Green Lantern comic books, and the B's are Batman comic books.)

W S S G S W W B G G G S G S G S B B

The value of the comic books in the eyes of each player is shown in the following table.

	Quintin	Ramon	Stephone	Tim
Each W is worth	$12	$9	$8	$5
Each S is worth	$7	$5	$7	$4
Each G is worth	$4	$5	$6	$4
Each B is worth	$6	$11	$14	$7

(a) Describe the placement of each player's markers. (Use Q_1, Q_2, Q_3 for Quintin's markers, R_1, R_2, R_3 for Ramon's markers, etc.)

(b) Describe the allocation of comic books to each player, and describe what comic books are left over.

(c) Suppose that the players agree to divide the leftover comic books using the method of sealed bids. Describe the outcome of this division.

66. Queenie, Roxy, and Sophie are dividing a set of 15 CDs— 6 Beach Boys CDs, 6 Grateful Dead CDs, and 3 opera CDs using the method of markers. Queenie loves The Beach Boys, but hates The Grateful Dead and opera. Roxy loves The Grateful Dead and The Beach Boys equally well, but hates opera. Sophie loves The Grateful Dead and opera equally well, but hates The Beach Boys. The CDs are lined up in an array as follows:

O O O GD GD GD BB BB BB GD GD GD O O O

(O represents the opera CDs, GD the Grateful Dead CDs, and BB the Beach Boys CDs.)

(a) Describe the placement of each player's markers. (Use Q_1, Q_2 for Queenie's markers, R_1, R_2 for Roxy's markers, etc.)

(b) Describe the allocation of CD's to each player, and describe what CDs are left over.

(c) Suppose that the players agree that each one gets to pick an extra CD from the leftover CDs. Suppose that Queenie picks first, Sophie picks second, and Roxy picks third. Describe which leftover CDs each one would pick.

67. Ana, Belle, and Chloe are dividing the array of 9 candy bars shown in the figure. The easy solution would be for each of them to get one of each type of candy bar, but that's boring, so they decide to give the method of markers a try. The players' value systems are as follows: (i) Ana likes all candy bars the same; (ii) Belle likes Nestle Crunch bars but hates Snickers and Reese's; (iii) Chloe likes Reese's twice as much as she likes Snickers or Nestle Crunch bars.

(a) Describe the placement of each player's markers. (Use A_1, A_2 for Ana's markers, B_1, B_2 for Belle's markers, and C_1, C_2 for Chloe's markers).

(***Hint:** For each player, compute the value of each piece as a fraction of the value of the booty first. This will help you figure out where the players would place their markers.*)

(b) Describe the allocation of candy bars to each player and which candy bars are left over.

(c) Suppose that the players decide to divide the left-over pieces by a random lottery where each player gets to choose one piece. Suppose that Belle gets to choose first, Chloe second and Ana last. Describe the division of the leftover pieces.

68. Arne, Bruno, Chloe, and Daphne are dividing the array of 12 candy bars shown in the figure using the method of markers. The players' value systems are as follows: (i) Arne values Nestle Crunch bars and Baby Ruth bars three times as much as Snickers, and values Snickers and Reese's the same; (ii) Bruno is allergic to Nestle Crunch Bars and can't eat them, values Snickers 1.5 times as much as Baby Ruth bars, and values Snickers and Reese's the same; (iii) Chloe values Snickers and Baby Ruth bars twice as much as Nestle Crunch bars, and values Reese's 1.5 times as much as Snickers; (iv) Daphne is allergic to Snickers and cannot eat them, values Reese's 1.5 times as much as Nestle Crunch bars, and values Reese's and Baby Ruth bars the same.

(a) Describe the placement of each player's markers.

(*Hint: For each player, compute the value of each piece as a fraction of the value of the booty first. This will help you figure out where the players would place their markers.*)

(b) Describe the allocation of candy bars to each player and which candy bars are left over.

(c) Suppose that the players decide to divide the left-over pieces by a random lottery where each player gets to choose one piece. Suppose that Chloe gets to choose first, Daphne second, Arne third, and Bruno last. Describe the final fair division after each player had a chance to pick one of the leftover pieces.

JOGGING

69. Every Friday night, Marty's Ice Cream Parlor sells "Kitchen Sink Sundaes" for $6.00 each. A KiSS consists of 12 mixed scoops of whatever flavors Marty wants to get rid of. The customer has no choice. Abe, Babe, and Cassandra decide to share a KiSS. Abe wants to eat half of it and pays $3.00 while Babe and Cassie pay $1.50 each. They decide to divide it by the lone-divider method. Abe spoons the sundae onto four plates (P, Q, R, and S) and says that he will be satisfied with any two of them.

(a) If both Babe and Cassie find only Q and R acceptable, discuss how to proceed.

(b) If Babe finds only Q and R acceptable, and Cassie finds only P and S acceptable, discuss how to proceed.

(c) If Babe and Cassie both find only R acceptable, discuss how to proceed.

70. Dishonest bidding doesn't pay. The purpose of this exercise is to show how players that bid dishonestly can end up losing their shirts. Four partners (Burly, Curly, Greedy, and Dandy) are dividing a million-dollar property using the lone-divider method. Using a map, Dandy divides the property into four parcels s_1, s_2, s_3, and s_4. The following table shows the value of the four parcels in the eyes of each partner.

	s_1	s_2	s_3	s_4
Dandy	$250,000	$250,000	$250,000	$250,000
Burly	$400,000	$200,000	$200,000	$200,000
Curly	$280,000	$320,000	$200,000	$200,000
Greedy	$320,000	$280,000	$280,000	$120,000

(a) Assuming all players bid honestly, describe the outcome of the fair division.

(b) Suppose that Burly and Curly both bid honestly, but Greedy gets greedy and decides to bid only for s_1 (figuring that doing that will get him s_1). Now there is a scenario under which Greedy ends up with a parcel worth only $220,000. Describe how this might happen.

71. Three players (P_1, P_2, and P_3) agree to divide the property shown using the last-diminisher method. The order of the players is P_1, P_2, P_3. The first player to play, P_1, makes a claim C as shown in the following figure.

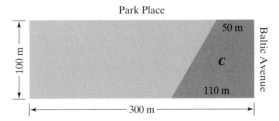

We know that both P_2's and P_3's value systems are the same and that they value the land uniformly.

(a) Give a geometric argument for why P_2 and P_3 would both pass in round 1 and P_1 would end up with C.

(b) Describe a possible cut that the divider in round 2 might make.

(c) Suppose that, after round 1 is over, P_2 and P_3 discover that the city requires that the next cut be made parallel to Park Place. Describe a possible cut that the divider in round 2 might make in this case.

(d) Repeat (c) for a cut that must be made parallel to Baltic Avenue.

72. Three players (P_1, P_2, and P_3) agree to divide the property shown using the last-diminisher method. The order of the players is P_1, P_2, P_3. The first player to play, P_1, makes a claim C as shown in the following figure.

Park Place

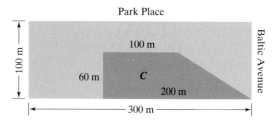

We know that both P_2's and P_3's value systems are the same and that they value the land uniformly.

(a) Give a geometric argument for why P_2 and P_3 would both pass in round 1 and P_1 would end up with C.

(b) Describe a possible cut that the divider in round 2 might make.

(c) Suppose that, after round 1 is over, P_2 and P_3 discover that the city requires that the next cut be made parallel to Baltic Avenue. Describe a possible cut that the divider in round 2 might make in this case.

73. Three players (P_1, P_2, and P_3) agree to divide the property shown using the last-diminisher method. The order of the players is P_1, P_2, P_3. The first player to play, P_1, makes a claim C as shown in the following figure.

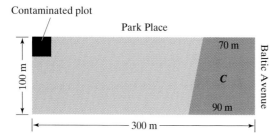

We know that both P_2's and P_3's value systems are the same and that they value the land uniformly, except for the square 20-meter-by-20-meter plot in the upper left corner of the property. This plot is contaminated by an old, underground gas station tank that will cost twice as much to remove and clean up as that square plot would otherwise be worth.

(a) Give an argument why P_2 and P_3 would both pass in round 1 and P_1 would end up with C.

(b) Suppose that after round 1 is over, P_2 and P_3 discover that the city requires that the next cut be made parallel to Baltic Avenue. Describe a possible cut that the divider in round 2 might make in this case.

74. Two players (A and B) wish to dissolve their partnership using the method of sealed bids. A bids x dollars and B bids y dollars, where $x < y$.

(a) Describe the first settlement.

(b) Express the surplus in terms of x and y.

(c) Describe the final settlement.

75. Consider the following variation of the divider-chooser method for two players. After the divider cuts the cake into two pieces, the chooser, unable to see either piece, picks his piece randomly by flipping a coin. The divider, of course, gets the other piece.

(a) Is this a fair-division scheme according to our definition? Explain your answer.

(b) Who would you rather be—divider or chooser? Explain.

76. Three partners (A, B, and C) own a business jointly. By mutual agreement they decide to split up—one of them will keep the business and the other two will get cash. They decide to do this using the method of sealed bids. Suppose that A bids x dollars, B bids y dollars, and C bids z dollars, with z being the highest bid.

(a) Describe the first settlement.

(b) Express the surplus in terms of x, y, and z.

(c) Describe the final settlement.

77. The National Basketball Association is expanding to Europe with three new teams—the Athens Olympians (A), Barcelona Toros (B), and Cologne Emperors (C). In a supplementary draft, each team will get to draft two of the top six European basketball players. The six players are ranked by each of the teams as follows.

Rank	Athens	Barcelona	Cologne
1	Allen	Evans	Carter
2	Bryant	Francis	Francis
3	Carter	Bryant	Evans
4	Duncan	Allen	Duncan
5	Evans	Duncan	Allen
6	Francis	Carter	Bryant

(a) Suppose the teams alternate turns in the draft in the order A-B-C-A-B-C. Describe which players get drafted by each team.

(b) Suppose the teams alternate turns in the draft in the order A-B-C-C-B-A. Describe which players get drafted by each team.

(c) In either draft, is there an alternate assignment of players that would benefit all three teams? Explain.

RUNNING

Exercises 78 and 79 show how the method of sealed bids can be used when some values are negative. If you offer to pay a bid (as for a purchase), the bid is listed as a positive amount. It follows that, if you offer to receive a bid (as for labor), the bid is listed as a negative amount. Regardless, the winning bid is the highest number.

78. Three women (Ruth, Sarah, and Tamara) share a house and wish to divide the chores: bathrooms, cooking, dishes, laundry, and vacuuming. For each chore, they privately write the least they are willing to receive monthly (their negative valuation) in return for doing that chore. The results are shown in the following table.

	Ruth	Sarah	Tamara
Clean bathrooms	$-20	$-30	$-40
Do cooking	$-50	$-10	$-25
Wash dishes	$-30	$-20	$-15
Mow the lawn	$-30	$-20	$-10
Vacuum and dust	$-20	$-40	$-15

Divide the chores using the method of sealed bids. Who does which chores? Who gets paid, and how much? Who pays, and how much?

79. Four roommates are going their separate ways after graduation and wish to divide up their jointly owned furniture (equal shares) and the moving chores by the method of sealed bids. Their bids (in dollars) on the items are shown in the following table.

	Quintin	Ramon	Stephone	Tim
Stereo	300	250	200	280
Couch	200	350	300	100
Table	250	200	240	80
Desk	150	150	200	220
Cleaning the rugs	-80	-70	-100	-60
Patching nail holes	-60	-30	-60	-40
Repairing the window	-60	-50	-80	-80

(a) What is each roommate's estimate of his part of the total?

(b) How much surplus cash is there?

(c) What is the final outcome?

(d) What percentage of the total value (of everything) does each roommate get using the roommate's own valuation?

(e) If Tim is dishonest and sneaks a peek at the bid lists of the other three roommates before filling out his own, how could he adjust his bids (in whole dollars) so as to get the same furniture as before, but no chores, and also maximize his cash receipts? Explain your reasoning.

80. Three players (A, B, and C) are going to share a cake consisting of three equal sections of chocolate, strawberry, and vanilla. The three figures below show how each player values each of the sections of cake (given as a percentage of the total value of the cake). Describe how you could cut this cake into three pieces so that each player can get a piece that he or she values at exactly 50% of the value of the cake.

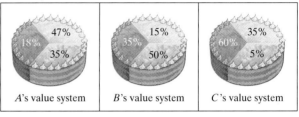

A's value system   *B*'s value system   *C*'s value system

81. (a) Suppose that four players divide a cake using the lone-chooser method. The chooser is C; the dividers are D_1, D_2, and D_3. Explain why, when properly played, the method guarantees to each player a share worth at least 25% of the cake.

(b) Suppose that N players divide a cake using the lone-chooser method. The chooser is C; the dividers are $D_1, D_2, \ldots, D_{N-1}$. Explain why, when properly played, the method guarantees to each player a fair share.

*Exercises 82 and 83 refer to a continuous fair-division method known as the Dubins-Spanier **moving knife method**. This method can be used to divide a single continuous item (say a cake) by having a "moving knife" that moves above the cake like the minute hand of a clock. An original cut is made at the "12:00 position" and then the "knife" starts moving clockwise around the cake until one of the players yells, "Stop!" At this point, the knife drops and the player yelling "Stop!" receives the piece cut by the moving knife. The "knife" starts moving again and play continues with the remaining players. The next player to yell "Stop" gets the next piece and so on.*

82. Suppose that Angela, Boris, and Carlos use the moving knife procedure to divide a $12 vanilla-strawberry cake. The value of each half of the cake in each player's eyes is shown in the following figure. An original cut is made from the center of the cake to the 12:00 position, and the knife starts rotating clockwise around the center of the cake.

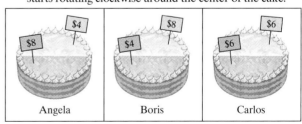

Angela   Boris   Carlos

(a) Determine which player is the first player to yell "Stop!", and describe the player's share.

(b) Determine which player is the second player to yell "Stop!", and describe the player's share.

(c) Describe the final division of the cake and the value of each piece to the player receiving it.

83. Suppose that Angela, Boris, Carlos, and Dale use the moving knife procedure to divide the $12 vanilla-strawberry cake. The value of each half of the cake in each player's eyes is shown in the following figure. An original cut is made from the center of the cake to the 12:00 position, and the knife starts rotating clockwise around the center of the cake.

Angela

Boris

Carlos

Dale

(a) Determine which player is the first player to yell "Stop!", and describe the player's share.

(b) Determine which player is the second player to yell "Stop!", and describe the player's share.

(c) Determine which player is the third player to yell "Stop!", and describe the player's share.

84. (a) Explain why, in the method of sealed bids, the surplus (after the original allocation and payments are made) is always positive or zero.

(b) Describe the circumstances under which the surplus is zero.

Projects and Papers

A. Envy-Free Fair Division

An *envy-free fair division* is a fair division in which each player ends up with a share that he or she feels is as good or better than that of any other player. Thus, in an envy-free fair division a player would never envy or covet another player's share. In the last decade, several important envy-free fair division methods have been developed.

Write a paper discussing the topic of envy-free fair division.

Notes: Some ideas of topics for your paper: (i) Discuss how envy-free fair division differs from the (proportional) type of fair division discussed in this chapter, (ii) describe a continuous envy-free fair division method for $N = 3$ players, (iii) give an outline of the Brams-Taylor method for continuous envy-free fair division for any number of players.

Suggested references for this paper are references 18, 9, and 1.

B. Fair Divisions with Unequal Shares

All the fair-division problems we discussed in this chapter were based on the assumption of *symmetry* (i.e., all players have equal rights in the division). Sometimes, players are not all equal and are entitled to larger or smaller shares than other players. This type of fair-division problem is called an *asymmetric* fair division (*asymmetric* means that the players

are not all equal in their rights). For example, Grandma's will may stipulate that her estate is to be divided as follows: Art is entitled to 25%, Betty is entitled to 35%, Carla is entitled to 30%, and Dave is entitled to 10%. (After all, it is her will, and if she wants to be difficult, she can!)

Write a paper discussing how some of the fair-division methods discussed in this chapter can be adapted for the case of asymmetric fair division. Discuss at least one discrete and one continuous asymmetric fair-division method.

C. The Mathematics of Forgiveness and Cooperation

It is generally acknowledged that as a species, humans are more selfless, cooperative, and forgiving to each other than any other animal species. On the surface, this appears to be a contradiction to the general Darwinian notion that only the ruthless get ahead. In recent years, scientists have been able to model the *cooperation* versus *noncooperation* question using a mathematical game called the *prisoner's dilemma*. Recent studies using the *prisoner's dilemma* have shown that (i) in the long run, cooperation makes sense mathematically; and (ii) the human impulse to cooperate may have a biochemical root.

Write a research paper describing these recent developments. This is a topic where mathematics, game theory, biology, and social psychology all interact, and you should touch on these interactions in your paper.

References and Further Readings

1. Brams, Steven, and Alan Taylor, "An Envy-Free Cake Division Protocol," *American Mathematical Monthly*, 102 (1995), 9–18.
2. Brams, Steven, and Alan Taylor, *Fair Division*. Cambridge, England: Cambridge University Press, 1996.
3. Brams, Steven, and Alan Taylor, *The Win-Win Solution: Guaranteeing Fair Shares to Everybody*. New York: W. W. Norton, 1999.
4. Dubins, L. E., "Group Decision Devices," *American Mathematical Monthly*, 84 (1977), 350–356.
5. Fink, A. M., "A Note on the Fair Division Problem," *Mathematics Magazine*, 37 (1964), 341–342.
6. Gardner, Martin, *aha! Insight*. New York: W. H. Freeman, 1978.
7. Hill, Theodore, "Determining a Fair Border," *American Mathematical Monthly*, 90 (1983), 438–442.
8. Hill, Theodore, "Mathematical Devices for Getting a Fair Share," *American Scientist*, 88 (2000), 325–331.
9. Hively, Will, "Dividing the Spoils," *Discover*, 16 (1995), 49–57.
10. Jones, Martin, "A Note on a Cake Cutting Algorithm of Banach and Knaster," *American Mathematical Monthly*, 104 (1997), 353–355.
11. Kaluza, Roman, *Through A Reporter's Eyes: The Life of Stefan Banach*. Boston, MA: Birkhauser, 1996.
12. Kuhn, Harold W., "On Games of Fair Division," *Essays in Mathematical Economics*, Martin Shubik, ed. Princeton, NJ: Princeton University Press, 1967, 29–37.
13. Olivastro, Dominic, "Preferred Shares," *The Sciences*, March–April 1992, 52–54.
14. Robertson, Jack, and William Webb, *Cake Cutting Algorithms: Be Fair If You Can*. Natick, MA: A. K. Peters, 1998.
15. Steinhaus, Hugo, *Mathematical Snapshots*, 3rd ed. New York: Dover, 1999.
16. Steinhaus, Hugo, "The Problem of Fair Division," *Econometrica*, 16 (1948), 101–104.
17. Stewart, Ian, "Fair Shares for All," *New Scientist*, 146 (June, 1995), 42–46.
18. Stewart, Ian, "Mathematical Recreations: Division without Envy," *Scientific American*, (1999), 110–111.
19. Stromquist, Walter, "How to Cut a Cake Fairly," *American Mathematical Monthly*, 87 (1980), 640–644.

4

The Mathematics of Apportionment

Making the Rounds

Representatives . . . shall be apportioned among the several States . . . according to their respective Numbers.

Article I, Section 2

In the stifling heat of the Philadelphia summer of 1787, delegates from the thirteen states met to draft a Constitution for a new nation. Except for Thomas Jefferson (then minister to France) and Patrick Henry (who refused to participate), all the main names of the American Revolution were there—George Washington, Ben Franklin, Alexander Hamilton, James Madison.

Without a doubt, the most important and heated debate at the Constitutional Convention concerned the makeup of the legislature. The small states wanted all states to have the same number of representatives; the larger states wanted some form of proportional representation. The Constitutional compromise, known as the Connecticut Plan, was a Senate, in which every state has two senators, and a House of Representatives, in which each state has a number of representatives that is a function of its population.

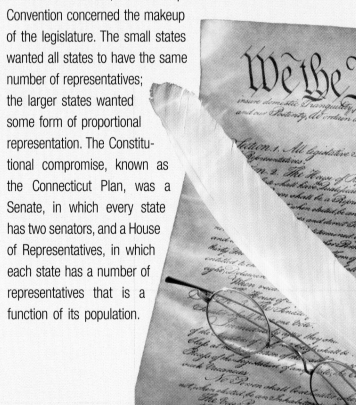

W hile the Constitution makes clear that seats in the House of Representatives are to be allocated to the states based on their populations (" . . . according to their respective Numbers"), it does not prescribe the formula for the calculations. Undoubtedly, the Founding Fathers felt that this was a relatively minor detail—a matter of simple arithmetic that could be easily figured out and agreed upon by reasonable people. Certainly it was not the kind of thing to clutter a Constitution with, or spend time arguing over in the heat of the summer. What the Founding Fathers did not realize is that Article 1, Section 2, set the Constitution of the United States into a collision course with a mathematical iceberg known today (but certainly not then) as *the apportionment problem*.

This chapter is an excursion into the mathematics behind the apportionment problem, and there are many interesting and surprising twists and turns to this excursion. We will start the chapter with an introduction to the basic concepts of apportionment: What is an apportionment problem? What is the nature of the problem? What are the mathematical issues we must deal with? Why should we care? In Sections 4.2 through 4.6, we will discuss four of the best-known apportionment methods that have been proposed over the years to solve apportionment problems. Surprisingly, the names associated with these methods are not names you would expect to find attached to mathematical procedures—Alexander Hamilton, Thomas Jefferson, John Quincy Adams, and Daniel Webster. This is a testament to the historical importance of the topic of this chapter. While our primary focus will be the mathematical aspects of apportionment, as we navigate the mathematics we will also get a glimpse of a little known but important chapter in American history.

4.1 Apportionment Problems

Obviously, the word *apportion* is the key word in this chapter. There are two critical elements in the dictionary definition of the word: (i) We are dividing and assigning things, and (ii) We are doing this on a proportional basis and in a planned, organized fashion.

ap·pôr·tion: to divide and assign in due and proper proportion or according to some plan.

Webster's New Twentieth Century Dictionary

We will start this section with a pair of examples that illustrate the nature of the problem we are dealing with. These examples will raise several questions, but not give any answers. Most of the answers will come later.

▶ **EXAMPLE 4.1** Kitchen Capitalism

Mom has a total of 50 identical pieces of candy (let's say caramels), which she is planning to divide among her five children (this is the *division* part). Like any good mom, she is intent on doing this fairly. Of course, the easiest thing to do would be to give each child 10 caramels—by most standards, that would be fair. Mom, however, is thinking of the long-term picture—she wants to teach her children about the value of work and about the relationship between work and reward. This leads her to the following idea: She announces to the kids that the candy is going to be divided at the end of the week in proportion to the amount of time each of them spends helping with the weekly kitchen chores—if you worked twice as long as your brother you get twice as much candy, and so on (this is the "due and proper proportion" part).

At the end of the week, the numbers are in. Table 4-1 shows the amount of work done by each child during the week. (Yes, mom did keep up-to-the-minute records!)

TABLE 4-1 Amount of Work (in minutes) per Child

Child	Alan	Betty	Connie	Doug	Ellie	Total
Minutes worked	150	78	173	204	295	900

According to the ground rules, Alan, who worked 150 out of a total of 900 minutes, is entitled to $16\frac{2}{3}\%$ of the 50 pieces of candy $[(150/900) = 16\frac{2}{3}\%]$, or $8\frac{1}{3}$ pieces. Here comes the problem: Since the pieces of candy are indivisible, it is impossible for Alan to get the exact share he is entitled to—he can get 8 pieces (and get shorted) or he can get 9 pieces (and someone else will get shorted). A similar problem occurs with each of the other children. Betty's exact fair share should be $4\frac{1}{3}$ pieces; Connie's should be $9\frac{11}{18}$ pieces; Doug's, $11\frac{1}{3}$ pieces; and Ellie's $16\frac{7}{18}$ pieces. (Be sure to double-check these figures!) Because none of these shares can be realized, an absolutely fair apportionment of the candy is going to be impossible. What should mom do?   ◀◀

Example 4.1 shows all the elements of an *apportionment problem*—there are objects to be divided (the pieces of candy) and there is a proportionality criterion for the division (number of minutes worked during the week). We will say that the pieces of candy are *apportioned* to the kids, and we will describe the final solution as an *apportionment* (Alan's *apportionment* is x pieces, Betty's *apportionment* is y pieces, etc.).

Let's now consider a seemingly different apportionment problem.

▶ **EXAMPLE 4.2** The Intergalactic Congress of Utopia

It is the year 2525, and all the planets in the Utopia galaxy have finally signed a peace treaty. Five of the planets (Alanos, Betta, Conii, Dugos, and Ellisium) decide to join forces and form an Intergalactic Federation. The Federation will be ruled by an Intergalactic Congress consisting of 50 "elders," and the 50 seats in the Intergalactic Congress are to be *apportioned* among the planets according to their respective populations.

The population data for each of the planets (in billions) are shown in Table 4-2. Based on these population figures, what is the correct *apportionment* of seats to each planet? (Good question. We will come back to it in the near future.)

TABLE 4-2 Intergalactic Federation: Population Figures (in billions) for 2525

Planet	Alanos	Betta	Conii	Dugos	Ellisium	Total
Population	150	78	173	204	295	900

Source: Intergalactic Census Bureau.

Example 4.2 is another example of an apportionment problem—here the objects being divided are the 50 seats in the Intergalactic Congress and the proportionality criterion is population. But there is more to it than that. The observant reader may have noticed that the numbers in Examples 4.1 and 4.2 are identical—it is only the setting that has changed. While the merits of the problem may be different, mathematically speaking, Examples 4.1 and 4.2 are one and the same apportionment problem. Solve either one and you have solved both!

Between the extremes of apportioning the seats in the Intergalactic Congress of Utopia (important, but too far away!) and apportioning the candy among the children in the house (closer to home, but the galaxy will not come to an end if the apportionments aren't done right!) fall many other real-life apportionment problems that are both important and relevant: apportioning nurses to shifts in a hospital, apportioning telephone calls to switchboards in a network, apportioning classrooms to departments in a university, and so on.

We will now introduce the most important apportionment example of this chapter. We will return to this example many times. While the story behind the example is obviously fiction, the issues it raises are real.

> EXAMPLE 4.3 The Congress of Parador

Parador is a small republic located in Central America and consisting of six states: Azucar, Bahia, Cafe, Diamante, Esmeralda, and Felicidad (*A, B, C, D, E,* and *F* for short). There are 250 seats in the Congress, which, according to the laws of Parador, are to be apportioned among the states in proportion to their respective populations. What is the "correct" apportionment?

Table 4-3 shows the population figures for the six states according to the most recent census.

TABLE 4-3 Republic of Parador (Populations by State)

State	A	B	C	D	E	F	Total
Population	1,646,000	6,936,000	154,000	2,091,000	685,000	988,000	12,500,000

The first step we will take to tackle this problem is to find a good *unit of measurement*. The most natural unit of measurement is the *ratio of people to seats*. In the case of Parador, this ratio turns out to be $12,500,000/250 = 50,000$. We call this ratio the **standard divisor** *SD*. (Typically, the standard divisor is not going to turn out to be such a nice whole number, but it is always going to be a rational number.)

The standard divisor $SD = 50,000$ tells us that in Parador, each seat in the Congress corresponds to 50,000 people. We can now use this yardstick to find the number of seats that each state *should get by the proportionality criterion*—all we have to do is divide the state's population by 50,000. For example, take state A. If we divide the population of A by the standard divisor, we get $1,646,000/50,000 = 32.92$. This number is called the **standard quota** (sometimes also known as the *exact* or *fair quota*) of state A. If seats in the Congress could be apportioned in fractional parts, then the fair and exact apportionment to A would be 32.92 seats, but of course, this is impossible—seats in the Congress are indivisible objects! Thus, the standard divisor of 32.92 is a standard that cannot meet, although in this case we might be able to come close to it: the next best solution might be to apportion to state A 33 seats. Hold that thought!

Using the standard divisor $SD = 50,000$, we can quickly find the standard quotas of each of the other states. These are shown in the second row of Table 4-4 (rounded to two decimal places). Notice that the sum of the standard quotas equals 250, the number of seats being apportioned.

TABLE 4-4 Republic of Parador: Standard Quotas for Each State ($SD = 50,000$)

State	A	B	C	D	E	F	Total
Population	1,646,000	6,936,000	154,000	2,091,000	685,000	988,000	12,500,000
Standard quota	32.92	138.72	3.08	41.82	13.70	19.76	250

Once we have computed the standard quotas, we get to the heart of the apportionment problem: How should we round these quotas into whole numbers? At first glance, this seems like a dumb question. After all, we all learned in school how to round decimals to whole numbers—round down if the fractional part is less than 0.5, round up otherwise. This kind of rounding is called *rounding to the nearest integer*, or simply *conventional rounding*. Unfortunately, conventional rounding will not work in this example, and Table 4-5 shows why not—we would end up giving out 251 seats in the Congress, and there are only 250 seats to give out! (Add an extra chair, you say? Sorry, this is not a banquet. The number of seats is fixed—we don't have the luxury of adding or subtracting seats!)

TABLE 4-5 Conventional Rounding of the Quotas

State	Population	Standard quota	Nearest integer
A	1,646,000	32.92	33
B	6,936,000	138.72	139
C	154,000	3.08	3
D	2,091,000	41.82	42
E	685,000	13.70	14
F	988,000	19.76	20
Total	12,500,000	250.00	**251**

So now that we know that conventional rounding of the standard quotas is not going to give us the solution to the apportionment of Parador's Congress, what do we try next? For now, we leave this example—a simple apportionment problem looking for a simple solution—with no answers (but, hopefully, with some gained insight!). Our search for a good solution to this problem will be the theme of our journey through the rest of this chapter.   ◀◀

Before moving on, we will revisit and expand some of the key concepts of this section.

The basic elements of every apportionment problem are as follows:

■ **The "states."** This is the term we will use to describe the *players* involved in the apportionment. [It is customary to use the legislature metaphor to define all apportionment concepts, but the concepts apply to any apportionment problem regardless of its context. Thus, in the original candy example (Example 4.1), the "states" are the children.] Unless they have specific names (Azucar, Bahia, etc.), we will let $A_1, A_2, \ldots, A_N$ denote the N states.

■ **The "seats."** This term describes the set of M *identical, indivisible objects* that are being divided among the N states. These "seats," of course, can be objects other than actual seats in a legislature. (In Example 4.1, the "seats" are the pieces of candy.) For convenience, we will assume that $M \geq N$. This ensures that every state can potentially get at least one seat. (This assumption does not imply that every state *must* get a seat! Such a requirement is not part of the general apportionment problem, although it is part of the constitutional requirements for the apportionment of the U.S. House of Representatives.)

■ **The "populations."** This is a set of N positive numbers (for simplicity we will assume them to be whole numbers) which are used as the basis for the apportionment of the seats to the states. (In Example 4.1, the "populations" were the minutes worked on the kitchen chores.) We will use $p_1, p_2, \ldots, p_N$ to denote the state's respective populations, and P to denote the total population ($P = p_1 + p_2 + \cdots + p_N$).

Two of the most important concepts of the chapter are the standard divisor and the standard quotas. We can now define these using the above terminology and notation.

■ **The standard divisor (SD).** This is the ratio of population to seats. It gives us a unit of measurement (SD people = 1 seat) for our apportionment calculations. Formally, $SD = P/M$.

■ **The standard quotas.** The standard quota of a state is the exact fractional number of seats that the state would get if fractional seats were allowed. We will use the notation $q_1, q_2, \ldots, q_N$ to denote the standard quotas of the respective states. To find a state's standard quota, we divide the state's population by the standard divisor: *quota = population/SD*. (In general, the standard quotas can be expressed as decimals or fractions—it would be almost a miracle if one of them turned out to be a whole number—and when using the decimal form, it's customary to round them to two or three decimal places.)

■ **Upper and lower quotas.** Associated with each standard quota are two other important numbers—the **lower quota** (the quota rounded down), and the **upper quota** (the quota rounded up). In the unlikely event that the quota is a whole number, the lower and upper quotas are the same. We will use L's to denote lower quotas and U's to denote upper quotas. For example, the standard quota $q_1 = 32.92$ has lower quota $L_1 = 32$ and upper quota $U_1 = 33$.

One of the important morals of Example 4.3 is that the intuitively obvious approach to solving an apportionment problem (take the standard quotas and round them off to either the lower or upper quotas using conventional rounding) is not a strategy we can rely on, since the final apportionment may not add up to M seats. In Example 4.3 we ended up apportioning more seats than we were supposed to. In other apportionment problems we could end up apportioning fewer seats than we are supposed to. Sometimes the stars are aligned and we end up apportioning the right number of seats. As a general method, this will not do.

Our main goal in this chapter is to discover a "good" apportionment method—a reliable procedure that (i) will always produce a valid apportionment (exactly M seats are apportioned), and (ii) will always produce a "fair" apportionment. In this quest, we will discuss several different methods and find out what is good and bad about each one.

Alexander Hamilton (1757–1804). It is doubtful that Hamilton was the first person to come up with the method that bears his name, but he was its most famous advocate and gets the credit (at least in the United States).

4.2 Hamilton's Method and the Quota Rule

Hamilton's method (also known as *Vinton's method* or the *method of largest remainders*) was used in the United States only between 1850 and 1900, but it is still used today to apportion the legislatures of Costa Rica, Namibia, and Sweden. While historically Hamilton's method did not come first, we will discuss it first because it is mathematically the simplest.

Hamilton's method can be described quite briefly: Every state gets at least its lower quota. As many states as possible get their upper quota, with the one with highest fractional part having first priority, the one with second highest fractional part second priority, and so on. A little more formally, it goes like this:

Hamilton's Method

- **Step 1.** Calculate each state's standard quota.

- **Step 2.** Give to each state (for the time being) its *lower quota*.

- **Step 3.** Give the surplus seats (one at a time) to the states with the largest fractional parts until there are no more surplus seats.

> **EXAMPLE 4.4** Parador's Congress (Hamilton's Method)

As promised, we are revisiting the Parador Congress apportionment problem. We are now going to find our first real solution to this problem—the solution given by Hamilton's method (sometimes, for the sake of brevity, we will call it the *Hamilton apportionment*). Table 4-6 shows all the details and speaks for itself. (Reminder to the reader: The standard quotas in column 3 of the table were computed in Example 4.3.)

		Step 1	Step 2	Fractional	Step 3	Hamilton
State	**Population**	**Quota**	**Lower quota**	**parts**	**Surplus**	**apportionment**
A	1,646,000	32.92	32	0.92	First	33
B	6,936,000	138.72	138	0.72	Last	139
C	154,000	3.08	3	0.08		3
D	2,091,000	41.82	41	0.82	Second	42
E	685,000	13.70	13	0.70		13
F	988,000	19.76	19	0.76	Third	20
Total	12,500,000	250.00	246	4.00	4	250

TABLE 4-6 Parador Congress Apportionment Under Hamilton's Method

At first glance, Hamilton's method appears to be quite fair. However, a careful look at Example 4.4 already shows some hints of possible unfairness. Compare the fates of state B, with a fractional part of 0.72, and state E, with a fractional part of 0.70. State B gets the last surplus seat; state E gets nothing! Sure enough, 0.72 is more than 0.70, so following the rules of Hamilton's method, we give priority to B. By the same token, B is a huge state, and as a percentage of its population the 0.72 represents an insignificant amount, whereas state E is a relatively small state, and its fractional part of 0.70 (for which it gets nothing) represents more than 5% of its population ($0.70/13.70 \approx 0.051 = 5.1\%$).

It could be reasonably argued that Hamilton's method has a major flaw in the way it relies entirely on the size of the fractional parts without consideration of what those fractional parts represent as a percent of the state's population. In so doing, Hamilton's method creates a systematic bias in favor of larger states over smaller ones. This is bad—a good apportionment method should be *population neutral*, meaning that it should not be biased in favor of large states over small ones or vice versa. In the next section we will see that bias in favor of large states is only one of the many serious flaws of Hamilton's method.

To be totally fair, Hamilton's method has two important things going for it: (1) It is very easy to understand, and (2) it satisfies an extremely important requirement for fairness called the *quota rule*.

> The idea of using the relative size of the fractional parts to determine the order of priority for the surplus seats is the basis of an apportionment method known as *Lowndes's method*. For more on Lowndes's method, see Exercises 52 and 53.

The Quota Rule

Suppose that the standard quota of state A is $q = 38.59$. This number represents the benchmark for what a fair apportionment of seats to A ought to be. Since giving A exactly 38.59 seats is impossible, it is reasonable to argue that A should get either its upper quota ($U = 39$ seats) or, in the worst of cases, its lower quota ($L = 38$ seats). Thus, no reasonable or fair apportionment should give A, say, 37 seats, or, for that matter, 40 seats. Either of these outcomes would be a clear violation of what seems like a fundamental requirement for fairness—*a state should not be apportioned a number of seats smaller than its lower quota or larger than its upper quota*. This rule is known as the *quota rule*.

> **The Quota Rule**
>
> No state should be apportioned a number of seats smaller than its lower quota or larger than its upper quota. (When a state is apportioned *a number smaller than its lower quota*, we call it a **lower-quota violation**; when a state is apportioned *a number larger than its upper quota*, we call it an **upper-quota violation**.)

An apportionment method that guarantees that every state will be apportioned either its lower quota or its upper quota is said to *satisfy the quota rule*. It is not hard to see that Hamilton's method satisfies the quota rule: Step 2 of Hamilton's method hands out to each state its lower quota. Right off the bat this guarantees that there will be no lower-quota violations. In Step 3, some states get one extra seat, some get none; no state can get more than one. This guarantees that there will be no upper-quota violations.

4.3 The Alabama and Other Paradoxes

The most serious (in fact, the fatal) flaw of Hamilton's method is commonly known as the **Alabama paradox**. In essence, the Alabama paradox occurs when an *increase in the total number of seats being apportioned, in and of itself, forces a state to lose one of its seats*. The best way to understand what this means is to look carefully at the following example.

> **EXAMPLE 4.5** More Seats Means Less Seats

The small country of Calavos consists of three states: Bama, Tecos, and Ilnos. With 200 seats in the House of Representatives, the apportionment of those seats under Hamilton's method is shown in Table 4-7.

TABLE 4-7 Hamilton Apportionment for $M = 200$ ($SD = 100$)					
State	**Population**	**Step 1**	**Step 2**	**Step 3**	**Apportionment**
Bama	940	9.4	9	1	**10**
Tecos	9030	90.3	90	0	90
Ilnos	10,030	100.3	100	0	100
Total	20,000	200.0	199	1	**200**

Now imagine that overnight the number of seats is increased to 201, but nothing else changes. Since there is one more seat to give out, the apportionment has to be recomputed. Table 4-8 shows the new apportionment (still using Hamilton's method) for a House with 201 seats. (You are strongly encouraged to verify these short calculations. Note that for $M = 200$ the SD is 100; for $M = 201$ the SD drops to 99.5.)

TABLE 4-8 Hamilton Apportionment for $M = 201$ ($SD \approx 99.5$)

State	Population	Step 1	Step 2	Step 3	Apportionment
Bama	940	9.45	9	0	**9**
Tecos	9030	90.75	90	1	91
Ilnos	10,030	100.80	100	1	101
Total	20,000	201.00	199	2	**201**

The shocking part of this story is the fate of Bama, the "little guy." When the House of Representatives had 200 seats Bama got 10 seats, but when the number of seats to be divided increased to 201 Bama's apportionment went down to 9 seats. How did this paradox occur? Notice the effect of the increase in M on the size of the decimal parts: In a House with 200 seats, Bama is at the head of the priority line for surplus seats, but when the number of seats goes up to 201, Bama gets shuffled to the back of the line. Welcome to the wacky arithmetic of Hamilton's method!   ◀◀

Example 4.5 illustrates the quirk of arithmetic behind the Alabama paradox: When we increase the number of seats to be apportioned, each state's standard quota goes up, but not by the same amount. As the fractional parts change, some states can move ahead of others in the priority order for the surplus seats. This can result in some state or states losing seats they already had.

But why is Alabama's name attached to this paradox? The answer is historical. In 1882, as different apportionment bills for the House of Representatives were being debated, it came to light that in a House with 299 seats, Alabama's apportionment would be 8 seats, whereas in a House with 300 seats, Alabama's apportionment would be 7 seats. The effect of adding that extra seat would be to give one more seat to Texas and one more seat to Illinois. To do this, a seat would have to be taken away from Alabama! The details are shown in Table 4-9.

TABLE 4-9 The Real Alabama Paradox (1882)

State	Standard quota with $M = 299$	Apportionment with $M = 299$	Standard quota with $M = 300$	Apportionment with $M = 300$
Alabama	7.646	**8**	7.671	**7**
Texas	9.64	9	9.672	10
Illinois	18.64	18	18.702	19

The final straw for Hamilton's method came during the 1901 apportionment debate. As part of the debate, House sizes of anywhere between 350 and 400 seats were considered. Within that range, the apportionment for the state of Maine went up and down like a roller coaster: For M between 350 and 356, Maine would get 4 seats, but for $M = 357$ Maine's apportionment goes down to 3 seats. Then up again to 4 seats for M between 358 and 381, then back down to 3 for $M = 382$, and so on. When a bill with $M = 357$ came out of the Census committee, all hell broke loose on the House floor. Fortunately, cooler heads prevailed and the bill never passed. Hamilton's method was never to be used again. (For a brief history of the apportionment controversies plaguing the United States House of Representatives over the years, see the Historical Note at the end of this chapter.)

If the Alabama paradox wasn't bad enough, Hamilton's method can fall victim to two other paradoxes called the *population paradox* and the *new-states paradox*. We will discuss these two paradoxes very briefly in the remainder of this section.

The Population Paradox

Sometime in the early 1900s, it was discovered that under Hamilton's method a state could potentially lose some seats because its population got too big! This phenomenon is known as the *population paradox*. To be more precise, the **population paradox** occurs when state *A loses* a seat to state *B* even though the population of *A grew at a higher rate* than the population of *B*. Too weird to be true, you say? Check out the next example.

 EXAMPLE 4.6 A Tale of Two Planets

In the year 2525 the five planets in the Utopia galaxy finally signed a peace treaty and agreed to form an Intergalactic Federation governed by an Intergalactic Congress. This is the story of the two apportionments that broke up the Federation.

Part I. The Apportionment of 2525. The first Intergalactic Congress was apportioned using Hamilton's method, based on the population figures (in billions) shown in the second column of Table 4-10. (This is the apportionment problem discussed in Example 4.2.) There were 50 seats apportioned.

TABLE 4-10 Intergalactic Congress: Apportionment of 2525					
Planet	**Population**	**Step 1**	**Step 2**	**Step 3**	**Apportionment**
Alanos	150	$8.\overline{3}$	8	0	8
Betta	**78**	$4.\overline{3}$	4	0	**4**
Conii	173	$9.6\overline{1}$	9	1	10
Dugos	204	$11.\overline{3}$	11	0	11
Ellisium	**295**	$16.3\overline{8}$	16	1	**17**
Total	900	50.00	48	2	50

Source: Intergalactic Census Bureau.

Let's go over the calculations. Since the total population of the galaxy is 900 billion, the standard divisor is $SD = 900/50 = 18$ billion. Dividing the planet populations by this standard divisor gives the standard quotas shown in the third column of Table 4-10. After the lower quotas are handed out (column 4), there are two surplus seats. The first surplus seat goes to Conii and the other one to Ellisium. The last column shows the apportionments. (Keep an eye on the apportionments of Betta and Ellisium—they are central to how this story unfolds.)

Part II. The Apportionment of 2535. After 10 years of peace, all was well in the Intergalactic Federation. The Intergalactic Census showed only a few changes in the planet's populations—an 8 billion increase in the population of Conii, and a 1 billion increase in the population of Ellisium. All other planets remained unchanged from 2525. Nonetheless, a new apportionment was required. Table 4-11 shows the details of the 2535 apportionment under Hamilton's method. Note that the total population increased to 909 billion, so the standard divisor for this apportionment was $909/50 = 18.18$. (You might want to check all the calculations in Table 4-11 for yourself!)

TABLE 4-11 Intergalactic Congress: Apportionment of 2535					
Planet	**Population**	**Step 1**	**Step 2**	**Step 3**	**Apportionment**
Alanos	150	8.25	8	0	8
Betta	**78**	4.29	4	1	**5**
Conii	181	9.96	9	1	10
Dugos	204	11.22	11	0	11
Ellisium	**296**	16.28	16	0	**16**
Total	909	50.00	48	2	50

Source: Intergalactic Census Bureau.

The one remarkable thing about the 2535 apportionment is that Ellisium *lost a seat* while its *population went up*, and that Betta *gained that seat* while its *population remained unchanged*! The epilogue to this story is that Ellisium used the 2535 apportionment as an excuse to invade Betta, and the years of peace and prosperity in Utopia were over. ≪

Example 4.6 is a rather trite story illustrating a fundamental paradox: *Under Hamilton's method, it is possible for a state with a positive population growth rate to lose one (or more) of its seats to another state with a smaller (or zero) population growth rate.* How can this population paradox happen? Go back to Example 4.6 and look at the effect that a small change of the standard divisor has on the relative sizes of the fractional parts. Once again, Hamilton's reliance on the fractional parts to allocate the surplus seats is its undoing. But wait, there is one more!

The New-States Paradox

In 1907, Oklahoma joined the Union. Prior to Oklahoma becoming a state, there were 386 seats in the House of Representatives. At the time, the fair apportionment to Oklahoma was five seats, so the size of the House of Representatives was changed from 386 to 391. The point of adding these five seats was to give Oklahoma its fair share of seats and leave the apportionments of the other states unchanged. However, when the new apportionments were calculated another paradox surfaced: Maine's apportionment went up from three to four seats, and

New York's went down from 38 to 37 seats. The perplexing fact that *the addition of a new state with its fair share of seats can, in and of itself, affect the apportionments of other states* is called the **new-states paradox**.

The following example gives a simple illustration of the new-states paradox. For a change of pace, we will discuss something other than legislatures.

EXAMPLE 4.7 Garbage Time

The Metro Garbage Company has a contract to provide garbage collection and recycling services in two districts of Metropolis, Northtown (with 10,450 homes) and the much larger Southtown (89,550 homes). The company runs 100 garbage trucks, which are apportioned under Hamilton's method according to the number of homes in the district. A quick calculation shows that the standard divisor is $SD = 1000$ homes, a nice round number which makes the rest of the calculations (shown in Table 4-12) easy. As a result of the apportionment, 10 garbage trucks are assigned to service Northtown and 90 garbage trucks to service Southtown.

TABLE 4-12 Metro Garbage Truck Apportionments

District	Homes serviced	Quota ($SD = 1000$)	Hamilton apportionment
Northtown	10,450	10.45	**10**
Southtown	89,550	89.55	**90**
Total	100,000	100.00	100

Now imagine that the Metro Garbage Company is bidding to expand its territory by adding the district of Newtown (5250 homes) to its service area. In its bid to the City Council the company promises to buy five additional garbage trucks for the Newtown run, so that its service to the other two districts is not affected. But when the new calculations (shown in Table 4-13) are carried out, there is a surprise: One of the garbage trucks assigned to Southtown has to be reassigned to Northtown! (You should check these calculations for yourself. Notice that the standard divisor has gone up a little and is now approximately 1002.38.)

TABLE 4-13 Revised Metro Garbage Truck Apportionments

District	Homes serviced	Quota ($SD \approx 1002.38$)	Hamilton apportionment
Northtown	10,450	10.42	**11**
Southtown	89,550	89.34	**89**
Newtown	5,250	5.24	5
Total	105,250	105.00	105

There are two key lessons we should take from this section: (i) In terms of fairness, Hamilton's method leaves a lot to be desired, and (ii) the critical flaw in Hamilton's method is the way it handles the surplus seats. Clearly, there must be a better apportionment method, so let's move on!

4.4 Jefferson's Method

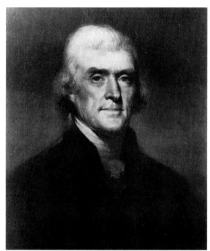

Thomas Jefferson (1743–1826). Jefferson's method (also known as the *method of greatest divisors*) was used in the United States from 1792 to 1840 and is still used to apportion the legislature in many countries, including Austria, Brazil, Finland, Germany, and the Netherlands.

The first apportionment method used in the United States House of Representatives was **Jefferson's method**. Jefferson's method is based on an approach very different from Hamilton's method, so there is some irony in the fact that we will explain the idea behind Jefferson's method by taking one more look at Hamilton's method.

Recall that under Hamilton's method we start by dividing every state's population by a fixed number (the standard divisor). This gives us the standard quotas. Step 2 is then to round *every* state's quota down. Notice that up to this point Hamilton's method uses a uniform policy for all states—every state is treated in exactly the same way. If you are looking for fairness, this is obviously good! But now comes the bad part (Step 3). We have some leftover seats which we need to distribute, but not enough for every state. Thus, we are forced to choose some states over others for preferential treatment. No matter how fair we try to be about it, there is no getting around that some states get that extra seat and others don't.

We saw in the previous section that it is in the handling of the surplus seats (Step 3) that Hamilton's method runs into trouble. So here is an interesting idea: Let's tweak things so that when the quotas are rounded down, *there are no surplus seats*! (The idea, in other words, is to get rid of Step 3.) But under Hamilton's method there is always going to be at least one surplus seat, so how do we work this bit of magic? The answer is by changing the divider, which then changes the quotas. The idea is that by using a *smaller* divider, we make the quotas bigger. If we hit it just right, when we round the slightly bigger quotas down ... poof! the surplus seats are gone. When we pull that bit of numerical magic, we have in fact implemented *Jefferson's method*.

The best way to get a good feel for the inner workings of Jefferson's method is through an example, so we will do that next.

▶ EXAMPLE 4.8 Parador's Congress (Jefferson's Method)

As promised, we are returning to the apportionment of the Parador Congress. First, a quick review of some the calculations we already did in Example 4.4. When we divide the populations by the standard divisor $SD = 50,000$, we get the standard quotas (third column of Table 4-14). When these quotas are rounded

TABLE 4-14 Rounding Down the Quotas: $SD = 50,000$ and $D = 49,500$					
State	Population	Standard quota ($SD = 50,000$)	Lower quota	Modified quota ($D = 49,500$)	Jefferson apportionment
A	1,646,000	32.92	32	33.25	33
B	6,936,000	**138.72**	138	140.12	**140**
C	154,000	3.08	3	3.11	3
D	2,091,000	41.82	41	42.24	42
E	685,000	13.70	13	13.84	13
F	988,000	19.76	19	19.96	19
Total	12,500,000	250.00	246		250

down, we end up with four surplus seats (fourth column of Table 4-14). Now here are some new calculations: When we divide the populations by the slightly small-er divisor $D = 49,500$ we get a slightly bigger set of quotas (fifth column of Table 4-14). When these *modified quotas* are rounded down (last column of Table 4-14), the surplus seats are gone, and we have a valid apportionment. It's nothing short of magical! And, of course, you are wondering, Where did that 49,500 come from? Let's call it a lucky guess for now.   ◀◀

Example 4.8 illustrates a key point: *Apportionments don't have to be based exclusively on the standard divisor*. If our intent is to *round the quotas down*, then $SD = 50,000$ is a bad choice (it forces us to deal with surplus seats), whereas $D = 49,500$ is a good choice (no surplus seats to worry about). One useful way to think about this change in divisors is as a *change in the units of measurement*. In-stead of using the standard unit 1 seat = 50,000 people, we chose to use the unit 1 seat = 49,500 people. There is nothing that says we can't do that!

Jefferson's method is but one of a group of apportionment methods based on the principle that the standard yardstick of 1 seat = SD people is not set in con-crete, and that, if necessary, we can change to a different yardstick: 1 seat = D people, where D is a suitably chosen number. The number D is called a **divisor** (sometimes we use the term **modified divisor** to emphasize the point that it is not the standard divisor), and apportionment methods that use modified divisors are called **divisor methods**. Different divisor methods are based on different philoso-phies of how the *modified quotas should be rounded to whole numbers*, but they all follow one script: When you are done rounding off the quotas, all M seats have been apportioned (no more, and no less). To be able to do this, you just need to find a suitable divisor D. For Jefferson's method, the rounding rule is to always round the quotas down. We will soon discuss two other divisor methods that use different rounding rules.

The formal description of Jefferson's method is surprisingly short. The devil is in the details.

Jefferson's Method

- **Step 1.** Find a "suitable" divisor D. [A suitable divisor is a divisor that produces an apportionment of exactly M seats when the *quotas* (popula-tions divided by D) are *rounded down*.]
- **Step 2.** Each state is apportioned its *lower quota*.

The hard part of implementing Jefferson's method is Step 1. How does one find a suitable divisor D? There are different ways to do it, and some are quite so-phisticated (see, for example, Project D), but we will use a basic, blue-collar ap-proach: educated trial and error. Our target is a set of lower quotas whose sum is M. For the sum of the *lower quotas* to equal M, we need to make the quotas *somewhat bigger than the standard quotas*. This can only be accomplished by choosing a divisor *somewhat smaller than SD*. (As the divisor goes down, the quota goes up, and vice versa.) We start by making an educated guess and choose a divisor D smaller than SD. If our first guess works, then we are lucky (or smart) and we are done. If we don't hit the target on our first shot, we adjust our aim. If the sum of the lower quotas is less than M, we know that the quotas are not quite big enough, and we need to make them bigger. This we do by *choosing an even smaller value for D*. On the other hand, if the sum of the lower quotas is more than M, we need to bring the quotas down by *choosing a bigger value for D*. Using this educated trial-and-error approach, each guess is closer to the target

In some very rare cases, there is no divisor D that works—see Exercise 58.

than the previous one, and usually in a matter of a few guesses we can find a divisor D that works. (In most problems there is a full range of values of D that works, so it's not like we have to hit a bulls eye!) A flowchart outlining the trial-and-error implementation of Jefferson's method is shown in Fig. 4-1.

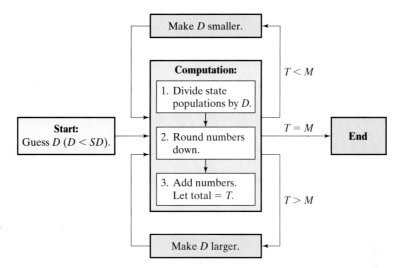

FIGURE 4-1 Flowchart for trial-and-error implementation of Jefferson's method

To illustrate the details of how to use the flowchart shown in Fig. 4-1, let's go back to Example 4.8. The first guess should be a divisor somewhat smaller than $SD = 50,000$. Suppose we start with $D = 49,000$. Using this divisor, we calculate the quotas, round them down, and add. We get a total of $T = 252$ seats. We overshot our target by two seats! We now refine our guess by choosing a *larger* divisor D (the point is to make the quotas *smaller*). A reasonable next guess (halfway between 50,000 and 49,000) is $D = 49,500$. We go through the computation and it works! (Note that $D = 49,450$ also works, as do many others.)

Jefferson's Method and the Quota Rule

By design, Jefferson's method is consistent in its approach to apportionment—the same formula applies to every state (find a suitable divisor, find the quota, round it down). By doing this, Jefferson's method is able to avoid the major pitfalls of Hamilton's method. There is, however, a serious problem with Jefferson's method, which can surface when we least expect it—*it can produce upper-quota violations!* In fact, take another look at the last column of Table 4-14 in Example 4.8, and look specifically at B's apportionment (140 seats) and standard quota (138.72). This is bad! To make matters worse, the upper-quota violations tend to consistently favor the larger states.

One of the most heated and contentious apportionment debates in U.S. history took place in 1832. The apportionment bill of 1832, based on Jefferson's method, gave New York 40 seats in the House of Representatives. When it came to light that New York's standard quota was only 38.59 seats, many delegates were horrified by such an unexpected violation of the quota rule, and the constitutionality of Jefferson's method came into question. Leading the move to dump Jefferson's method was Daniel Webster. The following is a brief excerpt from one of Daniel Webster's most famous speeches to the House of Representatives.

The House is to consist of 240 members. Now, the precise portion of power, out of the whole mass presented by the number of 240, to which New York would

be entitled according to her population, is 38.59; that is to say, she would be entitled to thirty-eight members, and would have a residuum or fraction; and even if a member were given her for that fraction, she would still have but thirty-nine. But the bill gives her forty ... for what is such a fortieth member given? Not for her absolute numbers, for her absolute numbers do not entitle her to thirty-nine. Not for the sake of apportioning her members to her numbers as near as may be because thirty-nine is a nearer apportionment of members to numbers than forty. But it is given, say the advocates of the bill, because the process [Jefferson's method] which has been adopted gives it. The answer is, no such process is enjoined by the Constitution.[1]

The 1832 apportionment bill passed nonetheless, but it was to be the last time the House of Representatives would use Jefferson's method. Two alternative methods were being proposed at the time as improvements over Jefferson's method—one by John Quincy Adams, the other by Daniel Webster himself. We will discuss them next.

4.5 Adams's Method

John Quincy Adams (1767–1848). Sixth President of the United States. Son of John Adams (second President) and distant cousin of Samuel Adams. Adams's method is also known as the *method of smallest divisors*.

Like Jefferson's method, Adams's method is a divisor method, but instead of rounding the quotas down, it rounds them up. For this to work, the quotas have to be made *smaller*, and this requires the use of a divisor D *larger* than the standard divisor SD. In essence, Adams's method is a mirror image of Jefferson's method.

> **Adams's Method**
>
> ■ **Step 1.** Find a "suitable" divisor D. [Here a suitable divisor is a divisor that produces an apportionment of exactly M seats when the *quotas* (populations divided by D) are *rounded up*.]
> ■ **Step 2.** Each state is apportioned its *upper quota*.

As before, we will illustrate Adams's method with the apportionment of Parador's Congress.

> **EXAMPLE 4.9 Parador's Congress (Adams's Method)**

By now you know the background story by heart. The populations are shown once again in column 2 of Table 4-15. The challenge, as is the case with any divisor method, is to find a suitable divisor D. We know that with Adams's method, D will have to be bigger than the standard divisor of 50,000. We start with the guess $D = 50,500$. (After all, $D = 49,500$ worked for Jefferson's method!) When the computation is carried out, the total is $T = 251$. (Columns 3 and 4 of Table 4-15 show the details.) This total is just one seat above our target of 250, so we need to make the quotas just a tad smaller. To do this, we must increase the divisor a little bit. We try $D = 50,700$. The new computations are shown in the last two columns

[1]Daniel Webster, *The Writings and Speeches of Daniel Webster*, Vol. VI (Boston: Little, Brown and Company, 1903).

TABLE 4-15 Rounding Up the Quotas: $D = 50,500$, and $D = 50,700$

State	Population	Quota ($D = 50,500$)	Upper quota ($D = 50,500$)	Quota ($D = 50,700$)	Adams's apportionment
A	1,646,000	32.59	33	32.47	33
B	6,936,000	137.35	138	136.80	137
C	154,000	3.05	4	3.04	4
D	2,091,000	41.41	42	41.24	42
E	685,000	13.56	14	13.51	14
F	988,000	19.56	20	19.49	20
Total	12,500,000		251		250

of Table 4-15. This divisor works! The apportionment under Adams's method is shown in the last column of Table 4-15.

Beyond illustrating how Adams's method can be implemented, Example 4.9 also highlights a serious weakness of this method—it can produce *lower-quota violations*! Once again, look at the apportionment of B (137 seats), and compare it with its standard quota (138.72). This is a different kind of violation, but just as serious as the one in Example 4.8—state B got 1.72 seats less than what it rightfully deserves! We can reasonably conclude that Adams's method is no better (or worse) than Jefferson's method—just different. (After considerable debate, the apportionment bill based on Adams's method never passed, and Adams's method was never used to apportion the House of Representatives.)

Whereas under Jefferson's method all violations of the quota rule are upper-quota violations, under Adams's method the violations of the quota rule are always lower-quota violations—see Exercise 55.

4.6 Webster's Method

What is the obvious compromise between rounding all the quotas down (Jefferson's method) and rounding all the quotas up (Adams's method)? What about rounding a quota down when the fractional part is less than 0.5, and rounding up otherwise?

It makes a lot of sense, but haven't we been there before and found it didn't quite work? Yes, and no. Yes, we have tried conventional rounding with the standard quotas and found it didn't work (see Example 4.3, Table 4-5). No, we haven't considered this approach using *modified quotas*. Now that we know that we can use different divisors to manipulate the quotas, it is always possible to find a suitable divisor that will make conventional rounding work. This is the idea behind Webster's method.

Daniel Webster (1782–1852). Lawyer, statesman, and Senator from Massachusetts. First proposed the apportionment method that bears his name in 1832. The method is also known as the *Webster-Willcox method* and *method of major fractions*.

Webster's Method

■ **Step 1.** Find a "suitable" divisor D. [Here a suitable divisor means a divisor that produces an apportionment of exactly M seats when the *quotas* (populations divided by D) are *rounded the conventional way*.]

■ **Step 2.** Find the apportionment of each state by rounding its quota the conventional way.

EXAMPLE 4.10 Webster's Method Meets Parador's Congress

We will now apportion Parador's Congress using Webster's method. Our first decision is to make a guess at the divisor D—should it be more than the standard divisor (50,000) or should it be less? Here we will use the standard quotas as a starting point. When we round off the standard quotas to the nearest integer we get a total of 251 (column 4 of Table 4-16). This number is too high (just by one seat), which tells us that we should try a divisor D a tad larger than the standard divisor. We try $D = 50,100$. Column 5 of Table 4-16 shows the respective modified quotas, and the last column shows these quotas rounded to the nearest integer. Now we have a valid apportionment! The last column of the table shows indeed the apportionment produced by Webster's method.

TABLE 4-16 Conventional Rounding: $D = 50,100$					
State	Population	Standard quota ($D = 50,000$)	Nearest integer	Quota ($D = 50,100$)	Webster's apportionment
A	1,646,000	32.92	33	32.85	33
B	6,936,000	138.72	139	138.44	138
C	154,000	3.08	3	3.07	3
D	2,091,000	41.82	42	41.74	42
E	685,000	13.70	14	13.67	14
F	988,000	19.76	20	19.72	20
Total	12,500,000	250.00	251		250

A flowchart illustrating how to find a suitable divisor D for Webster's method using educated trial and error is shown in Fig. 4-2. The most significant

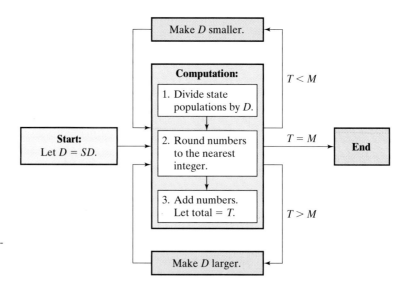

FIGURE 4-2 Flowchart for trial-and-error implementation of Webster's method

difference when we use trial and error to implement Webster's method as opposed to Jefferson's method (compare Figs. 4-1 and 4-2) is the choice of the starting value of D. With Webster's method *we always start with the standard divisor SD*. If we are lucky and SD happens to work, we are done! If the standard divisor SD doesn't quite do the job, we proceed with the trial-and-error approach as before until we find a divisor that works.

When the standard divisor works as a suitable divisor for Webster's method, every state gets an apportionment that is within 0.5 of its standard quota. This is as good an apportionment as one can hope for. If the standard divisor doesn't quite do the job, there will be at least one state with an apportionment that differs by more than 0.5 from its standard quota. In general, Webster's method tends to produce apportionments that don't stray too far from the standard quotas, although occasional violations of the quota rule (both lower- and upper-quota violations) are possible.

Webster's method has a lot going for it—it does not suffer from any paradoxes, and it shows no bias between small and large states. Its one flaw is that it can violate the quota rule, but such violations are rare in real-life apportionments. (Had Webster's method been used for every apportionment between 1790 and 2000, not a single violation of the quota rule would have shown up.) Surprisingly, Webster's method had a rather short tenure in the House of Representatives. It was used for the apportionment of 1842, then replaced by Hamilton's method, then reintroduced for the apportionments of 1901, 1911, and 1931, and then replaced again by the Huntington-Hill method, the apportionment method we currently use.

A detailed discussion of the Huntington-Hill method can be found in the Student Resource Manual (Mini-Excursion 1).

Conclusion

Apportionment is a problem of surprising mathematical depth—to say nothing of its political and historical complexities. Many of our Founding Fathers were deeply concerned with the issue, and the names of the apportionment methods we discussed bear witness to this fact.

In this chapter we covered apportionment methods named after Alexander Hamilton, Thomas Jefferson, John Quincy Adams, and Daniel Webster. These are some, but not all, of the apportionment methods that have been proposed and used over the years to apportion the U.S. House of Representatives. (Of the four, Adams's method was the only one never adopted.) Hamilton's method is known as a *quota method* and makes apportionment decisions sticking strictly to the use of the *standard divisor* and the *standard quotas*. (A different quota method called *Lowndes's method* is discussed in Exercises 52 and 53.) On the other hand, the methods proposed by Jefferson, Adams, and Webster are examples of *divisor methods*—methods based on the notion that divisors and quotas can be modified to work under different rounding philosophies. The only difference between divisor methods is in the rule used for rounding the quotas.

For the purposes of comparison it is useful to apply the different methods to the same apportionment problem, and this we did with the apportionment of Parador's Congress (Examples 4.4, 4.8, 4.9, and 4.10). The four solutions are summarized in Table 4-17. It should be noted that in this example each of the four methods produced a different solution, but this is not always going to happen. Often, two different methods applied to the same apportionment problem will produce identical apportionments.

TABLE 4-17 Parador's Congress: A Tale of Four Methods

State	Population	Standard quota	Solution 1 Hamilton	Solution 2 Jefferson	Solution 3 Adams	Solution 4 Webster
A	1,646,000	32.92	33	33	33	33
B	6,936,000	138.72	139	140	137	138
C	154,000	3.08	3	3	4	3
D	2,091,000	41.82	42	42	42	42
E	685,000	13.70	13	13	14	14
F	988,000	19.76	20	19	20	20
Total	12,500,000	250.00	250	250	250	250

While some of the methods are clearly better than others, none of them is perfect. Some produce paradoxes (such as the *Alabama paradox*, the *population paradox*, and the *new-states paradox*); others produce violations of the quota rule (*upper-quota* violations, *lower-quota* violations, or both). In addition, some methods show a systematic bias in favor of large states, others in favor of small states. Table 4-18 summarizes the potential problems with each method.

TABLE 4-18 Summary of the Four Methods

	Hamilton	Jefferson	Adams	Webster
Quota rule	No violations	Upper-quota violations possible	Lower-quota violations possible	Upper- and lower-quota violations possible
Alabama paradox	Possible	Not possible	Not possible	Not possible
Population paradox	Possible	Not possible	Not possible	Not possible
New-states paradox	Possible	Not possible	Not possible	Not possible
Bias in favor of	Large states	Large states	Small states	Neutral

From a purely mathematical point of view, an *ideal* apportionment method would be one that will not violate the quota rule, will not produce any paradoxes, and will not show bias between small and large states. For many years, the goal of mathematicians, social scientists, and politicians interested in the apportionment problem was to find such a method. In 1980, two mathematicians—Michel L. Balinski of the State University of New York at Stony Brook and H. Peyton Young of The Johns Hopkins University—made a surprising discovery: *An apportionment method that does not violate the quota rule and does not produce any paradoxes is a mathematical impossibility*. This fact is known as **Balinski and Young's impossibility theorem**. In essence, the theorem tells us that there are no perfect solutions to the riddle of apportionment—any apportionment method that satisfies the quota rule will produce paradoxes, and any apportionment method that does not produce paradoxes will produce violations of the quota rule.

That which cannot be done perfectly, must be done in a manner as near perfection as can be.

Daniel Webster

Much like Arrow's impossibility theorem in Chapter 1, Balinski and Young's impossibility theorem teaches us that there are inherent limitations to our democratic ideals—we believe in proportional representation and we believe in fairness, but we cannot consistently have both. Some problems do not have perfect solutions, and it is good to know that. We can then focus on the more realistic goal of finding, among all the imperfect solutions, the better ones.

Historical Note: A Brief History of Apportionment in the United States

The apportionment of the U.S. House of Representatives has an intriguing and convoluted history. As mandated by the Constitution, apportionments of the House are to take place every 10 years, following each census of the population. The real problem was that the Constitution left the method of apportionment and the number of seats to be apportioned essentially up to Congress. The only two restrictions stated in the constitution were that (i) "each State shall have at least one Representative," and (ii) "The number of Representatives shall not exceed one for every thirty thousand" (Article I, Section 2).

Following the 1790 Census, and after considerable and heated debate, Congress passed the first "act of apportionment" in 1792. The bill, sponsored by Alexander Hamilton (then Secretary of the Treasury), established a House of Representatives with $M = 120$ seats and apportioned under the method we now call *Hamilton's method*. In April of 1792, at the urging of then Secretary of State Thomas Jefferson, President George Washington vetoed the bill. (This was the first presidential veto in United States history.) Jefferson convinced Washington to support a different apportionment bill, based on a House of Representatives with $M = 105$ seats apportioned under the method we now call *Jefferson's method*. Unable to override the president's veto and facing a damaging political stalemate, Congress finally adopted Jefferson's bill. This is how the first House of Representatives came to be constituted.

Jefferson's method remained in use for five decades, up to and including the apportionment of 1832. The great controversy during the 1832 apportionment debate centered on New York's apportionment. Under Jefferson's method, New York would get 40 seats even though its *standard quota* was only 38.59. This apportionment exposed, in a very dramatic way, the critical weakness of Jefferson's method—it produces *upper-quota violations*. Two alternative apportionment bills were considered during the 1832 apportionment debate, one proposed by John Quincy Adams that would apportion the House using the method we now call *Adams's method*, and a second one, sponsored by Daniel Webster, that would do the same using the method we now call *Webster's method*. Both of these proposals were defeated and the original apportionment bill passed, but the 1832 apportionment was the last gasp for Jefferson's method.

Webster's method was adopted for the 1842 apportionment, but in 1852 Congress passed a law making Hamilton's method the "official" apportionment method for the House of Representatives. Since it is not unusual for Hamilton's method and Webster's method to produce exactly the same apportionment, an "unofficial" compromise was also adopted in 1852: Choose the number of seats in the House so that the apportionment is the same under either method. This was done again with the apportionment bills of 1852 and 1862.

In 1872, as a result of a power grab among states, an apportionment bill was passed that can only be described as a total mess—it was based on no particular method and produced an apportionment that was inconsistent with both Hamilton's method and Webster's method. The apportionment of 1872 was in violation of both the constitution (which requires that some method be used) and the 1852 law (which designated Hamilton's method as the method of choice).

In 1876 Rutherford B. Hayes defeated Samuel L. Tilden in one of the most controversial and disputed presidential elections in U.S. history. Hayes won in the Electoral College (despite having lost the popular vote) after Congress awarded him the disputed electoral votes from three southern states—Florida, Louisiana, and South Carolina. One of the many dark sidebars of the 1876 election was that, had the House of Representatives been legally apportioned, Tilden would have been the clear-cut winner in the Electoral College.

In 1882, the *Alabama paradox* first surfaced. In looking at possible apportionments for different House sizes, it was discovered that for $M = 299$ Alabama would get 8 seats, but if the House size were increased to $M = 300$ Alabama's apportionment would decrease to 7 seats. So how did Congress deal with this disturbing discovery? It essentially glossed it over, choosing a House with $M = 325$ seats, a number for which Hamilton's method and Webster's method would give the same apportionment. The same strategy was adopted in the apportionment bill of 1892.

In 1901, the *Alabama paradox* finally caught up with Congress. When the Census Bureau presented to Congress tables showing the possible apportionments under Hamilton's method for all House sizes between 350 and 400 seats, it was pointed out that two states—Maine and Colorado—were impacted by the Alabama paradox: For most values of M starting with $M = 350$, Maine would get 4 seats, but for $M = 357, 382, 386, 389,$ and 390, Maine's apportionment would go down to 3 seats. Colorado would get 3 seats for all possible values of M except $M = 357$, for which it would only get 2 seats. For $M = 357$, both Maine and Colorado lose, and, coincidentally, this just happened to be the House size that was proposed for the 1901 bill. Faster than you can say *we are being robbed*, the debate in Congress escalated into a frenzy of name-calling and accusations, with the end result being that the bill was defeated and Hamilton's method was scratched for good. The final apportionment of 1901 used Webster's method and a House with $M = 386$ seats.

Webster's method remained in use for the apportionments of 1901, 1911, and 1931 (no apportionment bill was passed following the 1920 Census, in direct violation of the Constitution).

In 1941, Congress passed a law that established a fixed size for the House of Representatives (435 seats) and a permanent method of apportionment known as the *method of equal proportions*, or *Huntington-Hill method*. The 1941 law (*Public Law 291, H.R. 2665, 55 Stat 761: An Act to Provide for Apportioning Representatives in Congress among the Several States by the Equal Proportions Method*) represented a realization by Congress that politics should be taken out of the apportionment debate, and that the apportionment of the House of Representatives should be purely a mathematical issue.

Are apportionment controversies then over? Not a chance. With a fixed-size House, one state's gain has to be another state's loss. In the 1990 apportionment, Montana was facing the prospect of losing one of its two seats—seats it had held in the House for 80 years. Not liking the message, Montana tried to kill the messenger. In 1991 Montana filed a lawsuit in federal District Court (*Montana v. United States Department of Commerce*) in which it argued that the Huntington-Hill method is unconstitutional and that either *Adams's method* or *Dean's method* (see Project A) should be used. (Under either one of these methods, Montana would keep its two seats.) A panel of three federal judges ruled by a 2-to-1 vote in favor of Montana. The case then went on appeal to the Supreme Court, which overturned the decision of the lower federal court and upheld the constitutionality of the Huntington-Hill method.

Key Concepts

Adams's method, **144**	Jefferson's method, **141**	quota rule, **135**
Alabama paradox, **136**	lower quota, **133**	standard divisor, **131**
apportionment method, **134**	lower-quota violation, **136**	standard quota, **132**
apportionment problem, **129**	modified divisor, **142**	upper quota, **133**
Balinski and Young's impossibility theorem, **148**	modified quota, **145**	upper-quota violation, **136**
	new-states paradox, **139**	Webster's method, **145**
Hamilton's method, **134**	population paradox, **138**	

Exercises

WALKING

A. Standard Divisors and Standard Quotas

1. The Bandana Republic is a small country consisting of four states (Apure, Barinas, Carabobo, and Dolores). There are $M = 160$ seats in the Bandana Congress. The populations of each state (in millions) are given in the following table.

State	Apure	Barinas	Carabobo	Dolores
Population	3.31	2.67	1.33	0.69

    (a) Find the standard divisor.

    (b) Using the standard divisor you found in (a), find each state's standard quota.

    (c) Using the standard quotas you found in (b), find each state's lower and upper quotas.

2. Consider the same Bandana Republic population figures as in Exercise 1, but in this exercise assume that the number of seats in the Bandana Congress is $M = 200$.

    (a) Find the standard divisor.

    (b) Using the standard divisor you found in (a), find each state's standard quota.

    (c) Using the standard quotas you found in (b), find each state's lower and upper quotas.

3. The Scotia Metropolitan Area Rapid Transit Service (SMARTS) operates 6 bus routes (*A, B, C, D, E,* and *F*) and 130 buses. The buses are apportioned among the

routes based on the average number of daily passengers per route, given in the following table.

Route	A	B	C	D	E	F
Average number of passengers	45,300	31,070	20,490	14,160	10,260	8,720

(a) Describe the "states" and the "seats" in this apportionment problem.

(b) Find the standard divisor. Explain what the standard divisor represents in this problem.

(c) Find the standard quotas.

4. The Placerville General Hospital has a nursing staff of 225 nurses working in four shifts: A (7:00 A.M. to 1:00 P.M.), B (1:00 P.M. to 7:00 P.M.), C (7:00 P.M. to 1:00 A.M.), and D (1:00 A.M. to 7:00 A.M.). The number of nurses apportioned to each shift is based on the average number of patients per shift, given in the following table.

Shift	A	B	C	D
Average number of patients	871	1029	610	190

(a) Describe the "states" and the "seats" in this apportionment problem.

(b) Find the standard divisor. Explain what the standard divisor represents in this problem.

(c) Find the standard quotas.

5. The Republic of Tropicana is a small country consisting of five states (A, B, C, D, and E). The total population of Tropicana is 23.8 million. According to the Tropicana constitution, the seats in the legislature are apportioned to the states according to their populations. The standard quota of each state is given in the following table.

State	A	B	C	D	E
Standard quota	40.50	29.70	23.65	14.60	10.55

(a) Find the number of seats in the Tropicana legislature.

(b) Find the standard divisor.

(c) Find the population of each state.

6. Tasmania State University (TSU) is made up of five different schools: Agriculture (A), Business (B), Education (E), Humanities (H), and Science (S). The total number of students at TSU is 12,500. The faculty positions at TSU are apportioned to the various schools based on the schools' respective enrollments. The standard quota for each school is given in the following table.

School	A	B	E	H	S
Standard quota	32.92	15.24	41.62	21.32	138.90

(a) Find the number of faculty positions at TSU.

(b) Find the standard divisor. What does the standard divisor represent in this problem?

(c) Find the number of students enrolled in each school.

7. According to the 2000 U.S. Census, 7.43% of the U.S. population lived in Texas. Compute Texas' standard quota in 2000 (rounded to two decimal places).

(***Hint:*** *Use the fact that the House of Representatives has 435 seats.*)

8. Based on the 2000 U.S. Census, California had a standard quota of 52.45. Estimate what percent of the U.S. population lived in California in 2000. Give your answer to the nearest tenth of a percent.

(***Hint:*** *Use the fact that the House of Representatives has 435 seats.*)

9. The Interplanetary Federation of Fraternia consists of six planets: Alpha Kappa, Beta Theta, Chi Omega, Delta Gamma, Epsilon Tau, and Phi Sigma (A, B, C, D, E, and F). The Federation is governed by the Inter Fraternia Congress, consisting of 200 seats apportioned among the planets according to their populations. The planet populations (given as percentages of the total population of Fraternia) are shown in the following table.

Planet	A	B	C	D	E	F
Percent of population	11.37	8.07	38.62	14.98	10.42	16.54

(a) Find the standard divisor (expressed as a percent of the population).

(b) Find the standard quota for each planet.

10. The small island nation of Margarita is made up of four islands: Aleta, Bonita, Corona, and Doritos. There are 125 seats in the Margarita Congress, which are apportioned among the islands according to their populations. The island populations (given as a percent of the total population of Margarita) are shown in the following table.

Island	Aleta	Bonita	Corona	Doritos
Percent of population	6.24	26.16	28.48	39.12

(a) Find the standard divisor (expressed as a percent of the population).

(b) Find the standard quota for each island.

B. Hamilton's Method

11. Find the apportionment of the Bandana Republic Congress as described in Exercise 1 ($M = 160$) under Hamilton's method.

12. Find the apportionment of the Bandana Republic Congress as described in Exercise 2 ($M = 200$) under Hamilton's method.

13. Find the apportionment of the SMARTS buses described in Exercise 3 under Hamilton's method.

14. Find the apportionment of the Placerville General Hospital nurses described in Exercise 4 under Hamilton's method.

15. Find the apportionment of the Republic of Tropicana legislature described in Exercise 5 under Hamilton's method.

16. Find the apportionment of the faculty at Tasmania State University described in Exercise 6 under Hamilton's method.

17. Find the apportionment of the Inter Fraternia Congress described in Exercise 9 under Hamilton's method.

18. Find the apportionment of the Margarita Congress described in Exercise 10 under Hamilton's method.

Exercises 19 through 21 are based on the following story: Mom found an open box of her children' favorite candy bars. She decides to apportion the candy bars among her three youngest children according to the number of minutes each child spent studying during the week (yes, she keeps accurate records of this!). Assume Bob studied a total of 54 minutes, Peter studied a total of 243 minutes, and Ron studied a total of 703 minutes.

19. (a) Suppose there were 11 candy bars in the box. Find the apportionment of the candy bars to the children under Hamilton's method.

    (b) Suppose that before mom hands out the candy bars the children decide to do a "little" more studying. Bob studies an additional 2 minutes (for a total of 56 minutes), Peter an additional 12 minutes (for a total of 255 minutes), and Ron an additional 86 minutes (for a total of 789 minutes). Using these new numbers, find the apportionment of the candy bars to the children under Hamilton's method.

    (c) The results of (a) and (b) illustrate one of the paradoxes of Hamilton's method. Which one? Explain.

20. (a) Suppose there were 10 candy bars in the box. Find the apportionment of the candy bars to the children under Hamilton's method.

    (b) Suppose that just before she hands out the candy bars, mom finds one extra candy bar. With 11 candy bars to hand out, she figures she better recompute the apportionment. Find the new apportionment of the candy bars under Hamilton's method.

    (c) The results of (a) and (b) illustrate one of the paradoxes of Hamilton's method. Which one? Explain.

21. (a) Suppose there were 11 candy bars in the box. Find the apportionment of the candy bars to the children under Hamilton's method. [Same problem as 19(a).]

    (b) Suppose there is another son, Jim, a college student not included in the original apportionment game. Unexpectedly Jim shows up and wants a piece of the action. He claims to have studied 580 minutes during the past week. Mom agrees to include him in the apportionment problem, but makes up for it by adding six candy more candy bars to the loot. (There are now 17 candy bars to be apportioned among the four boys.) Find the new apportionment of the candy bars under Hamilton's method.

    (c) The results of (a) and (b) illustrate one of the paradoxes of Hamilton's method. Which one? Explain.

22. A mother wishes to apportion 15 pieces of candy among her three daughters, Katie, Lilly, and Jaime, based on the number of minutes each child spent studying. The only information we have is that mom will use Hamilton's method and that Katie's standard quota is 6.53.

    (a) Explain why it is impossible for all three daughters to end up with five pieces of candy each.

    (b) Explain why it is impossible for Katie to end up with nine pieces of candy.

    (c) Explain why it is impossible for Lilly to end up with nine pieces of candy.

C. Jefferson's Method

23. Find the apportionment of the Bandana Republic Congress as described in Exercise 1 under Jefferson's method.

24. Find the apportionment of the Bandana Republic Congress as described in Exercise 2 under Jefferson's method.

25. Find the apportionment of the SMARTS buses described in Exercise 3 under Jefferson's method.

26. Find the apportionment of the Placerville General Hospital nurses described in Exercise 4 under Jefferson's method.

27. Find the apportionment of the Republic of Tropicana legislature described in Exercise 5 under Jefferson's method.

28. Find the apportionment of the faculty at Tasmania State University described in Exercise 6 under Jefferson's method.

29. Find the apportionment of the Inter Fraternia Congress described in Exercise 9 under Jefferson's method.

    (**Hint:** *Express the modified divisors in terms of percents of the total population.*)

30. Find the apportionment of the Margarita Congress described in Exercise 10 under Jefferson's method.

    (**Hint:** *Express the modified divisors in terms of percents of the total population.*)

31. Suppose that you just took a test on the apportionment chapter. One of the questions in the test was to calcu-

late what Texas's apportionment in the House of Representatives would be for the 2000 Census under Jefferson's method. Your friend claims that he did all the calculations and the answer is 31. It's immediately clear to you that your friend has the wrong answer. Explain how you know.

(**Hint:** *Do Exercise 7 first.*)

32. Under Jefferson's method, California's apportionment in the 2000 House of Representatives would have been 55 seats. What does this fact illustrate about Jefferson's method?

(**Hint:** *Do Exercise 8 first.*)

D. Adams's Method

33. Find the apportionment of the Bandana Republic Congress as described in Exercise 1 under Adams's method.

34. Find the apportionment of the Bandana Republic Congress as described in Exercise 2 under Adams's method.

35. Find the apportionment of the SMARTS buses described in Exercise 3 under Adams's method.

36. Find the apportionment of the Placerville General Hospital nurses described in Exercise 4 under Adams's method.

37. Find the apportionment of the Republic of Tropicana legislature described in Exercise 5 under Adams's method.

38. Find the apportionment of the faculty at Tasmania State University described in Exercise 6 under Adams's method.

39. Find the apportionment of the Inter Fraternia Congress described in Exercise 9 under Adams's method.

(**Hint:** *Express the modified divisors in terms of percents of the total population.*)

40. Find the apportionment of the Margarita Congress described in Exercise 10 under Adams's method.

(**Hint:** *Express the modified divisors in terms of percents of the total population.*)

41. Under Adams's method, California's apportionment in the 2000 House of Representatives would have been 50 seats even though California's standard quota is 52.45. What does this fact illustrate about Adams's method?

(**Hint:** *Do Exercise 8 first.*)

42. Suppose that you just took a test on the apportionment chapter. One of the questions in the test was to calculate what Texas's apportionment in the House of Representatives would be for the 2000 Census under Adams's method. Your friend claims that he did all the calculations and the answer is 34. It's immediately clear to you that your friend has the wrong answer. Explain how you know.

(**Hint:** *Do Exercise 7 first.*)

E. Webster's Method

43. Find the apportionment of the Bandana Republic Congress as described in Exercise 1 under Webster's method.

44. Find the apportionment of the Bandana Republic Congress as described in Exercise 2 under Webster's method.

45. Find the apportionment of the SMARTS buses described in Exercise 3 under Webster's method.

46. Find the apportionment of the Placerville General Hospital nurses described in Exercise 4 under Webster's method.

47. Find the apportionment of the Republic of Tropicana legislature discussed in Exercise 5 under Webster's method.

48. Find the apportionment of the faculty at Tasmania State University discussed in Exercise 6 under Webster's method.

49. Find the apportionment of the Inter Fraternia Congress discussed in Exercise 9 under Webster's method.

(**Hint:** *Express the modified divisors in terms of percents of the total population.*)

50. Find the apportionment of the Margarita Congress discussed in Exercise 10 under Webster's method.

(**Hint:** *Express the modified divisors in terms of percents of the total population.*)

JOGGING

51. Consider an apportionment problem with N states. The populations of the states are given by $P_1, P_2, \ldots, P_N$, and the respective quotas are given by $q_1, q_2, \ldots, q_N$. Describe in words what each of the following quantities represents.

(a) $q_1 + q_2 + \cdots + q_N$

(b) $\dfrac{P_1 + P_2 + \cdots + P_N}{q_1 + q_2 + \cdots + q_N}$

(c) $\dfrac{P_N}{P_1 + P_2 + \cdots + P_N}$

Lowndes's Method. *Exercises 52 and 53 refer to a variation of Hamilton's method known as Lowndes's method, first proposed in 1822 by South Carolina Representative William Lowndes. The basic difference between Hamilton and Lowndes's methods is that, in Lowndes's method, after each state is assigned the lower quota, the surplus seats are handed out in order of* **relative fractional parts**. *(The relative fractional part of a number is the fractional part divided by the integer part. For example, the relative fractional part of 41.82 is 0.82/41 = 0.02, and the relative fractional part of 3.08 is 0.08/3 = 0.027. Notice that while 41.82 would have priority over 3.08 under Hamilton's method, 3.08 has priority over 41.82 under Lowndes's method because 0.027 is greater than 0.02.)*

52. (a) Find the apportionment of Parador's Congress (Example 3) under Lowndes's method.

(b) Verify that the resulting apportionment is different from each of the apportionments shown in Table 4-17. In particular, list which states do better under Lowndes's method than under Hamilton's method.

53. Consider an apportionment problem with only two states, A and B. Suppose that state A has standard quota q_1 and state B has standard quota q_2 neither of which is a whole number. (Of course, $q_1 + q_2 = M$ must be a whole number.) Let f_1 represent the fractional part of q_1 and f_2 the fractional part of q_2.

(a) Find values q_1 and q_2 such that Lowndes's method and Hamilton's method result in the same apportionment.

(b) Find values q_1 and q_2 such that Lowndes's method and Hamilton's method result in different apportionments.

(c) Write an inequality involving q_1, q_2, f_1, and f_2 that would guarantee that Lowndes's method and Hamilton's method result in different apportionments.

54. Apportionments with only two states. Consider an apportionment problem with only two states, A and B. Suppose that state A has standard quota q_1 and state B has standard quota q_2, neither of which is a whole number. (Of course, $q_1 + q_2 = M$ must be a whole number.) Let f_1 represent the fractional part of q_1 and f_2 the fractional part of q_2.

(a) Explain why one of the fractional parts is bigger than or equal to 0.5 and the other is smaller than or equal to 0.5.

(b) Assuming neither fractional part is equal to 0.5, explain why Hamilton's method and Webster's method must result in the same apportionment.

(c) Explain why, in an apportionment problem involving only two states, Hamilton's method can never produce the Alabama paradox or the population paradox.

(d) Explain why, in an apportionment problem involving only two states, Webster's method can never violate the quota rule.

55. (a) Explain why, when Jefferson's method is used, any violations of the quota rule must be upper-quota violations.

(b) Explain why, when Adams's method is used, any violations of the quota rule must be lower-quota violations.

(c) Use parts (a) and (b) to justify why, in the case of an apportionment problem with just two states, neither Jefferson's nor Adams's method can possibly violate the quota rule.

56. For an arbitrary state X, let q represent its standard quota and s represent the number of seats apportioned to X under some unspecified apportionment method. Interpret in words the meaning of each of the following mathematical statements:

(a) $s - q \geq 1$

(b) $q - s \geq 1$

(c) $|s - q| \leq 0.5$

(d) $0.5 < |s - q| < 1$

57. This exercise is based on actual data taken from the 1880 census. In 1880, the population of Alabama was given at 1,262,505. With a House of Representatives consisting of $M = 300$ seats, the standard quota for Alabama was 7.671.

(a) Find the 1880 census population for the United States (rounded to the nearest person).

(b) Given that the standard quota for Texas was 9.672, find the population of Texas (to the nearest person).

58. The purpose of this exercise is to show that under rare circumstances, Jefferson's method may not work. A small country consists of four states with populations given in the following table, and there are $M = 49$ seats to be apportioned.

State	A	B	C	D
Population	500	1000	1500	2000

(a) Find the apportionment under Adams's method.

(b) Attempt to apportion the seats under Jefferson's method using the divisor $D = 100$. What happens if $D < 100$? What happens if $D > 100$?

(c) Explain why Jefferson's method will not work for this example.

59. The purpose of this exercise is to show that under rare circumstances, Adams's method may not work. A small country consists of four states with populations given in the following table, and there are $M = 51$ seats to be apportioned.

State	A	B	C	D
Population	500	1000	1500	2000

(a) Find the apportionment under Jefferson's method.

(b) Attempt to apportion the seats under Adams's method using the divisor $D = 100$. What happens if $D < 100$? What happens if $D > 100$?

(c) Explain why Adams's method will not work for this example.

60. Alternate version of Hamilton's method. Consider the following description of an apportionment method:

- **Step 1.** Find each state's standard quota.

- **Step 2.** Give each state (temporarily) its *upper quota* of seats. (You have now given away more seats than the number of seats available.)

- **Step 3.** Let K denote the number of extra seats you have given away in Step 2. Take away the K extra seats from the K states with the smallest fractional parts in their standard quotas.

Explain why this method produces exactly the same apportionment as Hamilton's method.

61. Consider the three apportionments shown in the following table. One of these apportionments was obtained under Jefferson's method, one under Adams's method, and one under Webster's method. Determine which method was used to obtain each apportionment and explain your reasoning.

State	A	B	C	D
Apportionment 1	5	2	1	17
Apportionment 2	5	2	2	16
Apportionment 3	5	1	1	18

RUNNING

62. The Hamilton-Jefferson hybrid method. The Hamilton-Jefferson hybrid method starts by giving each state its lower quota (as per Hamilton's method) and then apportioning the surplus seats using Jefferson's method.

(a) Use the Hamilton-Jefferson hybrid method to apportion $M = 22$ seats among four states according to the following populations: A (Population: 18,000), B (Population: 18,179), C (Population: 40,950), and D (Population: 122,871).

(b) Explain why the Hamilton-Jefferson hybrid method can produce apportionments that are different from both Hamilton and Jefferson apportionments.

(c) Explain why the Hamilton-Jefferson hybrid method can violate the quota rule.

63. The "Hamilton lottery" method. This method is a variation of Hamilton's method where the surplus seats are allocated to the players based on a lottery system. For the lottery each state gets a number of tickets proportional to the fractional part of its standard quota. The surplus seats are then allocated by drawing tickets in this lottery. (For example, if there are three states A, B, and C

with standard quotas of 9.47, 10.73, and 4.8, respectively, then prior to the lottery A would get 47 tickets, B would get 73 tickets, and C would get 80 tickets. The two surplus seats would then be assigned randomly by drawing two lottery tickets.)

(a) Describe how the "Hamilton lottery" method would work when applied to the Parador Congress apportionment problem discussed in the chapter (see Example 4.4). Assume an additional restriction: The total number of lottery tickets given out should be as small as possible. How many lottery tickets would each state get?

(b) Give an example of an apportionment problem where the "Hamilton lottery method" and Adams's method will produce different apportionments no matter what the results of the lottery are.

(c) Can the "Hamilton lottery" method violate the quota rule? If so, what kinds of violations does it produce?

64. Consider an apportionment problem with N states and M seats $(M \geq N)$. Let k denote the number of surplus seats in Step 3 of Hamilton's method.

(a) What is the *smallest* possible value of k? Give an example of an apportionment problem in which k has this value.

(b) What is the *largest* possible value of k? Give an example of an apportionment problem in which k has this value.

65. Explain why Jefferson's method cannot produce

(a) the Alabama paradox

(b) the new-states paradox

66. Explain why Adams's method cannot produce

(a) the Alabama paradox

(b) the new-states paradox

67. Explain why Webster's method cannot produce

(a) the Alabama paradox

(b) the new-states paradox

Projects and Papers

A. Dean's Method

Dean's method (also known as the *method of harmonic means*) was first proposed in 1832 by James Dean. [This is a different James Dean. Dean (1776–1849) was a Professor of Astronomy and Mathematics at Dartmouth College and the University of Vermont.] Dean's method is a divisor method similar to Webster's method but with a slightly different set of rules for rounding the quotas. In this project you are to prepare a report on Dean's method. Your report should include (i) a discussion of the **harmonic mean**, including some of its properties and applications; (ii) a description of Dean's method; (iii) a discussion of how Dean's method can be implemented using

trial and error; and (iv) a comparison of Dean's method with Webster's method with an example showing that the two methods can produce different apportionments.

You can find a good description of Dean's method in reference 3.

B. Apportionment Methods and the 2000 Presidential Election

After tremendous controversy over hanging chads and missed votes, the 2000 presidential election was decided in the Supreme Court, with George W. Bush getting the disputed electoral votes from Florida and thus beating Al

Gore by a margin of four electoral votes. Ignored in all the controversy was the significant role that the choice of apportionment method plays in a close presidential election. Would things have turned out differently if the House of Representatives had been apportioned under a different method?

In this project, you are asked to analyze and speculate how the election would have turned out had the House of Representatives (and thus the Electoral College) been apportioned under (i) Hamilton's method, (ii) Jefferson's method, and (iii) Webster's method.

Notes: (1) The 2000 Electoral College was based on the 1990 Census. (Remember to add 2 to the seats in the House to get each state's Electoral College votes!) You can go to *www.census.gov/main/www//cen1990.html* for the 1990 state population figures. (2) You can find an "apportionment calculator" that will do the calculations for you at this book's Web site: *http://prenhall.com/tannenbaum*. (Use the "Jump to ..." menu to jump to Chapter 4 and then click on "Applets.") (3) Give Florida's votes to Bush.

C. Apportionment Calculations via Spreadsheets

The power of a spreadsheet is its ability to be *dynamic*: Any and all calculations will automatically update after a change is made to the value in any given cell. This is particularly useful in implementing apportionment calculations that use educated trial and error, as well as when trying to rig up examples of apportionment problems that meet specific requirements.

This project has two parts.

Part 1. Implement each of the four apportionment methods discussed in the chapter using a spreadsheet. (Jefferson's, Adams's, and Webster's methods all fall under a similar template, so once you can implement one, the others are quite similar. Hamilton's method is a little different and, ironically, a bit harder than the other three to implement with a spreadsheet.) To check that everything works, do the different apportionments of Parador's Congress via the spreadsheets.

Part 2. Use your spreadsheets to create examples of the following situations:

   (i) An apportionment problem in which Hamilton's method and Jefferson's method produce exactly the same apportionments

   (ii) An apportionment problem in which Hamilton's method and Adams's method produce exactly the same apportionments

   (iii) An apportionment problem in which Hamilton's method and Webster's method produce exactly the same apportionments

   (iv) An apportionment problem in which Hamilton's method, Jefferson's method, Adams's method, and Webster's method all produce different apportionments

   (v) An apportionment problem in which Webster's method violates the quota rule

To do this project you should be reasonably familiar with Excel or a similar spreadsheet program.

D. Rank Index Implementations of Divisor Methods

From a computational point of view, the divisor methods of Jefferson, Adams, and Webster can be implemented using a priority system (called a **rank index**) that gives away seats one by one according to a table of values specifically constructed for this purpose. Each of the methods uses a slightly different set of rules to construct the rank index for the seats. In this project you are to describe how to construct the rank index for each of the three divisor methods studied in this chapter (Jefferson, Adams, and Webster). You should use a simple example to illustrate the procedure for each method.

A good starting point for this project is the American Mathematical Society apportionment Web site (reference 1). A more technical account can be found in reference 10.

E. The First Apportionment of the House of Representatives

The following table shows the United States population figures from the 1790 Census.

State	Population
Connecticut	236,841
Delaware	55,540
Georgia	70,835
Kentucky	68,705
Maryland	278,514
Massachusetts	475,327
New Hampshire	141,822
New Jersey	179,570
New York	331,589
North Carolina	353,523
Pennsylvania	432,879
Rhode Island	68,446
South Carolina	206,236
Vermont	85,533
Virginia	630,560
Total	3,615,920

For the first apportionment of 1792, two competing apportionment bills were considered: a bill to apportion 120 seats using Hamilton's method (sponsored by Hamilton), and a bill to apportion 105 seats using Jefferson's method (sponsored by Jefferson). Originally, Congress passed Hamilton's bill, but Washington was persuaded by Jefferson to veto it. Eventually, Congress approved Jefferson's bill, which became the basis the first mathematical apportionment of the House of Representatives.

This project has three parts.

Part 1. Historical Research. Summarize the arguments presented by Hamilton and Jefferson on behalf of their respective proposals. [Suggested sources: *The Papers of Alexander Hamilton, Vol. XI,* Harold C. Syrett (editor), and *The Works of Thomas Jefferson, Vol. VI,* Paul Leicester Ford (editor)].

Part 2. Mathematics. Calculate the apportionments for the House of Representatives under both proposals.

Part 3. Analysis. It has been argued by some scholars that there was more than mathematical merit behind Jefferson's thinking and Washington's support of it. Looking at the apportionments obtained in Part 2, explain why one could be suspicious of Jefferson and Washington's motives.

(***Hint:*** *Jefferson and Washington were both from the same state.*)

References and Further Readings

1. American Mathematical Society, "Apportionment Systems," *http://www.ams.org/featurecolumn/archive/apportionIII.html.*

2. Balinski, Michel L., and H. Peyton Young, "The Apportionment of Representation," *Fair Allocation: Proceedings of Symposia on Applied Mathematics,* 33 (1985), 1–29.

3. Balinski, Michel L., and H. Peyton Young, *Fair Representation; Meeting the Ideal of One Man, One Vote.* New Haven, CT: Yale University Press, 1982.

4. Balinski, Michel L., and H. Peyton Young, "The Quota Method of Apportionment," *American Mathematical Monthly,* 82 (1975), 701–730.

5. Brams, Steven, and Philip Straffin, Sr., "The Apportionment Problem," *Science,* 217 (1982), 437–438.

6. Census Bureau Web site, *http://www.census.gov.*

7. Eisner, Milton, *Methods of Congressional Apportionment,* COMAP Module #620.

8. Hoffman, Paul, *Archimedes' Revenge: The Joys and Perils of Mathematics.* New York: W. W. Norton & Co., 1988, chap. 13.

9. Huntington, E. V., "The Apportionment of Representatives in Congress." *Transactions of the American Mathematical Society,* 30 (1928), 85–110.

10. Huntington, E. V., "The Mathematical Theory of the Apportionment of Representatives," *Proceedings of the National Academy of Sciences, U.S.A.,* 7 (1921), 123–127.

11. Meder, Albert E., Jr., *Legislative Apportionment.* Boston: Houghton Mifflin Co., 1966.

12. Saari, D. G., "Apportionment Methods and the House of Representatives," *American Mathematical Monthly,* 85 (1978), 792–802.

13. Schmeckebier, L. F., *Congressional Apportionment.* Washington, DC: The Brookings Institution, 1941.

14. Steen, Lynn A., "The Arithmetic of Apportionment," *Science News,* 121 (May 8, 1982), 317–318.

15. Woodall, D. R., "How Proportional Is Proportional Representation?," *The Mathematical Intelligencer,* 8 (1986), 36–46.

16. Young, H. Peyton, "Dividing the House: Why Congress Should Reinstate an Old Reapportionment Formula," *The Brookings Institution,* Policy Brief #88, 2001.

part 2

Management Science

5

Euler Circuits

The Circuit Comes to Town

Sometimes great discoveries arise from the humblest and most unexpected of origins. Such is the case with the ideas we will explore in the next four chapters—the mathematical study of how things are interconnected. Our story begins in the 1700's in the medieval town of Königsberg, in Eastern Europe. At the time Königsberg was divided by a river into four separate sections, which were connected to each other by seven bridges. The old map of Königsberg shown here gives the layout of the city in 1736, the year a brilliant young mathematician named Leonhard Euler came passing through.

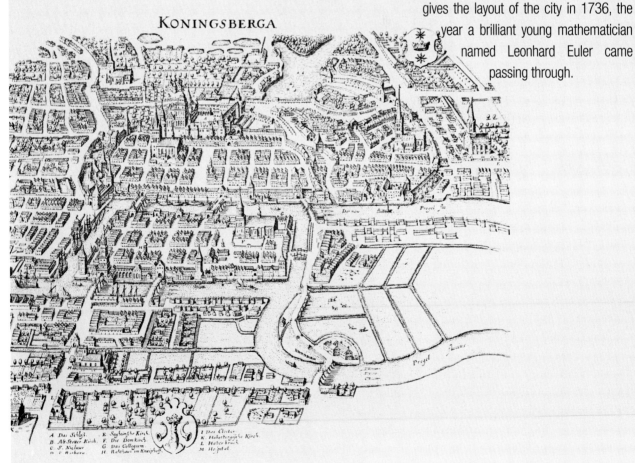

W hile visiting Königsberg, Euler was told of an innocent little puzzle of disarming simplicity: Is it possible for a person to take a walk around town in such a way that each of the seven bridges is crossed *once, but only once*? He heard the locals had tried, repeatedly and without success. Could he prove mathematically that this could not be done?

Euler, perhaps sensing that something important lay behind the frivolity of the puzzle, proceeded to solve it by demonstrating that indeed such a walk was impossible. But he actually did much more. In solving the puzzle of the Königsberg bridges, Euler laid the foundations for what was at the time a totally new type of geometry, which he called *geometris situs* ("the geometry of location"). From these modest beginnings, the basic ideas set forth by Euler eventually developed and matured into one of the most important and practical branches of modern mathematics, now known as *graph theory*. Modern applications of graph theory span practically every area of science and technology—from chemistry, biology, and computer science to psychology, sociology, and business management.

Over the next four chapters we will discuss the uses of graph theory as a tool for solving *management science* problems. These are problems in which the ultimate goal is to find efficient ways to organize and carry out complex tasks, usually tasks that involve a large number of variables, and in which it is not at all obvious how to make optimal decisions.

The theme of this chapter is the question of how to create *efficient routes* for the delivery of goods and services—such as mail delivery, garbage collection, police patrols, newspaper deliveries, and, most important, late-night pizza deliveries—along the streets of a city, town, or neighborhood. These types of management science problems are known as *Euler circuit problems* (named, of course, after the founding father of the field). Section 5.1 sets the table with a description of *routing problems* in general and the introduction of several examples of *Euler circuit problems* (which are solved at the end of the chapter). Sections 5.2 and 5.3 give an introduction to the *language and concepts* of *graph theory*—the mathematical toolkit that will help us tackle routing problems like the ones introduced in Section 5.1. Section 5.4 introduces the concept of *graph modeling*—the process by which a complicated real-life problem can be

Leonhard Euler (1707–1783). For more on Euler, see the biographical profile at the end of the chapter.

translated into the clean and precise language of graph theory. In Sections 5.5 and 5.6 we will learn about *Euler's theorems* and *Fleury's algorithm*—these provide the theoretical framework that will help us understand the structural aspects of Euler circuit problems. In Section 5.7 we will learn how to combine all of the preceding ideas to develop a strategy that allows us to solve Euler circuit problems.

5.1 Euler Circuit Problems

We will start this section with a brief discussion of routing problems. What is a *routing problem*? To put it in the most general way, routing problems are concerned with finding ways to route the delivery of *goods* and/or *services* to an assortment of *destinations*. The goods or services in question could be packages, mail, newspapers, pizzas, garbage collection, bus service, and so on. The delivery destinations could be homes, warehouses, distribution centers, terminals, and the like.

There are two basic questions that we are typically interested in when dealing with a routing problem. The first is called the *existence* question. The existence question is simple: Is an actual route possible? For most routing problems, the existence question is easy to answer, and the answer takes the form of a simple yes or no. When the answer to the existence question is yes, then a second question—the *optimization question*—comes into play. Of all the possible routes, which one is the *optimal route*? (*Optimal* here means "the best" when measured against some predetermined variable such as *cost*, *distance*, or *time*.) In most management science problems, the optimization question is where the action is.

In this chapter we will learn how to answer both the existence and optimization questions for a special class of routing problems known as **Euler circuit problems**. The common thread in all Euler circuit problems is what we might call, for lack of a better term, the *exhaustion requirement*—the requirement that the route must wind its way through ... *everywhere*. Thus, in an Euler circuit problem, by definition every single one of the streets (or bridges, or lanes, or highways) within a defined area (be it a town, an area of town, or a subdivision) must be covered by the route. We will refer to these types of routes as *exhaustive routes*. The most common services that typically require exhaustive routing are mail delivery, police patrols, garbage collection, street sweeping, and snow removal. More exotic examples can be census taking, precinct walking, electric meter reading, routing parades, tour buses, and so on.

To clarify some of the ideas we will introduce several examples of Euler circuit problems (just the problems for now—their solutions will come later in the chapter).

ex-haust: To treat completely; cover thoroughly.

American Heritage Dictionary

> ### EXAMPLE 5.1 Walking the 'Hood': Part 1

After a rash of burglaries, a private security guard is hired to patrol the streets of the Sunnyside neighborhood shown in Fig. 5-1. The security guard's assignment is to make an exhaustive patrol, on foot, through the entire neighborhood. Obviously, he doesn't want to walk any more than what is necessary. His starting point is the southeast corner across from the school (*S* in Fig. 5-1)—

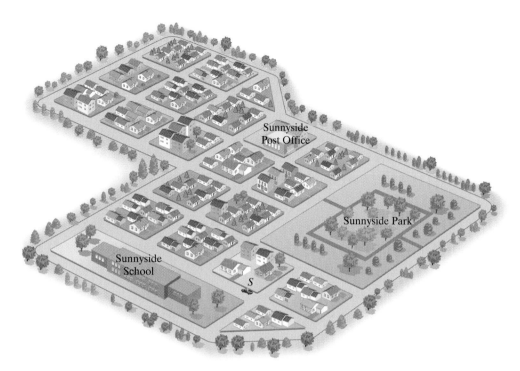

FIGURE 5-1 The Sunnyside neighborhood.

that's where he parks his car. (This is relevant because at the end of his patrol he needs to come back to *S* to pick up his car.) Being mathematically inclined, the security guard has two questions he would like to know the answer to: (i) is it possible to start and end at *S*, cover every block of the neighborhood, and pass through each block *just once?*, and (ii) if some of the blocks will have to be covered more than once, what is an *optimal* route that covers the entire neighborhood? (*Optimal* here means "with the minimal amount of walking.")   ◀◀

> ### EXAMPLE 5.2 Delivering the Mail: Part 1

A mail carrier has to deliver mail in the same Sunnyside neighborhood. The difference between the mail carrier's route and the security guard's route is that where there are homes on both sides of the street, the mail carrier must make *two* passes through each block (a separate pass for each side of the street); where there are homes on only one side of the street, the mail carrier needs to make only one pass through each block; and where there are no homes on either side of the street, the mail carrier does not have to walk at all. In addition, the mail carrier has no choice as to her starting and ending points—she has to start and end her route at the local post office. Much like the security guard, the mail carrier wants to find the optimal route that would allow her to cover the neighborhood with the least amount of walking. (Put yourself in her shoes and you would do the same—good weather or bad, she walks this route 300 days a year!)   ◀◀

> ## EXAMPLE 5.3 The Seven Bridges of Königsberg: Part 1

Basically, this is the true story with which we opened the chapter—with a little embellishment: A prize is offered to the first person who can find a way to walk across each one of the seven bridges of Königsberg without recrossing any and return to the original starting point. (For the reader's convenience we modernized the area map, now shown as Fig. 5-2.) A smaller prize is offered to anyone who can cross each of the seven bridges exactly once without necessarily returning to the original starting point. So far, no one has collected on either prize. Why?

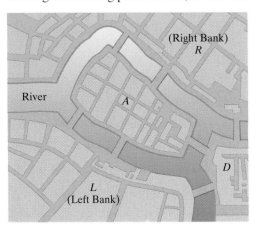

FIGURE 5-2

> ## EXAMPLE 5.4 The Bridges of Madison County: Part 1

This is a more modern version of Example 5.3. Madison County is a quaint old place, famous for its quaint old bridges. A beautiful river runs through the county, and there are four islands (*A, B, C,* and *D*) and 11 bridges joining the islands to both banks of the river (*R* and *L*) and one another (Fig. 5-3). A famous photographer is hired to take pictures of each of the 11 bridges for a national magazine. The photographer needs to drive across each bridge once for the photo shoot. Moreover, since there is a $25 toll (the locals call it a "maintenance tax") every time an out-of-town visitor drives across a bridge, the photographer wants to minimize the total cost of his trip and to recross bridges only if it is absolutely necessary. What is the optimal (cheapest) route for him to follow?

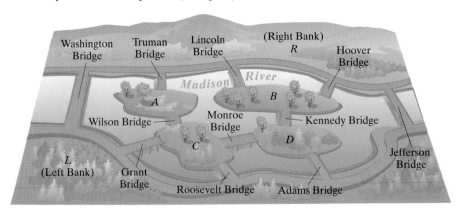

FIGURE 5-3

> ## EXAMPLE 5.5 Child's Play: Part 1

Figure 5-4 shows a few simple line drawings. The name of the game is to trace each drawing *without lifting the pencil or retracing any of the lines*. These kinds of tracings are called **unicursal tracings**. (When we end in the same place we started, we call it a *closed* unicursal tracing; when we start and end in different places, we call it an *open* unicursal tracing.) Which of the drawings in Fig. 5-4 can be traced with closed unicursal tracings? Which with only open ones? Which can't be traced (without cheating)? Some of us played such games in our childhood (in the good old days before X-Boxes and PlayStations) and may be able quickly to figure out the answers in the case of Figs. 5-4(a), (b), and (c). But what about slightly more complicated shapes, such as the one in Fig. 5-4(d)? How can we tell if a unicursal tracing (open or closed) is possible? Good question. We will answer it in Section 5.5.

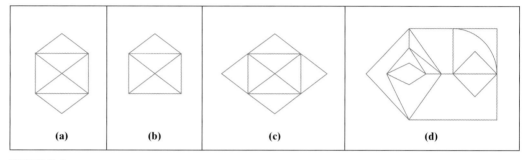

| (a) | (b) | (c) | (d) |

FIGURE 5-4

As the preceding examples illustrate, Euler circuit problems can come in a variety of forms. Fortunately, they can all be tackled by means of a single unifying mathematical concept—the concept of a *graph*. (A note of warning: These graphs have no relation to the graphs of functions you may have studied in algebra and calculus.)

5.2 Graphs

For starters, let's say that a **graph** is a picture consisting of *dots*, called **vertices**, and *lines*, called **edges**. The edges do not have to be straight lines, but they always have to connect two vertices. An edge connecting a vertex back with itself (which is allowed) is called a **loop**.

The foregoing is not to be taken as a precise definition of a graph, but rather as an informal description that will help us get by for the time being. To get a feel for what a graph is, let's look at a few examples.

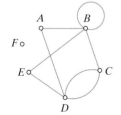

FIGURE 5-5

> ## EXAMPLE 5.6 Isolated Vertices, Loops, and Multiple Edges

Figure 5-5 shows an example of a typical graph. This graph has six vertices A, B, C, D, E, and F and eight edges. The edges can be described by giving the two vertices that are connected by the edge. Thus, the edges of this graph are AB, AD, BB, BC, BE, CD, CD, and DE.

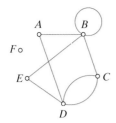

FIGURE 5-5 (repeated)

A convenient way to describe the vertices and edges of a graph is by using the notation of sets. In this graph the **vertex set** of the graph is $V = \{A, B, C, D, E, F\}$, and the **edge set** of the graph is $E = \{AB, AD, BB, BC, BE, CD, CD, \text{and } DE\}$. (To avoid confusion with vertices, it is a good idea to use a different font for the V and E that represent the vertex and edge sets.)

A few comments about the graph in Fig. 5-5: First, note that the point where edges BE and AD cross is *not* a vertex—it is just the crossing point of two edges. Second, note that vertex F is not connected to any other vertex. Such a vertex is called an **isolated vertex**. Third, note that this graph has a loop, namely the edge BB. Finally, note that it is permissible to have two edges connecting the same two vertices, as is the case with C and D. (This is the reason why CD is listed twice as an edge in the edge set.) When a graph has more than one edge connecting the same pair of vertices, it is said to have **multiple edges**. ⫷

EXAMPLE 5.7 The Dating Game

One of the reasons graphs are so useful is that they can tell a story in ways that simple words often can't. The graph in Fig. 5-6 is a simple illustration of an important idea, an idea we will return to in Section 5.4. (Note that Fig. 5-6 is considered a single graph, even though it consists of two separate, disconnected pieces. Such graphs are called *disconnected* graphs, and the individual pieces are called the *components* of the graph.)

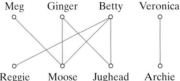

FIGURE 5-6

Imagine, if you will, that Archie, Betty, and so on are a group of teenagers, with their typical complex web of relationships, and imagine that, as part of a sociology project, we want to describe their recent dating patterns. We can do that very nicely with a graph like the one in Fig. 5-6, where the vertices of the graph represent the teenagers, and an edge connecting X to Y means that X and Y are currently dating. (You can draw some interesting conclusions from just looking at this graph—Veronica and Archie seem to be a couple, Betty is quite popular, and so on.)

Graphs such as the one in Fig. 5-6 but on a much grander scale are often used by social scientists to describe and analyze complex social networks. ⫷

EXAMPLE 5.8 A Graph with No Edges?

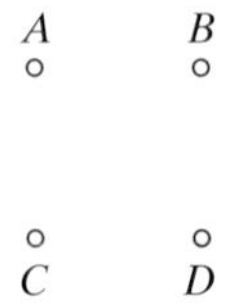

Yes, it's possible! The graph shown in Fig. 5-6 is legit, but if it is telling a story it is certainly a pretty boring one! This graph can be described by the vertex set $V = \{A, B, C, D\}$, and an *empty* edge set, $E = \{ \}$. Without edges, every vertex of the graph is an isolated vertex. We won't be seeing graphs like this too often, but when we do, it's good to know that this is still considered a graph. ⫷

$\begin{matrix} A & & B \\ \circ & & \circ \\ & & \\ \circ & & \circ \\ C & & D \end{matrix}$

FIGURE 5-7

> **EXAMPLE 5.9** Pictures Optional

The purpose of this example is to illustrate a subtle but important point. We tend to think that the graph *is the picture*, but a graph does not need a picture to define it—the graph is defined by just its vertex set and edge set. Thus, if we simply give the vertex set $\mathcal{V} = \{A, D, L, \text{and } R\}$ and the edge set $\mathcal{E} = \{AD, AL, AL, AR, AR, DL, DR\}$, we have defined a graph. What about a picture of the graph? What does it look like? Where do we place the vertices? How do we draw the edges? These questions are irrelevant—we can do it any way we want! Figure 5-8 shows three different renderings of the same graph. There are infinitely many more. [From a visual point of view, of course, they may not all be equal. In Fig. 5-8(a) the graph looks nice and clean; the graph in Fig. 5-8(b) looks . . . shall we say funky?]

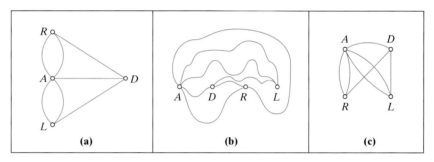

FIGURE 5-8

> **EXAMPLE 5.10** Take Me Out to the Ballpark

On any particular week of the baseball season one can look up the schedule for that week in a good newspaper or a television guide. A typical week's schedule for the National League East might look like this:

Monday: Pittsburgh at Montreal, New York at Philadelphia, Chicago at St. Louis

Tuesday: Pittsburgh at Montreal

Wednesday: New York at St. Louis, Philadelphia at Chicago

Thursday: Pittsburgh at St. Louis, New York at Montreal, Philadelphia at Chicago

Friday: Philadelphia at Montreal, Chicago at Pittsburgh

Saturday: Philadelphia at Pittsburgh, Chicago at New York, Montreal at St. Louis

Sunday: Philadelphia at Pittsburgh

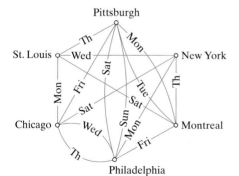

FIGURE 5-9

This schedule can just as well be told with a graph in which the vertices are the teams, and each game is an edge connecting the two teams playing that game. One possible version of this graph is shown in Fig. 5-9. (Note that the placement of the cities on the graph has nothing to do with their geographic location—all we are describing is the schedule.) One feature of this graph that makes it a little different from the previous examples is that the edges have labels. In Chapters 6 through 8 we will see many additional examples of graphs with labels on the edges.

With all of the preceding examples under our belt, we might be ready for a more formal definition of a graph.

> **Graphs**
>
> A **graph** is a *structure that defines pairwise relationships* within a set of objects. The objects are the vertices, and the pairwise relationships are the edges: *X is related to Y if and only if XY is an edge.*

It follows that any time we have a relationship between objects, whatever that relationship might be (date partner, friend, adversary, family member, etc.), *we can describe such a relationship by means of a graph.* This simple idea is the key reason for the tremendous usefulness of graph theory.

5.3 Graph Concepts and Terminology

Every branch of mathematics has its own peculiar jargon, and graph theory has more than its share. In this section we will introduce a few important concepts and terms that we will need later in the chapter. We will use Fig. 5-10 to illustrate these concepts.

- **Adjacent vertices.** Two vertices are said to be **adjacent** if there is an edge joining them. In this context, adjacency has nothing to do with physical proximity—a pair of vertices far apart but connected by an edge are adjacent; a pair of vertices near each other but not joined by and edge are not adjacent. In Fig. 5-10 vertices B and E are adjacent; C and D are not. Also, because of the loop at E, we can say that vertex E is adjacent to itself.

- **Adjacent edges.** Two edges are **adjacent** if they share a common vertex. In Fig. 5-10 edges AB and AD are adjacent; edges AB and DE are not.

- **Degree of a vertex.** The **degree of a vertex** is the number of edges at that vertex. When there is a loop at a vertex, the loop contributes twice toward the degree. We will use the notation $\deg(V)$ to denote the degree of vertex V. In Fig. 5-10, the degrees of the vertices are as follows: $\deg(A) = 3$, $\deg(B) = 5$, $\deg(C) = 3$, $\deg(D) = 2$, $\deg(E) = 4$, $\deg(F) = 3$, $\deg(G) = 1$, and $\deg(H) = 1$.

- **Odd and even vertices.** An **odd** vertex is a vertex of odd degree; an **even** vertex is a vertex of even degree. The graph in Fig. 5-10 has two even vertices (D and E) and six odd vertices (all the others).

- **Paths.** A **path** is a sequence of vertices with the property that each vertex in the sequence is *adjacent* to the next one. Thus, a path can also be thought of as describing a sequence of adjacent edges—a *trip*, if you will, along the edges of the graph. The key requirement in a path is that *an edge can be part of a path only once*—if you travel along an edge once, you can't travel through it again! (On the other hand, a vertex can appear on the path more than once.) The number of edges in the path is called the **length** of the path. The graph in Fig. 5-10 has many paths—here are just a few examples.

  - A, B, E, D. This is a path from vertex A to vertex D, consisting of edges AB, BE, and ED. The length of this path is 3.

  - A, B, C, A, D, E. This is a path of length 5 from A to E. The path visits vertex A twice, but no edge is repeated.

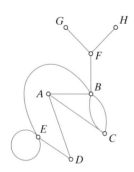

FIGURE 5-10

- *A, B, C, B, E.* This is also a path from *A* to *E*. This path is possible because there are two edges connecting *B* and *C*.

- *A, C, B, E, E, D.* This path of length 5 is possible because of the loop at *E*.

The following *are not* paths:

- *A, C, D, E.* There is no edge connecting *C* and *D*.

- *A, B, C, B, A, D.* The edge *AB* appears twice, so this is not a path.

- *A, B, C, B, E, E, D, A, C, B.* In this long string of vertices, everything is OK until the very end, when the edge *CB* appears for a third time. The first two instances are fine, because there are two edges connecting *B* and *C*. The third time, though, is one too many. One of the two edges would have to be retraveled.

- **Circuits.** A **circuit** has the same definition as a path, but has the additional requirement that the trip starts and ends at the same vertex. The following are a few of the circuits of the graph in Fig. 5-10.

  - *A, B, C, A.* This is a circuit of length 3. The edges of this circuit are *AB*, *BC*, and *CA*. Unlike a path, a circuit has no specific starting or ending point. We arbitrarily have to choose a starting point to describe the circuit on paper, but the circuit *A, B, C, A* can also be written as *B, C, A, B* or *C, A, B, C*.

  - *B, C, B.* When there are multiple edges, circuits of length 2 are possible. This is a circuit of length 2 and is the same circuit as *C, B, C*.

  - *E, E.* Yes, loops are circuits of length 1.

- **Connected graphs.** A graph is **connected** if, given any two vertices, there is a path joining them. A graph that is not connected is said to be **disconnected**. A disconnected graph is made up of separate connected **components**. (Figure 5-6 shows a disconnected graph with two components; Figure 5-7 shows a disconnected graph with four components, since each isolated vertex is a separate component).

- **Bridges.** Sometimes in a connected graph there is an edge such that if we were to erase it, the graph would become disconnected. Not surprisingly, such an edge is called a **bridge**. The graph in Fig. 5-10 has three bridges—*BF*, *FG*, and *FH*.

- **Euler paths.** An **Euler path** is a path that passes through *every* edge of a graph. Since it is a path, edges can only be traveled once. Thus, in an Euler path we travel along *every* edge of the graph *once and only once*. Note that a disconnected graph cannot have an Euler path, so a necessary condition for a graph to have an Euler path is that it be connected. At the same time, just because a graph is connected it doesn't mean that it will have an Euler path. The graph shown in Fig. 5-11(a) does not have an Euler path; the graph shown in Fig. 5-11(b) has several Euler paths. One of them is *L, A, R, D, A, R, D, L, A*. (Try to find another one that doesn't start at *L*.)

- **Euler circuits.** An **Euler circuit** is a circuit that passes through *every* edge of a graph. Essentially, an Euler circuit is a closed unicursal tracing of the edges of the graph. The graph in Fig. 5-11(c) has several Euler circuits. One of them is *L, A, R, D, A, R, D, L, A, L*. You should have no trouble finding others. Note that having an Euler circuit and having an Euler path are mutually exclusive—if a graph has an Euler circuit it cannot have an Euler path, and vice versa.

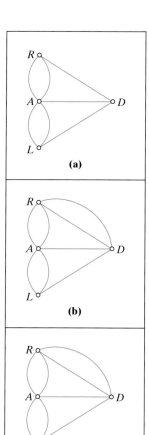

(a)

(b)

(c)

FIGURE 5-11

5.4 Graph Models

One of Euler's most important insights was the observation that certain types of problems can be conveniently rephrased as graph problems and that, in fact, graphs are just the right tool for describing many real-life situations. The notion of using a mathematical concept to describe and solve a real-life problem is one of the oldest and grandest traditions in mathematics. It is called *modeling*. Unwittingly, we have all done simple forms of modeling before, all the way back to elementary school. Every time we turn a word problem into an arithmetic calculation, an algebraic equation, or a geometric picture, we are modeling. We can now add to our repertoire one more tool for modeling: graph models.

In the next set of examples we are going to illustrate how we can use graphs to *model* some of the problems introduced in Section 5.1.

> **EXAMPLE 5.11** The Seven Bridges of Königsberg: Part 2

The Königsberg bridges question discussed in Example 5.3 asked whether it was possible to take a stroll through the old city of Königsberg and cross each of the seven bridges once and only once. To answer this question one obviously needs to take a look at the layout of the old city. A stylized map of the city of Königsberg is shown in Fig. 5-12(a). This map is not entirely accurate—the drawing is not to scale and the exact positions and angles of some of the bridges are changed. Does it matter?

A moment's reflection should convince us that many details on the original map are irrelevant to the question in point. The shape and size of the islands, the width of the river, the lengths of the bridges—none of these things really matter. So, then, what is it that does matter? Surprisingly little. *The only thing that truly matters to the solution of this problem is the relationship between land masses (islands and banks) and bridges.* Which land masses are connected to each other and by how many bridges? This information is captured by the red edges in Fig. 5-12(b). Thus, when we strip the map of all its superfluous information, we end up with the graph model shown in Fig. 5-12(c). The four vertices of the graph represent each of the four land masses; the edges represent the seven bridges. In this graph an *Euler circuit* would represent a stroll around the town that crosses each bridge once and ends back at the starting point; an *Euler path* would represent a stroll that crosses each bridge once but does not return to the starting point.

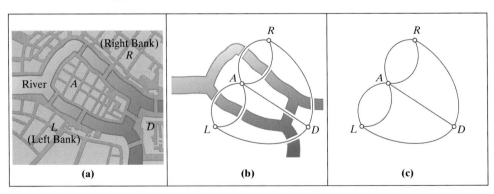

FIGURE 5-12

As big moments go this one may not seem like much, but Euler's idea to turn a puzzle about walking across bridges in a quaint medieval city into an abstract question about graphs was a "eureka" moment in the history of mathematics. (Here is a thought: Before you read on, take a break now, get yourself a cup of coffee, and try to mull over the significance of Example 5.11.)

> ### EXAMPLE 5.12 Walking the 'Hood': Part 2

In Example 5.1 we were introduced to the problem of the security guard who needs to walk the streets of the Sunnyside neighborhood [Fig. 5-13(a)]. The graph in Fig. 5-13(b)—where each edge represents a block of the neighborhood and each vertex an intersection—is a graph model of this problem. Does the graph have an Euler circuit? Euler path? Neither? (These are relevant questions that we will learn how to answer in the next section.) ◄◄

> ### EXAMPLE 5.13 Delivering the Mail: Part 2

Recall that unlike the security guard, the mail carrier (see Example 5.2) must make two passes through every block that has homes on both sides of the street (she has to physically place the mail in the mailboxes), make one pass through blocks that have homes on only one side of the street, and does not have to walk along blocks where there are no houses. In this situation an appropriate graph model requires two edges on the blocks that have homes on both sides of the street, one edge for the blocks that have homes on only one side of the street, and no edges for blocks having no homes on either side of the street. The graph that models this situation is shown in Fig. 5-13(c). ◄◄

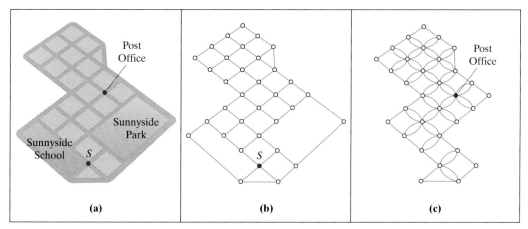

FIGURE 5-13 (a) The Sunnyside neighborhood. (b) A graph model for the security guard. (c) A graph model for the mail carrier.

5.5 Euler's Theorems

In this section we are going to develop the basic theory that will allow us to determine if a graph has an Euler circuit, an Euler path, or neither. This is important because, as we saw in the previous section, what are Euler circuit or Euler path questions in theory are real-life routing questions in practice. The three theorems we are going to see next (all due to Euler) are surprisingly simple and yet tremendously useful.

Euler's Circuit Theorem

- If a graph is *connected* and every vertex is *even*, then it has an Euler circuit (at least one, usually more).
- If a graph has *any* odd vertices, then it does not have an Euler circuit.

In practice, if we want to know if a graph has an Euler circuit or not, here is how we can use Euler's circuit theorem: First we make sure the graph is connected. (If it isn't, then no matter what else, an Euler circuit is impossible.) If the graph is connected, then we start checking the degrees of the vertices, one by one. As soon as we hit an odd vertex, we know that an Euler circuit is out of the question. If there are no odd vertices, then we know that the answer is yes—the graph does have an Euler circuit! (The theorem doesn't tell us how to find it—that will come soon.) Figure 5-14 illustrates the three possible scenarios. The graph in Fig. 5-14(a) cannot have an Euler circuit for the simple reason that it is disconnected. The graph in Fig. 5-14(b) is connected, but we can quickly spot odd vertices (*C* is one of them, there are others). This graph has no Euler circuits either. But the graph in Fig. 5-14(c) is connected and all the vertices are even. This graph does have Euler circuits.

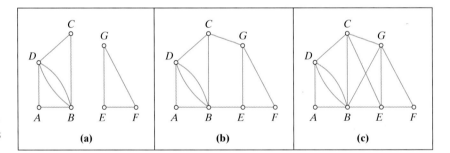

FIGURE 5-14 (a) Not connected. (b) Some vertices are odd. (c) All vertices are even.

The basic idea behind Euler's circuit theorem is that as we travel along an Euler circuit, every time we go through a vertex we use up two different edges at that vertex—one to come in and one to go out. We can keep doing this as long as the vertices are even. A single odd vertex means that at some point we are going to come into it and not be able to get out. An analogous theorem will work with Euler paths, but now we do need odd vertices for the starting and ending points of the path. All the other vertices have to be even. Thus, we have the following theorem.

Euler's Path Theorem

- If a graph is *connected* and has exactly *two* odd vertices, then it has an Euler path (at least one, usually more). Any such path must start at one of the odd vertices and end at the other one.
- If a graph has *more than two* odd vertices, then it cannot have an Euler path.

> **EXAMPLE 5.14** The Seven Bridges of Königsberg: Part 3

Back to the Königsberg bridges problem. In Example 5.11 we saw that the layout of the bridges in the old city can be modeled by the graph in Fig. 5-15(a). This graph has four odd vertices, and thus neither an Euler circuit nor an Euler path can exist. We now have an unequivocal answer to the puzzle: *There is no possible way anyone can walk across all of the bridges without having to recross some of them!* How many bridges will need to be recrossed? It depends. If we want to start and end in the same place, we must recross at least two of the bridges. One of the many possible routes is shown in Fig. 5-15(b). If we are allowed to start and end in different places, we can do it by recrossing just one of the bridges. One possible route is shown in Fig. 5-15(c).

FIGURE 5-15 (a) The original graph with four odd vertices. (b) A walk recrossing bridges *DL* and *RA*. (c) A walk recrossing bridge *DL* only.

> **EXAMPLE 5.15** Child's Play: Part 2

Figure 5-16 shows four graphs. These graphs correspond to the line drawings in Example 5.5 (Fig. 5-4). Using Euler's theorems, we can answer any questions about unicursal tracings.

The graph in Fig. 5-16(a) is connected and the vertices are all even. By Euler's circuit theorem we know that the graph has an Euler circuit, which implies that the original line drawing has a closed unicursal tracing.

The graph in Fig. 5-16(b) is connected and has exactly two odd vertices (*C* and *D*). By Euler's path theorem, the graph has an Euler path (open unicursal tracing). Moreover, we now know that the path has to start at *C* and end at *D*, or vice versa.

The graph in Fig. 5-16(c) has four odd vertices (*A*, *B*, *C*, and *D*), so it has neither an Euler path nor an Euler circuit.

The full power of Euler's theorems is best appreciated when the graphs get bigger. The graph in Fig. 5-16(d) is not extremely big, but we can no longer "eye-

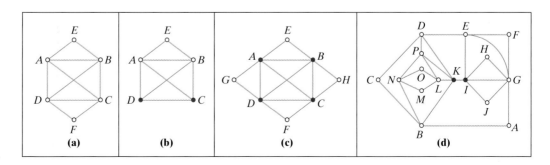

FIGURE 5-16 Odd vertires shown in red.

ball" an Euler circuit or path. On the other hand, a quick check of the degrees of the vertices shows that K and I are odd vertices and all the rest are even. We are now in business—an open unicursal tracing is possible, as long as we start it at K or I (and end it at the other one). Starting anyplace else will lead to a dead end. Good to know!   **◀◀**

Euler's circuit and path theorems tell us what happens when the graph has zero odd vertices or two or more odd vertices but leaves hanging the question of what to say about a graph with *just one* odd vertex. Didn't Euler consider this possibility? It turns out that he did, but he found that there is no such thing as a graph with just one odd vertex—such a graph is a mathematical impossibility! In fact, Euler proved a much more general fact: The number of odd vertices in a graph must be an *even* number.

Euler's Sum of Degrees Theorem

- The sum of the degrees of all the vertices of a graph equals twice the number of edges (and therefore is an even number).
- A graph always has an even number of *odd* vertices.

Euler's sum of degrees theorem is based on the following basic observation: Take any edge—let's call it XY. The edge contributes once to the degree of vertex X and once to the degree of vertex Y, so, in all, that edge makes a total contribution of 2 to the sum of the degrees. Thus when the degrees of all the vertices of a graph are added, the total is twice the number of edges. Since the total sum is an even number, it is impossible to have just one odd vertex, or three odd vertices, or five odd vertices, and so on. To put it in a slightly different way, *the odd vertices of a graph always come in twos.*

Table 5-1 is a summary of Euler's three theorems. It shows the relationship between the number of odd vertices in a *connected* graph and the existence of Euler paths or Euler circuits. (The assumption that the graph is connected is essential—a disconnected graph cannot have Euler paths or circuits regardless of what else is going on.)

TABLE 5-1

Number of odd vertices	Conclusion
0	Graph has Euler circuit
2	Graph has Euler path
$4, 6, 8, \ldots$	Graph has neither
$1, 3, 5, \ldots$	Better go back and double check! This is impossible!

5.6 Fleury's Algorithm

Euler's theorems help us answer the existence question for Euler circuits/paths: Does the graph have an Euler circuit, an Euler path, or neither? But when the graph has an Euler circuit or path, how do we find it? For small graphs, simple trial-and-error usually works fine, but real-life applications sometimes involve graphs with hundreds, or even thousands, of vertices. In these cases a trial-and-error approach is out of the question, and what is needed is a systematic strategy that tells us how to create an Euler circuit or path. In other words, we need an *algorithm*.

Algorithms

There are many types of problems that can be solved by simply following a set of procedural rules—very specific rules like *when you get to this point do this, … after you finish this, do that*, and so on. Given a specific problem X, an **algorithm** for solving X is a set of *procedural rules* that, when followed, always lead to some sort of "solution" to X. The problem X need not be a mathematics problem—algorithms are used, sometimes unwittingly, in all walks of life: directions to find someone's house, the instructions for assembling a new bike, or a recipe for baking an apple pie are all examples of real-life algorithms. A useful analogy is to think of the problem as a *dish* we want to prepare and the algorithm as a *recipe* for preparing that dish.

In many cases, there are several different algorithms for solving the same problem (there is more than one way to bake an apple pie); in other cases, the problem does not lend itself to an algorithmic solution. In mathematics, algorithms are either *formula* driven (you just apply the formula or formulas to the appropriate inputs) or *directive* driven (you must follow a specific set of directives). In this part of the book (Chapters 5 through 8) we will discuss many important algorithms of the latter type.

Algorithms may be complicated, but are rarely difficult. (There is a world of difference between complicated and difficult—accounting is complicated, calculus is difficult!) You don't have to be a brilliant and creative thinker to implement most algorithms—you just have to learn how to follow instructions carefully and methodically. For most of the algorithms we will discuss in this and the next three chapters, the key to success is simple: practice, practice, and more practice!

Fleury's Algorithm

We will now turn our attention to an algorithm that will allow us to find an *Euler circuit* in a connected graph with no odd vertices, or an *Euler path* in a connected graph with two odd vertices. Technically speaking, these are two separate algorithms, but in essence they are identical, so they can be described as one. (The algorithm we will give here is attributed to a Frenchman by the name of M. Fleury, who is alleged to have published a description of the algorithm in 1885. Other than his connection to this algorithm, little else is known about Monsieur Fleury.)

For a completely different algorithm, known as **Hierholzer's algorithm**, see Exercise 73.

The idea behind Fleury's algorithm can be paraphrased by that old piece of folk wisdom: *Don't burn your bridges behind you.* In graph theory the word *bridge* has a very specific meaning—it is the only edge connecting two separate sections (call them A and B) of a graph, as illustrated in Fig. 5-17. This means that if you are in A, you can only get to B by crossing the bridge. If you do that, and then want to get back to A, you will need to recross that same bridge. It follows that if you don't want to recross bridges, you better finish your business at A before you move on to B.

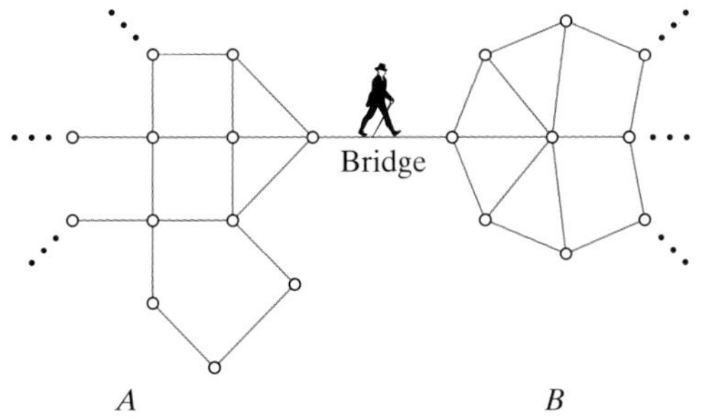

FIGURE 5-17 The bridge separates the two sections. Once you cross from A to B, the only way to get back to A is by recrossing the bridge.

Thus, Fleury's algorithm is based on a simple principle: To find an Euler circuit or an Euler path, *bridges are the last edges you want to cross*. Only do it if you have no choice! Simple enough, but there is a rub: The graph whose bridges we are supposed to avoid is not necessarily the original graph of the problem. Instead it is that part of the original graph which has yet to be traveled. The point is this: Once we travel along an edge, we are done with it! We will never cross it again, so from that point on, as far as we are concerned, it is as if that edge never existed. Our concerns lie only on how we are going to get around the *yet-to-be-traveled* part of the graph. Thus, when we talk about bridges that we want to leave as a last resort, we are really referring to *bridges of the to-be-traveled part of the graph*.

Fleury's Algorithm for Finding an Euler Circuit (Path)

- **Preliminaries.** Make sure that the graph is connected and either (i) has no odd vertices (circuit), or (ii) has two odd vertices (path).

- **Start.** Choose a starting vertex. [In case (i) this can be any vertex; in case (ii) it must be one of the two *odd* vertices.]

- **Intermediate steps.** At each step, if you have a choice, *don't choose a bridge of the yet-to-be-traveled part* of the graph. However, if you have only one choice, take it.

- **End.** When you can't travel any more, the circuit (path) is complete. [In case (i) you will be back at the starting vertex; in case (ii) you will end at the other odd vertex.]

The only complicated aspect of Fleury's algorithm is the bookkeeping. With each new step, the untraveled part of the graph changes and there may be new bridges formed. Thus, in implementing Fleury's algorithm it is critical to separate the *past* (the part of the graph that has already been traveled) from the *future* (the part of the graph that still needs to be traveled). While there are many different ways to accomplish this (you are certainly encouraged to come up with one of your own), a fairly reliable way goes like this: Start with *two* copies of the graph. Copy 1 is to keep track of the "future"; copy 2 is to keep track of the "past." Every time you travel along an edge, *erase* the edge from copy 1, but mark it (say in red) and label it with the appropriate number on copy 2. As you move forward, copy 1 gets smaller and copy 2 gets redder. At the end, copy 1 has disappeared; copy 2 shows the actual Euler circuit or path.

It's time to look at a couple of examples.

▶ EXAMPLE 5.16 Implementing Fleury's Algorithm

The graph in Fig. 5-18(a) is a very simple graph—it would easier to find an Euler circuit just by trial-and-error than using Fleury's algorithm. Nonetheless, we will do it using Fleury's algorithm. The real purpose of the example is to see the algorithm at work. Each step of the algorithm is explained in Figs. 5-18(b) through (h).

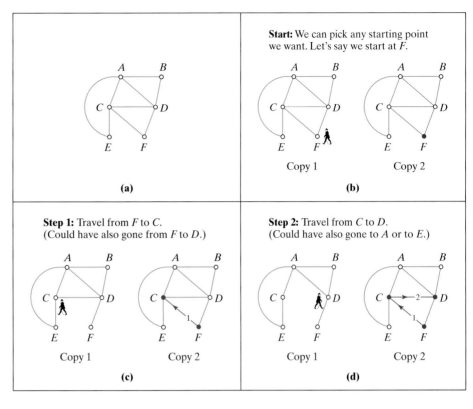

Start: We can pick any starting point we want. Let's say we start at F.

Copy 1   Copy 2

(b)

Step 1: Travel from F to C. (Could have also gone from F to D.)

Copy 1   Copy 2

(c)

Step 2: Travel from C to D. (Could have also gone to A or to E.)

Copy 1   Copy 2

(d)

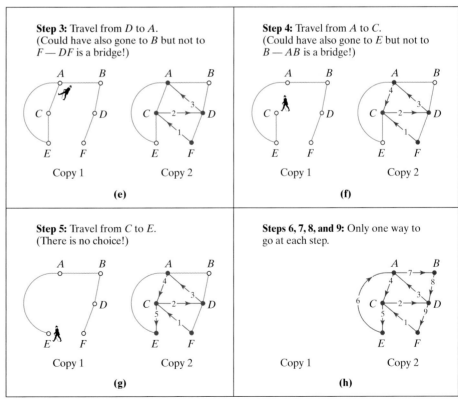

Step 3: Travel from D to A. (Could have also gone to B but not to F — DF is a bridge!)

Copy 1   Copy 2

(e)

Step 4: Travel from A to C. (Could have also gone to E but not to B — AB is a bridge!)

Copy 1   Copy 2

(f)

Step 5: Travel from C to E. (There is no choice!)

Copy 1   Copy 2

(g)

Steps 6, 7, 8, and 9: Only one way to go at each step.

Copy 1   Copy 2

(h)

FIGURE 5-18

EXAMPLE 5.17 Fleury's Algorithm For Euler Paths

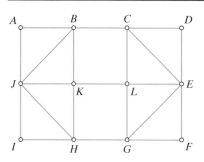

FIGURE 5-19

We will apply Fleury's algorithm to the graph in Fig. 5-19. Since it would be a little impractical to show each step of the algorithm with a separate picture as we did in Example 5.16, you are going to have to do some of the work. (If you haven't already done so, get some paper, a pencil, and an eraser.) Start by making two copies of the graph. Ready? Let's go!

- **Start.** This graph has two odd vertices, E and J. We can pick either one as the starting vertex. Let's start at J.

- **Step 1.** From J we have five choices, all of which are OK. We'll randomly pick K. (Erase JK on copy 1, and mark and label JK with a 1 on copy 2.)

- **Step 2.** From K we have three choices (B, L, or H). Any of these choices is OK. Say we choose B. (Now erase KB from copy 1 and mark and label KB with a 2 on copy 2.)

- **Step 3.** From B we have three choices (A, C, or J). Any of these choices is OK. Say we choose C. (Now erase BC from copy 1 and mark and label BC with a 3 on copy 2.)

- **Step 4.** From C we have three choices (D, E, or L). Any of these choices is OK. Say we choose L. (EML—that's shorthand for erase, mark, and label.)

- **Step 5.** From L we have three choices (E, G, or K). Any of these choices is OK. Say we choose K. (EML.)

- **Step 6.** From K we have only one choice—to H. Without further ado, we choose H. (EML)

- **Step 7.** From H we have three choices (G, I, or J). But for the first time, one of the choices is a bad choice. We should not choose G, as HG is a bridge of the yet-to-be-traveled part of the graph (see Fig. 5-20). Either of the other two choices is OK. Say we choose J. (EML)

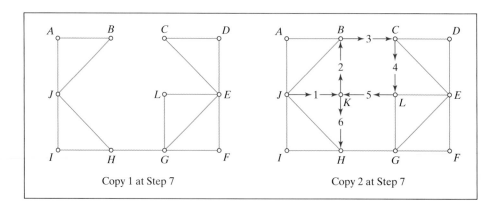

Copy 1 at Step 7   Copy 2 at Step 7

FIGURE 5-20

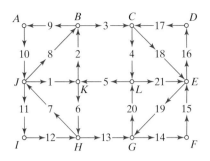

FIGURE 5-21

- **Step 8.** From J we have three choices (A, B, or I), but we should not choose I, as JI has just become a bridge. Either of the other two choices is OK. Say we choose B. (EML)

- **Step 9 through 13.** Each time we have only one choice. From B we have to go to A, then to J, I, H, and G.

- **Step 14 through 21.** Not to belabor the point, let's just cut to the chase. The rest of the path is given by G, F, E, D, C, E, G, L, E. There are many possible endings, and you should find a different one by yourself.

The completed Euler path (one of hundreds of possible ones) is shown in Fig. 5-21.

5.7 Eulerizing Graphs

In this section we will finally answer some of the routing problems raised at the beginning of the chapter. Their common thread the need to find *optimal exhaustive routes* in a connected graph. How is this done? Let's first refresh our memories of what this means. We will use the term *exhaustive route* to describe a route that travels along the edges of a graph and passes through *each and every edge* of the graph *at least once*. Such a route could be an Euler circuit (if the graph has no odd vertices), or an Euler path (if the graph has two odd vertices), but for graphs with more than two odd vertices, an exhaustive route will have to *recross* some of the edges. This follows from Euler's theorems.

We are interested in finding exhaustive routes that *recross* the fewest number of edges. Why? In many applications, each edge represents a unit of cost. The more edges along the route, the higher the cost of the route. In an exhaustive route, the first pass along an edge is a necessary expense, part of the requirements of the job. On the other hand, any additional pass along that edge represents a wasted expense (these extra passes are often described as *deadhead* travel). Thus, an exhaustive route that minimizes cost (*optimal exhaustive route*) is one with the fewest number of deadhead edges. (This is only true under the assumption that each edge equals one unit of cost.)

We are now going to see how the theory developed in the preceding sections will help us design optimal exhaustive routes for graphs with many (more than two) odd vertices. The key idea is that we can turn odd vertices into even vertices by adding "duplicate" edges in strategic places. This process is called **eulerizing** the graph.

> **EXAMPLE 5.18** Covering a 3 by 3 Street Grid

The graph in Fig. 5-22(a) represents a 3 block by 3 block street grid consisting of 24 blocks (count them if you don't believe it!). How can we find an optimal route that covers all the edges of the graph and ends back at the starting vertex?

Our first step is to identify the odd vertices. This graph has eight odd vertices (B, C, E, F, H, I, K, and L), shown in red. When we add a duplicate copy of edges BC, EF, HI, and KL, we get the graph in Fig. 5-22(b). This is a *eulerized* version of the original graph—its vertices are all even, so we know it has an Euler circuit. Moreover, it's clear we couldn't have done this with fewer than four duplicate edges. Figure 5-22(c) shows one of the many possible Euler circuits, with the

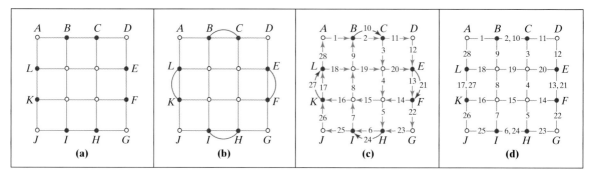

FIGURE 5-22

edges numbered in the order they are traveled. The Euler circuit described in Fig. 5-22(c) represents an exhaustive closed route along the edges of the original graph [Fig 5-22(d)] with the four duplicate edges (*BC*, *EF*, *HI*, and *KL*) indicating the deadhead blocks where a second pass is required. The total length of this route is 28 blocks (24 blocks in the grid plus 4 deadhead blocks), and this route is optimal—no matter how clever you are or how hard you try, if you want to travel along each block of the grid and start and end at the same vertex, you will have to pass through a minimum of 28 blocks! (There are many other ways to do it using just 28 blocks, but none with less than 28.) ◀◀

> ### EXAMPLE 5.19 Covering a 4 by 4 Street Grid

The graph in Fig. 5-23(a) represents a 4 block by 4 block street grid consisting of 40 blocks. The 12 odd vertices in the graph are shown in red. We want to eulerize the graph by adding the least number of edges. Figure 5-23(b) shows how *not to do it!* This graph violates the cardinal rule of eulerization—you can only duplicate edges that are part of the original graph. Edges *DF* and *NL* are new edges, not duplicates, so Fig. 5-23(b) is out! Figure 5-23(c) shows a legal eulerization, but it is not optimal, as it is obvious that we could have accomplished the same thing by adding fewer duplicate edges. Figure 5-23(d) shows an *optimal eulerization* of the original graph—one of several possible. Once we have an optimal eulerization, we have the blueprint for an optimal exhaustive closed route on the original

The final step is to find the actual route—see Exercise 30.

graph. Regardless of the specific details, we now know that the route will travel along 48 blocks—the 40 original blocks in the grid plus 8 deadhead blocks.

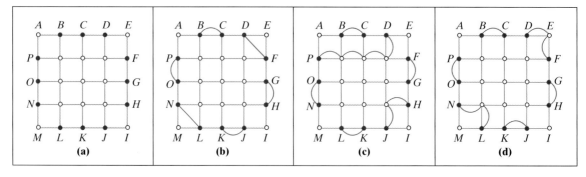

FIGURE 5-23 (a) The original graph. Odd vertices shown in red. (b) Bad move—*DF* and *NL* were not edges of the original graph! (c) This is a eulerization, but not an efficient one! (d) This is one of the many possible optimal eulerizations. ◀◀

In some situations we need to find an exhaustive route, but there is no requirement that it be closed—the route may start and end at different points. In these cases, we want to leave two odd vertices on the graph unchanged, and change the other odd vertices into even vertices by duplicating appropriate edges of the graph. This process is called a **semi-eulerization** of the graph. When we strategically choose how to do this so that the number of duplicate edges is as small as possible, we can obtain an *optimal exhaustive open route*. In this case the route will start at one of the two odd vertices and end at the other one.

▶ EXAMPLE 5.20 Parade Routes

Let's consider once again the 4 by 4 street grid shown in Fig. 5-23(a). Imagine that your job is to design a good route for a Fourth of July parade that must pass through each of the 40 blocks of the street grid. The key difference between this example and Example 5.19 is that when routing a parade, you do not want the parade to start and end in the same place. (In fact, for traffic control it is usually desirable to keep the starting and ending points of a parade as far from each other as possible.) The fire department has added one additional requirement to the parade route: The parade has to start at *B*. Your task, then, is to find a *semi-eulerization* of the graph that leaves *B* and one more odd vertex unchanged (preferably a vertex far from *B*), and changes all the other odd vertices into even vertices. (Why don't you give it a try before you read on?)

See Exercise 35(a).

Figure 5-24 shows two different semi-eulerizations. The semi-eulerization in Fig. 5-24(a) is *optimal* because it required only six duplicate edges, and this is as good as one can do. The optimal parade route could be found by finding an Euler path on the graph in Fig. 5-24(a). The only bad thing about this route is that the parade would end at *P*, a point a bit too close to the starting point (the police department is not happy about that!). A different semi-eulerization is shown in Fig. 5-24(b). The parade route in this case would not be optimal (it has seven dead-head blocks), but because it ends at *K*, it better satisfies the requirement that the starting and ending points be far apart.

See Exercise 35(b).

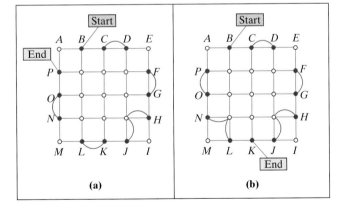

FIGURE 5-24 (a) An optimal semi-eulerization. (b) A good semi-eulerization (not optimal) with an ending point far from the starting point.

Example 5.20 illustrates the fact that sometimes in real-life situations cheapest may not be best, and that tradeoffs often have to be made between efficiency and other variables. Do we want a route that minimizes the number of deadhead blocks in the parade, or do we want a route one block longer but with starting and ending points that are far apart? The answer, of course, depends on how we prioritize these variables.

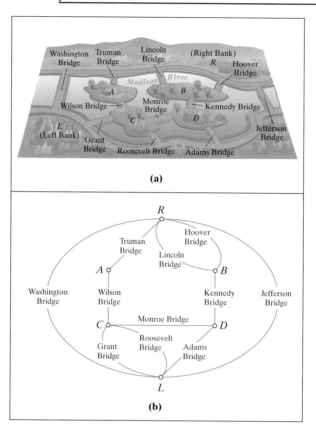

EXAMPLE 5.21 The Bridges of Madison County: Part 2

This is the conclusion of the routing problem first introduced in Example 5.4. A photographer needs to take photos of each of the 11 bridges in Madison County [Fig. 5-25(a)]. A graph model of the layout (vertices represent land masses, edges represent bridges) is shown in Fig. 5-25(b). The graph has four odd vertices (R, L, B, and D), so some bridges are definitely going to have to be recrossed. How many and which ones depends on the other parameters of the problem. (Recall that it costs $25 in toll fees to cross a bridge, so the baseline cost for crossing the 11 bridges is $275. Each recrossing is at an additional cost of $25.)

The following are a few of the possible scenarios one might have to consider. Each one requires a different eulerization and will result in a different route.

FIGURE 5-25

(a)

(b)

- The photographer needs to start and end his trip in the same place. This scenario requires an optimal eulerization of the graph in Fig. 5-25(b). This is not hard to do, and an optimal route can be found for a cost of $325.

See Exercise 62.

- The photographer has the freedom to choose any starting and ending points for his trip. In this case we can find an optimal semi-eulerization of the graph, requiring only one duplicate edge. Now an optimal route is possible at a cost of $300.

See Exercise 63(a).

- The photographer has to start his trip at B and end the trip at L. In this case we must find a semi-eulerization of the graph where B and L remain as odd vertices, R and D become even vertices. It is possible to do this with just two duplicate edges and thus find an optimal route that will cost $325.

See Exercise 63(b).

Even more complicated scenarios are possible, but we have to move on.

EXAMPLE 5.22 The Exhausted Patrol and the Grateful No Deadhead

This example brings us full circle to the first couple of examples of this chapter. In Example 5.1 we raised the question of finding an optimal exhaustive closed route for a security guard hired to patrol the streets of the Sunnyside subdivision. In Example 5.12 we created a graph model for this problem [Fig. 5-26(a)]. The graph

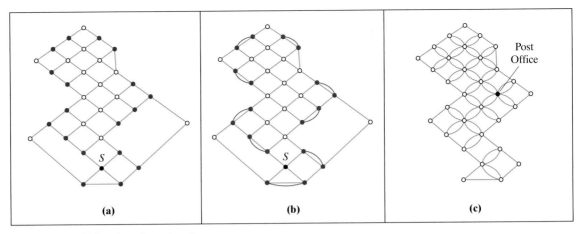

FIGURE 5-26 Odd vertices shown in red.

See Exercise 31.

model has 18 odd vertices, shown in red. We now know that the name of the game is to find an optimal eulerization of this graph. In this case the odd vertices pair up beautifully, and the optimal eulerization requires only nine duplicate edges. All the answers to the security guards questions can now be answered: An optimal route will require nine deadhead blocks. The actual route can be found using trial and error or Fleury's algorithm.

A slightly different problem is the one facing the mail carrier delivering mail along the streets of the Sunnyside subdivision. Much to the mail carrier's pleasant surprise, in the graph that models her situation [Fig. 5-26(c)] all the vertices are even (you just go ahead and check it out!). This means that the optimal route is an Euler circuit, which can be found once again using Fleury's algorithm (or trial-and-error if you prefer). Thanks to the lessons of this chapter, this mail carrier will not have to deadhead, for which she is extremely grateful!   ◀◀

See Exercise 32.

Conclusion

In this chapter we got our first introduction to three fundamental ideas. First, we learned about a simple but powerful concept for describing relationships within a set of objects—the concept of a *graph*. This idea can be traced back to Euler, some 270 years ago. Since then, the study of graphs has grown into one of the most important and useful branches of modern mathematics.

The second important idea of this chapter is the concept of a *graph model*. Every time we take a real-life problem and turn it into a mathematical problem, we are, in effect, modeling. Unwittingly, we have all done some form of mathematical modeling at one time or another—first using arithmetic and later using equations and functions to describe real-life situations. In this chapter we learned about a new type of modeling called graph modeling, in which we use graphs and the mathematical theory of graphs to solve certain types of routing problems.

By necessity, the routing problems that we solved in this chapter were fairly simplistic—crossing a few bridges, patrolling a small neighborhood, designing a parade route—what's all the fuss about? We should not be deceived by the simplicity of these examples—larger-scale variations on these themes have significant practical importance. In many big cities, where the efficient routing of municipal services (police patrols, garbage collection, etc.) is a major issue, the very theory that we developed in this chapter is being used on a large scale, the only difference being that many of the more tedious details are mechanized and carried out by a computer. (In New York City, for example, garbage collection, curb sweeping, snow removal, and other municipal services have been scheduled and organized using graph models since the 1970s, and the improved efficiency has yielded savings estimated in the tens of millions of dollars a year.)

The third important concept introduced in this chapter is that of an *algorithm*—a set of procedural rules that, when followed, provide solutions to certain types of problems. Perhaps without even realizing it, we had our first exposure to algorithms in elementary school, when we learned how to add, multiply, and divide numbers following precise and exacting procedural rules. In this chapter we learned about *Fleury's algorithm*, which helps us find an Euler circuit or an Euler path in a graph. In the next few chapters we will learn many other *graph algorithms*, some quite simple, others a bit more complicated. When it comes to algorithms of any kind, be they for doing arithmetic calculations or for finding circuits in graphs, there is one standard piece of advice that always applies: *Practice makes perfect.* (As is often the case, Yogi Berra said it better.)

In theory, there is no difference between theory and practice. In practice, there is.

Yogi Berra

Profile Leonhard Euler (1707–1783)

Leonhard Euler is universally recognized as one of the great, if not the greatest, mathematical geniuses in history. In the words of one of his biographers, "Euler was the Shakespeare of mathematics—universal, richly detailed and inexhaustible." In terms of sheer volume, Euler's mathematical production is staggering—his *Opera Omnia* (collected works) fills over 80 volumes and run over 25,000 pages. No other mathematician—in fact, no other scientist—in the history of humankind can come close to matching Euler in terms of creative output. Euler produced groundbreaking discoveries in practically every area of mathematics—analysis, number theory, algebra, geometry, and topology (which he started). Euler's theorems in graph theory (see Section 5.5) and his solution of the Königsberg bridge problem are just one note in the mathematical symphony that is his work.

Leonhard Euler was born in Basel, Switzerland, the son of a Protestant minister. From an early age, Euler exhibited two traits that set him apart from mere mortals—a prodigious memory and the ability to perform incredibly complicated calculations in his head. At the age of 14 Euler entered the University of Basel to study theology and follow in the footsteps of his father. The young man's incredible mathematical gifts soon came to the attention of Johann Bernoulli, professor of mathematics at Basel and one of the most famous and influential mathematicians of his generation. Under Bernoulli's mentorship Euler eventually dropped theology and decided to concentrate on the study of mathematics. At the age of 19, Euler graduated from the University of Basel, having completed a doctoral thesis in which he analyzed the works of Descartes and Newton.

By this time Euler was already a mathematician of some fame and was offered an academic position at the St. Petersburg Academy of Sciences in Russia—a remarkable and unprecedented offer for someone his age. For the next 14 years (1727 to 1741), Euler worked at the St. Petersburg Academy under the patronage of Catherine I. In addition to his groundbreaking work in pure mathematics, Euler was a practical man who worked on many important applied problems in physics, engineering, astronomy, and cartography. In

1733, at the age of 27, he was promoted to Professor of Mathematics at the St. Petersburg Academy. His first book, *Mechanica*, a two-volume text published in 1736, became the definitive work in mechanics for the next 50 years.

In 1738, due to an infection, Euler lost the sight in his right eye. Twenty-eight years later, he would lose the sight in his left eye, leaving him totally blind. Remarkably, neither event had a significant impact on his ability to produce copious amounts of mathematical work—he just dictated to his assistants and did most calculations in his head. In 1741 Euler accepted an offer from Frederick the Great of Prussia to become mathematics director of the Berlin Academy of Sciences. Euler worked in Berlin for the next 25 years (1741–1766), where he produced some of his greatest work,

including his most famous book—a multivolume textbook in elementary science called *Letters to a German Princess*. In 1766, mostly for political reasons, Euler returned to the St. Petersburg Academy of Sciences, where, despite his progressive blindness, he carried on with his prolific research for another 17 years.

The day he died, September 18, 1783, Euler worked in the morning on mathematical questions related to balloon flights, and in the afternoon he performed important calculations concerning the orbit of the planet Uranus—all just another day's work for him. In the late afternoon, while playing with one of his grandchildren, he suffered a massive stroke and died. His death marked the passing of one of the most creative men in the history of humankind.

Key Concepts

adjacent edges, **168**	Euler circuit, **169**	length (of a path or a circuit), **168**
adjacent vertices, **168**	Euler circuit problem, **162**	loop, **165**
algorithm, **175**	eulerization (of a graph), **179**	multiple edges, **166**
bridge (of a graph), **169**	Euler path, **169**	odd vertex, **168**
circuit, **169**	Euler's theorems, **172**	path, **168**
component (of a graph), **166**	even vertex, **168**	routing problems, **162**
connected graph, **169**	exhaustive route, **179**	semi-eulerization, **181**
degree (of a vertex), **168**	Fleury's algorithm, **175**	unicursal tracing, **165**
disconnected graph, **166**	graph, **165**	vertex, **165**
edge, **165**	graph model, **170**	vertex set, **166**
edge set, **166**	isolated vertex, **166**	

Exercises

WALKING

A. Graphs: Basic Concepts

1. For each of the following graphs, give the vertex set, the edge set, and the degree of each vertex.

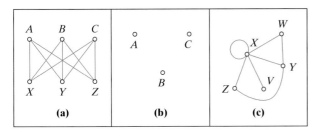

2. For each of the following graphs, give the vertex set, the edge set, and the degree of each vertex.

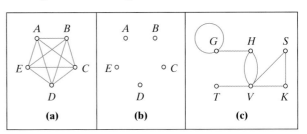

3. For the given vertex and edge sets, draw two different representations of the graph.

   (a) $\mathcal{V} = \{A, B, C, D\}$; $\mathcal{E} = \{AB, BC, BD, CD\}$

   (b) $\mathcal{V} = \{K, R, S, T, W\}$;
   $\mathcal{E} = \{RS, RT, TT, TS, SW, WW, WS\}$

4. For the given vertex and edge sets, draw two different representations of the graph.

   (a) $\mathcal{V} = \{L, M, N, P\}$;
   $\mathcal{E} = \{LP, MM, PN, MN, PM\}$

   (b) $\mathcal{V} = \{A, B, C, D, E\}$;
   $\mathcal{E} = \{AC, AE, BD, BE, CA, CD, CE, DE\}$

5. (a) Explain why the two figures represent the same graph.

   (b) Draw a third figure that represents the same graph.

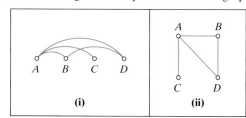

(i) (ii)

6. (a) Explain why the two figures represent the same graph.

   (b) Draw a third figure that represents the same graph.

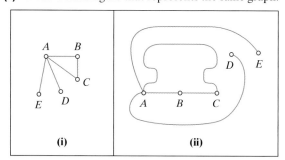

(i) (ii)

7. (a) Draw a connected graph with four vertices such that each vertex has degree 2.

   (b) Draw a disconnected graph with four vertices such that each vertex has degree 2.

   (c) Draw a graph with four vertices such that each vertex has degree 1.

8. (a) Draw a connected graph with eight vertices such that each vertex has degree 3.

   (b) Draw a disconnected graph with eight vertices such that each vertex has degree 3.

   (c) Draw a graph with eight vertices such that each vertex has degree 1.

9. Give an example of a graph with four vertices, each of degree 3 with

   (a) no loops and no multiple edges

   (b) loops but no multiple edges

   (c) multiple edges but no loops

(d) both multiple edges and loops

10. Give an example of a connected graph with five vertices, each of degree 4 with

    (a) no loops and no multiple edges.

    (b) loops but no multiple edges.

    (c) multiple edges but no loops.

    (d) both multiple edges and loops.

Exercises 11 and 12 refer to the graph shown in the figure.

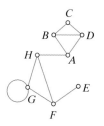

11. (a) Find a path from C to F passing through vertex B but not through vertex D.

    (b) Find a path from C to F passing through both vertex B and vertex D.

    (c) Find a path of length 4 from C to F.

    (d) Find a path of length 7 from C to F.

    (e) How many paths are there from C to A?

    (f) How many paths are there from H to F?

    (g) How many paths are there from C to F?

12. (a) Find a path from D to E passing through vertex G only once.

    (b) Find a path from D to E passing through vertex G twice.

    (c) Find a path of length 4 from D to E.

    (d) Find a path of length 8 from D to E.

    (e) How many paths are there from D to A?

    (f) How many paths are there from H to E?

    (g) How many paths are there from D to E?

13. Find all the circuits in the graph shown in the figure.

    (***Hint:** Organize your list by length, starting with circuits of length 1, then 2, and so on.*)

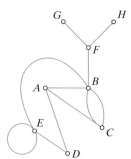

14. Find all the circuits in the graph shown in the figure.

(**Hint:** *Organize your list by length, starting with circuits of length 2, then 3, and so on.*)

B. Graph Models

15. Mr. Belding wishes to make a seating chart for one of his classes at Bayside High. (He wants to minimize the visiting among the students by separating friends as much as possible.) The students in the class are Zack, Screech, Kelly, Lisa, Jessie, Slater, and Tori. Zack is friends with everyone but Slater. Screech is friends with Zack and Slater. Kelly is friends with Zack, Lisa, Jessie, and Slater. Slater is friends with Screech, Kelly, Jessie, and Tori. Draw a graph that Mr. Belding might use to represent the friendship relationships among the students in the class.

16. The Kangaroo Lodge of Madison County has 10 members ($A, B, C, D, E, F, G, H, I,$ and J). The club has five working committees: the Rules Committee ($A, C, D, E, I,$ and J), the Public Relations Committee ($B, C, D, H, I,$ and J), the Guest Speaker Committee ($A, D, E, F,$ and H), the New Year's Eve Party Committee ($D, F, G, H,$ and I), and the Fund Raising Committee ($B, D, F, H,$ and J).

(a) Suppose we are interested in knowing which pairs of members are on the same committee. Draw a graph that models this situation.

(**Hint:** *Let the vertices of the graph represent the members.*)

(b) Suppose we are interested in knowing which committees have members in common. Draw a graph that models this situation.

(**Hint:** *Let the vertices of the graph represent the committees.*)

17. The following is a map of downtown Kingsburg, showing the Kings River running through the downtown area and the three islands (A, B, and C) connected to each other and both banks by seven bridges. The Chamber of Commerce wants to design a walking tour that crosses all the bridges. Draw a graph that models the layout of Kingsburg.

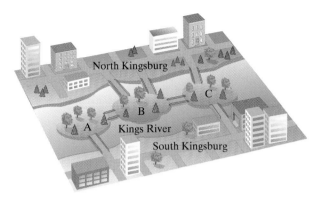

18. The following is a map of downtown Royalton, showing the Royalton River running through the downtown area and the three islands (A, B, and C) connected to each other and both banks by eight bridges. The Downtown Athletic Club wants to design the route for a marathon that passes through the downtown area and through each of the downtown bridges. Draw a graph that models the layout of Royalton.

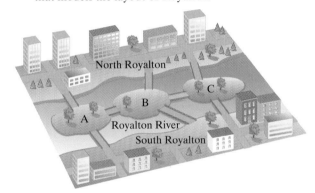

Exercises 19 and 20 refer to the Green Hills subdivision shown in the map.

19. A night watchman must walk the streets of the Green Hills subdivision. The night watchman needs to walk only once along each block. Draw a graph that models this situation.

20. A mail carrier must deliver mail on foot along the streets of the Green Hills subdivision. The mail carrier must make two passes on blocks that have houses on both sides of the street (once for each side of the street) and only one pass on blocks that have houses on only one side of the street. Draw a graph that models this situation.

C. Euler's Theorems

For each graph in Exercises 21 through 26, determine whether the graph has an Euler circuit, an Euler path, or neither of these. Explain your answer. (You do not have to show the actual path or circuit.)

21. (a)

(b)

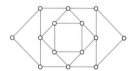

22. (a)

(b)

23. (a)

(b)

(c)

24. (a)

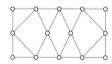

(b)

(c)

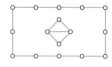

25. (a)

(b)

(c)

26. (a)

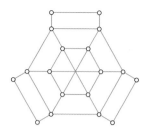

(b)

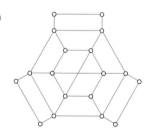

(c)

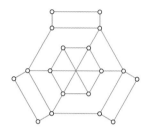

D. Finding Euler Circuits and Euler Paths

27. Find an Euler circuit for the graph. Show your answer by labeling the edges 1, 2, 3, and so on in the order in which they are traveled.

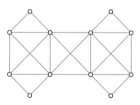

28. Find an Euler circuit for the graph. Show your answer by labeling the edges 1, 2, 3, and so on in the order in which they can be traveled.

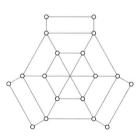

29. Find an Euler circuit for the graph. Use *M* as the starting and ending point of the circuit. Show your answer by labeling the edges 1, 2, 3, and so on in the order in which they are traveled.

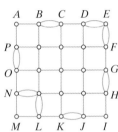

30. Find an Euler circuit for the graph. Use *B* as the starting and ending point of the circuit. Show your answer by labeling the edges 1, 2, 3, and so on in the order in which they are traveled.

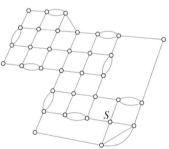

31. Find an Euler circuit for the graph. Use *S* as the starting and ending point of the circuit. Show your answer by labeling the edges 1, 2, 3, and so on in the order in which they are traveled.

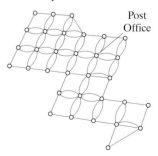

32. Find an Euler circuit for the graph. Use the post office as the starting and ending point of the circuit. Show your answer by labeling the edges 1, 2, 3, and so on in the order in which they are traveled.

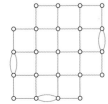

In Exercises 33 through 36, find an Euler path for each given graph. Show your answer by labeling the edges 1, 2, 3, and so on in the order in which they are traveled.

33.

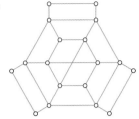

34.

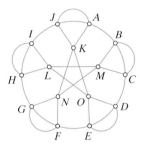

35. (a)

(b)

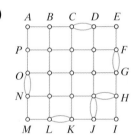

36.

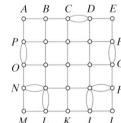

E. Unicursal Tracings

In Exercises 37 through 40, for each line drawing, find a unicursal tracing—if you can. (Show the unicursal tracing by labeling the edges 1, 2, 3, and so on in the order in which they are traced.) If the drawing does not have a unicursal tracing, explain why not.

37. (a)

38. (a)

(b)

(b)

(c)

(c)
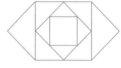

39. (a)

40. (a)

(b)

(b)

(c)

(c)

F. Eulerizations and Semi-Eulerizations

41. Find an optimal eulerization for each of the graphs.

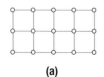

(a)

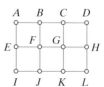

(b)

42. Find an optimal eulerization for each of the graphs.

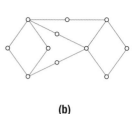

(a)   **(b)**

43. Find an optimal semi-eulerization for each of the graphs.

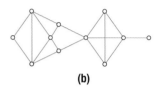

(a)   **(b)**

44. Find an optimal semi-eulerization for each of the graphs.

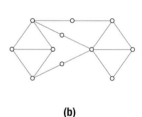

(a)   **(b)**

G. Miscellaneous

45. (a) Give an example of a connected graph with six vertices such that (i) every edge of the graph is a bridge, and (ii) there are exactly two vertices of degree 1.

(b) Give an example of a connected graph with six vertices such that (i) every edge of the graph is a bridge, and (ii) there are exactly three vertices of degree 1.

(c) Give an example of a connected graph with six vertices such that (i) every edge of the graph is a bridge, and (ii) there are exactly five vertices of degree 1.

46. If *G* is a connected graph with no bridges, how many vertices of degree 1 can *G* have? Explain your answer.

47. Suppose we want to trace the following graph with the fewest possible duplicate edges.

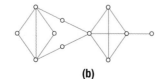

(a) Find an optimal semi-eulerization of the graph that starts at E and ends at H.

(b) Which edges of the graph will have to be retraced?

48. Suppose we want to trace the same graph as in Exercise 47, but now we want to start at B and end at K. Which edges of the graph will have to be retraced?

49. How many times would you have to lift your pencil to trace the following diagram of a tennis court, assuming that you do not trace over any line segment twice? Explain. (This is the problem facing a groundskeeper trying to mark the chalk lines of a tennis court.)

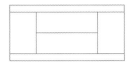

50. (a) Let G be a graph with five vertices and such that every vertex is adjacent to every other vertex, and having no loops or multiple edges. Does G have an Euler circuit? Explain.

(b) Let G be a graph with 10 vertices and such that every vertex is adjacent to every other vertex, and having no loops or multiple edges. Does G have an Euler circuit? Explain.

(c) Let G be a graph with 999 vertices and such that every vertex is adjacent to every other vertex, and having no loops or multiple edges. Does G have an Euler circuit? Explain.

Exercises 51 and 52 are a continuation of Exercises 19 and 20. You should do Exercise 19 before you try Exercise 51, and 20 before you try Exercise 52. A map of the Green Hills subdivision is shown again for your convenience.

51. A night watchman must walk the streets of the Green Hills subdivision and start and end the walk at the corner labeled A. The night watchman needs to walk only once along each block. Find an optimal route for the night watchman. Describe the route by labeling the edges $1, 2, 3, \ldots$ in the order in which they are traveled.

52. A mail carrier must deliver mail on foot along the streets of the Green Hills subdivision and start and end the walk at the post office. The mail carrier must walk along each block twice (once for each side of the street) except for blocks running along the edge of a park, where only one pass is needed. Find an optimal route for the mail carrier. Describe the route by labeling the edges $1, 2, 3, \ldots$ in the order in which they are traveled.

JOGGING

53. (a) Explain why in every graph the sum of the degrees of all the vertices equals twice the number of edges.

(b) Explain why every graph must have an even number of odd vertices.

54. Suppose a connected graph G has k odd vertices and you want to trace all of its edges. Assuming that you would not trace over any edges twice, what is the least number of times that you would have to lift your pencil? Explain.

55. **Regular graphs.** A graph is called *regular* if every vertex has the same degree. Let G be a connected regular graph with N vertices.

(a) Explain why if N is odd, G must have an Euler circuit.

(b) When N is even, G may or may not have an Euler circuit. Give examples of both situations.

56. **Complete bipartite graphs.** A complete bipartite graph is a graph having the property that the vertices of the graph can be divided into two groups A and B and each vertex in A is adjacent to each vertex in B, as shown in the figure. Two vertices in A are never adjacent, and neither are two vertices in B. Let m and n denote the number of vertices in A and B, respectively.

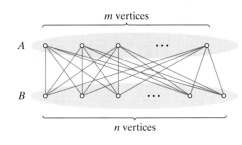

(a) Explain why if m and n are both even, then the complete bipartite graph has an Euler circuit.

(b) Explain why if $m = 2$ and n is odd, then the complete bipartite graph has an Euler path.

57. Explain why a connected graph having all even vertices cannot have any bridges.

58. The following is a map of downtown Kingsburg, showing the Kings River running through the downtown area and the three islands (*A*, *B*, and *C*) connected to each other and both banks by seven bridges. (See Exercise 17.) The Chamber of Commerce wants to design a walking tour that crosses all the bridges. Assume that each time you cross a bridge, it costs $5.00. Find an optimal (cheapest) route that crosses each of the bridges at least once.

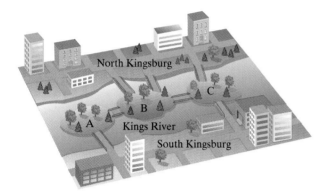

59. A policeman has to patrol along the streets of the subdivision shown in the following figure. The policeman wants to start his trip at the police station (located at *X*) and end the trip at his home (located at *Y*). He needs to cover each block of the subdivision at least once and at the same time he wants to duplicate the fewest possible number of blocks.

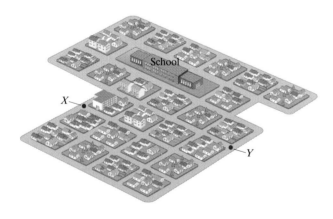

(a) How many blocks will he have to duplicate in an optimal trip through the subdivision?

(b) Describe an optimal trip through the subdivision. Label the edges 1, 2, 3, . . . in the order the policeman would travel them.

60. Consider the following game. You are given *N* vertices and required to build a graph by adding edges connecting these vertices. Each time you add an edge you must pay $1. You can stop when the graph is connected.

(a) Describe the strategy that will cost you the least money.

(b) What is the minimum amount of money needed to build the graph? (Give your answer in terms of *N*.)

61. Consider the following game. You are given *N* vertices and allowed to build a graph by adding edges connecting these vertices. For each edge you can add, you make $1. You are not allowed to add loops or multiple edges, and you must stop before the graph is connected (i.e., the graph you end up with must be disconnected).

(a) Describe the strategy that will give you the most money.

(b) What is the most money you can make building the graph? (Give your answer in terms of *N*.)

Exercises 62 and 63 refer to the problem of the bridges of Madison County discussed in Examples 5.4 and 5.21. A graph model for this problem is shown in the following figure.

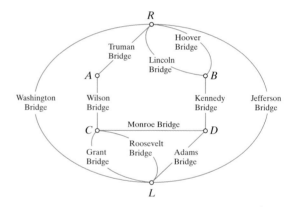

62. Describe an optimal route for the photographer if
(a) the route must start and end at *D* and the first bridge crossed must be the Adams Bridge
(b) the route must start and end at the same vertex, the first bridge crossed must be the Adams bridge, and the last bridge crossed must be the Grant Bridge

63. Describe an optimal route for the photographer if
(a) there are no restrictions on the starting and ending vertices
(b) the route must start at *B* and end at *L*

64. In his original paper on the Königsberg bridge problem (see reference 7), Euler posed the following question:

Let us take an example of two islands with four rivers forming the surrounding water. There are fifteen bridges marked a, b, c, d, etc., across the water around the islands and the adjoining rivers. The question is whether a journey can be arranged that will pass over all the bridges but not over any of them more than once.

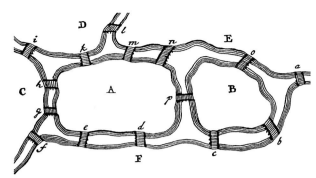

Give an answer to Euler's question. If the journey is possible, describe it. If it isn't, explain why not.

RUNNING

65. Suppose G is a connected graph with N vertices such that every vertex is even, and let k denote the number of bridges in G. Find all the possible values of k. Explain your answer.

66. Suppose G is a connected graph with $N - 2$ even vertices and 2 odd vertices, and let k denote the number of bridges in G. Find all the possible values of k. Explain your answer.

67. Let G denote a connected graph with N vertices ($N \geq 4$) such that every vertex has degree 2 or higher, and at least one of the vertices has degree 3 or higher.

  (a) Give an example of a graph G satisfying the preceding description and having exactly two circuits.

  (b) Explain why any graph satisfying the preceding description must have at least two circuits.

68. Suppose G and H are two graphs that have no common vertices and such that each graph has an Euler circuit. Let J be a (single) graph consisting of the graphs G, H, and one additional edge joining one of the vertices of G to one of the vertices of H. Explain why the graph J has no Euler circuit but does have an Euler path.

69. Suppose G is a disconnected graph with exactly two odd vertices. Explain why the two odd vertices must be in the same component of the graph.

70. Suppose G is a graph with N vertices ($N \geq 2$) with no loops and no multiple edges. Explain why G must have at least two vertices of the same degree.

71. Complements. The *complement* of a graph G is a graph having the same vertex set as G but consisting of exactly those edges that are not edges of G. (If AB is an edge in G, then it is *not* an edge in the complement; if AB is *not* an edge in G, then it is an edge in the complement.) Let N denote the number of vertices in G.

  (a) Suppose N is even ($N \geq 4$) and G has an Euler circuit. Explain why the complement of G cannot have an Euler circuit.

  (b) Suppose N is odd ($N \geq 5$) and G has an Euler circuit. Explain why the complement of G may or may not have an Euler circuit.

72. Kissing circuits. Two circuits in a graph are said to be *kissing* if they have no common edges but have at least one common vertex.

  (a) For the graph shown in the following figure, find a circuit kissing the circuit A, D, C, A (there is only one), and find two different circuits kissing the circuit A, B, D, A.

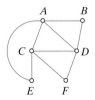

  (b) Suppose that G is a connected graph and every vertex in G is even. Let C be a circuit in G. Explain why the following statement is true: If C has no kissing circuits in G, then C must be an Euler circuit.

73. Hierholzer's algorithm. Another algorithm for finding an Euler circuit in a graph is known as *Hierholzer's algorithm*. The basic idea behind Hierholzer's algorithm is to start with an arbitrary circuit and then enlarge it by patching to it a *kissing circuit*. (For the definition of kissing circuits, see Exercise 72.) The formal description of Hierholzer's algorithm is as follows:

  ■ **Step 0.** Start with an arbitrary circuit C_0.

  ■ **Step 1.** Find a circuit C_0^* that kisses C_0 at a vertex v. The two circuits can be combined into the larger circuit C_1 (start at v, travel along C_0 back to v, then travel along C_0^* and end back at v). If there are no kissing circuits to C_0, then we are finished—C_0 is the Euler circuit. [See Exercise 72(b).]

  ■ **Step 2, 3, etc.** Repeat Step 1 until an Euler circuit is found.

  (a) Use Hierholzer's algorithm to find an Euler circuit for the graph shown in the figure (this is the graph in Example 5.13).

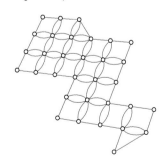

  (b) Suppose G is a connected graph with two odd vertices v and w and all other vertices even. Describe a modification of Hierholzer's algorithm for finding an Euler path in G.

Projects and Papers

A. Original Sources

Whenever possible, it is instructive to read about a great discovery from the original source. Euler's landmark paper in graph theory with his solution to the Königsberg bridge problem was published in 1736. Luckily, the paper was written in a very readable style and the full English translation is quite accessible—it appears as an article in *Scientific American* (reference 7) and in Newman's book *The World of Mathematics* (reference 9).

Write a summary/analysis of Euler's original paper. Include (i) a description of how Euler originally tackled the Königsberg bridge problem, (ii) a discussion of Euler's general conclusions, and (iii) a discussion of Euler's approach toward finding an Euler circuit/path.

B. Computer Representation of Graphs

In many real-life routing problems we have to deal with very large graphs—a graph could have thousands of vertices and tens of thousands of edges. In these cases, algorithms such as Fleury's algorithm (as well as others we will study in later chapters) are done by computer. Unlike humans, computers are not very good at interpreting pictures, so the first step in using computers to perform computations with graphs is to describe the graph in a way that the computer can understand it. The two most common ways to do so are by means of matrices.

In this project you are asked to write a short research paper describing the use of matrices to represent graphs. Explain (i) what is a **matrix**, (ii) what is the **adjacency matrix** of a graph, and (iii) what is the **incidence matrix** of a graph. Illustrate some of the graph concepts from this chapter (degrees of vertices, multiple edges, loops, etc.) in matrix terms. Include plenty of examples. You can find definitions and information on adjacency and incidence matrices of graphs in many graph theory books (see, for example, references 5 and 14).

C. The Chinese Postman Problem

A *weighted graph* is a graph in which the edges are assigned positive numbers called weights. The weights represent distances, times, or costs. Finding optimal routes that cover *all* the edges of a *weighted graph* is a problem known as the *Chinese postman problem*. (Chinese postman problems are a generalization of *Euler circuit* problems and, as a general rule, are much harder to solve, but most of the concepts developed in this chapter still apply.)

In this project you are asked to prepare a presentation on the Chinese postman problem for your class.

Some suggestions: (i) Give several examples to illustrate Chinese postman problems and how they differ from corresponding Euler circuit problems. (ii) Describe some possible real-life applications of Chinese postman problems. (iii) Discuss how to solve a Chinese postman problem in the simplest cases when the weighted graph has no odd vertices or has only two odd vertices (these cases can be solved using techniques learned in this chapter). (iv) Give a rough outline of how one might attempt to solve a Chinese postman problem for a graph with four odd vertices.

References and Further Readings

1. Beltrami, E., *Models for Public Systems Analysis*. New York: Academic Press, Inc., 1977.
2. Beltrami, E., and L. Bodin, "Networks and Vehicle Routing for Municipal Waste Collection," *Networks*, 4 (1974), 65–94.
3. Bogomolny, A., "Graphs," *http://www.cut-the-knot.com/do_you_know/graphs.shtml*.
4. Chartrand, Gary, *Graphs as Mathematical Models*. Belmont, CA: Wadsworth Publishing Co., 1977.
5. Chartrand, Gary, and Ortud R. Oellerman, *Applied and Algorithmic Graph Theory*. New York: McGraw-Hill, 1993.
6. Dunham, William, *Euler: The Master of Us All*. Washington, DC: The Mathematical Association of America, 1999.
7. Euler, Leonhard, "The Königsberg Bridges," trans. James Newman, *Scientific American*, 189 (1953), 66–70.
8. Minieka, E., *Optimization Algorithms for Networks and Graphs*. New York: Marcel Dekker, Inc., 1978.

9. Newman, James R., *The World of Mathematics*, vol. 1. New York: Simon & Schuster, 1956.

10. Roberts, Fred S., "Graph Theory and Its Applications to Problems of Society," *CBMS-NSF Monograph No. 29*. Philadelphia: Society for Industrial and Applied Mathematics, 1978, chap. 8.

11. Tucker, A. C., and L. Bodin, "A Model for Municipal Street-Sweeping Operations," in *Modules in Applied Mathematics*, Vol. 3, eds. W. Lucas, F. Roberts, and R. M. Thrall. New York: Springer-Verlag, 1983, 76–111.

12. Stein, S. K., *Mathematics: The Man-Made Universe* (3rd ed). New York: Dover, 2000.

13. Steinhaus, H., *Mathematical Snapshots* (3rd ed). New York: Dover, 1999.

14. West, Douglas, *Introduction to Graph Theory*. Upper Saddle River, NJ: Prentice Hall, 1996.

6

The Traveling Salesman Problem

Hamilton Joins the Circuit

To those of us who get a thrill out of planning for a big trip, the next decade holds a truly special treat: We humans are on our way to Mars. Why are we going to Mars, and what are we doing there?

For starters, Mars is the next great frontier for human exploration. Its promise lies in the fact that it is the most earthlike of all the planets in the solar system, the Martian soil is rich in chemicals and minerals, and, most important of all, there is strong evidence that there may be water just below the surface. The ultimate attraction, however, is the hope of finding life on Mars. Of all the planets in our solar system, Mars is the most likely place to show some evidence of life—probably primitive bacterial forms buried inside Martian rocks or under the Martian surface. But finding these tiny Martians—assuming they exist—raises many technical and logistical questions. What are the best places in Mars to explore? How do we get the equipment there? How long will it take to do the job? And, above all, how much will it cost? Once again, lurking behind the complexities of Mars exploration is an interesting and important routing problem.

Here are the details in a nutshell. Based on geological data already collected from earlier orbiting missions, NASA has identified a set of sites on the surface of Mars where the likelihood of finding evidence of past life is the highest (see Fig. 6-1). In January 2004, two rovers named *Spirit* and *Opportunity* successfully landed on Mars, returning incredible pictures and a lot of scientific data (so far, no evidence of life). But *Spirit* and *Opportunity* have limited mobility and were designed to only explore the area near their respective landing sites. A main goal for the next stage of Mars exploration is a *robotic sample-return mission*, in which a lander would land at one of the designated sites and release an unmanned rover controlled from Earth. The rover would then

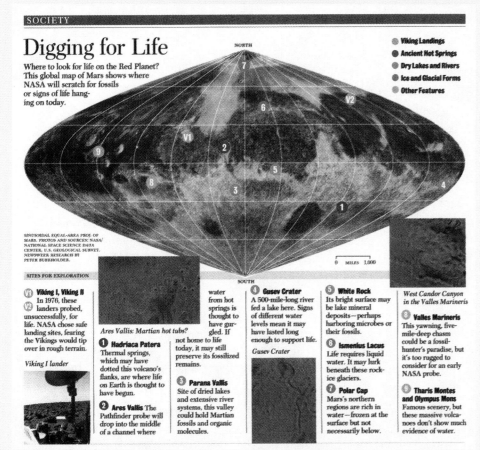

SOCIETY

Digging for Life

Where to look for life on the Red Planet? This global map of Mars shows where NASA will scratch for fossils or signs of life hanging on today.

- ⊙ **Viking Landings**
- ● **Ancient Hot Springs**
- ⊙ **Dry Lakes and Rivers**
- ⊙ **Ice and Glacial Forms**
- ⊙ **Other Features**

SINUSOIDAL EQUAL-AREA PROJ. OF MARS. PHOTOS AND SOURCES: NASA/ NATIONAL SPACE SCIENCE DATA CENTER, U.S. GEOLOGICAL SURVEY. NEWSWEEK RESEARCH BY PETER BURKHOLDER.

0   MILES   1,000

SITES FOR EXPLORATION

V1 V2 Viking I, Viking II In 1976, these landers probed, unsuccessfully, for life. NASA chose safe landing sites, fearing the Vikings would tip over in rough terrain.

Viking I lander

Ares Vallis: Martian hot tubs?

1 Hadriaca Patera Thermal springs, which may have dotted this volcano's flanks, are where life on Earth is thought to have begun.

2 Ares Vallis The Pathfinder probe will drop into the middle of a channel where water from hot springs is thought to have gurgled. If not home to life today, it may still preserve its fossilized remains.

3 Parana Vallis Site of dried lakes and extensive river systems, this valley could hold Martian fossils and organic molecules.

4 Gusev Crater A 500-mile-long river fed a lake here. Signs of different water levels mean it may have lasted long enough to support life.

Gusev Crater

5 White Rock Its bright surface may be lake mineral deposits—perhaps harboring microbes or their fossils.

6 Ismenius Lacus Life requires liquid water. It may lurk beneath these rock-ice glaciers.

7 Polar Cap Mars's northern regions are rich in water—frozen at the surface but not necessarily below.

West Candor Canyon in the Valles Marineris

8 Valles Marineris This yawning, five-mile-deep chasm could be a fossil-hunter's paradise, but it's too rugged to consider for an early NASA probe.

9 Tharis Montes and Olympus Mons Famous scenery, but these massive volcanoes don't show much evidence of water.

FIGURE 6-1

Source: NASA/National Space Science Data Center, U.S. Geological Survey. *Newsweek* research by Peter Burkholder. Copyright Newsweek, Inc. All rights reserved. Reprinted by permission.

travel to each of the other sites, collecting soil samples and performing experiments. After all the sites have been visited, the rover would return to the landing site where a return rocket would bring the samples back to Earth. The first of these sample-return missions is scheduled to be launched in 2014.

Figure 6-2 shows one of the many possible routes that the rover might take as it travels to each of the sites to look for life and collect soil samples. And there are hundreds of other possible ways to route the rover. Which one is the best?

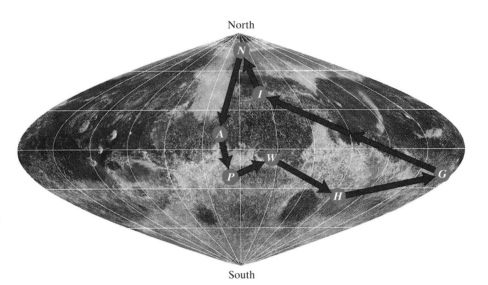

FIGURE 6-2 *A*: Ares Vallis (starting and ending point); *P*: Parana Vallis; *W*: White Rock; *H*: Hadriaca Patera; *G*: Gusev Crater; *I*: Ismenius Lacus; *N*: North Polar Cap.

Finding an optimal route for an unmanned Martian rover is an exotic illustration of a class of routing problems that go by the generic name of TSPs. The acronym TSP stands for *traveling salesman problem*, so called because the classic version of this problem is a traveling salesman trying to determine the cheapest route that passes through each of several cities, returning at the end of the trip to his hometown. The name has stuck as a generic name given to all sorts of equivalent problems, even if they have nothing to do with traveling salesmen. It is best to think of the "traveling salesman" as just a metaphor—a rover searching for the best route on Mars, a UPS driver trying to find the best way to deliver packages around town, or any ordinary Joe trying to plan the fastest route by which to run a bunch of errands on a Saturday morning.

The two broad goals of this chapter are (i) to develop an understanding of what a TSP is (Sections 6.1, 6.2, and 6.3), and (ii) to understand what it means to "solve" a TSP (Sections 6.4 through 6.8). As usual, to accomplish these goals a little theory is necessary, and several important graph concepts are introduced and discussed along the way—*Hamilton circuits* (Section 6.1), *complete graphs* (Section 6.2), *efficient algorithms* (Section 6.5), *optimal and approximate algorithms* (Section 6.6). In addition, some of the basic algorithms for solving TSPs are introduced and discussed in the second half of the chapter—the *brute-force* and *nearest-neighbor* algorithms in Section 6.5, the *repetitive nearest-neighbor* algorithm in Section 6.7, and the *cheapest-link* algorithm in Section 6.8.

6.1 Hamilton Circuits and Hamilton Paths

Hamilton paths and circuits are named after the Irish mathematician Sir William Rowan Hamilton (1805–1865). For more on Hamilton see the biographical profile at the end of this chapter.

A **Hamilton path** in a graph is a *path* that visits each *vertex* of the graph once and only once. Likewise, a **Hamilton circuit** in a graph is a *circuit* that visits each *vertex* of the graph once and only once (at the end, of course, the circuit must return to the starting vertex).

From the outset, it's important to understand that although their definitions sound almost identical, a Hamilton path (circuit) and an Euler path (circuit) are completely different concepts. In the former we require that every *vertex* of the graph be part of the path (circuit); in the latter we require that every *edge* of the graph be part of the path (circuit). The difference is a single word, but oh, what a difference it makes!

The main purpose of our first example is to highlight the differences between Hamilton circuits (paths) and Euler circuits (paths).

> **EXAMPLE 6.1 Hamilton Circuits (Paths) Versus Euler Circuits (Paths)**

How are Hamilton circuits (paths) related to Euler circuits (paths)? The various graphs in Fig. 6-3 show that there is essentially no relation between the two concepts.

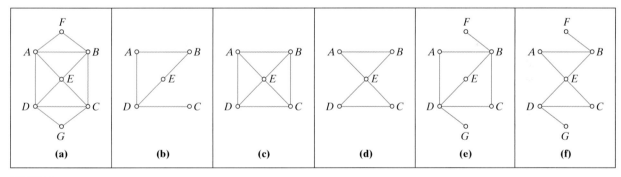

FIGURE 6-3

- Figure 6-3(a) shows a graph that (i) has Euler circuits (the vertices are all even), (ii) has Hamilton circuits. One such Hamilton circuit is *A, F, B, C, G, D, E, A*—there are plenty more. Note that once a graph has a Hamilton circuit, it automatically has a Hamilton path—the Hamilton circuit can always be truncated into a Hamilton path by dropping the last vertex of the circuit. For example, the Hamilton circuit *A, F, B, C, G, D, E, A* can be truncated into the Hamilton path *A, F, B, C, G, D, E*. (Contrast this with the mutually exclusive relationship between Euler circuits and paths—if a graph has an Euler circuit it cannot have an Euler path and vice versa.)

- Figure 6-3(b) shows a graph that (i) has no Euler circuits but does have Euler paths (for example *C, D, E, B, A, D*), (ii) has no Hamilton circuits (sooner or later you have to go to *C*, and then you are stuck) but does have Hamilton paths (for example, *A, B, E, D, C*). Aha, a graph can have a Hamilton path but no Hamilton circuit!

- Figure 6-3(c) shows a graph that (i) has neither Euler circuits nor paths (it has four odd vertices), (ii) has Hamilton circuits (for example A, B, C, D, E, A—there are plenty more), and consequently has Hamilton paths (for example, A, B, C, D, E).
- Figure 6-3(d) shows a graph that (i) has Euler circuits (the vertices are all even), (ii) has no Hamilton circuits (no matter what, you are going to have to go through E more than once!) but has Hamilton paths (for example, A, B, E, D, C).
- Figure 6-3(e) shows a graph that (i) has no Euler circuits but has Euler paths (F and G are the two odd vertices), (ii) has neither Hamilton circuits nor Hamilton paths.
- Figure 6-3(f) shows a graph that (i) has neither Euler circuits nor Euler paths (too many odd vertices), (ii) has neither Hamilton circuits nor Hamilton paths.

Table 6-1 summarizes the preceding observations.

See Exercise 9.

See Exercise 10.

TABLE 6-1 Existence of Euler and/or Hamilton Circuits/Paths

	Euler circuit	Euler path	Hamilton circuit	Hamilton path
Fig. 6-3(a)	Yes	No	Yes	Yes
Fig. 6-3(b)	No	Yes	No	Yes
Fig. 6-3(c)	No	No	Yes	Yes
Fig. 6-3(d)	Yes	No	No	Yes
Fig. 6-3(e)	No	Yes	No	No
Fig. 6-3(f)	No	No	No	No

The lesson of Example 6.1 is that the existence of an Euler path or circuit in a graph tells us nothing about the existence of a Hamilton path or circuit in that graph. This is important because it implies that Euler's circuit and path theorems from Chapter 5 are useless when it comes to Hamilton circuits and paths. But surely, there must be analogous "Hamilton circuit and path theorems" that we could use to determine if a graph has a Hamilton circuit, a Hamilton path, or neither. Surprisingly, no such theorems exist. Determining when a given graph does or does not have a Hamilton circuit or path can be very easy, but it also can be very hard—it all depends on the graph. [There are, however, nice theorems that identify special situations where a graph must have a Hamilton circuit. The best known of these theorems is *Dirac's theorem: If a connected graph has N vertices ($N > 2$) and all of then have degree bigger or equal to $N/2$, then the graph has a Hamilton circuit.*]

6.2 Complete Graphs

Sometimes the question, *Does the graph have a Hamilton circuit?* has an obvious *yes* answer, and the more relevant question turns out to be, *How many different Hamilton circuits does it have?* In this section we will answer this question for an important family of graphs called complete graphs.

A graph with N vertices in which *every* pair of distinct vertices is joined by an edge is called a **complete graph** on N vertices and denoted by the symbol K_N. Figure 6-4 shows the complete graphs on three, four, five, and six vertices, respectively.

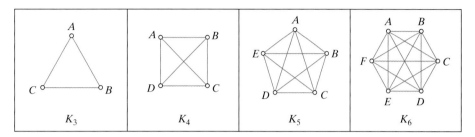

FIGURE 6-4

One of the key properties of K_N is that every vertex has degree $N - 1$. This implies that the sum of the degrees of all the vertices is $N(N - 1)$, and it follows from Euler's sum of degrees theorem (see page 174) that the number of edges in K_N is $N(N - 1)/2$. For a graph with N vertices and no multiple edges or loops $N(N - 1)/2$ is the maximum number of edges possible, and this maximum can only occur when the graph is K_N.

Number of Edges in K_N

- K_N has $N(N - 1)/2$ edges.
- Of all graphs with N vertices and no multiple edges or loops, K_N has the most edges.

Because K_N has a complete set of edges (every vertex is connected to every other vertex), it also has a complete set of Hamilton circuits—you can travel the vertices in *any* sequence you choose and you will not get stuck.

> **EXAMPLE 6.2** Hamilton Circuits in K_4

If we travel the four vertices of K_4 in an arbitrary order, we get a Hamilton path. For example, C, A, D, B is a Hamilton path [Fig. 6-5(a)]; D, C, A, B is another one [Fig. 6-5(b)], and so on. Each of these Hamilton paths can be closed into a Hamilton circuit—the path C, A, D, B begets the circuit C, A, D, B, C [Fig. 6-5(c)]; the path D, C, A, B begets the circuit D, C, A, B, D [Fig. 6-5(d)], and so on. It looks like we have an abundance of Hamilton circuits, but it is important to remember that the same Hamilton circuit can be written in many ways. For example, C, A, D, B, C is the same circuit as A, D, B, C, A [Fig. 6-5(c) describes either one]—the only difference is that in the first case we used C as the *reference point*; in the second case we used A as the reference point. There are two additional sequences that describe this same Hamilton circuit—D, B, C, A, D (with reference point D)

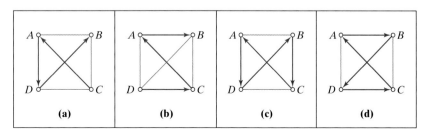

FIGURE 6-5

and B, C, A, D, B (with reference point B). Taking all of this into account, there are six different Hamilton circuits in K_4, as shown in Table 6-2 (the table also shows the four different ways each circuit can be written).

TABLE 6-2 The Hamilton Circuits of K_4

	Reference point is A	Reference point is B	Reference point is C	Reference point is D
1	A, B, C, D, A	B, C, D, A, B	C, D, A, B, C	D, A, B, C, D
2	A, B, D, C, A	B, D, C, A, B	C, A, B, D, C	D, C, A, B, D
3	A, C, B, D, A	B, D, A, C, B	C, B, D, A, C	D, A, C, B, D
4	A, C, D, B, A	B, A, C, D, B	C, D, B, A, C	D, B, A, C, D
5	A, D, B, C, A	B, C, A, D, B	C, A, D, B, C	D, B, C, A, D
6	A, D, C, B, A	B, A, D, C, B	C, B, A, D, C	D, C, B, A, D

EXAMPLE 6.3 Hamilton Circuits in K_5

Let's try to list all the Hamilton circuits in K_5. For simplicity, we will write each circuit just once, using a common reference point—say A. (As long as we are consistent, it doesn't really matter which reference point we pick.) Each of the Hamilton circuits will be described by a sequence that starts and ends with A, with the letters $B, C, D,$ and E sandwiched in between in some order. There are $4 \times 3 \times 2 \times 1 = 24$ different ways to shuffle the letters $B, C, D,$ and E, each producing a different Hamilton circuit. The complete list of the 24 Hamilton circuits in K_5 is shown in Table 6-3. The table is laid out so that each of the circuits in the

TABLE 6-3 The Hamilton Circuits of K_5 (Using A as the Reference Point)

1	A, B, C, D, E, A	13	A, E, D, C, B, A
2	A, B, C, E, D, A	14	A, D, E, C, B, A
3	A, B, D, C, E, A	15	A, E, C, D, B, A
4	A, B, D, E, C, A	16	A, C, E, D, B, A
5	A, B, E, C, D, A	17	A, D, C, E, B, A
6	A, B, E, D, C, A	18	A, C, D, E, B, A
7	A, C, B, D, E, A	19	A, E, D, B, C, A
8	A, C, B, E, D, A	20	A, D, E, B, C, A
9	A, C, D, B, E, A	21	A, E, B, D, C, A
10	A, C, E, B, D, A	22	A, D, B, E, C, A
11	A, D, B, C, E, A	23	A, E, C, B, D, A
12	A, D, C, B, E, A	24	A, E, B, C, D, A

table is directly opposite its *mirror-image circuit* (the circuit with vertices listed in reverse order). Although they are close relatives, a circuit and its mirror image are not considered the same circuit.

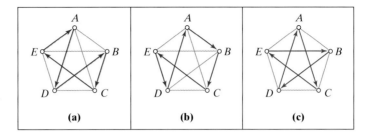

FIGURE 6-6 Three of the 24 Hamilton circuits in K_5. (Try to find them in Table 6.3.)

We can now generalize our observations in Examples 6.2 and 6.3. Suppose we have K_N, the complete graph with N vertices. We pick one of the vertices—call it A if you please—and make it the designated reference point for all the Hamilton circuits. Every Hamilton circuit will then take the form $A, *, *, \ldots, *, A$ (the *'s are wildcards that denote the remaining $N - 1$ vertices), and with each reordering of the *'s we get a different Hamilton circuit. This means that the question, "What is the number of Hamilton circuits in K_N?" boils down to the equivalent question, "How many different ways are there to rearrange the $(N - 1)$ vertices represented by the *'s?" The answer, as you may recall from Chapter 2, is given by the number $1 \times 2 \times 3 \times \cdots \times (N - 1)$, called the **factorial** of $(N - 1)$ and written as $(N - 1)!$ for short.

If this is your first exposure to factorials (or you need a refresher), it might be a good idea to stop here, read the brief discussion of factorials in Chapter 2, and then take a stab at Exercises 17 through 24.

Number of Hamilton Circuits in K_N

There are $(N - 1)!$ distinct Hamilton circuits in K_N.

TABLE 6-4 Number of Distinct Hamilton Circuits in K_N

N	(N − 1)!	N	(N − 1)!
2	1	12	39,916,800
3	2	13	479,001,600
4	6	14	6,227,020,800
5	24	15	87,178,291,200
6	120	16	1,307,674,368,000
7	720	17	20,922,789,888,000
8	5040	18	355,687,428,096,000
9	40,320	19	6,402,373,705,728,000
10	362,880	20	121,645,100,408,832,000
11	3,628,800		

Table 6-4 shows the number of Hamilton circuits in complete graphs with up to $N = 20$ vertices. The main point of Table 6-4 is to convince you that as we increase the number of vertices, the number of Hamilton circuits in the complete graph goes through the roof. Even a relatively small graph such as K_8 has over five thousand Hamilton circuits. Double the number of vertices to K_{16} and the number of Hamilton circuits exceeds 1.3 *trillion*. Double the number of vertices again to K_{32} and the number of Hamilton circuits—about *eight billion trillion trillion*—is so large that it defies ordinary human comprehension. (This incredible growth in the number of Hamilton circuits in K_N is not just an idle observation and it will become very relevant to us soon.)

6.3 Traveling Salesman Problems

The title *Traveling Salesman Problem* is catchy but a bit misleading, since most of the time the problems that fall under this heading have nothing to do with salespeople living out of a suitcase. The "traveling salesman" is a convenient metaphor for many different important real-life applications, all involving Hamilton circuits in complete graphs but only occasionally involving traveling salespeople. The next few examples illustrate a few of the many possible settings for a "traveling salesman" problem, starting, of course, with the traveling salesman's "traveling salesman" problem. (From here on we are following custom and dropping the quotes around *traveling salesman*, but remember—don't take the phrase too literally.)

> **EXAMPLE 6.4** A Tale of Five Cities: Part 1

Meet Willy, a traveling salesman. Willy has customers in five cities, which for the sake of brevity we will call $A, B, C, D,$ and E. Willie needs to schedule a sales trip that will start and end at A (that's Willy's hometown) and goes once through each of the other four cities. Other than starting and ending at A, there are no restrictions as to the sequence in which he visits the other cities.

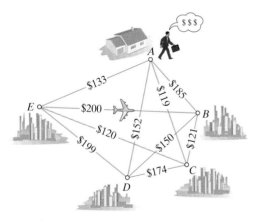

The graph in Fig. 6-7 shows the cost of a *one-way* airline ticket between each pair of cities. (We are making the assumption here that a one-way ticket costs the same regardless of whether you go from city X to city Y or vice versa. In real life this is not always true—today's airline prices have a logic of their own.) Naturally, Willy wants to cut down on his travel expenses as much as possible. What is the optimal (cheapest) Hamilton circuit for the graph in Fig. 6-7? We will return to this question soon.

FIGURE 6-7

> **EXAMPLE 6.5** Probing the Outer Reaches of Our Solar System

It is the year 2020. An expedition to explore the outer planetary moons in our solar system is about to be launched from planet Earth. The expedition is scheduled to visit Callisto, Ganymede, Io, Mimas, and Titan (the first three are moons of Jupiter; the last two, of Saturn), collect rock samples at each, and then return to Earth with the loot.

Figure 6-8 shows the mission time (in years) between any two moons. What is the optimal (shortest) Hamilton circuit for the graph in Fig. 6-8?

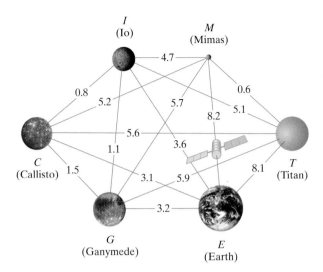

FIGURE 6-8

> ## EXAMPLE 6.6 Roving the Red Planet: Part 1

Figure 6-8 shows seven locations on Mars where NASA scientists believe there is a good chance of finding evidence of life. Imagine that you are in charge of planning a *sample-return* mission. First, you must land an unmanned rover in the Ares Vallis (A). Then you must direct the rover to travel to each site and collect and analyze soil samples. Finally, you must instruct the rover to return to the Ares Vallis landing site, where a return rocket will bring the best samples back to Earth. A trip like this will take several years and cost several billion dollars, so good planning is critical.

Figure 6-9 shows the estimated distances (in miles) that a rover would have to travel to get from one Martian site to another. What is the optimal (shortest) Hamilton circuit for the corresponding graph?

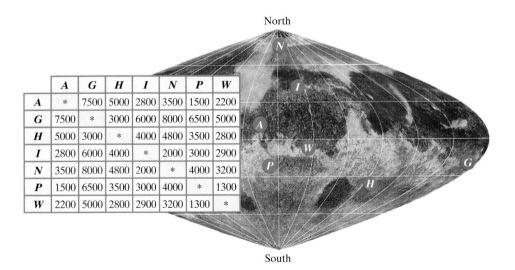

	A	**G**	**H**	**I**	**N**	**P**	**W**
A	*	7500	5000	2800	3500	1500	2200
G	7500	*	3000	6000	8000	6500	5000
H	5000	3000	*	4000	4800	3500	2800
I	2800	6000	4000	*	2000	3000	2900
N	3500	8000	4800	2000	*	4000	3200
P	1500	6500	3500	3000	4000	*	1300
W	2200	5000	2800	2900	3200	1300	*

FIGURE 6-9

Examples 6.4, 6.5, and 6.6 are variations on a single theme. In each case, we are presented with a complete graph whose edges have numbers attached to them. (For Example 6.6 it was easier to put the numbers in a table.) Any graph whose edges have numbers attached to them is called a **weighted graph**, and the numbers are called the **weights** of the edges. The graphs in Examples 6.4, 6.5, and 6.6 are called **complete weighted graphs**. In each example the weights of the graph represent a different variable. In Example 6.4 the weights represent *cost*, in Example 6.5 they represent *time*, and in Example 6.6 they represent *distance*. The problem we want to solve in each example is fundamentally the same: *Find an optimal Hamilton circuit (i.e., a Hamilton circuit with least total weight) for the given weighted graph.*

The following are just a few examples of real-life applications involving optimal Hamilton circuits in weighted graphs.

■ **Routing school buses.** A school bus picks up children in the morning and drops them off at the end of the day at designated stops. On a typical school bus route there may be 20 to 30 such stops, which can be represented by the vertices of a graph. The weight of the edge connecting bus-stop X to bus stop Y represents the time of travel between the two locations (with school buses, time of travel is always the most important variable). Since the bus repeats its route every day during the school year, finding an optimal route is crucial.

■ **Package deliveries.** Companies such as United Parcel Service (UPS) and Federal Express deal with this situation daily. Each truck has packages to deliver to a list of destinations. The travel time between any two delivery locations is known or can be estimated. The object is to deliver the packages to each of the delivery locations and return to the starting point in the least amount of time.

■ **Fabricating circuit boards.** In the process of fabricating integrated circuit boards, tens of thousands of tiny holes must be drilled in each board. This is done by using a stationary laser beam and rotating the board. To do this efficiently, the order in which the holes are drilled should be such that the entire drilling sequence is completed in the least amount of time. This makes for a very high tech TSP. In this TSP the vertices of the graph correspond to the holes on the circuit board and the weight of the edge connecting vertices X and Y is the time needed to rotate the board from drilling position X to drilling position Y.

■ **Scheduling jobs on a machine.** In many industries there are machines that perform multiple jobs. Think of the jobs as the vertices of a graph. After performing job X, the machine needs to be set up to perform another job. The amount of time required to reset the machine to perform job Y is the weight of the edge connecting vertices X and Y. The problem is to schedule the machine to run through all the jobs in a cycle such that the total amount of time is minimized.

■ **Running errands around town.** When we have a lot of errands to run, we like to follow the route that will take us to each of our destinations and then finally home in the shortest amount of time.

The word *weight* has a different meaning here from that in Chapter 2.

See Exercise 51.

6.4 Simple Strategies for Solving TSPs

> **EXAMPLE 6.7** A Tale of Five Cities: Part 2

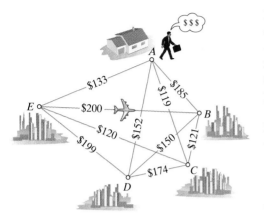

FIGURE 6.10

In Example 6.4 we met Willy the traveling salesman pondering his up-coming sales trip, dollar signs running through his head. (The one-way airfares between any two cities are shown again in Fig. 6-10.) Imagine now that Willy, unwilling or unable to work out the problem for him-self, decides to offer a reward of $50 to anyone who can find the best possible trip. Would it be worth $50 to you to work out this problem? (Yes.) So how would you do it? (At this point you should take a break from your reading, get a pencil and a piece of paper, and try to work out the problem on your own. Give yourself about 15 to 20 minutes.)

...

Welcome back.

If you are like most people, you probably followed one of two strategies in working through this problem.

Strategy 1 (Exhaustive Search)

Make a list of all possible Hamilton circuits. For each circuit in the list, cal-culate the total weight of the circuit. From all the circuits, choose the circuit with smallest total weight.

Table 6-5 shows a detailed implementation of this strategy. The 24 Hamilton circuits are split into two columns consisting of circuits and their mirror images. (Notice that since the one-way airfares are the same in either direction, a circuit and its mirror-image circuit will cost exactly the same. This is a useful shortcut that cuts the number of cal-culations in half.) The total costs for the circuits are shown in the middle column of the table. There are two optimal circuits, with a total cost of $676—$A, D, B, C, E, A$ and its mirror image A, E, C, B, D, A. (There are always going to be at least two opti-mal circuits, since the mirror image of an optimal circuit is also optimal. Either one of them can be used for a solu-tion.) The solution to Willy's travels is shown graphically in Fig. 6-11.

TABLE 6-5 Willy's Possible Routes and Their Costs

	Hamilton circuit	Total cost	Mirror-image circuit
1	A, B, C, D, E, A	$185 + 121 + 174 + 199 + 133 = 812$	A, E, D, C, B, A
2	A, B, C, E, D, A	$185 + 121 + 120 + 199 + 152 = 777$	A, D, E, C, B, A
3	A, B, D, C, E, A	$185 + 150 + 174 + 120 + 133 = 762$	A, E, C, D, B, A
4	A, B, D, E, C, A	$185 + 150 + 199 + 120 + 119 = 773$	A, C, E, D, B, A
5	A, B, E, C, D, A	$185 + 200 + 120 + 174 + 152 = 831$	A, D, C, E, B, A
6	A, B, E, D, C, A	$185 + 200 + 199 + 174 + 119 = 877$	A, C, D, E, B, A
7	A, C, B, D, E, A	$119 + 121 + 150 + 199 + 133 = 722$	A, E, D, B, C, A
8	A, C, B, E, D, A	$119 + 121 + 200 + 199 + 152 = 791$	A, D, E, B, C, A
9	A, C, D, B, E, A	$119 + 174 + 150 + 200 + 133 = 776$	A, E, B, D, C, A
10	A, C, E, B, D, A	$119 + 120 + 200 + 150 + 152 = 741$	A, D, B, E, C, A
11	A, D, B, C, E, A	$152 + 150 + 121 + 120 + 133 = 676$	A, E, C, B, D, A
12	A, D, C, B, E, A	$152 + 174 + 121 + 200 + 133 = 780$	A, E, B, C, D, A

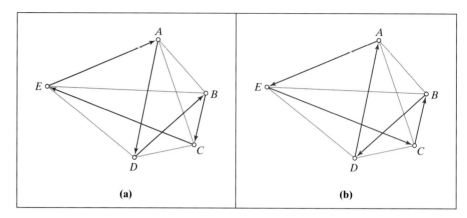

FIGURE 6-11 The two optimal circuits for Willy's trip. (Total travel cost = $676.)

(a)   (b)

All of the preceding could be reasonably done in somewhere between 10 and 20 minutes. Not a bad gig for $50!

Strategy 2 (Go Cheap)

Start from the home city. From there go to the city that is the cheapest to get to. From each new city go to the next new city that is cheapest to get to. When there are no more new cities to go to, go back home.

In Willy's case, this strategy works like this: Start at A. From A Willy goes to C (the cheapest place he can fly to from A). From C Willy goes to E (the cheapest city to fly to other than A). From E Willy goes to D, and from D he has little choice—the last new city to go to is B. From B Willy has to return home to A. The Hamilton circuit describing this trip is shown in Fig. 6-12. The total cost of the trip is $773. [$119 + $120 + $199 + $150 + $185 = $773.]

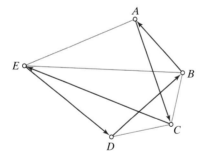

FIGURE 6-12

The go cheap strategy takes a lot less work than the exhaustive search strategy, but there is a hitch—the cost of the "solution" we get is $773, which is $97 more than the solution found under the exhaustive search strategy. Willy says this is not a solution and refuses to pay the $50. The go cheap strategy looks like a bust, but be patient—there is more to come.   ⋘

▷ EXAMPLE 6.8 A Tale of (Gasp!) Ten Cities: Part 1

Let's imagine now that Willy, who has done very well with his business, has expanded his sales territory to 10 cities (let's call them A through K). Willy wants us to help him once again find the cheapest trip that starts at A and goes to each of the other nine cities once. Flush with success and generosity, he is offering a whopping $200 as a reward for a solution to this problem. Should we accept the challenge?

The graph in Fig. 6-13 represents the 10 cities and the fares table shows the one-way fares between them. (With this many edges the graph would get pretty cluttered if we tried to show the weights on the graph itself. It is better to put them in a table.) We now know that a foolproof method for tackling this problem is the Exhaustive Search strategy. But before we plunge into it, let's think of what we are getting into. With 10 cities we have 9! = 362,880 Hamilton circuits. Assuming we could compute the cost of a new circuit every 30 seconds—and that's working fast—it would take about 3000 hours to do all 362,880 possible circuits. Shortcuts? Say you cut the work in half by skipping the mirror-image circuits. That's 1500 hours of work. If you worked nonstop, 24 hours a day, 7 days a week it would still take a couple of months!

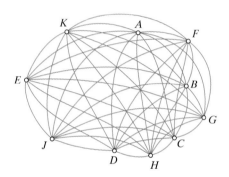

	A	B	C	D	E	F	G	H	J	K
A	*	185	119	152	133	321	297	277	412	381
B	185	*	121	150	200	404	458	492	379	427
C	119	121	*	174	120	332	439	348	245	443
D	152	150	174	*	199	495	480	500	454	489
E	133	200	120	199	*	315	463	204	396	487
F	321	404	332	495	315	*	356	211	369	222
G	297	458	439	480	463	356	*	471	241	235
H	277	492	348	500	204	211	471	*	283	478
J	412	379	245	454	396	369	241	283	*	304
K	381	427	443	489	487	222	235	478	304	*

FIGURE 6-13

Let's now try the Go Cheap strategy, the one that didn't work out so well in Example 6.7. This time we have to implement the strategy using the fare table in Fig. 6-13. The trip would start at A. Starting at the A-row of the table, we search for the smallest number in the row. It is 119, located under the C-column. This means that the trip should start with a flight from A to C. Now we go to the C-row in the table searching for the smallest number other than the 119 (A is already taken). That number is 120, located under the E-column. This means that the next segment of the trip is from C to E. Now we go to the E-row and locate the smallest number other than the numbers under the A and C columns (don't want to go back to either of those cities). That number is 199, located under the D-column. The process can be continued, one city at a time, and you are encouraged to finish this on your own—it shouldn't take more than a few minutes. The final result is the Hamilton circuit $A, C, E, D, B, J, G, K, F, H, A$, with a total cost of $2153.

So we now have a Hamilton circuit, but is it the optimal Hamilton circuit? If it isn't—is it at least close? Willy refuses to pay the $200 unless these questions can be answered to his satisfaction. We will settle this issue later in the chapter.

A good way to keep track of the cities that have been visited is to cross out their columns as we move on.

6.5 The Brute-Force and Nearest-Neighbor Algorithms

In this section we will look at the two strategies we informally developed in connection with Willy's sales trips and recast them in the form of precise *algorithms*. The *Exhaustive Search* strategy can be formalized into an algorithm generally known as the **brute-force algorithm**; the *Go Cheap* strategy can be formalized into an algorithm known as the **nearest-neighbor algorithm**. In both cases, the objective of the algorithm is to find an *optimal* (cheapest, shortest, fastest) Hamilton circuit in a complete weighted graph.

Algorithm 1: The Brute-Force Algorithm

- **Step 1.** Make a list of *all* the possible Hamilton circuits of the graph.
- **Step 2.** For each Hamilton circuit calculate its *total weight* (i.e., add the weights of all the edges in the circuit).
- **Step 3.** Choose an *optimal* circuit (there is always more than one optimal circuit to choose from!).

Algorithm 2: The Nearest-Neighbor Algorithm

- **Start:** Start at the designated starting vertex. If there is no designated starting vertex, pick any vertex.
- **First step:** From the starting vertex go to its *nearest neighbor* (i.e., the vertex for which the corresponding edge has the smallest weight).
- **Middle steps:** From each vertex go to its nearest neighbor, choosing only among *the vertices that haven't been yet visited*. (If there is more than one nearest neighbor, choose among them at random.) Keep doing this until all the vertices have been visited.
- **Last step:** From the last vertex return to the starting vertex.

Both of the preceding algorithms have some pros and some cons, which we will discuss next. The positive aspect of the brute-force algorithm is the fact that it is *optimal*. (An **optimal** algorithm is an algorithm that, when correctly implemented, is guaranteed to produce an optimal solution.) In the case of the brute-force algorithm, we know we are getting an optimal solution because we are choosing from among *all* possible Hamilton circuits.

The negative aspect of the brute-force algorithm is the amount of effort that goes into implementing the algorithm, which is (roughly) *proportional to the number of Hamilton circuits* that need to be checked. As we first saw in Table 6-4, as the number of vertices grows just a little, the number of Hamilton circuits in a complete graph grows at an incredibly fast rate.

What does this mean in practical terms? Let's start with human computation. To solve a TSP for a graph with 10 vertices (as real-life problems go, that's puny) the brute-force algorithm requires checking 362,880 Hamilton circuits. To do this by hand, even for a fast and clever human, it would take over 1000 hours. Thus, at $N = 10$ we are already beyond the limit of what can be considered reasonable human effort. A possible way to get around this difficulty, we might think, is to recruit a fast helper, such as a powerful computer. (In principle, this is a good idea, since a computer is the perfect tool to implement the brute-force algorithm—the algorithm is essentially a mindless exercise in arithmetic with a little bookkeeping

TABLE 6-6

N	SUPERHERO computation time
20	2 minutes
21	40 minutes
22	14 hours
23	13 days
24	10 months
25	20 years
26	500 years
27	13,000 years
28	350,000 years
29	9.8 million years
30	284 million years

thrown in, and both of these are things that computers are very good at.) So let's take a hopeful look at machine computation. Let's imagine, for the sake of argument, that we have unlimited free access to SUPERHERO—the fastest super-computer on the planet—which can compute *a quadrillion* Hamilton circuits per second. (One quadrillion equals a million billion. This is much faster than even the fastest current supercomputers can perform, but since we are just fantasizing, let's think big!) For N less than 20, SUPERHERO can run through the $(N - 1)!$ Hamilton circuits in a matter of seconds (or less). Things get more interesting when we start considering what happens beyond $N = 20$. Table 6-6 shows the approximate computation time required by SUPERHERO when implementing the brute-force algorithm for values of N ranging between 20 and 30.

The surprising lesson of Table 6-6 is that even with the world's best technology on our side, we very quickly reach the point beyond which using the brute-force algorithm is completely unrealistic (clearly for $N \geq 25$ and some might argue even for $N = 24$).

The brute-force algorithm is a classic example of what is formally known as an **inefficient algorithm**—an algorithm for which the number of steps needed to carry it out grows disproportionately with the size of the problem. The trouble with inefficient algorithms is that they are of limited practical use—they can realistically be carried out only when the problem is small.

The extraordinary computational effort required by the brute-force algorithm is a direct consequence of the way factorials grow—each time we increase the number of vertices of the graph from N to $N + 1$, *the amount of work required to carry out the brute-force algorithm increases by a factor of N.* For example, it takes 5 times as much work to go from 5 vertices to 6, 10 times more work to go from 10 vertices to 11, and 100 times as much work to go from 100 vertices to 101.

Fortunately, not all algorithms are inefficient. Let's discuss now the nearest-neighbor algorithm, in which we hop from vertex to vertex using a simple criterion: Choose the next available "nearest" vertex and go for it. Let's call the process of checking among the available vertices and finding the nearest one a single computation. Then, for a TSP with $N = 5$, we need to perform 5 computations. (Actually, the last step [go back to the starting vertex] is automatic—but for the sake of simplicity let's still call it a computation.) What happens when we double the number of vertices to $N = 10$? We now have to perform 10 computations. What if we jump to $N = 30$? Now we have to perform roughly 30 computations. Think about that—even if we are conservative and assume a couple of minutes per computation—a human being without a computer can implement the nearest-neighbor algorithm on a TSP with $N = 30$ vertices in an hour or less! Compare that with the last row of Table 6-6.

We can summarize the above observations by saying that the nearest-neighbor algorithm is an *efficient algorithm.* Roughly speaking, an **efficient algorithm** is an algorithm for which the amount of computational effort required to implement the algorithm grows in some reasonable proportion with the size of the input to the problem. (Granted, this is a pretty vague definition, with nettlesome fudge words such as *computational effort* and *reasonable proportion.* Unfortunately, a precise definition requires a more technical background that is beyond the scope of this book.)

The main problem with the nearest-neighbor algorithm is that it is *not an optimal algorithm.* If you go back to Example 6.7, you'll find that the nearest-neighbor circuit (*nearest-neighbor circuit* is short for "the circuit obtained using the nearest-neighbor algorithm") had a cost of $773, whereas the optimal Hamilton circuit had a cost of $676. This means that the nearest-neighbor circuit was off by approximately 14.35%. How did we come up with 14.35%? Here is the computation: $(773 - 676)/676 \approx 0.1435 = 14.35\%$. [In general, a relative measure of how "far" a given Hamilton circuit is from the optimal circuit is given by the **relative error** ε, given by $\varepsilon = (W - Opt)/Opt$, where W denotes the weight of the given circuit, and Opt denotes the weight of the optimal circuit. Of course, we can only find ε if we know the value of Opt.]

6.6 Approximate Algorithms

A really good algorithm for solving TSPs in general would have to be both *efficient* (like the nearest-neighbor algorithm) and *optimal* (like the brute-force algorithm). In practice this would ensure that we could find the optimal solution to any TSP. Unfortunately, nobody knows of such an algorithm. Moreover, *we don't even know why we don't know*. Is it because such an algorithm is actually a mathematical impossibility? Or is it because no one has yet been clever enough to find one?

Despite the efforts of some of the best mathematicians of our time, the answers to these questions have remained quite elusive. So far, no one has been able to come up with an efficient optimal algorithm for solving TSPs or, alternatively, to prove that such an algorithm does not exist. Because this question has profound implications in an area of computer science called *complexity theory*, it has become one of the most famous unsolved problems in modern mathematics.

In the meantime, we are faced with a quandary. In many real-world applications, it is necessary to find some sort of "solution" for TSPs involving graphs with hundreds and even thousands of vertices, and to do so in a reasonable time (using computers as helpers, for sure). Since the brute-force algorithm is out of the question, and since no efficient algorithm that guarantees an optimal solution is known, the only practical fallback strategy is to compromise—we give up on the expectation of an optimal solution and accept as a solution a Hamilton circuit that may be less than optimal. In exchange, we ask for quick results. Nowadays, this is the way that most real-life applications of TSPs are "solved."

We will use the term **approximate algorithm** to describe any algorithm that produces solutions that are, most of the time, reasonably close to the optimal solution. Sounds good, but what does "most of the time" mean? And how about "reasonably close"? Unfortunately, to answer these questions properly would take us beyond the scope of this book. We will have to accept the fact that in the area of analyzing algorithms we will be dealing with informal ideas rather than precise definitions. (A good introduction to the analysis of algorithms can be found in references 8 and 10.)

Let's return to the unfinished saga of Willy, the traveling salesman trying to find an optimal route for his 10-city sales trip (Example 6.8).

> **EXAMPLE 6.9** A Tale of Ten Cities: Part 2

When we left Willy a few pages back (Example 6.8), we had (a) decided not to try the brute-force algorithm (way too much work for just a $200 reward), and (b) used the nearest-neighbor algorithm to find the circuit $A, C, E, D, B, J, G, K, F, H$, A, with a total cost of $2153 (it took about 10 minutes). Is this an optimal circuit? If not, how close is it to the optimal circuit?

To answer these questions we would have to know the cost of the optimal circuit. It turns out that this cost is $1914. (This value was found using a computer program in just a few seconds, but this is deceptive. Somebody had to spend a lot of time writing the program, and someone [me] had to spend a fair amount of time entering the city-to-city costs. Even with a computer, it's a hassle.) We can conclude, with hindsight, that the nearest-neighbor circuit is off by a relative error of about 12.5%.

$(2153 - 1914)/1914 \approx 0.1249.$

The key point of Example 6.9 is that choosing an *approximate but efficient* algorithm over an *optimal but inefficient* algorithm is often a good tradeoff. As they say, "time is money." (This is a big departure from the conventional wisdom that math problems have either "right" or "wrong" answers. When dealing with algorithmic problems we need to consider not only the quality of the answer but also the amount of effort it takes to find it.)

In the next two sections, we will discuss two other approximate but efficient algorithms for "solving" TSPs: the *repetitive nearest-neighbor algorithm* and the *cheapest-link algorithm*.

6.7 The Repetitive Nearest-Neighbor Algorithm

As one might guess, the *repetitive nearest-neighbor algorithm* is a variation of the nearest-neighbor algorithm in which we repeat several times the entire nearest-neighbor circuit-building process. Why would we want to do this? The reason is that the nearest-neighbor circuit depends on the choice of the starting vertex. If we change the starting vertex, the nearest-neighbor circuit is likely to be different, and, if we are lucky, better. Since finding a nearest-neighbor circuit is an efficient process, it is not unreasonable to repeat the process several times, each time starting at a different vertex of the graph. In this way, we can obtain several different "nearest-neighbor solutions," from which we can then pick the best.

But what do we do with a Hamilton circuit that starts somewhere other than the vertex we really want to start at? That's not a problem. Remember that once we have a circuit, we can start the circuit anywhere we want. In fact, in an abstract sense, a circuit has no starting or ending point.

To illustrate how the repetitive nearest-neighbor algorithm works, let's return to the original five-city problem we last discussed in Example 6.5.

> **EXAMPLE 6.10** A Tale of Five Cities: Part 3

Once again, we are going to look at Willy's original five-city TSP [Fig. 6-14(a)] and this time try the repetitive nearest-neighbor algorithm. With luck we might find a better circuit than the original nearest-neighbor circuit.

In Example 6.7 we computed the nearest-neighbor circuit with A as the starting vertex, and we got A, C, E, D, B, A with a total cost of $773 [Fig. 6-14(b)]. But if we use B as the starting vertex, the nearest-neighbor circuit takes us from B to C, then to A, E, D, and back to B, with a total cost of $722, as shown in Fig. 6-14(c). Well, that is certainly an improvement! As for Willy—who must start and end his trip at A—this very same circuit would take the form A, E, D, B, C, A.

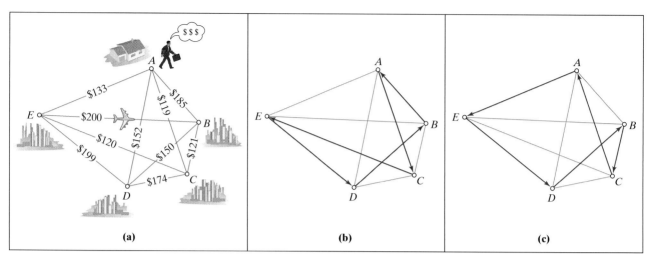

FIGURE 6-14

To verify these claims, try Exercise 31.

The process is once again repeated using C, D, and E as the starting vertices. When the starting point is C, we get the nearest-neighbor circuit C, A, E, D, B, C (total cost is \$722); when the starting point is D, we get the nearest-neighbor circuit D, B, C, A, E, D (total cost is \$722); and when the starting point is E, we get the nearest-neighbor circuit E, C, A, D, B, E (total cost is \$741). Among all the nearest-neighbor circuits we pick the cheapest one, shown in Fig. 6-14(c). ◀◀

A formal description of the **repetitive nearest-neighbor algorithm** is as follows.

Algorithm 3: The Repetitive Nearest-Neighbor Algorithm

- Let X be any vertex. Find the nearest-neighbor circuit using X as the starting vertex and calculate the total cost of the circuit.
- Repeat the process with each of the other vertices of the graph as the starting vertex.
- Of the nearest-neighbor circuits obtained, keep the best one. If there is a designated starting vertex, rewrite the circuit using that vertex as the reference point.

6.8 The Cheapest-Link Algorithm

The last—but not least—of the algorithms we will consider is known as the **cheapest-link algorithm**. The idea behind the cheapest-link algorithm is to piece together a Hamilton circuit by choosing its individual "links" (the edges in the circuit). It doesn't matter if in the intermediate stages the links are all over the place—if we are careful at the end they will all come together and form a Hamilton circuit.

This is how it works: Look at the entire graph and choose the cheapest edge of the graph, wherever that edge may be. Once this is done, choose the next cheapest edge of the graph, wherever that edge may be. (Don't worry if that edge is not adjacent to the first edge.) Continue this way, each time choosing the cheapest edge available, but following two rules: (i) Do not allow a circuit to form except at the very end, and (ii) do not allow three edges to come together at a vertex. A violation of either of these two rules will prevent forming a Hamilton circuit. Conversely, following these two rules guarantees that the end result will be a Hamilton circuit.

> **EXAMPLE 6.11** A Tale of Five Cities: Part 4

What better example to illustrate the cheapest-link algorithm than Willy's five-city TSP? Each step of the algorithm is illustrated in Fig. 6-15.

Among all the edges of the graph, the "cheapest link" is edge AC, with a cost of $119. For the purposes of recordkeeping, we will tag this edge as taken by marking it in red as shown in Fig. 6-15(a). (Nothing magical about red—you can choose any color you want.) The next step is to scan the entire graph again and choose the cheapest link available, in this case edge CE with a cost of $120. Again, we mark it in red, to indicate it is taken [Fig. 6-15(b)]. The next cheapest link available is edge BC ($121), but we should not choose BC—we would have three edges coming out of vertex C and this would prevent us from forming a circuit. This time we put a "do not use" mark on edge BC [Fig. 6-15(c)]. The next cheapest link available is AE ($133). But we can't take AE either—the vertices A, C, and E would form a small circuit, making it impossible to form a Hamilton circuit at the end. So we put a "do not use" mark on AE [Fig. 6-15(d)]. The next cheapest link available is BD ($150). Choosing BD would not violate either of the two rules, so we can add it to our budding circuit and mark it in red [Fig. 6-15(e)]. The next cheapest link available is AD ($152), and it works just fine [Fig. 6-15(f)]. At this point, we have only one way to close up the Hamilton circuit, edge BE, as shown in Fig. 6-15(g). The Hamilton circuit in red can now be described using any vertex as the reference

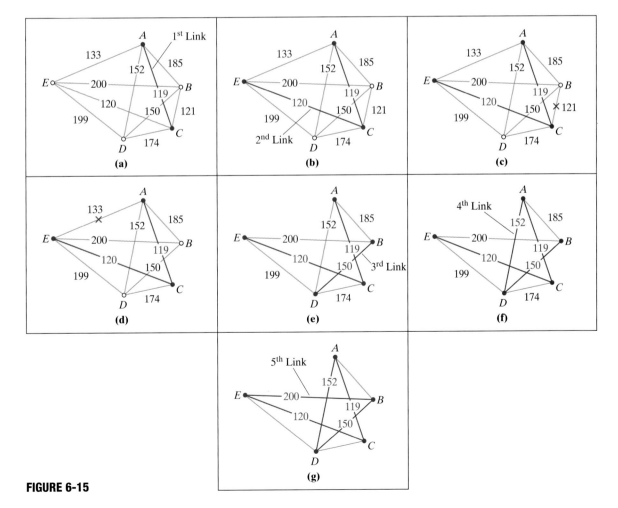

FIGURE 6-15

point. Since Willy lives at A, we describe it as A, C, E, B, D, A (or its mirror image). The total cost of this circuit is $741, which is a little better than the nearest-neighbor solution but not as good as the repetitive nearest-neighbor solution. ◀◀

A formal description of the cheapest-link algorithm is as follows.

Algorithm 4: The Cheapest-Link Algorithm

- **Step 1.** Pick the *cheapest link* (i.e., edge with smallest weight) available. (In case of a tie, pick one at random.) Mark it (say in red).
- **Step 2.** Pick the next cheapest link available and mark it.
- **Steps 3, 4, . . . , $N - 1$.** Continue picking and marking the cheapest un-marked link available that does not
  (a) close a circuit, or
  (b) create three edges coming out of a single vertex.
- **Step N.** Connect the last two vertices to close the red circuit.

For the last example of this chapter, we will return to the problem first described in the chapter opener—that of finding an optimal route for a rover exploring Mars.

▶ **EXAMPLE 6.12** Roving the Red Planet: Part 2

Figure 6-16(a) shows seven sites on Mars identified as particularly interesting sites for geological exploration. Our job is to find the shortest possible route for a rover that will land at A, visit all the sites and collect rock samples, and at the end, return to A. The distances between sites (in miles) are given in the graph shown in Fig. 6-16(b).

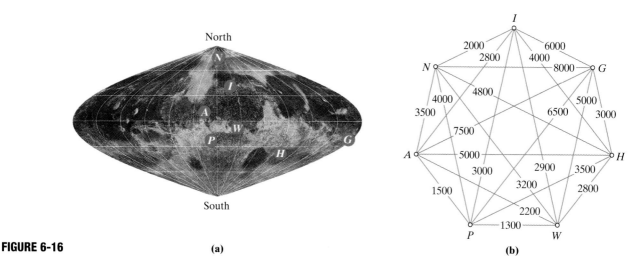

FIGURE 6-16   (a)   (b)

Let's look at some of the approaches one might use to tackle the rover routing problem.

Brute force: The brute-force algorithm would require us to check and compute $6! = 720$ different Hamilton circuits. We will pass on that idea for now.

Cheapest link: The cheapest-link algorithm is a reasonable algorithm to use—not trivial but not too hard either. A summary of the steps is shown in Table 6-7.

TABLE 6-7			
Step	**Cheapest edge available**	**Weight**	**Add to circuit?**
1	PW	1300	Yes
2	AP	1500	Yes
3	IN	2000	Yes
4	AW	2200	No
5 }	HW } tie	2800	Yes
6 }	AI }	2800	Yes
7	IW	2900	No
8 }	IP } tie	3000	No
9 }	GH }	3000	Yes
Last	GN only way to close circuit	8000	Yes

The Hamilton circuit obtained using this algorithm—A, P, W, H, G, N, I, A with a total length of 21,400 miles—is shown in Fig. 6-17.

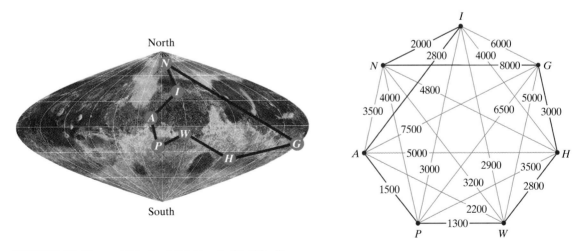

FIGURE 6-17 Cheapest-link circuit. Total length: 21,400 miles.

Nearest neighbor: The nearest-neighbor algorithm is the simplest of all the algorithms we learned. Starting from A we go to P, then to W, then to H, then to G, then to I, then to N, and finally back to A. The circuit obtained using this algorithm—A, P, W, H, G, I, N, A with a total length of 20,100 miles—is shown in Fig. 6-18. (We know that we can repeat this method using different starting points, but we won't bother with that at this time.)

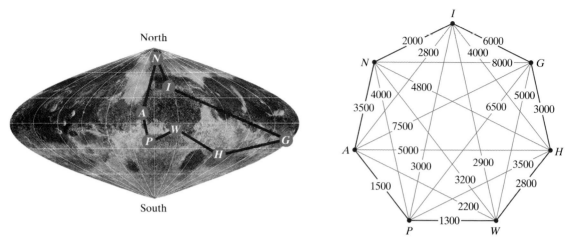

FIGURE 6-18 Nearest-neighbor circuit with starting vertex *A*. Total length: 20,100 miles.

The first surprise in Example 6.12 is the fact that the nearest-neighbor algorithm gave us a better solution than the cheapest-link algorithm. Sometimes the cheapest-link algorithm produces a better solution than the nearest-neighbor algorithm, but just as often, it's the other way around. The two algorithms are different but of equal standing—neither one can be said to be superior to the other one in terms of the quality of the solutions it produces.

The second surprise is that the nearest-neighbor circuit *A, P, W, H, G, I, N, A* turns out to be an *optimal* solution. (This can be verified using a computer and the brute-force algorithm.) Essentially, this means that in this particular example, the simplest of all methods happens to produce the optimal answer—a nice turn of events. Too bad we can't count on this happening on a more consistent basis!

Well, maybe next chapter.

Conclusion

The basic problem discussed in this chapter can be stated in fairly simple terms: How does one find an *optimal Hamilton circuit* in a *complete weighted graph*? (For historical reasons, these type of problems are known as *traveling salesman problems* [TSPs] even though most of the real-life applications of the problem have nothing to do with traveling salesmen.) Once the meaning of the technical terms *optimal Hamilton circuit* and *complete weighted graph* is clear, the problem does not sound all that difficult. Thus, the fact that this happens to be one of the most difficult problems in modern mathematics is quite remarkable.

How is this problem so difficult? After all, in this chapter we looked at several examples of TSPs and in most cases were able to find the optimal Hamilton circuit without too much trouble. But we haven't actually "solved" a general class of problems unless we have a method that produces a solution for any instance of the problem. In addition, when the method of solution is an algorithm, we have to worry about the computational effort that goes into producing a solution. A general algorithm for solving TSPs that would (i) always produce an optimal Hamilton circuit and (ii) do it within a reasonable amount of time has so far eluded the efforts of some of the world's best mathematicians.

The *nearest-neighbor* and *cheapest-link* algorithms are two fairly simple strategies for attacking TSPs. We have seen that both algorithms are *approximate algorithms*. This means that they are not likely to give us an optimal solution, although, as we saw in Example 6.12, on a lucky day, even that is possible. In most typical problems, however, we can expect either of these algorithms to give an approximate solution that is within a reasonable margin of error. With some problems the cheapest-link algorithm gives a better solution than the nearest-neighbor algorithm; with other problems it's the other way around. Thus, while they are not the same, neither is superior to the other.

Solving a TSP involving a large number of cities requires a combination of smart strategy and raw computational power—it is the perfect marriage of mathematics and computer science. Nowadays, research on TSPs—typically done by teams that include mathematicians and computer scientists—is carried out on several fronts. One major area of research interest is the quest for finding *optimal solutions* to larger and larger TSPs. (As of 2006, the record is an optimal Hamilton circuit connecting the 24,978 towns and cities in Sweden. For details, visit "The Traveling Salesman Problem Homepage" at *www.tsp.gatech.edu*.)

A second area that is attracting tremendous interest is that of finding improved and more efficient *approximate algorithms* for "solving" very large TSPs. Here progress is much faster, and many sophisticated approximate algorithms have been developed, including algorithms based on *recombinant DNA* techniques (see Project C and reference 1) as well as algorithms based on the collective intelligence of *ant colonies* (yes, that's ants, as in insects—see Project D and reference 5). Today, the best approximate algorithms can tackle TSPs with hundreds of thousands of vertices and produce solutions with relative errors *guaranteed* to be less than 1%.

On a purely theoretical level, the holy grail of TSP research still is the search for an *optimal and efficient general algorithm* that can solve all TSPs, or, alternatively, prove that such an algorithm does not exist. This conundrum is waiting for another Euler to come along.

A TSP with a little bite: What is the shortest Hamilton circuit that visits all 48 state capitals in the continental United States? The *optimal solution* shown has total length of approximately 12,000 miles. Even with a fast computer and special software, it could take hundreds of computer hours to find this solution using an *inefficient algorithm.*

For the same 48-city TSP, *approximate solutions* can be found quickly using *efficient algorithms.* The Hamilton circuit shown was found in a matter of seconds using the nearest neighbor algorithm (starting at Olympia, Washington). The total length of this approximate solution is 14,500 miles, roughly 20% longer than the optimal solution.

Profile   Sir William Rowan Hamilton (1805–1865)

Hamilton is one of the greatest and yet most tragic figures in the history of mathematics. A prodigy in languages, science, and mathematics at a very young age, he suffered from bouts of depression and fought the demon of alcohol for much of his adult life. In spite of many important mathematical discoveries, the story of his life is one of what might have been had the circumstances of his personal life turned out differently.

William Rowan Hamilton was born in Dublin, Ireland, exactly at midnight on August 3, 1805 (causing some confusion as to whether his birthday was August 3 or August 4). From a very young age he was taught by his uncle the Reverend James Hamilton, an eccentric clergyman and brilliant linguist. Under his uncle's tutelage, Hamilton learned Greek, Latin, and Hebrew by the age of 6, and by the time he was 13 he knew a dozen languages, including Persian, Arabic, and Sanskrit. At the age of 10 Hamilton became interested in mathematics, and by the age of 15 he was reading the works of Newton and Laplace. When he was just 17, Hamilton wrote his first two mathematical research papers, *On Contacts between Algebraic Curves and Surfaces*, and *Developments*. After reading the papers, John Brinkley, the Astronomer Royal of Ireland at that time, remarked, "This young man, I do not say will be, but is, the first [foremost] mathematician of his age."

In 1823, Hamilton entered Trinity College in Dublin, where he excelled in all subjects while continuing to publish groundbreaking research in mathematical physics. A year after entering college, Hamilton met and fell madly in love with a young lady from an upper-class family by the name of Catherine Disney, a watershed event in his life. Although Catherine might have reciprocated his affections, there was a problem. Hamilton was just a 19-year-old university student and did not come from a particularly wealthy family—according to the social norms of the day he was not a suitable match for Catherine. On the other hand, Hamilton was one of those mathematical geniuses that come around once in a generation. Not lacking for confidence, he was banking on his status and fame as a scientist to bring Catherine's parents around into eventually accepting him.

The following year Hamilton found out that Catherine's family had arranged for her marriage to another man. This was devastating news to him and he even considered suicide. Ironically, soon after Catherine was married, Hamilton's social status and financial position took a quantum leap forward. In 1827, not quite yet 22 years old and still an undergraduate, Hamilton won the appointment to the prestigious position of Astronomer Royal of Ireland as well as to a Professorship of Astronomy at Trinity College, an incredible personal triumph. Unfortunately, for him and Catherine it all happened a couple of years too late. By now Catherine was married and had a child.

Over the next 20 years, Hamilton's life was a checkered mixture of professional successes and personal setbacks. On the rebound, he soon entered into an ill-advised and unhappy marriage that bore him three children. He fought intermittent bouts of depression and started having serious problems with alcohol. At the same time, he pursued his research in mathematics and optics, where he produced many significant discoveries, including the formulas for conic refraction, which gained him much fame. In recognition of his many scientific accomplishments, Hamilton was knighted in 1835.

In 1843, Hamilton produced the mathematical discovery he is best known for, the *calculus of quaternions*. (Quaternions are essentially four-dimensional numbers that generalize the complex numbers and are ideally suited to describe mathematically many complex phenomena in physics.) Hamilton predicted that quaternions would revolutionize nineteenth-century physics, but he was a century ahead of his time—quaternions revolutionized *twentieth*-century physics, as they are a key element in quantum mechanics and Einstein's theory of relativity.

Hamilton's connection with graph theory and *Hamilton circuits* came late in his life and in a rather roundabout way. In 1857, Hamilton invented a board game he called the *Icosian game*, the purpose of which was to create a trip passing through each of 20 European cities once and only once. The cities were connected in the form of a *dodecahedral graph* (see Exercise 57). Hamilton sold the rights to the game to a London game dealer for 25 pounds, and the game was actually marketed throughout Europe, a precursor to some of the "connect-the-dots" games that we know today.

Key Concepts

approximate algorithm, **212**
algorithm, **210**
brute-force algorithm, **210**
cheapest-link algorithm, **216**
complete graph, **201**
complete weighted graph, **206**
efficient algorithm, **211**

factorial, **203**
Hamilton circuit, **199**
Hamilton path, **199**
inefficient algorithm, **211**
nearest-neighbor algorithm, **210**
optimal algorithm, **210**
relative error, **212**

repetitive nearest-neighbor
   algorithm, **214**
traveling salesman problem
   (TSP), **204**
weighted graph, **206**
weight, **206**

Exercises

WALKING

A. Hamilton Circuits and Hamilton Paths

1. For the following graph,

   (a) find three different Hamilton circuits.

   (b) find a Hamilton path that starts at A and ends at B.

   (c) find a Hamilton path that starts at D and ends at F.

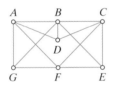

2. For the following graph,

   (a) find three different Hamilton circuits.

   (b) find a Hamilton path that starts at A and ends at B.

   (c) find a Hamilton path that starts at F and ends at I.

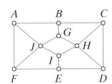

3. List all possible Hamilton circuits in the following graph.

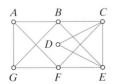

4. List all possible Hamilton circuits in the following graph.

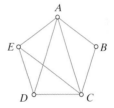

5. For the following graph,

   (a) find a Hamilton path that starts at A and ends at E.

   (b) find a Hamilton circuit that starts at A and ends with the edge EA.

   (c) find a Hamilton path that starts at A and ends at C.

   (d) find a Hamilton path that starts at F and ends at G.

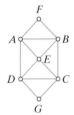

6. For the following graph,

   (a) find a Hamilton path that starts at A and ends at E.

   (b) find a Hamilton circuit that starts at A and ends with the edge EA.

   (c) find a Hamilton path that starts at A and ends at G.

   (d) find a Hamilton path that starts at F and ends at G.

7. For the following graph,
   (a) list all Hamilton circuits that start at vertex A.
   (b) list all Hamilton circuits that start at vertex D.
   (c) explain why your answers in (a) and (b) must have the same number of circuits.

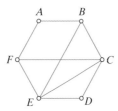

8. For the following graph,
   (a) list all Hamilton circuits that start at vertex A.
   (b) list all Hamilton circuits that start at vertex D.
   (c) explain why your answers in (a) and (b) must have the same number of circuits.

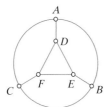

9. Explain why the following graph has neither Hamilton circuits nor Hamilton paths.

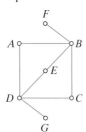

10. Explain why the following graph has no Hamilton circuit but does have a Hamilton path.

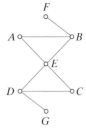

11. For the following graph,

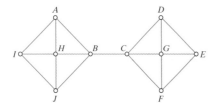

(a) find a Hamilton path that starts at A and ends at D.
(b) find a Hamilton path that starts at G and ends at H.
(c) explain why there is no Hamilton path that starts at B.
(d) explain why there is no Hamilton circuit.

12. For the following graph,

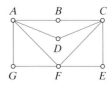

(a) find a Hamilton path that starts at D.
(b) find a Hamilton path that starts at B.
(c) explain why there is no Hamilton path that starts at A or C.
(d) explain why there is no Hamilton circuit.

13. For the following weighted graph,

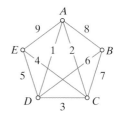

(a) find the weight of edge BD.
(b) find a Hamilton circuit that starts with edge BD, and give its weight.
(c) find a Hamilton circuit that ends with edge DB, and give its weight.

14. For the following weighted graph,

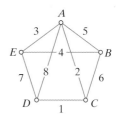

(a) find the weight of edge AD.
(b) find a Hamilton circuit that starts with edge AD, and give its weight.
(c) find a Hamilton circuit that ends with edge DA, and give its weight.

15. For the following weighted graph,

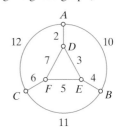

(a) find a Hamilton path that starts at A and ends at C, and give its weight.

(b) find a second Hamilton path that starts at A and ends at C, and give its weight.

(c) find the optimal (least weight) Hamilton path that starts at A and ends at C, and give its weight.

16. For the following weighted graph,

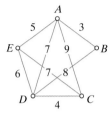

(a) find a Hamilton path that starts at B and ends at D, and give its weight.

(b) find a second Hamilton path that starts at B and ends at D, and give its weight.

(c) find the optimal (least weight) Hamilton path that starts at B and ends at D, and give its weight.

B. Factorials and Complete Graphs

In Exercise 17 through 20, you will need a calculator with a factorial key.

17. Using a calculator with a factorial key, compute each of the following (you may give your answer in scientific notation if necessary):

(a) 10!

(b) 20!

(c) 20!/10!

18. Using a calculator with a factorial key, compute each of the following (you may give your answer in scientific notation if necessary):

(a) 15!

(b) 30!

(c) 30!/15!

19. Using a calculator with a factorial key, compute each of the following (you may give your answer in scientific notation if necessary):

(a) 20!

(b) 40!

(c) The number of distinct Hamilton circuits in K_{40}

20. Using a calculator with a factorial key, compute each of the following (you may give your answer in scientific notation if necessary):

(a) 25!

(b) 50!

(c) The number of distinct Hamilton circuits in K_{25}

The purpose of Exercises 21 through 24 is for you to learn how to numerically manipulate factorials. (These exercises are repeates of Exercises 37–40 in Chapter 2.) You should answer these questions without using a calculator—otherwise, you are defeating the purpose of the exercise.)

21. (a) Given that 10! = 3,628,800, find 9!.

(b) Find 11!/10!.

(c) Find 11!/9!.

(d) Find 101!/99!.

22. (a) Given that 20! = 2,432,902,008,176,640,000, find 19!.

(b) Find 20!/19!.

(c) Find 201!/199!.

23. Find the value of each of the following. Give the answer in decimal form.

(a) $(9! + 11!)/10!$

(b) $(101! + 99!)/100!$

24. Find the value of each of the following. Give the answer in decimal form.

(a) $(19! + 21!)/20!$

  (***Hint:*** $1/20 = 0.05$)

(b) $(201! + 199!)/200!$

25. (a) How many edges are there in K_{20}?

(b) How many edges are there in K_{21}?

(c) If the number of edges in K_{50} is x, and the number of edges in K_{51} is y, what is the value of $y - x$?

26. (a) How many edges are there in K_{200}?

(b) How many edges are there in K_{201}?

(c) If the number of edges in K_{500} is x, and the number of edges in K_{501} is y, what is the value of $y - x$?

27. In each case, find the value of N.

(a) K_N has 120 distinct Hamilton circuits.

(b) K_N has 45 edges.

(c) K_N has 20,100 edges.

28. In each case, find the value of N.

(a) K_N has 720 distinct Hamilton circuits.

(b) K_N has 66 edges.

(c) K_N has 80,200 edges.

C. Brute-Force and Nearest-Neighbor Algorithms

29. For the weighted graph shown in the figure, (i) find the indicated circuit, and (ii) give its cost.

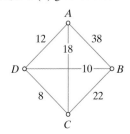

(a) An optimal Hamilton circuit (use the brute-force algorithm)

(b) The nearest-neighbor circuit for starting vertex A

(c) The nearest-neighbor circuit for starting vertex B

(d) The nearest-neighbor circuit for starting vertex C

30. For the weighted graph shown in the figure, (i) find the indicated circuit, and (ii) give its cost.

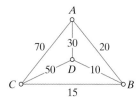

(a) An optimal Hamilton circuit (use the brute-force algorithm)

(b) The nearest-neighbor circuit for starting vertex A

(c) The nearest-neighbor circuit for starting vertex C

(d) The nearest-neighbor circuit for starting vertex D

31. For the weighted graph shown in the figure, (i) find the indicated circuit, and (ii) give its cost. (This is the graph discussed in Example 6.7.)

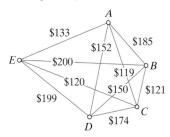

(a) The nearest-neighbor circuit for starting vertex B

(b) The nearest-neighbor circuit for starting vertex C

(c) The nearest-neighbor circuit for starting vertex D

(d) The nearest-neighbor circuit for starting vertex E

32. A delivery service must deliver packages at Buckman (B), Chatfield (C), Dayton (D), and Evansville (E), and then return to Arlington (A), the home base. The following graph shows the estimated travel times (in minutes) between the cities.

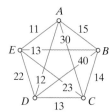

(a) Find the nearest-neighbor circuit for starting vertex A. What is the total travel time of this trip?

(b) Find the nearest-neighbor circuit for starting vertex D. Write the answer as it would be traveled if starting and ending at A.

(c) Suppose that the delivery truck's last stop before returning to A has to be at D. Find the optimal Hamilton circuit that satisfies this requirement. What is the total travel time of this trip?

33. The Brute Force Bandits is a rock band planning a five-city concert tour. The cities and the distances (in miles) between them are given in the weighted graph shown. The tour must start and end at A. The cost of the chartered bus the band is traveling in is $8 per mile.

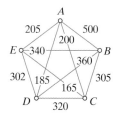

(a) Find the nearest-neighbor circuit for starting vertex A. What is the bus cost of this trip?

(b) Find the nearest-neighbor circuit for starting vertex B. Write the answer as it would be traveled by the band, starting and ending the trip at A. What is the bus cost of this trip?

(c) Suppose that the band decides to make B the first stop after A. Find the optimal Hamilton circuit that starts with the pair of cities A, B. What is the bus cost of this trip?

34. A space expedition is scheduled to visit the moons Callisto (C), Ganymede (G), Io (I), Mimas (M), and Titan (T) to collect rock samples at each and then return to Earth (E). The following graph summarizes the travel time (in years) between any two places. (This is the graph discussed in Example 6.5.)

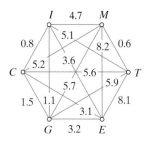

(a) Find the nearest-neighbor circuit for starting vertex E. What is the total travel time of this trip?

(b) Find the nearest-neighbor circuit for starting vertex T. Write the answer as it would be traveled by an expedition starting and ending at E.

(c) Suppose that scientific considerations require that the *last two* moons visited before the expedition returns to E should be T and C (in either order). Find the optimal Hamilton circuit that satisfies this requirement. What is the total travel time of this trip?

In Exercises 35 and 36, you are given the distance between the cities in a mileage chart, rather than in a weighted graph. The purpose of the exercises is for you to try to implement the nearest-neighbor algorithm using the mileage chart and without drawing a weighted graph. (Nobody can stop you from drawing the graph, but you would be doing a lot more work than is really necessary.)

35. Darren is a sales rep whose territory consists of the six cities shown in the mileage chart. Darren wants to visit costumers at each of the cities, starting and ending his trip in his home city of Atlanta. His travel costs (gas, insurance, etc.) average $0.75 per mile.

Mileage Chart

	Atlanta	Columbus	Kansas City	Minneapolis	Pierre	Tulsa
Atlanta	*	533	798	1068	1361	772
Columbus	533	*	656	713	1071	802
Kansas City	798	656	*	447	592	248
Minneapolis	1068	713	447	*	394	695
Pierre	1361	1071	592	394	*	760
Tulsa	772	802	248	695	760	*

(a) Find the nearest-neighbor circuit with Atlanta as the starting vertex. What is the total cost of this circuit?

(b) Find the nearest-neighbor circuit with Kansas City as the starting vertex. Write the circuit as it would be traveled by Darren, who must start and end the trip in Atlanta. What is the total cost of this circuit?

36. The Platonic Cowboys is a country western band based in Nashville. The Cowboys are planning a concert tour to the seven cities shown in the mileage chart.

Mileage Chart

	Boston	Dallas	Houston	Louisville	Nashville	Pittsburgh	St. Louis
Boston	*	1748	1804	941	1088	561	1141
Dallas	1748	*	243	819	660	1204	630
Houston	1804	243	*	928	769	1313	779
Louisville	941	819	928	*	168	388	263
Nashville	1088	660	769	168	*	553	299
Pittsburgh	561	1204	1313	388	553	*	588
St. Louis	1141	630	779	263	299	588	*

(a) Find the nearest-neighbor circuit with Nashville as the starting vertex. What is the total length of this circuit?

(b) Find the nearest-neighbor circuit with St. Louis as the starting vertex. What is the total length of this circuit?

D. Repetitive Nearest-Neighbor Algorithm

37. For the weighted graph in the figure, find the Hamilton circuit obtained by the repetitive nearest-neighbor algorithm. Write the circuit assuming that the starting and ending point is B.

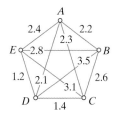

38. For the weighted graph in the figure, find the Hamilton circuit obtained by the repetitive nearest-neighbor algorithm. Write the circuit assuming that the starting and ending point is D.

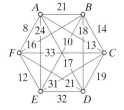

39. This exercise is a continuation of the Brute Force Bandits concert tour (Exercise 33). Find the Hamilton circuit obtained by the repetitive nearest-neighbor algorithm, and give the bus cost for this circuit.

40. This exercise is a continuation of the space expedition problem (Exercise 34). Find the Hamilton circuit obtained by the repetitive nearest-neighbor algorithm, and give the total travel time for this circuit.

41. This exercise is a continuation Darren's sales trip problem (Exercise 35). Find the Hamilton circuit obtained by the repetitive nearest-neighbor algorithm, and give the total mileage for this circuit.

42. This exercise is a continuation of The Platonic Cowboys concert tour (Exercise 36). Find the Hamilton circuit obtained by the repetitive nearest-neighbor algorithm, and give the total mileage for this circuit.

E. Cheapest-Link Algorithm

43. For the weighted graph in the figure, find the Hamilton circuit obtained by the cheapest-link algorithm, and give the total mileage for this circuit. Write the circuit assuming that the starting and ending point is B. (This is the graph in Exercise 37.)

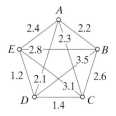

44. For the weighted graph in the figure, find the Hamilton circuit obtained by the cheapest-link algorithm, and give the total mileage for this circuit. Write the circuit assuming that the starting and ending point is *B*. (This is the graph in Exercise 38.)

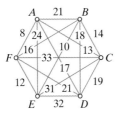

45. For the Brute Force Bandits concert tour discussed in Exercise 33, find the Hamilton circuit obtained by the cheapest-link algorithm, and give the bus cost for this circuit.

46. For the space expedition problem discussed in Exercise 34, find the Hamilton circuit obtained by the cheapest-link algorithm, and give the total travel time for this circuit.

47. For Darren's sales trip problem discussed in Exercise 35, find the Hamilton circuit obtained by the cheapest-link algorithm, and give the total mileage for this circuit.

48. For the Platonic Cowboys concert tour discussed in Exercise 36, find the Hamilton circuit obtained by the repetitive nearest-neighbor algorithm, and give the total mileage for this circuit.

F. Miscellaneous

49. For the following weighted graph, (i) find the indicated circuit, and (ii) give the weight of the circuit.

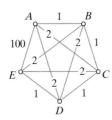

(a) The nearest-neighbor circuit for starting vertex *A*

(b) The Hamilton circuit obtained by the cheapest-link algorithm

(c) An optimal Hamilton circuit

50. Joe Fan wants to watch a game in each of the 30 major league baseball stadiums in North America. His plan is to fly to each team's home city, watch a game, and move on to the next city. Although Joe is nuts, he does want to do this in the cheapest possible way. Using *cheaptickets.com*, he plans to find the best fare from each city to every other city.

(a) Assuming it takes an average of 10 minutes to find the cheapest fare between two cities, how long is Joe going to be sitting in front of his computer (assume he is not taking any breaks)?

(b) After Joe looks up all the city-to-city fares at *cheaptickets.com*, he decides to use the brute-force algorithm to find the cheapest possible tour of the 30 cities. Explain why this is a really bad idea.

51. You have a busy day ahead of you. You must run the following errands (in no particular order): Go to the post office, deposit a check at the bank, pick up some French bread at the deli, visit a friend at the hospital, and get a haircut at Karl's Beauty Salon. You must start and end at home. Each block on the following map is exactly 1 mile.

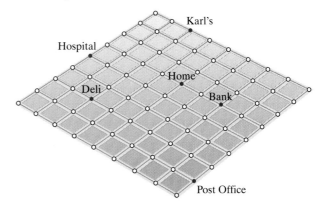

(a) Draw a weighted graph corresponding to this problem.

(b) Find the optimal (shortest) way to run all the errands. (Use any algorithm you think is appropriate.)

52. Rosa's Floral must deliver flowers to each of the five locations *A*, *B*, *C*, *D*, and *E* shown on the following map. The trip must start and end at the flower shop, located at *F*. Each block on the map is exactly 1 mile.

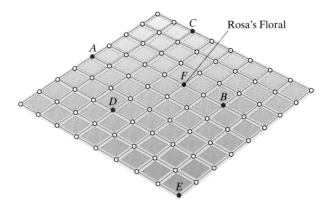

(a) Draw a weighted graph corresponding to this problem.

(b) Find the optimal (shortest) way to make all the deliveries. (Use any algorithm you think is appropriate.)

In Exercises 53 through 56, you are scheduling a dinner party for 6 people (A, B, C, D, E, and F). The guests are to be seated around a circular table, and you want to arrange the seating so that each guest is seated between two friends (i.e., the guests to the left and to the right are friends of the guest in between). You can assume that all friendships are mutual (when X is a friend of Y, Y is also a friend of X).

53. Suppose that you are told that all possible friendships can be deduced from the following information:

A is friends with *B* and *F*; *B* is friends with *A*, *C*, and *E*; *C* is friends with *B*, *D*, *E*, and *F*; *E* is friends with *B*, *C*, *D*, and *F*.

(a) Draw a "friendship graph" for the dinner guests.

(b) Find a possible seating arrangement for the party.

(c) Is there a possible seating arrangement in which *B* and *E* are seated next to each other? If there is, find it. If there isn't, explain why not.

54. Suppose that you are told that all possible friendships can be deduced from the following information:

A is friends with *B*, *C*, and *D*; *B* is friends with *A*, *C*, and *E*; *D* is friends with *A*, *E*, and *F*; *F* is friends with *C*, *D*, and *E*.

(a) Draw a "friendship graph" for the dinner guests.

(b) Find a possible seating arrangement for the party.

(c) Is there a possible seating arrangement in which *B* and *E* are seated next to each other? If there is, find it. If there isn't, explain why not.

55. Suppose that you are told that all possible friendships can be deduced from the following information:

A is friends with *C*, *D*, *E*, and *F*; *B* is friends with *C*, *D*, and *E*; *C* is friends with *A*, *B*, and *E*; *D* is friends with *A*, *B*, and *E*. Explain why it is impossible to have a seating arrangement in which each guest is seated between friends.

56. Suppose that you are told that all possible friendships can be deduced from the following information:

A is friends with *B*, *D*, and *F*; *C* is friends with *B*, *D*, and *F*; *E* is friends with *C* and *F*. Explain why it is impossible to have a seating arrangement in which each guest is seated between friends.

JOGGING

57. The following graph is called the *dodecahedral graph*, because it describes the relationship between the vertices and edges of a *dodecahedron*, a regular three-dimensional solid consisting of 12 faces all of which are regular pentagons. Find a Hamilton circuit in the graph. Show your answer by labeling the vertices in the order of travel.

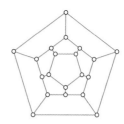

58. The following graph is called the *icosahedral graph*, because it describes the relationship between the vertices and edges of an *icosahedron*, a regular three-dimensional solid consisting of twenty faces all of which are equilateral triangles. Find a Hamilton circuit in the graph. Show your answer by labeling the vertices in the order of travel.

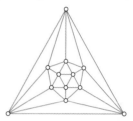

59. Find a Hamilton path in the following graph.

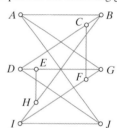

60. A 2 by 2 grid graph. The graph shown represents a street grid that is 2 blocks by 2 blocks. (Such graph is called a *2 by 2 grid graph.*) For convenience, the vertices are labeled by type: corner vertices C_1, C_2, C_3, and C_4, boundary vertices B_1, B_2, B_3, and B_4, and the interior vertex I.

(a) Find a Hamilton path in the graph that starts at I.

(b) Find a Hamilton path in the graph that starts at one of the corner vertices and ends at a different corner vertex.

(c) Find a Hamilton path that starts at one of the corner vertices and ends at I.

(d) Find (if you can) a Hamilton path that starts at one of the corner vertices and ends at one of the boundary vertices. If this is impossible, explain why.

61. Find (if you can) a Hamilton circuit in the 2 by 2 grid graph discussed in Exercise 60. If this is impossible, explain why.

62. A 3 by 3 grid graph. The graph shown represents a street grid that is 3 blocks by 3 blocks. The graph has four corner vertices (C_1, C_2, C_3, and C_4), eight boundary vertices (B_1 through B_8), and four interior vertices (I_1, I_2, I_3, and I_4).

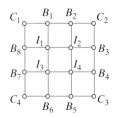

(a) Find a Hamilton circuit in the graph.

(b) Find a Hamilton path in the graph that starts at one of the corner vertices and ends at a different corner vertex.

(c) Find (if you can) a Hamilton path that starts at one of the corner vertices and ends at one of the interior vertices. If this is impossible, explain why.

(d) Given any two adjacent vertices of the graph, explain why there always is a Hamilton path that starts at one and ends at the other one.

63. A 3 by 4 grid graph. The graph that follows represents a street grid that is 3 blocks by 4 blocks.

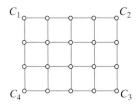

(a) Find a Hamilton circuit in the graph. (Make the circuit by labeling the vertices 1, 2, 3, … right on the graph.)

(b) Find a Hamilton path in the graph that starts at C_1 and ends at C_3.

(c) Find (if you can) a Hamilton path in the graph that starts at C_1 and ends at C_2. If this is impossible, explain why.

64. Explain why the cheapest edge in any graph is always part of the Hamilton circuit obtained using the nearest-neighbor algorithm.

65. Give an example of a complete weighted graph with six vertices such that the nearest-neighbor algorithm and the cheapest-link algorithm both give the optimal Hamilton circuit. Choose the weights of the edges to be all different.

66. (a) Give an example of a graph with four vertices in which the same circuit can be both an Euler circuit and a Hamilton circuit.

(b) Give an example of a graph with N vertices in which the same circuit can be both an Euler circuit and a Hamilton circuit. Explain why there is only one kind of graph for which this is possible.

67. (a) Explain why a graph that has a bridge cannot have a Hamilton circuit.

(b) Give an example of a graph with bridges that has a Hamilton path.

68. Explain why 21! is more than 100 billion times bigger than 10! (i.e., show that $21! > 10^{11} \times 10!$).

Exercises 69 and 70 refer to the following situation. Nick is a traveling salesman. His territory consists of the 11 cities shown on the mileage chart. Nick must organize a round trip that starts and ends in Dallas (that's his home) and visits each of the other 10 cities exactly once.

Mileage Chart

	Atlanta	Boston	Buffalo	Chicago	Columbus	Dallas	Denver	Houston	Kansas City	Louisville	Memphis
Atlanta	*	1037	859	674	533	795	1398	789	798	382	371
Boston	1037	*	446	963	735	1748	1949	1804	1391	941	1293
Buffalo	859	446	*	522	326	1346	1508	1460	966	532	899
Chicago	674	963	522	*	308	917	996	1067	499	292	530
Columbus	533	735	326	308	*	1028	1229	1137	656	209	576
Dallas	795	1748	1346	917	1028	*	781	243	489	819	452
Denver	1398	1949	1508	996	1229	781	*	1019	600	1120	1040
Houston	789	1804	1460	1067	1137	243	1019	*	710	928	561
Kansas City	798	1391	966	499	656	489	600	710	*	520	451
Louisville	382	941	532	292	209	819	1120	928	520	*	367
Memphis	371	1293	899	530	576	452	1040	561	451	367	*

69. Working directly from the mileage chart, find the nearest-neighbor circuit that starts at Dallas. Explain the procedure you used to find the answer.

70. Working directly from the mileage chart, find the Hamilton circuit obtained by the cheapest-link algorithm. Explain the procedure you used to find the answer.

71. Julie is the marketing manager for a small software company based in Boston. She is planning a sales trip to Michigan to visit customers in each of the nine cities shown in the mileage chart. She can fly from Boston to any one of the cities and fly out of any one of the cities back to Boston for the same price (call the arrival city A and the departure city D). Her plan is to pick up a rental car at A, drive to each of the other cities and drop off the rental car at the last city D. Slightly complicating the situation is the fact that Michigan has two separate peninsulas—an upper peninsula and a lower peninsula—and the only way to get from one to the other is through the Mackinaw Bridge connecting Cheboygan to Sault Ste. Marie. (There is a $2.50 toll to cross the bridge in either direction.)

Mileage Chart

	Detroit	Lansing	Grand Rapids	Flint	Cheboygan	Sault Ste. Marie	Marquette	Escanaba	Menominee
Detroit	*	90	158	68	280				
Lansing	90	*	68	56	221				
Grand Rapids	158	68	*	114	233				
Flint	68	56	114	*	215				
Cheboygan	280	221	233	215	*	78			
Sault Ste. Marie					78	*	164	174	227
Marquette						164	*	67	120
Escanaba						174	67	*	55
Menominee						227	120	55	*

(a) Suppose the rental car company charges 39 cents per mile plus a drop off fee of $250 if A and D are different cities (there is no charge if $A = D$). Find the optimal (cheapest) route and give the total cost.

(b) Suppose the rental car company charges 49 cents per mile but the car can be returned to any city without a drop off fee. Find the optimal route and give the total cost.

RUNNING

72. Complements. Let G be a graph with N vertices. The complement of G is the graph with the same vertices and the complementary set of edges. (Two distinct vertices that are adjacent in G are not adjacent in the complement, and, conversely, two distinct vertices not adjacent in G are adjacent in the complement.)

(a) If the number of edges in G is k, find the number of edges in the complement of G. (Assume no loops or multiple edges in either graph.)

(b) Explain why for $N = 4$ there is no graph such that both the graph and its complement have Hamilton circuits.

(c) Explain why for all $N \geq 5$ there are graphs such that both the graph and its complement have Hamilton circuits.

73. Complete bipartite graphs. A complete bipartite graph is a graph with the property that the vertices can be divided into two sets A and B and each vertex in set A is adjacent to each of the vertices in set B. There are no other edges! If there are m vertices in set A and n vertices in set B, the complete bipartite graph is written as $K_{m,n}$.

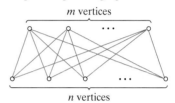

$K_{m,n}$

(a) For $n > 1$, the complete bipartite graphs of the form $K_{n,n}$ all have Hamilton circuits. Explain why.

(b) If the difference between m and n is exactly 1 (i.e., $|m - n| = 1$), the complete bipartite graph $K_{m,n}$ has a Hamilton path. Explain why.

(c) When the difference between m and n is more than 1, then the complete bipartite graph $K_{m,n}$ has no Hamilton path. Explain why.

74. m by n grid graphs. An m by n grid graph represents a rectangular street grid that is m blocks by n blocks, as indicated in the following figure. (You should try Exercises 60 through 63 before you try this one.)

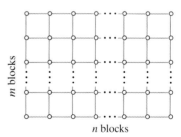

(a) If m and n are both odd, the m by n grid graph has a Hamilton circuit. Describe the circuit by drawing it on a generic graph.

(b) If either m or n is even and the other one is odd, then the m by n grid graph has a Hamilton circuit. Describe the circuit by drawing it on a generic graph.

(c) If m and n are both even, then the m by n grid graph does not have a Hamilton circuit. Explain why a Hamilton circuit is impossible.

75. The Petersen graph. The following graph is called the Petersen graph.

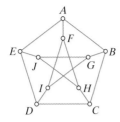

(a) Find a Hamilton path in the Petersen graph.

(b) Explain why the Petersen graph does not have a Hamilton circuit.

Projects and Papers

A. The Great Kaliningrad Circus

The Great Kaliningrad Circus has been signed for an extended 21-city tour of the United States, starting and ending in Miami, Florida. The 21 cities and the travel distances between cities are shown in the mileage chart (opposite page). The cost of transporting an entire circus the size of the Great Kaliningrad can be estimated to be about $1000 per mile, so determining the best, or near best, route for the circus tour is clearly a top priority for the tour manager, which, by the way, is you!

In this project, your job is to find the best route you can for the circus tour. (You are not asked to find the optimal route, just to find the best route you can!) You should prepare a presentation to the board of directors of the circus explaining what strategies you used to come up with your route, and why you think your route is a good one.

Note: The tools at your disposal to carry out this project are everything you learned in this chapter, your ingenuity, and your time (a wall-sized map of the United States and some pins may come in handy too!). This should be a low-tech project—you should not go to the Web and try to find some software that will do this job for you!

B. The Nearest-Insertion Algorithm

The *nearest-insertion algorithm* is another approximate algorithm used for tackling TSPs. The basic idea of the algorithm is to start with a subcircuit (a circuit that includes some, but not all, of the vertices) and enlarge it, one step at a time, by adding an extra vertex—the one that is closest to some ver-

tex in the circuit. By the time we have added all of the vertices, we have a full-fledged Hamilton circuit.

In this project, you should prepare a class presentation on the nearest-insertion algorithm. Your presentation should include a detailed description of the algorithm, at least two carefully worked out examples, and a comparison of the nearest-insertion and the nearest-neighbor algorithms.

C. Computing with DNA

DNA is the basic molecule of life—it encodes the genetic information that characterizes all living organisms. Due to the great advances in biochemistry of the last 20 years, scientists can now snip, splice, and recombine segments of DNA almost at will. In 1994, Leonard Adleman, a professor of computer science at the University of Southern California, was able to encode a graph representing seven cities into a set of DNA segments and to use the chemical reactions of the DNA fragments to uncover the existence of a Hamilton path in the graph. Basically, he was able to use the biochemistry of DNA to solve a graph theory problem. While the actual problem solved was insignificant, the idea was revolutionary, as it opened the door for the possibility of someday using DNA computers to solve problems beyond the reach of even the most powerful of today's electronic computers.

Write a research paper telling the story of Adleman's landmark discovery. How did he encode the graph into DNA? How did he extract the mathematical solution

Mileage Chart

	Atlanta	Boston	Buffalo	Chicago	Columbus	Dallas	Denver	Houston	Kansas City	Louisville	Memphis	Miami	Minneapolis	Nashville	New York	Omaha	Pierre	Pittsburgh	Raleigh	St. Louis	Tulsa
Atlanta	*	1037	859	674	533	795	1398	789	798	382	371	655	1068	242	841	986	1361	687	372	541	772
Boston	1037	*	446	963	735	1748	1949	1804	1391	941	1293	1504	1368	1088	206	1412	1726	561	685	1141	1537
Buffalo	859	446	*	522	326	1346	1508	1460	966	532	899	1409	927	700	372	971	1285	216	605	716	1112
Chicago	674	963	522	*	308	917	996	1067	499	292	530	1329	405	446	802	459	763	452	784	289	683
Columbus	533	735	326	308	*	1028	1229	1137	656	209	576	1160	713	377	542	750	1071	182	491	406	802
Dallas	795	1748	1346	917	1028	*	781	243	489	819	452	1300	936	660	1552	644	943	1204	1166	630	257
Denver	1398	1949	1508	996	1229	781	*	1019	600	1120	1040	2037	841	1156	1771	537	518	1411	1661	857	681
Houston	789	1804	1460	1067	1137	243	1019	*	710	928	561	1190	1157	769	1608	865	1186	1313	1160	779	478
Kansas City	798	1391	966	499	656	489	600	710	*	520	451	1448	447	556	1198	201	592	838	1061	257	248
Louisville	382	941	532	292	209	819	1120	928	520	*	367	1037	697	168	748	687	1055	388	541	263	659
Memphis	371	1293	899	530	576	452	1040	561	451	367	*	997	826	208	1100	652	1043	752	728	285	401
Miami	655	1504	1409	1329	1160	1300	2037	1190	1448	1037	997	*	1723	897	1308	1641	2016	1200	819	1196	1398
Minneapolis	1068	1368	927	405	713	936	841	1157	447	697	826	1723	*	826	1207	357	394	857	1189	552	695
Nashville	242	1088	700	446	377	660	1156	769	556	168	208	897	826	*	892	744	1119	553	521	299	609
New York	841	206	372	802	542	1552	1771	1608	1198	748	1100	1308	1207	892	*	1251	1565	368	489	948	1344
Omaha	986	1412	971	459	750	644	537	865	201	687	652	1641	357	744	1251	*	391	895	1214	449	387
Pierre	1361	1726	1285	763	1071	943	518	1186	592	1055	1043	2016	394	1119	1565	391	*	1215	1547	824	760
Pittsburgh	687	561	216	452	182	1204	1411	1313	838	388	752	1200	857	553	368	895	1215	*	445	588	984
Raleigh	372	685	605	784	491	1166	1661	1160	1061	541	728	819	1189	521	489	1214	1547	445	*	804	1129
St. Louis	541	1141	716	289	406	630	857	779	257	263	285	1196	552	299	948	449	824	588	804	*	396
Tulsa	772	1537	1112	683	802	257	681	478	248	659	401	1398	695	609	1344	387	760	984	1129	396	*

(Hamilton path) from the chemical solution? What other kinds of problems might be solved using DNA computing? What are the implications of Adelman's discovery for the future of computing?

D. Ant Colony Optimization

An individual ant is, by most standards, a dumb little creature, but collectively an entire ant colony can perform surprising feats of teamwork (such as lifting and carrying large leaves or branches) and self-organize to solve remarkably complex problems (finding the shortest route to a food source, optimizing foraging strategies, managing a smooth and steady traffic flow in congested ant highways). The ability of ants and other social insects to perform sophisticated group tasks goes by the name of *swarm intelligence*. Since ants don't talk to each other and don't have bosses telling them what to do, swarm intelligence is a decentralized, spontaneous type of intelligence that has many potential applications at the human scale. In recent years, computer scientists have been able to approach many difficult optimization problems (including TSPs) using *ant colony optimization* software (i.e., computer programs that use *virtual ants* to imitate the problem-solving strategies of real ants).

Write a research paper describing the concept of swarm intelligence and some of the recent developments in ant colony optimization methods and, in particular, the use of virtual ant algorithms for solving TSPs.

E. The Knight's Tour

The knight's tour is a mathematical problem with a long history and closely tied to the topic of this chapter. A legal move for a *knight* on a "chessboard" allows the knight to move another square that is two squares away in one direction (Left/Right or Up/Down) and one square away in the other direction. By a "chessboard" here we mean an array of squares such as the ones in an 8 by 8 traditional chessboard, but of arbitrary dimensions (m rows by n columns). A knight's tour is a trip by a knight starting at some specified square of the chessboard that passes through each of the other squares exactly once. If from the last square the knight can return to the starting square by one more move, then we have a closed knight's tour. Otherwise it's an open knight's tour. Depending on the dimensions of the chessboard and the starting position, a knight's tour may or may not be possible. (For example, in a 3 by 4 chessboard, closed tours are impossible but open tours are possible; in a 4 by 4 chessboard neither open nor closed tours are possible; in a 4 by 5 chessboard both open and closed tours are possible.) In this project, you are asked to prepare a presentation/paper discussing knight tours on chessboards of different sizes and the connection between knight tours and Hamilton circuits/paths on a graph.

You should have no trouble finding plenty of really good information on knight tours on the Web. This is a well-traveled topic.

References and Further Readings

1. Adleman, Leonard, "Computing with DNA," *Scientific American*, 279 (August 1998), 54–61.

2. Bell, E. T., "An Irish Tragedy: Hamilton," in *Men of Mathematics*. New York: Simon and Schuster, 1986, chap. 19.

3. Bellman, R., K. L. Cooke, and J. A. Lockett, *Algorithms, Graphs and Computers*. New York: Academic Press, Inc., 1970, chap. 8.

4. Berge, C., *The Theory of Graphs and Its Applications*. New York: Wiley, 1962.

5. Bonabeau, Eric, and G. Théraulaz, "Swarm Smarts," *Scientific American*, 282 (March 2000), 73–79.

6. Chartrand, Gary, *Graphs as Mathematical Models*. Belmont, CA: Wadsworth Publishing Co., Inc., 1977, chap. 3.

7. Devlin, Keith, *Mathematics: The New Golden Age*. London: Penguin Books, 1988, chap. 11.

8. Kolata, Gina, "Analysis of Algorithms: Coping With Hard Problems," *Science*, 186 (November 1974), 520–521.

9. Lawler, E. L., Lenstra, J. K., Rinooy Kan, A. H. G., and D. B. Shmoys, *The Traveling Salesman Problem*. New York: John Wiley & Sons, Inc., 1985.

10. Lewis, H. R., and C. H. Papadimitriou, "The Efficiency of Algorithms," *Scientific American*, 238 (January 1978), 96–109.

11. Michalewicz, Z., and D. B. Fogel, *How to Solve It: Modern Heuristics*. New York: Springer-Verlag, 2000.

12. Papadimitriou, Christos H., and Kenneth Steiglitz, *Combinatorial Optimization: Algorithms and Complexity*. New York: Dover, 1998, chap. 17.

13. Peterson, Ivars, *Islands of Truth: A Mathematical Mystery Cruise*. New York: W. H. Freeman & Co., 1990, chap. 6.

14. Wilson, Robin, and John J. Watkins, *Graphs: An Introductory Approach*. New York: John Wiley & Sons, Inc., 1990.

7

The Mathematics of Networks

The Cost of Being Connected

con-nec' tion:
(1) that which
connects or
unites; a tie; a
bond; means of
joining; (2) a line
of communication
from one point to
another.
Webster's New 20th
Century Dictionary

Transportation networks, communication networks, social networks, the Internet. Nowadays networks are everywhere, which is probably why it's so hard to find a good definition of the concept in the dictionary. Merriam-Webster's Online Dictionary lists many different definitions of the word *network*, which run the gamut, from "a radio or television company that produces programs for broadcast" to "a fabric or structure of cords or wires that cross at regular intervals and are knotted or secured at the crossings." Huh?

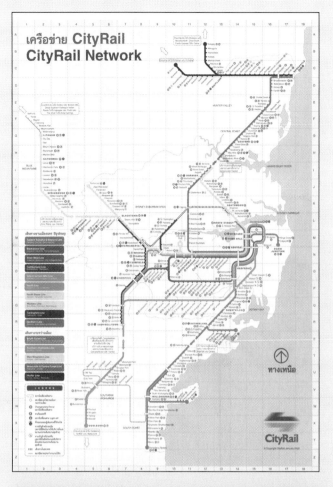

ortunately, our definition of a network is going to be really simple—essentially, *a network is a graph that is connected*. In this context the term is most commonly used when the graph models a real-life "network." Typically, the vertices of a network (sometimes called *nodes* or *terminals*) are "objects"—cities, subway stations, computer servers, radio telescopes, human beings, and so on. The edges of a network (which in this context are often called *links*) indicate *connections* among the objects—airline routes, rail lines, Internet connections, social connections, and so on.

Undoubtedly one of the largest, most complex, and most unpredictable networks around is the World Wide Web. The World Wide Web is a classic example of an *evolutionary* network—it follows no predetermined master plan and essentially evolves on its own without structure or centralized direction (yes, there are some rules of behavior but, as we all know, not that many). Most social animals, including humans, belong to many different overlapping evolutionary networks. (Think of how many different such networks you belong to—your family, your friends, *myspace.com*, and so on.)

At the opposite end of the spectrum from evolutionary networks are networks that are centrally planned and carefully designed to meet certain goals and objectives. Often these types of networks are very expensive to build, and one of the primary considerations when designing such networks is minimizing their cost. This certainly applies to networks of roads, fiber-optic cable lines, rail lines, power lines, and so on. In most of these cases the cost of the network is roughly proportional to its length (x dollars per mile), which means that the

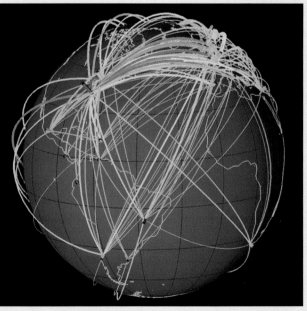

cheapest networks are going to be the shortest ones. Of course, there are other situations in which shortest may not necessarily be cheapest (freeways, for example—a bridge or an overpass can eat up more of the budget than many miles of roadbed).

The general theme of this chapter is the problem of finding *optimal networks* connecting a set of points. Within this context, an optimal network is a shortest or cheapest network. Thus, the design of an optimal network involves two basic goals: (i) to connect all the "terminals" (cities, train stations, power stations, etc.), and (ii) to make the total cost (length) of the network as small as possible. For obvious reasons, problems of this type are known as *minimum network problems*.

The backbone of a minimum network is a special type of graph called a *tree*. This chapter starts with a discussion of the properties of *trees* (Section 7.1) and *minimum spanning trees* (Section 7.2). Section 7.3 describes *Kruskal's algorithm*, an algorithm for finding the minimum spanning tree of any network. Sections 7.4 and 7.5 give an introduction to the geometry of *shortest networks*. (In contrast to the general theory of minimum spanning trees, which is very simple, the general theory of shortest networks is quite complicated, and Sections 7.4 and 7.5 are intended to offer just a brief glimpse of the overall picture.)

7.1 Trees

> **EXAMPLE 7.1** The Amazonian Cable Network

The Amazonia Telephone Company is contracted to provide telephone, cable, and Internet service to the seven small mining towns shown in Fig. 7-1(a). These towns are located deep in the heart of the Amazon jungle, which makes the project particularly difficult and expensive. Cell phone towers and wireless communication are out of the question in this setting, so the only practical option is to create a network of underground fiber-optic cable lines connecting the towns. In addition, it makes sense to bury the underground cable lines along the already existing roads connecting the towns. (How would you maintain and repair the lines otherwise?) The existing network of roads is also shown in Fig. 7-1(a). The weighted graph in Fig. 7-1(b) is a graph model describing the situation. The vertices of the graph represent the towns, the edges of the graph represent the existing roads, and the weight of each edge represents the cost (in millions of dollars) of putting a fiber-optic cable connection along that road.

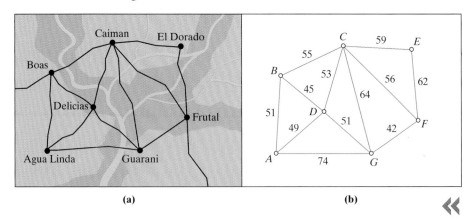

FIGURE 7-1   (a)   (b)

The basic question raised in Example 7.1 is: Find a network that (i) utilizes the existing network of roads; (ii) connects all the towns; and (iii) has the least cost.

When we reformulate these three requirements in the language of graphs, this is what we get:

(i) The network must be a **subgraph** of the original graph (in other words, its edges must come from the original graph).

(ii) The network must **span** the original graph (in other words, it must include *all* the vertices of the original graph).

(iii) The network must be **minimal** (in other words, the total weight of the network should be as small as possible).

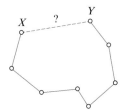

span: to extend, reach, or stretch across.

Webster's 20th Century Dictionary

FIGURE 7-2

The last requirement has an important corollary—a minimal network *cannot have any circuits*. Why not? Imagine that the solid edges in Fig. 7-2 represent already existing links in a mininal network. Why would you then build the link between *X* and *Y* and close the circuit. The edge *XY* would be a *redundant* link of the network. (In many real-life applications, other factors beyond cost—such as reliability, convenience, etc.—come into play. In these situations redundant links are quite important.)

It is now time for a few formal definitions.

■ A **network** is just another name for a connected graph. (This terminology is most commonly used when the graph models a real-life application.)

■ A network with no circuits is called a **tree**.

■ A **spanning tree** of a network is a *subgraph* that connects *all* the vertices of the network and has no circuits.

■ Among all spanning trees of a *weighted network*, one with least total weight is called a **minimum spanning tree** (MST) of the network.

EXAMPLE 7.2 Tree or Not?

Figure 7-3 shows eight different graphs. The graphs in Figs. 7-3(a) and (b) are disconnected, so they are not even networks, let alone trees. The graphs in Figs. 7-3(c) and (d) are networks that have circuits so neither of them is a tree. The graphs in Figs. 7-3(e) and (f) are networks with no circuits, so they are indeed trees. The structure of a family tree [Fig. 7-3(g)] and the structure formed by the bonds of some molecules [Fig. 7-3(h)] are also trees.

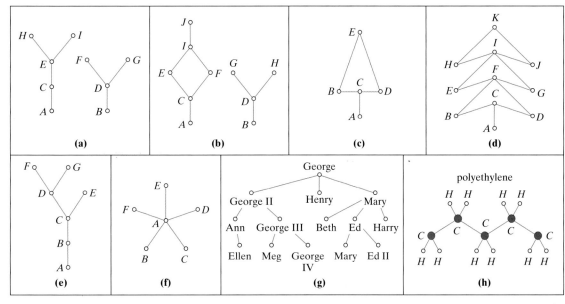

FIGURE 7-3

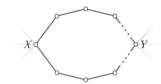

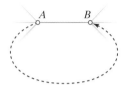

FIGURE 7-4 Two different paths joining *X* and *Y* make a circuit.

FIGURE 7-5 If *AB* is not a bridge, then it must be part of a circuit.

Properties of Trees

As graphs go, trees occupy an important niche between disconnected graphs and "overconnected" graphs. A tree is special by virtue of the fact that it is *barely connected*. This means several things:

- Given any two vertices *X* and *Y* of a tree, there is one and *only one* path joining *X* to *Y*. [Suppose there were *two different* paths joining *X* to *Y*. The graph would then have a circuit (Fig. 7-4).]
- Every edge of a tree is a *bridge*. [Suppose the edge *AB* is not a bridge. Then without *AB* the graph is still connected, so there would be an alternative path from *A* to *B*. This would imply that the edge *AB* is part of a circuit (Fig. 7-5).]
- Among all networks with *N* vertices, a tree is the one with the *fewest* number of edges.

The next example illustrates the special role that trees play in the creation of a network.

> ## EXAMPLE 7.3 Connect the Dots (and Then Stop)

Imagine the following "connect-the-dots" game: Start with eight isolated vertices. The object of the game is to create a network connecting the vertices by adding edges, one at a time. You are free to create any network you want. In this game, bridges are good and circuits are bad. (Imagine, for example, that for each bridge in your network you get a $10 reward, but for each circuit in your network you pay a $10 penalty.)

So grab a marker and start playing. We will let *M* denote the number of edges you have added at any point in time. In the early stages of the game ($M = 1, 2, \ldots, 6$) the graph is disconnected [Fig. 7-6(a) through (d)], so there is

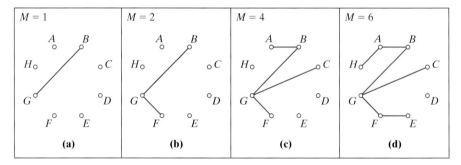

FIGURE 7-6 For small values of *M* the graph is disconnected.

no network. As soon as you get to $M = 7$ (and if you stayed away from forming any circuits) the graph becomes connected. Some of the possible configurations are shown in Fig. 7-7. Each of these networks is a tree, and thus each of the seven edges is a bridge. Stop here and you will come out $70 richer.

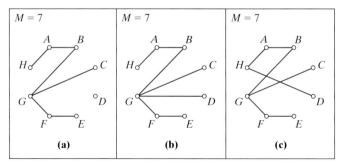

FIGURE 7-7 When $M = 7$, we have a tree—a network with no circuits.

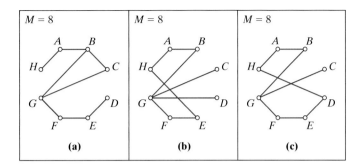

FIGURE 7-8 When $M = 8$, we have a network with a circuit.

Interestingly, this is as good as it will get. When $M = 8$, the graph will have a circuit—it just can't be avoided. In addition, the edges in that circuit will no longer be bridges. [The graph in Fig. 7-8(a) has a circuit and five bridges; the graph in Fig. 7-8(b) has a circuit and two bridges; and the graph in Fig. 7-8(c) has a circuit and only one bridge.] As the value of M increases, the number of circuits goes up (very quickly) and the number of bridges goes down [Figs. 7-9(a), (b), and (c)].

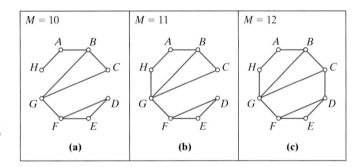

FIGURE 7-9 For larger values of M, we have a network with lots of circuits.

Our discussion of trees can be summarized into the following three key properties.

Property 1
- In a tree, there is one and only one path joining any two vertices.
- If there is one and only one path joining any two vertices of a graph, then the graph must be a tree.

Property 2
- In a tree, every edge is a bridge.
- If every edge of a graph is a bridge, then the graph must be a tree.

Property 3
- A tree with N vertices has $N - 1$ edges.
- If a *network* has N vertices and $N - 1$ edges, then it must be a tree.

FIGURE 7-10 A disconnected graph with 10 vertices and 9 edges.

Notice that a disconnected graph (i.e. not a network) can have N vertices and $N - 1$ edges, as shown in Fig. 7-10.

7.2 Spanning Trees

When we reformulate Property 3 in a slightly more general context, we get the following useful property:

> **Property 4**
>
> - If a network has N vertices and M edges, then $M \geq N - 1$.
>   [For convenience, we will refer to the difference $R = M - (N - 1)$ as the **redundancy** of the network.]
> - If $M = N - 1$, the network is a tree; if $M > N - 1$, the network has circuits and is not a tree. (In other words, a tree is a network with zero redundancy and a network with positive redundancy is not a tree.)

In the case of a network with positive redundancy there are many trees *within* the network that connect its vertices—these are what we call the *spanning trees* of the network.

> **EXAMPLE 7.4** Counting Spanning Trees: Part 1

The network in Fig. 7-11(a) has $N = 8$ vertices and $M = 8$ edges. The redundancy of the network is $R = 1$, so to find a spanning tree we will have to "discard" one edge. Five of these edges are bridges of the network, and they *will have to be part of any spanning tree*. The other three edges (BC, CG, and GB) form a circuit of length 3, and if we exclude any one of the three edges we will have a spanning tree. Thus, the network has three different spanning trees [Figs. 7-11(b), (c), and (d)].

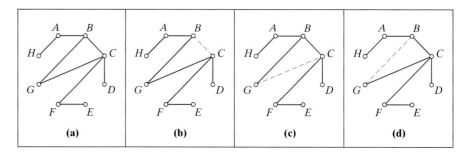

FIGURE 7-11

The network in Fig. 7-12(a) has $M = 9$ edges and $N = 8$ vertices. The redundancy of the network is $R = 2$, so to find a spanning tree we will have to "discard" two edges. Edges AB and AH are bridges of the network, so they will have to be part of any spanning tree. The other seven edges are split into two separate circuits (B, C, G, B of length 3 and C, D, E, F, C of length 4). A spanning tree can be found by "busting" each of the two circuits. This means excluding two edges in the network—any one of the edges of B, C, G, B, and any one of the edges of C, D, E, F, C. For example, if we exclude BC and CD we get the spanning tree shown in Fig. 7-12(b). We could also exclude BC and DE and get the spanning tree shown in Fig. 7-12(c). And so on. Given that there are $3 \times 4 = 12$ different ways to choose an edge from the circuit of length 3 and an edge from

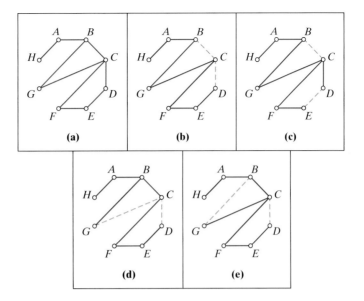

FIGURE 7-12

the circuit of length 4, we will not show all 12 spanning trees. [Figures 7-12(b) through (e) are just a sample.] ≪

EXAMPLE 7.5 Counting Spanning Trees: Part 2

The network in Fig. 7-13(a) is another network with $M = 9$ edges and $N = 8$ vertices. The difference between this network and the one in Fig. 7-12(a) is that here the circuits B, C, G, B and C, D, E, G, C have a common edge CG. Determining which pairs of edges can be excluded in this case is a bit more complicated.

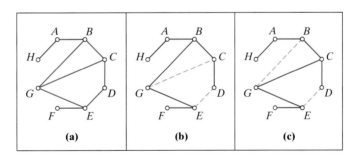

FIGURE 7-13

For a little practice counting spanning trees, see Exercises 15–18

If one of the excluded edges is the common edge CG, then the other excluded edge can be any other edge in a circuit. There are five choices (BC, CD, DE, EG, and GB), and each choice will result in a different spanning tree [one of these is shown in Fig. 7-13(b)]. The alternative scenario is to exclude two edges neither of which is the common edge CG. In this case one excluded edge has to be either BC or BG (to "bust" circuit B, C, G, B), and the other excluded edge has to be either CD, DE, or EG (to "bust" circuit C, D, E, G, C). There are $2 \times 3 = 6$ possible spanning trees that can be formed this way [one of these is shown in Fig. 7-13(c)]. Combining the two scenarios gives a total of 11 possible spanning trees for the network in Fig. 7-13(a). ≪

See Exercise 49.

> ### ▶ **EXAMPLE 7.6** Minimum Spanning Trees

Fig. 7-14(a) is an example of a weighted network with $N = 8$ vertices and $M = 11$ edges. For the sake of simplicity, let's assume the weights represent costs (\$). Our goal is to find the *minimum spanning tree* of the network—there is a \$100 reward if we can do it. This network has a redundancy of $R = 4$, and if nothing else, the lesson of Examples 7.4 and 7.5 is that it is likely to have quite a few spanning trees. If at all possible, we would like to find the minimum spanning tree without having to go through all possible spanning trees and compute their weights (we have gone down that road in Chapter 6 and know it's not the most efficient way to do business).

From what we learned in Examples 7.4 and 7.5, we can develop a reasonable strategy. We are going to have to "bust" all the circuits in the network, and we "bust" a circuit by choosing one of its edges and *excluding* it from the spanning tree. So, given a choice between an expensive edge and a cheaper edge, which one should we exclude? Obviously the more expensive one. Looking at circuit A, B, H, A, for example, it makes sense to exclude the expensive edge BH rather than the cheaper edges AB or AH. If we continue this way, excluding the most expensive edge left in each circuit, we end up with the spanning tree shown in Fig. 7-14(b).

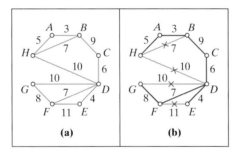

(a)   (b)

FIGURE 7-14

Intuitively, the spanning tree found in Fig. 7-14(b) looks like a good candidate for the minimum spanning tree of the network in Fig 7-14(a). But how can we be certain? And, more important, what assurances do we have that the process will work in more complicated graphs? These are the questions we will answer next.

7.3 Kruskal's Algorithm

There are several well-known algorithms for finding minimum spanning trees. In this section we will discuss one of the nicest of these, called **Kruskal's algorithm** (after the American mathematician Joseph Kruskal, who first proposed the algorithm in 1956).

Kruskal's algorithm is wickedly simple and almost identical to the *cheapest-link algorithm* we discussed in Chapter 6: We build the minimum spanning tree one edge at a time, choosing at each step the cheapest available edge. The only restriction to our choice of edges is that we should never choose an edge that creates a circuit. (On the other hand, having three or more edges coming out of a vertex is now OK.) What is truly remarkable about Kruskal's algorithm is that—unlike the cheapest-link algorithm—it always gives an optimal solution.

Before giving a formal description of Kruskal's algorithm, let's look at a simple example.

EXAMPLE 7.7 The Amazonian Cable Network: Part 2

In Example 7.1 we raised the following problem: What is the minimum spanning tree (MST) of the network shown in Fig. 7-15(b)? (Recall that the graph represented the network of roads connecting the seven towns, with the weights of each edge representing the costs [in millions] of laying a fiber-optic line along that road.)

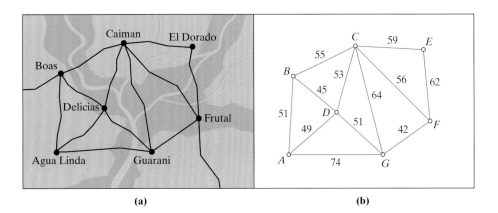

FIGURE 7-15   (a)   (b)

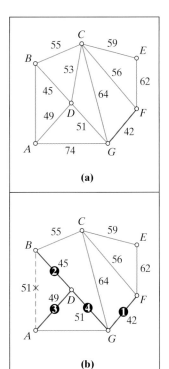

FIGURE 7-16

We will use Kruskal's algorithm to find the MST of the network shown in Fig. 7-15(b). Here are the details:

- **Step 1.** Among all the possible links, we choose the cheapest one, in this particular case GF (at a cost of $42 million). This link is going to be part of the MST, and we mark it in red (or any other color) as shown in Fig. 7-16(a). (Note that this does not have to be the first link actually built—we are putting the network together on paper only. On the ground, the schedule of construction is a different story, and there are many other factors that need to be considered.)

- **Step 2.** The next cheapest link available is BD at $45 million. We choose it for the MST and mark it in red.

- **Step 3.** The next cheapest link available is AD at $49 million. Again, we choose it for the MST and mark it in red.

- **Step 4.** For the next cheapest link there is a tie between AB and DG, both at $51 million. But we can rule out AB—it would create a circuit in the MST, and we can't have that! (For bookkeeping purposes it is a good idea to erase or cross out the link.) The link DG, on the other hand, is just fine, so we mark in red and make it part of the MST.
    Figure 7-16(b) shows how things look at this point.

- **Step 5.** The next cheapest link available is CD at $53 million. No problems here, so again, we mark it in red and make it part of the MST.

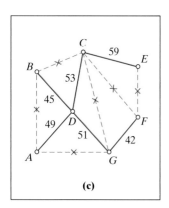

FIGURE 7-16 (continued)

Step 6. The next cheapest link available is *BC* at $55 million, but this link would create a circuit, so we cross it out. The next possible choice is *CF* at $56 million, but once again, this choice creates a circuit so we must cross it out. The next possible choice is *CE* at $59 million, and this is one we do choose. We mark it in red and make it part of the MST.

Step . . . Wait a second—we are finished! Even without looking at a picture, we can tell we are done—six links is exactly what is needed for an MST on seven vertices.

Figure 7-16(c) shows the MST in red. The total cost of the network is $299 million.

The following is a formal description of Kruskal's algorithm.

> **Kruskal's Algorithm**
>
> - **Step 1.** Pick the *cheapest edge* available. (In case of a tie, pick one at random.) Mark it (say in red).
> - **Step 2.** Pick the next cheapest link available and mark it.
> - **Steps 3, 4, . . . , $N - 1$.** Continue picking and marking the cheapest unmarked link available that does not create a circuit.

As algorithms go, Kruskal's algorithm is as good as it gets. First, it is easy to implement. Second, it is an *efficient* algorithm. As we increase the number of vertices and edges of the network, the amount of work grows more or less proportionally (roughly speaking, if finding the MST of a 30-city network takes you, say, 30 minutes, finding the MST of a 60-city network might take you 60 minutes). Last, but not least, Kruskal's algorithm is an *optimal* algorithm—it will always find a minimum spanning tree.

Thus, we have reached a surprisingly satisfying end to the MST story: No matter how complicated the network, we can find its minimum spanning tree by means of an algorithm that is easy to understand and implement, is *efficient*, and is also *optimal*. We couldn't ask for better karma.

7.4 The Shortest Network Connecting Three Points

Minimum spanning trees represent the optimal way to connect a set of points based on one key assumption—that all the connections should be along the links of the network. For example, in the Amazonian cable network problem, the optimal network was an MST because it didn't make sense to bury the fiber-optic lines through uncleared jungle—the only realistic placement of the underground fiber-optic lines was along the already existing roads (the links of the network). But what if, in a manner of speaking, we don't have to *follow the road*? What if we are free to create new vertices and links "outside" the original network? To clarify the distinction, let's look at a new type of cable network problem.

> ### EXAMPLE 7.8 The Outback Cable Network

The Amazonia Telephone Company has now relocated to Australia, and now calls itself the Outback Cable Network. The company has a new contract to create an underground fiber-optic network connecting the towns of Alcie Springs, Booker Creek, and Camoorea—three isolated towns located in the heart of the Australian outback [Fig. 7-17(a)]. By freakish coincidence (or good planning) the towns are equidistant, forming an equilateral triangle 500 miles on each side. The three towns are connected by three unpaved straight roads [Fig. 7-17(b)]. The question, once again, is what is the cheapest underground fiber-optic cable network connecting the three towns?

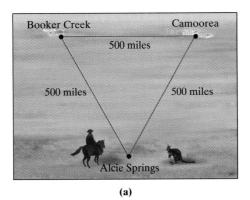

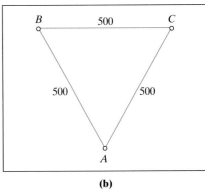

FIGURE 7-17   (a)   (b)

Obviously, the story has a lot of similarities to the Amazonian cable network problem we solved in Example 7.7, but there is one crucial difference, and this difference is based on where we can and cannot put the underground cable lines. The Amazon is heavy jungle, big rivers, and dangerous critters—you want to stay on the road (i.e., no shortcuts!). The Australian outback is flat scrub with few major roads to speak of—there is little or no disadvantage to going offroad (i.e., if you can find a shortcut, you take it!). Based on the flat and homogeneous nature of the terrain, we will assume that in the outback cable lines can be laid anywhere (along the road or offroad) and that there is a fixed cost of $100,000 per mile. Here, *cheapest* means "shortest," so the name of the game to design a network that is as short as possible. We shall call such a network the **shortest network** (SN).

The search for the shortest network often starts with a look at the minimum spanning tree. The MST can always be found using Kruskal's algorithm and it gives us a ceiling on the length of the shortest network. In this example the MST consists of two (any two) of the three sides of the equilateral triangle [Fig. 7-18(a)], and its length is 1000 miles.

It is not hard to find a network connecting the three towns shorter than the MST. The T-network in Fig. 7-18(b) is clearly shorter. The length of the segment *CJ* can be computed using properties of 30-60-90 triangles and is approximately 433 miles. It follows that the length of this network is approximately 933 miles.

We can do better. The Y-network shown in Fig. 7-18(c) is even shorter than the T-network. In this network there is a "Y"-junction at *S*, with three equal branches connecting *S* to each of *A*, *B*, and *C*. This network is approximately 866

A shortest network by definition is shorter than the MST or, in the worst of cases, equal to the MST—to have a shortest network longer than the MST is a contradiction in terms.

See Exercise 47. (For a primer on 30-60-90 triangles, see first Exercises 45 and 46.)

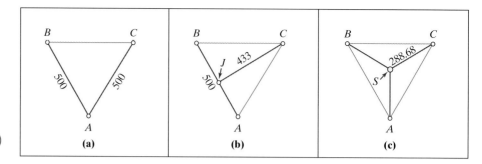

FIGURE 7-18 (a) The MST. (b) A shorter network with a T-junction. (c) The SN, with an interior Y-junction.

See Exercise 48.

miles long. A key feature of this network is the way the three branches come together at the junction point S, forming equal 120° angles.

Even without a formal mathematical argument, it is not hard to convince oneself that the Y-network in Fig. 7-18(c) is the shortest network connecting the three towns. An informal argument goes like this: Since the original layout of the towns is completely symmetric (it looks the same from each of the three towns), we would expect that the shortest network should also be completely symmetric. The only network that looks the same from each of the three towns is a "perfect" Y where all three segments have equal length. ◀◀

Before we move on, we need to discuss briefly the notion of a *junction point* in a network. (A **junction point** in a network is any point where two or more segments of the network come together.) The MST in Fig. 7-18(a) has a junction point at A, the network in Fig. 7-18(b) has a junction point at J, and the shortest network in Fig. 7-18(c) has a junction point at S. An important difference between them is that in the case of the MST, the junction point A is a *native* point—one of the original cities of the problem—but in the other two networks the junction points are *nonnative* points introduced as part of the solution.

There are three important terms that we will use in connection with junction points:

- In a network connecting a set of vertices, a junction point is said to be a **native** junction point if it is located at one of the vertices.
- A nonnative junction point located somewhere other than at one of the original vertices is called an **interior** junction point of the network.
- An interior junction point consisting of three line segments coming together forming equal 120° angles (a "perfect" Y-junction if you will) is called a **Steiner point** of the network.

Steiner points are named after the Swiss mathematician Jakob Steiner (1796–1863).

▶ EXAMPLE 7.9 The Third Trans-Pacific Cable

This is a true story. In 1989, a consortium of several of the world's biggest telephone companies (among them AT&T, MCI, Sprint, and British Telephone) completed the Third Trans-Pacific Cable (TPC-3) line, a network of submarine fiber-optic lines linking Japan and Guam to the United States (via Hawaii). The approximate straight-line distances (in miles) between the three endpoints of the network (Chikura, Japan; Tanguisson Point, Guam; and Oahu, Hawaii) are shown in Fig. 7-19. By and large, laying submarine cable has a fixed cost per mile (somewhere between $50,000 and $70,000), so *cheapest* means "shortest" and the problem is once again to find the shortest network connecting the three endpoints.

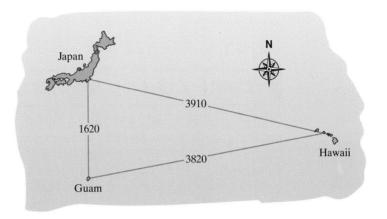

FIGURE 7-19

Given what we learned in Example 7.8, a reasonable guess is that the shortest network is going to require an interior junction point, somewhere inside the Japan–Guam–Hawaii triangle. But where? Figure 7-20(a) shows the answer: The junction point S is located in such a way that three branches of the network coming out of S form equal 120° angles—in other words, S is the *Steiner point of the triangle.*

Figure 7-20(b) shows a stamp issued by the Japanese Postal Service to commemorate the completion of TCP-3. The length of the shortest network is 5180 miles, but this is a theoretical length only, based on straight-line distances. With submarine cable one has to add as much as 10% to the straight-line lengths

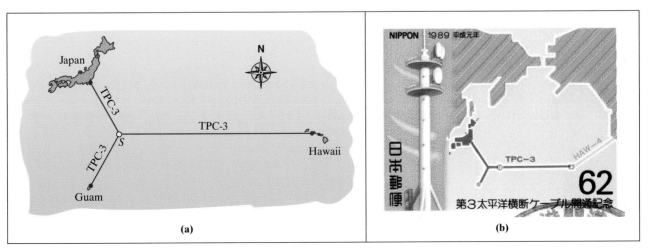

FIGURE 7-20

because of the uneven nature of the ocean floor. (The actual length of submarine cable used in TPC-3 is about 5690 miles.) ◀◀

From Examples 7.8 and 7.9 we are tempted to conclude that the key to finding the shortest network connecting three points (cities) *A*, *B*, and *C* is to find the *Steiner point S* inside triangle *ABC* as the interior junction point of the network (i.e., the three segments of the network *AS*, *BS*, and *CS* form equal 120° angles at *S*). This is true as long as we can find a Steiner point inside the triangle, but, as we will see in the next example, *not every triangle has a Steiner point!*

▶ EXAMPLE 7.10 A High-Speed Rail Network

Off and on, there is talk of building a high-speed rail connection between Los Angeles and Las Vegas. To make it more interesting, let's add a third city—Salt Lake City—to the mix. The straight-line distances between the three cities are shown in Fig. 7-21(a). The mathematical question again is, What is the shortest network connecting these three cities? (High-speed trains run on expensive special tracks that have to be built expressly for them, and it is critical to make the network as short as possible.)

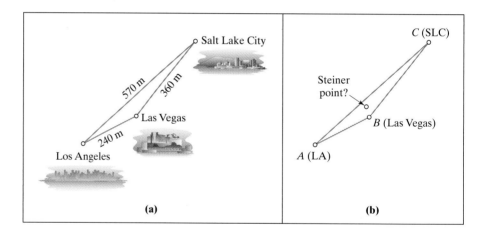

(a)   (b)

FIGURE 7-21

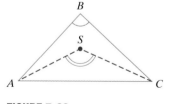

FIGURE 7-22

The first idea that comes to mind is that we should locate the Steiner point of the triangle *ABC* [Fig. 7-21(b)]. Imagine that there is such a point somewhere inside the triangle *ABC* and call it *S*. The first observation is a simple but important general property of triangles illustrated in Fig. 7-22: *For any triangle ABC and interior point S, angle ASC must be bigger than angle ABC.* It follows that for angle *ASC* to measure 120°, the measure of angle *ABC* must be *less* than 120°.

Unfortunately (or fortunately—depends on how you look at it) the angle *ABC* in this example measures about 155°, and therefore there is no Steiner junction point inside the triangle. Without a Steiner junction point, how do we find the shortest network? The answer turns out to be surprisingly simple: In this situ-

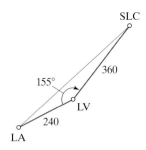

FIGURE 7-23 The SN connecting LA, LV (Las Vegas), and SLC (Salt Lake City).

ation the shortest network consists of the two shortest sides of the triangle, as shown in Fig. 7-23. Please notice that this shortest network happens to be the *minimum spanning tree* as well!

Two important lessons are to be drawn from Example 7.10. First, if a triangle has an angle whose measure is bigger or equal to 120°, then the triangle has no Steiner point. Second, when a triangle has no Steiner point, the shortest network connecting the vertices of the triangle consists of the two shorter sides of the triangle, and this happens to also be the *minimum spanning tree* of the triangle.

When we put it all together, combining the lessons of Examples 7.8, 7.9, and 7.10, the shortest network connecting three points is given by the following general solution:

The Shortest Network Connecting Three Points

■ If one of the angles of the triangle is 120° or more, the shortest network linking the three vertices consists of the two shortest sides of the triangle [Fig. 7-24(a)]. In this situation, *the shortest network coincides with the minimum spanning tree.*

■ If all three angles of the triangle are less than 120°, the shortest network is obtained by finding a *Steiner point S* inside the triangle and joining *S* to each of the vertices [Fig. 7-24(b)].

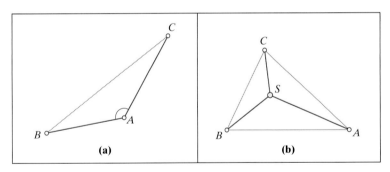

FIGURE 7-24 The SN connecting *A, B*, and *C*: (a) One angle is greater than or equal to 120°. (b) All three angles are less than 120°.

Finding the Steiner Point: Torricelli's Construction

Evangelista Torricelli (1608–1647). For more on Torricelli's story, see the biographical profile at the end of the chapter.

Finding the shortest network connecting three points is a problem with a long and interesting history going back some 400 years. In the early 1600s the Italian Evangelista Torricelli came up with a remarkably simple and elegant method for locating a Steiner junction point inside a triangle. All you need to carry out Torricelli's construction is a straightedge and a compass; all you need to understand why it works is a few facts from high school geometry. Here it goes:

Torricelli's Construction

Suppose A, B, and C form a triangle such that all three angles of the triangle are less than 120° [Fig. 7–25(a)].

- **Step 1.** Choose any of the three sides of the triangle (say BC) and construct an equilateral triangle BCX, so that X and A are on opposite sides of BC [Fig. 7-25(b)].
- **Step 2.** Circumscribe a circle around equilateral triangle BCX [Fig. 7-25(c)].
- **Step 3.** Joint X to A with a straight line [Fig. 7-25(d)]. The point of intersection of the line segment XA with the circle is the Steiner point!

Why does the construction work? It is sufficient to prove that under this construction, angles BSA and BSC both measure 120°. Then angle CSA will also measure 120°, proving that S is indeed the Steiner point.

Fact 1: Angle BSX measures 60°. *Reasons:* (a) Angle BCX measures 60° since BCX is an equilateral triangle by construction; (b) Angles BSX and BCX have equal measures because the two angles are inscribed in the same circle and are opposite to the same chord BX.

Fact 2: Angle CSX measures 60°. *Reason:* Use an analogous argument as for Fact 1.

Fact 3: Angle BSC measures 120°. *Reason:* Facts 1 and 2.

Fact 4: Angle BSA measures 120°. *Reason:* Fact 1.

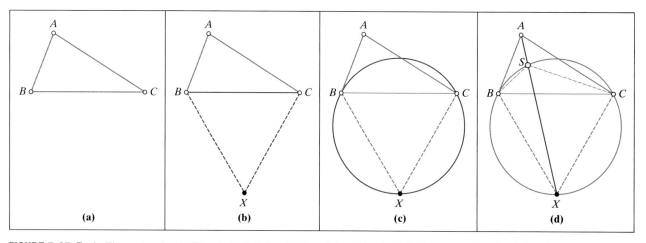

FIGURE 7-25 Torricelli's construction: (a) Triangle ABC. (b) Construct equilateral triangle BXC. (c) Circumscribe triangle BXC in a circle. (d) S is the point of intersection of the circle and the line segment joining X and A.

7.5 Shortest Networks for Four or More Points

When it comes to finding shortest networks, things get really interesting when we have to connect four points.

EXAMPLE 7.11 Four-City Networks: Part 1

Imagine four cities (A, B, C, and D) that need to be connected by an underground fiber-optic cable network. Suppose that the cities sit on the vertices of a square 500 miles on each side, as shown in Fig. 7-26(a). What does the optimal network connecting these cities look like? It depends on the situation.

If we *don't want to create any interior junction points in the network* (either because we don't want to venture off the prescribed paths—as in the jungle scenario—or because the cost of creating a new junction is too high), then the answer is a *minimum spanning tree*, such as the one shown in Fig. 7-26(b). The length of the MST is 1500 miles.

On the other hand, if *interior junction points* are allowed, somewhat shorter networks are possible. One obvious improvement is the network shown in Fig. 7-26(c), with an X-junction located at O, the center of the square. The length of this network is approximately 1414 miles.

We can shorten the network even more if we place not one but two interior junction points inside the square. It's not immediately obvious that this is a good move, but once we recognize that two junction points might be better than one, then it's not hard to see that the best option is to make the two interior junction points Steiner points. There are two different networks possible with two Steiner points inside the square, and they are shown in Figs. 7-26(d) and (e). These two networks are essentially equal (one is a rotated version of the other) and clearly have the same length—approximately 1366 miles. It is impossible to shorten these any further—the two networks in Figs. 7-26(d) and (e) are the shortest networks connecting the four cities.

See Exercise 62.

See Exercise 70.

See Exercise 63.

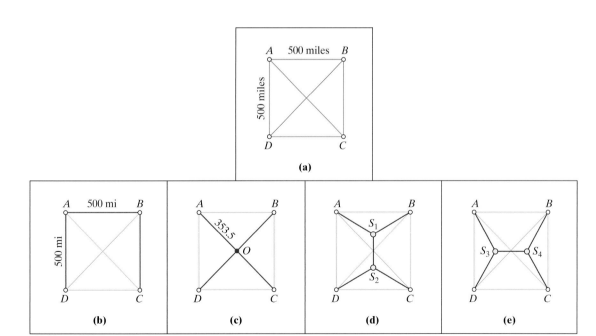

FIGURE 7-26 (a) Four cities located at the vertices of a square. (b) An MST (1500 miles). (c) A shorter network (approximately 1414 miles). (d) An SN (approximately 1366 miles). Junction points S_1 and S_2 are both Steiner points. (e) An equivalent SN, with Steiner points S_3 and S_4.

> ### EXAMPLE 7.12 Four-City Networks: Part 2

Let's repeat what we did in Example 7.11, but this time imagine that the four cities are located at the vertices of a rectangle, as shown in Fig. 7-27(a). By now, we have some experience on our side, so we can cut to the chase. We know that the MST is 1000 miles long [Fig. 7-27(b)]. That's the easy part.

See Exercise 64.

For the shortest network, let's think about what happened in the previous example—a square, after all, is just a special case of a rectangle. An obvious candidate would be a network with two interior Steiner junction points. There are two such networks, shown in Figs. 7-27(c) and 7-27(d), but this time they are not equivalent. The length of the network in Fig. 7-27(c) is approximately 993 miles, while the length of the network in Fig. 7-27(d) is approximately 920 miles—a pretty significant difference. Given these lengths, it is obvious that the network in Fig. 7-27(c) cannot be the shortest network, but what about Fig. 7-27(d)? If there is any justice in this mathematical world, this network fits the pattern and ought to be the shortest. In fact, it is!

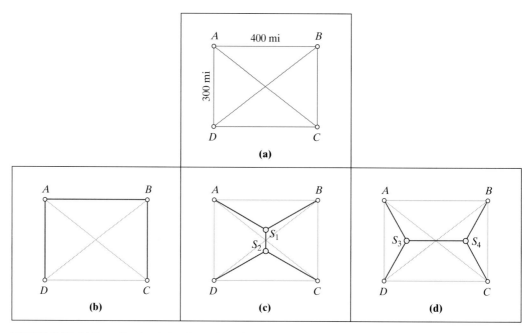

FIGURE 7-27 (a) Four cities located at the vertices of a rectangle. (b) An MST (1000 miles). (c) A network with two Steiner junction points (approximately 993 miles.) (d) The SN (approximately 920 miles).

> ### EXAMPLE 7.13 Four-City Networks: Part 3

Let's look at four cities once more. This time, imagine that the cities are located at the vertices of a skinny trapezoid, as shown in Fig. 7-28(a). The minimum spanning tree is shown in Fig. 7-28(b), and it is 600 miles long.

What about the shortest network? Based on our experience with Examples 7.11 and 7.12, we are fairly certain that we should be looking for a network with a

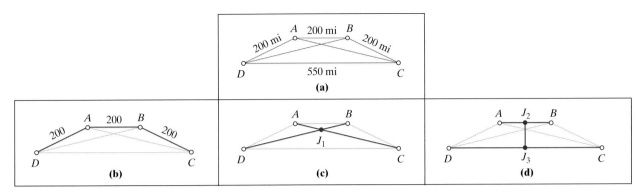

FIGURE 7-28 (a) Four cities located at the vertices of a trapezoid. (b) The MST (600 miles). (c) A network with an X-junction at J_1 is longer than the MST (774.6 miles). (d) A network with a couple of T-junction points at J_2 and J_3 is even worse (846.8 miles).

couple of interior Steiner junction points. After a little trial-and-error, however, we realize that such a layout is impossible! The trapezoid is too skinny, or, to put it in a slightly more formal way, the angles at A and B are greater than $120°$. Since no Steiner points can be placed inside the trapezoid, the shortest network, whatever it is, will have to be one without Steiner junction points.

Well, we say to ourselves, if not Steiner junction points, how about other kinds of interior junction points? How about X-junctions [Fig 7-28(c)], or T-junctions [Fig 7-28(d)], or Y-junctions where the angles are not all $120°$? As reasonable as this idea sounds, a remarkable thing happens: *The only possible interior junction points in a shortest network are Steiner points.* For convenience, we will call this the *interior junction rule* for shortest networks.   ◀◀

> **The Interior Junction Rule for Shortest Networks**
>
> In a shortest network the *interior junction points* are all *Steiner points.*

The interior junction rule is an important and powerful piece of information in building shortest networks, and we will come back to it soon. Meanwhile, what does it tell us about the situation of Example 7.13? It tells us that the shortest network cannot have any interior junction points. Steiner junction points are impossible because of the geometry; other types of junction points do not work because of the interior junction rule. But we also know that the shortest network without interior junction points is the minimum spanning tree! Conclusion: For the four cities of Example 7.13, *the shortest network is the minimum spanning tree*!

▶ EXAMPLE 7.14 Four-City Networks: Part 4

For the last time, let's look at four cities A, B, C, and D. This time, the cities sit as shown in Fig. 7-29(a). The MST is shown in Fig. 7-29(b), and its length is 1000 miles.

We know by now that in searching for the shortest network, the MST is a good starting point—it might even be the shortest network. We also know that if the shortest network is not the MST it will have interior junction points, which by the interior junction rule will have to be Steiner points. Because of the layout of these cities, it is geometrically impossible to build a network with two interior Steiner points. On the other hand, there are three possible networks with a single interior Steiner point [Figs. 7-29(c), (d), and (e)].

See Exercise 71.

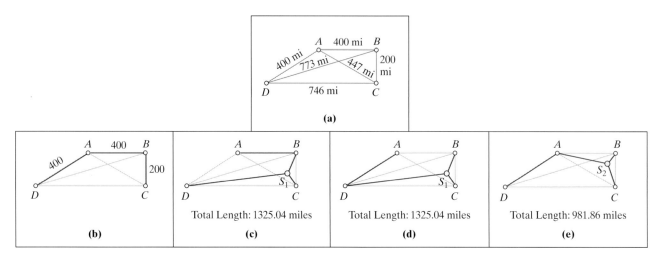

FIGURE 7-29 (a) The four cities. (b) First contender: the MST (1000 miles). (b) Second contender (1325 miles). (c) Third contender (1325 miles). (e) Fourth contender (982 miles) is the SN.

We have now narrowed the list of contenders for the shortest network title to four: the MST or one of the networks shown in Figs. 7-29(c), (d), or (e). All we have to do now to figure which one is the SN is to compute their lengths. The length of the MST is 1000 miles (we knew that!). The networks shown in Figs. 7-29(c) and (d) are both approximately 1325 miles long—that certainly rules them out! The network shown in Fig. 7-29(e) is approximately 982 miles long, which makes it the champ!

There are several lessons that we can draw from the four examples we discussed in this section. The first and clearest lesson is that finding the shortest network is far from easy—even for four cities it's a tough nut to crack. The second lesson is that we should not use the first lesson as an excuse and give up—we can learn a lot about a problem by just poking around. The third lesson is—don't forget the MST! The MST is always a candidate for the title of shortest network, and even when it isn't it gives us a good baseline for the SN. The fourth lesson is that the shortest networks seem to have an affinity for Steiner points—whenever the SN has interior points, it likes them to be Steiner points.

What happens when the number of cities grows—five, six, one hundred? How do we find the shortest network? Here, mathematicians face a situation analogous to the one discussed in Chapter 6—no optimal and efficient algorithm is known, and there are serious doubts that such an algorithm actually exists. At this point, the best we can do is to take advantage of the following rule, which we will informally call the **shortest network rule**.

The Shortest Network Rule

The shortest network connecting a set of points is either

- a *minimum spanning tree* (no interior junction points) or
- a *Steiner tree*. [A **Steiner tree** is a network with no circuits (i.e. a tree) such that all interior junction points are Steiner points.]

This means that we can always find the shortest network by rummaging through all possible Steiner trees, finding the shortest one among them, and comparing it with the minimum spanning tree. The shorter of these two has to be the shortest network. This sounds like a good idea, but, once again, the problem is the explosive growth of the number of possible Steiner trees. With as few as 10 cities, the possible number of Steiner trees we would have to rummage through could be in the millions; with 20 cities, it could be in the billions.

What's the alternative? Just as in Chapter 6, if we are willing to settle for an *approximate solution* (in other words, if we are willing to accept a short network—not necessarily *the shortest*), we can tackle the problem no matter how large the number of cities. Some excellent approximate algorithms for finding short networks are known, and one of them happens to be Kruskal's algorithm for finding a minimum spanning tree. (Mathematicians Frank Hwang of AT&T Bell Labs and Ding-Xhu Du of the University of Minnesota showed in 1990 that the percentage difference in length between the minimum spanning tree and the shortest network *is never more than 13.4%*—see reference 8. Thus, the MST can be used as a decent approximation to the shortest network, with a relative error

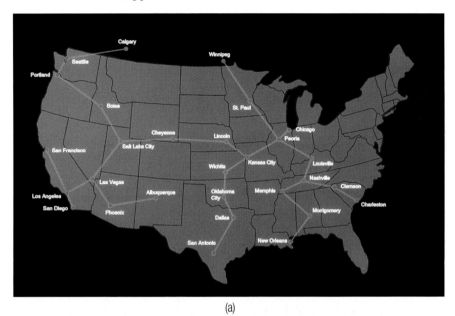

(a)

(b)

FIGURE 7-30 (a) The minimum spanning tree connecting 29 cities in the United States and Canada. Total length: approximately 7600 miles. This MST was found using Kruskal's algorithm in a matter of a few seconds. (b) The shortest network connecting the same 29 cities. (The network has 13 interior Steiner points, shown in yellow.) Total length: approximately 7400 miles (a savings of about 3%). This SN was found by a computer using a sophisticated algorithm developed by E. J. Cockayne and D. E. Hewgill of the University of Victoria.

guaranteed to be under 13.4% but usually much less than that.) Better approximations to the shortest network can be found using sophisticated algorithms that can produce networks that are within 1% of the SN. These are good enough for most real-life applications.

Solution to Old Puzzle: How Short a Shortcut?

By GINA KOLATA

Two mathematicians have solved an old problem in the design of networks that has enormous practical importance but has baffled some of the sharpest minds in the business.

Dr. Frank Hwang of A.T.&T. Bell Laboratories in Murray Hill, N.J., and Dr. Ding Xhu Du, a postdoctoral student at Princeton University, announced at a meeting of theoretical computer scientists last week that they had found a precise limit to the design of paths connecting three or more points.

Designers of things like computer circuits, long-distance telephone lines, or mail routings place great stake in finding the shortest path to connect a set of points, like the cities an airline will travel to or the switching stations for long-distance telephone lines. Dr. Hwang and Dr. Du proved, without using any calculations, that an old trick of adding extra points to a network to make the path shorter can reduce the length of the path by no more than 13.4 percent.

"This problem has been open for 22 years," said Dr. Ronald L. Graham, a research director at A.T.&T. Bell Laboratories who spent years trying in vain to solve it. "The problem is of tremendous interest at Bell Laboratories, for obvious reasons."

In 1968, two other mathematicians guessed the proper answer, but until now no one had proved or disproved their conjecture.

In the tradition of Dr. Paul Erdos, an eccentric Hungarian mathematician who offers prizes for solutions to hard problems that interest him, Dr. Graham offered $500 for the solution to this one. He said he is about to mail his check.

Add Point, Reduce Path

The problem has its roots in an observation made by mathematicians in 1640. They noticed that if you want to find a path connecting three points that form the vertices of an equilateral triangle, the best thing to do is to add an extra point in the middle of the set. By connecting these four points, you can get a shorter route around the original three than if you had simply drawn lines between those three points in the first place.

Later, mathematicians discovered many more examples of this paradoxical phenomenon. It turned out to be generally true that if you want to find the shortest path connecting a set of points, you can shorten the path by adding an extra point or two.

But it was not always clear where to add the extra points, and even more important, it was not always clear how much you would gain. In 1977, Dr. Graham and his colleagues, Dr. Michael Garey and Dr. David Johnson of Bell Laboratories, proved that there was no feasible way to find where to place the extra points, and since there are so many possibilities, trial and error is out of the question.

The Shorter Route

The length of string needed to join the three corners of a triangle is shorter if a point is added in the middle, as in the figure at right.

Source: F.K. Hwang

These researchers and others found ways to guess where an extra point or so might be beneficial, but they were always left with the nagging question of whether this was the best they could do.

The conjecture, made in 1968 by Dr. Henry Pollack, who was formerly a research director at Bell Laboratories, and Dr. Edgar Gilbert, a Bell mathematician, was that the best you could ever do by adding points was to reduce the length of the path by the square root of three divided by two, or about 13.4 percent. This was exact-

Relief for designers of computer and telephone networks.

ly the amount the path was reduced in the original problem with three vertices of an equilateral triangle.

Knowing Your Limits

In the years since, researchers proved that the conjecture was true if you wanted to connect four points, then they proved it for five.

"Everyone who contributed devised some special trick," Dr. Hwang said. "But still, if you wanted to generalize to the next number of points, you couldn't do it."

Dr. Du and Dr. Hwang's proof relied on converting the problem into one involving mathematics of continuously moving variables. The idea was to suppose that there is an arrangement for which added points make you do worse than the square root of three over two. Then, Dr. Du and Dr. Hwang showed, you can slide the points in your network around and see what happens to your attempt to make the connecting path shorter. They proved that there was no way you could slide points around to do any better than a reduction in the path length of the square root of three over two.

Dr. Graham said the proof should allow network designers to relax as they search for the shortest routes. "If you can never save more than the square route of three over two, it may not be worth all the effort it takes to look for the best solution," he said.

Conclusion

In this chapter we discussed the problem of creating an optimal network connecting a set of points, where *optimal* usually means "cheapest" or "shortest" (sometimes both). In practice, the points represent geographical locations (cities, telephone centrals, pumping stations, etc.), and the links can be rail lines, fiber-optic cable lines, power lines, pipelines, and so on. Depending on the circumstances, we considered two different ways of doing this.

The MST version. In the first half of the chapter, we were required to build the network in such a way that no junction points other than the original locations were allowed, in which case the optimal network turned out to be a *minimum spanning tree*. Finding a minimum spanning tree was the great success story of the chapter—Kruskal's algorithm is a simple *optimal* and *efficient* algorithm for finding MSTs.

The SN version. In the second half of the chapter, we considered what on the surface appears to be a minor change—allowing the network to have interior junction points. In this case, the problem became one of finding the *shortest network* connecting the points. Here the situation got considerably more complicated, but we did learn a few things:

- Sometimes the minimum spanning tree and the shortest network are one and the same, but more often than not they are different. In the latter case, the shortest network is obviously shorter than the minimum spanning tree, but not by all that much. The percentage difference between the minimum spanning tree and the shortest network is always under 13.4%, and in most problems it is a lot less than that.

- In a shortest network, the interior junction points that are added must be *Steiner* points—junction points formed by three lines joining at 120° angles. It follows that *if a shortest network has interior junction points, it must be a Steiner tree, and if it doesn't, then it must be the minimum spanning tree.* (A *Steiner tree* is a network without circuits where all interior junction points are Steiner points.)

There is a way to find some of these Steiner trees using soap-film "computers"—see the appendix to this chapter for details.

- In general, the number of minimum spanning trees for a given set of points is small (usually just one) and easy to find using Kruskal's algorithm, but the number of possible Steiner trees for the same set of points is usually very large and difficult to find. With 7 points, for example, the possible number of Steiner trees is in the thousands, and with 10, it's in the millions.

As a generic problem, the shortest network problem has a lot of similarities to the traveling salesman problem: There is no known optimal and efficient algorithm that can solve all instances of the problem, and most researchers in this area suspect that no such algorithm exists. In practice the situation is not as bad as it sounds. For most real-life applications, we can find approximate solutions that are very close to the optimal (with margins of error that are less than 1%) using sophisticated new algorithms. In most cases this is good enough. In a few cases it isn't, and mathematicians are constantly striving to find even better algorithms.

Profile Evangelista Torricelli (1608–1647)

We usually first hear about Torricelli in a physics class. He was, after all, the man who invented the barometer (originally called *Torricelli's tube*), the man who discovered the laws of atmospheric pressure (called *Torricelli's laws*), and the man who gave us the equation describing the flow of a liquid discharging through an opening in the bottom of a container (known as *Torricelli's equation*). Somewhat overshadowed by his fame as a physicist is the fact that Torricelli was also one of the most gifted mathematicians of his time. In spite of a short life span, he produced many important contributions to geometry, including the elegant method described earlier in the chapter for finding the shortest network connecting the vertices of a triangle (*Torricelli's construction*).

Evangelista Torricelli was born in Faenza, Italy, the son of a textile artisan of modest means. As a young man he was educated at the Jesuit school in Faenza and tutored by his uncle, a Catholic monk. At the age of 18 he was sent by his uncle to continue his studies in Rome. His tutor in Rome was Father Benedetto Castelli, a well-known scholar who had studied under Galileo. Castelli recognized that he had a brilliant student on his hands and tutored Torricelli in most of the mathematics and physics that were known at that time. Torricelli completed his studies by writing a treatise on the path of projectiles using Galileo's recently discovered laws of motion. Castelli forwarded Torricelli's completed manuscript on projectiles to Galileo, who was so impressed with the mathematical sophistication used by Torricelli in this work that he offered him a position as his secretary and personal assistant.

By this time, Galileo was in serious trouble with the Catholic Church for writing a book supporting the theory that it's the Earth that moves around the Sun, and not the other way around. This was considered heresy, and Galileo was tried and placed under house arrest at his own estate near Florence. Torricelli, as much as he admired Galileo, was afraid of getting into trouble himself (guilt by association was common practice in those days), so he postponed accepting Galileo's offer for some time. For the next few years, Torricelli supported himself by working as an assistant to several professors at the University of Rome and by grinding lenses for telescopes, a delicate craft at which he was a master. (Torricelli became known as the best telescope maker of his time and also developed an ingenious new type of microscope using tiny drops of crystal.)

In 1641 Torricelli finally decided to join Galileo and work for him, but by this time Galileo was blind due to an eye infection and in pretty poor health. Galileo died three months after Torricelli joined him. Upon Galileo's death, Torricelli was invited by the Grand Duke Ferdinand II of Tuscany to succeed Galileo as court mathematician and professor of mathematics at the University of Florence. Torricelli remained in this position and lived in the ducal palace in Florence until his death in 1647.

Having a prestigious academic post and being financially secure allowed Torricelli to concentrate on his scientific work. During this period, Torricelli conducted the experiment he is best remembered for, filling a long glass tube with mercury and turning it upside down (carefully keeping any air from getting in) into a dish also containing mercury. The level of mercury always fell a certain amount, creating a vacuum inside the tube. This single experiment secured Torricelli's name in history, as it disproved the prevailing doctrine of the time that vacuums cannot be sustained in nature. The experiment also paved the way to his discovery of the barometer.

Torricelli died in Florence at the age of 39 due to an illness of undetermined nature. A man best remembered for his practical contributions to our understanding of our physical world, he was most proud of his theoretical work in mathematics. Once, when challenged because of some inaccuracies in his calculations on the paths of cannonballs, he replied, "I write for philosophers, not bombardiers."

Key Concepts

Exercises

WALKING

A. Trees

1. For each graph, determine whether the graph is a tree or not. If not, explain why not.

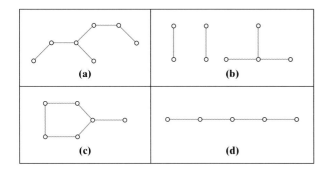

2. For each graph, determine whether the graph is a tree or not. If not, explain why not.

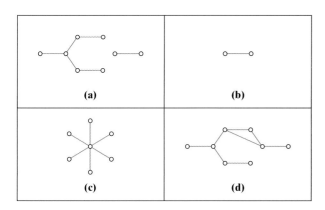

In Exercises 3 through 10, assume that G is a graph with no loops or multiple edges, and choose the option that best applies: (I) G is definitely a tree (explain why); (II) G is definitely not a tree (explain why); (III) G may or may not be a tree (in this case give two examples of graphs that fit the description—one a tree and the other one not).

3. (a) *G* has 8 vertices and 10 edges.

   (b) *G* has 8 vertices and 7 edges.

   (c) *G* has 8 vertices and 7 bridges.

4. (a) *G* has 10 vertices and 11 edges and is a connected graph.

(b) *G* has 10 vertices and 9 edges.

(c) *G* has 10 vertices and 9 edges, and there is a path from any vertex to any other vertex.

5. (a) *G* has 8 vertices, and there is exactly one path from any vertex to any other vertex.

   (b) *G* has 8 vertices and no bridges.

   (c) *G* has 8 vertices, is connected, and every edge in *G* is a bridge.

6. (a) *G* has 10 vertices (*A* through *J*), and there is exactly one path from *A* to *J*.

   (b) *G* has 10 vertices (*A* through *J*), and there are two different paths from *A* to *J*.

   (c) *G* has 10 vertices and 9 bridges.

7. (a) *G* has 8 vertices and no circuits.

   (b) *G* has 8 vertices, 7 edges, and exactly one circuit.

   (c) *G* has 8 vertices, 7 edges, and no circuits.

8. (a) *G* has 10 vertices, and there is a Hamilton circuit in *G*.

   (b) *G* has 10 vertices, and there is a Hamilton path in *G*.

   (c) *G* has 10 vertices, has no circuits, and there is a Hamilton path in *G*.

9. (a) *G* is connected and has 8 vertices, and every vertex has even degree.

   (b) *G* is connected and has 8 vertices. There are 2 vertices of odd degree, and all other vertices have even degree.

   (c) *G* is connected and has 8 vertices. The degree of every vertex is either 1 or 3.

10. (a) *G* is connected and has 10 vertices. Every vertex has degree 9.

   (b) *G* is connected and has 10 vertices. One of the vertices has degree 9, and all other vertices have degree less than 9.

   (c) *G* is connected and has 10 vertices, and every vertex has degree 2.

B. Spanning Trees

11. Find a spanning tree for each of the following networks.

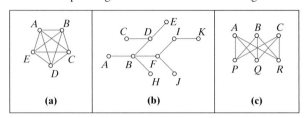

(a) (b) (c)

12. Find a spanning tree for each of the following networks.

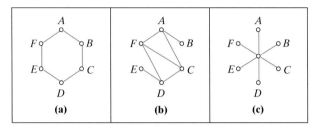

(a) (b) (c)

13. Find all the spanning trees of each of the following networks.

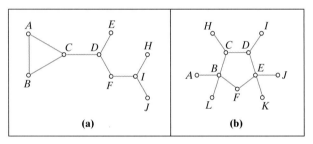

(a) (b)

14. Find all the spanning trees of each of the following networks.

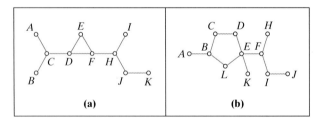

(a) (b)

15. For each of the following networks, find the number of different spanning trees.

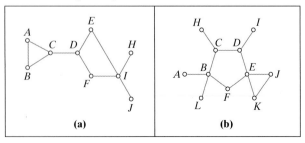

(a) (b)

16. For each of the following networks, find the number of different spanning trees.

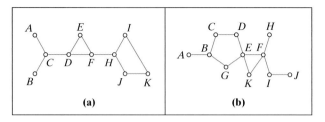

(a) (b)

17. For each of the following networks, find the number of different spanning trees.

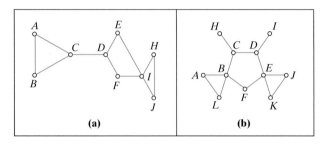

(a) (b)

18. For each of the following networks, find the number of different spanning trees.

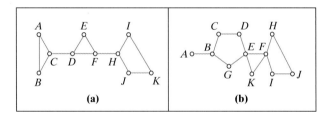

(a) (b)

C. Minimum Spanning Trees and Kruskal's Algorithm

19. For the weighted network shown in the figure,

(a) find the MST using Kruskal's algorithm.

(b) give the weight of the MST found in (a).

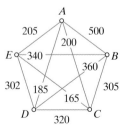

20. For the weighted network shown in the figure,
(a) find the MST using Kruskal's algorithm.
(b) give the weight of the MST found in (a).

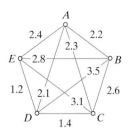

21. For the weighted network shown in the figure,
   (a) find the MST using Kruskal's algorithm.
   (b) give the weight of the MST found in (a).

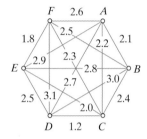

22. For the weighted network shown in the figure,
   (a) find the MST using Kruskal's algorithm.
   (b) give the weight of the MST found in (a).

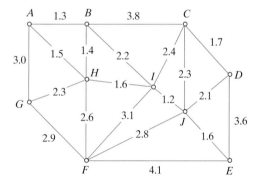

23. The following mileage chart shows the distances between Atlanta, Columbus, Kansas City, Minneapolis, Pierre, and Tulsa. Working directly from the mileage chart, implement Kruskal's algorithm and find the minimum spanning tree connecting the cities.

Mileage Chart

	Atlanta	Columbus	Kansas City	Minneapolis	Pierre	Tulsa
Atlanta	*	533	798	1068	1361	772
Columbus	533	*	656	713	1071	802
Kansas City	798	656	*	447	592	248
Minneapolis	1068	713	447	*	394	695
Pierre	1361	1071	592	394	*	760
Tulsa	772	802	248	695	760	*

24. The following mileage chart shows the distances between Boston, Dallas, Houston, Louisville, Nashville, Pittsburgh, and St. Louis. Working directly from the mileage chart, implement Kruskal's algorithm and find the minimum spanning tree connecting the cities.

Mileage Chart

	Boston	Dallas	Houston	Louisville	Nashville	Pittsburgh	St. Louis
Boston	*	1748	1804	941	1088	561	1141
Dallas	1748	*	243	819	660	1204	630
Houston	1804	243	*	928	769	1313	779
Louisville	941	819	928	*	168	388	263
Nashville	1088	660	769	168	*	553	299
Pittsburgh	561	1204	1313	388	553	*	588
St. Louis	1141	630	779	263	299	588	*

25. The 3 by 4 grid shown in the figure represents a network of streets (3 blocks by 4 blocks) in a small subdivision. For landscaping purposes, it is necessary to get water to each of the corners by laying down a system of pipes along the streets. The cost of laying down the pipes is $40,000 per mile, and each block of the grid is exactly half a mile long. Without using Kruskal's algorithm, find the cost of the cheapest network of pipes connecting all the corners of the subdivision. Explain your answer.

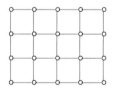

26. The 4 by 5 grid shown in the figure represents a network of streets (4 blocks by 5 blocks) in a small subdivision. For landscaping purposes, it is necessary to get water to each of the corners by laying down a system of pipes along the streets. The cost of laying down the pipes is $40,000 per mile, and each block of the grid is exactly half a mile long. Without using Kruskal's algorithm, find the cost of the cheapest network of pipes connecting all the corners of the subdivision. Explain your answer.

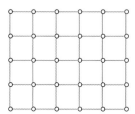

D. Steiner Points and Shortest Networks

27. Find the length of the shortest network connecting the three cities A, B, and C shown in each figure. All distances are rounded to the nearest mile. (Figures are not drawn to scale.)

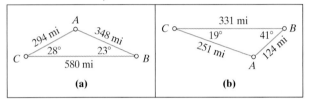

(a)　　　　　　　　　**(b)**

28. Find the length of the shortest network connecting the three cities A, B, and C shown in each figure. All distances are rounded to the nearest mile. (Figures are not drawn to scale.)

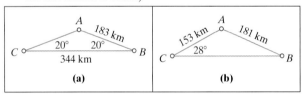

(a)　　　　　　　　　**(b)**

29. Find the length of the shortest network connecting the three cities A, B, and C shown in each figure. All distances are rounded to the nearest kilometer. (Figures are not drawn to scale.)

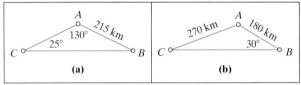

(a)　　　　　　　　　**(b)**

30. Find the length of the shortest network connecting the three cities A, B, and C shown in each figure. All distances are rounded to the nearest kilometer. (Figures are not drawn to scale.)

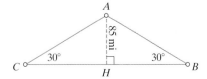

(a)　　　　　　　　　**(b)**

31. Find the length of the shortest network (rounded to the nearest mile) connecting the three cities A, B, and C shown in the figure. (Figure not drawn to scale.) Explain your answer.

(**Hint:** *You will need to use properties of 30-60-90 triangles.. For a primer on 30-60-90 triangles, go to Exercises 45 and 46.*)

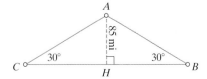

32. Find the length of the shortest network (rounded to the nearest kilometer) connecting the three cities A, B, and C shown in the figure. (Figure not drawn to scale.) Explain your answer.

(**Hint:** *You will need to use the Pythagorean theorem.*)

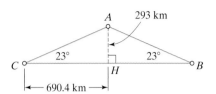

33. Assume that one of three points—X, Y, or Z—is a Steiner point of triangle ABC. The distances (rounded to the nearest mile) of each of these three points to the vertices of the triangle are given in the table. Determine which of the three points is the Steiner point. Explain your answer.

Distance in miles

	X	Y	Z
A	54	72	41
B	61	94	87
C	125	77	104

34. Assume that one of three points—X, Y, or Z—is a Steiner point of triangle ABC. The distances (rounded to the nearest mile) of each of these three points to the vertices of the triangle are given in the table. Determine which of the three points is the Steiner point. Explain your answer.

Distance in miles

	X	Y	Z
A	380	390	700
B	620	680	260
C	300	190	550

Exercises 35 and 36 refer to five cities (A, B, C, D, and E) located as shown in the figure. Assume that $AC = AB$, $DC = DB$, and $< CDB$ measures $120°$.

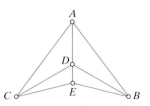

35. (a) Which is larger, $CD + DB$ or $CE + ED + EB$? Explain.

(b) What is the shortest network connecting cities C, D, and B? Explain.

(c) What is the shortest network connecting cities C, E, and B? Explain.

36. (a) Which is larger, $CA + AB$ or $DC + DA + DB$? Explain.

(b) Which is larger, $EC + EA + EB$ or $DC + DA + DB$? Explain.

(c) What is the shortest network connecting cities A, B, and C? Explain.

E. Miscellaneous

37. Consider a network with M edges. Let k denote the number of bridges in the network.

(a) If $M = 5$, list all the possible values of k.

(b) If $M = 123$, describe the set of all possible values of k.

38. Consider a network with M edges. Let k denote the number of bridges in the network.

(a) If $M = 7$, list all the possible values of k.

(b) If $M = 2481$, describe the set of all possible values of k.

39. Consider a network with N vertices and M edges.

(a) If $N = 123$ and $M = 122$, find the number of circuits in the network.

(b) If $N = 123$ and $M = 123$, find the number of circuits in the network.

(c) If $N = 123$, $M = 123$, and there is a circuit of length 5, find the number of bridges in the network.

40. Consider a network with N vertices and M edges.

(a) If $N = 2481$ and $M = 2480$, find the number of circuits in the network.

(b) If $N = 2481$ and $M = 2481$, find the number of circuits in the network.

(c) If $N = 2481$, $M = 2481$, and there is a circuit of length 10, find the number of bridges in the network.

41. Give an example of a graph with $N = 11$ vertices and $M = 10$ edges having

(a) exactly one circuit.

(b) exactly two circuits.

(c) exactly three circuits.

42. Give an example of a graph with $N = 14$ vertices and $M = 13$ edges having

(a) exactly one circuit.

(b) exactly two circuits.

(c) exactly three circuits.

43. The triangle ABC shown in the figure is an isosceles right triangle ($AB = BC$), and S is the Steiner point of the triangle. (Figure not drawn to scale.) Find the measure of angle BAS.

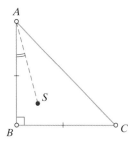

44. The triangle ABC shown in the figure is an isosceles triangle ($AB = AC$), and the measure of angle BAC is $36°$. (Figure not drawn to scale.) If S is the Steiner point of the triangle, find the measure of angle ABS.

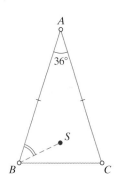

30-60-90° Triangles. *The purpose of Exercises 45 and 46 is to review the relationship as relating the lengths of the three sides of a 30-60-90 triangle. We will denote the lengths of the three sides of the 30-60-90 triangle by s (for short leg), l (for long leg), and h (for hypotenuse). (Use the approximation $\sqrt{3} \approx 1.73$ in your calculations.)*

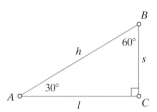

45. (a) Express h in terms of s.

(**Hint:** *A 30-60-90 triangle is half of an equilateral triangle.*)

(b) Express l in terms of s.
[***Hint:*** *Use (a) and the Pythagorean theorem.*]

(c) If $s = 21.0$ cm, find h and l (rounded to the nearest tenth of a centimeter).

(d) If $h = 30.4$ cm, find s and l (rounded to the nearest tenth of a centimeter).

46. (a) Express s in terms of h.
[***Hint:*** *Try Exercise 45(a) first.*]

(b) Express l in terms of h.
[***Hint:*** *Try Exercise 45(b) first.*]

(c) Express h in terms of l.

(d) If $l = 173.0$ cm, find s and h (rounded to the nearest centimeter).

Exercises 47 and 48 refer to the Outback Cable Network discussed in Example 7.8.

47. Show that the length of the T-network shown in the figure (rounded to the nearest mile) is 933 miles.

(***Hint:*** *Triangle CAJ is a 30-60-90 triangle.*)

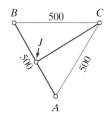

48. Show that the length of the Y-network shown in the figure (rounded to the nearest mile) is 866 miles.

(***Hint:*** *Triangle ABC can be broken up into six congruent 30-60-90 triangles.*)

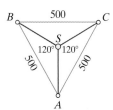

JOGGING

49. Find the number of different spanning trees of the network shown in the figure.

(***Hint:*** *Try Exercise 17 first.*)

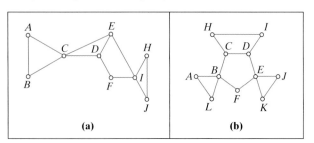

(a) **(b)**

50. (a) Let G be a tree with N vertices. Find the sum of the degrees of all the vertices in G.

(b) Explain why a tree must have at least two vertices of degree 1. (A vertex of degree 1 in a tree is called a *leaf*.)

51. Regular trees are rare. A graph is called *regular* if every vertex has the same degree.

(a) Give an example of a regular tree with $N = 2$ vertices.

(b) Explain why for $N \geq 3$, a tree with N vertices *cannot* be regular.

[***Hint:*** *Use the result of Exercise 50(b).*]

52. (a) Give an example of a tree with six vertices such that the degrees of the vertices are $1, 1, 2, 2, 2$, and 2.

(b) Give an example of a tree with N vertices such that the degrees of the vertices are $1, 1, 2, 2, 2, \ldots, 2$.

(c) Give an example of a tree with five vertices such that the degrees of the vertices are $1, 1, 1, 1, 4$.

(d) Give an example of a tree with N vertices such that the degrees of the vertices are $1, 1, 1, \ldots, 1, N - 1$.

53. Redundancy. Recall that the *redundancy* of a network with N vertices and M edges is given by $R = M - (N - 1)$.

(a) Describe all networks with redundancy $R = 0$.

(b) Explain why a network has redundancy $R = 1$ has exactly one circuit.

(c) Explain why if there are no loops or multiple edges, the maximum value of the redundancy of a network is given by $R = (N^2 - 3N + 2)/2$.

54. Suppose that G is a network with N vertices and M edges, with $M \geq N$. Explain why the number of bridges in G is smaller than or equal to $(M - 3)$.

55. Cayley's theorem. Cayley's theorem says that the number of spanning trees in a complete graph with N vertices is given by N^{N-2}.

(a) Verify this result for the cases $N = 3$ and $N = 4$ by finding all spanning trees for complete graphs with three and four vertices.

(b) Which is larger, the number of Hamilton circuits or the number of spanning trees in a complete graph with N vertices? Explain.

56. A highway system connecting nine cities— $C_1, C_2, C_3, \ldots, C_9$—is to be built. Use Kruskal's algorithm to find a minimum spanning tree for this problem. The accompanying table shows the cost (in millions of dollars) of putting a highway between any two cities.

	C_1	C_2	C_3	C_4	C_5	C_6	C_7	C_8	C_9
C_1		1.3	3.4	6.6	2.6	3.5	5.7	1.1	3.8
C_2	1.3		2.4	7.9	1.7	2.3	7.0	2.4	3.9
C_3	3.4	2.4		9.9	3.4	1.0	9.1	4.4	6.5
C_4	6.6	7.9	9.9		8.2	9.7	0.9	5.5	4.9
C_5	2.6	1.7	3.4	8.2		4.8	7.4	3.7	3.5
C_6	3.5	2.3	1.0	9.7	4.8		8.9	4.4	5.8
C_7	5.7	7.0	9.1	0.9	7.4	8.9		4.7	3.9
C_8	1.1	2.4	4.4	5.5	3.7	4.4	4.7		2.8
C_9	3.8	3.9	6.5	4.9	3.5	5.8	3.9	2.8	

57. The four cities (A, B, C, and D), shown in the figure must be connected into a telephone network. The cost of laying down telephone cable connecting any two of the cities is given (in millions of dollars) by the weights of the edges in the graph. In addition, in any city that serves as a junction point of the network, expensive switching equipment must be installed. The cost of installing this equipment is given (in millions of dollars) by the numbers inside the circles. Find the minimum-cost telephone network connecting these four cities.

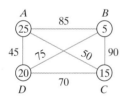

58. Five cities (A, B, C, D, and E) are located as shown on the following figure. The five cities need to be connected by a railroad, and the cost of building the railroad system connecting any two cities is proportional to the distance between the two cities. Find the length of the railroad network of minimum cost (assuming that no additional junction points can be added).

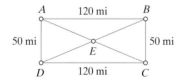

59. Middletown is designing a new high-speed rail system to connect Downtown, the Airport, and the Coliseum. The elevated tracks for the rail system will run along the existing roads, which run on a grid as shown in the figure. In addition, a switching station (junction point) will be placed somewhere on the grid.

(a) Determine the optimal location for the switching station.

(b) Find the optimal network connecting the three locations and give the cost of the network. (It costs one million dollars per mile of elevated track and half a million dollars for the switching station.)

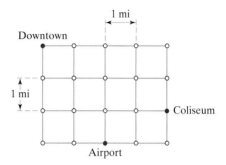

60. Happyville is building a new high-speed rail system to connect the Airport, the Midtown Mall, the University, and Slugger's Ballpark. The elevated tracks for the rail system will run along the existing roads, which run on a grid as shown in the figure. In addition, two switching stations (junction points) will be placed at two different locations on the grid.

(a) Determine the optimal locations for the two switching stations.

(b) Find the optimal network connecting the four locations and give the cost of the network. (It costs two million dollars per mile of elevated track plus half a million dollars for each switching station.)

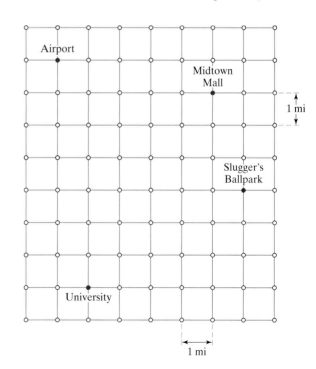

61. Consider triangle ABC having an equilateral triangle EFG inside, as shown in the following figure.

(a) Find the measure of angles BFA, AEC, and CGB.

(b) Explain why triangle ABC has a Steiner point S.

(c) Explain why the Steiner point of triangle ABC lies inside triangle EFG.

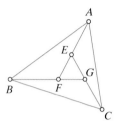

62. Show that the length of the network shown in the figure (rounded to the nearest mile) is 1414 miles.

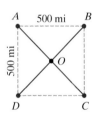

63. Show that the length of the shortest network connecting the vertices of the square shown in the figure (rounded to the nearest mile) is 1366 miles. (S_1 and S_2 are Steiner points.)

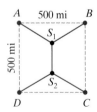

64. (a) Show that the length of the network shown in (a) is $400\sqrt{3} + 300 \approx 993$ miles. (S_1 and S_2 are Steiner points.)

(b) Show that the length of the network shown in (b) is $300\sqrt{3} + 400 \approx 919.6$ miles. (S_1 and S_2 are Steiner points.)

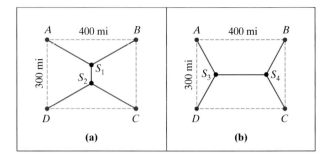

65. Boruvka's algorithm. In 1926, the Czech mathematician Otakar Boruvka gave the following algorithm for finding the minimum spanning tree of a network.

- **Start:** Start with a "blank" copy of the network (no edges, just the vertices).

- **Step 1.** For each vertex, choose an edge incident to that vertex that (i) has least weight and (ii) does not create any cycles. Add these edges (to the blank copy). At this point you will have a collection of disconnected subtrees.

- **Steps 2, 3, etc.** For each subtree formed in the previous step, choose an edge incident to that subtree that (i) has least weight and (ii) does not create any cycles. Continue doing this until the individual subtrees are all connected into a network. This will be the MST.

(a) For the network in the figure, show the subtrees formed at Step 1 of Boruvka's algorithm.

(b) For the network in the figure, show the subtrees formed at Step 2 of Boruvka's algorithm.

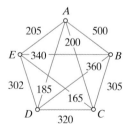

66. Suppose you are in charge of designing the shortest fiber-optic cable network connecting the Ohio cities of Cincinnati, Toledo, and Canton. At the junction point of the network, some very expensive special equipment needs to be installed, and this equipment must be serviced by technical staff living close to the junction point. Using the map of Ohio provided, identify the city closest to the location of the Steiner point. You may use Torricelli's method or any other method you deem appropriate. (You will need a pencil, a ruler, and a compass.)

RUNNING

67. Show that if a tree has a vertex of degree K, then there are at least K vertices in the tree of degree 1.

68. A disconnected graph such that each of its components is a tree is called a *forest*.

    (a) If G is a forest with N vertices, M edges, and K components, express K in terms of N and M.

    (b) What is the minimum number of vertices of degree 1 in a forest with N vertices and K components? (See Exercise 50.)

    (c) What is the maximum number of vertices of degree 1 in a forest with N vertices and K components?

69. (a) Suppose that there is an edge in a network that must be included in any spanning tree. Give an algorithm for finding the minimum spanning tree that includes a given edge.
    (***Hint:*** *Modify Kruskal's algorithm.*)

    (b) Consider the mileage chart given in Exercise 56. Suppose that C_3 and C_4 are the two largest cities in the area and that the chamber of commerce insists that a section of highway directly connecting them must be built (or heads will roll). Find the minimum spanning tree that includes the section of highway between C_3 and C_4.

70. In the figure, A, B, C, and D are the vertices of a square. In (i), S_1 and S_2 are Steiner points. In (ii), X and Y are two arbitrary points that are not Steiner points. Show that the network in (i) is shorter than the network in (ii). (You can assume any facts concerning shortest networks for three points.)

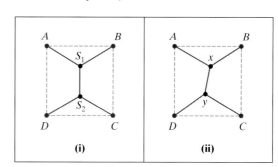

(i)   (ii)

71. Suppose that $A, B, C,$ and D are four towns located as in the figure. Show that it is impossible have a network with two Steiner points inside the trapezoid connecting the four cities.

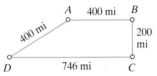

72. Suppose that $A, B, C,$ and D are four towns located at the vertices of a rectangle with length b and height a as shown in the figure. Assume that $a < b$. Determine the conditions that must be satisfied by a and b so that the length of the MST shown in (ii) is less than the length of the Steiner tree shown in (iii).

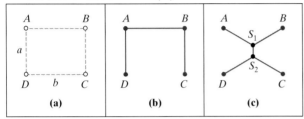

73. Three cities, $A, B,$ and $C,$ are located at the vertices of an isosceles right triangle, as shown in the figure.

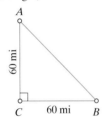

(a) Compute the length of the minimum spanning tree connecting the three cities. Call it W.

(b) Compute the length of the shortest network connecting the three cities. Call it Z.

(c) Compute the relative percentage difference between the two answers; that is, compute $(W - Z)/Z$ and express it as a percent.

74. Six cities, A through F, are located at the vertices of a regular hexagon. A seventh city H is located at the center of the hexagon, as shown in the figure.

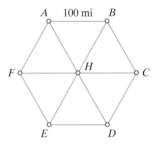

(a) Find the shortest network connecting the seven cities.

(b) Compute the length of the shortest network found in (a).

75. Consider the 1 by 3 square grid shown in the figure.

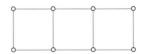

(a) Find the shortest network connecting the eight vertices in the grid.

(b) Compute the length of the shortest network found in (a). (Assume the squares in the grid are 1 by 1.)

76. Consider the 3 by 3 square grid shown in the figure.

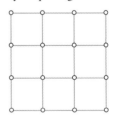

(a) Find the shortest network connecting the sixteen vertices in the grid.

(b) Compute the length of the shortest network found in (a). (Assume the squares in the grid are 1 by 1.)

Projects and Papers

A. The Kruskal-Steiner Fiber-Optic Cable Network Project

Kruskal-Steiner Corporation is a new telecommunications company providing phone/Internet service to selected cities in the United States. Kruskal-Steiner plans to build a major fiber-optic cable network connecting the 21 cities shown in the mileage chart. The network construction costs are estimated at $100,000 per mile, so this is serious stuff.

You have been hired as a consultant for this project. Your job is to design a network connecting the 21 cities and sell your design to the board of directors. Your contract states that your consulting fee of $10,000 will be paid only if your network is better (shorter) than the *minimum spanning tree* (MST) connecting the 21 cities. In addition, you will get a bonus of $1000 for every mile that your network saves over the MST, so it's in your best interest to try to improve on the MST as much as you can.

Mileage Chart

	Atlanta	Boston	Buffalo	Chicago	Columbus	Dallas	Denver	Houston	Kansas City	Louisville	Memphis	Miami	Minneapolis	Nashville	New York	Omaha	Pierre	Pittsburgh	Raleigh	St. Louis	Tulsa
Atlanta	*	1037	859	674	533	795	1398	789	798	382	371	655	1068	242	841	986	1361	687	372	541	772
Boston	1037	*	446	963	735	1748	1949	1804	1391	941	1293	1504	1368	1088	206	1412	1726	561	685	1141	1537
Buffalo	859	446	*	522	326	1346	1508	1460	966	532	899	1409	927	700	372	971	1285	216	605	716	1112
Chicago	674	963	522	*	308	917	996	1067	499	292	530	1329	405	446	802	459	763	452	784	289	683
Columbus	533	735	326	308	*	1028	1229	1137	656	209	576	1160	713	377	542	750	1071	182	491	406	802
Dallas	795	1748	1346	917	1028	*	781	243	489	819	452	1300	936	660	1552	644	943	1204	1166	630	257
Denver	1398	1949	1508	996	1229	781	*	1019	600	1120	1040	2037	841	1156	1771	537	518	1411	1661	857	681
Houston	789	1804	1460	1067	1137	243	1019	*	710	928	561	1190	1157	769	1608	865	1186	1313	1160	779	478
Kansas City	798	1391	966	499	656	489	600	710	*	520	451	1448	447	556	1198	201	592	838	1061	257	248
Louisville	382	941	532	292	209	819	1120	928	520	*	367	1037	697	168	748	687	1055	388	541	263	659
Memphis	371	1293	899	530	576	452	1040	561	451	367	*	997	826	208	1100	652	1043	752	728	285	401
Miami	655	1504	1409	1329	1160	1300	2037	1190	1448	1037	997	*	1723	897	1308	1641	2016	1200	819	1196	1398
Minneapolis	1068	1368	927	405	713	936	841	1157	447	697	826	1723	*	826	1207	357	394	857	1189	552	695
Nashville	242	1088	700	446	377	660	1156	769	556	168	208	897	826	*	892	744	1119	553	521	299	609
New York	841	206	372	802	542	1552	1771	1608	1198	748	1100	1308	1207	892	*	1251	1565	368	489	948	1344
Omaha	986	1412	971	459	750	644	537	865	201	687	652	1641	357	744	1251	*	391	895	1214	449	387
Pierre	1361	1726	1285	763	1071	943	518	1186	592	1055	1043	2016	394	1119	1565	391	*	1215	1547	824	760
Pittsburgh	687	561	216	452	182	1204	1411	1313	838	388	752	1200	857	553	368	895	1215	*	445	588	984
Raleigh	372	685	605	784	491	1166	1661	1160	1061	541	728	819	1189	521	489	1214	1547	445	*	804	1129
St. Louis	541	1141	716	289	406	630	857	779	257	263	285	1196	552	299	948	449	824	588	804	*	396
Tulsa	772	1537	1112	683	802	257	681	478	248	659	401	1398	695	609	1344	387	760	984	1129	396	*

For this project you are asked to prepare a presentation to the board. In your presentation you should (i) describe the MST and compute the total cost of its construction, (ii) describe your proposed network and show where it provides improvements over the MST, and (iii) estimate the length of your network and the savings it generates over the MST.

Notes: (1) You should start by getting two accurate, good-quality maps of the United States. (2) Use Kruskal's algorithm to first find the MST, drawing it on one of the maps. (To implement Kruskal's algorithm directly on the map, you may find it convenient to first sort the distances in the mileage chart from smallest to biggest, and use this sorted list to build the MST.) (3) Looking at the angles formed by the branches of the MST in the map, try to identify places where shortcuts are possible by introducing an interior Steiner junction point. (4) Estimate the length of each of the shortcuts and then the total length of your network. [You will need the following formula: If *l* denotes the length of the shortest network connecting the vertices of a triangle having angles that measure less than 120° each and sides of lengths *a*, *b*, and *c*, then

$$l = \sqrt{\frac{a^2 + b^2 + c^2}{2} + \frac{\sqrt{3}}{2}\sqrt{2a^2b^2 + 2a^2c^2 + 2b^2c^2 - (a^4 + b^4 + c^4)}}]$$

(5) The last, and most important step: Calculate how much money you made for your efforts ($10,000 plus $1000 for every mile you saved over the MST).

B. Validating Torricelli's Construction

In this chapter we introduced Torricelli's method for finding a Steiner point inside a triangle but did not give a justification as to why the Steiner point gives the shortest network.

Torricelli proved the following three facts about a triangle *ABC* having all angles less than 120°:

Fact 1. There is a unique point *S* inside the triangle with the property that the angles *ASB*, *ASC*, and *BSC* are all 120° (we call it the *Steiner point* of the triangle).

Fact 2. For any point *P* different from *S*, $\overline{PA} + \overline{PB} + \overline{PC} > \overline{SA} + \overline{SB} + \overline{SC}$. (This implies that the network obtained using *S* as the junction point is the shortest network connecting the three vertices of the triangle.)

Fact 3. $\overline{SA} + \overline{SB} + \overline{SC} = \overline{AX}$, where *X* is the vertex of the equilateral triangle built on side *BC* and on the opposite side of *A*, as shown in the figure.

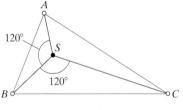

Fact 1: *S* is unique

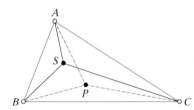

Fact 2: $\overline{PA} + \overline{PB} + \overline{PC} > \overline{SA} + \overline{SB} + \overline{SC}$

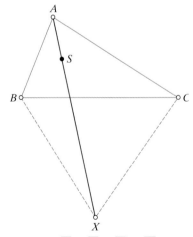

Fact 3: $\overline{SA} + \overline{SB} + \overline{SC} = \overline{AX}$

In this project you are asked to give a complete mathematical proof of each of the facts.

Notes: (1) All of the foregoing can be proved using standard facts from Euclidean geometry. (2) There are several different ways you can approach these proofs, and you can use other well-known theorems in Euclidean geometry (such as *Viviani's theorem* and *Ptolemy's theorem*) in your proofs.

C. Prim's Algorithm

Prim's algorithm is another well-known algorithm for finding the minimum spanning tree of a weighted graph. Like Kruskal's algorithm, Prim's algorithm is an optimal and efficient algorithm for finding MSTs.

Prepare a classroom presentation on Prim's algorithm. (i) Describe the algorithm. (ii) Illustrate how the algorithm works using the following two examples: (a) the weighted graph in Exercise 22 and (b) the mileage chart in Exercise 24. (iii) Compare Prim's algorithm to Kruskal's algorithm and discuss their differences.

D. Minimizing with Soap Film

Occasionally, we can enlist nature's aid to help us solve reasonably interesting math problems. One of the better-known examples of this is the use of soap-film solutions to solve *minimal surface* and *shortest distance* problems.

For this project, prepare a classroom presentation on soap-film solutions and how they can be used to solve certain optimization problems. Discuss (i) the types of optimization problems in which soap-film solutions can be used, and (ii) the basic laws of physics that explain why and how soap-film solutions work in these problems.

Appendix The Soap-Bubble Solution

Every child knows about the magic of soap bubbles. Take a wire or plastic ring, dip it into a soap-and-water solution, blow through the ring, and presto—beautiful iridescent geometric shapes magically materialize to delight and inspire our fantasy. Adults are not averse to a puff or two themselves.

What's special about soapy water that makes this happen? A very simplistic understanding of the forces of nature that create soap bubbles will help us understand how these same forces can be used to find (imagine, of all things) Steiner trees connecting a given set of points.

Take a liquid (any liquid), and put it into a container. When the liquid is at rest, there are two categories of molecules: those that are on the surface and those that are below the surface. The molecules below the surface are surrounded on all sides by other molecules like themselves and are therefore in perfect balance—the forces of attraction between molecules all cancel each other out. The molecules on the surface, however, are only partly surrounded by other like molecules and are therefore unbalanced. For this reason, an

A bevy of Steiner junctions: Just as a rolling ball seeks the bottom of a hill, soap films seek configurations of minimal energy. Soap bubbles trapped between glass plates always meet in sets of three, their boundaries forming 120° angles at the junction point.

additional force called **surface tension** comes into play for these molecules. As a result of this surface tension, the surface layer of any liquid behaves exactly as if it were made of a very thin, elastic material. The amount of elasticity of this surface layer depends on the structure of the molecules in the liquid. Soap or detergent molecules are particularly well suited to create an extremely elastic surface layer. (A good soap-film solution can be obtained by adding a small amount of dishwashing liquid to water stirring gently to minimize surface bubbles, and, if necessary, adding a small amount of glycerin to make the soap film a little more stable.)

The connection between the preceding brief lesson in soapy solutions and the material in this chapter is made through one of the fundamental principles of physics: A physical system will remain in a certain configuration only if it cannot easily change to another configuration that uses less energy. Because of its extreme elasticity, the surface layer of a soapy solution has no trouble changing its shape until it feels perfectly comfortable (i.e., at a position of relatively minimal energy). When the energy is proportional to the distance, minimal energy results in minimal distance—ergo, Steiner trees.

Suppose that we have a set of points $(A_1, A_2, \ldots, A_N)$ for which we want to find a shortest network. We can find a Steiner tree that connects these points by means of an ingenious device that we will call a soap-bubble computer. To begin with, we draw the points $A_1, A_2, \ldots, A_N$ to exact scale on a piece of paper. (As much as possible, choose the scale so that points are neither too close to each other nor too far apart—somewhere from 1 to 4 inches should do just fine.) We now take two sheets of Plexiglas or Lucite and, using the paper map as a template, drill small holes on both sheets of Plexiglas at the exact locations of the points. Then we put thin metal or plastic pegs through the holes in such a way that the two sheets are held about an inch apart.

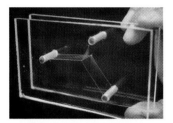
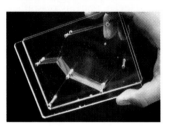

When we dip our device into a soap-and-water solution and pull it out, the soap-bubble computer goes to work. The film layer formed between the plates connects the various pegs. For a while it moves, seeking a configuration of minimal energy. Very shortly thereafter, it settles into a Steiner tree.

It is a bit of a disappointment that the Steiner tree we get is not necessarily the shortest network. The reasons for this are beyond the scope of our discussion. (The interested reader is referred to the excellent technical discussion of soap-film computers in reference 1.) At the same time, we should be thankful for what nature has provided: a simple device that can compute, in seconds, what might take us hours to do with pencil and paper.

References and Further Readings

1. Almgren, Fred J., Jr., and Jean E. Taylor, "The Geometry of Soap Films and Soap Bubbles," *Scientific American*, 235 (July 1976), 82–93.

2. Barabási, Albert-László, *Linked: The New Science of Networks*. Cambridge, MA: Perseus Publishing, 2002.

3. Barabási, Albert-László, and Eric Bonabeau, "Scale-Free Networks," *Scientific American*, 288 (May 2003), 60–69.

4. Bern, M., and R. L. Graham, "The Shortest Network Problem," *Scientific American*, 260 (January 1989), 84–89.

5. Buchanan, Mark, *Nexus: Small Worlds and the Groundbreaking Theory of Networks*. New York: W. W. Norton, 2003.

6. Chung, F., M. Gardner, and R. Graham, "Steiner Trees on a Checkerboard," *Mathematics Magazine*, 62 (April 1984), 83–96.

7. Cockayne, E. J., and D. E. Hewgill, "Exact Computation of Steiner Minimal Trees in the Plane," *Information Processing Letters*, 22 (1986), 151–156.

8. Du, D.-Z., and F. K. Hwang, "The Steiner Ratio Conjecture of Gilbert and Pollack is True," *Proceedings of the National Academy of Sciences, U.S.A.*, 87 (December 1990), 9464–9466.

9. Gardner, Martin, "Mathematical Games: Casting a Net on a Checkerboard and Other Puzzles of the Forest," *Scientific American,* 254 (June 1986), 16–23.

10. Gilbert, E. N., and H. O. Pollack, "Steiner Minimal Trees," *SIAM Journal of Applied Mathematics*, 16 (1968), 1–29.

11. Graham, R. L., and P. Hell, "On the History of the Minimum Spanning Tree Problem," *Annals of the History of Computing*, 7 (January 1985), 43–57.

12. Gross, Jonathan, and Jay Yellen, *Graph Theory*. Boca Raton, FL: CRC Press, 1999, chap. 4.

13. Hayes, Brian, "Graph Theory in Practice: Part I," *American Scientist*, 88 (January–February 2000), 9–13.

14. Hayes, Brian, "Graph Theory in Practice: Part II," *American Scientist*, 88 (March–April 2000), 104–109.

15. Hwang, F. K., D. S. Richards, and P., Winter, *The Steiner Tree Problem*. Amsterdam: North Holland, 1992.

16. Robinson, P., "Evangelista Torricelli," *Mathematical Gazette*, 78 (1994), 37–47.

17. Wilson, Robin J., and John J. Watkins, *Graphs: An Introductory Approach*. New York: John Wiley & Sons, 1990, chap 10.

8

The Mathematics of Scheduling

Directed Graphs and Critical Paths

Waste neither time nor money, but make the best use of both.

Ben Franklin.

How long does it take to build a house? Here is a deceptively simple question that defies an easy answer. Some of the factors involved are obvious: the size of the house, the type of construction, the number of workers, the kinds of tools and machinery used. Less obvious, but equally important, is another variable: the ability to organize and coordinate the timing of people, equipment, and work so that things get done in a timely way. For better or for worse, this last issue boils down to a mathematics problem, just one example in a large family of problems that fall under an area of graph theory known as *combinatorial scheduling*. Discussing some of the basic ideas behind combinatorial scheduling will be the theme of this chapter.

Let's get back to our original question. According to the National Association of Home Builders, it takes 1092 man-hours to build the average American house. Since we are not being picky about details, we'll just call it 1100 hours. Essentially this means that it would take a single construction worker (assuming that this worker can do every single job required for building a house), about 1100 hours of labor to finish this hypothetical average American house. Let's now turn the question on its head. If we had 1100 equally capable workers, could we get the same house built in one hour? Of course not! In fact, we could put tens of thousands of workers on the job and we still could not get the house built in one hour. Some inherent physical limitations to the speed with which a house can be built are outside of the builder's control. Some jobs cannot be speeded up beyond a certain point, regardless of how many workers one puts on that job. Even more significantly, some jobs can only be started only after certain other jobs have been completed. (Roofing, for example, can be started only after framing has been completed.)

Combinatorial scheduling involves such questions as: How fast could a house be built if one cared only about speed? (To the best of our knowledge, the record is a bit under 24 hours.) How fast could we build the house if we had 10 equally able construction workers at our disposal at all times? What if we had only three workers? What if we needed to finish the entire project within a given time frame—say, 30 days? How many workers should we hire then? The same questions could equally well be asked if we replaced "building a house" with many other types of projects—from preparing a banquet to launching a space shuttle.

This chapter starts with an introduction to the *key concepts* and terminology of combinatorial scheduling (Section 8.1). Typically, scheduling problems are modeled using a special type of graph called a *directed graph* (or digraph for short)—we discuss digraphs and some of their basic properties in Section 8.2. In Section 8.3 we will discuss the general rules for creating schedules—the key concept here is that of a *priority list*. The two most commonly used algorithms for "solving" a scheduling problem are the *decreasing-time* and the *critical-path algorithms*—we discuss these in Sections 8.4, 8.5, and 8.6. In Section 8.7 we briefly discuss scheduling tasks that have no precedence relations among them (*independent tasks*).

8.1 The Basic Elements of Scheduling

We will now introduce the principal characters in any scheduling story.

- **The processors.** Every job requires workers. We will use the term *processors* to describe the "workers" who carry out the work. While the word *processor* may sound a little cold and impersonal, it does underscore an important point: Processors need not be human beings. In scheduling, a processor could just as well be a robot, a computer, an automated teller machine, and so on. For the purposes of our discussion, we will use N to represent the number of processors and $P_1, P_2, P_3, \ldots,$ P_N to denote the processors themselves. We will assume throughout the chapter that $N \geq 2$ (for $N = 1$ scheduling is trivial and not very interesting).

- **The tasks.** In every complex project there are individual pieces of work, often called "jobs" or "tasks." We will need to be a little more precise than that, however. We will define a *task* as an indivisible unit of work that (either by nature or by choice) cannot be broken up into smaller units. Thus, by definition a task cannot be shared—it is always *carried out by a single processor*. In general, we will use capital letters $A, B, C, \ldots,$ to represent the tasks, although in specific situations it is convenient to use abbreviations (such as *WE* for "wiring the electrical system," *PL* for "plumbing," etc.).

  At a particular moment in time a task can be in one of four possible states:
  - *completed*,
  - *in execution* (the task is being carried out by one of the processors),
  - *ready* (the task has not been started but could be started at this time), or
  - *ineligible* (the task cannot be started because some of the prerequisites for the task have not yet been completed).

A task can be: *ineligible* (upper left), *ready* (upper right), *in execution* (lower left), and *completed* (lower right).

■ **The processing times.** The *processing time* for a given task X is the amount of time, without interruption, required by *one processor* to execute X. When dealing with human processors, there are many variables (ability, attitude, work ethic, etc.) that can affect the processing time of a task, and this adds another layer of complexity to an already complex situation. On the other hand, if we assume a "robotic" interpretation of the processors (either because they are indeed machines or because they are human beings trained to work in a very standardized and uniform way), then scheduling becomes somewhat more manageable.

To keep things simple we will work under the following three assumptions:

■ Any processor can execute any task (call this the *versatility* assumption).

■ The processing time for a task is the same regardless of which processor is executing the task (call this the *uniformity* assumption).

■ Once a processor starts a task, it will complete it without interruption (call this the *perseverance* assumption).

Under the preceding assumptions, the concept of *processing time* for a task makes good sense—and we can conveniently incorporate this information by including it inside parentheses next to the name of the task. Thus, the notation $X(5)$ tells us that the task called X has a processing time of 5 units (be it minutes, hours, days, or any other unit of time) *regardless of which processor is assigned to execute the task*.

■ **The precedence relations.** Precedence relations are formal restrictions on the order in which the tasks can be executed, much like those course prerequisites in the school catalog that tell you that you can't take course Y until you have completed course X. In the case of tasks, these prerequisites are called *precedence relations*. A typical precedence relation is of the form *task X precedes task Y*, and it means that task Y cannot be started until task X has been completed. A precedence relation can be conveniently abbreviated by writing $X \rightarrow Y$, or described graphically as shown in Fig. 8-1(a). A single scheduling problem can have hundreds or even thousands of precedence relations, each adding another restriction on the scheduler's freedom.

At the same time, it also happens fairly often that there are no restrictions on the order of execution between two tasks in a project. When a pair of tasks X and Y have no precedence requirements between them (neither $X \rightarrow Y$ nor $Y \rightarrow X$), we say that the tasks are **independent**. When two tasks are independent, either one can be started before the other one, or they can both be started at the same time—some people put their shoes on before their shirt, others put their shirt on before their shoes, and, occasionally, some of us have been known to put our shoes and shirt on at the same time. Graphically, we can tell that two tasks are independent if there are no arrows connecting them [Fig. 8-1(b)].

FIGURE 8-1 (a) X precedes Y. (b) X and Y are independent tasks. (c) When $X \rightarrow Y$ and $Y \rightarrow Z$, then $X \rightarrow Z$ is implied. (d) These tasks cannot be scheduled because of the cyclical nature of the precedence relations.

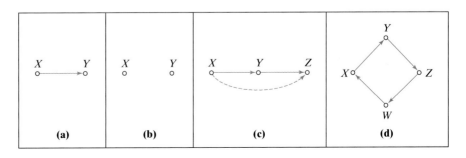

Two final comments about precedence relations are in order. First, precedence relations are *transitive*: If $X \rightarrow Y$ and $Y \rightarrow Z$, then it must be true that $X \rightarrow Z$. In a sense, the last precedence relation is implied by the first two, and it is really unnecessary to mention it [Fig. 8-1(c)]. Thus, we will make a distinction between two types of precedence relations: *basic* and *implicit*. Basic precedence relations are the ones that come with the problem and that we must follow in the process of creating a schedule. If we do this, the implicit precedence relations will be taken care of automatically.

The second observation is that *we cannot have a set of precedence relations that form a cycle!* Imagine having to schedule the tasks shown in Fig. 8-1(d): X precedes Y, which precedes Z, which precedes W, which in turn precedes X. Clearly, this is logically impossible. From here on, we will assume that there are no cycles of precedence relations among the tasks.

Processors, tasks, processing times, and *precedence relations* are the basic ingredients that make up a scheduling problem. They constitute, in a manner of speaking, the hand that is dealt to us. But how do we play such a hand? To get a small inkling of what's to come, let's look at the following simple example.

> ### EXAMPLE 8.1 Repairing a Wreck

Imagine that you just wrecked your brand new car, but thank heavens you are OK, and the insurance company will pick up the tab. You take the car to the best garage in town, operated by the Tappet brothers Click and Clack (we'll just call them P_1 and P_2). The repairs on the car can be broken into four different tasks: (*A*) exterior body work (4 hours), (*B*) engine repairs (5 hours), (*C*) painting and exterior finish work (7 hours), and (*D*) repair transmission (3 hours). The only precedence relation for this set of tasks is that the painting and exterior finish work cannot be started until the exterior body work has been completed ($A \rightarrow C$). The two brothers always work together on a repair project, but each takes on a different task (so they won't argue with each other). Under these assumptions, how should the different tasks be scheduled? Who should do what and when?

Even in this simple situation, many different schedules are possible. Figure 8-2 shows several possibilities, each one illustrated by means of a timeline. Figure 8-2(a) shows a schedule that is very inefficient. All the short tasks are assigned to one processor (P_1) and all the long tasks to the other processor (P_2)—obviously not a very clever strategy. Under this schedule, the project **finishing time** (the duration of the project from the start of the first task to the completion of the last task) is 12 hours. (We will use *Fin* to denote the project finishing time, so for this project we can write *Fin* = 12 hours.)

Figure 8-2(b) shows what looks like a much better schedule, but it violates the precedence relation $A \rightarrow C$ (as much as we would love to, we cannot start task C until task A is completed). On the other hand, if we force P_2 to be idle for one hour, waiting for the green light to start task C, we get a perfectly good schedule, shown in Fig. 8-2(c). Under this schedule the finishing time of the project is *Fin* = 11 hours.

The schedule shown in Fig. 8-2(c) is an improvement over the first schedule. Can we do even better? No! No matter how clever we are and no matter how

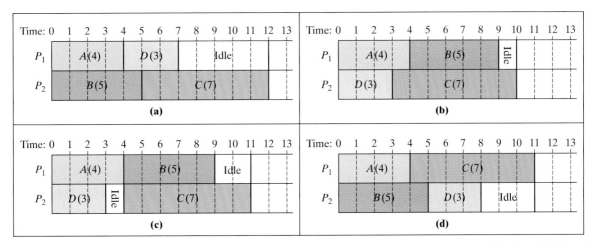

FIGURE 8-2 (a) A legal schedule with *Fin* = 12 hours. (b) An *illegal* schedule (the precedence relation $A \rightarrow C$ is violated). (c) An optimal schedule (*Opt* = 11 hours). (d) A different optimal schedule.

many processors we have at our disposal, the precedence relation $A(4) \rightarrow C(7)$ implies that 11 hours is a minimum barrier that we cannot break—it takes 4 hours to complete A, 7 hours to complete C, and *we cannot start C until A is completed*! Thus, the schedule shown in Fig. 8-2(c) is an **optimal schedule** and the finishing time of *Fin* = 11 hours is the **optimal finishing time**. (From now on we will use *Opt* instead of *Fin* when we are referring to the optimal finishing time.) Figure 8-2(d) shows a different optimal schedule with finishing time *Opt* = 11 hours. ◀◀

As scheduling problems go, Example 8.1 was a fairly simple one. But even from this simple example, we can draw some useful lessons. First, notice that even though we had only four tasks and two processors, we were able to create several different schedules. The four we looked at were just a sampler—there are other possible schedules that we didn't bother to discuss. Imagine what would happen if we had hundreds of tasks and dozens of processors—the number of possible schedules to consider would be overwhelming. In looking for a good, or even optimal, schedule, we are going to need a systematic way to sort through the many possibilities. In other words, we are going to need some good *scheduling algorithms*.

The second useful thing we learned in Example 8.1 is that when it comes to the finishing time of a project, there is an *absolute minimum* time that no schedule can break, no matter how good an algorithm we use or how many processors we put to work. In Example 8.1, this absolute minimum was 11 hours, and, as luck would have it, we easily found a schedule [actually two—Figs. 8-2(c) and (d)] with a finishing time to match it. Every project, no matter how simple or complicated, has such an absolute minimum (called the *critical time*) that depends on the processing times and precedence relations for the tasks and not on the number of processors used. We will return to the concept of critical time in Section 8.5.

To set the stage for a more formal discussion of scheduling algorithms, we will introduce the most important example of this chapter. While couched in what seems like science fiction terms, the situation it describes is not totally farfetched.

> **EXAMPLE 8.2** Building That Dream Home on Mars: Part 1

It is the year 2050, and several human colonies have already been established on Mars. Imagine that you accept a job offer to work in one of these colonies. What will you do about housing?

Like everyone else on Mars, you will be provided with a living pod called a Martian Habitat Unit (MHU). MHUs are shipped to Mars in the form of prefabricated kits that have to be assembled on the spot—an elaborate and unpleasant job if you are going to do it yourself. A better option is to hire special workers who will do all of the assembly for you. On Mars, these workers come in the form of robots called *HUBRIs* (Habitat Unit Building Robot I), which can be rented by the hour at the local Rent-a-Robot outlet.

The assembly of an MHU consists of 15 separate tasks, and there are 17 different precedence relations among these tasks that must be followed. The tasks, their respective processing times, and the precedence relations are all shown in Table 8-1.

TABLE 8-1

Task	Symbol (processing time)	Precedence relations
Assemble pad	$AP(7)$	
Assemble flooring	$AF(5)$	
Assemble wall units	$AW(6)$	
Assemble dome frame	$AD(8)$	
Install floors	$IF(5)$	$AP \rightarrow IF, AF \rightarrow IF$
Install interior walls	$IW(7)$	$IF \rightarrow IW, AW \rightarrow IW$
Install dome frame	$ID(5)$	$AD \rightarrow ID, IW \rightarrow ID$
Install plumbing	$PL(4)$	$IF \rightarrow PL$
Install atomic power plant	$IP(4)$	$IW \rightarrow IP$
Install pressurization unit	$PU(3)$	$IP \rightarrow PU, ID \rightarrow PU$
Install heating units	$HU(4)$	$IP \rightarrow HU$
Install commode	$IC(1)$	$PL \rightarrow IC, HU \rightarrow IC$
Complete interior finish work	$FW(6)$	$IC \rightarrow FW$
Pressurize dome	$PD(3)$	$HU \rightarrow PD$
Install entertainment unit	$EU(2)$	$PU \rightarrow EU, HU \rightarrow EU$

Here are some of the basic questions we will want to address: How can we get your MHU built quickly? How many *HUBRIs* should you rent to do the job? How do we create a suitable work schedule that will get the job done? (A *HUBRI* will do whatever it is told, but someone has to tell it what to do and when.) These are all tough questions to answer, but we will be able to do it later in the chapter.

8.2 Directed Graphs (Digraphs)

A **directed graph**, or **digraph** for short, is a graph in which the edges have a direction associated with them, typically indicated by an arrowhead. Digraphs are particularly useful when we want to describe asymmetric relationships (X related to Y does not imply that Y must be related to X).

The classic example of an asymmetric relationship is romantic love: Just because X is in love with Y, there is no guarantee that Y reciprocates that love. Given two individuals X and Y and some asymmetric relationship (say love), we have four possible scenarios: Neither loves the other [Fig. 8-3(a)], X loves Y but Y does not love X [Fig. 8-3(b)], Y loves X but X does not love Y [Fig. 8-3(c)], and they love each other [Fig. 8-3(d)].

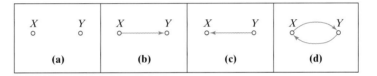

FIGURE 8-3

To distinguish digraphs from ordinary graphs, we use slightly different terminology. In a digraph, instead of talking about edges we talk about **arcs**. Every arc is defined by its *starting vertex* and its *ending vertex*, and we respect that order when we write the arc. Thus, if we write XY, we are describing the arc in Fig. 8-3(b) as opposed to the arc YX shown in Fig. 8-3(c). A list of all the arcs in a digraph is called the **arc-set** of the digraph. [The digraph in Fig. 8-3(d) has arc-set $\mathcal{A} = \{XY, YX\}$.]

If XY is an arc in the digraph, we say that vertex X is **incident to** vertex Y, or, equivalently, that Y is **incident from** X). The arc YZ is said to be **adjacent** to the arc XY if the starting point of YZ is the ending point of XY. (Essentially, this means one can go from X to Z by way of Y). In a digraph, a **path** from vertex X to vertex W consists of a sequence of arcs $XY, YZ, ZU, \ldots, VW$ such that each arc is adjacent to the one before it and no arc appears more than once in the sequence—it is essentially a trip from X to W along the arcs in the digraph. The best way to describe the path is by listing the vertices in the order of travel: X, Y, Z, and so on.

When the path starts and ends at the same vertex, we call it a **cycle** of the digraph. Just like circuits in a regular graph, cycles in digraphs can be written in more than one way—the cycle X, Y, Z, X is the same as the cycles Y, Z, X, Y and Z, X, Y, Z.

In a digraph, the notion of the degree of a vertex is replaced by the concepts of *indegree* and *outdegree*. The **outdegree** of X is the number of arcs that have X as their *starting point* (outgoing arcs); the **indegree** of X is the number of arcs that have X as their *ending point* (incoming arcs).

The following example illustrates some of the aforementioned concepts.

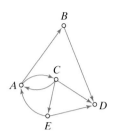

FIGURE 8-4

> ## EXAMPLE 8.3 Digraph Basics

The digraph in Fig. 8-4 has vertex-set $\mathcal{V} = \{A, B, C, D, E\}$ and arc-set $\mathcal{A} = \{AB, AC, BD, CA, CD, CE, EA, ED\}$. In this digraph, A is *incident to* B and C, but not to E. By the same token, A is *incident from* E as well as from C. The indegree of vertex A is 2, and so is the outdegree. The indegree of vertex C is 1, and the outdegree is 3. We leave it to the reader to find the indegrees and outdegrees of each of the other vertices of the graph.

In this digraph, there are several paths from A to D, such as A, C, D; $A, C, E,$ D; A, B, D; and even A, C, A, B, D. On the other hand, A, E, D is not a path from A to D (because you can't travel from A to E). There are two cycles in this digraph: A, C, E, A (which can also be written as C, E, A, C and E, A, C, E), and $A,$ C, A.   ≪

While love is not to be minimized as a subject of study, there are many other equally important applications of digraphs:

- **Traffic flow.** In most cities some streets are one-way streets and others are two-way streets. In this situation, digraphs allow us to visualize the flow of traffic through the city's streets. The vertices are intersections, and the *arcs* represent one-way streets. (To represent a two-way street, we use two arcs, one for each direction.)

- **Telephone traffic.** To track and analyze the traffic of telephone calls through their network, telephone companies use digraphs called "call digraphs." In these digraphs the vertices are telephone numbers, and an arc from X to Y indicates that a call was initiated from telephone number X to telephone number Y.

- **Tournaments.** Digraphs are frequently used to describe certain types of tournaments, with the vertices representing the teams (or individual players) and the arcs representing the outcomes of the games played in the tournament (the arc XY indicates that X defeated Y). Tournament digraphs can be used in any sport where the games cannot end in a tie (basketball, tennis, etc.).

- **Organization charts.** In any large organization (a corporation, the military, a university, etc.) it is important to have a well-defined chain of command. The best way to describe the chain of command is by means of a digraph often called an *organization chart*. In this digraph the *vertices* are the individuals in the organization, and an *arc* from X to Y indicates that X is Y's immediate boss (i.e., Y takes orders directly from X).

As you probably guessed by now, digraphs are also used in scheduling. There is no better way to visualize the tasks, processing times, and precedence relations in a project than by means of a digraph where the vertices represent the tasks (with their processing times indicated in parentheses) and the arcs represent the precedence relations.

> ## EXAMPLE 8.4 Building That Dream Home on Mars: Part 2

Let's return to the scheduling problem first discussed in Example 8.2. We can take the tasks and precedence relations given in Table 8-1 and create a digraph like the one shown in Fig. 8-5(a). It is helpful to try to place the vertices of the di-

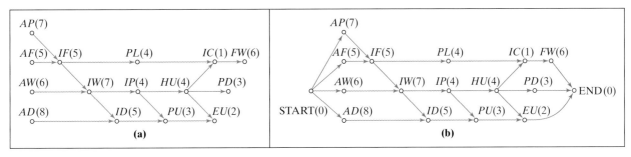

FIGURE 8-5

graph so that the arcs point from left to right, and this can usually be done with a little trial-and-error. After this is done, it is customary to add two fictitious tasks called START and END, where START indicates the imaginary task of getting the project started (cutting the red ribbon, so to speak), and END indicates the imaginary task of declaring the project complete (pop the champagne). By giving these fictitious tasks zero processing time, we avoid affecting the time calculations for the project. This modified digraph, shown in Fig. 8-5(b), is called the **project digraph**. The project digraph allows us to better visualize the execution of the project as a flow, moving from left to right.   ◀◀

8.3 Scheduling with Priority Lists

The project digraph is the basic graph model used to package all the information in a scheduling problem, but there is nothing in the project digraph itself that specifically tells us how to create a schedule. We are going to need something else, some set of instructions that indicates the order in which tasks should be executed. We can accomplish this by the simple act of prioritizing the tasks in some specified order, called a *priority list.*

A **priority list** is a list of all the tasks prioritized in the order we prefer to execute them. If task X comes before task Y in the priority list, then X gets priority over Y. This means that when it comes to a choice between the two, *X is executed ahead of Y*. However, if X is not yet *ready* for execution, *we skip over it and move on to the first ready task after X in the priority list*. (If there are no ready tasks after X in the priority list, the processors must sit idle and wait until a task becomes ready.)

The process of scheduling tasks using a priority list and following these basic rules is known as the **priority-list model** for scheduling. The priority-list model is a completely general model for scheduling—every priority list produces a schedule, and any schedule can be created from some (usually more than one) "parent" priority list. The trick is going to be to figure out *which* priority lists give us good schedules and which don't. We will come back to this topic in Sections 8.4 and 8.5.

Since each time we change the order of the tasks we get a different priority list, there are as many priority lists as there are ways to order the tasks. For three tasks, there are six possible priority lists; for 4 tasks, there are 24 priority lists; for 10 tasks, there are more than 3 million priority lists; and for 100 tasks, there are more priority lists than there are molecules in the universe.

Clearly, a shortage of priority lists is not going to be our problem. If this sounds familiar, it's because we already have seen the concept behind this type of explosive

For a brief review of factorials, see
Section 2.4.

growth before. Like sequential coalitions (Chapter 2) and Hamilton circuits (Chapter 6), the number of priority lists follows a factorial growth rule: *The number of possible priority lists in a project with M tasks, is* $M! = 1 \times 2 \times 3 \times \cdots \times M$.

Before we proceed, we will illustrate how the priority-list model for scheduling works with a couple of small but important examples. Even with such small examples, there is a lot to keep track of, and you are well advised to have pencil and paper in front of you as you follow the details.

> ### EXAMPLE 8.5 Preparing for Launch: Part 1

Immediately preceding the launch of a satellite into space, last-minute system checks need to be performed by the on-board computers, and it is important to complete these system checks as quickly as possible—for both cost and safety reasons. Suppose that there are five system checks required: $A(6)$, $B(5)$, $C(7)$, $D(2)$, and $E(5)$, with the numbers in parentheses representing the hours it takes one computer to perform that system check. In addition, there are precedence relations: D cannot be started until both A and B have been finished, and E cannot be started until C has been finished. The project digraph is shown in Fig. 8-6.

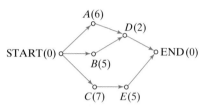

FIGURE 8-6

Let's assume that there are two identical computers on board (P_1 and P_2) that will carry out the individual system checks. How do we use the priority-list model to create a schedule for these two processors? For starters, we will need a priority list. Suppose that the priority list is given by listing the system checks in alphabetical order.

Priority list: $A(6)$, $B(5)$, $C(7)$, $D(2)$, $E(5)$

Let's look at the evolution of the project under the priority-list model. (We will use T to indicate the elapsed time in hours.)

- **$T = 0$ (START).** $A(6)$, $B(5)$, and $C(7)$ are the only *ready* tasks. Following the priority list, we assign $A(6)$ to P_1 and $B(5)$ to P_2.
- **$T = 5$.** P_1 is still *busy* with $A(6)$; P_2 has just *completed* $B(5)$. $C(7)$ is the only available *ready* task. We assign $C(7)$ to P_2.
- **$T = 6$.** P_1 has just *completed* $A(6)$; P_2 is *busy* with $C(7)$. $D(2)$ has just become a *ready* task (A and B have been completed). We assign $D(2)$ to P_1.
- **$T = 8$.** P_1 has just *completed* $D(2)$; P_2 is still *busy* with $C(7)$. There are no *ready* tasks at this time for P_1, so P_1 has to sit *idle*.
- **$T = 12$.** P_1 is *idle*; P_2 has just *completed* $C(7)$. Both processors are *ready* for work. $E(5)$ is the only ready task, so we assign $E(5)$ to P_1, P_2 sits *idle*. (Note that in this situation, we could have just as well assigned $E(5)$ to P_2 and let P_1 sit idle. The processors don't get tired and don't care if they are working or idle, so the choice is random.)
- **$T = 17$ (END).** P_1 has just *completed* $E(5)$, and therefore the project is completed. A timeline for this schedule is shown in Fig. 8-7.

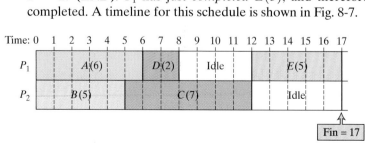

FIGURE 8-7

The project finishing time is *Fin* = 17 hours. Is this a good schedule? Given the excessive amount of idle time (a total of 9 hours), one might suspect this is a rather bad schedule. How could we improve it? We might try changing the priority list.

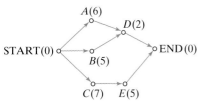

FIGURE 8-8

> **EXAMPLE 8.6** Preparing for Launch: Part 2

We are going to schedule the same project with the same processors, but with a different priority list. (The project digraph is shown again in Fig. 8-8. When scheduling, it's really useful to have the project digraph right in front of you.)

Let's try this time a reverse alphabetical order for the priority list. Why? Why not—at this point we are just shooting in the dark! (Don't worry, we are just feeling our way around—we will become a lot more enlightened later in the chapter.)

Priority list: $E(5), D(2), C(7), B(5), A(6)$

- **$T = 0$ (START).** $C(7)$, $B(5)$, and $A(6)$ are the only *ready* tasks. Following the priority list, we assign $C(7)$ to P_1 and $B(5)$ to P_2.

- **$T = 5$.** P_1 is still *busy* with $C(7)$; P_2 has just *completed* $B(5)$. $A(6)$ is the only available ready task. We assign $A(6)$ to P_2.

- **$T = 7$.** P_1 has just *completed* $C(7)$; P_2 is *busy* with $A(6)$. $E(5)$ has just become a *ready* task, and we assign it to P_1.

- **$T = 11$.** P_2 has just *completed* $A(6)$; P_1 is *busy* with $E(5)$. $D(2)$ has just become a *ready* task, and we assign it to P_2.

- **$T = 12$.** P_1 has just *completed* $E(5)$; P_2 is *busy* with $D(2)$. There are no tasks left, so P_1 sits idle.

- **$T = 13$ (END).** P_2 has just *completed* the last task, $D(2)$. Project is completed.

The timeline for this schedule is shown in Fig. 8-9. The project finishing time is *Fin* = 13 hours.

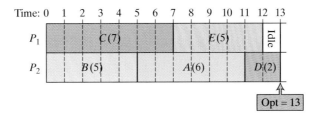

FIGURE 8-9

It's easy to see that this schedule is a lot better than the one obtained in Example 8.5. In fact, we were pretty lucky—this schedule turns out to be optimal for two processors. Two processors cannot finish this project in less than 13 hours because there is a total of 25 hours worth of work (the sum of all processing times), which implies that in the best of cases it would take 12.5 hours to finish the project. But since the processing times are all whole numbers and tasks cannot be split, the finishing time cannot be less than 13 hours! Thus, *Opt* = 13 hours. ◀◀

Thirteen hours is still a long time for the computers to go over their system checks. Since that's the best we can do with two computers, the only way to speed things up is to add a third computer to the "workforce." Adding another computer to the satellite can be quite expensive, but hopefully it will speed things up enough to make it worth it. Let's see.

> ## EXAMPLE 8.7 Preparing for Launch: Part 3

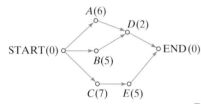

FIGURE 8-10

We will now schedule the same project using $N = 3$ processors (P_1, P_2, P_3). For the reader's convenience the project digraph is shown again in Fig. 8-10. We will use the "good" priority list we found in the previous example.

Priority list: $E(5), D(2), C(7), B(5), A(6)$

- ■ **$T = 0$ (START).** $C(7), B(5)$, and $A(6)$ are the *ready* tasks. We assign $C(7)$ to P_1, $B(5)$ to P_2, and $A(6)$ to P_3.

- ■ **$T = 5$.** P_1 is *busy* with $C(7)$; P_2 has just *completed* $B(5)$; and P_3 is *busy* with $A(6)$. There are no available ready tasks for P_2 [$E(5)$ can't be started until $C(7)$ is done, and $D(2)$ can't be started until $A(6)$ is done], so P_2 sits idle until further notice.

- ■ **$T = 6$.** P_3 has just *completed* $A(6)$; P_2 is *idle*; and P_1 is still *busy* with $C(7)$. $D(2)$ has just become a *ready* task. We randomly assign $D(2)$ to P_2 and let P_3 be idle, since there are no other ready tasks. [Note that we could have just as well assigned $D(2)$ to P_3 and let P_2 be idle.]

- ■ **$T = 7$.** P_1 has just *completed* $C(7)$ and $E(5)$ has just become a *ready* task, so we assign it to P_1. There are no other tasks to assign, so P_3 continues to sit idle.

- ■ **$T = 8$.** P_2 has just *completed* $D(2)$. There are no other tasks to assign, so P_2 and P_3 both sit *idle*.

- ■ **$T = 12$ (END).** P_1 has just *completed* the last task, $E(5)$, so the project is completed.

Strange things can happen when an additional processor is added to a project—see, for example, Exercise 67.

The timeline for this schedule is shown in Fig. 8-11. The project finishing time is *Fin* = 12 hours, a pathetically small improvement over the two-processor schedule found in Example 8.6. The cost of adding a third processor doesn't seem to justify the benefit.

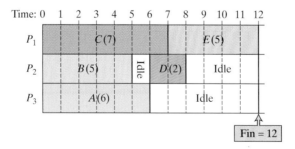

FIGURE 8-11

The previous three examples give us a general sense of how to create a schedule from a project digraph *and* a priority list. We will now formalize the ground rules of the **priority-list model** for scheduling.

At any particular moment in time throughout a project, a processor can be either *busy* or *idle* and a task can be *ineligible, ready, in execution*, or *completed*. Depending on the various combinations of these, there are three different scenarios to consider:

- **All processors are busy.** In this case, there is nothing we can do but wait.

- **One processor is free.** In this case, we scan the priority list from left to right, looking for the first *ready* task in the priority list, which we assign to that processor. (Remember that for a task to be *ready*, all the tasks that are incident to it in the project digraph must have been completed.) If there are no ready tasks at that moment, the processor must stay idle until things change.

- **More than one processor is free.** In this case, the *first ready* task on the priority list is given to one free processor, the second ready task is given to another free processor, and so on. If there are more free processors than ready tasks, some of the processors will remain idle until one or more tasks become ready. Since the processors are identical and tireless, the choice of which processor is assigned which task is completely arbitrary.

It's fair to say that the basic idea behind the priority-list model is not difficult, but there is a lot of bookkeeping involved, and that becomes critical when the number of tasks is large. At each stage of the schedule we need to keep track of the status of each task—which tasks are *ready* for processing, which tasks are *in execution*, which tasks have been *completed*, which tasks are still *ineligible*. One convenient recordkeeping strategy goes like this: On the priority list itself *ready* tasks are circled in red [Fig. 8-12(a)]. When a ready task is picked up by a processor and goes into *execution*, put a single red slash through the red circle [Fig. 8-12(b)]. When a task that has been in execution is completed, put a second red slash through the circle [Fig. 8-12(c)]. At this point, it is also important to check the project digraph to see if any new tasks have all of a sudden become eligible. Tasks that are *ineligible* remain unmarked [Fig. 8-12(d)].

FIGURE 8-12 "Road" signs on a priority list. (a) Task *X* is *ready*. (b) Task *X* is *in execution*. (c) Task *X* is *completed*. (d) Task *X* is *ineligible*.

As they say, the devil is in the details, so a slightly more substantive example will help us put everything together—the project digraph, the priority list model and the bookkeeping strategy.

▶ EXAMPLE 8.8 Building That Dream Home on Mars: Part 3

This is the third act of the Martian Habitat Unit building project. We are finally ready to start the project of assembling that MHU, and, like any good scheduler, we will first work the entire schedule out with pencil and paper. Let's start with the assumption that maybe we can get by with just two robots (P_1 and P_2). For the reader's convenience, the project digraph is shown again in Fig. 8-13.

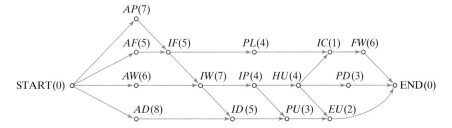

FIGURE 8-13

Let's start with a random priority list.

Priority list: $\widehat{AD(8)}$, $\widehat{AW(6)}$, $\widehat{AF(5)}$, $IF(5)$, $\widehat{AP(7)}$, $IW(7)$, $ID(5)$, $IP(4)$, $PL(4)$, $PU(3)$, $HU(4)$, $IC(1)$, $PD(3)$, $EU(2)$, $FW(6)$ (Ready tasks are circled in red.)

- **$T = 0$ (START).** Status of processors: P_1 starts AD; P_2 starts AW.
  Priority list: $\widehat{AD}$, $\widehat{AW}$, $\widehat{AF}$, IF, $\widehat{AP}$, $IW, ID, IP, PL, PU, HU, IC, PD, EU, FW$.

- **$T = 6$.** P_1 busy (executing AD); P_2 completed AW and starts AF.
  Priority list: $\widehat{AD}$, $\widehat{AW}$, $\widehat{AF}$, IF, $\widehat{AP}$, $IW, ID, IP, PL, PU, HU, IC, PD, EU, FW$.

- **$T = 8$.** P_1 completed AD and starts AP; P_2 is busy (executing AF).
  Priority list: $\widehat{AD}$, $\widehat{AW}$, $\widehat{AF}$, IF, $\widehat{AP}$, $IW, ID, IP, PL, PU, HU, IC, PD, EU, FW$.

- **$T = 11$.** P_1 busy (executing AP); P_2 completed AF, but since there are no ready tasks, it remains idle.
  Priority list: $\widehat{AD}$, $\widehat{AW}$, $\widehat{AF}$, IF, $\widehat{AP}$, $IW, ID, IP, PL, PU, HU, IC, PD, EU, FW$.

- **$T = 15$.** P_1 completed AP. IF becomes a ready task and goes to P_1; P_2 stays idle.
  Priority list: AD, $\widehat{AW}$, $\widehat{AF}$, $\widehat{IF}$, $\widehat{AP}$, $\widehat{IW}$, $ID, IP, PL, PU, HU, IC, PD, EU, FW$.

At this point, we will let you take over and finish the schedule. (Remember—the object is to learn how to keep track of the status of each task, and the only way to do this is with practice.)

After a fair amount of work, we obtain the final schedule shown in Fig. 8-14, with project finishing time $Fin = 44$ hours.

See Exercise 63.

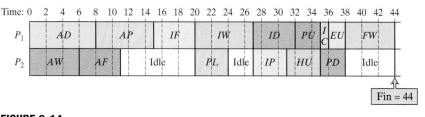

FIGURE 8-14

FIGURE 8-15

Scheduling under the priority-list model can be thought of as a two-part process: (1) Choose a priority list, and (2) use the priority list and follow the ground rules of the model to come up with a schedule (Fig. 8.15). As we saw in the previous example, the second part is long and tedious, but purely mechanical—it can be done by anyone (or anything) that is able to follow a set of instructions, be it a meticulous student or a properly programmed computer. We will use the term *scheduler* to describe the entity (be it student or machine) that takes a priority list as input and produces the schedule as output.

Ironically, it is the seemingly easiest part of this process—choosing a priority list—that is actually the most interesting. Among all the priority lists there is one or more that give a schedule with the optimal finishing time (we will call these *optimal priority lists*). How do we find an optimal priority list? Short of that, how do we find "good" priority lists, that is, priority lists that give schedules with finishing times reasonably close to the optimal? These are both important questions, and some answers are coming up next.

8.4 The Decreasing-Time Algorithm

Our first attempt to find a good priority list is to formalize what is a seemingly sensible and often used strategy: *Do the longer jobs first and leave the shorter jobs for last*. In terms of priority lists, this strategy is implemented by creating a priority list where the tasks are listed in decreasing order of processing times—longest first, second longest next, and so on. (When there are two or more tasks with equal processing times, we order them randomly.)

A priority list where the tasks are listed in decreasing order of processing times is called, not surprisingly, a **decreasing-time priority list**, and the process of creating a schedule using a decreasing-time priority list is called the **decreasing-time algorithm**.

> **EXAMPLE 8.9** Building That Dream Home on Mars: Part 4

Figure 8-16 shows, once again, the project digraph for the Martian Habitat Unit building project. To use the decreasing-time algorithm we first prioritize the 15 tasks in a decreasing-time priority list.

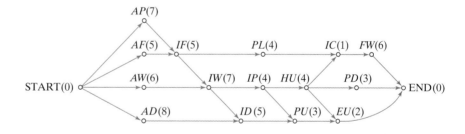

FIGURE 8-16

The details of the process are shown in Table 8-2, but you may want to try re-creating the schedule on your own—see Exercise 64.

Decreasing-time priority list: $AD(8)$, $AP(7)$, $IW(7)$, $AW(6)$, $FW(6)$, $AF(5)$, $IF(5)$, $ID(5)$, $IP(4)$, $PL(4)$, $HU(4)$, $PU(3)$, $PD(3)$, $EU(2)$, $IC(1)$

Using the decreasing-time algorithm with $N = 2$ processors, we get the schedule shown in Fig. 8-17, with project finishing time $Fin = 42$ hours.

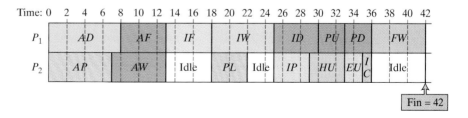

FIGURE 8-17

When looking at the finishing time under the decreasing-time algorithm, one can't help but feel disappointed. The sensible idea of prioritizing the longer jobs ahead of the shorter jobs turned out to be somewhat of a dud—at least in this

TABLE 8-2

Step	Time	Priority-List Status	Schedule Status
1	$T = 0$	~~AD~~(8) ~~AP~~(7) IW(7) ~~AW~~(6) FW(6) ~~AF~~(5) IF(5) ID(5) IP(4) PL(4) HU(4) PU(3) PD(3) EU(2) IC(1)	Time: 0 2 4 6 8 10 12 14 16 18 20 22 24 26 28 30 32 34 36 38 40 42 P_1: AD P_2: AP
2	$T = 7$	~~AD~~(8) ~~AP~~(7) IW(7) ~~AW~~(6) FW(6) ~~AF~~(5) IF(5) ID(5) IP(4) PL(4) HU(4) PU(3) PD(3) EU(2) IC(1)	Time: 0 2 4 6 8 10 12 14 16 18 20 22 24 26 28 30 32 34 36 38 40 42 P_1: AD P_2: AP AW
3	$T = 8$	~~AD~~(8) ~~AP~~(7) IW(7) ~~AW~~(6) FW(6) ~~AF~~(5) IF(5) ID(5) IP(4) PL(4) HU(4) PU(3) PD(3) EU(2) IC(1)	Time: 0 2 4 6 8 10 12 14 16 18 20 22 24 26 28 30 32 34 36 38 40 42 P_1: AD AF P_2: AP AW
4	$T = 13$	~~AD~~(8) ~~AP~~(7) IW(7) ~~AW~~(6) FW(6) ~~AF~~(5) ~~IF~~(5) ID(5) IP(4) PL(4) HU(4) PU(3) PD(3) EU(2) IC(1)	Time: 0 2 4 6 8 10 12 14 16 18 20 22 24 26 28 30 32 34 36 38 40 42 P_1: AD AF IF P_2: AP AW
5	$T = 18$	~~AD~~(8) ~~AP~~(7) ~~IW~~(7) ~~AW~~(6) FW(6) ~~AF~~(5) ~~IF~~(5) ID(5) IP(4) ~~PL~~(4) HU(4) PU(3) PD(3) EU(2) IC(1)	Time: 0 2 4 6 8 10 12 14 16 18 20 22 24 26 28 30 32 34 36 38 40 42 P_1: AD AF IF IW P_2: AP AW Idle PL
6	$T = 22$	~~AD~~(8) ~~AP~~(7) ~~IW~~(7) ~~AW~~(6) FW(6) ~~AF~~(5) ~~IF~~(5) ID(5) IP(4) ~~PL~~(4) HU(4) PU(3) PD(3) EU(2) IC(1)	Time: 0 2 4 6 8 10 12 14 16 18 20 22 24 26 28 30 32 34 36 38 40 42 P_1: AD AF IF IW P_2: AP AW Idle PL
7	$T = 25$	~~AD~~(8) ~~AP~~(7) ~~IW~~(7) ~~AW~~(6) FW(6) ~~AF~~(5) ~~IF~~(5) ~~ID~~(5) ~~IP~~(4) ~~PL~~(4) HU(4) PU(3) PD(3) EU(2) IC(1)	Time: 0 2 4 6 8 10 12 14 16 18 20 22 24 26 28 30 32 34 36 38 40 42 P_1: AD AF IF IW ID P_2: AP AW Idle PL Idle IP
8	$T = 29$	~~AD~~(8) ~~AP~~(7) ~~IW~~(7) ~~AW~~(6) FW(6) ~~AF~~(5) ~~IF~~(5) ~~ID~~(5) ~~IP~~(4) ~~PL~~(4) ~~HU~~(4) PU(3) PD(3) EU(2) IC(1)	Time: 0 2 4 6 8 10 12 14 16 18 20 22 24 26 28 30 32 34 36 38 40 42 P_1: AD AF IF IW ID P_2: AP AW Idle PL Idle IP HU
9	$T = 30$	~~AD~~(8) ~~AP~~(7) ~~IW~~(7) ~~AW~~(6) FW(6) ~~AF~~(5) ~~IF~~(5) ~~ID~~(5) ~~IP~~(4) ~~PL~~(4) ~~HU~~(4) ~~PU~~(3) PD(3) EU(2) IC(1)	Time: 0 2 4 6 8 10 12 14 16 18 20 22 24 26 28 30 32 34 36 38 40 42 P_1: AD AF IF IW ID PU P_2: AP AW Idle PL Idle IP HU
10	$T = 33$	~~AD~~(8) ~~AP~~(7) ~~IW~~(7) ~~AW~~(6) FW(6) ~~AF~~(5) ~~IF~~(5) ~~ID~~(5) ~~IP~~(4) ~~PL~~(4) ~~HU~~(4) ~~PU~~(3) PD(3) ~~EU~~(2) IC(1)	Time: 0 2 4 6 8 10 12 14 16 18 20 22 24 26 28 30 32 34 36 38 40 42 P_1: AD AF IF IW ID PU PD P_2: AP AW Idle PL Idle IP HU EU
11	$T = 35$	~~AD~~(8) ~~AP~~(7) ~~IW~~(7) ~~AW~~(6) FW(6) ~~AF~~(5) ~~IF~~(5) ~~ID~~(5) ~~IP~~(4) ~~PL~~(4) ~~HU~~(4) ~~PU~~(3) ~~PD~~(3) ~~EU~~(2) ~~IC~~(1)	Time: 0 2 4 6 8 10 12 14 16 18 20 22 24 26 28 30 32 34 36 38 40 42 P_1: AD AF IF IW ID PU PD P_2: AP AW Idle PL Idle IP HU EU IC
12	$T = 36$	~~AD~~(8) ~~AP~~(7) ~~IW~~(7) ~~AW~~(6) ~~FW~~(6) ~~AF~~(5) ~~IF~~(5) ~~ID~~(5) ~~IP~~(4) ~~PL~~(4) ~~HU~~(4) ~~PU~~(3) ~~PD~~(3) ~~EU~~(2) ~~IC~~(1)	Time: 0 2 4 6 8 10 12 14 16 18 20 22 24 26 28 30 32 34 36 38 40 42 P_1: AD AF IF IW ID PU PD FW P_2: AP AW Idle PL Idle IP HU EU IC
13	$T = 42$	~~AD~~(8) ~~AP~~(7) ~~IW~~(7) ~~AW~~(6) ~~FW~~(6) ~~AF~~(5) ~~IF~~(5) ~~ID~~(5) ~~IP~~(4) ~~PL~~(4) ~~HU~~(4) ~~PU~~(3) ~~PD~~(3) ~~EU~~(2) ~~IC~~(1)	Time: 0 2 4 6 8 10 12 14 16 18 20 22 24 26 28 30 32 34 36 38 40 42 P_1: AD AF IF IW ID PU PD FW P_2: AP AW Idle PL Idle IP HU EU IC Idle

example! What went wrong? If we work our way backward from the end, we can see that we made a bad choice at $T = 33$ hours. At this point there were three ready tasks [$PD(3)$, $EU(2)$, and $IC(1)$], and both processors were available. Based on the decreasing-time priority list, we chose the two "longer" tasks, $PD(3)$ and $EU(2)$, ahead of the short task, $IC(1)$. This was a bad move! $IC(1)$ is a much more *critical* task than the other two because we can't start task $FW(6)$ until we finish task $IC(1)$. Had we looked ahead at some of the tasks incident from these three, we might have noticed this.

An even more blatant example of the weakness of the decreasing-time algorithm occurs at the very start of this schedule when the algorithm fails to take into account the fact that task $AF(5)$ should have a very high priority. Why? $AF(5)$ is one of the two tasks that must be finished before $IF(5)$ can be started, and $IF(5)$ must be finished before $IW(7)$ can be started, which must be finished before $IP(4)$ and $ID(5)$ can be started, and so on down the line.

When there is a long path of tasks in the project digraph, it seems clear that the first task along that path should be started as early as possible. This idea leads to the following informal rule: *The greater the total amount of work that lies ahead of a task, the sooner that task should be started.*

8.5 Critical Paths

To formalize the preceding ideas, we will introduce the concepts of *critical paths* and *critical times*.

Critical Paths and Critical Times

- For a given vertex X of a project digraph, the **critical path for** X is the path *from X to END* with *longest* processing time. (The *processing time* of a path is defined to be the sum of the processing times of all the vertices in the path.) When we add the processing times of all the tasks along the critical path for a vertex X, we get the **critical time for** X. (By definition, the critical time of END is 0.)

- The path with longest processing time from START to END is called the **critical path** for the project, and the total processing time for this critical path is called the **critical time** for the project.

> **EXAMPLE 8.10** Building That Dream Home on Mars: Part 5

Figure 8-18 shows the project digraph for the Martian Habitat Unit building project. We will find critical paths and critical times for several vertices of the project digraph.

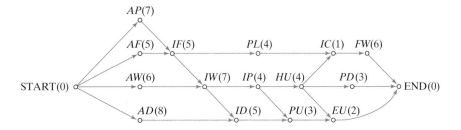

FIGURE 8-18

Let's start with a relatively easy case—the vertex *HU*. A quick look at Fig. 8-18 should convince you that there are only three paths from *HU* to END, namely

- *HU, IC, FW*, END, with processing time 4 + 1 + 6 = 11 hours,
- *HU, PD*, END, with processing time 4 + 3 = 7 hours, and
- *HU, EU*, END, with processing time 4 + 2 = 6 hours.

Of the three paths, the first one has the longest processing time, so *HU, IC, FW*, END is the *critical path* for vertex *HU*. The *critical time* for *HU* is 11 hours.

Next, let's find the critical path for vertex *AD*. There is only one path from *AD* to END, namely *AD, ID, PU, EU*, END, which makes the decision especially easy. Since this is the only path, it is automatically the longest path and therefore the *critical path for AD*. The *critical time* for *AD* is 8 + 5 + 3 + 2 = 18 hours.

To find the critical path for the project, we need to find the path from START to END with longest processing time. Since there are dozens of paths from START to END, let's just eyeball the project digraph for a few seconds and take our best guess....

OK, if you guessed START, *AP, IF, IW, IP, HU, IC, FW*, END, you have good eyes. This is indeed the *critical path*. It follows that the *critical time* for the Martian Habitat Unit building project is 34 hours. ⟪

We will soon discuss the special role that the critical time and the critical path play in scheduling, but before we do so, let's address the issue of how to find critical paths. In a large project digraph there may be thousands of paths from START to END, and the "eyeballing" approach we used in the preceding example is not likely to work. What we need here is an efficient algorithm, and fortunately there is one—it is called the *backflow algorithm*.

The Backflow Algorithm

- **Step 1.** Find the critical time for every vertex of the project digraph. This is done by starting at END and working backward toward START according to the following rule: *critical time for X = processing time of X plus largest critical time among the vertices incident from X.* The general idea is illustrated in Fig. 8-19. [To help with the recordkeeping, it is suggested that you write the critical time of the vertex in square brackets [ ] to distinguish it from the processing time in parentheses ( ).]

- **Step 2.** Once we have the critical time for every vertex in the project digraph, critical paths are found by just following the *path along largest critical times*. In other words, the critical path for any vertex *X* (and that includes START) is obtained by starting at *X* and moving to the adjacent vertex with largest critical time, and from there to the adjacent vertex with largest critical time, and so on.

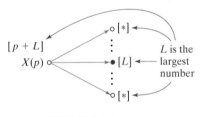

FIGURE 8-19

While the backflow algorithm sounds a little complicated when described in words, it is actually pretty easy to implement in practice, as we will show in the next example.

 EXAMPLE 8.11 Building That Dream Home on Mars: Part 6

We are now going to use the backflow algorithm to find the critical time for each of the vertices of the Martian Habitat Unit project digraph (Fig. 8–20).

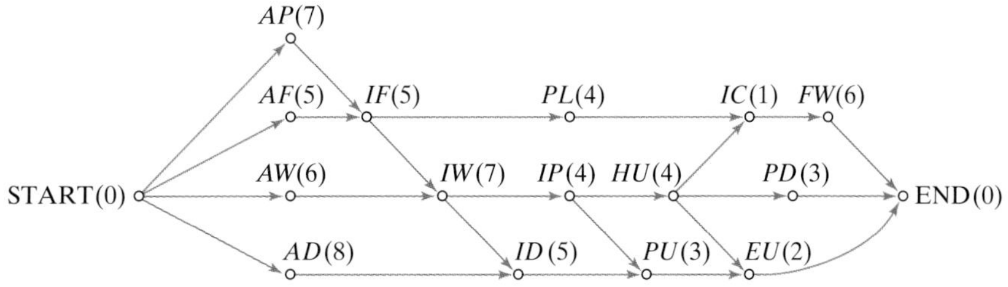

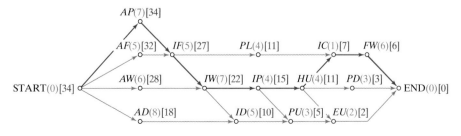

FIGURE 8-20

- **Step 1**
  - Start at END. The critical time of END is 0, so we add a [0] next to END(0).
  - The backflow now moves to the three vertices that are incident to END, namely, $FW(6)$, $PD(3)$, and $EU(2)$. In each case the critical time is the processing time plus 0, so the critical times are $FW[6]$, $PD[3]$, and $EU[2]$. We add this information to the project digraph.
  - From $FW[6]$, the backflow moves to $IC(1)$. The vertex $IC(1)$ is incident only to $FW[6]$, so the critical time for IC is $1 + 6 = 7$. We add a [7] next to IC in the project digraph.
  - The backflow now moves to $HU(4)$, $PL(4)$, and $PU(3)$. There are three vertices $HU(4)$ is incident to ($IC[7]$, $PD[3]$, and $EU[2]$). Of the three, the one with the largest critical time is $IC[7]$. This means that the critical time for HU is $4 + 7 = 11$. $PL(4)$ is only incident to $IC[7]$, so its critical time is $4 + 7 = 11$. $PU(3)$ is only incident to $EU[2]$, so its critical time is $3 + 2 = 5$. Add [11], [11], and [5] next to HU, PL, and PU, respectively.
  - The backflow now moves to $IP(4)$ and $ID(5)$. $IP(4)$ is incident to $HU[11]$ and $PU[5]$, so the critical time for IP is $4 + 11 = 15$. $ID(5)$ is only incident to $PU(5)$, so its critical time is $5 + 5 = 10$. We add [15] next to IP, and [10] next to ID.
  - The backflow now moves to $IW(7)$. The critical time for IW is $7 + 15 = 22$. (Please verify that this is correct!)
  - The backflow now moves to $IF(5)$. The critical time for IF is $5 + 22 = 27$. (Ditto.)
  - The backflow now moves to $AP(7)$, $AF(5)$, $AW(6)$, and $AD(8)$. Their respective critical times are $7 + 27 = 34, 5 + 27 = 32, 6 + 22 = 28$, and $8 + 10 = 18$.
  - Finally, the backflow reaches START(0). We still follow the same rule—the critical time is $0 + 34 = 34$. This is the critical time for the project!
- **Step 2** The critical time for every vertex of the project digraph is shown in Fig. 8-21. We can now find the critical path by following the trail of largest critical times: START, $AP, IF, IW, IP, HU, IC, FW$, END.

FIGURE 8-21

A word of caution: The project's critical time is not necessarily the same as the project's *optimal completion time*. The optimal completion time depends on how many processors are working on the project; the critical time has nothing to do with the number of processors at work.

Why are the critical path and critical time of a project of special significance? We saw earlier in the chapter that for every project there is a theoretical time barrier below which a project cannot be completed, regardless of how clever the scheduler is or how many processors are used. Well, guess what? This theoretical barrier *is the project's critical time*.

If a project is to be completed in the optimal completion time, it is absolutely essential that all the tasks in the critical path be done at the earliest possible time. Any delay in starting up one of the tasks in the critical path will necessarily delay the finishing time of the entire project. (By the way, this is why this path is called *critical*.)

Unfortunately, it is not always possible to schedule the tasks on the critical path one after the other, bang, bang, bang without delay. For one thing, processors are not always free when we need them. (Remember that a processor cannot stop in the middle of one task to start a new task.) Another reason is the problem of uncompleted predecessor tasks. We cannot concern ourselves only with tasks along the critical path and disregard other tasks that might affect them through precedence relations. There is a whole web of interrelationships that we need to worry about. Optimal scheduling is extremely complex.

8.6 The Critical-Path Algorithm

The concept of critical paths can be used to create very good (although not necessarily optimal) schedules. The idea is to use *critical times* rather than processing times to prioritize the tasks. The priority list we obtain when we write the tasks in decreasing order of critical times (with ties broken randomly) is called the **critical-time priority list**, and the process of creating a schedule using the critical-time priority list is called the **critical-path algorithm**.

Critical-Path Algorithm

- **Step 1 (Find critical times).** Using the backflow algorithm, find the *critical time* for every task in the project.
- **Step 2 (Create priority list).** Using the critical times obtained in Step 1, create the *critical-time priority list*.
- **Step 3 (Create schedule).** Using the critical-time priority list obtained in Step 2, create the *schedule*.

There are, of course, plenty of small details that need to be attended to when carrying out the critical-path algorithm, especially in Steps 1 and 3. Fortunately, everything that needs to be done we now know how to do.

> **EXAMPLE 8.12** Building That Dream Home on Mars: Part 7

We will now schedule the Martian Habitat Unit building project with $N = 2$ processors using the critical-path algorithm.

We took care of Step 1 in Example 8.11. The critical times for each task are shown in red in Fig. 8-22.

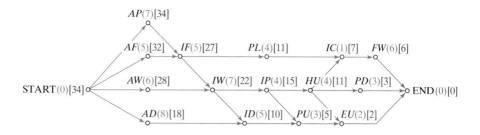

FIGURE 8-22

Step 2 follows directly from Step 1. The critical-time priority list for the project is $AP[34]$, $AF[32]$, $AW[28]$, $IF[27]$, $IW[22]$, $AD[18]$, $IP[15]$, $PL[11]$, $HU[11]$, $ID[10]$, $IC[7]$, $FW[6]$, $PU[5]$, $PD[3]$, $EU[2]$.

See Exercise 65.

Step 3 is a lot of busywork—the details are left to the reader.

The timeline for the resulting schedule is given in Fig. 8-23. The project finishing time is $Fin = 36$ hours. This is a very good schedule, but it is not an optimal schedule. (Figure 8-24 shows the timeline for an optimal schedule with finishing time $Opt = 35$ hours.)

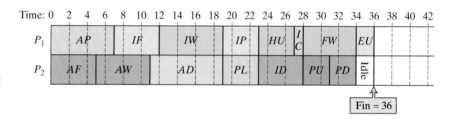

FIGURE 8-23 Timeline for the MHU building project under the critical-path algorithm ($N = 2$).

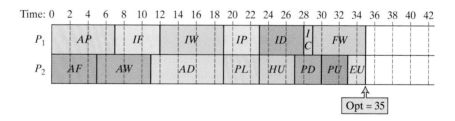

FIGURE 8-24 Timeline for an optimal schedule for the MHU building project ($N = 2$).

The critical-path algorithm is an excellent *approximate* algorithm for scheduling a project, but as Example 8.12 shows, it does not always give an optimal schedule. In this regard, scheduling problems are like traveling salesman problems (Chapter 6) and shortest network problems (Chapter 7)—there are *efficient approximate* algorithms for scheduling, but no *efficient optimal* algorithm is currently known. Of the standard scheduling algorithms, the critical-path algorithm is by far the most commonly used. Other, more sophisticated algorithms have been developed in the last 30 years, and under specialized circumstances they can outperform the critical-path algorithm, but as an all-purpose algorithm for scheduling, the critical-path algorithm is hard to beat.

8.7 Scheduling with Independent Tasks

In this section, we will briefly discuss what happens to scheduling problems in the special case when there are no precedence relations to worry about. This situation arises whenever we are scheduling tasks that are all independent—for example, scheduling a group of typists in a steno pool to type a bunch of reports of various lengths.

It is tempting to think that without precedence relations hanging over one's head, scheduling becomes a simple problem, and one should be able to find optimal schedules without difficulty, but appearances are deceiving. *There are no efficient optimal algorithms known for scheduling, even when the tasks are all independent.*

While, in a theoretical sense, we are not much better able to schedule independent tasks than to schedule tasks with precedence relations, from a purely practical point of view, there are a few differences. For one thing, there is no getting around the fact that the nuts-and-bolts details of creating a schedule using a priority list become tremendously simplified when there are no precedence relations to mess with. In this case, we just assign the tasks to the processors as they become free in exactly the order given by the priority list. Second, without precedence relations, the critical-path time of a task equals its processing time. This means that the *critical-time list* and *decreasing-time list* are exactly the same list, and, therefore, the decreasing-time algorithm and the critical-path algorithm become one and the same. Before we go on, let's look at a couple of examples of scheduling with independent tasks.

> ## EXAMPLE 8.13 Preparing for Lunch: Part 1

Imagine that you and your two best friends are cooking a nine-course luncheon as part of a charity event. We will consider each of the nine courses an independent task, to be done by just one of the chefs (P_1, P_2, or P_3). The nine courses are $A(70)$, $B(90)$, $C(100)$, $D(70)$, $E(80)$, $F(20)$, $G(20)$, $H(80)$, and $I(10)$, with their processing times given in minutes. Let's first use an alphabetical priority list.

Priority list: $A(70)$, $B(90)$, $C(100)$, $D(70)$, $E(80)$, $F(20)$, $G(20)$, $H(80)$, $I(10)$

Since there are no precedence relations, there are no ineligible tasks, and all tasks start out as ready tasks. As soon as a processor is free, it picks up the next available task in the priority list. From the bookkeeping point of view, this is a piece of cake. We leave it to the reader to verify that the resulting schedule is the one in Fig. 8-25, with finishing time *Fin* = 220 minutes. It is obvious from the figure that this is not a very good schedule.

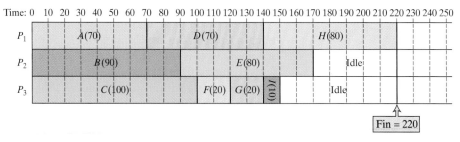

FIGURE 8-25

If we use the decreasing-time priority list (which in this case is also the critical-time priority list), we are bound to get a much better schedule.

Decreasing-time list: $C(100)$, $B(90)$, $E(80)$, $H(80)$, $A(70)$, $D(70)$, $F(20)$, $G(20)$, $I(10)$

The resulting schedule is shown in Fig. 8-26 and it is clearly an optimal schedule, since there is no idle time for any of the processors throughout the project. The optimal finishing time for the project is $Opt = 180$ minutes.

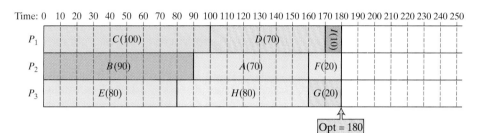

FIGURE 8-26

In Example 8.13, the critical-path algorithm gave us the optimal schedule, but, unfortunately, this need not always be the case.

EXAMPLE 8.14 Preparing for Lunch: Part 2

After the success of your last banquet, you and your two friends are asked to prepare another banquet. This time it will be a seven-course meal. The courses are all independent tasks, and their processing times (in minutes) are $A(50)$, $B(30)$, $C(40)$, $D(30)$, $E(50)$, $F(30)$, and $G(40)$.

The decreasing-time priority list is $A(50)$, $E(50)$, $C(40)$, $G(40)$, $B(30)$, $D(30)$, and $F(30)$. The resulting schedule, shown in Fig. 8-27, has project finishing time

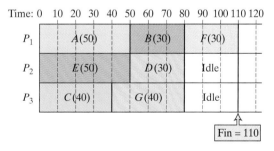

FIGURE 8-27

$Fin = 110$ minutes. An optimal schedule with finishing time $Opt = 90$ minutes is shown in Fig. 8-28.

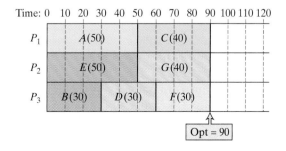

FIGURE 8-28

The Relative Error

If we know the optimal finishing time *Opt*, we can measure how "close" a particular schedule is to being optimal using the concept of *relative error*. For a project with finishing time *Fin*, the **relative error** (denoted by ε) is given by $\varepsilon = \dfrac{Fin - Opt}{Opt}$.

In Example 8.14, we found that the critical-path algorithm results in the project finishing time *Fin* = 110 minutes, and we also found that *Opt* = 90 minutes. Thus, the relative error for the schedule obtained using the critical-path algorithm is $\varepsilon = (110 - 90)/90 = 20/90 \approx 0.2222$.

It is customary to express the relative error in terms of percents, which in this case translates into $\varepsilon = 22.22\%$. Essentially, we interpret this to mean that in Example 8.14, the schedule produced by the critical-path algorithm is 22.22% longer than the optimal schedule. Interestingly enough, we couldn't do any worse than this—even if we tried.

For more on Graham, see the biographical profile at the end of this chapter.

In 1969, the American mathematician Ron Graham showed that for independent tasks, the critical-path algorithm will produce schedules with relative errors that fall within a specified, narrow range. Specifically, Graham proved that the relative error falls somewhere between 0 and $\dfrac{N - 1}{3N}$, where N is the number of processors. Table 8-3 shows the range of the relative error for a few values of N.

TABLE 8-3 Range of the Relative Error ε	
N	**Range of the relative error**
2	Between 0% and $\dfrac{2 - 1}{3 \times 2} = \dfrac{1}{6} \approx 16.66\%$
3	Between 0% and $\dfrac{3 - 1}{3 \times 3} = \dfrac{2}{9} \approx 22.22\%$
4	Between 0% and $\dfrac{4 - 1}{3 \times 4} = \dfrac{3}{12} = 25\%$
5	Between 0% and $\dfrac{5 - 1}{3 \times 5} = \dfrac{4}{15} \approx 26.66\%$
6	Between 0% and $\dfrac{6 - 1}{3 \times 6} = \dfrac{5}{18} \approx 27.78\%$
$\vdots$	$\vdots$
100	Between 0% and $\dfrac{100 - 1}{3 \times 100} = \dfrac{99}{300} = 33\%$

See Exercise 61.

Table 8-3 gives us a good sense of what is happening: As the number of processors increases, so does the range of the relative error, but the rate of increase slows down dramatically by the time we get to $N = 5$. Moreover, it is not hard to show that the range of the relative error will always fall between 0 and $33\frac{1}{3}\%$. Graham's discovery essentially reassures us that when the tasks are independent, the finishing time obtained using the critical-path algorithm is never going to be too far off from the optimal finishing time—no matter how many tasks need to be scheduled or how many processors are available to carry them out.

Conclusion

In one form or another, the scheduling of human (and nonhuman) activity is a pervasive and fundamental problem of modern life. At its most informal, it is part and parcel of the way we organize our everyday living (so much so that we are often scheduling things without realizing we are doing so). In its more formal incarnation, the systematic scheduling of a set of activities for the purposes of saving either time or money is a critical issue in management science. Business, industry, government, education—wherever there is a big project, there is a schedule behind it.

By now, it should not surprise us that at their very core, scheduling problems are mathematical in nature and that the mathematics of scheduling can range from the simple to the extremely complex. In this chapter we focused on a very specific type of scheduling problem in which we are given a set of *tasks*, a set of *precedence relations* among the tasks, and a set of identical *processors*. The objective is to schedule the tasks by properly assigning tasks to processors so that the *project finishing time* is as small as possible.

To tackle these scheduling problems systematically, we first developed a graph model of the problem, called the *project digraph*, and a general framework by means of which we can create, compare, and analyze schedules, called the *priority-list model*. Within the priority-list model, many strategies can be followed (with each strategy leading to the creation of a specific priority list). In the chapter, we considered two basic strategies for creating schedules. The first was the *decreasing-time algorithm*, a strategy that intuitively makes a lot of sense but that in practice often results in inefficient schedules. The second strategy, called the *critical-path algorithm*, is generally a big improvement over the decreasing-time algorithm, but it falls short of the ideal goal of guaranteeing an optimal schedule. The critical-path algorithm is by far the best known and most widely used algorithm for scheduling in business and industry.

When scheduling with independent tasks, the decreasing-time algorithm and the critical-path algorithm become one and the same, and the project finishing times they generate will never be off by much from the optimal finishing times.

Although several other, more sophisticated strategies for scheduling have been discovered by mathematicians in the last 40 years, no optimal, efficient scheduling algorithm is presently known, and the general feeling among the experts is that there is little likelihood that such an algorithm actually exists. Efficient scheduling, nonetheless, will always remain a significant human goal—another task, we must execute, in that grand cosmic project we call organized life.

Science is organized knowledge.
Wisdom is organized life.

Immanuel Kant

Profile Ronald L. Graham (1935–)

Ron Graham is, by all accounts, the consummate juggler (in every sense of the word). For a period of almost 40 years, as a mathematician, director, vice president, and head scientist at Bell Labs, AT&T's legendary research arm, Graham was famous for his ability to keep an eclectic array of activities all going at once— doing research in several areas of mathematics, administering a major research institute, lecturing and teaching all over the world, teaching himself Chinese, and *really* juggling. (Graham is a past president of the International Jugglers Association, the inventor of many new juggling routines, and an expert on the mathematics of juggling.)

Ronald L. Graham was born in 1935 in Taft, California. Graham discovered his natural talent and love of mathematics at an early age, and by the time he was 15, without even finishing high school, he won a full scholarship to the University of Chicago. After brief stints at Chicago and the University of California, Berkeley, Graham joined the Air Force and was stationed in Fairbanks, Alaska, where he served as a communication specialist while completing his undergraduate studies at the University of Alaska. After his tour of duty with the Air Force, Graham returned to UC Berkeley, where he completed a Ph.D. in mathematics in 1962. While a student at Berkeley, Graham, an accomplished gymnast and trampolinist, supported himself by performing with a small circus troupe called the Bouncing Bears, which performed acrobatic stunts at schools and fairs.

Graham joined Bell Labs in 1962. For the next 37 years he held a long list of scientific and administrative roles: researcher, head of the Mathematics Division, director of the Mathematics Center, vice president of research, and head scientist. Soon after joining Bell Labs, Graham was approached by engineers working on the computer system that was being developed to manage the antiballistic missile (ABM) defense program. One of the major issues engineers were facing was how to schedule the computers to identify and lock onto incoming enemy missiles in an efficient way. This turned out to be a classic scheduling problem of the kind discussed in this chapter (the computers are the *processors*; identifying and destroying the incoming missiles are the *tasks*). To work on these problems, Graham developed much of the theory of scheduling as we now it today. In so doing he also created a new field in the study of algorithms known as *worst-case analysis* (see Project A).

Graham retired from Bell Labs in 1999. Not one to sit idle, he promptly accepted an endowed chair as the Irwin and Joan Jacobs Professor of Computer Science and Information Science at the University of California, San Diego. Graham continues to teach, carry on with his mathematics research, lecture around the world, juggle, and, in general, keep going through life at a pace that amazes everyone else. When once asked how he managed to keep so many different things going, Graham's reply was, "There are 168 hours in a week."

Key Concepts

adjacent (arcs), **281**
arc, **281**
arc-set, **281**
backflow algorithm, **292**
critical path, **291**
critical-path algorithm, **294**
critical-time priority list, **294**
critical time, **291**
cycle, **281**
decreasing-time algorithm, **289**
decreasing-time priority list, **289**

digraph, **281**
finishing time (*Fin*), **278**
incident from, **281**
incident to, **281**
indegree, **281**
independent tasks, **277**
optimal finishing time (*Opt*), **279**
optimal schedule, **279**
outdegree, **281**
path, **281**
precedence relation, **277**

priority list, **283**
priority-list model, **283**
processing time, **277**
processor, **276**
project digraph, **283**
relative error (ε), **298**
task (ineligible, ready, in execution, completed), **276**
vertex-set, **282**

Exercises

A. Directed Graphs

1. For the digraph shown in the figure,

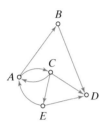

(a) give the vertex-set and arc-set.

(b) find the indegree of each vertex.

(c) find the outdegree of each vertex.

2. For the digraph shown in the figure,

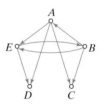

(a) give the vertex-set and arc-set.

(b) find the indegree of each vertex.

(c) find the outdegree of each vertex.

3. For the digraph shown in the figure,

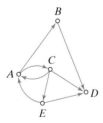

(a) find all vertices that are incident *to A*.

(b) find all vertices that are incident *from A*.

(c) find all vertices that are incident *to D*.

(d) find all vertices that are incident *from D*.

(e) find all the arcs adjacent to *AC*.

(f) find all the arcs adjacent to *CD*.

4. For the digraph shown in the figure,

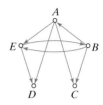

(a) find all vertices that are incident *to A*.

(b) find all vertices that are incident *from A*.

(c) find all vertices that are incident *to D*.

(d) find all vertices that are incident *from D*.

(e) find all the arcs adjacent to *AE*.

(f) find all the arcs adjacent to *BC*.

5. Draw a digraph with the given vertex and arc-set.

(a) $\mathcal{V} = \{A, B, C, D\}$, $\mathcal{A} = \{AB, AC, AD, BD, DB\}$

(b) $\mathcal{V} = \{A, B, C, D, E\}$, $\mathcal{A} = \{AC, AE, BD, BE, CD, DC, ED\}$

6. Draw a digraph with the given vertex and arc-set.

(a) $\mathcal{V} = \{A, B, C, D\}$, $\mathcal{A} = \{AB, AC, AD, BC, BD, DB, DC\}$

(b) $\mathcal{V} = \{V, W, X, Y, Z\}$, $\mathcal{A} = \{VW, VZ, WZ, XV, XY, XZ, YW, ZY, ZW\}$

7. Draw a digraph with the given vertex and arc-set.

(a) $\mathcal{V} = \{A, B, C, D, E\}$, $\mathcal{A} = \{BA, BE, CE, EB, EC, ED\}$

(b) $\mathcal{V} = \{W, X, Y, Z\}$, with W incident to X and Y; X incident to Y and Z; Y incident to Z and W; Z incident to W and X

8. Draw a digraph with the given vertex and arc-set.

(a) $\mathcal{V} = \{A, B, C, D, E\}$, $\mathcal{A} = \{AB, AE, CB, CD, DB, DE, EB, EC\}$

(b) $\mathcal{V} = \{W, X, Y, Z\}$, with every vertex incident to every other vertex

9. Consider the digraph with vertex-set $\mathcal{V} = \{A, B, C, D, E\}$ and arc-set $\mathcal{A} = \{AB, AE, CB, CE, DB, EA, EB, EC\}$. Without drawing the digraph, determine each of the following.

(a) The outdegree of A

(b) The indegree of A

(c) The outdegree of D

(d) The indegree of D

10. Consider the digraph with vertex-set $V = \{V, W, X, Y, Z\}$ and arc-set $A = \{VW, VZ, WZ, XY, XZ, YW, ZY, ZW\}$. Without drawing the digraph, determine each of the following.

(a) The outdegree of V

(b) The indegree of V

(c) The outdegree of Z

(d) The indegree of Z

11. Consider the digraph shown in the following figure.

(a) Find a path from vertex A to vertex F.

(b) Find a Hamilton path from vertex A to vertex F.
(Note: A Hamilton path is a path that passes through every vertex of the graph once.)

(c) Find a cycle in the digraph.

(d) Explain why vertex F cannot be part of any cycle.

(e) Explain why vertex A cannot be part of any cycle.

(f) Find all the cycles in this digraph.

12. Consider the digraph shown in the following figure.

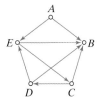

(a) Find a path from vertex A to vertex D.

(b) Explain why the path you found in (a) is the only possible path from vertex A to vertex D.

(c) Find a cycle in the digraph.

(d) Explain why vertex A cannot be part of a cycle.

(e) Explain why vertex B cannot be part of a cycle.

(f) Find all the cycles in this digraph.

13. A city has several one-way streets as well as two-way streets. The White Pine subdivision is a rectangular area six blocks long and two blocks wide. Streets alternate between one way and two way, as shown in the following figure. Draw a digraph that represents the traffic flow in this neighborhood.

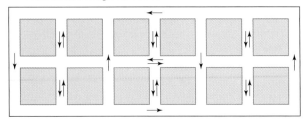

14. A mathematics textbook for liberal arts students consists of 10 chapters. While many of the chapters are independent of the others, some chapters require that previous chapters be covered first. The accompanying diagram illustrates the dependence. Draw a digraph that represents the dependence/independence relation among the chapters in the book.

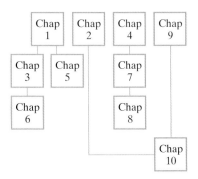

15. The digraph in the following figure is a *respect* digraph. That is, the vertices of the digraph represent members of a group, and an arc XY represents the fact that X respects Y.

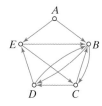

(a) If you had to choose one person to be the leader of the group, whom would you pick? Explain.

(b) Who would be the worst choice to be the leader of the group? Explain.

16. The digraph in the following figure is an example of a *tournament* digraph. In this example the vertices of the digraph represent five volleyball teams in a round-robin tournament (i.e., every team plays every other team). An arc XY represents the fact that X defeated Y in the tournament.
*(**Note:** There are no ties in volleyball.)*

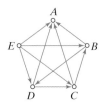

(a) Which team won the tournament? Explain.

(b) Which team came in last in the tournament? Explain.

B. Project Digraphs

17. Draw a project digraph for a project consisting of the eight tasks given in the following table.

Task	Processing time	Tasks that must be completed before the task can start
A	3	
B	10	C, F, G
C	2	A
D	4	G
E	5	C
F	8	A, H
G	7	H
H	5	

18. Draw a project digraph for a project consisting of the eight tasks given in the following table.

Task	Length of task	Tasks that must be completed before the task can start
A	5	C
B	5	C, D
C	5	
D	2	G
E	15	A, B
F	6	D
G	2	
H	2	G

19. Eight computer programs need to be executed. One of the programs requires 10 minutes to complete, 2 programs require 7 minutes each to complete, 2 more require 12 minutes each to complete, and 3 of the programs require 20 minutes each to complete. Moreover, none of the 20-minute programs can be started until both of the 7-minute programs have been completed, and the 10-minute program cannot be started until both of the 12-minute programs have been completed. Draw a project digraph for this scheduling problem.

20. Ten computer programs need to be executed. Three of the programs require 4 minutes each to complete, 3 more require 7 minutes each to complete, and 4 of the programs require 15 minutes each to complete. Moreover, none of the 15-minute programs can be started until all of the 4-minute programs have been completed. Draw a project digraph for this scheduling problem.

21. Apartments Unlimited is an apartment maintenance company that refurbishes apartments before new tenants move in. The following table shows the tasks performed when refurbishing a one-bedroom apartment, the processing time required for each task (measured in 15-minute units), and the precedence relations between tasks. Draw a project digraph for refurbishing a one-bedroom apartment.

Tasks	Symbol/time	Precedence relations
Bathrooms (clean)	$B(8)$	$P \rightarrow B$
Carpets (shampoo)	$C(4)$	$S \rightarrow C, W \rightarrow C$
Filters (replace)	$F(1)$	
General cleaning	$G(8)$	$B \rightarrow G, F \rightarrow G, K \rightarrow G$
Kitchen (clean)	$K(12)$	$P \rightarrow K$
Fix Lighting	$L(1)$	
Paint	$P(32)$	$L \rightarrow P$
Smoke detectors	$S(1)$	$G \rightarrow S$
Windows (wash)	$W(4)$	$G \rightarrow W$

22. A ballroom is to be set up for a large wedding reception. The following table shows the tasks to be carried out, their processing times (in hours) based on one person doing that task, and the precedence relations between the tasks. Draw a project digraph for setting up the wedding reception.

Tasks	Symbol/time	Precedence relations
Set up tables and chairs	$TC(1.5)$	
Set tablecloths and napkins	$TN(0.5)$	$TC \rightarrow TN$
Make flower arrangements	$FA(2.2)$	
Unpack crystal, china, and flatware	$CF(1.2)$	$PT \rightarrow TD$
Put place settings on table	$PT(1.8)$	$TN \rightarrow PT, CF \rightarrow PT$
Arrange table decorations	$TD(0.7)$	$FA \rightarrow TD$
Set up the sound system	$SS(1.4)$	
Set up the bar	$SB(0.8)$	$TC \rightarrow SB$

C. Schedules and Priority Lists

Exercises 23 through 26 refer to a project consisting of 11 tasks (A through K) with the following processing times (in hours): A(10), B(7), C(11), D(8), E(9), F(5), G(3), H(6), I(4), J(7), K(5).

23. **(a)** A schedule with $N = 3$ processors produces finishing time $Fin = 31$ hours. What is the total idle time for all the processors?

   (b) Explain why a schedule with $N = 3$ processors must have finishing time $Fin \geq 25$ hours.

24. **(a)** A schedule with $N = 5$ processors has finishing time $Fin = 19$ hours. What is the total idle time for all the processors?

   (b) Explain why a schedule with $N = 5$ processors must have finishing time $Fin \geq 15$ hours.

25. Explain why a schedule with $N = 6$ processors must have a finishing time $Fin \geq 13$ hours.

26. **(a)** Explain why a schedule with $N = 10$ processors must have finishing time $Fin \geq 11$ hours.

   (b) Explain why it doesn't make sense to put more than 10 processors on this project.

Exercises 27 through 32 refer to the following project digraph.

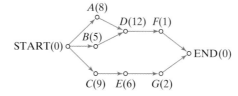

27. Using the priority list D, C, A, E, B, G, F, schedule the project with $N = 2$ processors. Show the project timeline and give its finishing time.

28. Using the priority list G, F, E, D, C, B, A, schedule the project with $N = 2$ processors. Show the project timeline and give its finishing time.

29. Using the priority list D, C, A, E, B, G, F, schedule the project with $N = 3$ processors. Show the project timeline and give its finishing time.

30. Using the priority list G, F, E, D, C, B, A, schedule the project with $N = 3$ processors. Show the project timeline and give its finishing time.

31. Give five different priority lists that will produce the following schedule.

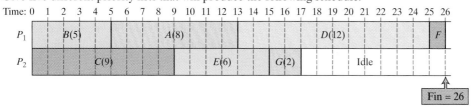

32. Give five different priority lists that will produce the following schedule.

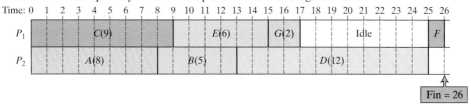

Exercises 33 through 36 refer to the Apartments Unlimited scheduling problem given in Exercise 21.

33. Explain why the following schedule for refurbishing an apartment with one worker is illegal. See figure below.

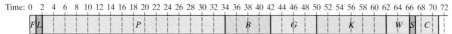

34. Explain why the following schedule for refurbishing an apartment with two workers is illegal. See figure below.

Time: 0 2 4 6 8 10 12 14 16 18 20 22 24 26 28 30 32 34 36 38 40 42 44 46 48 50 52 54

35. Using the priority list $B, C, F, G, K, L, P, S, W$, create a schedule for refurbishing an apartment with two workers. Show the project timeline and give its finishing time.

36. Using the priority list $W, C, G, S, K, B, L, P, F$, create a schedule for refurbishing an apartment with two workers. Show the project timeline and give its finishing time.

D. The Decreasing-Time Algorithm

Exercises 37 and 38 refer to the following project digraph.

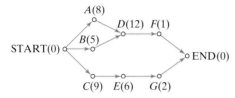

37. Using the decreasing-time algorithm, schedule the project with $N = 2$ processors. Show the timeline for the project and give the project finishing time.

38. Using the decreasing-time algorithm, schedule the project with $N = 3$ processors. Show the timeline for the project and give the project finishing time.

Exercises 39 and 40 refer to the Apartments Unlimited scheduling problem given in Exercise 21.

39. Using the decreasing-time algorithm, schedule the project with $N = 3$ workers. Show the timeline for the project and give the project finishing time.

40. Using the decreasing-time algorithm, schedule the project with $N = 4$ workers. Show the timeline for the project and give the project finishing time.

Exercises 41 through 44 refer to a copy center that must copy 13 court transcripts for a major trial. The times required (in hours) for the 13 jobs in decreasing order are A(12), B(7), C(7), D(6), E(6), F(5), G(5), H(5), I(5), J(4), K(4), L(3), M(3). The precedence relations are that both of the seven-hour jobs must be completed before any of the five-hour jobs can be started.

41. (a) Schedule the jobs on two copiers using the decreasing-time algorithm. Show the timeline and give the project finishing time.

(b) Find an optimal schedule for two copiers.

42. (a) Schedule the jobs on three copiers using the decreasing-time algorithm. Show the timeline and give the project finishing time.

(b) Find an optimal schedule for three copiers and the optimal finishing time *Opt*.

43. (a) Schedule the jobs on six copiers using the decreasing-time algorithm. Show the timeline and give the project finishing time.

(b) Find an optimal schedule for six copiers and the optimal finishing time *Opt*.

(c) Explain why it doesn't make sense to assign more than six copiers to this project.

44. (a) Schedule the jobs on five copiers using the decreasing-time algorithm.

(b) Explain why it is impossible to schedule the jobs on five copiers with a completion time that is less than 15 hours.

(c) Find an optimal schedule for five copiers and the optimal finishing time *Opt*.

E. Critical Paths and the Critical-Path Algorithm

Exercises 45 and 46 refer to the following project digraph.

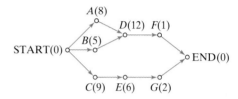

45. (a) Use the backflow algorithm to find the critical time for each vertex.

(b) Find the critical path for the project.

(c) Schedule the project with $N = 2$ processors using the critical-path algorithm. Show the timeline and give the project finishing time.

(d) Explain why the schedule obtained in (c) is optimal.

46. (a) Schedule the project with $N = 3$ processors using the critical-path algorithm. Show the timeline and give the project finishing time.

(b) Explain why the schedule obtained in (a) is optimal.

Exercises 47 and 48 refer to the following project digraph.

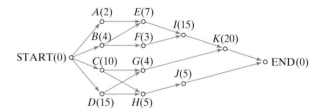

47. Schedule the project with $N = 3$ processors using the critical-path algorithm. Show the timeline and give the project finishing time.

48. (a) Use the backflow algorithm to find the critical time for each vertex.

(b) Find the critical path for the project.

(c) Schedule the project with $N = 2$ processors using the critical-path algorithm. Show the timeline and give the project finishing time.

Exercises 49 and 50 refer to the project discussed in Exercise 18. The tasks, processing times, and precedence relations are shown again in the following table.

Task	Length of task	Tasks that must be completed before the task can start
A	5	C
B	5	C, D
C	5	
D	2	G
E	15	A, B
F	6	D
G	2	
H	2	G

49. Schedule the project with $N = 3$ processors using the critical-path algorithm. Show the timeline and give the project finishing time.

50. Schedule the project with $N = 2$ processors using the critical-path algorithm. Show the timeline and give the project finishing time.

F. Scheduling with Independent Tasks

51. Consider eight independent tasks with processing times (in hours) given by 1, 2, 3, 4, 5, 6, 7, and 8.
    (a) Schedule these tasks with $N = 2$ processors using the critical-path algorithm. Show the timeline and give the project finishing time F.
    (b) Find the optimal finishing time Opt for $N = 2$ processors.
    (c) Give the relative error ε for the schedule found in (a) expressed as a percent.

52. Consider 16 independent tasks with processing times (in hours) given by 1, 2, 3, ..., 15, 16.
    (a) Schedule these tasks with $N = 2$ processors using the critical-path algorithm. Show the timeline and give the project finishing time.
    (b) Find the optimal finishing time Opt for $N = 2$ processors.
    (c) Give the relative error ε for the schedule found in (a) expressed as a percent.

53. Consider six independent tasks with processing times (in hours) given by 1, 2, 3, 4, 5, and 6.
    (a) Find the finishing time Fin for $N = 3$ processors using the critical-path algorithm.
    (b) Find the optimal finishing time Opt for $N = 3$ processors.

54. Consider 18 independent tasks with processing times (in hours) given by 1, 2, 3, ..., 17, 18.
    (a) Find the finishing time Fin for $N = 3$ processors using the critical-path algorithm.
    (b) Find the optimal finishing time Opt for $N = 3$ processors.

55. Consider the following set of independent tasks: $A(4)$, $B(4)$, $C(5)$, $D(6)$, $E(7)$, $F(4)$, $G(5)$, $H(6)$, $I(7)$.
    (a) Schedule these tasks with $N = 4$ processors using the critical-path algorithm. Show the timeline and give the project finishing time Fin.
    (b) Find an optimal schedule for $N = 4$ processors. Show the timeline and give the optimal finishing time Opt.
    (c) Give the relative error ε for the schedule found in (a).

56. Consider the following set of independent tasks: $A(4)$, $B(3)$, $C(2)$, $D(8)$, $E(5)$, $F(3)$, $G(5)$.
    (a) Schedule these tasks with $N = 3$ processors using the critical-path algorithm. Show the timeline and give the project finishing time Fin.
    (b) Find an optimal schedule for $N = 3$ processors. Show the timeline and give the optimal finishing time Opt.
    (c) Give the relative error ε for the schedule found in (a).

57. (a) Consider nine independent tasks with processing times (in hours) given by 1, 1, 2, 3, 5, 8, 13, 21, and 34. Schedule these tasks with $N = 2$ processors using the critical-path algorithm. Show the timeline and give the project finishing time.
    (b) Consider 10 independent tasks with processing times (in hours) given by 1, 1, 2, 3, 5, 8, 13, 21, 34, and 55. Schedule these tasks with $N = 2$ processors using the critical-path algorithm. Show the timeline and give the project finishing time.

58. (a) Consider 11 independent tasks with processing times (in hours) given by 1, 1, 2, 3, 5, 8, 13, 21, 34, 55, and 89. Schedule these tasks with $N = 2$ processors using the critical-path algorithm. Show the timeline and give the project finishing time.
    (b) Consider 12 independent tasks with processing times (in hours) given by 1, 1, 2, 3, 5, 8, 13, 21, 34, 55 89, and 144. Schedule these tasks using the decreasing-time algorithm with two processors. Show the timeline and give the project finishing time.

G. Miscellaneous

59. Explain why, in any digraph, the sum of all the indegrees must equal the sum of all the outdegrees.

60. **Symmetric and totally asymmetric digraphs.** A digraph is called **symmetric** if, whenever there is an arc from vertex X to vertex Y, there is *also* an arc from vertex Y to vertex X. A digraph is called **totally asymmetric** if, whenever there is an arc from vertex X to vertex Y, there *is not* an arc from vertex Y to vertex X. For each of the following, state whether the digraph is symmetric, totally asymmetric, or neither.

(a) A digraph representing the streets of a town in which all streets are one-way streets

(b) A digraph representing the streets of a town in which all streets are two-way streets

(c) A digraph representing the streets of a town in which there are both one-way and two-way streets

(d) A digraph in which the vertices represent a bunch of men, and there is an arc from vertex X to vertex Y if X is a brother of Y

(e) A digraph in which the vertices represent a bunch of men, and there is an arc from vertex X to vertex Y if X is the father of Y

61. When scheduling independent tasks with N processors using the critical-path algorithm, the relative error ε is guaranteed to be between 0 and $(N - 1)/(3N)$. (See the discussion in Section 8.7.)

   (a) Calculate the range of values of the relative error ε when scheduling independent tasks with $N = 7$, $N = 8$, $N = 9$, and $N = 10$ processors.

   (b) Explain why, regardless of the number of processors used, when scheduling independent tasks with the critical-path algorithm, the relative error ε is less than $33\frac{1}{3}\%$.

62. Consider the project digraph shown in the figure.

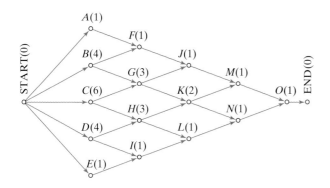

   (a) Schedule the project with $N = 2$ processors using the critical-path algorithm. Show the timeline and give the project finishing time.

   (b) If the tasks were independent, could you do better than the schedule found in (a)? Explain your answer.

Exercises 63 through 66 refer to Martian Habitat Unit building project as discussed in Examples 8.8, 8.9, and 8.12.

63. Find a schedule for building a Martian Habitat Unit with $N = 2$ processors using the priority list $AD(8)$, $AW(6)$, $AF(5)$, $IF(5)$, $AP(7)$, $IW(7)$, $ID(5)$, $IP(4)$, $PL(4)$, $PU(3)$, $HU(4)$, $IC(1)$, $PD(3)$, $EU(2)$, $FW(6)$. Show the timeline and give the project finishing time. (See Example 8.8.)

64. Find a schedule for building a Martian Habitat Unit with $N = 2$ processors using the decreasing-time algorithm. (See Example 8.9. Do the work on your own, and then compare with the step-by-step details shown in Table 8-2.)

65. Find a schedule for building a Martian Habitat Unit with $N = 2$ processors using the critical-path algorithm. Show the timeline and give the project finishing time. (See Example 8.12.)

66. Find a schedule for building a Martian Habitat Unit with $N = 3$ processors using the critical-path algorithm. Show the timeline and give the project finishing time.

JOGGING

Exercises 67, 68, and 69 refer to the following example, discovered by Ron Graham in 1969. (See reference 6.) (Assume that the processing times are given in hours.)

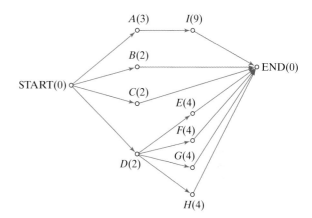

67. (a) Using the priority list A, B, C, D, E, F, G, H, I, schedule the project with $N = 3$ processors.

   (b) Using the same priority list as in (a), schedule the project with $N = 4$ processors.

   (c) There is a paradox in the results obtained in (a) and (b). Describe the paradox.

68. (a) Using the priority list A, B, C, D, E, F, G, H, I, schedule the project with $N = 3$ processors. [This is the same question as in 67(a). If you already answered 67(a), you can recycle the answer.]

   (b) Imagine that the processors are fine-tuned so that the processing times for all the tasks are shortened by 1 hour [i.e., the tasks now are $A(2)$, $B(1)$,..., $H(3)$]. Using the same priority list as in (a), schedule this project with $N = 3$ processors.

   (c) There is a paradox in the results obtained in (a) and (b). Describe the paradox.

69. (a) Using the priority list $A, B, C, D, E, F, G, H, I$, schedule the project with $N = 3$ processors. [Same question again as 67(a) and 68(a). Getting your money's worth on this one!]

(b) Suppose that the precedence relations $D \rightarrow E$ and $D \rightarrow F$ are no longer required. All other precedence relations and processing times are as in the original project. Using the same priority list as in (a), schedule this project with $N = 3$ processors.

(c) There is a paradox in the results obtained in (a) and (b). Describe the paradox.

70. In 1961, T. C. Hu of the University of California showed that in any scheduling problem in which all the tasks have equal processing times and in which the original project digraph (without the START and END vertices) is a tree, the critical-path algorithm will give an optimal schedule. Using this result, find an optimal schedule for the scheduling problem with the following project digraph, using three processors. Assume that each task takes three days. (Notice that we have omitted the START and END vertices.)

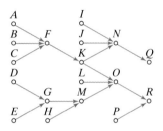

71. Let W represent the sum of the processing times of all the tasks, N be the number of processors, and Fin the finishing time for a project.

(a) Explain the meaning of the inequality $Fin \geq W/N$ and why it is true for any schedule.

(b) Under what circumstances is $Fin = W/N$?

(c) What does the value $N \times Fin - W$ represent?

72. For each schedule find a priority list that generates the schedule.

(Note: There is more than one possible answer.)

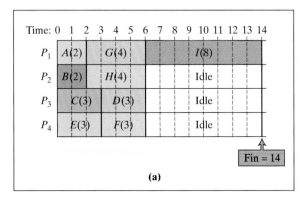

(a)

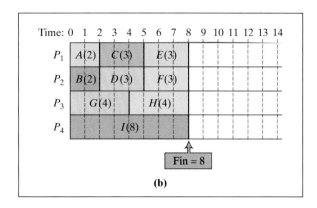

(b)

RUNNING

73. You have $N = 2$ processors to process M independent tasks with processing times $1, 2, 3, \ldots, M$. Find the optimal schedule and give the optimal completion time Opt in terms of M.

*(**Hint:** Consider four separate cases based on the remainder when N is divided by 4.)*

74. You have $N = 3$ processors to process M independent tasks with processing times $1, 2, 3, \ldots, M$. Find the optimal schedule and give the optimal completion time Opt in terms of M.

*(**Hint:** Consider six separate cases based on the remainder when N is divided by 6.)*

75. (a) You have $N = 2$ processors to process M independent tasks with processing times given by the first M Fibonacci numbers: $1, 1, 2, 3, 5, 8, 13, 21, \ldots, F_M$. [Each Fibonacci number (after the first two) equals the sum of the preceding two Fibonacci numbers. We will study Fibonacci numbers in all their glory in Chapter 9.] Find an optimal schedule for these tasks and determine the optimal completion time Opt.

*(**Hint:** Use the following property of Fibonacci numbers: $1 + 1 + 2 + 3 + 5 + 8 + \cdots + F_K = F_{K+2} - 1$.)*

(b) Explain why when scheduling the same tasks as in (a) with $N \geq 3$ processors, $Opt = F_M$.

Projects and Papers

A. Worst-Case Analysis of Scheduling Algorithms

We saw in this chapter that when creating schedules there is a wide range of possibilities, from *very bad schedules to optimal* (best possible) schedules. Our focus was on creating good (if possible optimal) schedules, but the study of bad schedules is also of considerable importance in many applications. In the late 1960s and early 1970s, Ronald Graham of AT&T Bell Labs pioneered the study of worst-case scenarios in scheduling, a field known as *worst-case analysis*. As the name suggests, the issue in worst-case analysis is to analyze how bad a schedule can get. This research was motivated by a deadly serious question—how would the performance of the antiballistic missile defense system of the United States be affected by a failure in the computer programs that run the system?

In his research on worst-case analysis in scheduling, Graham made the critical discovery that there is a limit on how bad a schedule can be—no matter how stupidly a schedule is put together, the project finishing time *Fin* must satisfy the inequality $Fin \leq \left(2 - \dfrac{1}{N}\right)$. *Opt*, where N is the number of processors and *Opt* is the optimal completion time of the project.

Prepare a presentation explaining Graham's worst-case analysis result and its implications. Give examples comparing optimal schedules and the worst possible schedules for projects involving $N = 2$ and $N = 3$ processors.

B. Dijkstra's Shortest-Path Algorithm

Just like there are weighted graphs, there are also *weighted digraphs*. A weighted digraph is a digraph in which each arc has a weight (distance, cost, time) associated with it. Weighted digraphs are used to model many important optimization problems in vehicle routing, pipeline flows, and so on. (Note that a weighted digraph is not the same as a project digraph—in the project digraph the vertices rather than the edges are the ones that have weights.) A fundamental question when working with weighted digraphs is finding the shortest path from a given vertex X to a given vertex Y. (Of course, there may be no path at all from X to Y, in which case the question is moot.) The classic algorithm for finding the shortest path between two vertices in a digraph is known as *Dijkstra's algorithm*, named after the Dutch computer scientist Edsger Dijkstra, who first proposed the algorithm in 1959.

Prepare a presentation describing Dijkstra's algorithm. Describe the algorithm carefully and illustrate how the algorithm works with a couple of examples. Discuss why Dijkstra's algorithm is important. Is it efficient? Optimal? Discuss some possible applications of the algorithm.

C. Tournaments

In the language of graph theory, a *tournament* is a digraph whose underlying graph is a complete graph. In other words, to create a tournament you can start with K_N (the complete graph on N vertices) and then change each edge into an arc by putting an arbitrary direction on it. The reason these graphs are called tournaments is that they can be used to describe the results of a tournament in which every player plays against every other player (no ties allowed).

Write a paper on the mathematics of tournaments (as defined previously). If you studied Chapter 1, you should touch on the connection between tournaments and the results of elections decided under the *method of pairwise comparisons*.

References and Further Readings

1. Baker, K. R., *Introduction to Sequencing and Scheduling*. New York: John Wiley & Sons, Inc., 1974.

2. Brucker, Peter, *Scheduling Algorithms*. New York: Springer-Verlag, 2001.

3. Coffman, E. G., *Computer and Jobshop Scheduling Theory*. New York: John Wiley & Sons, Inc., 1976, chaps. 2 and 5.

4. Conway, R. W., W. L. Maxwell, and L. W. Miller, *Theory of Scheduling*. New York: Dover, 2003.

5. Dieffenbach, R. M., "Combinatorial Scheduling," *Mathematics Teacher*, 83 (1990), 269–273.

6. Garey, M. R., R. L. Graham, and D. S. Johnson, "Performance Guarantees for Scheduling Algorithms," *Operations Research*, 26 (1978), 3–21.

7. Graham, R. L., "Bounds on Multiprocessing Timing Anomalies," *SIAM Journal of Applied Mathematics*, 17 (1969), 263–269.

8. Graham, R. L., "The Combinatorial Mathematics of Scheduling," *Scientific American*, 238 (1978), 124–132.

9. Graham, R. L., "Combinatorial Scheduling Theory," in *Mathematics Today*, ed. L. Steen. New York: Springer-Verlag, Inc., 1978, 183–211.

10. Graham, R. L., E. L. Lawler, J. K. Lenstra, and A. H. G. Rinnooy Kan, "Optimization and Approximation in Deterministic Sequencing and Scheduling: A Survey," *Annals of Discrete Mathematics*, 5 (1979), 287–326.

11. Gross, Jonathan, and Jay Yellen, *Graph Theory and Its Applications*. Boca Raton, FL: CRC Press, 1999, chap. 11.

12. Hillier, F. S., and G. J. Lieberman, *Introduction to Operations Research*, 3rd ed. San Francisco: Holden-Day, Inc., 1980, chap. 6.

13. Kerzner, Harold, *Project Management: A Systems Approach to Planning, Scheduling, and Controlling*, 7th ed. New York: John Wiley & Sons, Inc., 2001, chap. 12.

14. Roberts, Fred S., *Graph Theory and Its Applications to Problems of Society*, CBMS/NSF Monograph No. 29. Philadelphia: Society for Industrial and Applied Mathematics, 1978.

15. Wilson, Robin, *Introduction to Graph Theory*, 4th ed. Harlow, England: Addison-Wesley Longman Ltd., 1996.

part 3

Growth and Symmetry

9

Spiral Growth in Nature

Fibonacci Numbers and the Golden Ratio

In nature's portfolio of architectural works, the shell of the chambered nautilus holds a place of special distinction. The spiral-shaped shell, with its revolving interior stairwell of ever-growing chambers, is more than a splendid piece of natural architecture—it is also a work of remarkable mathematical ingenuity.

Humans have imitated nature's wondrous spiral designs in their own architecture for centuries, all the while trying to understand the magic behind them. What are the physical laws that govern spiral growth in nature? Why do so many different and unusual mathematical concepts come into play? How do the mathematical concepts and physical laws mesh together? What is the source of the intrinsic beauty of nature's spirals? Trying to answer these questions is the goal of this chapter.

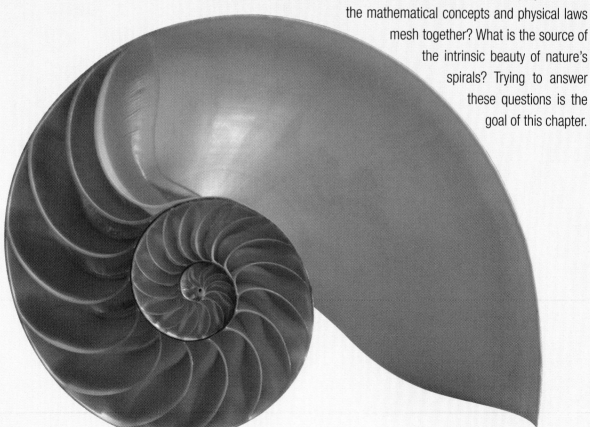

More than 2000 years ago, the ancient Greeks got us off to a great head start in our quest to understand nature with two great contributions: Euclidean geometry and irrational numbers. The next important mathematical connection came 800 years ago, with the serendipitous discovery, by a medieval scholar named Leonardo Fibonacci, of an amazing group of numbers now called Fibonacci numbers. (For more on Fibonacci, see the biographical profile at the end of this chapter.) Next came the discovery, by the French philosopher and mathematician René Descartes in 1638, that an equation he had studied for purely theoretical reasons ($r = ae^\theta$) is the very same equation that describes the spirals generated by seashells. Since then, other surprising connections between spiral-growing organisms and seemingly abstract mathematical concepts

have been discovered, some within the last 25 years. Exactly why and how these concepts play such a crucial role in the development of natural forms is not yet fully understood—not by humans, that is—a humbling reminder that nature still is the oldest and wisest of teachers.

The first three sections of this chapter cover three different mathematical concepts—the *Fibonacci numbers* (Section 9.1), the *golden ratio* (Section 9.2), and *gnomons* (Section 9.3). All three of these concepts are deep and interesting topics on their own, but they are also some of the key ideas necessary for understanding spiral growth in nature. A glimpse of the connection between these concepts and spiral growing organisms is given in Section 9.4.

9.1 Fibonacci Numbers

Table 9-1 shows the first twelve numbers in a famous family of numbers called the **Fibonacci numbers**. It doesn't take long to see that these numbers are generated by a simple pattern: Each number is *the sum of the two numbers before it*. For obvious reasons, the rule does not apply to the first two numbers in the list, which stand on their own. The . . . at the end indicate that the list of Fibonacci numbers goes on indefinitely.

TABLE 9-1
1, 1, 2, 3, 5, 8, 13, 21, 34, 55, 89, 144, . . .

Fibonacci Numbers in Nature

What is so special about Fibonacci numbers? Among other things, it is the way they show up in the most unexpected places. Consistently, we find Fibonacci numbers when we count the number of petals in certain varieties of flowers: 3 (lily, trillium, iris); 5 (buttercup, columbine larkspur); 8 (cosmo, clematis rue anemone); 13 (yellow daisy, marigold); 21 (English daisy, aster); 34 (oxeye daisy); 55 (Coral Gerber daisy), and so on. Fibonacci numbers also appear consistently in conifers, seeds, and fruits. The bracts in a pine cone, for example, spiral in two different directions in 8 and 13 rows; the scales in a pineapple spiral in three different directions in 8, 13, and 21 rows; the seeds in the center of a sunflower spiral in 55 and 89 rows.

$F_6 = 8$

$F_7 = 13$

$F_8 = 21$

$F_9 = 34$

$F_{10} = 55$

Why is nature so fond of Fibonacci numbers? Surprisingly, the answer lies in the rich tapestry of mathematical properties that lurk right below their simple definition. We will explore just a few of these properties throughout the chapter.

The Mathematics of Fibonacci Numbers

The Fibonacci numbers form what mathematicians call an *infinite sequence*—an ordered list of numbers that goes on forever. We will study several types of infinite sequences in Chapter 10, but for now, we will focus on *the Fibonacci sequence*.

> **The Fibonacci Sequence**
>
> 1, 1, 2, 3, 5, 8, 13, 21, 34, 55, 89, 144, 233, 377, . . .

As with any other sequence, the *terms* of the Fibonacci sequence are ordered from left to right: The *first* term is 1, the *second* term is also 1, the *third* term is 2, . . . , the *tenth* term is 55, and so on. In mathematical notation we express this by using the letter F (for Fibonacci) followed by a *subscript* that indicates the *position* of the term in the sequence. In other words, $F_1 = 1$, $F_2 = 1$, $F_3 = 2$, $F_4 = 3, \ldots, F_{10} = 55$, and so on. A generic Fibonacci number is usually written as F_N (where N represents a generic position). If we want to describe the Fibonacci number that comes before F_N, we write F_{N-1}; the Fibonacci number two places before F_N is F_{N-2}, and so on. Clearly, this notation allows us to describe relations among the Fibonacci numbers in a clear and concise way that would be hard to match by just using words.

The rule that generates Fibonacci numbers—*a Fibonacci number equals the sum of the two preceding Fibonacci numbers*—is called a **recursive rule** because it defines a number in the sequence using other (earlier) numbers in the sequence.

This is a good time to try Exercises 1 through 4 (to get better acquainted with the subscript notation) and Exercises 15 through 18 (to understand why the notation is so useful).

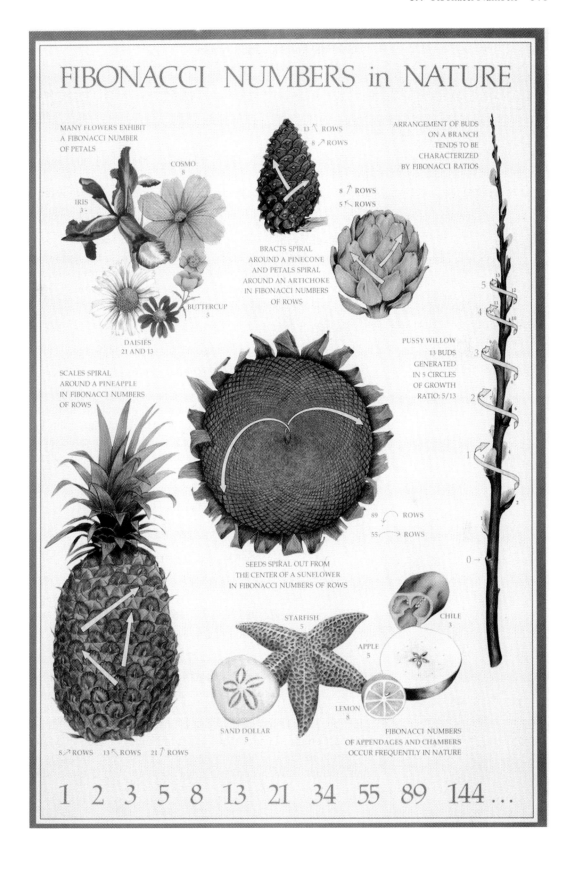

FIBONACCI NUMBERS in NATURE

MANY FLOWERS EXHIBIT A FIBONACCI NUMBER OF PETALS

COSMO 8

IRIS 3

BUTTERCUP 5

DAISIES 21 AND 13

13 ↖ ROWS
8 ↗ ROWS

BRACTS SPIRAL AROUND A PINECONE AND PETALS SPIRAL AROUND AN ARTICHOKE IN FIBONACCI NUMBERS OF ROWS

8 ↗ ROWS
5 ↖ ROWS

ARRANGEMENT OF BUDS ON A BRANCH TENDS TO BE CHARACTERIZED BY FIBONACCI RATIOS

PUSSY WILLOW
13 BUDS GENERATED IN 5 CIRCLES OF GROWTH
RATIO: 5/13

SCALES SPIRAL AROUND A PINEAPPLE IN FIBONACCI NUMBERS OF ROWS

89 ↻ ROWS
55 → ROWS

SEEDS SPIRAL OUT FROM THE CENTER OF A SUNFLOWER IN FIBONACCI NUMBERS OF ROWS

STARFISH 5

CHILE 3

APPLE 5

LEMON 8

SAND DOLLAR 5

8 ↗ ROWS   13 ↖ ROWS   21 ↗ ROWS

FIBONACCI NUMBERS OF APPENDAGES AND CHAMBERS OCCUR FREQUENTLY IN NATURE

1 2 3 5 8 13 21 34 55 89 144 ...

Using the "*F*-notation" notation, the preceding recursive rule can be expressed by a simple and elegant formula:

$$\underbrace{F_N}_{\substack{\text{a generic}\\\text{Fibonacci}\\\text{number}}} = \underbrace{F_{N-1}}_{\substack{\text{the Fibonacci}\\\text{number right}\\\text{before it}}} + \underbrace{F_{N-2}}_{\substack{\text{the Fibonacci}\\\text{number two positions}\\\text{before it}}}$$

There is one thing still missing. The preceding rule cannot be applied to the first two Fibonacci numbers, F_1 (there are no Fibonacci numbers before it) and F_2 (there is only one Fibonacci number before it—the rule requires two), so for a complete description, we must "anchor" the rule by giving the values of the first two Fibonacci numbers, namely $F_1 = 1$ and $F_2 = 1$. We will call these initial anchoring values the **seeds** of the sequence.

The *seeds* $F_1 = 1$, and $F_2 = 1$ combined with the *recursive rule* $F_N = F_{N-1} + F_{N-2}$ give us a full definition of the Fibonacci numbers. Since this is not the only definition, we will describe it as *the recursive definition*.

> The recursive rule
> $F_N = F_{N-1} + F_{N-2}$ does not, in and of itself, define the Fibonacci numbers. For example, when the same rule is applied with seeds 1 and 3, it generates a different sequence of numbers known as the *Lucas numbers*. For details, see Exercise 52.

Fibonacci Numbers (Recursive Definition)

- $F_1 = 1, F_2 = 1$ (the *seeds*)
- $F_N = F_{N-1} + F_{N-2}$ (the *recursive rule*)

Using the preceding recursive definition, we can, in theory, compute any Fibonacci number, but in practice this is easier said than done. How would you find, on your own and without cheating, F_{100}? If you have a couple of hours to spare, you can do it (get a cup of cup of coffee and a good calculator first). Begin with $F_1 = 1$, $F_2 = 1$ and start cranking: $F_2 + F_1$ gives F_3; $F_3 + F_2$ gives F_4; and so on. If all goes well, you will eventually get to $F_{97} = 83,621,143,489,848,422,977$ and $F_{98} = 135,301,852,344,706,746,049$. Then comes

$$F_{99} = 135,301,852,344,706,746,049 + 83,621,143,489,848,422,977$$
$$= 218,922,995,834,555,169,026$$

and finally

$$F_{100} = 218,922,995,834,555,169,026 + 135,301,852,344,706,746,049$$
$$= 354,224,848,179,261,915,075$$

The moral of the preceding story is that the recursive definition will eventually lead to F_{100} (or F_{500}, or any other large Fibonacci number, for that matter), but it is going to take a ridiculous amount of work.

The practical limitations of the recursive definition lead naturally to the question, Is there an *explicit* (direct) formula for computing Fibonacci numbers? Indeed, Leonhard Euler (the very same Euler of chapter 5 fame) discovered such a formula in 1736. (The formula was rediscovered 100 years later by the Frenchman Jacques Binet, who somehow ended up getting the credit.)

Binet's Formula

$$F_N = \left(\frac{1}{\sqrt{5}}\right)\left[\left(\frac{1 + \sqrt{5}}{2}\right)^N - \left(\frac{1 - \sqrt{5}}{2}\right)^N\right]$$

Binet's formula looks quite intimidating at first, but once we break it down, it's straightforward. Three constants (all irrational) are at the heart of Binet's formula: $(1 + \sqrt{5})/2$, $(1 - \sqrt{5})/2$, and $\sqrt{5}$. If we substitute these three constants by the respective letters a, b, and c, we get a friendlier looking version of the formula: $F_N = \dfrac{a^N - b^N}{c}$.

Binet's formula is an example of an **explicit formula**—it allows us to calculate a Fibonacci number without needing to calculate all the preceding Fibonacci numbers. From a practical point of view this is a big improvement over the recursive definition, and with a good calculator (scientific or graphing) or a computer, you can use Binet's formula to find large Fibonacci numbers with just a few keystrokes.

See Exercises 19 through 22, and especially 56(b).

9.2 The Golden Ratio

Before we continue our discussion of Fibonacci numbers, let's take a brief detour.

The Equation $x^2 = x + 1$

Here is a problem that looks like it came straight out of your typical Algebra 2 book: *Solve the quadratic equation $x^2 = x + 1$.* You probably know how to do this without help, but just to be sure, let's go through the steps. First, move all terms to the left-hand side to get $x^2 - x - 1 = 0$. Next, use the *quadratic formula*. [Just in case you forgot: The solutions of $ax^2 + bx + c = 0$ are given by $x = \left(-b \pm \sqrt{b^2 - 4ac}\right)/2a$.] For $x^2 - x - 1 = 0$, the quadratic formula gives the two solutions $(1 + \sqrt{5})/2$ and $(1 - \sqrt{5})/2$. (It would be a great idea if you worked out the details right now!)

For similar problems that require the quadratic formula, see Exercises 29 and 30.

Notice that the two solutions we just found happen to be the two numbers that show up in the numerator of Binet's formula. Both of these numbers are irrational and therefore have infinite, nonrepeating decimal expansions. For working purposes we will round them to three decimal places: $(1 + \sqrt{5})/2 \approx 1.618$ and $(1 - \sqrt{5})/2 \approx -0.618$. It is worth noting that while $(1 + \sqrt{5})/2$ is positive and $(1 - \sqrt{5})/2$ is negative, their decimal parts are identical.

See Exercise 55.

The Golden Ratio

We will now focus our attention on the number $(1 + \sqrt{5})/2$—one of the most remarkable and famous numbers in all of mathematics. The appreciation (bordering on veneration) accorded to this number goes all the way back to the ancient Greek philosophers. Over the years the number has taken many different names—the *divine proportion*, the *golden number*, the *golden section*, and in modern times the **golden ratio**. The modern tradition is to denote this number by the Greek letter ϕ (phi), and from here on we will assume that ϕ represents one thing and one thing only: the irrational number $(1 + \sqrt{5})/2$. (With apologies to those in the Greek system!)

The fact that ϕ is a solution of the equation $x^2 = x + 1$ means that $\phi^2 = \phi + 1$. We can get a great deal of mileage out of this identity. For example, if we multiply both sides of the identity by ϕ, we get $\phi^3 = \phi^2 + \phi$. If we now replace ϕ^2 by $\phi + 1$, we get the identity $\phi^3 = (\phi + 1) + \phi = 2\phi + 1$. We can repeat this process to come up with the identity $\phi^4 = 3\phi + 2$. [Here are the steps in a nutshell: (i) multiply both sides of $\phi^3 = 2\phi + 1$ by ϕ to get $\phi^4 = 2\phi^2 + \phi$; (ii) replace ϕ^2 by $\phi + 1$ to get $\phi^4 = 2(\phi + 1) + \phi = 3\phi + 2$.]

Continuing this way, we can express every power of ϕ in terms of ϕ:

$$\phi^5 = 5\phi + 3; \quad \phi^6 = 8\phi + 5; \quad \phi^7 = 13\phi + 8, \text{ and so on}$$

The general formula that expresses powers of ϕ in terms of ϕ is

Powers of the Golden Ratio

$$\phi^N = F_N\phi + F_{N-1}$$

See Exercise 56.

In some ways you may think of the preceding formula as the opposite of Binet's formula. Whereas Binet's formula uses powers of the golden ratio to calculate Fibonacci numbers, this formula uses Fibonacci numbers to calculate powers of the golden ratio.

The next connection between the Fibonacci numbers and the golden ratio is possibly the most surprising one. Let's grab a calculator and take a look at the ratios F_N/F_{N-1} (i.e., Fibonacci number divided by the previous Fibonacci number). Table 9-2 shows the first 16 values of the ratio F_N/F_{N-1}.

TABLE 9-2 Ratios of Consecutive Fibonacci Numbers

N	F_N	F_N/F_{N-1}	N	F_N	F_N/F_{N-1}	N	F_N	F_N/F_{N-1}
2	1	$1/1 = 2.0$	7	13	$13/8 = 1.625$	12	144	$144/89 = 1.61797\ldots$
3	2	$2/1 = 2$	8	21	$21/13 = 1.61538\ldots$	13	233	$233/144 = 1.61805\ldots$
4	3	$3/2 = 1.5$	9	34	$34/21 = 1.61904\ldots$	14	377	$377/233 = 1.61802\ldots$
5	5	$5/3 = 1.66666\ldots$	10	55	$55/34 = 1.61764\ldots$	15	610	$610/377 = 1.61803\ldots$
6	8	$8/5 = 1.6$	11	89	$89/55 = 1.61818\ldots$	16	987	$987/610 = 1.61803\ldots$

Clearly, something interesting is happening here. As the Fibonacci numbers get bigger, the ratio of consecutive Fibonacci numbers appears to settle down to a fixed value, and, most remarkable of all, that *fixed value turns out to be the golden ratio*! This observation can be expressed as $F_N/F_{N-1} \approx \phi$. Moreover, the larger the value of F_N the closer the ratio F_N/F_{N-1} gets to the golden ratio.

9.3 Gnomons

The most common usage of the word *gnomon* is to describe the pin of a sundial—the part that casts the shadow that shows the time of day. The original Greek meaning of the word *gnomon* is "one who knows," so it's not surprising that the word should find its way into the vocabulary of mathematics.

In this section, we will discuss a different meaning for the word *gnomon*. Before we do so, we will take a brief detour to review a fundamental concept of high school geometry—*similarity*.

We know from geometry that two objects are said to be **similar** if one is a scaled version of the other. When a slide projector takes the image in a slide and blows it up onto a screen, it creates a *similar* but larger image. When a photocopy machine reduces the image on a sheet of paper, it creates a *similar* but smaller image.

The following important facts about similarity of basic two-dimensional figures will come in handy later in the chapter:

■ **Triangles:** Two triangles are similar if and only if the measures of their respective angles are the same. Alternatively, two triangles are similar if and only if their sides are proportional. In other words, if Triangle 1 has sides of length a, b, and c, then Triangle 2 is similar to Triangle 1 if and only if its sides have length $k \cdot a$, $k \cdot b$, and $k \cdot c$ for some positive constant k (Fig. 9-1).

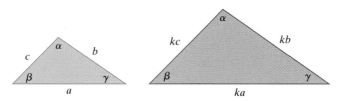

FIGURE 9-1 Similar triangles ($k = 1.6$).

■ **Squares:** Two squares are always similar.

■ **Rectangles:** Two rectangles are similar if their corresponding sides are proportional (Fig. 9-2).

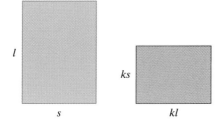

FIGURE 9-2 Similar rectangles ($k = 0.75$).

■ **Circles and disks:** Two circles are always similar. Any circular disk (a circle plus all of its interior) is similar to any other circular disk.

■ **Circular rings:** Two circular rings are similar if and only if their inner and outer radii are proportional (Fig. 9-3).

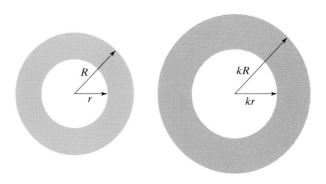

FIGURE 9-3 Similar rings ($k = 4/3$).

We will now return to the main topic of this section—gnomons. In geometry, a **gnomon** G to a figure A is a connected figure which, when suitably *attached* to A, produces a new figure similar to A. By "attached," we mean that the two figures are coupled into one figure without any overlap. Informally, we will describe it this way: G is a gnomon to A if *G&A is similar to A* (Fig. 9-4). Here the symbol & should be taken to mean, "attached in some suitable way."

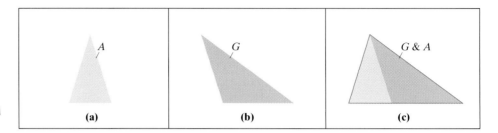

FIGURE 9-4 (a) The original object A. (b) The gnomon G. (c) $G\&A$ is similar to A.

> ### EXAMPLE 9.1 Gnomons of Squares

Consider the square S in Fig. 9-5(a). The L-shaped figure G in Fig. 9-5(b) is a gnomon to the square—when G is attached to S as shown in Fig. 9-5(c), we get the square S'.

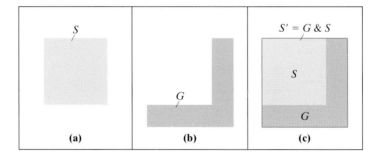

FIGURE 9-5 (a) A square S. (b) The gnomon G. (c) $G\&S$ form a larger square.

Note that the wording is *not* reversible. The square S *is not* a gnomon to the L-shaped figure G, since there is no way to attach the two to form an L-shaped figure similar to G. ≪

> ### EXAMPLE 9.2 Gnomons of Circular Disks

Consider the circular disk C with radius r in Fig. 9-6(a). The O-ring G in Fig. 9-6(b) with inner radius r is a gnomon to C. Clearly, $G\&C$ form the circular disk C' shown in Fig. 9-6(c). Since all circular disks are similar, C' is similar to C.

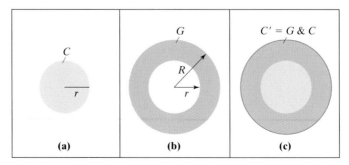

FIGURE 9-6 (a) A circular disk C. (b) The gnomon G. (c) $G\&C$ form a larger circular disk.

EXAMPLE 9.3 Gnomons of Rectangles

Consider a rectangle R of height h and base b as shown in Fig. 9-7(a). The L-shaped G shown in Fig. 9-7(b) can clearly be attached to R to form the larger rectangle R' shown in Fig. 9-7(c). This does not, in and of itself, guarantee that G is a gnomon to R. The rectangle R' [with height $(h + x)$ and base $(b + y)$] is similar to R if and only if their corresponding sides are proportional. This requires that $\dfrac{b}{h} = \dfrac{(b + y)}{(h + x)}$. With a little algebraic manipulation, this can be simplified to $\dfrac{b}{h} = \dfrac{y}{x}$.

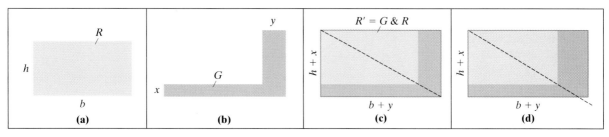

FIGURE 9-7 (a) A rectangle R. (b) A candidate for gnomon G. (c) $G\&R$ is similar to R. (d) $G\&R$ is not similar to R.

There is a simple geometric way to determine if the L-shaped G is a gnomon to R—just extend the diagonal of R in $G\&R$. If the extended diagonal passes through the outside corner of G, then G is a gnomon [Fig. 9-7(c)]; if it doesn't, then it isn't [Fig. 9-7(d)]. ◀◀

EXAMPLE 9.4 A Golden Triangle

In this example, we are going to do things a little bit backward. Let's start with an isosceles triangle T, with vertices B, C, and D whose angles measure $72°$, $72°$, and $36°$, respectively, as shown in Fig. 9-8(a). On side CD we mark the point A so that BA is congruent to BC [Fig. 9-8(b)]. (A is the point of intersection of side CD and the circle of radius BC and center B.) Since T' is an isosceles triangle, angle BAC measures $72°$ and it follows that angle ABC measures $36°$. This implies that triangle T' has equal angles as triangle T, and thus, they are similar triangles.

"So what?" you may ask. Where is the gnomon to triangle T? We don't have one yet! But we *do* have a gnomon to triangle T'—it is triangle BAD, labeled G' in Fig. 9-8(c). After all, $G'\&T'$ is a triangle similar to T'. Note that gnomon G' is an isosceles triangle with angles that measure $36°$, $36°$, and $108°$.

FIGURE 9-8 (a) A 72–72–36° isosceles triangle T. (b) T' is similar to T. (c) $G'\&T' = T$.

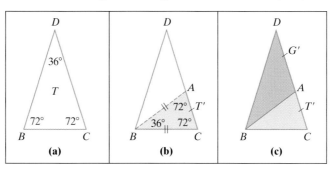

◀◀

We now know how to find a gnomon not only to triangle T' but also to any 72–72–36° triangle, including the original triangle T: Attach a 36–36–108° triangle to one of the longer sides [Fig. 9-9(a)]. If we repeat this process indefinitely, we get a spiralling series of ever increasing 72–72–36° triangles [Fig. 9-9(b)]. It's not too far-fetched to use a family analogy: Triangles T and G are the *parents*, with T having the *dominant* genes; the *offspring* of their union looks just like T (but bigger). The offspring then has offspring of its own (looking exactly like grandfather T), and so on ad infinitum.

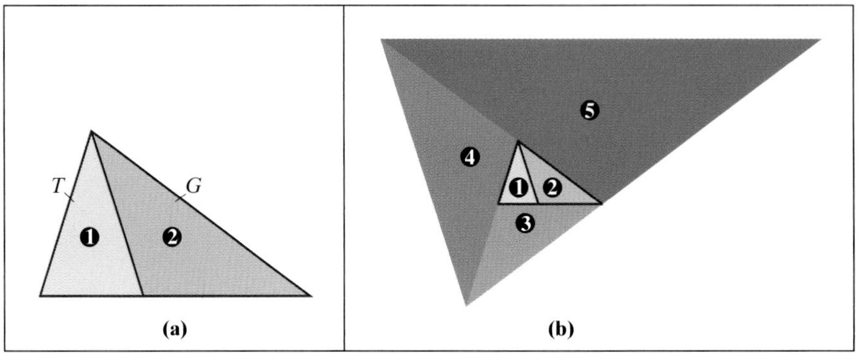

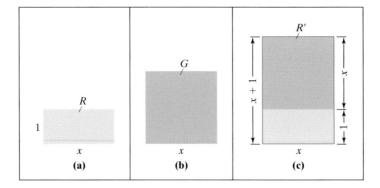

FIGURE 9-9 The process of adding a 36–36–108° gnomon G to a 72–72–36° triangle T can be repeated indefinitely, producing a spiralling chain of ever-increasing similar triangles.

Example 9.4 is of special interest to us for two reasons. First, this is the first time we have an example where the figure and its gnomon are of the same type (isosceles triangles). Second, the isosceles triangles in this story (72–72–36° and 36–36–108°) have a property that makes them unique: In both cases, the ratio of their sides (longer side over shorter side) is the golden ratio. These are the only two isosceles triangles with this property, and for this reason they are called **golden triangles**.

See Exercise 65.

▶ EXAMPLE 9.5 Rectangles Having Square Gnomons

In this example, we start a rectangle R with sides of length 1 (short side) and x (long side) [Fig. 9-10(a)].

For reasons that we will discuss soon, we want to rig things up so that R has a square gnomon G [Fig. 9-10(b)]. There is only one way this can happen—the square has to have sides of length x, and the rectangle $R' = G\&R$ [Fig. 9-10(c)] must be similar to R. This requires that

$$\frac{\text{(long side of } R')}{\text{(long side of } R)} = \frac{\text{(short side of } R')}{\text{(short side of } R)}, \text{ or } \frac{(x+1)}{x} = \frac{x}{1}$$

FIGURE 9-10

Now here comes the interesting part: The equation $\dfrac{(x+1)}{x} = \dfrac{x}{1}$ is equivalent to the quadratic equation $x^2 = x + 1$, and we saw that the only positive solution to the latter is the golden ratio ϕ. (Being the length of the longer side of the rectangle, x must be positive, and in this case is bigger than 1.) It follows that the only way that our original rectangle R has a square gnomon is if it has its short side of length 1 and its long side of length $\phi = (1 + \sqrt{5})/2$. ⟪

Our choice of a rectangle R in Example 9.5 with short side of length 1 is not a restriction—with the proper choice of units any rectangle can be described this way. This means that the conclusion of Example 9.5 applies to any rectangle having sides whose ratio is ϕ: *Any rectangle whose sides are in the proportion of the golden ratio has a square gnomon, and, conversely, if R is a rectangle having a square gnomon, the ratio of its sides (long side/short side) equals the golden ratio ϕ.*

Golden and Fibonacci Rectangles

A rectangle whose sides are in the proportion of the golden ratio is called a **golden rectangle**. In other words, a golden rectangle is a rectangle with sides l (long side) and s (short side) satisfying $l/s = \phi$. A close relative to a golden rectangle is a **Fibonacci rectangle**—a rectangle whose sides are consecutive Fibonacci numbers.

> **EXAMPLE 9.6** Fibonacci Is Almost Golden

Figure 9-11 shows an assortment of rectangles (Please note that the rectangles are not drawn to the same scale). Some are golden, some are close.

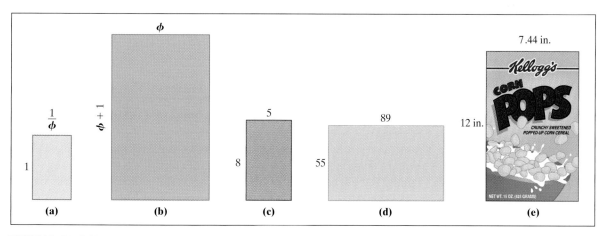

FIGURE 9-11 Golden and "almost golden" rectangles. To the naked eye, they all appear similar.

- The rectangle in Fig. 9-11(a) has $l = 1$ and $s = 1/\phi$. Since $l/s = 1/(1/\phi) = \phi$, this is a golden rectangle.
- The rectangle in Fig. 9-11(b) has $l = \phi + 1$ and $s = \phi$. Here $l/s = (\phi + 1)/\phi$. Since $\phi + 1 = \phi^2$ this is another golden rectangle.

■ The rectangle in Fig. 9-11(c) has $l = 8$ and $s = 5$. This is a Fibonacci rectangle, since 5 and 8 are consecutive Fibonacci numbers. The ratio of the sides is $l/s = 8/5 - 1.6$, so this is certainly not a golden rectangle. On the other hand, the ratio 1.6 is reasonably close to $\phi = 1.618\ldots$, so we will think of this rectangle as "almost golden."

■ The rectangle in Fig. 9-11(d) with $l = 89$ and $s = 55$ is another Fibonacci rectangle. Since $89/55 = 1.61818\ldots$, in theory this rectangle is not a golden rectangle. In practice, this rectangle is as good as golden—the ratio of the sides is the same as the golden ratio up to three decimal places.

■ The rectangle in Fig. 9-11(e) shows the front of a box of Corn Pops. The dimensions are 12 in by 7.44 in. This rectangle is neither a golden nor a Fibonacci rectangle. On the other hand, the ratio of the sides ($12/7.44 \approx 1.613$) is very close to the golden ratio. It is safe to say that, sitting on a supermarket shelf, that box of Corn Pops looks temptingly golden. ◀◀

From a design perspective, golden (and almost golden) rectangles have a special appeal, and they show up in many everyday objects, from posters to cereal boxes. In some sense, golden rectangles strike the perfect middle ground between being too "skinny" and being too "squarish." A prevalent theory, known as the *golden ratio hypothesis*, is that human beings have an innate aesthetic bias in favor of golden rectangles, which, so the theory goes, appeal to our natural sense of beauty and proportion.

The golden ratio hypothesis, originated with the experiments of the German psychologist Gustav Fechner in 1876. For more on this topic take a look at Project E.

9.4 Spiral Growth in Nature

In nature, where form usually follows function, the perfect balance of a golden rectangle shows up in spiral-growing organisms, often in the form of consecutive Fibonacci numbers. To see how this connection works, consider the following example, which serves as a model for certain natural growth processes.

> **EXAMPLE 9.7** Stacking Squares on Fibonacci Rectangles

Start with a 1-by-1 square [square ❶ in Fig. 9-12(a)]. Attach to it a 1-by-1 square [square ❷ in Fig. 9-12(b)]. Squares ❶ and ❷ together form a 2-by-1 Fibonacci rectangle. We will call this the "second-generation" shape. For the third generation, tack on a 2-by-2 square [square ❸ in Fig. 9-12(c)]. The "third-generation" shape (❶, ❷, and ❸ together) is the 2-by-3 Fibonacci rectangle in Fig. 9-12(c). Next, tack onto it a 3-by-3 square [square ❹ in Fig. 9-12(d)], giving a 5-by-3 Fibonacci rectangle. Then tack on a 5-by-5 square [square ❺ in Fig. 9-12(e)], resulting in an 8-by-5 Fibonacci rectangle. You get the picture—we can keep doing this as long as we want.

We might imagine these growing Fibonacci rectangles as a living organism. At each step, the organism grows by adding a square (a very simple, basic shape). The interesting feature of this growth is that as the Fibonacci rectangles

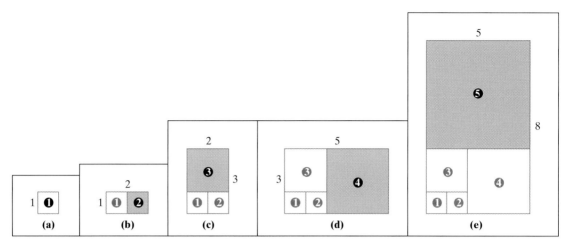

FIGURE 9-12 Fibonacci rectangles beget Fibonacci rectangles.

grow larger, they become very close to golden rectangles, and as such, they become essentially similar to each other. This kind of growth—getting bigger while maintaining the same overall shape—is characteristic of the way many natural organisms grow.   ≪

The next example is a simple variation of Example 9.7.

> ### EXAMPLE 9.8 The Growth of a "Chambered" Fibonacci Rectangle

Let's revisit the growth process of the previous example, except that now let's create within each of the squares being added an interior "chamber" in the form of a quarter-circle. We need to be a little more careful about how we attach the chambered square in each successive generation, but other than that, we can repeat the sequence of steps in Example 9.7 to get the sequence of shapes shown in Fig. 9-13. These figures depict the consecutive generations in the evolution of the

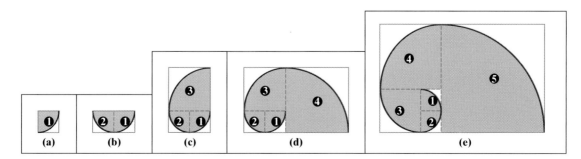

FIGURE 9-13 The growth of a "chambered" Fibonacci rectangle.

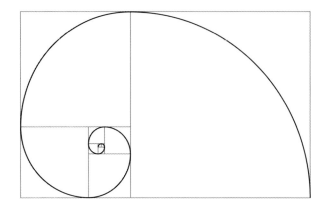

FIGURE 9-14 A Fibonacci spiral after 10 "generations."

chambered Fibonacci rectangle. The outer spiral formed by the circular arcs is often called a **Fibonacci spiral**, shown in Fig. 9-14.   ◀◀

Gnomonic Growth

Natural organisms grow in essentially two different ways. Humans, most animals, and many plants grow following what can informally be described as an *all-around growth* rule. In this type of growth, all living parts of the organism grow simultaneously—but not necessarily at the same rate. One characteristic of this type of growth is that there is no obvious way to distinguish between the newer and the older parts of the organism. In fact, the distinction between new and old parts does not make much sense. The historical record (so to speak) of the organism's growth is lost. By the time the child becomes an adult, no identifiable traces of the child (as an organism) remain—that's why we need photographs!

Contrast this with the kind of growth exemplified by the shell of the chambered nautilus, a ram's horn, or the trunk of a redwood tree. These organisms grow following a *one-sided* or *asymmetric growth* rule, meaning that the organism has a part added to it (either by its own or outside forces) in such a way that the old organism together with the added part form the new organism. At any stage of the growth process, we can see not only the present form of the organism, but also the organism's entire past. All the previous stages of growth are the building blocks that make up the present structure.

The other important aspect of natural growth is the principle of *self-similarity*: Organisms like to maintain their overall shape as they grow. This is where gnomons come into the picture. For the organism to retain its shape as it grows, the new growth must be a *gnomon* of the entire organism. We will call this kind of growth process **gnomonic growth**.

We have already seen abstract mathematical examples of gnomonic growth (Examples 9.7 and 9.8). Here is a pair of more realistic examples.

FIGURE 9-15 The growth rings in a redwood tree—an example of circular gnomonic growth.

▶ EXAMPLE 9.9 Circular Gnomonic Growth

We know from Example 9.2 that the gnomon to a circular disk is an O-ring with an inner radius equal to the radius of the circle. We can thus have circular gnomonic growth (Fig. 9-15) by the regular addition of O-rings. O-rings added one layer at a time to a starting circular structure preserve the circular shape throughout the structure's growth. When carried to three dimensions, this is a good model for the way the trunk of a redwood tree grows. And this is why we can "read" the history of a felled redwood tree by studying its rings. ◀◀

▶ EXAMPLE 9.10 Spiral Gnomonic Growth

FIGURE 9-16 Gnomonic growth of a chambered nautilus.

For more on logarithmic spirals, see Project C.

Figure 9-16 shows a diagram of a cross section of the chambered nautilus. The chambered nautilus builds its shell in stages, each time adding another chamber to the already existing shell. At every stage of its growth, the shape of the chambered nautilus shell remains the same—the beautiful and distinctive spiral shown in the photograph. This is a classic example of gnomonic growth—each new chamber added to the shell is a gnomon of the entire shell. The gnomonic growth of the shell proceeds, in essence, as follows: Starting with its initial shell (a tiny spiral similar in all respects to the adult spiral shape), the animal builds a chamber (by producing a special secretion around its body that calcifies and hardens). The resulting, slightly enlarged spiral shell is similar to the original one. The process then repeats itself over many stages, each one a season in the growth of the animal. Each new chamber adds a gnomon to the shell, so that the shell grows and yet remains similar to itself. This process is a real-life variation of the mathematical spiral-building process discussed in Example 9.8. The curve generated by the outer edge of a nautilus shell's cross section is called a *logarithmic spiral*.

More complex examples of gnomonic growth occur in sunflowers, daisies, pineapples, pine cones, and so on. Here, the rules that govern growth are somewhat more involved, but Fibonacci numbers and the golden ratio once again play a prominent role. The interested reader is encouraged to follow up on this topic by way of Project A.

Conclusion

Some of the most beautiful shapes in nature arise from a basic principle of design: *form follows function*. The beauty of natural shapes is a result of their inherent elegance and efficiency, and imitating nature's designs has helped humans design and build beautiful and efficient structures of their own.

Left: Sunflower. Right: Dome design (Pier Paolo Nervi, architect).

In this chapter, we examined a special type of growth—gnomonic growth—where an organism grows by the addition of gnomons, thereby preserving its basic shape even as it grows. Many beautiful spiral-shaped organisms, from seashells to flowers, exhibit this type of growth.

To us, understanding the basic principles behind spiral growth was relevant because it introduced us to some wonderful mathematical concepts that have been known and studied in their own right for centuries—*Fibonacci numbers*, *the golden ratio*, *golden rectangles*, and *gnomons*.

To humans, these abstract mathematical concepts have been, by and large, intellectual curiosities. To nature—the consummate artist and builder—they are the building tools for some of its most beautiful creations. Whatever lesson one draws from this, it should include something about the inherent value of mathematical ideas.

Profile Leonardo Fibonacci (circa 1175–1250)

Leonardo Fibonacci was born in Pisa, Italy sometime around 1175 (there are no reliable records of Leonardo's exact date of birth). His actual family name was Bonaccio, so the name Fibonacci, which he preferred and by which he is now known, is a sort of nickname derived from the Latin for *filius Bonacci*, meaning "son of Bonaccio." Pisa, best known nowadays as the home of the famous leaning tower, was at the time an independent and prosperous city-state with important trade routes throughout the Mediterranean. Leonardo's father, Guglielmo Bonaccio, was a civil servant/diplomat who served as the agent and representative for Pisan business interests in North Africa. When Leonardo was a young boy he moved with his father to North Africa, where he spent most of his youth. Thus, rather than receiving a traditional European education, Leonardo was educated in the Moorish tradition, a happenstance that might very well have affected the entire course of Western civilization.

As his father wanted him to be a merchant, a significant part of Leonardo's education was the study of mathematics and accounting. Under the Moors, Leonardo learned arithmetic and algebra using the Hindu-Arabic decimal system of numeration—the system we all use today but unknown in Europe at the time. Leonardo was not the first European to learn the Hindu-Arabic system of numbers and their arithmetic, but he was the first one to truly grasp their extraordinary potential.

Sometime around the year 1200, when he was about 25, Leonardo returned to Pisa. For the next 25 years, he dedicated himself to bringing the Hindu-Arabic methods of mathematics to Europe. During this period he published at least four books (these are the ones for which manuscripts still exist—it is believed that there were two other books for which the manuscripts are lost). His first book, *Liber Abaci (The Book of Calculation)*, was published in 1202. (A long-awaited English translation of *Liber Abaci* was published in 2002 [reference 17].)

It is not too farfetched to claim that the publication of *Liber Abaci* is one of those little-known events that changed the course of history. In *Liber Abaci*, Fibonacci described the workings of the Hindu-Arabic number system and richly illustrated it with marvellous problems and examples. The advantages of this place-value-based decimal system for accounting and bookkeeping were so obvious that in a very short time it became the de facto standard for conducting business, the Roman system was abandoned, and the rest is history.

One of the many examples that Fibonacci introduced in *Liber Abaci* was a simple rabbit-breeding problem: *A certain man put a pair of rabbits in a place surrounded on all sides by a wall. How many pairs of rabbits can be produced from that pair in a year if it is supposed that every month each pair begets a new pair which from the second month on becomes productive?* The solution to this famous problem generates the numbers $1, 1, 2, 3, 5, 8, 13, 21, 34, 55, 89 \ldots$, which we now call *Fibonacci numbers*. To Fibonacci, the rabbit problem was just another example among the many discussed in his book—little did he know that the numbers generated in the solution would be his ticket to immortality. It is a great irony that the man who brought the Western world its modern system for doing mathematics is best remembered now for the reproductive habits of a bunch of mythical bunnies.

By the time Fibonacci published his last book, *Liber Quadratorum*, he was a famous man, but surprisingly little is known about the latter part of his life or about the date and circumstances of his death, believed to be between 1240 and 1250.

Key Concepts

Binet's formula, **316**
explicit (definition, formula), **317**
Fibonacci number, **313**
Fibonacci rectangle, **323**
Fibonacci sequence, **314**

Fibonacci spiral, **326**
gnomon, **320**
gnomonic growth, **326**
golden ratio (ϕ), **317**
golden rectangle, **323**

golden triangle, **322**
recursive (definition, rule), **314**
similar (shapes, objects), **318**

Exercises

WALKING

A. Fibonacci Numbers

1. Compute the value of each of the following.
   (a) F_{10}
   (b) $F_{10} + 2$
   (c) F_{10+2}
   (d) $F_{10}/2$
   (e) $F_{10/2}$

2. Compute the value of each of the following.
   (a) F_{12}
   (b) $F_{12} - 1$
   (c) F_{12-1}
   (d) $F_{12}/4$
   (e) $F_{12/4}$

3. Compute the value of each of the following.
   (a) $F_1 + F_2 + F_3 + F_4 + F_5$
   (b) $F_{1+2+3+4+5}$
   (c) $F_3 \times F_4$
   (d) $F_{3\times4}$
   (e) F_{F_4}

4. Compute the value of each of the following.
   (a) $F_1 + F_3 + F_5 + F_7$
   (b) $F_{1+3+5+7}$
   (c) F_{10}/F_5
   (d) F_{10/F_5}
   (e) F_{F_7}

5. Describe in words what each of the expressions represents.
   (a) $3F_N + 1$
   (b) $3F_{N+1}$

(c) $F_{3N} + 1$

(d) F_{3N+1}

6. Describe in words what each of the expressions represents.

(a) $F_{2N} - 3$

(b) F_{2N-3}

(c) $2F_N - 3$

(d) $2F_{N-3}$

7. Given that $F_{36} = 14{,}930{,}352$ and $F_{37} = 24{,}157{,}817$,

(a) find F_{38}.

(b) find F_{35}.

8. Given that $F_{31} = 1{,}346{,}269$ and $F_{33} = 3{,}524{,}578$,

(a) find F_{32}.

(b) find F_{34}.

9. Which of the two rules (I or II) is equivalent to the recursive rule $F_N = F_{N-1} + F_{N-2}, N \geq 3$? Explain.

(I) $F_{N+2} = F_{N+1} + F_N, N > 0$

(II) $F_N = F_{N+1} + F_{N+2}, N > 0$

10. Which of the two rules (I or II) is equivalent to the recursive rule $F_N = F_{N-1} + F_{N-2}, N \geq 3$? Explain.

(I) $F_{N-1} - F_N = F_{N+1}, N > 1$

(II) $F_{N+1} - F_N = F_{N-1}, N > 1$

11. Write each of the following integers as the sum of *distinct* Fibonacci numbers.

(a) 47

(b) 48

(c) 207

(d) 210

12. Write each of the following integers as the sum of *distinct* Fibonacci numbers.

(a) 52

(b) 53

(c) 107

(d) 112

13. Consider the following sequence of equations involving Fibonacci numbers.

$$1 + 2 = 3$$
$$1 + 2 + 5 = 8$$
$$1 + 2 + 5 + 13 = 21$$
$$1 + 2 + 5 + 13 + 34 = 55$$

(a) Write down a reasonable choice for the fifth equation in this sequence.

(b) Find the subscript that will make the following equation true.

$$F_1 + F_3 + F_5 + \ldots + F_{21} = F_?$$

(c) Find the subscript that will make the following equation true (assume N is odd).

$$F_1 + F_3 + F_5 + \ldots + F_N = F_?$$

14. Consider the following sequence of equations involving Fibonacci numbers.

$$2(2) - 3 = 1$$
$$2(3) - 5 = 1$$
$$2(5) - 8 = 2$$
$$2(8) - 13 = 3$$

(a) Write down a reasonable choice for the fifth equation in this sequence.

(b) Find the subscript that will make the following equation true.

$$2(F_?) - F_{15} = F_{12}$$

(c) Find the subscript that will make the following equation true.

$$2(F_{N+2}) - F_{N+3} = F_?$$

15. **Fact:** *If we add any Fibonacci number to the Fibonacci number three positions before it, we get twice the Fibonacci number before it.*

(a) Verify this fact for four different Fibonacci numbers of your choice.

(b) Using F_N as a generic Fibonacci number, write this fact as a mathematical formula.

16. **Fact:** *If we make a list of any 10 consecutive Fibonacci numbers, the sum of all these numbers divided by 11 is always equal to the seventh number on the list.*

(a) Verify this fact for the list consisting of the first 10 Fibonacci numbers.

(b) Using F_N as the first Fibonacci number on the list, write this fact as a mathematical formula.

17. **Fact:** *If we make a list of any four consecutive Fibonacci numbers, twice the third one minus the fourth one is always equal to the first one.*

(a) Verify this fact for the list that starts with F_4.

(b) Using F_N as the first Fibonacci number on the list, write this fact as a mathematical formula.

18. **Fact:** *If we make a list of any four consecutive Fibonacci numbers, the first one times the fourth one is always equal to the third one squared minus the second one squared.*

(a) Verify this fact for the list that starts with F_8.

(b) Using F_N as the first Fibonacci number on the list, write this fact as a mathematical formula.

B. The Golden Ratio

For Exercises 19 through 22 you will need a scientific or graphing calculator (or a computer with a mathematics software package).

19. Calculate each of the following to five decimal places.

(a) $21\left(\dfrac{1 + \sqrt{5}}{2}\right) + 13$

(b) $\left(\dfrac{1 + \sqrt{5}}{2}\right)^{8}$

(c) $\dfrac{\left(\dfrac{1 + \sqrt{5}}{2}\right)^{8} - \left(\dfrac{1 - \sqrt{5}}{2}\right)^{8}}{\sqrt{5}}$

20. Calculate each of the following to five decimal places.

(a) $55\left(\dfrac{1 + \sqrt{5}}{2}\right) + 34$

(b) $\left(\dfrac{1 + \sqrt{5}}{2}\right)^{10}$

(c) $\dfrac{\left(\dfrac{1 + \sqrt{5}}{2}\right)^{10} - \left(\dfrac{1 - \sqrt{5}}{2}\right)^{10}}{\sqrt{5}}$

21. Using $\phi = (1 + \sqrt{5})/2$, calculate each of the following, rounded to the nearest integer.

(a) $\phi^{8}/\sqrt{5}$

(b) $\phi^{9}/\sqrt{5}$

(c) Without using a calculator, first try to guess the value of $\phi^{7}/\sqrt{5}$ rounded to the nearest integer. Verify your guess with a calculator.

22. Using $\phi = (1 + \sqrt{5})/2$, calculate each of the following, rounded to the nearest integer.

(a) $\phi^{10}/\sqrt{5}$

(b) $\phi^{11}/\sqrt{5}$

(c) Without using a calculator, first try to guess the value of $\phi^{12}/\sqrt{5}$ rounded to the nearest integer. Verify your guess with a calculator.

23. Find the values of the integers a and b in each of the equations.

(a) $\phi^{9} = a\phi + b$

(b) $\phi^{9} = a\sqrt{5} + b$

24. Find the values of the integers a and b in each of the equations.

(a) $\phi^{12} = a\phi + b$

(b) $\phi^{12} = a\sqrt{5} + b$

25. Given that $F_{499} \approx 8.6168 \times 10^{103}$, find an approximate value for F_{500} in scientific notation.

(**Hint:** $F_{N}/F_{N-1} \approx \phi$.)

26. Given that $F_{1002} \approx 1.138 \times 10^{209}$, find an approximate value for F_{1000} in scientific notation.

(**Hint:** $F_{N}/F_{N-1} \approx \phi$.)

27. The *Fibonacci sequence of order 2* is the sequence of numbers 1, 2, 5, 12, 29, 70,.... Each term in this sequence (from the third term on) equals two times the term before it plus the term two places before it; in other words, $A_{N} = 2A_{N-1} + A_{N-2} (N \geq 3)$.

(a) Compute A_{7}.

(b) Use your calculator to compute to five decimal places the ratio A_{N}/A_{N-1} for $N = 7$.

(c) Use your calculator to compute to five decimal places the ratio A_{N}/A_{N-1} for $N = 11$.

(d) Guess the value (to five decimal places) of the ratio A_{N}/A_{N-1} when $N > 11$.

28. The *Fibonacci sequence of order 3* is the sequence of numbers 1, 3, 10, 33, 109,.... Each term in this sequence (from the third term on) equals three times the term before it plus the term two places before it; in other words, $A_{N} = 3A_{N-1} + A_{N-2} (N \geq 3)$.

(a) Compute A_{6}.

(b) Use your calculator to compute to five decimal places the ratio A_{N}/A_{N-1} for $N = 6$.

(c) Guess the value (to five decimal places) of the ratio A_{N}/A_{N-1} when $N > 6$.

C. Fibonacci Numbers and Quadratic Equations

Exercises 29 through 34 involve quadratic equations with coefficients that are Fibonacci numbers. [The solutions to any quadratic equation of the form $ax^2 + bx + c = 0$ are given by $x = \left(-b \pm \sqrt{b^2 - 4ac}\right)/2a$. This is the well-known quadratic formula.]

29. Find the solutions of each equation. Use a calculator to approximate the solutions to five decimal places.

(a) $x^2 = 2x + 1$

(b) $3x^2 = 5x + 8$

30. Find the solutions of each equation. Use a calculator to approximate the solutions to five decimal places.

(a) $x^2 = 3x + 1$

(b) $8x^2 = 5x + 3$

31. Consider the quadratic equation $55x^2 = 34x + 21$.

(a) Without using the quadratic formula, find one of the solutions to this equation.

(**Hint:** *One of the solutions is a small integer.*)

(b) Without using the quadratic formula, find the other solution to the equation.

(**Hint:** *The sum of the two solutions of the quadratic equation $ax^2 + bx + c = 0$ equals $-b/a$.*)

32. Consider the quadratic equation $21x^2 = 34x + 55$.

(a) Without using the quadratic formula, find one of the solutions to this equation.

(**Hint:** *One of the solutions is a small negative integer.*)

(b) Without using the quadratic formula, find the other solution to the equation.

(**Hint:** *The sum of the two solutions of the quadratic equation $ax^2 + bx + c = 0$ equals $-b/a$.*)

33. Consider the quadratic equation $F_N x^2 = F_{N-1} x + F_{N-2}$.
    (a) Explain why $x = 1$ is one of the solutions to the equation.
    (b) Explain why the other solution is given by $x = (F_{N-1}/F_N) - 1$.
    [**Hint:** See the hint for Exercise 31(b).]

34. Consider the quadratic equation $F_N x^2 = F_{N+1} x + F_{N+2}$.
    (a) Explain why $x = -1$ is one of the solutions to the equation.
    (b) Explain why the other solution is given by $x = (F_{N+1}/F_N) + 1$.
    [**Hint:** See the hint for Exercise 32(b).]

D. Similarity

35. R and R' are similar rectangles. Suppose that the width of R is a and the width of R' is $3a$.
    (a) If the perimeter of R is 41.5 in, what is the perimeter of R'?
    (b) If the area of R is 105 sq in, what is the area of R'?

36. O and O' are similar O-rings. The inner radius of O is 5 ft and the inner radius of O' is 15 ft.
    (a) If the circumference of the outer circle of O is 14π, what is the circumference of the outer circle of O'?
    (b) Suppose it takes 1.5 gallons of paint to paint the O-ring O. If the paint is used at the same rate, how much paint is needed to paint the O-ring O'?

37. T and T' in the figure are similar triangles. (The triangles are not drawn to scale.)

    (a) If the perimeter of T is 13 in, what is the perimeter of T' (in meters)?
    (b) If the area of T is 20 sq in, what is the area of T' (in square meters)?

38. P and P' in the figure are similar polygons.

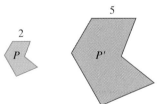

    (a) If the perimeter of P is 10, what is the perimeter of P'?
    (b) If the area of P is 30, what is the area of P'?

39. R is a rectangle with sides of length x and 3. R' is a rectangle with sides of length $8 - x$ and 5. Determine the value(s) of x that make R and R' similar.

40. R is a rectangle with sides of length x and 2. R' is a rectangle with sides of length $x + 1$ and 6. Determine all value(s) of x that make R and R' similar.

E. Gnomons

41. Find the value of c so that the shaded rectangle is a gnomon to the white 3 by 9 rectangle. (Figure is not drawn to scale.)

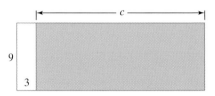

42. Find the value of x so that the shaded figure is a gnomon to the white rectangle. (Figure is not drawn to scale.)

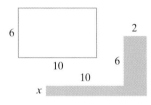

43. Find the value of x so that the shaded "rectangular ring" is a gnomon to the white rectangle. (Figure is not drawn to scale.)

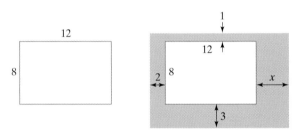

44. Find the value of x so that the shaded figure is a gnomon to the white rectangle. (Figure is not drawn to scale.)

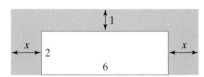

45. Rectangle A is 10 by 20. Rectangle B is a gnomon to rectangle A. What are the dimensions of rectangle B?

46. Find the value of x so that the shaded "rectangular ring" is a gnomon to the white rectangle. (Figure is not drawn to scale.)

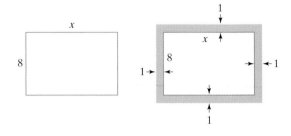

47. In the figure, triangle BCA is a 36–36–108° triangle with sides of length ϕ and 1. Suppose triangle ACD is a gnomon to triangle BCA.

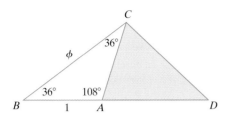

(a) Find the measure of the angles of triangle ACD.

(b) Find the length of the three sides of triangle ACD.

48. Find the values of x and y so that the shaded triangle is a gnomon to the white triangle.

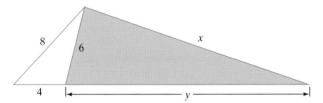

49. Find the values of x and y so that the shaded figure is a gnomon to the white triangle.

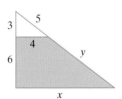

50. Find the values of x and y so that the shaded triangle is a gnomon to the white triangle.

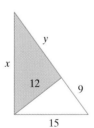

JOGGING

51. Consider the following sequence of numbers: 5, 5, 10, 15, 25, 40, 65, ... If A_N is the Nth term of this sequence, write A_N in terms of F_N.

52. The *Lucas numbers* L_N are defined by the recursive rule $L_N = L_{N-1} + L_{N-2}$ ($N \geq 3$) and the seeds $L_1 = 1, L_2 = 3$. (The Lucas sequence has the same recursive rule as the Fibonacci sequence but different seeds.) The first few terms of the Lucas sequence are 1, 3, 4, 7, 11, 18, 29, 47, ...

(a) Show that $L_N = 2F_{N+1} - F_N$.

(**Hint:** *Show that the expression $2F_{N+1} - F_N$ satisfies the recursive rule and the definition of the seeds of the Lucas sequence.*)

(b) Show that an explicit formula for the Lucas numbers is given by $L_N = \left(\dfrac{1 + \sqrt{5}}{2}\right)^N + \left(\dfrac{1 - \sqrt{5}}{2}\right)^N$.

[**Hint:** *Use the result of (a) combined with Binet's formula.*]

53. Let $T_N = 7F_{N+1} + 4F_N$.

(a) Find the values of T_1 through T_8.

(b) Show that the numbers T_N satisfy the same recursive rule as the Fibonacci numbers: $T_N = T_{N-1} + T_{N-2}$.

(c) Give a complete recursive definition (recursive rule and seeds) of the sequence of numbers T_N.

54. (This exercise is a generalization of the previous exercise. You should do Exercise 53 before you try this one.) Let $T_N = aF_{N+1} + bF_N$, where a and b are fixed integers.

(a) Find T_1 through T_5.

(b) Show that the numbers T_N satisfy the same recursive rule as the Fibonacci numbers: $T_N = T_{N-1} + T_{N-2}$.

(c) Consider the sequence of numbers 5, 11, 16, 27, 43, ... Find the values of a and b such that these numbers fit the description $T_N = aF_{N+1} + bF_N$.

55. Explain why the irrational numbers $(1 + \sqrt{5})/2$ and $(1 - \sqrt{5})/2$ have exactly the same decimal expansion.

(**Hint:** *What do you get when you add the two numbers?*)

56. (a) Explain what happens to the values of $\left(\dfrac{1 - \sqrt{5}}{2}\right)^N$ as N gets larger.

(**Hint:** *Get a calculator and experiment with $N = 6$, 7, 8, ... until you get the picture.*)

(b) Using your answer in (a), justify the following assertion: $F_N \approx \phi^N/\sqrt{5}$. [This is a simplified version of Binet's formula. An even more useful version is $F_N = [\phi^N/\sqrt{5}]$, where the symbol $[x]$ means *round x to the nearest integer*.]

(c) Using (b), explain why $F_N/F_{N-1} \approx \phi$.

57. A rectangle R has sides of length $l = 144\phi + 89$ and $s = 89\phi + 55$. Explain why R is a golden rectangle.

58. Find a positive solution of the equation $x^8 - 21x - 13 = 0$.

59. A rectangle has dimensions l (long side) by s (short side). If the rectangle is a gnomon to itself, find the value of the ratio l/s.

60. Explain why the L-shaped figure shown cannot have a square gnomon.

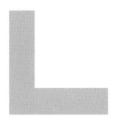

61. Find the values of x, y, and z so that the shaded figure has an area eight times the area of the white triangle and at the same time is a gnomon to the white triangle. (Figure not drawn to scale.)

62. Find the values of x and y so that the shaded triangle is a gnomon to the white triangle. (Figure not drawn to scale.)

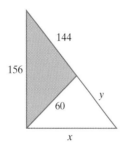

63. Find the values of x and y so that the shaded L-shaped figure has an area of 75 and at the same time is a gnomon to the white rectangle. (Figure not drawn to scale.)

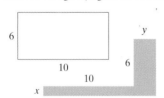

64. Let $ABCD$ be an arbitrary rectangle as shown in the following figure. Let AE be perpendicular to the diagonal BD and EF perpendicular to AB as shown. Show that the rectangle $BCEF$ is a gnomon to the rectangle $ADEF$.

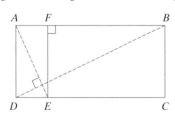

65. In the figure, triangle BCD is a 72–72–36° triangle with base of length 1 and longer side of length x. (Using this choice of values, the ratio of the longer side to the shorter side is $x/1 = x$.)

(a) Show that $x = \phi$.

  (***Hint:*** *Triangle ACB is similar to triangle BCD.*)

(b) What are the interior angles of triangle DAB?

(c) Show that in the isosceles triangle DAB, the ratio of the longer to the shorter side is also ϕ.

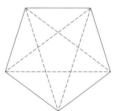

66. The regular pentagon in the following figure has sides of length 1. Show that the length of any one of its diagonals is ϕ.

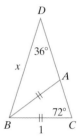

67. (a) A regular decagon (10 sides) is inscribed in a circle of radius 1. Find the perimeter in terms of ϕ.

(b) Repeat (a) with radius r. Find the perimeter in terms of ϕ and r.

68. (a) Without using a calculator, find the exact value of
$$\sqrt{1 + \sqrt{1 + \sqrt{1 + \phi}}}.$$
  (***Hint:*** $\phi^2 = \phi + 1$.)

(b) Without using a calculator, find the exact value of
$$1 + \cfrac{1}{1 + \cfrac{1}{1 + \cfrac{1}{\phi}}}.$$

RUNNING

69. You are designing a straight path 2 ft wide using rectangular paving stones with dimensions 1 ft by 2 ft. How many different designs are possible for a path of length

(a) 4 ft?

(b) 8 ft?

(c) N ft?

  (***Hint:*** *Give the answer in terms of Fibonacci numbers.*)

70. Show that $(F_1)^2 + (F_2)^2 + (F_3)^2 + \cdots + (F_N)^2 = F_N \cdot F_{N+1}$.

(**Hint:** *See Figure 9-12.*)

71. Show that $F_1 + F_2 + F_3 + \cdots + F_N = F_{N+2} - 1$.

72. Show that $F_1 + F_3 + F_5 + \cdots + F_N = F_{N+1}$. (Note that on the left side of the equation we are adding the Fibonacci numbers with odd subscripts up to N.)

73. Show that every positive integer greater than 2 can be written as the sum of distinct Fibonacci numbers.

74. Show that the sum of any 10 consecutive Fibonacci numbers is a multiple of 11.

75. Suppose that T_N represents a sequence of numbers satisfying the "Fibonacci recursive rule" $T_N = T_{N-1} + T_{N-2}$, with seeds $T_1 = c$ and $T_2 = d$.

(**a**) Show that there are constants a and b such that $T_N = aF_{N+1} + bF_N$.

(**Hint:** *See Exercises 53 and 54.*)

(**b**) Show that the ratios T_{N+1}/T_N get closer and closer to ϕ as N increases.

76. (**a**) Show that the ratios F_{N+2}/F_N get closer and closer to ϕ^2 as N increases.

(**b**) What number do the ratios F_{N+3}/F_N approximate as N increases? Explain.

77. In the figure, $ABCD$ is a square and the three triangles I, II, and III have equal areas. Show that $x/y = \phi$.

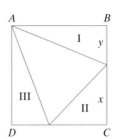

78. During the time of the Greeks the star pentagram was a symbol of the Brotherhood of Pythagoras. A typical diagonal of the large outside regular pentagon is broken up into three segments of lengths x, y, and z, as shown in the following figure.

(**a**) Show that $x/y = \phi$, $(x + y)/z = \phi$, and $(x + y + z)/(x + y) = \phi$.

(**b**) Show that if $y = 1$, then $x = \phi$, $(x + y) = \phi^2$, and $x + y + z = \phi^3$.

Projects and Papers

A. Fibonacci Numbers, the Golden Ratio, and Phyllotaxis

In this chapter we mentioned that Fibonacci numbers and the golden ratio often show up in both the plant and animal worlds. In this project, you are asked to expand on this topic.

1. Give several detailed examples of the appearance of Fibonacci numbers and the golden ratio in the plant world. Include examples of branch formation in plants, leaf arrangements around stems, and seed arrangements on circular seedheads (such as sunflower heads).

2. Discuss the concept of *phyllotaxis*. What is it, and what are some of the mathematical theories behind it?

(*Notes:* The literature on Fibonacci numbers, the golden ratio, and phyllotaxis is extensive. A search on the Web should provide plenty of information. Two excellent Web sites on this subject are Ron Knott's site at the University of Surrey, England (http://www.mcs.surrey.ac.uk/Personal/R.Knott/Fibonacci), and the site *Phyllotaxis: An Interactive Site for the Mathematical Study of Plant Pattern Formation* (http://www.math.smith.edu/~phyllo/) based at Smith College.

B. The Golden Ratio in Art, Architecture, and Music

It is often claimed that from the time of the ancient Greeks through the Renaissance to modern times, artists, architects, and musicians have been fascinated by the golden ratio. Choose one of the three fields (art, architecture, or music) and write a paper discussing the history of the golden ratio in that field. Describe famous works of art, architecture, or music in which the golden ratio is alleged to have been used. How? Who were the artists, architects, and composers?

Be forewarned that there are plenty of conjectures, unsubstantiated historical facts, controversies, claims, and counterclaims surrounding some of the alleged uses of the golden ratio and Fibonacci numbers. Whenever appropriate, you should present both sides to a story.

C. Logarithmic Spirals

In this project, you are to investigate logarithmic spirals. You should explain what a logarithmic spiral is and how it is related to Fibonacci numbers and the golden ratio, and you should reference several examples of where they occur in the natural world.

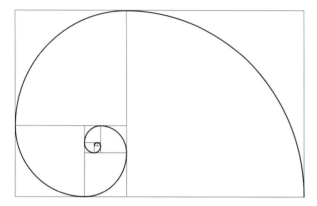

D. Figurate Numbers

The *triangular* numbers are numbers in the sequence 1, 3, 6, 10, 15, ... (The *N*th **triangular** number T_N is given by the sum $1 + 2 + 3 + \cdots + N$.) In a similar way, *square, pentagonal*, and *hexagonal* numbers can be defined. In this project, you are to investigate these types of numbers (called *figurate* numbers), give some of their more interesting properties, and discuss the relationship between these numbers and *gnomons*.

E. The Golden Ratio Hypothesis

A long-held belief among those who study how humans perceive the outside world (mostly psychologists and psychobiologists) is that the *golden ratio* plays a special and prominent role in the human interpretation of "beauty." Shapes and objects whose proportions are close to the golden ratio are believed to be more pleasing to human sensibilities than those that are not. This theory, generally known as *the golden ratio hypothesis*, originated with the experiments of the famous psychologist Gustav Fechner in the late 1800s. In a classic experiment, Fechner showed rectangles of various proportions to hundreds of subjects and asked them to choose. His results showed that the rectangles that were close to the proportions of the golden ratio were overwhelmingly preferred over the rest. Since Fechner's original experiment, there has been a lot of controversy about the golden ratio hypothesis, and many modern experiments have cast serious doubts about its validity.

Write a paper describing the history of the golden ratio hypothesis. Start with a description of Fechner's original experiment. Follow up with other experiments duplicating Fechner's results and some of the more recent experiments that seem to disprove the golden ratio hypothesis. Conclude with your own analysis.

References and Further Readings

1. Adam, John, *Mathematics in Nature: Modeling Patterns in the Natural World*. Princeton, NJ: Princeton University Press, 2003.

2. Ball, Philip, *The Self-made Tapestry: Pattern Formation in Nature*. New York: Oxford University Press, 2001.

3. Conway, J. H., and R. K. Guy, *The Book of Numbers*. New York: Springer-Verlag, 1996.

4. Douady, S., and Y. Couder, "Phyllotaxis as a Self-Organized Growth Process," in *Growth Patterns in Physical Sciences and Biology*, eds. J. M. Garcia-Ruiz et al. New York: Plenum Press, 1983.

5. Erickson, R. O., "The Geometry of Phyllotaxis," in *The Growth and Functioning of Leaves*, eds. J. E. Dale and F. L. Milthrope. New York: Cambridge University Press, 1983.

6. Gardner, Martin, "About Phi, an Irrational Number That Has Some Remarkable Geometrical Expressions," *Scientific American*, 201 (August 1959), 128–134.

7. Gardner, Martin, "The Multiple Fascinations of the Fibonacci Sequence," *Scientific American*, 220 (March 1969), 116–120.

8. Gazalé, M. J., *Gnomon: From Pharaohs to Fractals*. Princeton, NJ: Princeton University Press, 1999.

9. Gullberg, Jan, *Mathematics: From the Birth of Numbers*. New York: W. W. Norton, 1997.

10. Herz-Fischler, Roger, *A Mathematical History of the Golden Number*. New York: Dover, 1998.

11. Jean, R. V., *Mathematical Approach to Pattern and Form in Plant Growth*. New York: John Wiley & Sons, Inc., 1984.

12. Livio, Mario, *The Golden Ratio: The Story of Phi, the World's Most Astonishing Number*. New York: Random House, 2002.

13. May, Mike, "Did Mozart Use the Golden Section?" *American Scientist*, 84 (March–April 1996), 118.

14. Markovsky, George, "Misconceptions About the Golden Ratio," *College Mathematics Journal*, 23 (1992), 2–19.

15. Neill, William, and Pat Murphy, *By Nature's Design*. San Francisco, CA: Chronicle Books, 1993.

16. Putz, John F., "The Golden Section and the Piano Sonatas of Mozart," *Mathematics Magazine*, 68 (1995), 275–282.

17. Sigler, Laurence, *Fibonacci's Liber Abaci*. New York: Springer-Verlag, 2002.

18. Stewart, Ian, *Life's Other Secret: The New Mathematics of the Living World*. New York: John Wiley & Sons, 1999, chap. 6.

19. Stewart, Ian, "Daisy, Daisy, Give Me Your Answer, Do," *Scientific American* (January 1995), 96–99.

20. Thompson, D'Arcy, *On Growth and Form*. New York: Dover, 1992, chaps. 11, 13, and 14.

10

The Mathematics of Population Growth

There Is Strength in Numbers

> " . . . you, be ye fruitful and multiply."
>
> Genesis 9:7

The connection between the study of population growth and mathematics goes back to the very beginnings of civilization. One of the reasons that humans invented the first numbering systems was their need to handle the rudiments of counting populations—how many sheep in the flock, how many people in the tribe, and so on. By biblical times, simple models of population growth were being used to measure crop production and even to estimate the yields of future crops. Today, mathematical models of population growth are a fundamental tool in our efforts to understand the rise and fall of endangered wildlife populations, fishery stocks, agricultural pests, infectious diseases, radioactive waste, and so on. Entire modern disciplines, such as *mathematical ecology, population biology*, and *epidemiologiy*, are built around the mathematics of population growth.

I n its modern usage, the term *population growth* has a very broad meaning, due primarily to the way that both words—*population* and *growth*—are interpreted nowadays. The Latin root of *population* is *populus* (which means "people"), so that in its original interpretation the word refers to human populations. Over time, this scope has been expanded to apply to any collection of objects (animate or inanimate), and it is not uncommon to use the term *population,* whether speaking of penguins or dollars in a bank account.

Second, we normally think of the word *growth* as being applied to things that get bigger, but in this chapter we will ascribe a slightly more technical meaning to it. *Growth* can mean *positive growth* (i.e., a population getting bigger) as well as *negative growth*—usually called *decay*—which applies to populations whose numbers are getting smaller. This is convenient because in many models populations fluctuate and often we don't know ahead of time how a population is going to change: Is it going to have positive growth or negative growth? By allowing *growth* to mean either, we need not concern ourselves with making the distinction.

As the role of mathematics in the study of population growth has expanded, so has its complexity, and the mathematical tools in use today to study population models can be quite sophisticated. The overall mathematical principles involved, however, are reasonably simple. (The devil is always in the details!) We will start the chapter with an introduction to the general principles behind mathematical models of population growth (Section 10.1), followed by a discussion of three important models—the *linear growth* model (Section 10.2), the *exponential growth* model (Section 10.3), and the *logistic growth* model (Section 10.4).

10.1 The Dynamics of Population Growth

The growth of a population is a *dynamical process*, meaning that it represents a situation that changes over time. Mathematicians distinguish between two kinds of population growth: *continuous* and *discrete*. In

continuous growth the dynamics of change are in effect all the time—every hour, every minute, every second, there is change. The classic example of this kind of growth is represcntcd by money left in an account drawing interest on a continuous basis (see Project A). We will not study continuous growth in this chapter because the mathematics involved (calculus) is beyond the scope of this book.

The second type of growth, **discrete growth**, is the most common and natural way by which populations change. We can think of it as a *stop-and-go* type of situation. For a while nothing happens; then there is a sudden change in the population. We call such a change a **transition**. Then for a while nothing happens again; then another transition takes place, and so on. This discrete, stop-and-go model of population growth can be used to simulate many real-life situations: animal populations that follow consistent breeding patterns (*breeding seasons*), money in an account that draws interest at regular intervals, the balance on a loan amortized with regular monthly payments, and so on. In all of these examples, the critical feature is the existence of transitions that occur at regular intervals, and the length of time between transitions will not make much of a difference. Thus, the same discrete model can be used when the period between transitions is measured in years, hours, or seconds.

The basic problem of population growth is to predict the future: What will happen to a given population over time? What are the long-term patterns of growth (if any)? In the case of discrete growth models we deal with these questions by finding the rules that govern the transitions. We will call these the **transition rules**. After all, if we have a way to figure out how the population changes each time there is a transition, then (with a little help from mathematics) we can usually figure out how the population changes after many transitions.

The ebb and flow of a particular population over time can be conveniently thought of as a list of numbers called the **population sequence**. Every population sequence starts with an initial population P_0 (called the *seed* of the population sequence) and continues with P_1, P_2, and so on, where P_N is the size of the population in the Nth generation. The reason a population sequence starts with P_0 rather than P_1 is convenience—with this notation the subscripts match the generations: P_1 denotes the population in the first generation, P_2 the population in the second generation, and so on. Figure 10-1 is a schematic illustration of how a population sequence is generated.

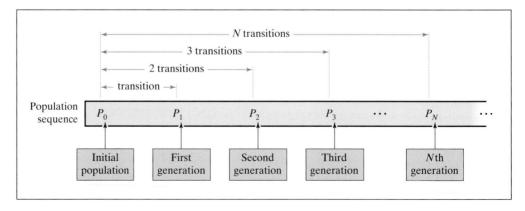

FIGURE 10-1 A generic population sequence. P_N is the population in the Nth generation.

A very convenient way to describe the population sequence is by means of a *time-series graph*. In a time-series graph the horizontal axis usually represents time (with the tick marks generally corresponding to the transitions), and the vertical axis usually represents the size of the population. A time-series graph can consist of just marks (such as dots) indicating the population size at each generation or of dots joined by lines, which sometimes helps the visual effect. The former is called a **scatter plot**, the latter a **line graph**. Figure 10-2 shows a scatter plot and a line graph for the same population sequence.

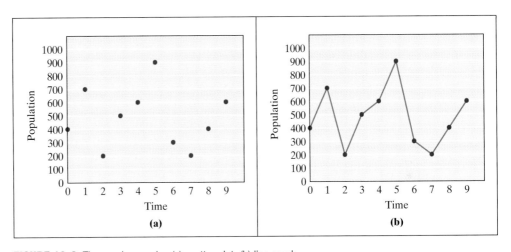

FIGURE 10-2 Time-series graphs: (a) scatter plot; (b) line graph.

Just to get our feet wet, we will start with a classic example of rabbit growth. This famous example was introduced by Leonardo Fibonacci in 1202.

EXAMPLE 10.1 Fibonacci's Rabbits

For details on the historical origins of this example, see the biographical profile of Fibonacci at the end of Chapter 9.

A man puts one pair of rabbits in a certain place entirely surrounded by a wall. How many pairs of rabbits can be produced from that pair in a year if the nature of these rabbits is such that every month each pair bears a new pair which from the second month on becomes productive?

Fibonacci, *Liber Abaci*

We will use P_N to denote the number of male–female pairs in the Nth month, starting with the seed $P_0 = 1$ (the starting pair) and assuming that this starting pair is not "productive" until the second month. Thus, in the first month we still have just the original pair ($P_1 = 1$), but by the second month we have the original pair plus a new pair ($P_2 = 2$). In the third month the new pair is still too young to breed, while the original pair bears another new pair, so $P_3 = 2 + 1 = 3$. Of these three pairs of rabbits two are going to bear new pairs in the fourth month, so $P_4 = 3 + 2 = 5$. Continuing this way, we realize that in any given month N the number of pairs P_N equals the number of pairs in the preceding month (P_{N-1}) plus the number of new pairs (one for every pair that has been around for at least two months, i.e., P_{N-2}). Thus, the transition rule for the growth of the rabbit population is $P_N = P_{N-1} + P_{N-2}$.

	Initial population	First generation	Second generation	Third generation	Fourth generation	Fifth generation	Sixth generation
Elapsed time	0	1 month	2 months	3 months	4 months	5 months	6 months
New pairs	1	0	1	1	2	3	5
Old pairs	0	1	1	2	3	5	8
Total	$P_0 = 1$	$P_1 = 1$	$P_2 = 2$	$P_3 = 3$	$P_4 = 5$	$P_5 = 8$	$P_6 = 13$

FIGURE 10-3 Fibonacci's rabbits: P_N is the number of male–female pairs in the Nth generation.

Figure 10-3 visually illustrates the transition rule. The blue arrows mean that each productive pair begets a new pair; the red arrows mean that every pair (young or old) is a productive pair by the next month.

The numbers $1, 1, 2, 3, 5, 8, \ldots$ that describe the growth of the Fibonacci rabbit population (in pairs) are the very Fibonacci numbers we discussed in Chapter 9 (and the reason why these numbers are called Fibonacci numbers). There is one small bookkeeping issue you should be careful with—the population sequence starts at P_0 whereas the Fibonacci sequence starts with F_1. In other words, $P_0 = F_1, P_1 = F_2, P_2 = F_3, \ldots, P_N = F_{N+1}$. Accordingly, the answer to the original question raised by Fibonacci is that after one year (12 months) the number of rabbit pairs is $P_{12} = F_{13} = 233$.

You may wonder how long the rabbit population can continue growing under the model we developed in Example 10.1. According to this model, after 15 months the population would be $P_{15} = F_{16} = 987$ pairs—that's a lot of rabbits to have in an enclosed space! Further down the line old rabbits are going to start dying, and that will have to be accounted for as well.

Clearly, the growth described by the model given by the transition rule $P_N = P_{N-1} + P_{N-2}$ is not sustainable over the long haul, and even in the short term represents only a crude and simplistic approximation of reality. Real-life rabbits live and breed by considerably more complicated rules, which we could never hope to capture in a simple equation, and, in general, this is true about most other types of equations that attempt to model the growth of natural populations. Is there any use, then, for simplistic mathematical models of how populations grow? The answer is yes! We can make excellent predictions about the growth of a population over time, even when we don't have a completely realistic set of transition rules. The secret is to capture the variables that are really influential in determining how the population grows, put them into a few transition rules that describe how the variables interact, and forget about the small details.

In Section 10.4 we will see a much better model for the growth of animal populations living in a confined space.

10.2 The Linear Growth Model

The linear growth model is the simplest of all models of population growth. In this model, in each generation the population increases (or decreases) by a fixed amount called the *common difference*. The easiest way to see how the model works is with an example.

EXAMPLE 10.2 How Much Garbage Can We Take?

The city of Cleansburg is considering a new law that would restrict the monthly amount of garbage allowed to be dumped in the local landfill to a maximum of 120 tons a month. There is concern among local officials that, unless this restriction on dumping is imposed, the landfill will reach its maximum capacity of 20,000 tons in a few years. Currently, there are 8000 tons of garbage already in the landfill. Assuming that the law is passed and the landfill collects exactly 120 tons of garbage each month, how much garbage will there be in the landfill five years from now? How long before the landfill reaches its 20,000-ton capacity?

The population in this example is the garbage in the landfill, and since we only care about monthly totals, we define the transitions as taking place once a month. The key assumption about this population growth problem is that the monthly garbage at the landfill grows by 120 tons a month.

From a starting population of $P_0 = 8000$ tons, the next few terms of the population sequence are $P_1 = 8000 + 120 = 8120$, $P_2 = 8120 + 120 = 8240$, $P_3 = 8240 + 120 = 8360$, and so on. After five years of this garbage we will have had 60 transitions, each representing an increase of 120 tons, giving a population of $P_{60} = 8000 + 60(120) = 15{,}200$ tons.

To determine the number of transitions it would take for the landfill to reach its 20,000-ton maximum, we set up the equation $8000 + 120N = 20{,}000$, which has as a solution $N = 100$. This means that it will take 100 monthly transitions (eight years and four months) for the landfill to reach its maximum capacity. Based on this information, local officials should start making plans soon for a new landfill.

The line graph in Fig. 10-4(a) illustrates the growth of the garbage in the dump over the five years in question, and, by the way, so does the line graph in Fig. 10-4(b). These two line graphs illustrate why linear growth is called linear growth—no matter how we scale it, stretch it or slide it, those "red dots" will always line up.

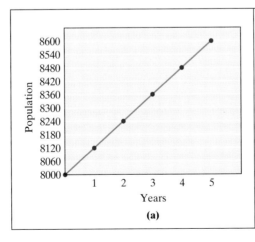
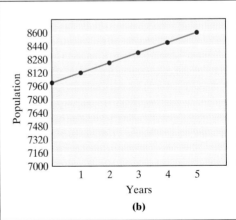

FIGURE 10-4

Example 10.2 is a typical example of the general linear growth model, whose basic characteristic is that in each transition a constant amount—call it d—is added to the previous population. A population that grows according to a linear growth

model produces a type of sequence commonly known as an **arithmetic sequence**. (Technically speaking, the arithmetic sequence is just the numerical description of a population growing according to a linear growth model—informally, linear growth and arithmetic sequences can be considered synonymous.) The number d is called the **common difference** for the arithmetic sequence, because any two consecutive values of the arithmetic sequence will always differ by the amount d.

A population that grows according to a linear growth model is typically described by two constants: the initial population P_0 (the *seed* of the population sequence) and the common difference d. With these two values we can build up the population sequence using the **recursive formula** $P_N = P_{N-1} + d$. The rule is called *recursive* because each term of the population sequence is defined in terms of the previous term. As we saw in Chapter 9, the major drawback of the recursive description is the fact that to calculate any one value of the population sequence we essentially have to know the previous value, which requires us to know the value before that, and so on all the way down the line. This can turn out to be quite an inconvenience.

Fortunately, the same population sequence (linear growth with seed P_0 and common difference d) can also be described by the **explicit formula** $P_N = P_0 + N \times d$. We call it an *explicit formula* because this formula allows us to calculate any term of the population sequence directly—no previous terms of the sequence (except the seed, of course) are needed. Figure 10-5 illustrates the argument why $P_N = P_0 + N \times d$: To get to P_N the starting population P_0 goes through N transitions, each of which consists of adding d.

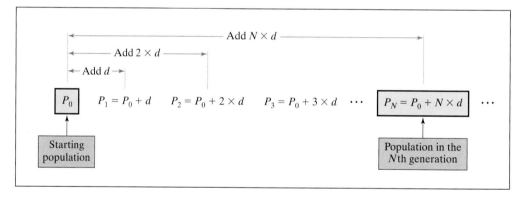

FIGURE 10-5 Population growth based on a linear growth model.

The linear growth model with seed P_0 and common difference d is described by either of the following two formulas:

Linear Growth

- $P_N = P_{N-1} + d$ (*recursive formula*)
- $P_N = P_0 + N \times d$ (*explicit formula*)

> ## EXAMPLE 10.3 A Frappuchino in Every Corner?

Cosmic Java is the name of a new national chain of coffee shops, competing with you know who. Cosmic Java opened in 2005 with 370 coffee shops and plans to open 60 new coffee shops each year after that. Here are some questions of some relevance to the company's long-term strategic plans: (i) How many Cosmic Java coffee shops will there be in the year 2020? (ii) In what year will Cosmic Java reach the milestone of having 2000 coffee shops?

The Cosmic Java story illustrates a classic example of linear growth. The parameters of this model are an initial population of $P_0 = 370$ (which also sets the baseline year as 2005) and a common difference $d = 60$.

Want to know the number of Cosmic Java coffee shops in 2020? No problem. This is asking for the value of the population 15 years after the starting year, and this is given by P_{15}. Using the explicit formula, we get

$$P_{15} = 370 + 60(15) = 1270$$

To determine how many years it will take for the number of coffee shops to exceed 2000, we solve the inequality

$$P_N = 370 + 60N > 2000$$

which gives

$$N > 1630/60 \approx 27.17$$

This means that somewhere between the year 2032 (27 years after 2005) and 2033 Cosmic Java will reach its milestone of having 2000 coffee shops. We can hardly wait! ◀◀

The explicit formula for linear growth requires that we know the common difference d and the seed P_0. Sometimes these two numbers are not given to us directly, but we can derive them if we know two terms of the population sequence (any two will work). Example 10.4 illustrates how to do this.

▶ EXAMPLE 10.4 Two Points Determine a Line

A population grows according to a linear growth model. Unfortunately, all we have to work with is $P_9 = 1324$ and $P_{25} = 2684$.

To find the explicit formula for this population sequence, we need to find P_0 and d, which we do by solving the system of two equations and two unknowns

$$\begin{cases} P_0 + 25d & = 2684 \\ P_0 + 9d & = 1324 \end{cases}$$

To solve the system we subtract the second equation from the first and get

$$16d = 1360$$

and thus

$$d = 85$$

Replacing d by 85 in either of the two equations gives $P_0 = 559$. Now that we have d and P_0 we are in business—our population sequence is given by the explicit formula

$$P_N = 559 + 85N$$

 ◀◀

The Arithmetic Sum Formula

Given a population that grows according to a linear growth model, we often need to know what is the sum of the first so many terms of the population sequence—say the sum of the first 10 terms, or the first 100 terms, or the first 500 terms. Of course, in the case of the sum of the first 10 terms we can simply write the terms down and add them the long way, but this is not a reasonable approach when dealing with the addition of hundreds of terms. The next example illustrates how we can find such sums in an efficient way.

> ### EXAMPLE 10.5 The Cost of Building Up Inventory

Jane Doe is a small tractor manufacturer. In order to unveil its radically new 2008 model, the company goes into a 72-week production schedule. The plan is to manufacture 30 tractors each week for the next 72 weeks. After they are manufactured, the tractors are stored in one of several warehouses for the remainder of the time between production and release date. If the storage costs are $10 per tractor per week, what is the total storage cost to the company over the 72-week period?

In this problem the weekly storage costs for weeks 1 through 72 are given by the sequence

$$P_0 = \$0 \text{ (week 0)}, P_1 = \$300 \text{ (week 1)}, P_2 = \$600 \text{ (week 2)},$$

$$\cdots,$$

$$P_{71} = \$21,300 \text{ (week 71)}, P_{72} = \$21,600 \text{ (week 72)}$$

The weekly storage costs form an arithmetic sequence with initial term $P_0 = 0$ and common difference $d = 300$. (Starting the count at $0 is a little contrived here, but it is consistent with the way we look at population growth.) The new wrinkle in this problem is that to find the total storage cost over the 72-week period we will have to compute the rather long sum $S = 300 + 600 + 900 + \cdots + 21,300 + 21,600$. We definitely don't want to add these 72 numbers the old-fashioned way—even with a calculator this is a horrible thought!

Luckily, there is a wonderful arithmetic trick that will do the job. The key idea is to write out the sum twice, once forward and once backward.

$$S = \quad 300 + \quad 600 + \quad 900 + \cdots + 21600 \quad \text{(forward)}$$
$$S = 21,600 + 21,300 + 21,000 + \cdots + \quad 300 \quad \text{(backward)}$$

Once they are lined up this way, we notice that each of the 72 columns on the right-hand side of the equal signs adds up to 21,900. Thus, the grand total of both lines is $72 \times 21,900$. This is twice the sum S we are trying to compute, so dividing by two gives S.

$$S = \frac{72 \times 21,900}{2} = \$788,400$$

> **EXAMPLE 10.6** Missing the Last Term? No Problem.

Consider an arithmetic sequence $A_0 = 5$, $A_1 = 12$, $A_2 = 19,\ldots$. (Note the change in the choice of letters. We use A's rather than P's to emphasize that we are working with an arithmetic sequence.) We would like to find the sum of the first 1000 terms of this arithmetic sequence; in other words, find

$$S = \underbrace{5 + 12 + 19 + \cdots}_{1000 \text{ terms}}$$

We would like to use the same trick we used in Example 10.5, but to do that we will first need to find the *last term* of the sum. We can find the last term using the explicit formula for linear growth: $A_{999} = 5 + 999 \times 7 = 6998$. (A common mistake is to think that this last term is given by A_{1000}, but since the count starts with A_0 the last term is really A_{999}.)

Now that we have the first and last terms of the sum, we can use the trick we used in Example 10.5.

$$S = \quad 5 + \quad 12 + \quad 19 + \cdots + 6998 \quad \text{(forward)}$$
$$S = 6998 + 6991 + 6984 + \cdots + \quad 5 \quad \text{(backward)}$$

Note how each column adds up to 7003 (the sum of the *first* and *last* terms of S). Adding all 1000 columns gives a grand total of $2 \times S = 1000 \times 7003 = 7{,}003{,}000$. It follows that

$$S = \frac{1000 \times 7003}{2} = 3{,}501{,}500 \qquad \blacktriangleleft\!\blacktriangleleft$$

The approach used in Examples 10.5 and 10.6 works with *any* arithmetic sequence $A_0, A_1, A_2,\ldots$. We can add up any number of consecutive terms easily with the following formula, which we will informally refer to as the *arithmetic sum formula*.

Arithmetic Sum Formula

$$A_0 + A_1 + \cdots + A_{N-1} = \frac{(A_0 + A_{N-1})N}{2}$$

Before we go on, a couple of comments about the arithmetic sum formula are in order. First, it is an extremely useful and important formula. It's not known who first discovered it, but it goes back a long way, and the first known reference to it can be found in Book IX of Euclid's *Elements*, written around 300 B.C. At a minimum the formula has been known for 2300 years, possibly much longer. Second, why did we write the last term of the sum as A_{N-1} rather than A_N? (Surely, A_N seems a lot more user friendly.) The reason is that the sum starts with A_0, which makes the number of terms and the subscript for the last term offset by 1. In short, the Nth term is A_{N-1}, and if you want A_N you need to go to the $(N + 1)$st term.

You may find the following informal version of the arithmetic sum formula more to your liking: *To find the sum of terms of an arithmetic sequence, add the first term and the last term, multiply the result by the number of terms, and then divide by 2.* (Please remember that this rule applies *only* when you are adding *consecutive* terms of an *arithmetic sequence*.)

> ## EXAMPLE 10.7 Don't Know How Many Terms? No Problem.

Suppose we need to compute the arithmetic sum

$$4 + 13 + 22 + 31 + 40 + \cdots + 922$$

Here the first and last terms are given—the missing piece of the puzzle is the number of terms in this sum. We can find the number of terms (N) by setting $922 = A_{N-1}$ and using the explicit formula for linear growth (the common difference is $d = 9$):

$$A_{N-1} = 922 = 4 + 9(N - 1)$$

From the equation we get $9(N - 1) = 918$, which gives $N - 1 = 102$ and $N = 103$. It follows that

$$4 + 13 + 22 + 31 + 40 + \cdots + 922 = \frac{(4 + 922) \times 103}{2} = 47{,}689 \quad \text{«}$$

10.3 The Exponential Growth Model

In everyday life we often express rates of change in terms of percents. "I just got an 8% raise," "Test scores up by 3.7%," "Going out of business sale—45% off on all merchandise." The notion of percentage change plays an important part in the exponential growth model, and for this reason, before we start our discussion of exponential growth in earnest, we will take a brief look at the use of percentages to calculate increases and decreases.

> ## EXAMPLE 10.8 Tacking on Percentage Increases

A firm manufactures an item at a cost of C dollars. The item is marked up 10% and sold to a distributor. The distributor then marks the item up 20% (based on the price he or she paid) and sells the item to a retailer. The retailer marks that price up 50% and sells the item to the public. By what percent has the item been marked up over its original cost?

- Original cost of item: C
- Cost to distributor after 10% markup: $D = 110\%$ of $C = (1.1)C$
- Cost to retailer after 20% markup:
  $R = 120\%$ of $D = (1.2)D = (1.2)(1.1)C = (1.32)C$
- Price (P) to the public after 50% markup:
  $P = 150\%$ of $R = (1.5)R = (1.5)(1.32)C = (1.98)C$

  Conclusion: The markup over the original cost is 98%. «

> ## EXAMPLE 10.9 Combining Markups and Markdowns

A retailer buys an item for C dollars and marks it up 80%. To make room for new merchandise, the item is put on sale for 40% off the marked price. After a while, an additional 10% markdown is taken off the sale price. What is the net percentage markup on this item?

- Original cost of item: C
- Price after 80% markup: $P = 180\%$ of $C = (1.8)C$
- Sale price after 40% markdown:
  $S = 60\%$ of $P = (0.6)P = (0.6)(1.8)C = (1.08)C$
- Sale price after additional 10% markdown:
  $F = 90\%$ of $S = (0.9)S = (0.9)(0.6)(1.8)C = (0.972)C$

Surprisingly, when we put all the markups and markdowns together, the retailer is selling the item for 97.2% of its original cost; in other words, a 2.8% net markdown!   ≪

The main points of Examples 10.8 and 10.9 can be summarized as follows:

- To *increase* a number C by $x\%$, we multiply C by the number $(1 + x/100)$. (The $x/100$ represents $x\%$ in decimal form; adding the 1 represents the fact that we are *increasing* the original number C.)
- To *decrease* a number C by $x\%$, we multiply C by the number $(1 - x/100)$.

Note the difference between linear and exponential growth—in linear growth we *add* a fixed constant; in exponential growth we *multiply* by a fixed constant.

Let's return now to the exponential growth model. The exponential growth model is arguably the most important and widely used (some would say overused) model of population growth. Exponential growth is based on the idea of a constant growth rate—in each transition the population changes by a fixed *factor* called the *common ratio*.

▶ EXAMPLE 10.10 The Power of Compounding

The sum of $1000 is deposited in a retirement account that pays 10% *annual* interest—interest is paid once a year at the end of the year. (A 10% interest rate is a little unrealistic these days, but it's a nice number to use for starters.) How much money is there in the account after 25 years if the interest is left in the account? How much money is there in the account after N years?

Table 10-1 will help us get started.

TABLE 10-1

	Account balance at beginning of year	Interest earned for the year	Account balance at end of year
Year 1	$1000	$100	$(1.1) \times 1000 = \$1100$
Year 2	$1100	$110	$(1.1) \times 1100 = \$1210$
Year 3	$1210	$121	$(1.1) \times 1210 = \$1331$
⋮	⋮	⋮	⋮
Year 24	?	?	?
Year 25	?	?	?

The critical observation is that the account balance at the end of any given year is obtained by taking a 10% increase on the account balance at the start of that year. [(Year Ending $) = 1.1 × (Year Starting $).] This means that the account balance at the end of year 1 is $1000 × (1.1). (To best see what is going on, we will not carry out the multiplication.) In year 2 the account starts with $1000 × (1.1) and at the end of the year grows by a factor of 1.1 to $1000 × (1.1)^2$. Repeating the argument as many times as we have to [the account balance at the end of the third year is $1000 × (1.1)^3$, and so on], we conclude that the account balance at the end of the 25th year is

$$\$1000 \times (1.1)^{25} = \$10{,}834.71$$

In general, the account balance at the end of the Nth year is given by

$$P_N = \$1000 \times (1.1)^N$$

Figure 10-6(a) plots the growth of the money in the account for the first eight years. Figure 10-6(b) plots the growth of the money in the account for the first 30 years.

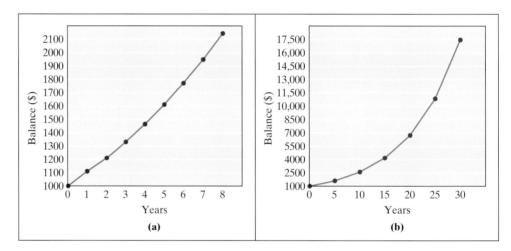

FIGURE 10-6 Cumulative growth of $1000 at 10% interest compounded annually.

Example 10.10 is a classic example of exponential growth. The money draws interest, then the money plus the interest draw interest, and so on. While the most familiar examples of exponential growth have to do with the growth of money, the exponential growth model is useful in the study of biological populations as well. The key property of exponential growth is *a constant rate of growth*. This implies that each transition consists of multiplying the size of the population by *a constant factor determined by the growth rate but not equal to the growth rate!* (In Example 10.10 the 10% growth rate corresponds to a constant factor of 1.1.)

A sequence defined by repeated multiplication—every term in the sequence after the first is obtained by multiplying the preceding term by a fixed amount r—is called a **geometric sequence**. The constant factor r is called the **common ratio** of the geometric sequence—it is the ratio of two successive terms in the sequence. (To ensure that the population sequence does not have negative numbers, we will restrict the values of the common ratio r to positive numbers, although no such restriction is necessary when dealing with geometric sequences in general.)

As in the case of linear growth, the exponential growth model can be described by both recursive and explicit formulas. The recursive formula tells us how to get a term P_N of the population sequence from the preceding term P_{N-1}: Multiply P_{N-1} by the common ratio r. The explicit formula tells us how to get a term P_N of the population sequence using just the seed P_0 and the common ratio r: multiply P_0 by the Nth power of r.

Exponential Growth ($r > 0$)

- $P_N = r \cdot P_{N-1}$ (*recursive formula*)
- $P_N = P_0 \cdot r^N$ (*explicit formula*)

When $r > 1$, the terms of the sequence get bigger and we have real growth (positive growth), but when $0 < r < 1$, the terms of the sequence get smaller and we have a situation known as *exponential decay*.

▶ **EXAMPLE 10.11** Eradicating the Gamma Virus

The numbers in this example are made up and the Gamma virus is fictitious, but the example illustrates some of the important issues that public health officials must deal with when tracking the spread and control of disease.

Thanks to improved vaccines and good public health policy, the number of reported cases of the Gamma virus has been dropping by 70% a year since 2004, when there were 1 million reported cases of the virus. If the present rate continues, how many reported cases of the virus can we predict by the year 2010? How long will it take to eradicate the virus?

In this example we are dealing with exponential decay. We are tracking the number of recorded cases of the Gamma virus starting with the year 2004. The initial population is given by $P_0 = 1,000,000$ and the common ratio corresponding to a 70% *annual decrease* is given by $r = 1 - 0.7 = 0.3$.

The number of reported cases of the Gamma virus in the year 2010 is given by $P_6 = 1,000,000 \times (0.3)^6 = 729$. By 2011 this number will drop to about 220 cases ($729 \times 0.3 = 218.7$), by 2012 to about 66 cases, by 2013 to about 20 cases, and by 2014 to 6 cases.

The story of the eradication of the Gamma virus is best illustrated by the line graph in Fig. 10-7. This kind of graph is typical of exponential decay.

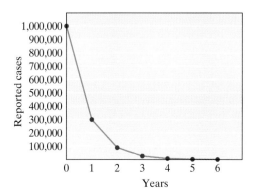

FIGURE 10-7 Exponential growth with $r = 0.3$.

Let's now return to the topic of investing money in an interest-bearing account—the classic application of exponential growth. The most basic scenario is the case of *annual compounding* (the interest is added to the account once a year and only at the end of the year). As long as the money is left untouched in the account, the "population" (money in the account) grows according to the exponential growth model—the seed is given by the original investment (called the *principal*), and the common ratio r is derived from the interest rate (r equals 1 plus the interest rate written as a decimal).

If we let P_0 denote the principal, i the annual interest rate (expressed as a decimal), and N the number of years the money is left in the account, the explicit formula for exponential growth becomes the following *annual compounding formula*:

Annual Compounding Formula

$$\$P_N = P_0 \cdot (1 + i)^N$$

▶ EXAMPLE 10.12 Funky Numbers? No Problem

Suppose you invest \$367.51 in an account that pays an annual interest rate of $4\frac{1}{2}\%$ a year and you let the money grow. How much money will there be in the account at the end of 7 years? What if you leave the money in the account for $7\frac{1}{2}$ years?

Here $P_0 = 367.51$ and $i = 0.045$ ($4\frac{1}{2}\%$ written in decimal form).

Other than the fact that the numbers in this example are a bit funky, this is a straightforward example of annual compounding. Using the annual compounding formula, the amount of money in the account after 7 years is $P_7 = \$367.51 \times (1.045)^7 = \$500.13$ (rounded to the nearest penny). Because the interest is compounded only at the end of each year, after $7\frac{1}{2}$ years the money in the account is still \$500.13. ◀◀

With a little tweaking, the exponential growth model can be applied to investments where the interest is compounded over shorter periods, such as monthly or daily compounding. This means that the transitions happen more frequently and thus the money grows faster—good news for the investor!

▶ EXAMPLE 10.13 Monthly Compounding Is Good

This example is a variation of Example 10.10. We are going to start with a principal of \$1000 invested in an account that pays 10% *annual* interest, but this time the interest is *compounded monthly*. If we don't touch the money, how much money will there be in the account at the end of 5 years? How much money will there be in the account after $5\frac{1}{2}$ years? What about 25 years?

The key observation here is that while the interest rate is given as an annual rate, the interest is compounded monthly (i.e., the transitions occur once a month). Thus, to find the effective interest rate for the compounding period, we need to divide 10% by 12. This gives us the **periodic interest rate** $p = 0.10/12$. Adding 1 to this periodic interest rate gives the common ratio $r = 1 + (0.1/12) = 12.1/12$.

After 5 years (60 monthly transitions) the money in the account (rounded to the nearest penny) is given by

$$P_{60} = \$1000 \times \left(\frac{12.1}{12}\right)^{60} = \$1645.31$$

(A note of caution: Avoid rounding off as much as you can. For example, if you round off the fraction 12.1/12 to 1.0083 or even 1.00833, you get a significantly different answer—check it out!)

After $5\frac{1}{2}$ years (66 monthly transitions) the money in the account (rounded to the nearest penny) is given by

$$P_{66} = \$1000 \times \left(\frac{12.1}{12}\right)^{66} = \$1729.31$$

After 25 years the money in the account (rounded to the nearest penny) is given by

$$P_{300} = \$1000 \times \left(\frac{12.1}{12}\right)^{300} = \$12{,}056.95$$

⟪

Had you rounded 12.1/12 to 1.0083 and used this value to compute P_{300} you would have ended up with only $11,937.96!

If we compare the last result in Example 10.13 with the result in Example 10.10 (same principal and same interest rate but with annual compounding), we can see that the frequency with which the interest is compounded makes a big difference on the value of an investment.

▶ **EXAMPLE 10.14 Daily Compounding Is Even Better**

Let's take Example 10.13 one step further. Let's take $1000 and invest it at 10% annual interest compounded *daily*. How much money will there be in the account at the end of the 5 years? $5\frac{1}{2}$ years? 25 years?

In this case, the period between transitions is one day, and thus, there are 365 transitions in a year. (We are not going to worry about leap years—banks don't pay or charge any extra interest on leap years.) The periodic interest rate is $p = 0.10/365$, and thus $r = 1 + (0.10/365) = 365.10/365$.

The money in the account at the end of 5 years (i.e., after $5 \times 365 = 1825$ compounding periods) rounded to the nearest penny is

$$P_{1825} = \$1000\left(\frac{365.10}{365}\right)^{1825} = \$1648.61$$

When the money is left in the account for $5\frac{1}{2}$ years we need to round "half a year" down to 182 days (banks always round in their favor). Thus, at the end of $5\frac{1}{2}$ years ($5 \times 365 + 182 = 2007$ compounding days), the money in the account rounded to the nearest penny is

$$P_{2007} = \$1000\left(\frac{365.10}{365}\right)^{2007} = \$1732.89$$

After 25 years (9125 compounding days), the money in the account is

$$P_{9125} = \$1000\left(\frac{365.10}{365}\right)^{9125} = \$12{,}178.32$$

⟪

We can generalize the results of Examples 10.13 and 10.14 to get a *general compounding formula* for computing the growth of P_0 left in an account that pays an annual interest rate i compounded k times a year.

General Compounding Formula

$$\$P_N = P_0\left(1 + \frac{i}{k}\right)^{Nk}$$

> ## EXAMPLE 10.15 Shopping Interest Rates

You have an undisclosed amount of money to invest in a savings account. Bank *A* offers 10% annual interest *compounded yearly*, bank *B* offers 9.75% annual interest *compounded monthly*, and bank *C* offers 9.5% annual interest *compounded daily*. Which bank offers the best deal?

Note that the problem does not indicate the amount of money we invest or the length of time we plan to leave the money in the account. The answer to the problem depends only on *i* (the annual interest rate) and *k* (the number of compounding periods in a year). The way to compare these different accounts is to use a common yardstick—for example, how much does $1 grow in 1 year?

At bank *A*, offering 10% interest compounded once a year, in 1 year $1 becomes $1.10.

At bank *B*, offering 9.75% annual interest compounded monthly, in one year $1 becomes

$$\$\left(1 + \frac{0.0975}{12}\right)^{12} \approx \$1.102$$

And with bank *C*, offering 9.5% annual interest compounded daily, in one year $1 becomes

$$\$\left(1 + \frac{0.095}{365}\right)^{365} \approx \$1.0996$$

We can now see that bank *B* offers the best deal and bank *C* offers the worst deal: The 9.75% rate compounded monthly represents a 10.2% increase after a year while the 9.5% interest rate compounded daily represents a 9.96% annual increase. ≪

Banks use the term **annual yield** to describe the percentage increase of an investment over a one-year period. When you invest $1000 and after a year you have $1099.60 you made a profit of 9.96%—this is your annual yield.

> The differences among the three banks may appear insignificant when we look at the effect over one year, but they become quite significant when we invest over a longer period—see Exercise 39.

The Geometric Sum Formula

Suppose you want to find the sum of the first 100 (or 1000) terms of a geometric sequence. Just like with arithmetic sequences, there is a nice handy formula that will allow us to do it without having to actually add the terms. We will call this formula the *geometric sum formula*. Here we will give the formula first and then explain how it can be derived. (To simplify the notation we are using *a* instead of P_0 for the seed.)

Geometric Sum Formula

$$a + ar + ar^2 + \cdots + ar^{N-1} = \frac{a(r^N - 1)}{r - 1}$$

The geometric sum formula works for all values of the common ratio r except $r = 1$. (For $r = 1$ the formula breaks down because we get a zero denominator on the right-hand side. Fortunately, the case $r = 1$ is trivial, since then every term in the sum is a, and the sum equals $N \cdot a$.)

The geometric sum formula can be derived using high school algebra combined with the following clever observation: Call the sum on the left-hand side S. If we multiply $S = a + ar + ar^2 + \cdots + ar^{N-1}$ by r, we get $r \cdot S = ar + ar^2 + ar^3 + \cdots + ar^{N-1} + ar^N$, which matches S in all of its terms except the last. If we subtract $S = a + ar + ar^2 + \cdots + ar^{N-1}$ from $r \cdot S = ar + ar^2 + ar^3 + \cdots + ar^{N-1} + ar^N$, we get a lot of cancellations and end up with $r \cdot S - S = ar^N - a$. A little more algebra gives us $(r - 1)S = a(r^N - 1)$,

See Exercise 69.

and from this we get $S = \dfrac{a(r^N - 1)}{r - 1}$.

Like its arithmetic counterpart, the geometric sum formula is an important tool in many real-life applications. We will illustrate a couple of these applications in the next two examples.

▶ **EXAMPLE 10.16** The *X-Virus*

At the emerging stages, the spread of many infectious diseases—such as HIV and the West Nile virus—often follows an exponential growth model. Let's consider the case of an imaginary infectious disease called the *X-virus*, for which no vaccine is known. The first appearance of the *X-virus* occurred in 2005 (year 0), when a total of 5000 cases of the disease were recorded in the United States. Epidemiologists estimate that until a vaccine is developed, the virus will spread at a 40% *annual rate of growth*, and it is expected that it will take at least 10 years until an effective vaccine becomes available. Under these assumptions, how many estimated cases of the *X-virus* will occur in the United States over the 10-year period 2005–2014?

We can track the spread of the virus by looking at the number of new cases of the virus reported each year. This is a geometric sequence with $P_0 = a = 5000$ (the seed) and common ratio $r = 1.4$ (40% annual growth):

5000 cases in 2005

$5000 \times 1.4 = 7000$ new cases in 2006

$5000 \times (1.4)^2 = 9800$ new cases in 2007

$\vdots$

$5000 \times (1.4)^9$ new cases in 2014

It follows that the total number of cases over the 10-year period is given by the sum

$$5000 + 5000 \times 1.4 + 5000 \times (1.4)^2 + \cdots + 5000 \times (1.4)^9$$

Using the geometric sum formula, this sum (rounded to the nearest whole number) equals

$$\frac{5000 \times [(1.4)^{10} - 1]}{(1.4 - 1)} \approx 349{,}068$$

Our computation shows that about 350,000 people will contract the *X-virus* over the next 10 years. And what would happen if, due to budgetary or technical problems, it takes 15 years to develop a vaccine? All we have to do is change N to 15 in the geometric sum formula:

$$\frac{5000 \times [(1.4)^{15} - 1]}{(1.4 - 1)} \approx 1{,}932{,}101$$

These are sobering numbers—the exponential growth model predicts that if the development of a vaccine is delayed from 2009 to 2014, the number of *X-virus* cases would grow from 350,000 to almost 2 million! ◀◀

▶ EXAMPLE 10.17 Setting Up a College Fund

A mother decides to set up a college trust fund for her newborn child. The plan is to deposit $100 a month for the next 18 years (i.e., 216 deposits) into a savings account that pays 6% annual interest *compounded monthly*. How much money will there be in the account at the end of 18 years?

This is a problem of exponential growth with a twist: Each $100 deposit grows at the same monthly rate $\left[r = 1 + \left(\dfrac{0.06}{12} \right) = 1.005 \right]$, but the number of periods it compounds is different for each deposit. Let's make a list.

- First deposit of $100 draws interest compounded for 216 months, producing $100(1.005)^{216}$
- Second deposit of $100 draws interest compounded for 215 months, producing $100(1.005)^{215}$
- Third deposit of $100 draws interest compounded for 214 months, producing $100(1.005)^{214}$
  ⋮
- Two-hundred-sixteenth deposit of $100 draws interest for 1 month, producing $100(1.005)$

The total amount in the account at the end of 18 years will be

$$100(1.005)^{216} + 100(1.005)^{215} + \cdots + 100(1.005)$$

This is a geometric sum (written backward) with $a = 100(1.005) = 100.50$, $r = 1.005$, and $N = 216$. By the geometric sum formula, this total equals

$$\$\frac{(100.50)[(1.005)^{216} - 1]}{1.005 - 1} \approx \$38{,}929$$

Act II. At this point, mom realizes that by the time Junior goes off to college this sum will cover maybe a year of college expenses. She wonders if she should double the monthly deposits and make it $200 a month. How much money would then be in the trust fund at the end of 18 years?

The net effect of doubling the monthly payments is that the total in the trust fund is doubled to $77,858. In fact, there is a general principle at work here: If she were to increase the monthly payments by a factor of $x\%$, the total in the trust fund would also increase by $x\%$. This follows from the observation that in the geometric sum formula $S = \dfrac{a(r^N - 1)}{r - 1}$, the only variable affected by a change in the monthly payments is a—the other two variables (r and N) are not affected.

The plan is looking better, but mom still wonders if it is going to be enough (what a mom!).

See Exercise 68.

Act III. The last scheme she considers is to make monthly deposits of $400, but to only do it for 9 years (in other words, double the amount again but cut the number of payments in half). We are now talking of $400 payments for 108 months. After making the necessary adjustments, the geometric sum formula gives (rounded to the nearest penny)

$$\$\frac{400(1.005)\left[(1.005)^{108} - 1\right]}{1.005 - 1} \approx \$57{,}381.44$$

This is the balance in the trust fund at the end of 9 years. At this point mom stops making additional deposits, but the money remains in the trust fund for another 9 years and continues to accrue interest at a 6% annual rate compounded monthly. This is another compounding problem with $P_0 = \$57{,}381.40$, $N = 9$, $i = 0.06$, and $k = 12$.

Using the general compounding formula, we get the final answer:

$$\$57{,}381.44\left(1 + \frac{0.06}{12}\right)^{108} \approx \$98{,}334.54$$

Now that makes for a healthy-looking trust fund!   ◀◀

10.4 The Logistic Growth Model

When dealing with animal populations, the two models we have studied so far are mostly inadequate. As we now know, *linear growth* models situations where the *amount* of growth between transitions is constant. This model might work for populations of inanimate objects (garbage, consumer goods, sales figures, and so on) but fails completely when some form of breeding must be taken into account. *Exponential growth*, on the other hand, represents the case in which there is *unrestrained* breeding (e.g., money left to compound in a bank account, and sometimes in the early stages of an actual animal population). In population biology, however, it is generally the case that the rate of growth of an animal population is not constant and often depends on the density of the population. Small populations have plenty of room to spread out and grow, and thus their growth rates tend to be high. As the population density increases there is less room to grow and there is more competition for resources—the growth rate tends to taper off. Sometimes the population density is so high that resources become scarce or depleted, leading to population decay and even to extinction.

The effects of population density on growth rates were studied in the 1950s by behavioral psychologist John B. Calhoun. Calhoun's now classic studies showed that when rats were placed in a closed environment, their behavior and growth rates were normal as long as the rats were not too crowded. When their environment became too crowded, the rats started to exhibit abnormal behaviors, such as infertility and cannibalism, which effectively put a brake on the population growth rate. In extreme cases, the entire rat population became extinct.

Calhoun's experiments with rats are but one classic illustration of a general principle in population biology: *A population's growth rate is negatively impacted by the population's density*. This principle is particularly important in cases where the population is confined to a limited environment.

Population biologists call such an environment the **habitat**. The habitat might be a cage (as in Calhoun's rat experiments), a lake (as for a population of fish), a garden (as for a population of snails), and, of course, Earth itself (everyone's habitat).

Of the many mathematical models that attempt to deal with a variable growth rate in a fixed habitat, the simplest is a model first proposed in 1838 by the Belgian mathematician Pierre François Verhulst. Verhulst called his model the **logistic growth model**. To put it very informally, the key idea in the logistic growth model is that the rate of growth of the population is directly proportional to the amount of "elbow room" available in the population's habitat. Thus, lots of elbow room means a high growth rate, little elbow room means a low growth rate (possibly less than 1, which, as we know, means that the population is actually decreasing), and if there is no elbow room at all the population becomes extinct.

Population biologists use the term **carrying capacity** to describe the total saturation point of an habitat. The carrying capacity (denoted by C) is a ceiling for a population living in a given habitat—by definition the population size will never exceed C ($P_N \leq C$). For a population of size P_N, we will think of the difference $C - P_N$ as the absolute amount of elbow room remaining for this population. An even more convenient way to think of the elbow room is in relative terms: The ratio $\dfrac{C - P_N}{C}$ represents the amount of elbow room available expressed as a fraction (percentage) of the carrying capacity.

When the relative elbow room for a population is very high (close to 100%), the population can breed without constraints, and the classic exponential growth equation $P_{N+1} = r \cdot P_N$ does a good job of describing the growth of the population. (Here the growth rate r is a constant called the **growth parameter** that depends only on the type of population we are dealing with.) The key idea behind the logistic growth model is that this growth rate decreases in direct proportion to the amount of elbow room—if the elbow room is 40%, then the growth rate is $(0.4) \cdot r$, and so on. Thus,

$$\text{growth rate for period } N = r \cdot \left(\frac{C - P_N}{C} \right)$$

and from this, we get the following transition rule for the logistic growth model:

$$P_{N+1} = r \cdot \left(\frac{C - P_N}{C} \right) \cdot P_N$$

This transition rule can be simplified considerably by switching the notation and using relative population figures $p_N = P_N/C$ to replace the absolute population figures P_N. The lowercase population figures p_N represent the fraction (or percentage) of the carrying capacity taken up by the population. We will call these values the *p*-**values** of the population sequence. Note that these *p*-values will always fall between 0 (zero population) and 1 (complete saturation of the habitat).

To express the preceding transition rule in term of *p*-values, we divide both sides by C and get

$$\frac{P_{N+1}}{C} = r \cdot \left(\frac{C - P_N}{C} \right) \cdot \frac{P_N}{C} = r \cdot \left(1 - \frac{P_N}{C} \right) \cdot \frac{P_N}{C}$$

Substituting p_{N+1} and p_N for P_{N+1}/C and P_N/C, respectively, gives the following equation, called the *logistic equation*.

The Logistic Equation

$$p_{N+1} = r \cdot (1 - p_N) \cdot p_N$$

The values of r must be restricted to be between 0 and 4 because for r bigger than 4 the *p-values* can fall outside the 0 to 1 range.

In the examples that follow, we will look at the growth pattern of an imaginary population using the logistic equation. In each case, all we need to get started is the seed p_0 and the growth parameter r. The logistic equation and a good calculator—or better yet, a spreadsheet—will do the rest. A note of warning: The calculations shown in the examples that follow were done with a computer and carried to 16 decimal places before being rounded off to 3 or 4 decimal places. Thus, they may not match exactly the numbers you get out of a calculator.

▶ EXAMPLE 10.18 A Stable Equilibrium

Fish farming is big business these days, so you decide to give it a try. You have access to a large natural pond in which you plan to set up a rainbow trout hatchery. The carrying capacity of the pond is $C = 10{,}000$ fish, and the growth parameter of this type of rainbow trout is $r = 2.5$. We will use the logistic equation to model the growth of the fish population in your pond.

You start by seeding the pond with an initial population of 2000 rainbow trout (that is, 20% of the pond's carrying capacity, or $p_0 = 0.2$). After the first year (trout have an annual hatching season) the population is given by

$$p_1 = 2.5 \times (1 - 0.2) \times (0.2) = 0.4$$

The population of the pond has doubled, and things are looking good! Unfortunately, most of the fish are small fry and not ready to be sent to market. After the second year the population of the pond is given by

$$p_2 = 2.5 \times (1 - 0.4) \times (0.4) = 0.6$$

The population is no longer doubling but the hatchery is still doing well. You are looking forward to even better yields after the third year. But on the third year you get a big surprise:

$$p_3 = 2.5 \times (1 - 0.6) \times (0.6) = 0.6$$

Stubbornly, you wait for better luck the next year, but

$$p_4 = 2.5 \times (1 - 0.6) \times (0.6) = 0.6$$

From the second year on, the hatchery is stuck at 60% of the pond capacity—nothing is going to change unless external forces come into play. We describe this situation as one where the population is at a *stable equilibrium*. Figure 10-8 shows a line graph of the pond's fish population for the first four years. ◀◀

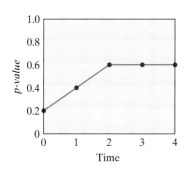

FIGURE 10-8 $r = 2.5$, $p_0 = 0.2$.

▶ EXAMPLE 10.19 An Attracting Point

Consider the same setting as in Example 10.18 (same pond and the same variety of rainbow trout with $r = 2.5$) but suppose you initially seed the pond with 3000 rainbow trout (30% of the pond's carrying capacity). How will the fish population grow if we start with $p_0 = 0.3$?

The first six years of population growth are as follows:

$$p_1 = 2.5 \times (1 - 0.3) \times (0.3) = 0.525$$

$$p_2 = 2.5 \times (1 - 0.525) \times (0.525) \approx 0.6234$$

$$p_3 = 2.5 \times (1 - 0.6234) \times (0.6234) \approx 0.5869$$

$$p_4 = 2.5 \times (1 - 0.5869) \times (0.5869) \approx 0.6061$$

$$p_5 = 2.5 \times (1 - 0.6061) \times (0.6061) \approx 0.5968$$

$$p_6 = 2.5 \times (1 - 0.5968) \times (0.5968) \approx 0.6016$$

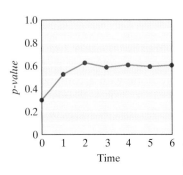

FIGURE 10-9 $r = 2.5, p_0 = 0.3$.

Clearly, something different is happening here. The trout population appears to be fluctuating—up, down, up again, back down—but always hovering near the value of 0.6. We leave it to the reader to verify that as one continues with the population sequence, the p-values inch closer and closer to 0.6 in an oscillating (up, down, up, down, . . .) manner. The value 0.6 is called an *attracting point* of the population sequence. Figure 10-9 shows a line graph of the pond's fish population for the first six years.

EXAMPLE 10.20 Complementary Seeds

In Example 10.19 we seeded the pond at 30% of its carrying capacity ($p_0 = 0.3$). If we seed the pond with the complementary seed $p_0 = 1 - 0.3 = 0.7$, we end up with the same populations.

$$p_1 = 2.5 \times 0.3 \times 0.7 = 2.5 \times 0.7 \times 0.3 = 0.525$$

$$p_2 = 2.5 \times (1 - 0.525) \times (0.525) \approx 0.6234$$

and so on.

Example 10.20 points to a simple but useful general rule about logistic growth—the seeds p_0 and $(1 - p_0)$ always produce the same population sequence. This follows because in the expression $p_1 = r \cdot (1 - p_0) \cdot p_0$, p_0 and $(1 - p_0)$ play interchangeable roles—if you change p_0 to $(1 - p_0)$, then you are also changing $(1 - p_0)$ to p_0. Nothing gained, nothing lost! Once the values of the p_1's are the same, the rest of the p-values follow suit. The moral of this observation is that you should never seed your pond at higher than 50% of its carrying capacity.

EXAMPLE 10.21 The Two-Cycle Pattern

You decided that farming rainbow trout is too difficult. You are moving on to raising something easier—goldfish. The particular variety of goldfish you will grow has growth parameter $r = 3.1$.

Suppose you start by seeding a tank at 20% of its carrying capacity ($p_0 = 0.2$). Following the logistic growth model, the first 16 p-values of the goldfish population are

For the sake of brevity, the details are left to the reader.

$p_0 = 0.2,$	$p_1 = 0.496,$	$p_2 \approx 0.775,$	$p_3 \approx 0.541,$
$p_4 \approx 0.770,$	$p_5 \approx 0.549,$	$p_6 \approx 0.767,$	$p_7 \approx 0.553,$
$p_8 \approx 0.766,$	$p_9 \approx 0.555,$	$p_{10} \approx 0.766,$	$p_{11} \approx 0.556,$
$p_{12} \approx 0.765,$	$p_{13} \approx 0.557,$	$p_{14} \approx 0.765,$	$p_{15} \approx 0.557,$ . . .

An interesting pattern emerges here. After a few breeding seasons, the population settles into a two-cycle, alternating between a high-population period at 0.765 and a low-population period at 0.557. Figure 10-10 convincingly illustrates the oscillating nature of the population sequence.

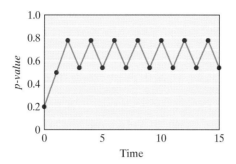

FIGURE 10-10
$r = 3.1$, $p_0 = 0.2$.

Example 10.21 describes a situation not unusual in population biology—animal populations that alternate cyclically between two different levels of population density. Even more complex cyclical patterns are possible when we increase the growth parameter just a little.

> ### EXAMPLE 10.22 A Four-Cycle Pattern

You are now out of the fish-farming business and have acquired an interest in entomology—the study of insects. Let's apply the logistic growth model to study the population growth of a type of flour-beetle with growth parameter $r = 3.5$. The seed will be $p_0 = 0.44$. (There is no particular significance to the choice of the seed—you can change the seed and you will still get an interesting population sequence.)

Following are a few specially selected p-values. We leave it to the reader to verify these numbers and fill in the missing details.

$$p_0 = 0.440, \quad p_1 \approx 0.862, \quad p_2 \approx 0.415, \quad p_3 \approx 0.850,$$
$$p_4 \approx 0.446, \quad p_5 \approx 0.865, \quad \ldots \quad \quad p_{20} \approx 0.497,$$
$$p_{21} \approx 0.875, \quad p_{22} \approx 0.383, \quad p_{23} \approx 0.827, \quad p_{24} \approx 0.501,$$
$$p_{25} \approx 0.875, \quad \ldots$$

It took a while, but we can now see a pattern: Since $p_{25} = p_{21}$, the population will repeat itself in a four-period cycle ($p_{26} = p_{22}$, $p_{27} = p_{23}$, $p_{28} = p_{24}$, $p_{29} = p_{25} = p_{21}$, etc.), an interesting and surprising turn of events. Figure 10-11 shows the line graph of the first 26 p-values.

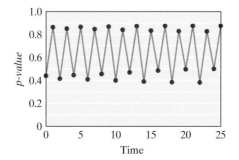

FIGURE 10-11
$r = 3.5$, $p_0 = 0.44$.

The cyclical behavior exhibited in Example 10.22 is not unusual, and many insect populations follow cyclical patterns of various lengths—7-year locusts, 17-year cicadas, and so on.

In the logistic growth model, the highest allowed value of the growth parameter r is $r = 4$. Example 10.23 illustrates what happens in this case.

▶ EXAMPLE 10.23 A Glimpse of Chaos

Let's start with $p_0 = 0.2$ and look at the p-values generated by the logistic equation when $r = 4$. Following are the first 20 p-values.

$$p_0 = 0.2000, \quad p_1 = 0.6400, \quad p_2 \approx 0.9216, \quad p_3 \approx 0.2890,$$

$$p_4 \approx 0.8219, \quad p_5 \approx 0.5854, \quad p_6 \approx 0.9708, \quad p_7 \approx 0.1133,$$

$$p_8 \approx 0.4020, \quad p_9 \approx 0.9616, \quad p_{10} \approx 0.1478, \quad p_{11} \approx 0.5039,$$

$$p_{12} \approx 0.9999, \quad p_{13} \approx 0.0002, \quad p_{14} \approx 0.0010, \quad p_{15} \approx 0.0039,$$

$$p_{16} \approx 0.0157, \quad p_{17} \approx 0.0617, \quad p_{18} \approx 0.2317, \quad p_{19} \approx 0.7121$$

Figure 10-12 is a line graph showing these first 20 p-values.

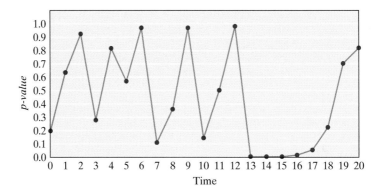

FIGURE 10-12
$r = 4.0$, $p_0 = 0.2$.

The surprise here is the absence of any apparent pattern. In fact, no matter how much further we continue computing p-values, we will find no predictable pattern—to an outside observer the p-values for this population sequence appear to be quite erratic and seemingly random. Of course, we know better—they are all coming from the logistic equation.

The logistic growth model exhibits many interesting surprises. In addition to Exercises 45 through 54 at the end of the chapter, you are encouraged to experiment on your own much like we did in the preceding examples: Choose a p_0 between 0 and 0.5, choose an r between 3 and 4, and fire up your calculator!

An excellent nontechnical account of the surprising patterns produced by the logistic growth model can be found in references 1 and 2. More technical accounts of the logistic equation can be found in references 3, 8, and 9.

Conclusion

In this chapter we studied three simple but important models of population growth.

In the *linear model*, the population is described by an *arithmetic sequence* of the form $P_0, P_0 + d, P_0 + 2d, P_0 + 3d, \ldots$. In each transition period the population grows by the addition of a fixed amount d called the *common difference*. Linear growth is most common in situations where there is no "breeding" such as populations of inanimate objects—commodities, resources, garbage, and so on.

In the *exponential model*, the population is described by a *geometric sequence* of the form $P_0, P_0r, P_0r^2, P_0r^3, \ldots$. In each transition period the population grows by multiplication by a positive constant r called the *common ratio*. When $r > 1$ the population grows exponentially, but when $0 < r < 1$ the exponential growth turns into exponential decay. (When $r = 1$, the population is constant.) Exponential growth is typical of situations in which there is some type of "breeding" in the population and the amount of breeding is directly proportional to the size of the population—money drawing interest in a bank account, the early stages of spread of an epidemic, the decay of radioactive materials, and so on.

The *logistic model* of population growth is described by the logistic equation $p_{N+1} = r(1 - p_N)p_N$. This model is used to describe the growth of biological populations whose growth rate is in direct proportion to the amount of space available in the population's habitat. When confined to a single-species habitat, many animal populations (including human populations) are governed by the logistic model or simple variations of it.

Most serious studies of population growth involve models with much more complicated mathematical descriptions, but to us, that is neither here nor there. Ultimately, the details are not as important as the overall picture: a realization that mathematics can be useful even in its most simplistic forms to describe and predict the rise and fall of populations in many fields—from the human realms of industry, finance, and public health to the natural world of population biology and ecology.

Profile Sir Robert May (1936–)

Robert May was born in Sydney, Australia, in 1936. A man of many talents and interests, May could very well be the most versatile scientist of his generation. After starting his undergraduate studies in chemical engineering, May switched to physics for his doctoral work. He received a Ph.D. in theoretical physics from the University of Sydney in 1959, and in 1962, at the age of 26, he became a professor of physics there. Soon after, he changed direction again, dropping a successful career in physics research to start a new career in population ecology. An ordinary Joe changing careers too many times is called a "flake," but May was no ordinary Joe. By the mid-1970s May had become the world's foremost authority in mathematical ecology, and in 1973 he was appointed Professor of Biology at Princeton University, where he remained until 1988.

While at Princeton, May produced his pioneering work in mathematical models of population growth. The conventional wisdom at the time among ecologists was that to model a complex ecosystem one needs a complicated mathematical model, and the more complexity shown in the system the more sophisticated the equations and variables of the model need to be. The flip side of this was the notion that a simple mathematical model could not possibly reproduce complicated patterns of behavior. In a groundbreaking paper entitled "Simple Mathematical Models with Very Complicated Dynamics" published in the scientific journal *Nature* in 1976 (reference 10), May showed that the extremely simple (by mathematical ecology standards) *logistic growth model* could exhibit surprisingly exotic patterns, such as periodicity, bifurcation, and seemingly random fluctuation. May's 1976 paper paved the way for the development of a new and revolutionary branch of mathematics called *chaos theory* (for details, see reference 2).

After Princeton, May's career took another turn, as he became an important public figure and world advocate for science, the environment, and public policy. Between 1988 and 1995, May served as a Royal Society Research Professor of Zoology at the University of Oxford, England. In 1995 he was appointed Chief Scientific Advisor to the Government of the United Kingdom, and in 2000 he was appointed president of The Royal Society of London, one of the most distinguished scientific posts in the world. May is the recipient of many important prizes and awards, including the prestigious MacArthur "genius" award (1984), the Linnean Medal for Zoology (1991), and the Crafoord Prize in Biosciences (1996). He was awarded the honorary titles of Knight of the United Kingdom (1996), Companion of the Order of Australia (1998), and in 2001 was made a Lord of the United Kingdom for his many contributions to science and society.

Key Concepts

Exercises

WALKING

A. Linear Growth and Arithmetic Sequences

1. Consider a population that grows according to the recursive rule $P_N = P_{N-1} + 125$, with initial population $P_0 = 80$.

    (a) Find P_1, P_2, and P_3.

    (b) Find P_{100}.

    (c) Give an explicit description of the population sequence.

2. Consider a population that grows according to the recursive rule $P_N = P_{N-1} + 23$, with initial population $P_0 = 57$.

    (a) Find P_1, P_2, and P_3.

    (b) Find P_{200}.

    (c) Give an explicit description of the population sequence.

3. Consider a population that grows according to a linear growth model. The initial population is $P_0 = 75$, and the common difference is $d = 5$.

    (a) Find P_{30}.

    (b) How many generations will it take for the population to reach 1000?

    (c) How many generations will it take for the population to reach 1002?

4. Consider a population that decays according to a linear model. The initial population is $P_0 = 520$, and the common difference is $d = -20$.

    (a) Find P_{24}.

    (b) How many generations will it take for the population to reach 10?

    (c) How many generations will it take for the population to become extinct?

5. Consider a population that grows according to a linear growth model. The initial population is $P_0 = 8$, and the population in the 10th generation is $P_{10} = 38$.

(a) Find the common difference d.

(b) Find P_{50}.

(c) Give an explicit description of the population sequence.

6. Consider a population that grows according to a linear growth model. The population in the fifth generation is $P_5 = 37$, and the population in the seventh generation is $P_7 = 47$.

(a) Find the common difference d.

(b) Find the initial population P_0.

(c) Give an explicit description of the population sequence.

7. An arithmetic sequence has $A_1 = 11$ and $A_2 = -4$.

(a) Find A_3.

(b) Find A_0.

(c) How many terms in the sequence are bigger than 30? Explain.

8. An arithmetic sequence has $A_5 = 20$ and $A_6 = -5$.

(a) Find A_7.

(b) Find A_4.

(c) How many terms in the sequence are bigger than 50? Explain.

9. Mr. G. Q. is a snappy dresser and has an incredible collection of neckties. Each month, he buys himself five new neckties. Let P_0 represent the number of neckties he starts out with and P_N be the number of neckties in his collection at the end of the Nth month. Assume that he started out with just three neckties and that he never throws neckties away.

(a) Give a recursive description for P_N.

(b) Give an explicit description for P_N.

(c) Find P_{300}.

10. A nuclear power plant produces 12 lb of radioactive waste every month. The radioactive waste must be stored in a special storage tank. On January 1, 2000, there were 25 lb of radioactive waste in the tank. Let P_N represent the amount of radioactive waste (in pounds) in the storage tank after N months.

(a) Give a recursive description for P_N.

(b) Give an explicit description for P_N.

(c) If the maximum capacity of the storage tank is 500 lb, when will the tank reach its maximum capacity?

11. (a) Find $\underbrace{2 + 5 + 5 + \cdots + 5}_{100 \text{ terms}}$.

(b) Find $\underbrace{2 + 7 + 12 + \cdots}_{100 \text{ terms}}$.

12. (a) Find $\underbrace{21 + 7 + 7 + \cdots + 7}_{57 \text{ terms}}$.

(b) Find $\underbrace{21 + 28 + 35 + \cdots}_{57 \text{ terms}}$.

13. (a) The first two terms of an arithmetic sequence are 12 and 15. The number 309 is which term of the arithmetic sequence?

(b) Find $12 + 15 + 18 + \cdots + 309$.

14. (a) An arithmetic sequence has first term 1 and common difference 9. The number 2701 is which term of the arithmetic sequence?

(b) Find $1 + 10 + 19 + \cdots + 2701$.

15. Consider a population that grows according to a linear growth model. The initial population is $P_0 = 23$, and the common difference is $d = 7$.

(a) Find $P_0 + P_1 + P_2 + \cdots + P_{999}$.

(b) Find $P_{100} + P_{101} + \cdots + P_{999}$.

16. Consider a population that grows according to a linear growth model. The initial population is $P_0 = 7$, and the population in the first generation is $P_1 = 11$.

(a) Find $P_0 + P_1 + P_2 + \cdots + P_{500}$.

(b) Find $P_{100} + P_{101} + \cdots + P_{500}$.

17. The city of Lightsville currently has 137 streetlights. As part of an urban renewal program, the city council has decided to install and have operational 2 additional streetlights at the end of each week for the next 52 weeks. Each streetlight costs $1 to operate for 1 week.

(a) How many streetlights will the city have at the end of 38 weeks?

(b) How many streetlights will the city have at the end of N weeks? (Assume $N \leq 52$.)

(c) What is the cost of operating the original 137 lights for 52 weeks?

(d) What is the additional cost for operating the newly installed lights for the 52-week period during which they are being installed?

18. A manufacturer currently has on hand 387 widgets. During the next 2 years, the manufacturer will be increasing his inventory by 37 widgets per week. (Assume that there are exactly 52 weeks in one year.) Each widget costs 10 cents a week to store.

(a) How many widgets will the manufacturer have on hand after 20 weeks?

(b) How many widgets will the manufacturer have on hand after N weeks? (Assume $N \leq 104$.)

(c) What is the cost of storing the original 387 widgets for 2 years (104 weeks)?

(d) What is the additional cost of storing the increased inventory of widgets for the next 2 years?

B. Exponential Growth and Geometric Sequences

19. A population grows according to an exponential growth model. The initial population is $P_0 = 11$ and the common ratio is $r = 1.25$.

  (a) Find P_1.

  (b) Find P_9.

  (c) Give an explicit formula for P_N.

20. A population grows according to an exponential growth model, with $P_0 = 8$ and $P_1 = 12$.

  (a) Find the common ratio r.

  (b) Find P_9.

  (c) Give an explicit formula for P_N.

21. A population grows according to the recursive rule $P_N = 4P_{N-1}$, with initial population $P_0 = 5$.

  (a) Find P_1, P_2, and P_3.

  (b) Give an explicit formula for P_N.

  (c) How many generations will it take for the population to reach 1 million?

22. A population *decays* according to an exponential model, with $P_0 = 3072$ and common ratio $r = 0.75$.

  (a) Find P_5.

  (b) Give an explicit formula for P_N.

  (c) How many generations will it take for the population to fall below 200?

23. Crime in Happyville is on the rise. Each year the number of crimes committed increases by 50%. Assume that there were 200 crimes committed in 2000, and let P_N denote the number of crimes committed in the year $2000 + N$.

  (a) Give a recursive description of P_N.

  (b) Give an explicit description of P_N.

  (c) If the trend continues, approximately how many crimes will be committed in Happyville in the year 2010?

24. Since 2000, when 100,000 cases were reported, each year the number of new cases of equine-flu has decreased by 20%. Let P_N denote the number of new cases of equine flu in the year $2000 + N$.

  (a) Give a recursive description of P_N.

  (b) Give an explicit description of P_N.

  (c) If the trend continues, approximately how many new cases of equine flu will be reported in the year 2015?

25. Consider the geometric sequence with first term $P_0 = 3$ and common ratio $r = 2$.

  (a) Find P_{100}.

  (b) Give an explicit formula for P_N.

  (c) Find $P_0 + P_1 + \cdots + P_{100}$.

  (d) Find $P_{50} + P_{51} + \cdots + P_{100}$.

26. Consider the geometric sequence with first four terms 1, 3, 9, and 27.

  (a) Find P_{100}.

  (b) Give an explicit formula for P_N.

  (c) Find $P_0 + P_1 + \cdots + P_{100}$.

  (d) Find $P_{50} + P_{51} + \cdots + P_{100}$.

C. Financial Applications

27. You have a coupon worth 15% off any item (including sale items) in a store. The particular item you want is on sale at 30% off the marked price of $100. The store policy allows you to use your coupon before the 30% discount or after the 30% discount. (You can take 15% off the marked price first and then take 30% off the resulting price, or you can take 30% off the marked price first and then take 15% off the resulting price.)

  (a) What is the dollar amount of the discount in each case?

  (b) What is the total percentage discount in each case?

  (c) Suppose the article costs P dollars (instead of $100). What is the percentage discount in each case?

28. A membership store gives a 10% discount on all purchases to its members. If the store marks each item up 50% (based on its cost), what is the markup actually realized by the store when an item is sold to a member?

29. Suppose you deposit $3250 in a savings account that pays 9% annual interest, with interest credited to the account at the end of each year. Assuming that no withdrawals are made, how much money will be in the account after four years?

30. Suppose you deposit $1237.50 in a savings account that pays 8.25% annual interest, with interest credited to the account at the end of each year. Assuming that no withdrawals are made, how much money will be in the account after three years?

31. Suppose you deposited $3420 on January 1, 2006, in an account paying $6\frac{5}{8}$% annual interest, with interest credited to the account on December 31 of each year. On January 1, 2009, the interest rate drops to $5\frac{3}{4}$%. What will be the balance in your account on January 1, 2013?

32. Suppose you deposited $2500 on January 1, 2004, in an account paying $5\frac{3}{8}$% annual interest, with interest credited to the account on December 31 of each year. On January 1, 2008, the interest rate drops to $4\frac{3}{4}$%. What will be the balance in your account on January 1, 2011?

33. Suppose you deposited $3420 on January 1, 2003, in a savings account paying $6\frac{5}{8}$% annual interest, with interest credited to the account on December 31 of each year. On January 1, 2005, you withdrew $1500, and on January 1, 2006, you withdrew $1000. If you make no other withdrawals, what will be the balance in your account on January 1, 2009?

34. Suppose you deposited $2500 on January 1, 2003, in a savings account paying $5\frac{3}{8}\%$ annual interest, with interest credited to the account on December 31 of each year. On January 1, 2006, you withdrew $850. If you make no other withdrawals, what will be the balance in your account on January 1, 2011?

35. Suppose $5000 is deposited in a savings account that pays 12% annual interest compounded monthly.

(a) What is the monthly interest rate on this account?

(b) Assuming that no withdrawals are made, how much money will be in the account after five years?

(c) What is the annual yield on this account?

36. Suppose $874.83 is deposited in a savings account that pays $7\frac{3}{4}\%$ annual interest compounded daily.

(a) What is the daily interest rate on this account?

(b) Assuming that no withdrawals are made, how much money will be in the account after two years?

(c) What is the annual yield on this account?

37. You have some money to invest. The Great Bulldog Bank offers accounts that pay 6% annual interest compounded yearly. The First Northern Bank offers accounts that pay 5.75% annual interest compounded monthly. The Bank of Wonderland offers 5.5% annual interest compounded daily. What is the annual yield for each bank?

38. Complete the following table.

Annual interest rate	Compounded	Annual yield
12%	Yearly	12%
12%	Semiannually	?
12%	Quarterly	?
12%	Monthly	?
12%	Daily	?
12%	Hourly	?

39. You have a small inheritance ($1000) that you wish to invest in a long term investment (25 years). You are considering three options: Bank A offers savings accounts with an annual yield of 10%, bank B offers savings accounts with an annual yield of 10.2%, and bank C offers savings accounts with an annual yield of 9.96%. Assuming you make no withdrawals, how much money will there be in the savings account after 25 years if you deposit $1000 in

(a) bank A?

(b) bank B?

(c) bank C?

You might as well assume these banks are in Timbuktu—American banks don't offer anything close to these annual yields.

40. Your bank is offering a special promotion for its preferred customers. If you buy a $500 certificate of deposit (CD), at the end of the year you can cash the CD for $555. What is the annual yield of this investment?

41. You decide to open a Christmas Club account at a bank that pays 6% annual interest compounded monthly. You deposit $100 on the first of January and on the first of each succeeding month through November. How much will you have in your account on the first of December?

42. You decide to save money to buy a car by opening a special account at a bank that pays 8% annual interest compounded monthly. You deposit $300 on the first of each month for 36 months. How much will you have in your account at the end of the 36th month?

43. You are interested in buying a car five years from now, and you estimate that the future cost of the car will be $10,000. You decide to deposit enough money today so that five years from now you will have the $10,000. How much money do you need to deposit in a savings account that offers a 10% interest rate compounded

(a) annually?

(b) quarterly?

(c) monthly?

44. You have $1000 to invest. Suppose you find an investment that guarantees an $8\frac{1}{2}\%$ annual yield, with the interest paid once a year at the end of the year. How many years will it take for you to at least double your original investment?

D. Logistic Growth Model

45. A population grows according to the logistic growth model, with growth parameter $r = 0.8$. Starting with an initial population given by $p_0 = 0.3$,

(a) find p_1.

(b) find p_2.

(c) determine what percent of the habitat's carrying capacity is taken up by the third generation.

46. A population grows according to the logistic growth model, with growth parameter $r = 0.6$. Starting with an initial population given by $p_0 = 0.7$,

(a) find p_1.

(b) find p_2.

(c) determine what percent of the habitat's carrying capacity is taken up by the third generation.

47. For the population discussed in Exercise 45 ($r = 0.8$, $p_0 = 0.3$),

(a) find the values of p_1 through p_{10}.

(b) what does the logistic growth model predict in the long term for this population?

48. For the population discussed in Exercise 46 ($r = 0.6$, $p_0 = 0.7$),

(a) find the values of p_1 through p_{10}.

(b) what does the logistic growth model predict in the long term for this population?

49. A population grows according to the logistic growth model, with growth parameter $r = 1.8$. Starting with an initial population given by $p_0 = 0.4$,

    (a) find the values of p_1 through p_{10}.

    (b) what does the logistic growth model predict in the long term for this population?

50. A population grows according to the logistic growth model, with growth parameter $r = 1.5$. Starting with an initial population given by $p_0 = 0.8$,

    (a) find the values of p_1 through p_{10}.

    (b) what does the logistic growth model predict in the long term for this population?

51. A population grows according to the logistic growth model, with growth parameter $r = 2.8$. Starting with an initial population given by $p_0 = 0.15$,

    (a) find the values of p_1 through p_{10}.

    (b) what does the logistic growth model predict in the long term for this population?

52. A population grows according to the logistic growth model, with growth parameter $r = 2.5$. Starting with an initial population given by $p_0 = 0.2$,

    (a) find the values of p_1 through p_{10}.

    (b) what does the logistic growth model predict in the long term for this population?

53. A population grows according to the logistic growth model, with growth parameter $r = 3.25$. Starting with an initial population given by $p_0 = 0.2$,

    (a) find the values of p_1 through p_{10}.

    (b) what does the logistic growth model predict in the long term for this population?

54. A population grows according to the logistic growth model, with growth parameter $r = 3.51$. Starting with an initial population given by $p_0 = 0.4$,

    (a) find the values of p_1 through p_{10}.

    (b) what does the logistic growth model predict in the long term for this population?

E. Miscellaneous

55. Each of the following sequences follows either a linear, exponential, or a logistic growth model. For each sequence, determine which model applies (if more than one applies, then indicate all the ones that apply).

    (a) $2, 4, 8, 16, 32, \ldots$

    (b) $2, 4, 6, 8, 10, \ldots$

    (c) $0.8, 0.4, 0.6, 0.6, 0.6, \ldots$

    (d) $0.81, 0.27, 0.09, 0.03, 0.01, \ldots$

    (e) $0.49512, 0.81242, 0.49528, 0.81243, 0.49528, \ldots$

    (f) $0.9, 0.75, 0.6, 0.45, 0.3, \ldots$

    (g) $0.7, 0.7, 0.7, 0.7, 0.7, \ldots$

56. Each of the following graphs describes a population that grows according to a linear, exponential, or logistic model. For each line graph, determine which model applies.

(a)

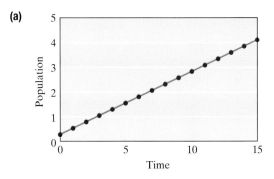

(b)

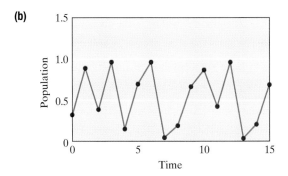

(c)

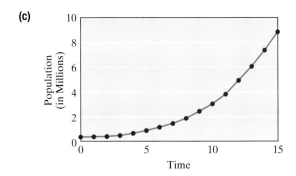

(d)

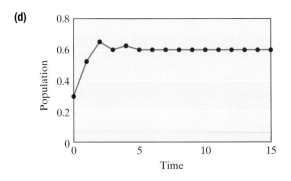

(e)

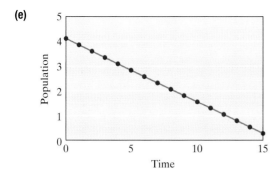

(f)

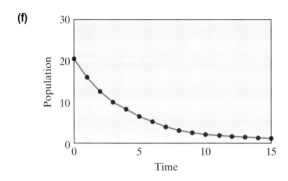

57. A population of laboratory rats grows according to the following transition rule: $P_N = P_{N-1} + 2P_{N-2}$. The initial population is $P_0 = 6$, and the population in the first generation is $P_1 = 10$.

(a) Find P_2 and P_3.

(b) Explain why there is always an even number of rats.

58. A population of guinea pigs grows according to the following transition rule: $P_N = 2P_{N-1} - P_{N-2}$. The initial population is $P_0 = 3$, and the population in the first generation is $P_1 = 5$.

(a) Find P_2 and P_3.

(b) Explain why there is always an odd number of guinea pigs.

(c) Give an explicit description of the population sequence.

JOGGING

59. You buy a $500 certificate of deposit (CD), and at the end of two years you cash it for $561.80. What is the annual yield of this investment?

60. What annual interest rate compounded semiannually gives an annual yield of 21%?

61. **(a)** Find a right triangle whose sides are consecutive terms of an arithmetic sequence with common difference $d = 2$.

(b) Find a right triangle whose sides are consecutive terms of a geometric sequence with common ratio r.

62. How much should a retailer mark up her goods so that when she has a 25% off sale, the resulting prices will still reflect a 50% markup (on her cost)?

63. For three consecutive years the tuition at Tasmania State University increased by 10%, 15%, and 10%, respectively. What was the total percentage increase overall during the three-year period?

64. **(a)** The *Happyville Gazette* wants to sign a one-year contract with a Web service provider to have each edition (including back issues) of its newspaper available online. The contract specifies the cost of storage will be 2 cents per edition per day. Determine the cost of this one-year contract.

(b) Suppose that the *Gazette* also needs to hire a Web designer to format the online newspaper. The Web designer charges $3000 for the first month, and his fee increases by 2% every month thereafter. Determine the yearly cost of hiring the Web designer.

65. Before Annie set off for college, Daddy Warbucks offered her a choice between the following two incentive programs:

- **Option 1.** A $100 reward for every A she gets in a college course

- **Option 2.** One cent for her first A, 2 cents for the second A, 4 cents for the third A, 8 cents for the fourth A, and so on

Annie chose Option 1. After getting a total of 30 A's in her college career, Annie is happy with her reward of $100 \times 30 = \$3000$. Unfortunately, Annie did not get an A in math. Help her figure out how much she would have made had she chosen option 2.

66. Compute the sum

$$1 + 1 + 2 + \frac{1}{2} + 4 + \frac{1}{4} + 8 + \frac{1}{8} + \cdots + 4096 + \frac{1}{4096}$$

67. Suppose that $P_0, P_1, P_2, \ldots, P_N$ are the terms of a geometric sequence. Suppose, moreover, that the sequence satisfies the recursive rule $P_N = P_{N-1} + P_{N-2}$, for $N \geq 2$. Find the common ratio r.

68. The purpose of this exercise is to fill in all the missing details of Example 10.17.

(a) In Example 10.17 it is claimed that

$$100(1.005)^{216} + 100(1.005)^{215} + \cdots + 100(1.005) =$$
$$\frac{100(1.005)(1.005^{216} - 1)}{0.005}$$

Use the geometric sum formula to explain why the preceding statement is true. (What is a? What is r?)

(b) If $S = a(1.005)^{216} + a(1.005)^{215} + \cdots + a(1.005)$, then express the sum
$b(1.005)^{216} + b(1.005)^{215} + \cdots + b(1.005)$
in terms of a, b, and S.

(c) If you deposit $400 each month in a savings account that pays 6% annual interest compounded monthly and leave all the money in the account, at the end of nine years you will have a total of

$\$\dfrac{400(1.005)(1.005^{108} - 1)}{0.005}$. Fill in the details.

69. Show that the geometric sum formula

$$a + ar + ar^2 + \cdots + ar^{N-1} = \frac{a(r^N - 1)}{r - 1} \text{ holds for all}$$

a and all $r \neq 1$.

70. Show that the sum of the first N terms of an arithmetic sequence with first term c and common difference d is

$$\frac{N}{2}[2c + (N - 1)d].$$

71. Give an example of a geometric sequence in which P_0, P_1, P_2, and P_3 are integers, and all the terms from P_4 on are fractions.

72. Consider a population that grows according to the logistic growth model with initial population given by $p_0 = 0.7$. What growth parameter r would keep the population constant?

73. Suppose that you are in charge of stocking a lake with a certain type of alligator with a growth parameter $r = 0.8$. Assuming that the population of alligators grows according to the logistic growth model, is it possible for you to stock the lake so that the alligator population is constant? Explain.

74. Consider a population that grows according to the logistic growth model with growth parameter $r(r > 1)$. Find p_0 in terms of r so that the population is constant.

75. Suppose the habitat of a population of snails has a carrying capacity of $C = 20,000$ and the current population is 5000. Suppose also that the growth parameter for this particular type of snail is $r = 3.0$. What does the logistic growth model predict for this population after four transition periods?

RUNNING

76. You are purchasing a home for $120,000 and are shopping for a loan. You have a total of $31,000 to put down, including the closing costs of $1000 and any loan fee that might be charged. Bank A offers a 10%-annual-interest loan amortized over 30 years with 360 equal monthly payments. There is no loan fee. Bank B offers a 9.5%-annual-interest loan amortized over 30 years with 360 equal monthly payments. There is a 3% loan fee (i.e., a one-time up-front charge of 3% of the loan). Which loan is better?

77. A friend of yours sells his car to a college student and takes a personal note (cosigned by the student's rich uncle) for $1200 with no interest, payable at $100 per month for 12 months. Your friend immediately approaches you and offers to sell you this note. How much should you pay for the note if you want an annual yield of 12% on your investment?

78. The purpose of this exercise is to understand why we assume that, under the logistic growth model, the growth parameter r is between 0 and 4.

(a) What does the logistic equation give for p_{N+1} if $p_N = 0.5$ and $r > 4$? Is this a problem?

(b) What does the logistic equation predict for future generations if $p_N = 0.5$ and $r = 4$?

(c) If $0 \leq p \leq 1$, what is the largest possible value of $(1 - p)p$?

(d) Explain why, if $0 < p_0 < 1$ and $0 < r < 4$, then $0 < p_N < 1$, for every positive integer N.

79. Suppose $r > 3$. Using the logistic growth model, find a population p_0 such that $p_0 = p_2 = p_4 \ldots$, but $p_0 \neq p_1$.

80. Show that if $P_0, P_1, P_2, \ldots$ is an arithmetic sequence, then $2^{P_0}, 2^{P_1}, 2^{P_2}, \ldots$ must be a geometric sequence.

Projects and Papers

A. The Many Faces of e

Just as the golden ratio ϕ is tied to spiral growth in nature (see Chapter 9), another famous irrational number—the number e—is tied to many continuous models of population growth. (For example, if $1 is deposited in an account which earns 100% interest and is compounded *continuously*, the $1 will grow to e [approximately $2.72] after one year.) The irrational number e has many remarkable mathematical properties. In this project you are asked to present and discuss five of the more interesting mathematical properties of e. For each property, give a historical background (if possible), a simple mathematical explanation (avoid technical details if you can), and a real-life application (if possible).

B. The Malthusian Doctrine

In 1798, Thomas Malthus wrote his famous *Essay on the Principle of Population*. In this essay, Malthus put forth the principle that population grows according to an exponential growth model, whereas food and resources grow according to a linear growth model. Based on this doctrine, Malthus predicted that humankind was doomed to a future where the supply of food and other resources would be unable to keep pace with the needs of the world's population.

Write an analysis paper detailing some of the consequences of Malthus's doctrine. Does the doctrine apply in a modern technological world? Can the doctrine be the explanation for the famines in sub-Saharan Africa? Discuss the many possible criticisms that can be leveled against Malthus's doctrine. To what extent do you agree with Malthus's doctrine?

C. The Logistic Equation and the United States Population

The logistic growth model, first discovered by Verhulst, was rediscovered in 1920 by the American population ecologists Raymond Pearl and Lowell Reed. Pearl and Reed compared the population data for the United States between 1790 and 1920 with what would be predicted using a logistic equation and found that the numbers produced by the equation and the real data matched quite well.

In this project, you are to discuss and analyze Pearl and Reed's 1920s paper (reference 12). Here are some suggested questions you may want to discuss: Is the logistic model a good model to use with human populations? What might be

a reasonable estimate for the carrying capacity of the United States? What happens with Pearl and Reed's model when you expand the census population data all the way to the latest population figures available?
(Note: Current and historical U.S. population data can be found at www.census.gov.)

D. World Population Growth

How do demographers model world population? Is this different from how they model, say, the population of the United States? How does this process compare with that used by biologists in determining the size of a future salmon spawn?

In this project, you will compare and contrast the process demographers use to model human population growth with that which biologists use to model animal populations.

References and Further Readings

1. Cipra, Barry, "Beetlemania: Chaos in Ecology," in *What's Happening in the Mathematical Sciences 1998–1999*. Providence, RI: American Mathematical Society, 1999.

2. Gleick, James, *Chaos: Making a New Science*. New York: Viking Penguin, Inc., 1987, chap. 3.

3. Gordon, W. B., "Period Three Trajectories of the Logistic Map," *Mathematics Magazine*, 69 (1996), 118–120.

4. Hoppensteadt, Frank, *Mathematical Methods of Population Biology*. Cambridge: Cambridge University Press, 1982.

5. Hoppensteadt, Frank, *Mathematical Theories of Populations: Demographics, Genetics and Epidemics*. Philadelphia: Society for Industrial and Applied Mathematics, 1975.

6. Hoppensteadt, Frank, and Charles Peskin, *Mathematics in Medicine and the Life Sciences*. New York: Springer-Verlag, 1992.

7. Kingsland, Sharon E., *Modeling Nature: Episodes in the History of Population Ecology*. Chicago: University of Chicago Press, 1985.

8. Maor, Eli, *e: The Story of a Number*. Princeton, NJ: Princeton University Press, 1998.

9. May, Robert M., "Biological Populations with Nonoverlapping Generations: Stable Points, Stable Cycles and Chaos," *Science*, 186 (1974), 645–647.

10. May, Robert M., "Simple Mathematical Models with Very Complicated Dynamics," *Nature*, 261 (1976), 459–467.

11. May, Robert M., and George F. Oster, "Bifurcations and Dynamic Complexity in Simple Ecological Models," *American Naturalist*, 110 (1976), 573–599.

12. Pearl, Raymond, and Lowell J. Reed, "On the Rate of Growth of the Population of the United States since 1790 and Its Mathematical Representation," *Proceedings of the National Academy of Sciences USA*, 6 (June 1920), 275–288.

13. Sigler, Laurence, *Fibonacci's Liber Abaci*. New York: Springer-Verlag, 2002.

14. Smith, J. Maynard, *Mathematical Ideas in Biology*. Cambridge: Cambridge University Press, 1968.

15. Swerdlow, Joel, "Population," *National Geographic* (October 1998), 2–35.

11

Symmetry

Mirror, Mirror, Off the Wall . . .

> Symmetry is a vast subject, significant in art and nature. Mathematics lies at its root, and it would be hard to find a better one on which to demonstrate the working of the mathematical intellect.
>
> Hermann Weyl

It is said that Eskimos have dozens of different words for ice. Ice is, after all, a universal theme in the Eskimo's world. By the same token one would expect our own vocabulary to have dozens of different words to describe the notion of symmetry—symmetry is as pervasive to our world as ice is to the Eskimos'. Surprisingly, just the opposite is the case. In everyday conversation we use a single word—*symmetry*—to cover an incredibly diverse set of situations and ideas.

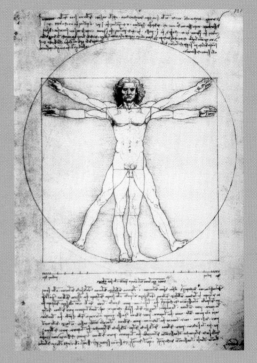

xactly what is symmetry? The answer depends very much on the context of the question. In everyday language, *symmetry* is often taken to mean *mirror* or *bilateral symmetry*, the left-right symmetry exhibited (at least on the outside) by the human body. But in everyday language the word *symmetry* also has an aesthetic connotation—a snowflake is often used as the shining example of symmetry because it is so well proportioned and beautiful. This is often how the word is used in art and architecture. Along similar lines, the word *symmetry* is used in music, poetry, and literature to describe particular elements of style. In all of these contexts, symmetry is perceived as something good—even when we can't put our finger on exactly what it is.

At its very heart, however, symmetry is a mathematical notion, and to grasp the full meaning and relevance of symmetry in our everyday world it is helpful to have some understanding of the mathematics behind it. In this chapter we will take a look at symmetry from a mathematical perspective. For technical reasons (i.e., to keep the theory reasonably simple) we will focus on symmetry in the context of two-dimensional objects and shapes. (Since we live in a three-dimensional world, the objects we discuss will be only idealized versions of reality.)

The chapter starts by introducing the concept of a *rigid motion* (Section 11.1), a fundamental idea behind the mathematical definition of symmetry. There are four basic types of rigid motions of two-dimensional objects living in a plane, and these are discussed in the next four sections—*reflections* in Section 11.2, *rotations* in Section 11.3, *translations* in Section 11.4, and *glide reflections* in Section 11.5. In Section 11.6 we use the rigid motions to formalize the mathematical interpretation of symmetry and to develop the concept of the *symmetry type* of a shape or object. In the last section (Section 11.7) we introduce *border* and *wallpaper patterns* and discuss their classification in terms of symmetry types.

11.1 Rigid Motions

As are many other core concepts, symmetry is rather hard to define, and we will not even attempt a proper definition until Section 11.6. We will start our discussion with just an informal stab at the mathematical (or geometric if you prefer) interpretation of symmetry.

EXAMPLE 11.1 Symmetries of a Triangle

Figure 11-1 shows three triangles: (a) a scalene triangle (all three sides are different), (b) an isosceles triangle, and (c) an equilateral triangle. In terms of symmetry, how do these triangles differ? Which one is the most symmetric? Least symmetric? (What do you think?)

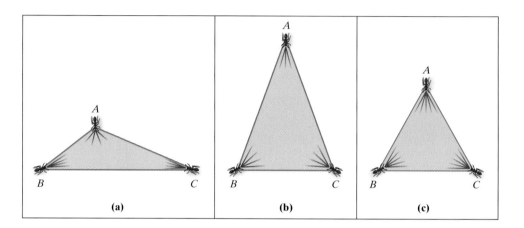

FIGURE 11-1

Even without a formal understanding of what symmetry is, most people would answer that the equilateral triangle in (c) is the most symmetric, and the scalene triangle in (a) is the least symmetric. This is in fact correct, but why? Think of an imaginary observer—say a tiny (but very observant) ant—standing at the vertices of each of the triangles, looking toward the opposite side. In the case of the scalene triangle (a), the view from each vertex is different. In the case of the isosceles triangle (b), the view from vertices B and C is the same, but the view from vertex A is different. In the case of the equilateral triangle (c), the view is the same from each of the three vertices.

Let's say, for starters, that *symmetry* is a property of an object that looks the same to an observer standing at different vantage points. This is still pretty vague but a start nonetheless. Now instead of talking about an observer moving around to different vantage points think of the object itself moving—forget the observer. For example, saying that the isosceles triangle (b) looks the same to an observer whether he stands at vertex B or vertex C is equivalent to saying that triangle (b) itself can be moved so that vertices B and C swap locations and the triangle as a whole looks exactly as before. Thus, we can think of symmetry as a property related to an object that can be moved in such a way that when all the moving is done, the object looks exactly as it did before.   «

Given the preceding observations, it is not surprising that to understand symmetry we need to understand the various ways in which we can "move" an object. This is our lead-in to what is a key concept in this chapter—the notion of a *rigid motion*.

The act of taking an object and moving it from some starting position to some ending position *without altering its shape or size* is called a **rigid motion** (and sometimes an *isometry*). If, in the process of moving the object, we stretch it, tear it, or generally alter its shape or size, then the motion is *not* a rigid motion. Since in a rigid motion the size and shape of an object are not altered, dis-

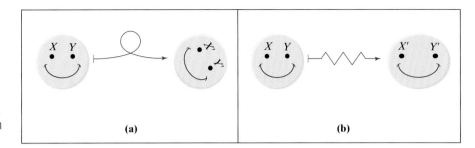

FIGURE 11-2 (a) A rigid motion preserves distances between points. (b) If the shape is altered, the motion is not rigid.

tances between points are preserved: *The distance between any two points X and Y in the starting position is the same as the distance between the same two points in the ending position.* Figure 11-2 illustrates the idea behind the concept of a rigid motion.

In defining rigid motions we are completely result oriented. We are only concerned with the *net effect* of the motion—where the object started and where the object ended. What happens during the "trip" is irrelevant. This implies that a rigid motion is completely defined by the starting and ending positions of the object being moved, and two rigid motions that move an object from a starting position A to an ending position B are **equivalent rigid motions**—never mind the details of how they go about it.

Because rigid motions are defined strictly in terms of their net effect, there is a surprisingly small number of scenarios. In the case of two-dimensional objects in a plane, there are only four possibilities: Every rigid motion is equivalent to a *reflection*, a *rotation*, a *translation*, or a *glide reflection*. We will call these four types of rigid motions the **basic rigid motions of the plane**.

A rigid motion of the plane—let's call it $\mathcal{M}$—moves each point in the plane from its starting position P to an ending position P', also in the plane. (From here on we will use script letters such as $\mathcal{M}$ and $\mathcal{N}$ to denote rigid motions—this should eliminate any possible confusion between the point M and the rigid motion $\mathcal{M}$.) We will call the point P' the **image** of the point P under the rigid motion $\mathcal{M}$ and describe this informally by saying that $\mathcal{M}$ *moves P to P'*. (We will also stick to the convention that the image point has the same label as the original point but with a prime symbol added.) It may happen that a point P is moved back to itself under $\mathcal{M}$, in which case we call P a **fixed point** of the rigid motion $\mathcal{M}$.

We will now discuss each of the basic rigid motions in the plane in a little more detail.

In three-dimensional space there are six basic rigid motions: reflection, rotation, translation, and glide reflection plus *rotary reflection* and *screw displacement*—see Project B.

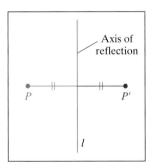

FIGURE 11-3

11.2 Reflections

A **reflection** in the plane is a rigid motion that moves an object into a new position that is a mirror image of the starting position. In two dimensions, the "mirror" is a line called the **axis** of reflection.

From a purely geometric point of view a reflection can be defined by showing how it moves a generic point P in the plane. This is shown in Fig. 11-3: The image of any point P is found by drawing a line through P perpendicular to the axis l and finding the point P' on the opposite side of l at the same distance as P from l. Points on the axis itself are fixed points of the reflection.

▶ EXAMPLE 11.2 Reflections of a Triangle

Figure 11-4 shows three cases of reflection of a triangle ABC. In all cases the reflected triangle $A'B'C'$ is shown in red. In (a) the axis of reflection l does not intersect the triangle ABC. In (b) the axis of reflection l cuts through the triangle ABC—here the points where l intersects the triangle are fixed points of the triangle. In (c) the reflected triangle $A'B'C'$ falls on top of the original triangle ABC. The vertex B is a fixed point of the triangle, but the vertices A and C swap positions under the reflection.

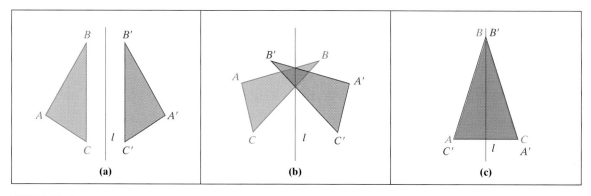

(a)   (b)   (c)

FIGURE 11-4

The following is a list of some basic but useful facts about reflections.

> A reflection is completely determined by its axis l.

The preceding fact follows directly from the observation that if we know the axis of reflection l, then we know the image of any point P.

> A reflection is completely determined by a single point-image pair P and P' (as long as P is not a fixed point).

The preceding fact follows from the observation that if we have a point P and its image P', then we can find the axis of the reflection—it is the perpendicular bisector of the segment joining the two points.

> A reflection is an *improper* rigid motion.

The preceding fact represents one of the most important properties of a reflection. It basically means that reflections change orientations—when reflected, a left hand becomes a right hand [Fig. 11-5(a)], and the hands of a reflected clock move counterclockwise instead of clockwise [Fig. 11-5(b)]. Any rigid motion that reverses left-right and clockwise-counterclockwise orientations is called an **improper** rigid motion.

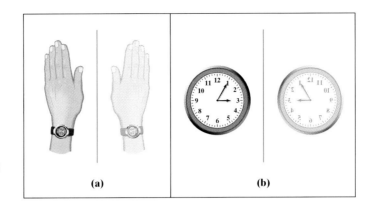

(a)   (b)

FIGURE 11-5 Reflections are *improper*—the image of a left hand is a right hand, and the image of a clock is a "counterclock."

If the same reflection is applied twice, every point ends up exactly where it started.

The preceding fact essentially says that if P' is the image of P under a reflection, then $(P')' = P$ (the image of the image is the original point). Thus, the net effect of applying the same reflection twice is the same as *not having moved the object at all.*

This leads us to an interesting semantic question: Is not moving an object at all itself a rigid motion? On the one hand, it seems rather absurd to say yes. If we are talking about motion, then there should be some kind of movement, however small. On the other hand, we are equally compelled to argue that the result of combining two (or more) consecutive rigid motions should itself be a rigid motion regardless of what the net effect is. If this is the case, then combining two consecutive reflections with the same axis (which produces the same result as no motion at all) should be a rigid motion.

The latter is the mathematically correct way to look at it, and we will adopt the convention that one of the possible rigid motions of an object is equivalent to not moving it at all. This "nothing doing" rigid motion is called the **identity motion**. From here on, when we speak of moving an object under some generic rigid motion we must allow for the possibility that it is the identity motion and that in fact the object has not moved at all.

11.3 Rotations

Informally, a rotation in the plane is a rigid motion that pivots or swings an object around a fixed point O. A rotation is defined by giving the **rotocenter** (the point O that acts as the center of the rotation), and the **angle of rotation** (actually the *measure* of an angle indicating the amount of rotation). The angle of rotation is given in either degrees or radians (we will use degrees, but converting degrees to radians or radians to degrees is easy if you just remember that 180 degrees equals π radians). In addition, it is necessary to specify the direction (clockwise or counterclockwise) associated with the angle of rotation.

Figure 11-6 illustrates geometrically how a *clockwise* rotation with rotocenter O and angle of rotation α moves a point P to the point P'.

FIGURE 11-6

EXAMPLE 11.3 Rotations of a Triangle

Figure 11-7 illustrates three cases of rotation of a triangle ABC. In all cases the reflected triangle $A'B'C'$ is shown in red. In (a) the rotocenter O lies outside the triangle ABC. The 90° clockwise rotation moved the triangle from the "12 o'clock position" to the "3 o'clock position." (Note that a 90° counterclockwise rotation would have moved the triangle ABC to the "9 o'clock position".) In (b) the rotocenter O is at the center of the triangle ABC. The 180° rotation turns the triangle "upside down." For obvious reasons, a 180° rotation is often called a *half-turn*. (With half-turns the result is the same whether we rotate clockwise or counterclockwise, so it is unnecessary to specify a direction.) In (c) the 360° rotation moves every point back to its original position—from the rigid motion point of view it's as if the triangle had not moved.

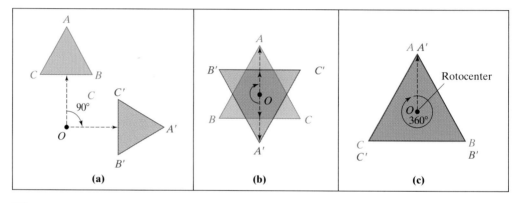

FIGURE 11-7

The last observation leads to the following reformulation of the well-known fact (at least to gymnasts, figure skaters, skateboarders, and assorted spinners) that a 360° rotation puts you back exactly where you started.

A 360° rotation is equivalent to the identity motion.

The preceding fact has several useful consequences. First, any rotation by an angle more than 360° is equivalent to another rotation with the same center by an angle between 0° and 360°—all we have to do is divide the angle by 360 and take the remainder. For example, as a rigid motion, a clockwise rotation by 759° is equivalent to a clockwise rotation by an angle of 39°, because 759 divided by 360 gives a quotient of 2 and a remainder of 39. Second, any rotation specified in a clockwise orientation can just as well be specified in a counterclockwise orientation.

A common misconception is to confuse a 180° rotation with a reflection, but the two are very different. A rotation, unlike a reflection, is always a **proper** rigid motion—in other words, it preserves left-right and clockwise-counterclockwise orientations within the rotated object.

A rotation is a *proper* rigid motion.

If you are given a point P and its image P' and told that the rigid motion that moves P to P' is a rotation, can you determine which rotation? The answer is *no*—any point located on the perpendicular bisector of the segment PP' can be the rotocenter of some rotation that moves P to P', as shown in Fig. 11-8(a). Given a second pair of points Q and Q', we can identify the rotocenter O of the rotation: It is the point where the perpendicular bisectors of PP' and QQ' meet, as shown in Fig. 11-8(b). [In the special case where PP' and QQ' happen to have the same perpendicular bisector, as in Fig. 11-8(c), the rotocenter O is the intersection of PQ and $P'Q'$.] Once we have identified the rotocenter O, the angle of rotation α is given by the measure of angle POP' (or for that matter QOQ'—they are the same).

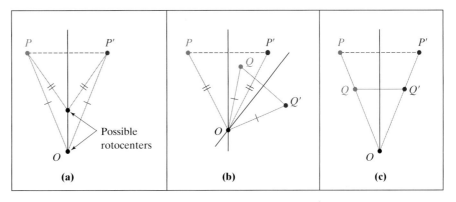

FIGURE 11-8 Identifying the rotocenter requires at least two pairs of points P, P' and Q, Q'.

A rotation is completely determined by *two* point-image pairs P, P' and Q, Q'.

The rotocenter is a fixed point of a rotation, and, with one exception, it is the only fixed point. The exception is the case when the rotation is the identity motion (i.e., any rotation equivalent to a 360° rotation). In this case every point is a fixed point.

A rotation that is not the identity motion has only one fixed point—the rotocenter O.

Wrestlers, ballroom dancers, and those who have sat on a swivel chair know that the effect of a clockwise rotation with rotocenter O and angle of rotation α can be undone by a counterclockwise rotation with the same rotocenter and angle.

Combining a clockwise rotation with rotocenter O and angle α with a counterclockwise rotation with the same rotocenter and angle gives the identity rigid motion.

11.4 Translations

A **translation** consists of essentially sliding an object in a specified *direction* and by a specified amount (the *length* of the translation). The two pieces of information (direction and length of the translation) are combined in the form of a **vector of translation** (usually denoted by v). The vector of translation is represented by an arrow—the arrow points in the direction of translation and the length of the arrow is the length of the translation.

> ### ▶ EXAMPLE 11.4 Translation of a Triangle

Figure 11-9 illustrates the translation of a triangle ABC. There are three "different" arrows shown in the figure but they all have the same length and direction, so they describe the same vector of translation v. As long as the arrow points in the proper direction and has the right length, the placement of the arrow in the plane is immaterial.

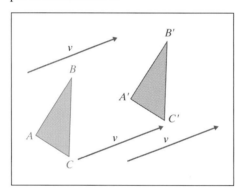

FIGURE 11-9   ◀◀

If we are given a point P and its image P' under a translation, the arrow joining P to P' gives v, the vector of the translation. Once we know the vector v, we know where the translation moves any other point.

> A translation is completely determined by a single point-image pair P and P'.

Translations have no fixed points and preserve the left-right or clockwise-counterclockwise orientations in the plane.

> A translation has no fixed points.

> A translation is a *proper* rigid motion.

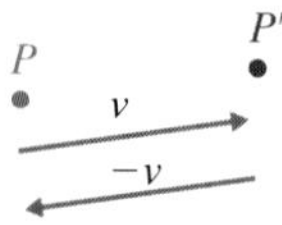

Finally, the effect of a translation with vector v can be undone by a translation of the same length but in the opposite direction. This vector, shown in Fig. 11-10, can be conveniently described as $-v$.

FIGURE 11-10

> Combining a translation with vector v and a translation with vector $-v$ gives the identity rigid motion.

11.5 Glide Reflections

A **glide reflection** is a compound rigid motion obtained by combining a translation (the glide) with a reflection with axis parallel to the direction of translation. Thus, a glide reflection is described by two things: the vector of the translation v and the axis of the reflection l, and these two *must* be parallel.

▶ EXAMPLE 11.5 Glide Reflection of a Triangle

Figure 11-11 illustrates the result of applying the glide reflection with vector v and axis l to the triangle ABC. We can do this in two different ways, but the final result will be the same. In Fig. 11-11(a) the translation is applied first, moving triangle ABC to the intermediate position $A*B*C*$. The reflection is then applied to $A*B*C*$, giving the final position $A'B'C'$. If we apply the reflection first, then the triangle ABC gets moved to the intermediate position $A*B*C*$ [Fig. 11-11(b)] and then translated to the final position $A'B'C'$.

Notice that any point and its image under the glide reflection [for example, A and A' in Fig. 11-11(a)] are on opposite sides but equidistant from the axis l. This implies that the midpoint of the segment joining a point and its image under a glide reflection *must* fall on the axis l.

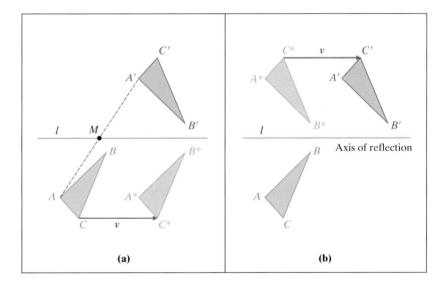

(a) (b)

FIGURE 11-11

If we are given a point P and its image P' and told that the rigid motion that moves P to P' is a glide reflection, we do not have enough information to determine the glide reflection but we do know that the axis l must pass through the midpoint M of the line segment PP'. If we are given a second point-image pair Q and Q', then the axis of the reflection must also pass through the midpoint N of the line segment QQ' [Fig. 11-12(a)], and we can then identify the axis l as the line passing trough M and N. Once we find the axis of reflection l, we can find the image of one of the points—say P'—under the reflection. This gives the intermediate point $P*$, and the vector that moves P to $P*$ is the vector of translation v, as shown in Fig. 11-12(b). [In the event that the midpoints of PP' and QQ' are the same point M, as shown in Fig. 11-12(c), we can still find

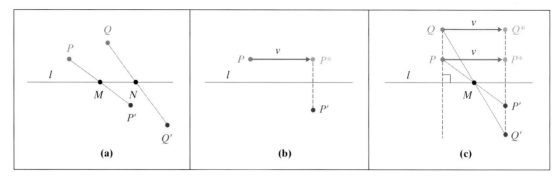

FIGURE 11-12

the axis *l* by drawing a line perpendicular to the line *PQ* passing through the common midpoint *M*.]

> A glide reflection is completely determined by *two* point-image pairs *P*, *P'* and *Q*, *Q'*.

A glide reflection has no fixed points (we can thank the translation piece for that) and is an *improper* rigid motion (we can thank the reflection piece for that).

> A glide reflection has no fixed points.

> A glide reflection is an improper rigid motion.

Finally, the effect of a glide reflection with vector of translation *v* and axis of reflection *l* can be undone by a glide reflection with the same axis of reflection and vector of translation −*v*.

> Combining a glide reflection with vector *v* and axis *l* with a glide reflection with vector −*v* and axis *l* gives the identity rigid motion.

For the reader's convenience, Table 11-1 gives a summary of the key properties of the four basic rigid motions in the plane discussed in Sections 11.2 through 11.5.

TABLE 11-1

Rigid motion	Specified by	Proper/improper	Fixed points	Number of point-image pairs needed
Reflection	axis of reflection *l*	improper	all points on *l*	one
Rotation (other than 360°)	rotocenter *O* and angle *α*	proper	*O* only	two
Translation	vector of translation *v*	proper	none	one
Glide reflection	vector of translation *v* and axis of reflection *l*	improper	none	two
Identity (360° rotation)		proper	all points	

11.6 Symmetry as a Rigid Motion

With an understanding of rigid motions and their classification, we will be able to look at the notion of symmetry in a much more precise way. Here, finally, is a good definition of **symmetry**, one that probably would not have made much sense at the start of this chapter:

> **Symmetry**
>
> A *symmetry* of an object (or shape) is a rigid motion that moves the object back onto itself.

In other words, in a symmetry one cannot tell, at the end of the motion, that the object has been moved. It is important to note that this does not necessarily force the rigid motion to be the identity motion. Individual points may be moved to different positions, even though the whole object is moved back into itself. And of course, the identity motion is itself a symmetry, one possessed by every object and that from now on we will call simply the **identity**.

For two-dimensional objects in the plane, there are only four possible types of rigid motions, and therefore only four possible types of symmetry: *reflection symmetry, rotation symmetry, translation symmetry*, and *glide reflection symmetry*.

▶ EXAMPLE 11.6 The Symmetries of a Square

What are the possible rigid motions that move the square in Fig. 11-13(a) back onto itself? First, there are *reflection symmetries*. For example, if we use the line l_1 in Fig. 11-13(b) as the axis of reflection, the square falls back into itself with points A and B interchanging places and C and D interchanging places. It is not hard to think of three other reflection symmetries, with axes l_2, l_3, and l_4 as shown in Fig. 11-13(b). Are there any other symmetries? Yes—the square has *rotation symmetries* as well. Using the center of the square O as the rotocenter, we can rotate the square by an angle of 90°. This moves the upper-left corner A to the upper-right corner B, B to the lower-right corner C, C to the lower-left corner D, and D to the upper-right corner A. Likewise, rotations with rotocenter O and angles of 180°, 270°, and 360°, respectively, are also symmetries of the square. Notice that the 360° rotation is just the identity symmetry.

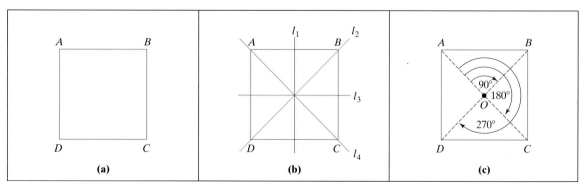

FIGURE 11-13 (a) The original square, (b) reflection symmetries (axes l_1, l_2, l_3, and l_4), (c) rotation symmetries (rotocenter O and angles of 90°, 180°, 270°, and 360°, respectively).

All in all, we have easily found eight symmetries for the square in Fig. 11-13(a): Four of them are reflections, the other four are rotations. Could there be more? What if we combined one of the reflections with one of the rotations? A symmetry combined with another symmetry, after all, has to be itself a symmetry. It turns out that the eight symmetries we listed are all there are—no matter how we combine them we always end up with one of the eight.

See Exercise 74.

EXAMPLE 11.7 The Symmetries of a Propeller

Let's now consider the symmetries of the shape shown in Fig. 11-14(a)—a two-dimensional version of a boat propeller (or a ceiling fan if you prefer) with four blades. Once again we have a shape with four reflection symmetries [the axes of reflection are l_1, l_2, l_3, and l_4 as shown in Fig. 11-14(b)] and four rotation symmetries [with rotocenter O and angles of 90°, 180°, and 270°, and 360°, respectively as shown in Fig. 11-14(c)]. And, just as with the square, there are no other possible symmetries.

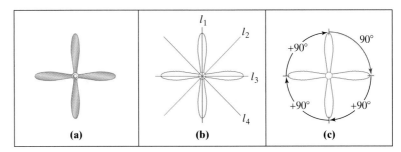

FIGURE 11-14

An important lesson lurks behind Examples 11.6 and 11.7: *Two different-looking objects can have exactly the same set of symmetries*. A good way to think about this is that the square and the propeller, while certainly different objects, are members of the same "symmetry family", as they carry exactly the same symmetry genes.

Formally, we will say that two objects or shapes are of the same **symmetry type** if they have exactly the same set of symmetries. The symmetry type for the square, the propeller, and each of the objects shown in Fig. 11-15 is called D_4 (shorthand for four reflections and four rotations).

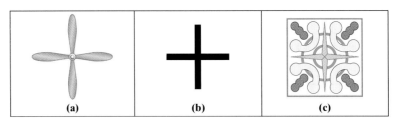

FIGURE 11-15 Objects with symmetry type D_4.

EXAMPLE 11.8 The Symmetry Type Z_4

Let's consider now the propeller shown in Fig. 11-16(a). This object is only slightly different from the one in Example 11.7, but from the symmetry point of view the difference is significant—here we still have the four rotation symmetries (90°, 180°, 270°, and 360°) but there are no reflection symmetries! [This follows from

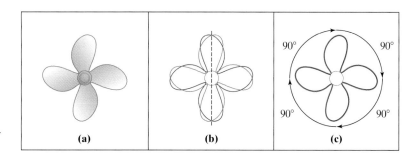

FIGURE 11-16 A propeller with symmetry type Z_4 (four rotation symmetries, no reflection symmetries).

the fact that the individual blades of the propeller have no reflection symmetry. As can be seen in Fig. 11-16(b), a vertical reflection is not a symmetry, and neither are any of the other reflections.] This object belongs to a new symmetry family called Z_4 (shorthand for the symmetry type of objects having four rotations only).   ◄◄

▶ EXAMPLE 11.9 The Symmetry Type Z_2

Here is one last propeller example. Every once in a while a propeller looks like the one in Fig. 11-17(a), which is kind of a cross between Figs. 11-16(a) and 11-15(a)—only opposite blades are the same. This figure has no reflection symmetries, and a 90° rotation won't work either [Fig. 11-16(b)]. The only symmetries of this shape are a 180° rotation (turn it upside down and it looks the same!) and the 360° rotation (the identity) are possible as symmetries of this propeller.

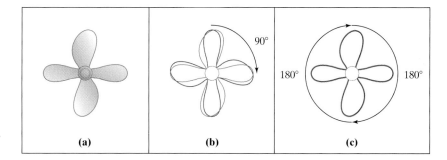

FIGURE 11-17 A propeller with symmetry type Z_2 (two rotation symmetries, no reflection symmetries).

Any object having only two rotation symmetries (the identity and a 180° rotation symmetry) is said to be of symmetry type Z_2. Figure 11-18 shows a few additional examples of shapes and objects with symmetry type Z_2.

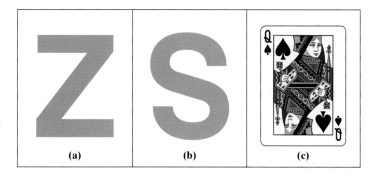

FIGURE 11-18 Objects with symmetry type Z_2. (a) The letter Z, (b) the letter S (in some fonts but not in others), and (c) the Queen of Spades (and many other cards in the deck).   ◄◄

EXAMPLE 11.10 The Symmetry Type D_1

One of the most common symmetry types occurring in nature is that of objects having a single reflection symmetry plus a single rotation symmetry (the identity). This symmetry type is called D_1. Figure 11-19 shows several examples of shapes and objects having symmetry type D_1. Notice that it doesn't matter if the axis of reflection is vertical, horizontal, or slanted.

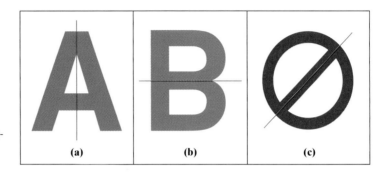

FIGURE 11-19 Objects with symmetry type D_1 (one reflection symmetry plus the identity symmetry).

(a)　　(b)　　(c)

EXAMPLE 11.11 The Symmetry Type Z_1

Many objects and shapes are informally considered to have no symmetry at all, but this is a little misleading, since *every object has at least the identity symmetry*. Objects whose only symmetry is the identity are said to have symmetry type Z_1. Figure 11-20 shows a few examples of objects of symmetry type Z_1—there are plenty of such objects around.

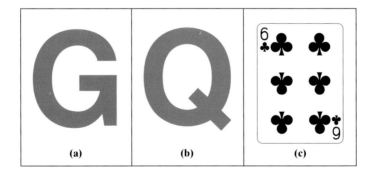

FIGURE 11-20 Objects with symmetry type Z_1 (only symmetry is the identity symmetry). Why doesn't the six of clubs have a half-turn symmetry? (The answer is given by the two middle clubs.)

(a)　　(b)　　(c)

EXAMPLE 11.12 Objects with Lots of Symmetry

In everyday language, certain objects and shapes are said to be "highly symmetric" when they have lots of rotation and reflection symmetries. Figures 11-21(a) and 11-21(b) show two very different looking snowflakes, but from the symmetry point of view they are the same: All snowflakes have six reflection symmetries and six rotation symmetries. Their symmetry type is D_6. (Try to find the six axes

FIGURE 11-21 Left and *center left:* snowflakes (symmetry type D_6). *Center right:* decorative ceramic plate (symmetry type D_9). *Right:* dome design (symmetry type D_{36}).

of reflection symmetry and the six angles of rotation symmetry.) Figure 11-21(c) shows a decorative ceramic plate. It has nine reflection symmetries and nine rotation symmetries, and, as you may have guessed, its symmetry type is called D_9. Finally, in Fig 11-21(d) we have an architectural blueprint of the dome of the Sports Palace in Rome, Italy. The design has 36 reflection and 36 rotation symmetries (symmetry type D_{36}). ◀◀

In each of the objects in Example 11.12, the number of reflections matches the number of rotations. This was also true in Examples 11.6, 11.7, and 11.10. Co-incidence? Not at all. When a finite object or shape has *both* reflection and rotation symmetries, the number of rotation symmetries (which includes the identity) has to match the number of reflection symmetries! Any finite object or shape with exactly N reflection symmetries and N rotation symmetries is said to have symmetry type D_N.

> ## EXAMPLE 11.13 The Symmetry Type D_∞

If we are looking for a two-dimensional shape that has *as much symmetry as possible*, we don't have to look past the wheels of a car. The wheel works so wonderfully well as a means of locomotion because of the infinitely many rotation and reflection symmetries of the circle. In a circle, a rotation with center at the center of the circle and by any angle whatsoever is a symmetry, and any line passing through the center of the circle can be used as an axis of reflection symmetry. We call the symmetry type of the circle D_∞ (the ∞ is the mathematical symbol for "infinity"). ◀◀

> ## EXAMPLE 11.14 Shapes with Rotations, but No Reflections

The *D* comes from *dihedral symmetry*—the technical mathematical term used to describe this group of symmetry types.

We now know that if a finite two-dimensional shape has rotations *and* reflections, then it must have exactly the same number of each. In this case, the shape belongs to the *D* family of symmetries, specifically, it has symmetry type D_N. However, we also saw in Examples 11.8, 11.9, and 11.11 shapes that have rotations, *but no* reflections. In

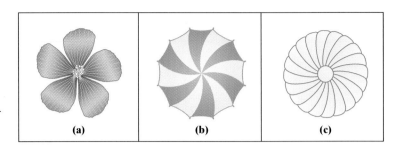

FIGURE 11-22 (a) Hibiscus (symmetry type Z_5), (b) top view of a parasol (symmetry type Z_6), and (c) turbine (symmetry type Z_{20}).

this case, we used the notation Z_N to describe the symmetry type, with the subscript N indicating the actual number of rotations. Figure 11–22 shows examples of objects having symmetry type Z_N for different values of N. ◀◀

We are now in a position to classify the possible symmetries of any *finite* two-dimensional shape or object. (The word *finite* is in there for a reason, which will become clear in the next section.) The possibilities boil down to a surprisingly short list of symmetry types:

- ■ **D_N.** This is the symmetry type of shapes with both rotation and reflection symmetries. The subscript N ($N = 1, 2, 3$, etc.) denotes the number of reflection symmetries, which is always equal to the number of rotation symmetries. (The rotations are an automatic consequence of the reflections—an object can't have reflection symmetries without having an equal number of rotation symmetries.)

- ■ **Z_N.** This is the symmetry type of shapes with rotation symmetries only. The subscript N ($N = 1, 2, 3$, etc.) denotes the number of rotation symmetries.

- ■ **D_∞.** This is the symmetry type of a circle, the only two-dimensional shape with an infinite number of rotations and reflections.

11.7 Patterns

Well, we've come a long way, but we have yet to see examples of shapes having translation or glide reflection symmetry. If we think of objects and shapes as being finite, then translation symmetry is impossible: A slide will always move such an object to a position different from its original position! But if we broaden our horizons and consider infinite "shapes," then translation and glide reflection symmetries are indeed possible.

We will formally define a **pattern** as an infinite "shape" consisting of an infinitely repeating basic design called the **motif** of the pattern. The reason we have "shape" in quotation marks is that a pattern is really an abstraction—in the real world there are no infinite objects as such, although the idea of an infinitely repeating motif is familiar to us from wallpaper, carpet designs, pottery, and so on.

Just like finite shapes, *patterns* can be classified by their symmetries. The classification of patterns according to their symmetry type is of fundamental importance in the study of molecular and crystal organization in chemistry, so it is not surprising that some of the first people to seriously investigate the symmetry

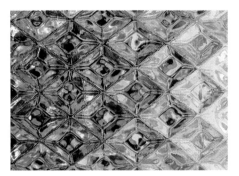  

FIGURE 11-23 *Left:* crystal structure. *Center:* Oriental rug (Izmir, 18th century). *Right:* vase (Corinthian, 7th century B.C.).

See reference 13.

types of patterns were crystallographers. Archeologists and anthropologists have also found that analyzing the symmetry types used by a particular culture in their textiles and pottery helps them gain a better understanding of that culture.

We will briefly discuss the symmetry types of *border* and *wallpaper* patterns. A comprehensive study of patterns is beyond the scope of this book, so we will not go into as much detail as we did with finite shapes.

Border Patterns

Border patterns are *linear* patterns where a basic *motif* repeats itself indefinitely in a linear direction, as in a frieze, a ribbon, or the border design of a pot or a basket.

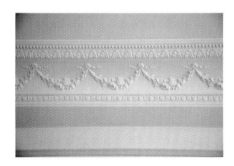  

FIGURE 11-24 *Left:* decorative frieze. *Center:* ribbon. *Right:* flower pot.

The most common direction in a border pattern (what we will call the *direction of the pattern*) is horizontal, but in general a border pattern can be in any direction (vertical, slanted 45°, etc.). (For typesetting in a book, it is much more convenient to display a border pattern horizontally, so you will only see examples of horizontal border patterns.)

We will now discuss what kinds of symmetries are possible in a border pattern. Fortunately, the number of possibilities is quite small.

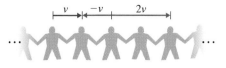

FIGURE 11-25

- **Translations.** A border pattern always has translation symmetries—they come with the territory. There is a *basic* translation symmetry v (v moves each copy of the motif one unit to the right), the opposite translation $-v$, and any multiple of v or $-v$ (Fig. 11-25).

■ **Reflections.** A border pattern can have (i) no reflection symmetry [Fig. 11-26(a)], (ii) *horizontal* reflection symmetry only, [Fig. 11-26(b)], (iii) *vertical* reflection symmetries [Fig. 11-26(c)] only, or (iv) both *horizontal* and *vertical* reflection symmetries [Fig. 11-26(d)]. In this last case the border pattern automatically picks up a half-turn symmetry as well. In terms of reflection symmetries, these are the only four possible scenarios.

See Exercise 67.

(a)   (b)   (c)   (d)

FIGURE 11-26

■ **Rotations.** Like with any other object, the identity (i.e., a 360° rotation) is a rotation symmetry of every border pattern, so every border pattern has at least one rotation symmetry. The only other possible rotation symmetry of a border pattern is a half-turn (180° rotation). Clearly, no other angle of rotation can take a horizontal pattern and move it back onto itself. Thus, in terms of rotation symmetry there are two kinds of border patterns: those whose only rotation symmetry is the identity [Fig. 11-27(a)], and those having half-turn symmetry in addition to the identity [Fig. 11-27(b)].

(a)   (b)

FIGURE 11-27

■ **Glide reflections.** A border pattern can have a glide reflection symmetry, but there is only one way this can happen: The axis of reflection *has* to be a line along the center of the pattern, and the *reflection part of the glide reflection is not by itself a symmetry of the pattern*. This means that a border pattern having horizontal reflection symmetry such as the one shown in Fig. 11-28(a) *does not* have glide reflection symmetry. On the other hand, the border pattern shown in Fig. 11-28(b) does not have horizontal reflection symmetry (the footprints do not fall back onto other footprints), but a glide by the vector w combined with a reflection along the axis l result in an honest-to-goodness glide reflection symmetry. (An important property of the glide reflection symmetry is that the vector w is always half the length of the basic translation symmetry v. This implies that two consecutive glide reflection symmetries are equivalent to the basic translation symmetry.) The border pattern in Fig. 11-28(c) has a vertical reflection symmetry as well as a glide reflection symmetry. In these cases a half-turn symmetry (rotocenter O) comes free in the bargain.

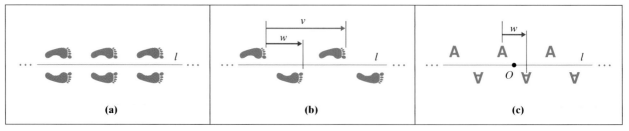

(a)   (b)   (c)

FIGURE 11-28

Combining the preceding observations, we get the following list of possible symmetries of a border pattern. (For simplicity we assume that the border pattern is in a horizontal direction.)

1. The *identity symmetry:* All border patterns have it.

2. *Translation symmetry:* All border patterns have it. There is a basic translation v, the opposite translation $-v$, and any multiples of these (see Fig. 11-25).

3. *Horizontal reflection symmetry:* Some patterns have it, some don't. There is only one possible horizontal axis of reflection and it must run through the middle of the pattern [see Fig. 11-26(b)].

4. *Vertical reflection symmetry:* Some patterns have it, some don't. The vertical axes of reflection (i.e., the axes are perpendicular to the direction of the pattern) can run through the middle of a motif or between two motifs [see Fig. 11-26(c)].

5. *Half-turns:* Some patterns have them, some don't. The rotocenter must be located at the center of a motif or between two motifs [see Fig. 11-27(b)].

6. *Glide reflections:* Some patterns have them, some don't. Neither the reflection nor the glide are symmetries on their own. The length of the glide w is half that of the basic translation v. The axis of the reflection runs through the middle of the pattern [see Fig. 11-28(b)].

Based on the various possible combinations of these symmetries, border patterns can be classified into just *seven different symmetry families*, which we list next. (Since all border patterns have identity symmetry and translation symmetry, we will only make reference to the other possible symmetries.)

■ **11.** This symmetry type represents border patterns that have no symmetries other than the identity and translation symmetry. The pattern in Fig. 11-26(a) is an example of this symmetry type.

■ **1m.** This symmetry type represents border patterns with only horizontal reflection symmetry. The pattern in Fig. 11-26(b) is an example of this symmetry type.

■ **m1.** This symmetry type represents border patterns with only vertical reflection symmetry. The pattern in Fig. 11-26(c) is an example of this symmetry type.

■ **mm.** This symmetry type represents border patterns with both horizontal and vertical reflection symmetry. When both of these symmetries are present, then there is also half-turn symmetry. The pattern in Fig. 11-26(d) is an example of this symmetry type.

■ **12.** This symmetry type represents border patterns with only half-turn symmetry. The pattern in Fig. 11-27(b) is an example of this symmetry type.

■ **1g.** This symmetry type represents border patterns with only glide reflection symmetry. The pattern in Fig. 11-28(b) is an example of this symmetry type.

■ **mg.** This symmetry type represents border patterns with vertical reflection and glide reflection symmetry. When both of these symmetries are present, then there is also half-turn symmetry. The pattern in Fig. 11-28(c) is an example of this symmetry type.

Table 11-2 shows the symmetries (besides the identity) for each of the seven symmetry types.

TABLE 11-2

Symmetry type	Translation	Horizontal reflection	Vertical reflection	Half-turn	Glide reflection
11	Yes	No	No	No	No
1m	Yes	Yes	No	No	No
m1	Yes	No	Yes	No	No
mm	Yes	Yes	Yes	Yes	No
12	Yes	No	No	Yes	No
1g	Yes	No	No	No	Yes
mg	Yes	No	Yes	Yes	Yes

Wallpaper Patterns

Wallpaper patterns are patterns that fill the plane by repeating a *motif* indefinitely along several (two or more) nonparallel directions. Typical examples of such patterns can be found in wallpaper (of course), carpets, textiles, and so on.

With wallpaper patterns things get a bit more complicated, so we will skip the details. The possible symmetries of a wallpaper pattern are as follows:

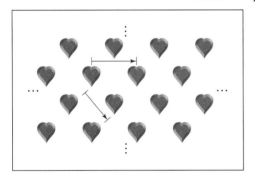

FIGURE 11-29

- **Translations.** Every wallpaper pattern has translation symmetry in at least two different (nonparallel) directions (Fig. 11-29).
- **Reflections.** A wallpaper pattern can have (i) no reflections, (ii) reflections in only one direction, (iii) reflections in two nonparallel directions, (iv) reflections in three nonparallel directions, (v) reflections in four nonparallel directions, and (vi) reflections in six nonparallel directions. There are no other possibilities. (Examples are shown in the chapter appendix.) Note that particularly conspicuous in its absence is the case of reflections in exactly five different directions.
- **Rotations.** In terms of rotation symmetries, a wallpaper pattern can have (i) the identity only, (ii) two rotations (identity and 180°), (iii) three rotations (identity, 120°, and 240°), (iv) four rotations (identity, 90°, 180°, and 270°), and (v) six rotations (identity, 60°, 120°, 180°, 240°, and 300°). There are no other possibilities. (Examples are shown in the chapter appendix.) Once again, note that a wallpaper pattern cannot have exactly five different rotations.
- **Glide reflections.** A wallpaper pattern can have (i) no glide reflections, (ii) glide reflections in only one direction, (iii) glide reflections in two nonparallel directions, (iv) glide reflections in three nonparallel directions, (v) glide reflections in four nonparallel directions, and (vi) glide reflections in six nonparallel directions. There are no other possibilities. (Examples are shown in the chapter appendix.)

In the early 1900s, it was shown mathematically that there are *only 17 possible symmetry types for wallpaper patterns.* This is quite a surprising fact—it means that the hundreds and thousands of wallpaper patterns one can find at a decorating store all fall into just 17 different symmetry families.

The 17 symmetry types of wallpaper patterns are listed and illustrated in the chapter appendix.

Conclusion

Symmetry, as wide or as narrow as you may define its meaning, is one idea by which man through the ages has tried to comprehend and create order, beauty and perfection.

Hermann Weyl

Real-life tangible physical objects as well as abstract shapes from geometry, art, and ornamental design are often judged and measured by a yardstick that can be both mathematical and aesthetic: How much symmetry and what types of symmetry does the object have? In this chapter we acquired the tools that allow us to answer both of these questions in the case of two-dimensional objects or shapes.

Most of the time, when we talk about a shape we assume we are dealing with a finite shape (there is some box big enough to put it in). In this case there are really only two possible scenarios: The shape has rotation symmetries only (a *Z-something* kind of shape), or it has both rotation and reflection symmetries in equal amounts (a *D-something* kind of shape). It is quite remarkable that there are no other possibilities. Nowhere in the universe of finite two-dimensional shapes does there exist, for example, a shape with three reflection symmetries and five rotation symmetries—it is mathematically impossible!

Patterns—infinite shapes consisting of an infinitely repeating motif—are even more surprising in terms of their possible symmetry types. *Border patterns*—commonly found in ribbons, baskets and pottery—are patterns where the motif is repeated indefinitely in a single direction. With border patterns there are *only seven possible symmetry types*, and these are listed in Table 11-2. *Wallpaper patterns*—such as those found in wallpapers and textiles—are patterns where a motif repeats itself in more than one direction and fills the plane. With wallpaper patterns there are only 17 possible symmetry types, and these are listed in the appendix at the end of this chapter.

Profile Sir Roger Penrose (1931–)

Much like the tiling of a bathroom floor, a mathematical *tiling* of the plane is an arrangement of tiles (of one or several shapes) that fully cover the plane without overlaps or gaps. A tiling of the plane is an abstract concept—unlike a bathroom floor, which has a finite area, a plane goes on forever. And yet, there is an easy method for tiling a plane: Find a basic design (a single tile or a combination of several tiles)

that can be fitted snugly with other copies of itself, and just repeat the same design, mindlessly marching off to infinity in all directions. This kind of tiling, called a *periodic tiling*, is essentially equivalent to a wallpaper pattern. With the right *motif* (the basic design that repeats), incredibly beautiful and exotic periodic tilings are possible (witness the amazing tilings found in some of M. C. Escher's graphic designs). Because tiling a plane is an infinite task, it stands to reason that the only way it can be done is by means of a periodic tiling, and this was the conventional wisdom for centuries. Artists, designers, chemists, and physicists all banked on this assumption. Enter Roger Penrose. In 1974, Penrose, a mathematical physicist already famous for his contributions to cosmology (he proved mathematically that black holes can exist) found an *aperiodic* tiling of the plane—a method for tiling the plane not based on the periodic repetition of one motif. Using only two basic tile shapes, Penrose was able to show that it is possible to tile a plane with infinitely changing patterns (see Project C for more details).

Roger Penrose was born in Colchester, England. His father, a famous medical geneticist, moved the family to Canada during World War II, and Roger spent his early school years in London, Ontario. After the war, the family returned to the other London, where Penrose began his university studies, earning his B.Sc. degree in mathematics from University College, and a Ph.D. in mathematical physics from Cambridge University.

Sir Roger (he was knighted in 1994) has done groundbreaking research in mathematics, mathematical physics, cosmology, and neuroscience and has held appointments at many academic institutions, including Cambridge University, Princeton University, and the University of Texas. He is currently the Rouse Ball Professor of Mathematics at the University of Oxford, England. In addition to his serious professional research in mathematics and physics, Penrose is famous for his contributions to recreational mathematics. In fact, his original interest in tilings of the plane was strictly recreational (he just loves challenging puzzles). Ironically, his

discovery of aperiodic tilings turned out to have amazing and unexpected practical implications. In 1982, following Penrose's lead, chemists working at the National Institute of Standards and Technology discovered the existence of *quasicrystals*, crystal structures with nonrepetitive patterns analogous to Penrose's aperiodic tilings. The discovery of quasicrystals turned the world of crystallography and materials science on its head. The aperiodic structure of quasicrystals defies all the traditional rules of crystal formation in nature and has led to the discovery of heretofore unknown properties of solid matter. As new alloys based on quasicrystalline structure are being developed (the latest is a nonscratch coating for frying pans), it is hard to imagine that we owe it all to one man's singular love of patterns and puzzles.

Key Concepts

angle of rotation, **377**	glide reflection symmetry, **390**	rotation, **377**
axis (of reflection, of symmetry), **375**	identity (rigid motion), **377**	rotation symmetry, **383**
	image, **375**	rotocenter, **377**
basic rigid motions of the plane, **375**	improper (rigid motion), **376**	symmetry, **383**
bilateral symmetry, **373**	motif, **388**	symmetry type, **384**
border pattern (one-dimensional pattern), **389**	pattern, **388**	translation, **380**
	proper (rigid motion), **378**	translation symmetry, **389**
equivalent rigid motion, **375**	reflection, **375**	vector (of translation), **380**
fixed point, **375**	reflection symmetry, **383**	wallpaper pattern (two-dimensional pattern), **392**
glide reflection, **381**	rigid motion (isometry), **374**	

Exercises

WALKING

A. Reflections

1. Which point in the following figure is the image of *P* under

(a) the reflection with axis l_1?

(b) the reflection with axis l_2?

(c) the reflection with axis l_3?

(d) the reflection with axis l_4?

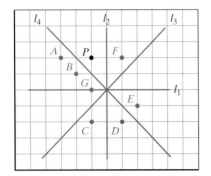

2. Which point in the following figure is the image of *P* under

(a) the reflection with axis l_1?

(b) the reflection with axis l_2?

(c) the reflection with axis l_3?

(d) the reflection with axis l_4?

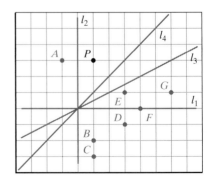

3. Given a reflection with axis *l* as shown in the following figure, find

   (a) the image of *S* under the reflection.

   (b) the image of quadrilateral *PQRS* under the reflection.

   (c) the fixed point of the reflection closest to *Q*.

   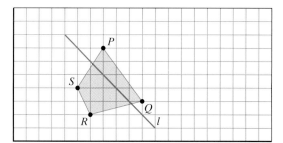

4. Given a reflection with axis *l* as shown in the following figure, find

   (a) the image of *P* under the reflection.

   (b) the image of triangle *PQR* under the reflection.

   (c) the fixed point of the reflection closest to *P*.

   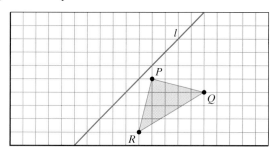

5. Given a reflection that sends the point *P* to the point *P′* as shown in the following figure, find

   (a) the axis of reflection.

   (b) the image of *S* under the reflection.

   (c) the image of quadrilateral *PQRS* under the reflection.

   (d) a point on the quadrilateral *PQRS* that is a fixed point of the reflection.

   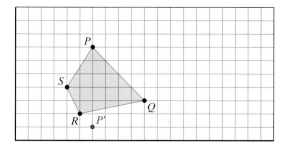

6. Given a reflection that sends the point *P* to the point *P′* as shown in the following figure, find

   (a) the axis of reflection.

   (b) the image of *S* under the reflection.

   (c) the image of quadrilateral *PQRS* under the reflection.

   (d) a point on the quadrilateral *PQRS* that is a fixed point of the reflection.

   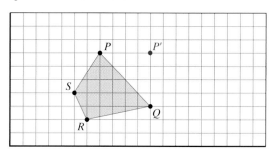

7. Given a reflection that sends the point *P* to the point *P′* as shown in the following figure, find

   (a) the axis of reflection.

   (b) the image of triangle *PQR* under the reflection.

   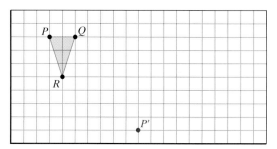

8. Given a reflection that sends the point *R* to the point *R′* as shown in the following figure, find

   (a) the axis of reflection.

   (b) the image of quadrilateral *PQRS* under the reflection.

   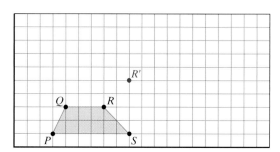

9. Consider a reflection for which *A* and *B* in the following figure are fixed points. Find the image of the shaded region under the reflection.

   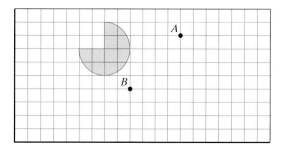

10. Consider a reflection for which A and B in the following figure are fixed points. Find the image of the shaded region under the reflection.

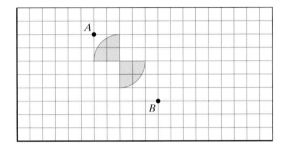

B. Rotations

Exercises 11 and 12 refer to the following figure.

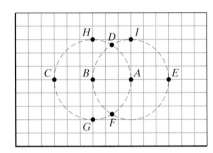

11. Which point in the figure is

   (a) the image of B under a 90° clockwise rotation with rotocenter A?

   (b) the image of B under a 180° rotation with rotocenter A?

   (c) the image of A under a 90° clockwise rotation with rotocenter B?

   (d) the image of D under a 60° clockwise rotation with rotocenter B?

   (e) the image of D under a 120° clockwise rotation with rotocenter B?

   (f) the image of D under a 120° counterclockwise rotation with rotocenter B?

   (g) the image of I under a 3690° clockwise rotation with rotocenter A?

   (h) the image of I under a 7530° clockwise rotation with rotocenter A?

12. Which point in the figure is

   (a) the image of C under a 90° clockwise rotation with rotocenter B?

   (b) the image of C under a 90° counterclockwise rotation with rotocenter B?

   (c) the image of H under a 90° clockwise rotation with rotocenter B?

   (d) the image of F under a 60° clockwise rotation with rotocenter A?

   (e) the image of F under a 120° clockwise rotation with rotocenter B?

   (f) the image of I under a 90° clockwise rotation with rotocenter H?

   (g) the image of G under a 3870° counterclockwise rotation with rotocenter B?

   (h) the image of F under a 5550° counterclockwise rotation with rotocenter B?

13. In each of the following, give an answer between 0° and 360°.

   (a) A clockwise rotation by an angle of 250° is equivalent to a counterclockwise rotation by an angle of _____.

   (b) A clockwise rotation by an angle of 710° is equivalent to a clockwise rotation by an angle of _____.

   (c) A counterclockwise rotation by an angle of 710° is equivalent to a clockwise rotation by an angle of _____.

   (d) A clockwise rotation by an angle of 3681° is equivalent to a clockwise rotation by an angle of _____.

14. In each of the following, give an answer between 0° and 360°.

   (a) A clockwise rotation by an angle of 500° is equivalent to a clockwise rotation by an angle of _____.

   (b) A clockwise rotation by an angle of 500° is equivalent to a counterclockwise rotation by an angle of _____.

   (c) A clockwise rotation by an angle of 5000° is equivalent to a clockwise rotation by an angle of _____.

   (d) A clockwise rotation by an angle of 50,000° is equivalent to a clockwise rotation by an angle of _____.

15. Given a rotation that moves the point B to the point B' and the point C to the point C' as shown in the followng figure, find

   (a) the rotocenter.

   (b) the image of triangle ABC under the rotation.

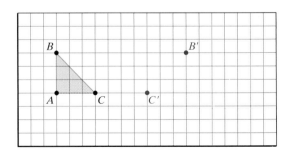

16. Given a rotation that moves the point A to the point A' and the point B to the point B' as shown in the following figure, find

(a) the rotocenter.

(b) the image of triangle ABC under the rotation.

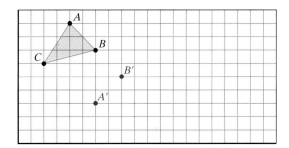

17. Given a rotation that moves the point P to the point P' and the point S to the point S' as shown in the following figure, find

(a) the rotocenter.

(b) the angle of rotation.

(c) the image of quadrilateral $PQRS$ under the rotation.

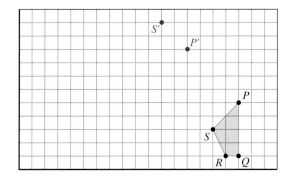

18. Given a rotation that moves the point Q to the point Q' and the point R to the point R' as shown in the following figure, find

(a) the rotocenter.

(b) the angle of rotation.

(c) the image of quadrilateral $PQRS$ under the rotation.

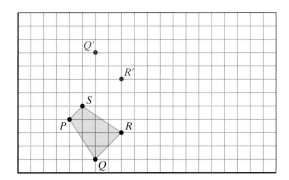

19. Given a 90° clockwise rotation that moves the point B to the point B' as shown in the following figure, find

(a) the rotocenter.

(b) the image of triangle ABC under the rotation.

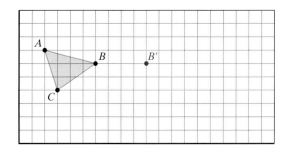

20. Given a half-turn (180° rotation) that moves the point A to the point A' as shown in the following figure, find

(a) the rotocenter.

(b) the image of the shaded region under the rotation.

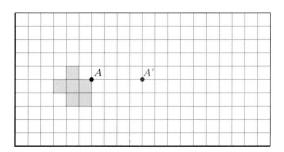

C. Translations

21. Which point in the following figure is the image of P under

(a) the translation with vector v_1?

(b) the translation with vector v_2?

(c) the translation with vector v_3?

(d) the translation with vector v_4?

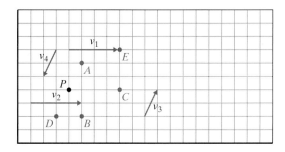

22. Which point in the following figure is the image of *P* under

   (a) the translation with vector v_1?

   (b) the translation with vector v_2?

   (c) the translation with vector v_3?

   (d) the translation with vector v_4?

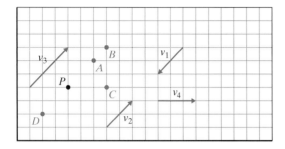

23. Given a translation that sends the point *E* to the point *E'* as shown in the following figure, find

   (a) the image of *A* under the translation.

   (b) the image of figure *ABCDE* under the translation.

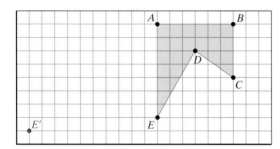

24. Given a translation that sends the point *Q* to the point *Q'* as shown in the following figure, find

   (a) the image of *P* under the translation.

   (b) the image of figure *PQRS* under the translation.

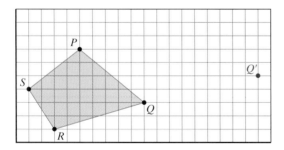

25. The grid graph shown in the following figure represents a street grid where each block of the grid is exactly 1 mile long. A taxi needs to go from point *A* to point *A'*. Describe the taxi's trip in terms of one or more translations. For each translation describe the length and direction of the vector of translation.

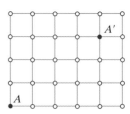

26. The grid graph shown in the following figure represents a street grid where each block of the grid is exactly 1 mile long. A taxi needs to go from point *A* to point *A'*. Describe the taxi's trip in terms of one or more translations. For each translation describe the length and direction of the vector of translation.

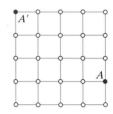

D. Glide Reflections

27. Given a glide reflection with vector *v* and axis *l* as shown in the following figure, find the image of the triangle *ABC* under the glide reflection.

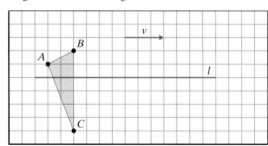

28. Given a glide reflection with vector *v* and axis *l* as shown in the following figure, find the image of the quadrilateral *ABCD* under the glide reflection.

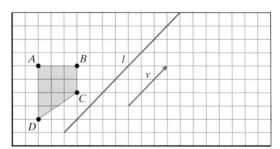

29. Given a glide reflection that sends the point *B* to the point *B'* and the point *D* to the point *D'* as shown in the following figure, find

   (a) the axis of the glide reflection.

   (b) the image of *A* under the glide reflection.

(c)   the image of the shaded figure under the glide reflection.

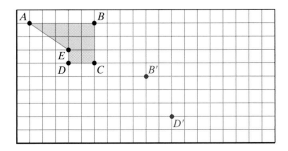

30. Given a glide reflection that sends the point A to the point A' and the point C to the point C' as shown in the following figure, find

(a)   the axis of the glide reflection.

(b)   the image of B under the glide reflection.

(c)   the image of the shaded figure under the glide reflection.

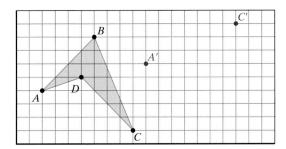

31. Given a glide reflection that sends the point B to the point B' and the point C to the point C' as shown in the following figure, find

(a)   the axis of the glide reflection.

(b)   the image of the shaded figure under the glide reflection.

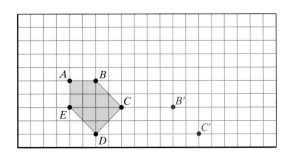

32. Given a glide reflection that sends the point P to the point P' and the point Q to the point Q' as shown in the following figure, find

(a)   the axis of the glide reflection.

(b)   the image of the shaded figure under the glide reflection.

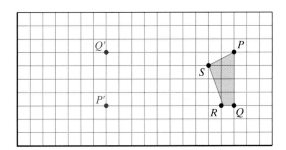

33. Given a glide reflection that sends the point P to the point P' and the point Q to the point Q' as shown in the following figure, find

(a)   the axis of the glide reflection.

(b)   the image of the shaded figure under the glide reflection.

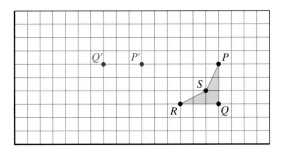

34. Given a glide reflection with translation vector v that sends the point I to the point I' as shown in the following figure, find the image of the shaded figure under the glide reflection.

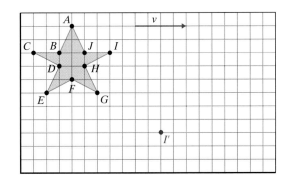

E. Symmetries of Finite Shapes

In Exercises 35 through 38, list all the symmetries of each figure. Describe each symmetry by giving specifics—the axes of reflection, the centers and angles of rotation, and so on.

35.

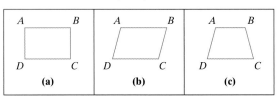

36.

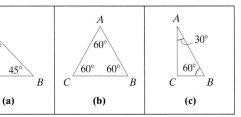

37.

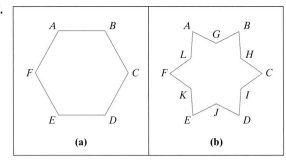

38.

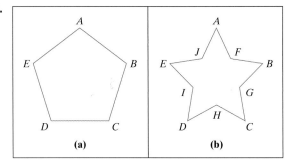

39. For each of the figures in Exercise 35, give its symmetry type.

40. For each of the figures in Exercise 36, give its symmetry type.

41. For each of the figures in Exercise 37, give its symmetry type.

42. For each of the figures in Exercise 38, give its symmetry type.

43. Find the symmetry type for each of the following letters.

  (a) A

  (b) D

  (c) L

  (d) Z

  (e) Ω

  (f) Φ

44. Find the symmetry type for each of the following letters.

  (a) T

  (b) E

  (c) N

  (d) H

  (e) Σ

  (f) ζ

45. Give an example of a capital letter of the alphabet that has symmetry type

  (a) Z_1.

  (b) D_1.

  (c) Z_2.

  (d) D_2.

46. Give an example of a numeral that has symmetry type

  (a) Z_1.

  (b) D_1.

  (c) Z_2.

  (d) D_2.

47. Give an example of

  (a) a natural object (plant, animal, mineral) that has symmetry type D_5. Explain your answer.

  (b) a human-made object (logo, gadget, consumer product, etc.) that has symmetry type D_5. Explain your answer.

  (c) a natural object (plant, animal, mineral) that has symmetry type Z_1. Explain your answer.

  (d) a human-made object (logo, gadget, consumer product, etc.) that has symmetry type Z_1. Explain your answer.

48. Give an example of

  (a) a natural object (plant, animal, mineral) that has symmetry type D_6. Explain your answer.

  (b) a human-made object (logo, gadget, consumer product, etc.) that has symmetry type D_6. Explain your answer.

  (c) a natural object (plant, animal, mineral) that has symmetry type Z_2. Explain your answer.

  (d) a human-made object (logo, gadget, consumer product, etc.) that has symmetry type Z_2. Explain your answer.

F. Symmetries of Border Patterns

49. For each of the following border patterns, give its symmetry type using the standard crystallography notation ($mm, mg, m1, 1m, 1g, 12, 11$).

(a) ... **A A A A A** ...
(b) ... **D D D D D** ...
(c) ... **Z Z Z Z Z** ...
(d) ... **L L L L L** ...

50. For each of the following border patterns, give its symmetry type using the standard crystallography notation ($mm, mg, m1, 1m, 1g, 12, 11$).

(a) ... **J J J J J** ...
(b) ... **T T T T T** ...
(c) ... **C C C C C** ...
(d) ... **N N N N N** ...

51. For each of the following border patterns, give its symmetry type using the standard crystallography notation ($mm, mg, m1, 1m, 1g, 12, 11$).

(a) ... **qpqpqpqp** ...
(b) ... **pdpdpdpd** ...
(c) ... **pbpbpbpb** ...
(d) ... **pqbdpqbd** ...

52. For each of the following border patterns, give its symmetry type using the standard crystallography notation ($mm, mg, m1, 1m, 1g, 12, 11$).

(a) ... **qbqbqbqb** ...
(b) ... **qdqdqdqd** ...
(c) ... **dbdbdbdb** ...
(d) ... **qpdbqpdb** ...

53. If a border pattern consists of repeating a motif of symmetry type Z_2, what is the symmetry type of the border pattern?

54. If a border pattern consists of repeating a motif of symmetry type Z_1, what is the symmetry type of the border pattern?

G. Miscellaneous

55. Explain why any proper rigid motion that has a fixed point must be equivalent to a rotation.

56. Explain why any rigid motion other than the identity that has two or more fixed points must be equivalent to a reflection.

*Exercises 57 through 66 refer to the product of two rigid motions. Given two rigid motions M and N, we can combine the two rigid motions by first applying M and then applying N to the result. The rigid motion defined by combining M and N in that order is called the **product** of M and N. Note that the product of N and M (where we apply N first and M second) is not the same as the product of M and N.*

57. Find the image of P under the product of

(a) the reflection with axis l_1 and the reflection with axis l_2.

(b) the reflection with axis l_2 and the reflection with axis l_1.

(c) the reflection with axis l_2 and the reflection with axis l_3.

(d) the reflection with axis l_3 and the reflection with axis l_2.

(e) the reflection with axis l_1 and the reflection with axis l_4.

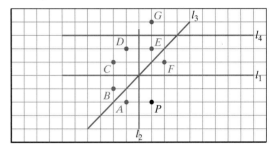

58. Find the image of P under the product of

(a) the reflection with axis l and the 90° clockwise rotation with rotocenter A.

(b) the 90° clockwise rotation with rotocenter A and the reflection with axis l.

(c) the reflection with axis l and the 180° rotation with rotocenter A.

(d) the 180° rotation with rotocenter A followed by the reflection with axis l.

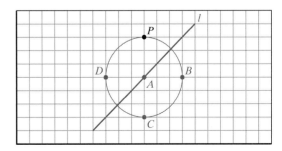

59. In each case, state whether the rigid motion M is proper or improper.

(a) M is the product of a proper rigid motion and an improper rigid motion.

(b) $\mathcal{M}$ is the product of an improper rigid motion and an improper rigid motion.

(c) $\mathcal{M}$ is the product of a reflection and a rotation.

(d) $\mathcal{M}$ is the product of a reflection and a reflection.

60. In each case, state whether the rigid motion $\mathcal{M}$ has (i) no fixed points, (ii) exactly one fixed point, or (iii) infinitely many fixed points.

(a) $\mathcal{M}$ is the product of a reflection with axis l_1 and a reflection with axis l_2. Assume the lines l_1 and l_2 intersect at a point C.

(b) $\mathcal{M}$ is the product of a reflection with axis l_1 and a reflection with axis l_3. Assume the lines l_1 and l_3 are parallel.

61. Suppose that a rigid motion $\mathcal{M}$ is the product of a reflection with axis l_1 and a reflection with axis l_2, where l_1 and l_2 intersect at a point C. Explain why $\mathcal{M}$ must be a rotation with center C.

[***Hint:*** *See Exercises 59(d) and 60(a).*]

62. Suppose that the rigid motion $\mathcal{M}$ is the product of the reflection with axis l_1 and the reflection with axis l_3, where l_1 and l_3 are parallel. Explain why $\mathcal{M}$ must be a translation.

[***Hint:*** *See Exercises 59(d) and 60(b).*]

JOGGING

63. Suppose that lines l_1 and l_2 intersect at C and that the angle between them as shown in the following figure is α.

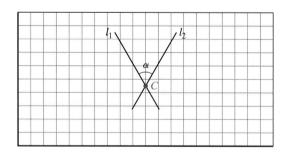

(a) Give the rotocenter, angle, and direction of rotation of the product of the reflection with axis l_1 and the reflection with axis l_2.

(b) Give the rotocenter, angle, and direction of rotation of the product of the reflection with axis l_2 and the reflection with axis l_1.

64. Suppose that lines l_1 and l_3 are parallel and that the distance between them as shown in the figure is d.

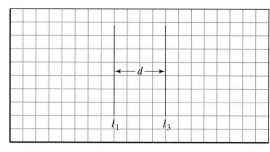

(a) Give the length and direction of the vector of the product of the reflection with axis l_1 and the reflection with axis l_3.

(b) Give the length and direction of the vector of the product of the reflection with axis l_3 and the reflection with axis l_1.

65. Translation 1 moves point P to point P'; translation 2 moves point Q to point Q', as shown in the figure.

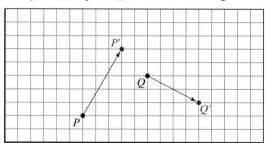

(a) Find the images of P and Q under the product of translation 1 and translation 2.

(b) Show that the product of translation 1 and translation 2 is a translation. Give a geometric description of the vector of the translation.

66. (a) Given a glide reflection with axis l and vector v as shown in the figure, find the image of the triangle ABC under the product of the glide reflection with itself.

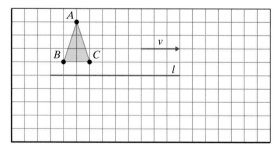

(b) Show that the product of a glide reflection with itself is a translation. Describe the direction and amount of the translation in terms of the direction and amount of the original glide.

67. (a) Explain why a border pattern cannot have a reflection symmetry along an axis forming 45° with the direction of the pattern.

(b) Explain why a border pattern can have only horizontal and/or vertical reflection symmetry.

68. Suppose $\mathcal{M}$ is a translation by the vector v and $\mathcal{N}$ is a 90° clockwise rotation with rotocenter O, as shown in the following figure.

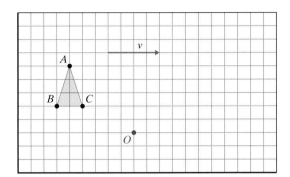

(a) Find the image of triangle ABC under the product of $\mathcal{M}$ and $\mathcal{N}$.

(b) Explain why the product of $\mathcal{M}$ and $\mathcal{N}$ is a rotation, and find the rotocenter and angle of the rotation.

69. Suppose $\mathcal{M}$ is a 90° clockwise rotation with rotocenter O, and $\mathcal{N}$ is a glide reflection with axis l and vector v, as shown in the following figure.

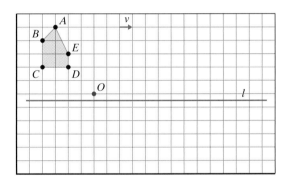

(a) Find the image of the shaded figure under the product of $\mathcal{M}$ and $\mathcal{N}$.

(b) Explain why the product of $\mathcal{M}$ and $\mathcal{N}$ is a reflection, and find the axis of the reflection.

70. Using copies of the symbol ♥ (and rotated versions of it), construct border patterns of symmetry type

(a) *mg*.

(b) *m*1.

(c) 1*m*.

(d) 11.

(e) 12.

(f) 1*g*.

(g) *mm*.

71. A rigid motion $\mathcal{M}$ moves the triangle PQR into the triangle $P'Q'R'$ as shown in the figure. Explain why the rigid motion $\mathcal{M}$ must be a glide reflection.

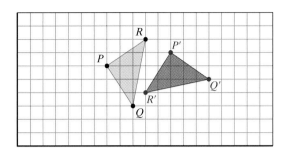

72. A *palindrome* is a word that is the same when read forward or backward. MOM is a palindrome and so is ANNA. (For simplicity, we will assume all letters are capitals.)

(a) Explain why if a word has vertical reflection symmetry, then it must be a palindrome.

(b) Give an example of a palindrome (other than ANNA) that doesn't have vertical reflection symmetry.

(c) If a palindrome has vertical reflection symmetry, what can you say about the symmetries of the individual letters in the word?

(d) Find a palindrome with 180° rotational symmetry.

RUNNING

73. Let the six symmetries of the equilateral triangle ABC shown in the figure be denoted as follows: r_1: reflection with axis l_1, r_2: reflection with axis l_2, r_3: reflection with axis l_3, R_1: 120° clockwise rotation with rotocenter O, R_2: 240° clockwise rotation with rotocenter O, I: the identity symmetry.

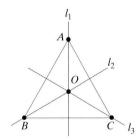

Complete the following table by entering, in each row and column of the table, the symmetry that results when applying first the symmetry in the row followed by the symmetry in the column. (For example, the entry in row r_1 column r_2 is R_1 because the reflection r_1 followed by the reflection r_2 equals the rotation R_1.)

	r_1	r_2	r_3	R_1	R_2	I
r_1		R_1				
r_2						
r_3						
R_1						
R_2						
I						

74. Let the eight symmetries of the square $ABCD$ shown in the figure be denoted as follows: r_1: reflection with axis l_1, r_2: reflection with axis l_2, r_3: reflection with axis l_3, r_4: reflection with axis l_4, R_1: 90° clockwise rotation with rotocenter O, R_2: 180° clockwise rotation with rotocenter O, R_3: 270° clockwise rotation with rotocenter O, I: the identity symmetry.

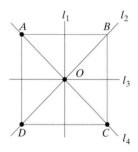

Complete the following table by entering, in each row and column of the table, the symmetry that results when applying first the symmetry in the row followed by the symmetry in the column.

	r_1	r_2	r_3	r_4	R_1	R_2	R_3	I
r_1								
r_2								
r_3								
r_4								
R_1								
R_2								
R_3								
I								

75. Find the symmetry type of the wallpaper pattern. (***Hint:*** *Use the flowchart on p. 408.*)

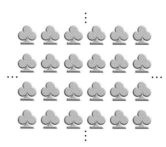

76. Find the symmetry type of the wallpaper pattern. (***Hint:*** *Use the flowchart on p. 408.*)

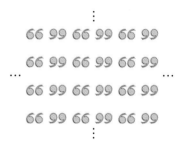

77. Find the symmetry type of the wallpaper pattern. (***Hint:*** *Use the flowchart on p. 408.*)

78. Find the symmetry type of the wallpaper pattern. (***Hint:*** *Use the flowchart on p. 408.*)

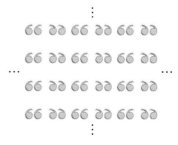

79. Find the symmetry type of the wallpaper pattern.

(**Hint:** *Use the flowchart on p. 408.*)

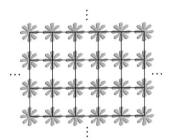

80. Find the symmetry type of the wallpaper pattern.

(**Hint:** *Use the flowchart on p. 408.*)

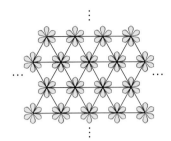

Projects and Papers

A. Patterns Everywhere

Border patterns can be found in many objects from the real world—ribbons, wallpaper borders, and architectural friezes. Even ceramic pots and woven baskets exhibit border patterns (when the pattern goes around in a circle it can be unraveled as if it were going on a straight line). Likewise, wallpaper patterns can be found in wallpapers, textiles, neckties, rugs, and gift-wrapping paper. Patterns are truly everywhere.

This project consists of two separate subprojects.

Part 1. Find examples from the real world of each of the seven possible border-pattern symmetry types. Do not use photographs from a book or designs you have just downloaded from some Web site. This part is not too hard, and it is a warm-up for Part 2, the real challenge.

Part 2. Find examples from the real world of each of the 17 wallpaper-pattern symmetry types. The same rules apply as for Part 1. Use the flowchart on p. 408 to classify the wallpaper patterns.

Notes: (1) Your best bet is to look at wallpaper patterns and borders at a wallpaper store and/or gift-wrapping paper and ribbons at a paper store. You will have to do some digging—a few of the wallpaper-pattern symmetry types are hard to find. (2) For ideas, you may want to visit Steve Edwards's excellent Web site *Tiling Plane and Fancy* at *http://www2.spsu.edu/math/tile/index.htm*.

B. Three-Dimensional Rigid Motions

For two-dimensional objects, we have seen that every rigid motion is of one of four basic types. For three-dimensional objects moving in three-dimensional space, there are *six* pos-

sible types of rigid motion. Specifically, every rigid motion in three-dimensional space is equivalent to a **reflection**, a **rotation**, a **translation**, a **glide reflection**, a **rotary reflection**, or a **screw displacement**.

Prepare a presentation on the six possible types of rigid motions in three-dimensional space. For each one give a precise definition of the rigid motion, describe its most important properties and give illustrations as well as real-world examples.

C. Penrose Tilings

In the mid-1970s, the British mathematician Roger Penrose made a truly remarkable discovery—it is possible to cover the plane using *aperiodic tilings*, something like a wallpaper pattern with a constantly changing motif. (A brief biographical profile of Penrose can be found at the end of the chapter.) One of the simplest and most surprising aperiodic tilings discovered by Penrose is based on just two shapes—the two rhombi shown in the figure.

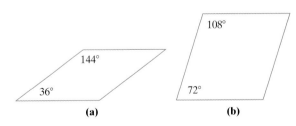

Prepare a presentation discussing and describing Penrose's tilings based on figures (a) and (b). Include in your presentation the connection between figures (a) and (b) and the golden ratio, as well as the connection between Penrose tilings and quasicrystals.

Appendix The 17 Wallpaper Symmetry Types

Symmetry Type	Translation (2 or more directions)	Rotations (given by the smallest angle)					Reflections (Number of directions)					Glide Reflections (Number of directions)					Example
		identity	2-fold	4-fold	3-fold	6-fold	1	2	4	3	6	1	2	4	3	6	
		D E G R E E S															
		0	180	90	120	60											
p1	✓	✓															
pm	✓	✓					✓										
pg	✓	✓										✓					
cm	✓	✓					✓					✓					
p2	✓	✓	✓														
pmg	✓	✓	✓				✓					✓					
pmm	✓	✓	✓					✓									
pgg	✓	✓	✓										✓				
cmm	✓	✓	✓					✓					✓				

Symmetry Type	Translation (2 or more directions)	identity 0°	2-fold 180°	4-fold 90°	3-fold 120°	6-fold 60°	Refl 1	Refl 2	Refl 4	Refl 3	Refl 6	Glide 1	Glide 2	Glide 4	Glide 3	Glide 6	Example
		\[Rotations (given by the smallest angle)\]					\[Reflections (Number of directions)\]					\[Glide Reflections (Number of directions)\]					
p4	✓	✓	✓[a]	✓[b]													
p4m	✓	✓	✓[a]	✓[b]					✓				✓				
p4g	✓	✓	✓[a]	✓[b]				✓						✓			
p3	✓	✓			✓												
p3m1	✓	✓			✓[c]					✓					✓		
p31m	✓	✓			✓[d]					✓					✓		
p6	✓	✓	✓		✓	✓											
p6m	✓	✓	✓		✓	✓					✓	✓				✓	

a, b : Different rotocenters.

c : All rotocenters on axes of reflection.

d : Not all rotocenters on axes of reflection.

Flowchart for Classifying Wallpaper Patterns

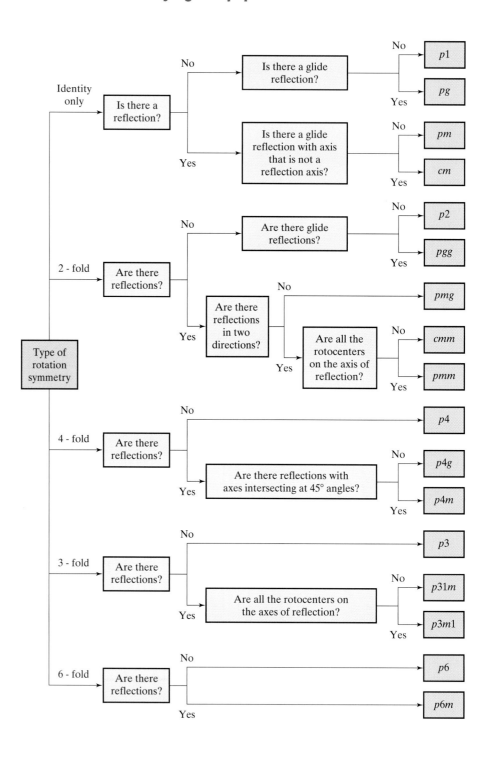

References and Further Readings

1. Bunch, Bryan, *Reality's Mirror: Exploring the Mathematics of Symmetry*. New York: John Wiley & Sons, Inc., 1989.

2. Coxeter, H. S. M., *Introduction to Geometry*, 2nd ed. New York: John Wiley & Sons, Inc., 1969.

3. Crowe, Donald W., "Symmetry, Rigid Motions, and Patterns," *UMAP Journal*, 8 (1987), 206–236.

4. Field, M., and M. Golubitsky, *Symmetry in Chaos*. New York: Oxford University Press, 1992.

5. Gardner, Martin, *The New Ambidextrous Universe: Symmetry and Asymmetry from Mirror Reflections to Superstrings*, 3rd ed. New York: W. H. Freeman & Co., 1990.

6. Grünbaum, Branko, and G. C. Shephard, *Tilings and Patterns: An Introduction*. New York: W. H. Freeman & Co., 1989.

7. Hargittai, I., and M. Hargittai, *Symmetry: A Unifying Concept*. Bolinas, CA: Shelter Pubns., 1994.

8. Hofstadter, Douglas R., *Gödel, Escher, Bach: An Eternal Golden Braid*. New York: Vintage Books, 1980.

9. Martin, George E., *Transformation Geometry: An Introduction to Symmetry*. New York: Springer-Verlag, 1994.

10. Rose, Bruce, and Robert D. Stafford, "An Elementary Course in Mathematical Symmetry," *American Mathematical Monthly*, 88 (1981), 59–64.

11. Schattsneider, Doris, *Visions of Symmetry: Notebooks, Periodic Drawings, and Related Work of M. C. Escher*. New York: W. H. Freeman & Co., 1990.

12. Shubnikov, A. V., and V. A. Koptsik, *Symmetry in Science and Art*. New York: Plenum Publishing Corp., 1974.

13. Washburn, Dorothy K., and Donald W. Crowe, *Symmetries of Culture*. Seattle, WA: University of Washington Press, 1988.

14. Weyl, Hermann, *Symmetry*. Princeton, NJ: Princeton University Press, 1952.

15. Wigner, Eugene, *Symmetries and Reflections*. Bloomington: Indiana University Press, 1967.

12

The Geometry of Fractal Shapes

Fractally Speaking

The wild and wooly images on these pages—what are they? Abstract works of art? Images from nature? Computer-generated art? All of the above. In spite of their varied origins (see the Photo Credits on p. 444), these images have one important thing in common: They are all images of special shapes known as *fractals*.

The purpose of this chapter is to introduce the basic ideas behind fractal geometry—a geometry very different from the traditional Euclidean geometry of triangles, squares, and circles—and to show the connection between abstract geometric fractals and the many fractal-like structures that can be found in nature.

The chapter is organized around a series of landmark examples of fractal geometry: the *Koch snowflake* in Section 12.1; the *Sierpinski gasket* in Section 12.2; the *chaos game* in Section 12.3; the *twisted Sierpinski gasket* in Section 12.4; the *Mandelbrot set* in Section 12.5. Each of these examples has something new to add to our fractal excursion, and as we explore these examples we will also learn something about the general ideas of fractal geometry.

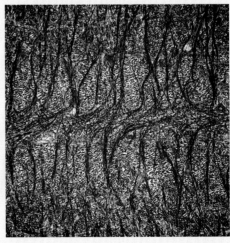

12.1 The Koch Snowflake

Our first example of a *geometric fractal* is a shape known as the **Koch snowflake**. We describe the Koch snowflake as a geometric fractal because it is created by means of a recursive sequence of geometric steps:

The Koch Snowflake (Recursive Construction)

- **Start.** Start with a solid *equilateral* triangle [Fig. 12-1(a)]. The size of the triangle is irrelevant, so for simplicity we will say that the sides of the triangle are of length 1.

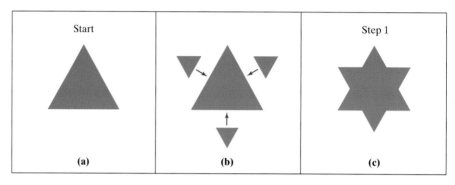

FIGURE 12-1

- **Step 1.** To the middle third of each of the sides of the original triangle add an equilateral triangle with sides of length $1/3$, as shown in Fig. 12-1(b). The result is the 12-sided "star of David" shown in Fig. 12-1(c).

- **Step 2.** To the middle third of each of the 12 sides of the star in Step 1 add an equilateral triangle with sides of length one-third the length of that side. The result is a "snowflake" with $12 \times 4 = 48$ sides, each of length $(1/3)^2 = 1/9$, as shown in Fig. 12-2(a). (Each of the sides "crinkles" into four new sides; each new side has length $1/3$ the previous side.)

- **Step 3.** Apply *Procedure KS* to the "snowflake" in Step 2. This gives the more elaborate "snowflake" shown in Fig. 12-2(b). Without counting we can figure out that this snowflake has $48 \times 4 = 192$ sides, each of length $(1/3)^3 = 1/27$.

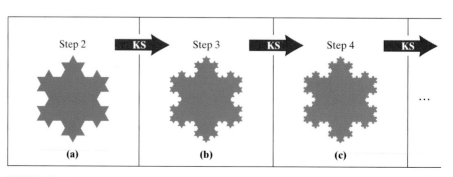

FIGURE 12-2

- **Step 4.** Apply *Procedure KS* to the "snowflake" in Step 3. This gives the "snowflake" shown in Fig. 12-2(c). (You definitely don't want to do this by hand—there are 192 tiny little equilateral triangles that are being added!)

- **Step 5, 6 etc.** Apply *Procedure KS* to the "snowflake" obtained in the previous step.

At each step of this process we create a new "snowflake," but after a while it's hard to tell that there is any change. There is a small difference between the snowflakes in Figs. 12-2(b) and (c)—a few more steps and the new triangles being added become tiny specks. Soon enough, the images become *visually stable*: To the naked eye there is no difference between one snowflake and the next. For all practical purposes we are seeing the ultimate destination of this trip: the **Koch snowflake** itself (Fig. 12-3).

FIGURE 12-3

Because the Koch snowflake is constructed in an infinite sequence of steps, an actual rendering of it is impossible—Fig. 12-3 is only an approximation. This is no reason to be concerned or upset—the Koch snowflake is an abstract shape which we can study and explore by looking at imperfect pictures of it (much like we do in high school geometry when we study circles even though it is impossible to draw a perfect circle).

The construction of the Koch snowflake is an example of a *recursive process*, a process in which the same set of rules (what we called *Procedure KS*) is applied repeatedly in an infinite feedback loop—the output at one step becomes the input at the next step. One main advantage of recursive descriptions is that they allow for very simple and efficient definitions, even when the objects being defined are quite complicated. The Koch snowflake, for example, is a fairly complicated shape, but we can define it in two lines using a form of shorthand we will call a **replacement rule**—a rule that specifies how to substitute one piece for another.

Recursive processes were also discussed in Chapter 1 (recursive ranking methods), Chapter 9 (Fibonacci numbers), and Chapter 10 (population growth models).

The Koch Snowflake

- **Start:** Start with a solid equilateral triangle ▲.
- **Replacement Rule:** Whenever you see a boundary line segment, apply *Procedure KS* to it.

The Koch Curve and Self-Similarity

If we only consider the boundary of the Koch snowflake and forget about the interior, we get an infinitely jagged curve known as the **Koch curve** (or sometimes the *snowflake curve*) shown in Fig. 12-4(a). Clearly, Fig.12-4(a) is just a rough rendering of the Koch curve, so our natural curiosity pushes us to take a closer look. We'll just randomly pick a small section of the Koch curve and magnify it [Fig. 12-4(b)]. The surprise (or not!) is that we see nothing new—the small detail looks just like the rough detail. Magnifying further is not going to be much help. Figure 12-4(c) shows a detail of the Koch curve after magnifying it by a factor of almost 100.

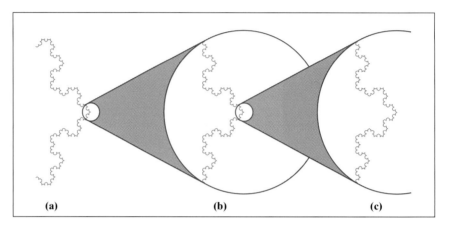

FIGURE 12-4 (a) A small piece of the Koch curve. (b) The piece in (a) magnified ×9. (c) The same piece magnified ×81.

This seemingly remarkable characteristic of the Koch curve of looking the same at different scales is called **self-similarity** (or *symmetry of scale*). As the name suggests, it implies that a shape is similar to a part of itself, which in turn is similar to an even smaller part, and so on. In the case of the Koch curve the self-similarity has three important properties: (i) it is *infinite* (the self-similarity takes place over infinitely many levels of magnification), (ii) it is *universal* (the same self-similarity occurs along every part of the Koch curve), and (iii) it is *exact* (we see the exact same image at every level of magnification).

Self-similarity is not nearly as rare as it first appears. There are many natural shapes that exhibit some form of self-similarity, even if in nature self-similarity can never be infinite or exact. A classic example of natural self-similarity can be found in, of all places, a head of cauliflower. A head of cauliflower is made of individual pieces called florets. Each floret looks much like a smaller head of cauliflower with even smaller florets, and so on. The "so on" here is good only for three or four steps—at some point there are no more florets—so the self-similarity is *finite*. Moreover, the florets look very much like the head of cauliflower but they are not exact clones of it. In these cases we describe the self-similarity as *approximate* (as opposed to *exact*) self-similarity.

In spite of their differences (one is an abstract human-made geometric shape with infinite and exact self-similarity, the other a natural object with finite and approximate self-similarity), the Koch snowflake and the head of cauliflower have a lot in common in terms of structure and form. Even though you can't make soup with it, one might informally say that the Koch snowflake is a two-dimensional mathematical blueprint for the structure of cauliflower.

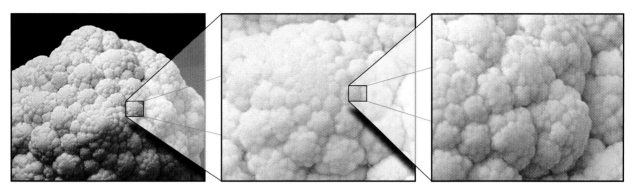

Approximate self-similarity in a head of cauliflower. At various levels of magnification we see … more or less the same thing.

Perimeter and Area of the Koch Snowflake

Perhaps the most surprising property of the Koch snowflake is that it has an infinitely long boundary and yet it encompasses a finite area—a notion that seems to defy common sense.

To compute the boundary of the Koch snowflake, let's look at the boundary of the figures obtained in Steps 1 and 2 of the construction (Fig. 12-5). At each step we replace a side by four sides that are 1/3 as long. Thus, at any given step the perimeter is 4/3 times the perimeter at the preceding step. This implies that the perimeters keep growing with each step, and growing very fast indeed. After infinitely many steps the perimeter is infinite.

See Exercises 1 and 2.

FIGURE 12-5 At each step of the construction, the length of the boundary increases by a factor of 4/3.

■ The Koch snowflake has infinite perimeter.

To compute the exact area of the Koch snowflake is considerably more difficult, but, as we can see from Fig. 12-6, the Koch snowflake fits inside the circle that circumscribes the original equilateral triangle and, thus, its area is reasonably small.

Start   Step 1   Step 2   Step 3   Step 4

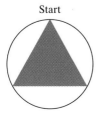    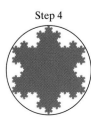

FIGURE 12-6

Having a very large boundary packed within a small area (or volume) is an important characteristic of many self-similar shapes in nature (long boundaries improve functionality while small volumes keep energy costs down). The vascular system of veins and arteries in the human body is a perfect example of the trade-off that nature makes between length and volume: While veins, arteries and capillaries take up only a small fraction of the body's volume, their reach, by necessity, is enormous. Laid end to end, the veins, arteries, and capillaries of a single human being would extend over 40,000 miles.

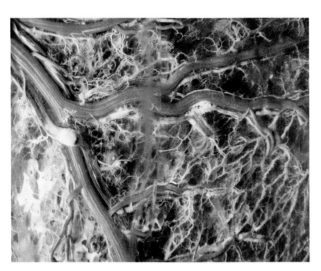
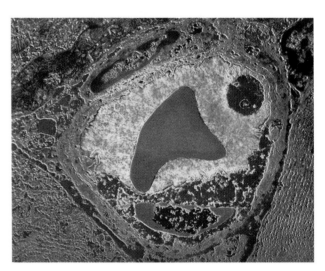

Left: The vascular network of the human body—forty thousand miles of veins, arteries, and capillaries packed inside small quarters. *Right:* Cross section of a blood capillary, with a single red blood cell in the center (magnification: 7070 times).

In what follows, we will give an outline of the argument that shows the exact computation of the area of the Koch snowflake, leaving the technical details as exercises for the reader. In fact, the reader who wishes to do so may skip the forthcoming explanation without prejudice.

For simplicity we will call the area of the starting triangle A. Using A as the starting point we can calculate the area of the figures at the early steps of the construction—all we need to know is the number of new triangles being added and the area of each new triangle. A careful look at Fig. 12-7 will show us the pattern: To get from Start to Step 1 we are adding three new triangles each of area $A/9$. (When the sides of a triangle are scaled by a factor of s the area is scaled by a factor of s^2. Here $s = 1/3$.) Thus, the total area of the star in Step 1 is $A + 3 \cdot (A/9) = A + (A/3)$. To get to Step 2 we add an additional 12 small triangles, each of area $A/81$. The total new area added is $12 \cdot (A/81)$. We will find it convenient to rewrite this in the form $(A/3) \cdot (4/9)$. Continuing this way (and after a good look at Fig. 12-7), we can see that in the Nth step of the construction we add $3(4^{N-1})$ new triangles, each having an area of $(1/9)^N \cdot A$, for a total added area of $(A/3) \cdot (4/9)^{N-1}$.

The total area at the Nth step of the construction is A plus the sum of all the new areas added up to that point:

$$A + \frac{A}{3} + \frac{A}{3}\left(\frac{4}{9}\right) + \frac{A}{3}\left(\frac{4}{9}\right)^2 + \cdots + \frac{A}{3}\left(\frac{4}{9}\right)^{N-1}$$

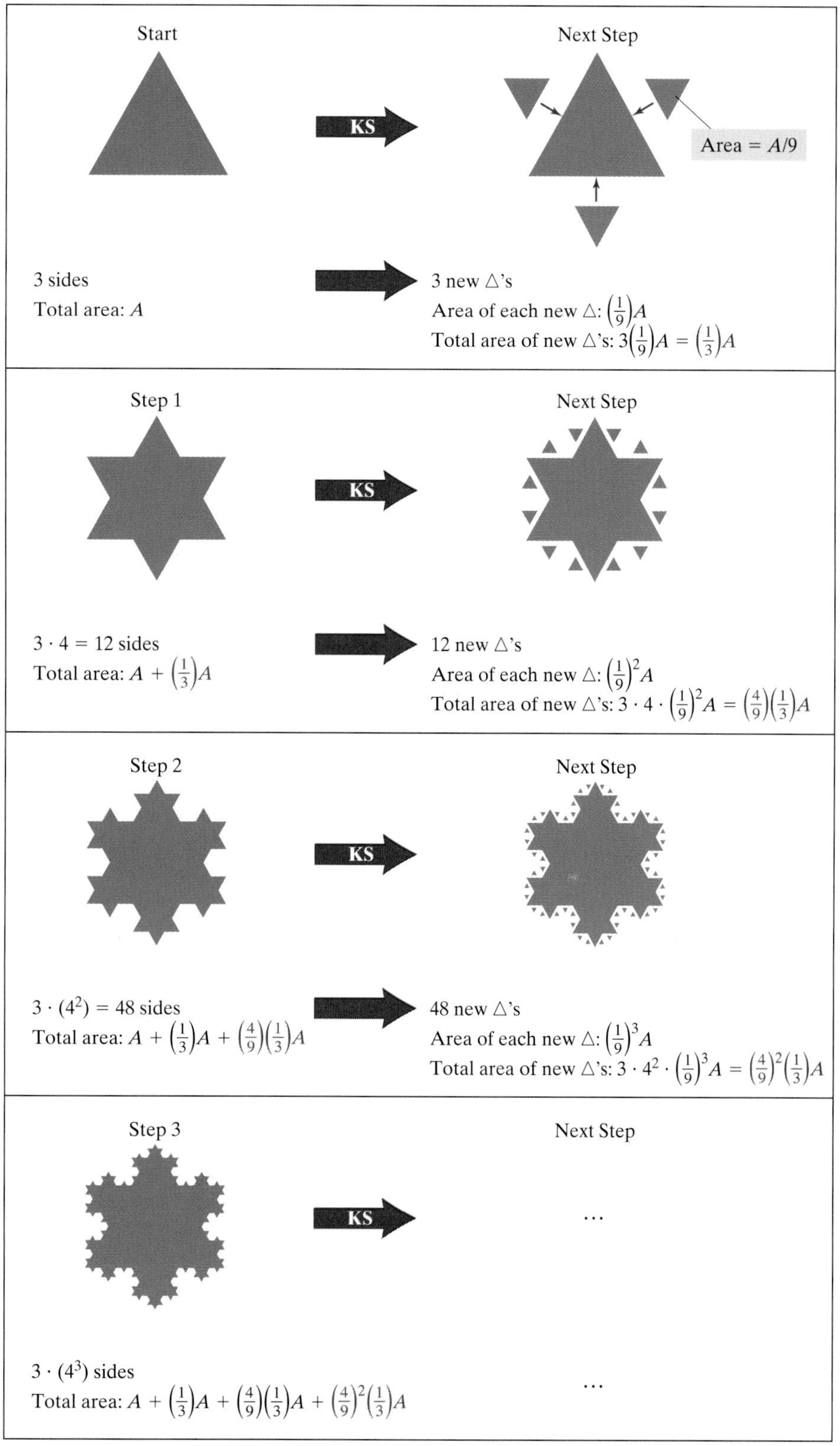
Start
KS
Next Step
Area = $A/9$
3 sides
Total area: A
3 new $\triangle$'s
Area of each new $\triangle$: $\left(\frac{1}{9}\right)A$
Total area of new $\triangle$'s: $3\left(\frac{1}{9}\right)A = \left(\frac{1}{3}\right)A$
Step 1
KS
Next Step
$3 \cdot 4 = 12$ sides
Total area: $A + \left(\frac{1}{3}\right)A$
12 new $\triangle$'s
Area of each new $\triangle$: $\left(\frac{1}{9}\right)^2 A$
Total area of new $\triangle$'s: $3 \cdot 4 \cdot \left(\frac{1}{9}\right)^2 A = \left(\frac{4}{9}\right)\left(\frac{1}{3}\right)A$
Step 2
KS
Next Step
$3 \cdot (4^2) = 48$ sides
Total area: $A + \left(\frac{1}{3}\right)A + \left(\frac{4}{9}\right)\left(\frac{1}{3}\right)A$
48 new $\triangle$'s
Area of each new $\triangle$: $\left(\frac{1}{9}\right)^3 A$
Total area of new $\triangle$'s: $3 \cdot 4^2 \cdot \left(\frac{1}{9}\right)^3 A = \left(\frac{4}{9}\right)^2\left(\frac{1}{3}\right)A$
Step 3
KS
Next Step
...
$3 \cdot (4^3)$ sides
Total area: $A + \left(\frac{1}{3}\right)A + \left(\frac{4}{9}\right)\left(\frac{1}{3}\right)A + \left(\frac{4}{9}\right)^2\left(\frac{1}{3}\right)A$
...

FIGURE 12-7

See Exercise 61.

If we disregard the first A, the remaining terms of the preceding sum are terms in a geometric sequence with starting term $A/3$ and common ratio $r = 4/9$. We discussed geometric sequences and their sums in Chapter 10. Using the geometric sum formula (see p. 354), the preceding expression becomes

$$A + \frac{3}{5}A\left[1 - \left(\frac{4}{9}\right)^N\right]$$

Now, as N becomes larger and larger, $(4/9)^N$ becomes smaller and smaller (this happens whenever the base is between 0 and 1), and the expression inside the square brackets becomes closer and closer to 1. It follows that the entire expression becomes closer and closer to $A + (3/5)A = (8/5)A = (1.6)A$. This implies that after infinitely many steps of adding smaller and smaller areas, we get the amazing result that the area of the Koch snowflake is 1.6 times the area of the starting equilateral triangle.

> ■ The area of the Koch snowflake is 1.6 times the area of the starting equilateral triangle.

12.2 The Sierpinski Gasket

With the insight gained by our study of the Koch snowflake, we will now look at another well-known geometric fractal called the **Sierpinski gasket**, named after the Polish mathematician Waclaw Sierpinski (1882–1969).

Just like with the Koch snowflake, the construction of the Sierpinski gasket starts with a solid triangle, but this time, instead of repeatedly *adding* smaller and smaller versions of the original triangle, we will *remove* them according to the following procedure:

For convenience, we will call this process "cutting the middle" out of the triangle, and more formally, *Procedure SG*.

The Sierpinski Gasket (Recursive Construction)

- **Start.** Start with *any* solid triangle ABC [Fig. 12-8(a)]. (Often an equilateral triangle or a right triangle is used, but here we chose a random triangle to underscore the fact that it can be a triangle of arbitrary shape.)

- **Step 1.** Remove the triangle connecting the midpoints of the sides of the solid triangle. This gives the shape shown in Fig. 12-8(b)—consisting of three solid triangles each a half-scale version of the original and a hole where the middle triangle used to be.

- **Step 2.** To each of the three triangles in Fig. 12-8(b) apply *Procedure SG*. The result is the "gasket" shown in Fig. 12-8(c) consisting of $3^2 = 9$ triangles each at one-fourth the scale of the original triangle, plus three small holes of the same size and one larger hole in the middle.

- **Step 3.** To each of the nine triangles in Fig. 12-8(c) apply *Procedure SG*. The result is the "gasket" shown in Fig. 12-8(d) consisting of $3^3 = 27$

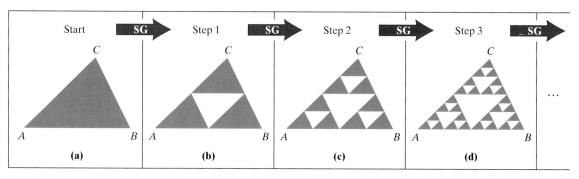

FIGURE 12-8

triangles each at one-eighth the scale of the original triangle, nine small holes of the same size, three medium-sized holes, and one large hole in the middle.

- **Step 4, 5, etc.** Apply *Procedure SG* to each triangle in the "gasket" obtained in the previous step.

After a few more steps the figure becomes visually stable—the naked eye cannot tell the difference between the gasket obtained at one step and the gasket obtained at the next step. At this point we have a good rendering of the Sierpinski gasket itself. In your mind's eye you can think of Fig. 12-9 as a picture of the Sierpinski gasket (in reality it is the gasket obtained at Step 7 of the construction process).

FIGURE 12-9

The Sierpinski gasket is clearly a fairly complicated geometric shape, and yet it can be defined in two lines using the following recursive *replacement rule*.

The Sierpinski Gasket

- **Start:** Start with an arbitrary solid triangle ▲.
- **Replacement rule:** Whenever you see a ▲ apply *Procedure SG* to it.

Looking at Fig. 12-9, we might think that the Sierpinski gasket is made of a huge number of tiny triangles, but this is just an optical illusion—there are no

Some folks describe it as Sierpinski dust!

solid triangles in the Sierpinski gasket, just specks of the original triangle surrounded by a sea of white triangular holes. If we were to magnify any one of those small specks, we would see more of the same—specks surrounded by white triangles (Fig. 12-10). This, of course, is another example of *self-similarity*. As with the Koch curve, the self-similarity of the Sierpinski gasket is (i) infinite, (ii) universal, and (iii) exact.

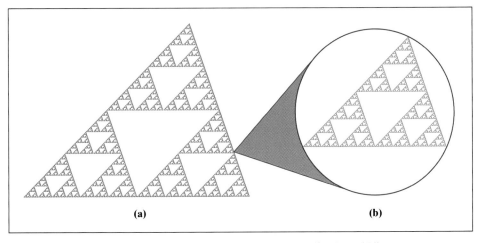

FIGURE 12-10 (a) Sierpinski gasket. (b) Sierpinski gasket detail (magnification ×256).

As a geometric object existing in the plane, the Sierpinski gasket should have an area, but it turns out that its area is infinitely small, smaller than any positive quantity. Paradoxical as it may first seem, the mathematical formulation of this fact is that the Sierpinski gasket has *zero area*. At the same time, the boundary of the "gaskets" obtained at each step of the construction grows without bound, and this implies that the Sierpinski gasket has an infinitely long boundary. Thus, we have the following remarkable combination of facts:

See Exercises 21 and 22.

- The Sierpinski gasket has zero area but infinitely long boundary.

There is a surprising cutting-edge application of the Sierpinski gasket to one of the most vexing problems of modern life: the inconsistent reception on your cell phone. A really efficient cell phone antenna must have a very large boundary (more boundary means stronger reception), a very small area (less area means more energy efficiency), and exact self-similarity (self-similarity means that the antenna works equally well at every frequency of the radio spectrum). What better choice than a design based on a Sierpinski gasket? Fractal antennas based on this concept are now used not only in cell phones but also in wireless modems and GPS receivers (see Project D).

In nature, designs that resemble the Sierpinski gasket can be found on the shells of certain families of sea snails and volutes. Exactly why this happens is not fully understood, but the shells sure are beautiful!

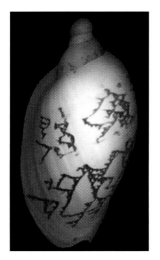
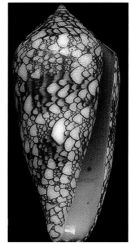

Left: Livonia mamilla. *Center:* Conus canonicus. *Right:* Conus ammiralis.

12.3 The Chaos Game

This example involves the laws of chance. We start with an arbitrary triangle with vertices *A, B,* and *C* and an honest die [Fig. 12-11(a)]. Before we start we assign two of the six possible outcomes of rolling the die to each of the vertices of the triangle. For example, if we roll a 1 or a 2, then *A* is the "winner"; if we roll a 3 or a 4, then *B* is the "winner"; and if we roll a 5 or a 6, then *C* is the "winner." We are now ready to play the game.

- ■ **Start.** Roll the die. Start at the "winning" vertex. Say we roll a 5. We then start at vertex *C* [Fig. 12-11(b)].

- ■ **Step 1.** Roll the die again. Say we roll a 2, so the winner is vertex *A*. We now *move to the point M_1 halfway between the previous position C and the winning vertex A.* Mark a point at the new position M_1 [Fig. 12-11(c)].

- ■ **Step 2.** Roll the die again, and *move to the point halfway between the last position M_1 and the winning vertex.* [Say we roll a 3—the move then is to M_2 halfway between M_1 and *B* as shown in Fig. 12-11(d).] Mark a point at the new position M_2.

- ■ **Step 3, 4, etc.** Continue rolling the die, each time *moving to a point halfway between the last position and the winning vertex* and marking that point.

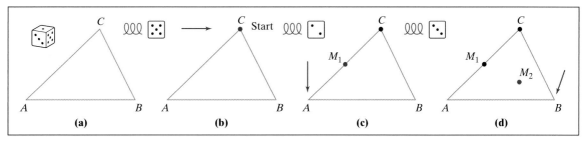

FIGURE 12-11 The Chaos Game: Always move from the previous position towards the chosen vertex, and stop halfway.

Figure 12-12(a) shows the pattern of points after 50 rolls of the die—just a bunch of scattered dots. Figure 12-12(b) shows the pattern of points after 500 rolls; Figure 12-12(c) shows the pattern of points after 5000 rolls. The last figure is unmistakable: a Sierpinski gasket! The longer we play the chaos game, the closer we get to a Sierpinski gasket. After 100,000 rolls of the die, it would be impossible to tell the difference between the two. (For a nice Java applet that lets you play the chaos game, go to *www.cevis.uni-bremen.de/fractals/nsfpe/Chaos_Game/Chaos1.html*)

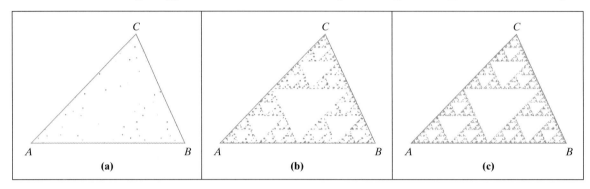

FIGURE 12-12 The "footprint" of the chaos game after (a) 50 rolls of the die, (b) 500 rolls of the die, and (c) 5000 rolls of the die.

This is a truly surprising turn of events. The chaos game is ruled by the laws of chance, and thus we would expect that essentially a random pattern of points would be generated. Instead, we get an approximation to the Sierpinski gasket, and the longer we play the chaos game, the better the approximation gets. An important implication of this is that it is possible to generate self-similar shapes using the laws of chance.

12.4 The Twisted Sierpinski Gasket

Our next example is a simple variation of the original Sierpinski gasket. For lack of a better name, we will call it the *twisted Sierpinski gasket*.

The construction starts out exactly like the one for the regular Sierpinski gasket, with a solid triangle [Fig. 12-13(a)] from which we cut out the middle triangle, whose vertices we will call *M*, *N*, and *L* [Fig. 12-13(b)]. The next move (which we will call the "twist") is new. Each of the points *M*, *N*, and *L* is moved a small amount in a random direction—as if jolted by an earthquake—to new positions *M'*, *N'*, and *L'*. One possible resulting shape is shown in Fig. 12-13(c).

For convenience, we will use the term *Procedure TSG* to describe the combination of the two moves ("cut" and then "twist").

■ **Cut.** Cut the middle out of a triangle [Fig. 12-13(b)].

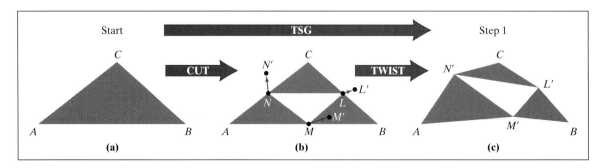

FIGURE 12-13 The two moves in *procedure TSG*: the *cut* and the *twist*.

■ **Twist.** Translate each of the midpoints of the sides by a small random amount and in a random direction. [Fig. 12-13(c)].

When we repeat *procedure TSG* in an infinite recursive process, we get the twisted Sierpinski gasket.

The Twisted Sierpinski Gasket (Recursive Construction)

■ **Start.** Start with an arbitrary solid triangle.

■ **Step 1.** Apply *procedure TSG* to the starting triangle. This gives the "twisted gasket" shown in Fig. 12-14(b), with three twisted triangles and a (twisted) hole in the middle.

■ **Step 2.** To each of the three triangles in Fig. 12-14(b) apply *procedure TSG*. The result is the "twisted gasket" shown in Fig. 12-14(c), consisting of nine twisted triangles and four holes of various sizes.

■ **Step 3, 4, etc.** Apply *procedure TSG* to each triangle in the "twisted gasket" obtained in the previous step.

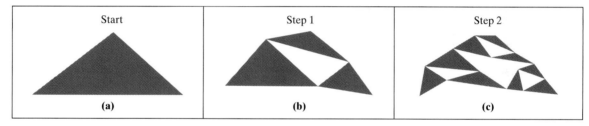

FIGURE 12-14 First two steps in the construction of the twisted Sierpiski gasket.

FIGURE 12-15

Figure 12-15 shows an example of a twisted Sierpinski gasket at Step 7 of the construction. Even without touch-up, we can see that this image has the unmistakable look of a mountain. Add a few of the standard tools of computer graphics—color, lighting, and shading—and we can get a very realistic-looking mountain indeed.

Left: Photo of a real mountain range. *Right:* Computer generated mountain range.

The construction of the twisted Sierpinski gasket can be also described by a two-line recursive replacement rule:

Twisted Sierpinski Gasket

- **Start:** Start with an arbitrary solid triangle.
- **Replacement rule:** Wherever you see a solid triangle, apply *procedure TSG* to it.

The twisted Sierpinski gasket has infinite and universal self-similarity, but due to the randomness of the twisting process the self-similarity is only *approximate*—when we magnify any part of the gasket we see similar, but not identical images, as illustrated in Fig. 12-16. It is the approximate self-similarity that gives this shape that natural look that the original Sierpinski gasket lacks, and it illustrates why randomness is an important element in creating artificial imitations of natural-looking shapes.

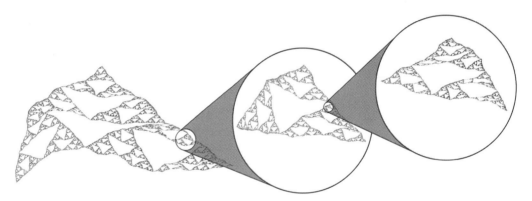

FIGURE 12-16 Approximate self-similarity in the twisted Sierpinski gasket.

By changing the shape of the starting triangle, we can change the shape of the mountain, and by changing the rules for how large we allow the random displacements to be, we can change the mountain's texture. Both of these are computer-generated images.

12.5 The Mandelbrot Set

For more on Mandelbrot, see the biographical profile at the end of this chapter.

In this section we will introduce one of the most interesting and beautiful geometric shapes ever created by the human hand, an object called the **Mandelbrot set** after the Polish-born mathematician Benoit Mandelbrot. Some of the mathematics behind the Mandelbrot set goes a bit beyond the level of this book, so we will describe the overall idea and try not to get bogged down in the details.

We will start this section with a brief visual tour. The Mandelbrot set is the shape shown in Fig. 12-17(a), a strange-looking blob of black. Using a strategy we will describe later, the different regions outside the Mandelbrot set can be colored according to their mathematical properties, and when this is done [Fig. 12-17(b)] the Mandelbrot set comes to life like a switched-on neon sign.

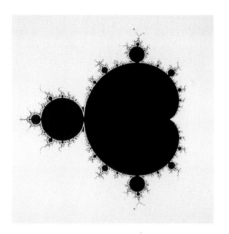 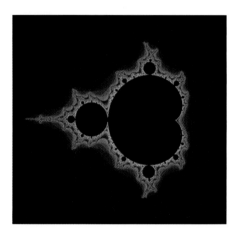

FIGURE 12-17 (a) The Mandelbrot set (black) on a white background. (b) The Mandelbrot set comes to life when color is added to the outside.

So Natr'alists observe, A Flea Hath smaller Fleas that on him prey and these have smaller Fleas to bite'em And so proceed, ad infinitum.

Jonathan Swift

In the wild imagination of some, the Mandelbrot set looks like some sort of bug—an exotic extraterrestrial flea. The "flea" is made up of a heart-shaped body (called a *cardioid*), a head, and an antenna coming out of the middle of the head. A careful look at Fig. 12-17 shows that the flea has many smaller fleas that prey on it, but we can only begin to understand the full extent of the infestation when we look at Fig. 12-18(a)—a finely detailed close-up of the boundary of the Mandelbrot set. When we magnify the view around the boundary even further, we can see that these secondary fleas have fleas that prey on them [Figs. 12-18(b) and (c)], and further magnification would show this repeats itself ad infinitum. Clearly, Jonathan Swift was onto something!

It is clear from Fig. 12-18 that the Mandelbrot set has some strange form of *infinite* and *approximate* self-similarity—at infinitely many levels of magnification we see the same theme—like-looking fleas surrounded by smaller fleas. At the same time, we can see that there is tremendous variation in the regions surrounding the individual fleas. The images we see are a peek into a psychedelic coral reef—a world of strange "urchins" and "seahorse tails" in Fig. 12-18(b), "anemone" and "starfish" in Fig. 12-18(c). Further magnification shows an even

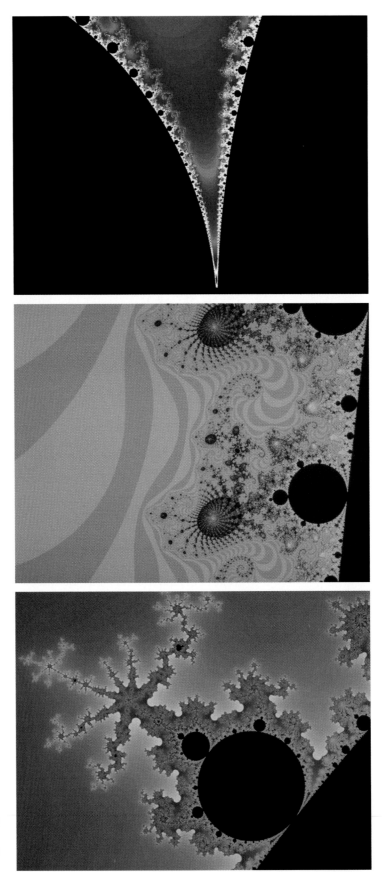

FIGURE 12-18 (a) Detailed close-up of a small region on the boundary. (b) An even tighter close-up of the boundary (c) A close-up of one of the secondary fleas.

more exotic and beautiful landscape. Figure 12-19(a) is a close-up of one of the seahorse tails in Fig. 12-18(b). A further close-up of a section of Fig. 12-19(a) is shown in Fig. 12-19(b), and an even further magnification of it is seen in Fig. 12-19(c), revealing a tiny copy of the Mandelbrot set surrounded by a beautiful arrangement of swirls, spirals, and seahorse tails. [The magnification for Fig. 12-19(c) is approximately 10,000 times the original.]

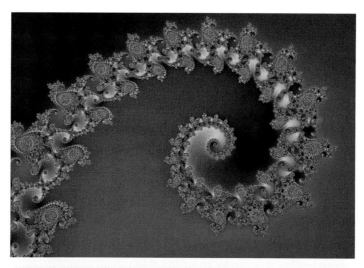

FIGURE 12-19 (a) A close-up of one of the "seahorse tails" in Fig. 12-18(b) (×100 magnification). (b) A further close-up of a section of Fig. 12-19(a) shows the fine detail and more seahorse tails (×200 magnification). (c) At infinitely many levels of magnification, old themes show up in new settings (×10,000 magnification).

Anywhere we choose to look at these pictures, we will find (if we magnify enough) copies of the original Mandelbrot set, always surrounded by an infinitely changing, but always stunning, background. The infinite, approximate self-similarity of the Mandelbrot set manages to blend infinite repetition and infinite variety, creating a landscape as consistently exotic and diverse as nature itself.

Complex Numbers and Mandelbrot Sequences

How does this magnificent mix of beauty and complexity called the Mandelbrot set come about? Incredibly, the Mandelbrot set itself can be described mathematically by a recursive process involving simple computations with *complex numbers*.

You may recall having seen complex numbers in high school algebra. Among other things, complex numbers allow us to take square roots of negative numbers and solve quadratic equations of any kind. The basic building block for complex numbers is the number $i = \sqrt{-1}$. Starting with i we can build all other complex numbers, such as $3 + 2i$, $-0.125 + 0.75i$, and even $1 + 0i = 1$, or $-0.75 + 0i = -0.75$. Complex numbers can be added, multiplied, divided, squared, and so on. (For a quick review of the basic operations with complex numbers see Exercises 41 through 44.)

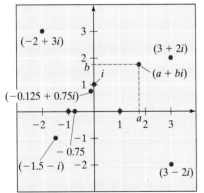

For our purposes, the most important fact about complex numbers is that they have a geometric interpretation: The complex number $(a + bi)$ can be identified with the point (a, b) in a Cartesian coordinate system, as shown in Fig. 12-20. This identification means that every complex number can be thought of as a point in the plane and that operations with complex numbers have geometric interpretations (see Exercises 45 and 46).

The key concept in the construction of the Mandelbrot set is that of a *Mandelbrot sequence*. A **Mandelbrot sequence** (with seed s) is an infinite sequence of complex numbers that starts with an arbitrary complex number s and then each successive term in the sequence is obtained recursively by *adding the seed s to the previous term squared*. Figure 12-21 shows a schematic illustration of how a generic Mandelbrot sequence is generated. [Notice that the only complex number operations involved are (i) squaring a complex number, and (ii) adding two complex numbers. If you know how to do these two things, you can compute Mandelbrot sequences.]

FIGURE 12-20 Every complex number is a point in the Cartesian plane; every point in the Cartesian plane is a complex number.

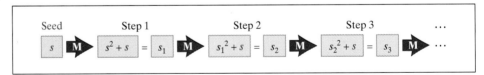

FIGURE 12-21 A Mandelbrot sequence with seed s.

Much like the Koch snowflake and the Sierpinski gasket, a Mandelbrot sequence can be defined by means of a recursive replacement rule:

Mandelbrot Sequence

- **Start:** Choose an arbitrary complex number s, called the **seed** of the Mandelbrot sequence. Set the seed s to be the initial term of the sequence ($s_0 = s$).

- **Procedure M:** To find the next term in the sequence, square the preceding term and add the seed ($s_{N+1} = s_N^2 + s$).

The next set of examples will illustrate the different patterns of growth exhibited by Mandelbrot sequences as we vary the seeds. These different patterns of growth are going to tell us how to generate the Mandelbrot set itself and the incredible images that we saw in our visual tour. The idea goes basically like this: Each point in the Cartesian plane is a complex number and thus the seed of some Mandelbrot sequence. The pattern of growth of that Mandelbrot sequence determines whether the seed is inside the Mandelbrot set (black point) or outside (nonblack point). In the case of nonblack points, the color assigned to the point is also determined by the pattern of growth of the corresponding Mandelbrot sequence. Let's try it.

> ### EXAMPLE 12.1 Escaping Mandelbrot Sequences

Figure 12-22 shows the first few terms of the Mandelbrot sequence with seed $s = 1$. (Since integers and decimals are also complex numbers, they make perfectly acceptable seeds.) The pattern of growth of this Mandelbrot sequence is clear—the terms are getting larger and larger, and they are doing so very quickly. Geometrically, it means that the points in the Cartesian plane that represent the numbers in this sequence are getting further and further away from the origin. For this reason we call this sequence an *escaping* Mandelbrot sequence.

Seed   Step 1   Step 2   Step 3   Step 4

$s = 1$ **M** $s_1 = 1^2 + 1$ $= 2$ **M** $s_2 = 2^2 + 1$ $= 5$ **M** $s_3 = 5^2 + 1$ $= 26$ **M** $s_4 = 26^2 + 1$ $= 677$ $\cdots$

FIGURE 12-22 Mandelbrot sequence with seed $s = 1$ (*escaping very quickly*).

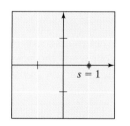

FIGURE 12-23 The seed is painted blue.

In general, when the points that represent the terms of a Mandelbrot sequence move further and further away from the origin, we will say that the Mandelbrot sequence is **escaping**. The basic rule that defines the Mandelbrot set is that seeds of escaping Mandelbrot sequences are *not* in the Mandelbrot set and must be assigned some color other than black. While there is no specific rule that tells us what color should be assigned, the overall color palette is based on how fast the sequence is escaping. The typical approach is to use "hot" colors such as reds, yellows, and oranges for seeds that escape slowly, and "cool" colors such as blues and purples for seeds that escape quickly. The seed $s = 1$, for example, escapes very quickly, and the corresponding point in the Cartesian plane is painted blue [Fig 12-23]. ◀◀

> ### EXAMPLE 12.2 Periodic Mandelbrot Sequences

Figure 12-24 shows the first few terms of the Mandelbrot sequence with seed $s = -1$. The pattern that emerges here is also clear—the numbers in the sequence alternate between 0 and -1. In this case we say that the Mandelbrot sequence is *periodic*.

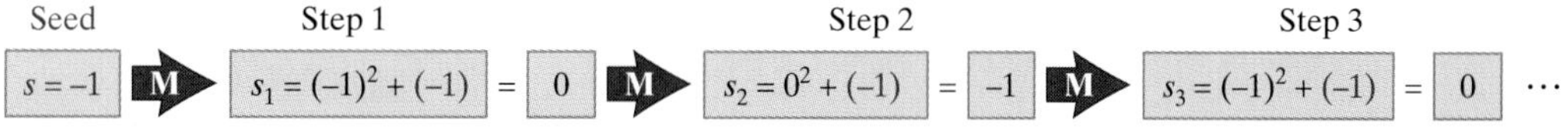

Seed   Step 1   Step 2   Step 3

$s = -1$ **M** $s_1 = (-1)^2 + (-1)$ $= 0$ **M** $s_2 = 0^2 + (-1)$ $= -1$ **M** $s_3 = (-1)^2 + (-1)$ $= 0$ $\cdots$

FIGURE 12-24 Mandelbrot sequence with seed $s = -1$ (*periodic*).

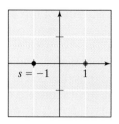

FIGURE 12-25 The seed is in the Mandelbrot set (black point).

In general, a Mandelbrot sequence is said to be **periodic** if at some point the numbers in the sequence start repeating themselves in a cycle. When the Mandelbrot sequence is periodic, the seed is a point of the Mandelbrot set, and thus, it is assigned the color black (Fig. 12-25). ≪

> **EXAMPLE 12.3** Attracted Mandelbrot Sequences

Figure 12-26 shows the first few terms in the Mandelbrot sequence with seed $s = -0.75$. Here the growth pattern is not obvious, and additional terms of the sequence are needed. Further computation (a calculator will definitely come in handy) shows that as we go further and further out in this sequence, the terms get closer and closer to the value -0.5. In this case we will say that the sequence is *attracted* to the value -0.5.

FIGURE 12-26 Mandelbrot sequence with seed $s = -0.75$ (*attracted*).

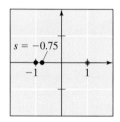

FIGURE 12-27 The seed is in the Mandelbrot set (black point).

In general, when the terms in a Mandelbrot sequence get closer and closer to a fixed complex number a, we say that a is an **attractor** for the sequence, or, equivalently, that the sequence is **attracted** to a. Just as with periodic sequences, when a Mandelbrot sequence is attracted, the seed s is in the Mandelbrot set and colored black (Fig. 12-27). ≪

So far, all of our examples have been based on rational number seeds (mostly to keep things simple), but the truly interesting cases occur when the seeds are complex numbers. The next two examples deal with complex number seeds.

> **EXAMPLE 12.4** A Periodic Mandelbrot Sequence with Complex Terms

In this example we will examine the growth of the Mandelbrot sequence with seed $s = i$. Starting with $s = i$ (and using the fact that $i^2 = -1$), we get $s_1 = i^2 + i = -1 + i$. If we now square s_1 and add i, we get $s_2 = (-1 + i)^2 + i = -i$, and repeating the process gives $s_3 = (-i)^2 + i = -1 + i$. At this point we notice that $s_3 = s_1$, which implies $s_4 = s_2$, $s_5 = s_1$, and so on (Fig. 12-28). This, of course, means that this Mandelbrot sequence is *periodic*, with its terms alternating between the complex numbers $-1 + i$ (odd terms) and $-i$ (even terms).

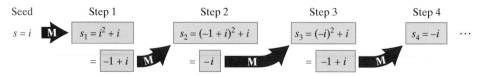

FIGURE 12-28 Mandelbrot sequence with seed $s = i$ (*periodic*).

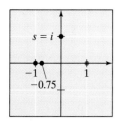

FIGURE 12-29 The seed is in the Mandelbrot set (black point).

The key conclusion from the preceding computations is that the seed i is a black point inside the Mandelbrot set (Fig. 12-29). ◀◀

The next example illustrates the case of a Mandelbrot sequence with three complex attractors.

> **EXAMPLE 12.5** A Mandelbrot Sequence
> with Three Complex Attractors

In this example we will examine the growth of the Mandelbrot sequence with seed $s = -0.125 + 0.75i$. The first few terms of the Mandelbrot sequence are shown. We leave it to the enterprising reader to check these calculations. (You will need a scientific calculator that handles complex numbers—if you don't have one, you can download a free virtual calculator for doing complex number arithmetic from *www.calc3D.com* or other similar websites.)

$s_0 = -0.125 + 0.75i$ $\qquad$ $s_1 = -0.671875 + 0.5625i$ $\qquad$ $s_2 = 0.0100098 - 0.00585938i$

$s_3 = -0.124934 + 0.749883i$ $\qquad$ $s_4 = -0.671716 + 0.562628i$ $\qquad$ $s_5 = 0.00965136 - 0.00585206i$

$s_6 = -0.124941 + 0.749887i$ $\qquad$ $s_7 = -0.67172 + 0.562617i$ $\qquad$ $s_8 = 0.00967074 - 0.00584195i$

We can see that the terms in this Mandelbrot sequence are complex numbers that essentially cycle around in sets of three and are approaching three different attractors. Since this Mandelbrot sequence is attracted, the seed $s = -0.125 + 0.75i$ represents another point in the Mandelbrot set. ◀◀

The Mandelbrot Set

Given all the previous examples and discussion, a formal definition of the Mandelbrot set using seeds of Mandelbrot sequences sounds incredibly simple: If the Mandelbrot sequence is *periodic* or *attracted*, the seed is a point of the Mandelbrot set and assigned the color black; if the Mandelbrot sequence is *escaping*, the seed is a point outside the Mandelbrot set and assigned a color that depends on the speed at which the sequence is escaping (hot colors for slowly escaping sequences, cool colors for fast escaping sequences). There are a few technical details that we omitted, but essentially these are the key ideas behind the amazing pictures that we saw in Figs. 12-18 and 12-19. In addition, of course, a computer is needed to carry out the computations and generate the images.

Because the Mandelbrot set provides a bounty of aesthetic returns for a relatively small mathematical investment, it has become one of the most popular mathematical playthings of our time. There are now literally hundreds of software programs available that allow one to explore the beautiful landscapes surrounding the Mandelbrot set, and many of these programs are freeware. (You can find plenty of these by going to Google and entering the search term "Mandelbrot set.")

Conclusion

The study of *fractals* and their geometry has become one of the hottest mathematical topics of the last 20 years. It is a part of mathematics that combines complex and interesting theories, beautiful graphics, and extreme relevance to the real world. In this chapter we only scratched the surface of this deep and rich topic.

The word **fractal** (from the Latin *fractus*, meaning "broken up or fragmented") was coined by Benoit Mandelbrot in the mid-1970s to describe objects as diverse as the Koch curve, the Sierpinski gasket, the twisted Sierpinski gasket, and the Mandelbrot set, as well as many shapes in nature, such as clouds, mountains, trees, rivers, a head of cauliflower, and the vascular system in the human body.

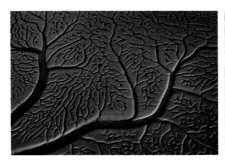  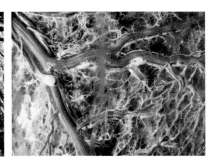

These objects share one key characteristic—they all have some form of self-similarity. (Self-similarity is not the only defining characteristic of a fractal—others, such as *fractional dimension*, are discussed in Project A.) There is a striking visual difference between the fractal geometry of self-similar shapes and the traditional geometry of lines, circles, spheres, and so on. This visual difference is most apparent when we compare the look and texture of natural objects (mountains, trees, coastlines, etc.) with that of human-made objects such as bridges, machines, buildings, and so on.

Geometry as we have known it in the past was developed by the Greeks about 2000 years ago and passed on to us essentially unchanged. It was (and still is) a great triumph of the human mind, and it has allowed us to develop much of our technology, engineering, architecture, and so on. As a tool and a language for modeling and representing nature, however, Greek geometry has by and large been a failure. The discovery of fractal geometry seems to have given science the right mathematical language to overcome this failure, and thus it promises to be one of the great achievements of twentieth-century mathematics. Today, fractal geometry is used to study the patterns of clouds and how they affect the weather, to diagnose the pattern of contractions of a human heart, to design more efficient antennas, and to create some of the otherworldly computer graphics that animate many of the latest science fiction movies.

Profile Benoit Mandelbrot (1924–)

With the publication in 1983 of his classic book *The Fractal Geometry of Nature*, Benoit Mandelbrot became somewhat of a scientific celebrity. Later, with images of the Mandelbrot set and other exotic fractals becoming part of our popular culture (through screensavers, T-shirt designs, and television ads), the Mandelbrot name became an icon of the computer age.

Born in Warsaw, Poland in 1924, Mandelbrot's family moved to France when he was 11 years old. These were diffi-

cult times—the beginning of World War II—and Mandelbrot received little formal education during his formative years. One of his uncles was a mathematics professor, and Mandelbrot learned some mathematics from him, but by and large, the mathematics that Mandelbrot learned in his youth was self-taught. To Mandelbrot, this turned out to be an asset rather than a liability, as he credits much of his refined mathematical intuition and his ability to think "outside the box" to the unstructured nature of his early education.

After the war, Mandelbrot was able to receive a first-rate university education: an undergraduate degree in mathematics from the École Polytechnique in Paris in 1947, an M.S. degree in Aeronautics from the California Institute of Technology in 1948, and a Ph.D. in Mathematics from the University of Paris in 1952. After a brief stint doing research and teaching in Europe, Mandelbrot accepted an appointment in the research division of IBM and moved to the United States in 1958. For the next 30 years Mandelbrot worked at IBM's T. J. Watson Research Center in New York State, first as a Research Fellow and eventually as an IBM Fellow, the most prestigious research position within the company. It was during his years at IBM that Mandelbrot developed and refined his theories on fractal geometry.

Upon his retirement from IBM in 1987, Mandelbrot accepted an endowed chair at Yale University, where he is currently the Abraham Robinson Professor of Mathematical Sciences.

Key Concepts

Exercises

WALKING

A. The Koch Snowflake and Variations

Exercises 1 through 6 refer to the Koch snowflake.

1. If the starting equilateral triangle has sides of length 1 in., fill in the missing entries in the following table.

	Start	Step 1	Step 2	Step 3	Step 4	. . .	Step 30	Step N
Number of sides	3	12						
Length of each side	1 in.	1/3 in.						
Length of boundary	3 in.	4 in.					(*) mi	

*Give your answer rounded to the nearest mile. Use 1 mile = 63,360 inches.

2. If the starting equilateral triangle has sides of length 6 cm, fill in the missing entries in the following table.

	Start	Step 1	Step 2	Step 3	Step 4	. . .	Step 40	Step N
Number of sides	3	12						
Length of each side	6 cm	2 cm						
Length of boundary	18 cm	24 cm					(*) km	

*Give your answer rounded to the nearest kilometer. Use 1 km = 100,000 cm.

3. If the starting equilateral triangle has an area of 24 square inches, fill in the missing entries in the following table.

   (***Hint:** Take a look at Fig. 12-7.*)

	Start	Step 1	Step 2	Step 3	Step 4
Area	24 in.^{2}				

4. If the starting equilateral triangle has sides of length 1 in, fill in the missing entries in the following table. Leave all your answers in radical form.

   (***Hints:** (1) An equilateral triangle with sides of length 1 has area $\sqrt{3}/4$. (2) Try Exercise 3 first.*)

	Start	Step 1	Step 2	Step 3	Step 4
Area	$\left(\sqrt{3}/4\right)$ in.^{2}				

5. Find the area of the Koch snowflake if the starting equilateral triangle has an area of 24 square inches.

6. Find the area of the Koch snowflake if the starting equilateral triangle has an area of $\sqrt{3}/4$ square inches.

*Exercises 7 through 10 refer to the **Koch square snowflake**, a geometric fractal defined by the following recursive construction:*

Koch Square Snowflake

- **Start.** *Start with a solid square.*
- **Step 1.** *Divide each side of the square into three equal segments and attach to the middle segment of each side a solid square with sides of length one-third the length of the sides of the square.*
- **Step 2.** *Divide each side of the figure obtained in the previous step into three equal segments and attach to the middle segment of each side a solid square with sides of length one third the length of the previous side. (Call this Procedure KSS.)*
- **Step 3, 4, etc.** *Apply Procedure KSS to the figure obtained in the previous step.*

The first two steps of the construction of the Koch square snowflake are illustrated in the following figure.

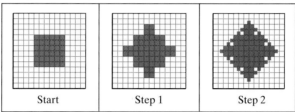

| | Start | Step 1 | Step 2 |

7. If the starting square has sides of length 1 in., fill in the missing entries in the following table.

	Start	**Step 1**	**Step 2**	**Step 3**	**. . .**	**Step 50**	**Step N**
Number of sides	4	20					
Length of each side	1 in.	1/3 in.					
Length of boundary	4 in.	20/3 in.				(*) mi	

*Give your answer rounded to the nearest mile. Use 1 mile = 63,360 inches.

8. If the starting square has sides of length a, fill in the missing entries in the following table.

	Start	**Step 1**	**Step 2**	**Step 3**	**Step 4**	**. . .**	**Step N**
Number of sides	4	20					
Length of each side	a	$a/3$					
Length of boundary	$4a$	$(20/3)a$					

9. If the starting square has sides of length 1, fill in the missing entries in the following table.

	Start	**Step 1**	**Step 2**	**Step 3**	**Step 4**
Area	1	13/9			

10. If the starting square has area A, fill in the missing entries in the following table.

	Start	**Step 1**	**Step 2**	**Step 3**	**Step 4**
Area	A	$(13/9)A$			

*Exercises 11 through 14 refer to the **Koch antisnowflake**. The Koch antisnowflake is obtained by reversing the process for constructing the Koch snowflake—instead of adding a solid equilateral triangle to the middle of each side of the figure, we cut out the equilateral triangle. The Koch antisnowflake is defined by the following recursive replacement rule.*

Koch Antisnowflake

- **Start:** Start with a solid equilateral triangle △.
- **Replacement rule:** Whenever you see a boundary line segment ___outside___ , replace it with ___outside___ ∨.

The first two steps of the construction of the Koch antisnowflake are illustrated in the following figure.

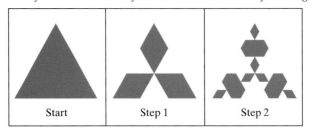

| Start | Step 1 | Step 2 |

11. If the starting equilateral triangle has sides of length a, fill in the missing entries in the following table.

	Start	Step 1	Step 2	Step 3	Step 4	. . .	Step 40	Step N
Number of sides	3							
Length of each side	a	$a/3$						
Length of boundary	$3a$							

12. If the starting equilateral triangle has sides of length 1 in., fill in the missing entries in the following table.

	Start	Step 1	Step 2	Step 3	Step 4	. . .	Step 30	Step N
Number of sides	3	12						
Length of each side	1 in.	1/3 in.						
Length of boundary	3 in.	4 in.					(*) mi	

*Give your answer rounded to the nearest mile. Use 1 mile = 63,360 inches.

13. If the area of the starting equilateral triangle is 81 in^2, fill in the missing entries in the following table.

   (***Hint:*** *The construction of the Koch antisnowflake parallels the construction of the Koch snowflake with triangles being removed instead of added.*)

	Start	Step 1	Step 2	Step 3	Step 4	Step 5
Area	81 in.^{2}	54 in.^{2}				

14. If the area of the starting equilateral triangle is A, fill in the missing entries in the following table.

	Start	Step 1	Step 2	Step 3	Step 4	Step 5
Area	A	$(2/3)A$				

*Exercises 15 through 18 refer to the **quadratic Koch island**. The quadratic Koch island is a curve defined by the following recursive replacement rule:*

Quadratic Koch Island

- **Start:** Start with a square ☐.
- **Replacement rule:** Replace a horizontal segment with a ⌐‿⌐ and a vertical segment with a ⌐‿.

15. Carefully draw the figures at Steps 1 and 2 of the construction of the quadratic Koch island.
   (***Hint:*** *Use graph paper and start with a 16-by-16 square.*)

16. If the starting square has sides of length 1, fill in the missing entries in the following table.

	Start	Step 1	Step 2	Step 3	Step 4	. . .	Step N
Number of sides	4	28					
Length of boundary	4	8					

17. (a) If the area enclosed by the starting square is A, fill in the missing entries in the following table.

	Start	Step 1	Step 2	Step 3	Step 4
Area enclosed by	A				

(b) Find the area of the quadratic Koch island.

18. Explain why the quadratic Koch island has infinite perimeter.
(***Hint:** Do Exercise 16 first.*)

B. The Sierpinski Gasket and Variations

Exercises 19 through 24 refer to the Sierpinski gasket.

19. If the area of the starting triangle is 1, fill in the missing entries in the following table.

	Start	Step 1	Step 2	Step 3	Step 4	. . .	Step N
Area	1	3/4					

20. If the starting triangle has perimeter P, fill in the missing entries in the following table.

	Start	Step 1	Step 2	Step 3	Step 4	. . .	Step N
Number of triangles	1	3					
Perimeter of each triangle	P	1/2					
Length of boundary	P	3/2					

21. Explain why the area of the Sierpinski gasket is infinitely small, smaller than any positive quantity.
(***Hints:** (1) Do Exercise 19 first. (2) The area of the starting triangle is irrelevant, so you can assume for simplicity that it is 1. (3) You should use the fact that if b is between 0 and 1, b^N gets closer and closer to 0 as N gets larger.*)

22. Explain why the perimeter of the Sierpinski gasket is infinite.
(***Hint:** Do Exercise 20 first.*)

23. Suppose the starting triangle for the Sierpinski gasket is a right triangle whose legs have length 6 in. and 8 in., respectively. Fill in the missing entries in the following table.

	Start	Step 1	Step 2	Step 3	Step 4	. . .	Step N
Number of triangles	1	3					
Perimeter of each triangle							
Length of boundary							

24. Suppose the starting triangle for the Sierpinski gasket is a right triangle whose legs have length 6 in. and 8 in., respectively. Complete the following table.

	Start	Step 1	Step 2	Step 3	Step 4	. . .	Step N
Area							

*Exercises 25 through 28 refer to the **Sierpinski carpet**. The Sierpinski carpet is a square version of the Sierpinski gasket and is defined by the following recursive construction.*

Sierpinski Carpet

- **Start.** Start with a solid square.
- **Step 1.** Subdivide the square into nine equal sub-squares and remove the central subsquare.
- **Step 2.** Subdivide each of the remaining squares into nine subsquares and remove the central subsquare. (Call the procedure of subdividing a square into nine subsquares and removing the middle square *Procedure SC*.)
- **Step 3, 4, etc.** Apply *procedure SC* to each square of the "carpet" obtained in the previous step.

The first two steps of the construction of the Sierpinski carpet are illustrated in the following figure.

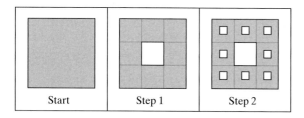

| | Start | Step 1 | Step 2 |

25. Using graph paper, carefully draw the "carpet" obtained at Step 3 of the construction.

26. If the starting square has sides of length 1 in., fill in the missing entries in the following table.

	Start	Step 1	Step 2	Step 3	...	Step N
Area	1	8/9				

27. Suppose the starting square has sides of length 1.

   (a) Complete the following table. (Enter your answers as fractions.)

	Start	Step 1	Step 2	Step 3	Step 4
Perimeter	4	16/3			

   (b) Suppose the perimeter of the "carpet" at step N is L. Express the perimeter of the "carpet" at step $N + 1$ in terms of L.

28. Complete the following table.

	Start	Step 1	Step 2	Step 3	Step 4	Step 5
Number of square holes	0	1	9			

*Exercises 29 through 32 refer the **Sierpinski ternary gasket**. The Sierpinski ternary gasket is a variation of the Sierpinski gasket defined by the following recursive replacement rule.*

Sierpinski Ternary Gasket

- Start with a solid triangle ▲.
- Whenever you see a ▲, replace it with a ▲.

The first two steps of the construction of the Sierpinski ternary gasket are shown in the following figure.

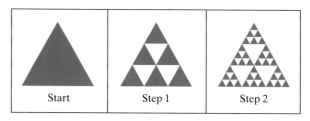

| | Start | Step 1 | Step 2 |

29. If the area of the starting triangle is A, fill in the missing entries in the following table.

	Start	Step 1	Step 2	Step 3	Step 4	...	Step N
Number of triangles	1	6					
Area of each triangle	A	$\frac{1}{9}A$					
Area of gasket	A	$\frac{2}{3}A$					

30. If the starting triangle has perimeter P, fill in the missing entries in the following table.

	Start	Step 1	Step 2	Step 3	Step 4	...	Step N
Number of triangles	1	6					
Perimeter of each triangle	P						
Perimeter of gasket	P						

31. Explain why the area of the ternary Sierpinski gasket is infinitely small, smaller than any positive quantity.

   (***Hint:*** *Do Exercises 21 and 29 first.*)

32. Explain why the perimeter of the ternary Sierpinski gasket is infinite.

(**Hint:** *Do Exercise 30 first.*)

C. The Chaos Game and Variations

Exercises 33 through 36 refer to the chaos game as described in the chapter. Start with an isosceles right triangle ABC with AB = AC = 32, as shown in the figure. You should use graph paper with 10 squares per inch or make a copy of the figure and work directly on it. Assume that vertex A corresponds to numbers 1 and 2, vertex B to numbers 3 and 4, and vertex C to numbers 5 and 6.

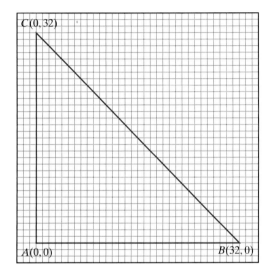

33. Suppose that an honest die is rolled 6 times and that the outcomes are 3, 1, 6, 4, 5, and 5. Carefully draw the points P_1 through P_6 corresponding to these outcomes.

(**Note:** *Each of the points P_1 through P_6 falls on a grid point of the graph. You should be able to identify the location of each point without using a ruler.*)

34. Suppose that an honest die is rolled 6 times and that the outcomes are 2, 6, 1, 4, 3, and 6. Carefully draw the points P_1 through P_6 corresponding to these outcomes.

(**Note:** *Each of the points P_1 through P_6 falls on a grid point of the graph. You should be able to identify the location of each point without using a ruler.*)

35. Using a rectangular coordinate system with A at $(0, 0)$, B at $(32, 0)$, and C at $(0, 32)$, complete the following table.

Number rolled	3	1	2	3	5	5
Point	P_1	P_2	P_3	P_4	P_5	P_6
Coordinates	$(32, 0)$	$(16, 0)$				

36. Using a rectangular coordinate system with A at $(0, 0)$, B at $(32, 0)$, and C at $(0, 32)$, complete the following table.

Number rolled	2	6	5	1	3	6
Point	P_1	P_2	P_3	P_4	P_5	P_6
Coordinates	$(0, 0)$	$(0, 16)$				

Exercises 37 through 40 refer to a variation of the chaos game discussed in Section 12.3. When played a large number of times, the set of points generated by this game approximates a Sierpinski carpet. (See Exercises 25 through 28.) Here we start with a square ABCD, such as the one shown in the figure. We need to identify each of the four vertices of the square with four equally likely random outcomes. An easy way to do this is to roll a fair die. We will say that A is the "winner" if we roll a 1; B is if we roll a 2; C is if we roll a 3; and D is if we roll a 4. If we roll a 5 or 6, we disregard the roll and roll again.

Each roll of the die generates a point inside or on the boundary of the square according to the following rules.

- **Start.** Roll the die. Mark the "winning" vertex and call it P_1.

- **Step 1.** Roll the die again. From P_1 move two-thirds of the way toward the next winning vertex. Mark this point and call it P_2.

- **Step 2, 3, etc.** Continue rolling the die, each time moving to a point two-thirds of the way between the previous position to the winning vertex.

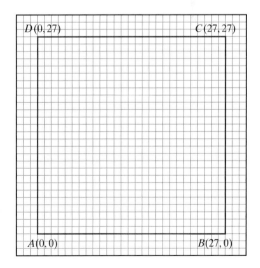

37. Using graph paper, carefully locate the points P_1, P_2, P_3, and P_4 corresponding to

(a) the sequence of rolls 4, 2, 1, 2.

(b) the sequence of rolls 3, 2, 1, 2.

(c) the sequence of rolls 3, 3, 1, 1.

38. Using graph paper, carefully locate the points $P_1, P_2, P_3,$ and P_4 corresponding to

(a) the sequence of rolls $2, 2, 4, 4$.

(b) the sequence of rolls $2, 3, 4, 1$.

(c) the sequence of rolls $1, 3, 4, 1$.

39. Using a rectangular coordinate system with A at $(0, 0), B$ at $(27, 0), C$ at $(27, 27),$ and D at $(0, 27),$ find the coordinates of the points $P_1, P_2, P_3,$ and P_4 corresponding to

(a) the sequence of rolls $4, 2, 1, 2$.

(b) the sequence of rolls $3, 1, 1, 3$.

(c) the sequence of rolls $1, 3, 4, 2$.

40. Using a rectangular coordinate system with A at $(0, 0), B$ at $(27, 0), C$ at $(27, 27),$ and D at $(0, 27),$ find the coordinates of the points $P_1, P_2, P_3,$ and P_4 corresponding to

(a) the sequence of rolls $2, 3, 4, 1$.

(b) the sequence of rolls $4, 2, 2, 4$.

(c) the sequence of rolls $3, 1, 2, 4$.

D. Operations with Complex Numbers

Exercises 41 through 46 are a review of complex number arithmetic. Recall that (i) to add two complex numbers you simply add the real parts and the imaginary parts [e.g., $(2 + 3i) + (5 + 2i) = 7 + 5i$]; (ii) To multiply two complex numbers you multiply them as is they were polynomials and use the fact that $i^2 = -1$ [e.g., $(2 + 3i)(5 + 2i) = 10 + 4i + 15i + 6i^2 = 4 + 19i$]. Finally, if you know how to multiply two complex numbers, then you also know how to square them, since $(a + bi)^2 = (a + bi)(a + bi)$.

41. Simplify each expression:

(a) $(1 + i)^2 + (1 + i)$

(b) $(1 - i)^2 + (1 - i)$

(c) $(-1 + i)^2 + (-1 + i)$

42. Simplify each expression:

(a) $(2 + 3i)^2 + (2 + 3i)$

(b) $(2 - 3i)^2 + (2 - 3i)$

(c) $(-2 + 3i)^2 + (3 - 2i)$

43. Simplify each expression. (Write your answers in decimal form and round to three significant digits.)

(a) $(-0.25 + 0.25i)^2 + (-0.25 + 0.25i)$

(b) $(-0.25 - 0.25i)^2 + (-0.25 - 0.25i)$

44. Simplify each expression.

(a) $(-0.25 + 0.125i)^2 + (-0.25 + 0.125i)$

(b) $(-0.2 + 0.8i)^2 + (-0.2 + 0.8i)$

45. (a) Plot the points corresponding to the complex numbers $(1 + i), i(1 + i), i^2(1 + i),$ and $i^3(1 + i)$.

(b) Plot the points corresponding to the complex numbers $(3 - 2i), i(3 - 2i), i^2(3 - 2i),$ and $i^3(3 - 2i)$.

(c) What geometric effect does multiplication by i have on a complex number?

46. (a) Plot the points corresponding to the complex numbers $(1 + i), -i(1 + i), -i^2(1 + i),$ and $-i^3(1 + i)$.

(b) Plot the points corresponding to the complex numbers $(0.8 + 1.2i), -i(0.8 + 1.2i), (-i)^2(0.8 + 1.2i),$ and $(-i)^3(0.8 + 1.2i)$.

(c) What geometric effect does multiplication by $-i$ have on a complex number?

E. Mandelbrot Sequences

Exercises 47 through 54 refer to Mandelbrot sequences as discussed in the chapter.

47. Consider the Mandelbrot sequence with seed $s = -2$.

(a) Find $s_1, s_2, s_3,$ and s_4.

(b) Find s_{100}.

(c) Is this Mandelbrot sequence *escaping, periodic,* or *attracted*? Explain.

48. Consider the Mandelbrot sequence with seed $s = 2$.

(a) Find $s_1, s_2, s_3,$ and s_4.

(b) Is this Mandelbrot sequence *escaping, periodic,* or *attracted*? Explain.

49. Consider the Mandelbrot sequence with seed $s = -0.5$.

(a) Using a calculator, find s_1 through s_5, rounded to four decimal places.

(b) Suppose you are given $s_N = -0.366$. Using a calculator, find s_{N+1}, rounded to four decimal places.

(c) Is this Mandelbrot sequence *escaping, periodic,* or *attracted*? Explain.

50. Consider the Mandelbrot sequence with seed $s = -0.25$.

(a) Using a calculator, find s_1 through s_{10}, rounded to six decimal places.

(b) Suppose you are given $s_N = -0.207107$. Using a calculator, find s_{N+1}, rounded to six decimal places.

(c) Is this Mandelbrot sequence *escaping, periodic,* or *attracted*? Explain.

51. Consider the Mandelbrot sequence with seed $s = 1/2$.

(a) Find $s_1, s_2,$ and s_3 without using a calculator. Give the answers in fractional form.

(b) Suppose that $s_N > 1$. Explain why this implies that $s_{N+1} > s_N$.

(c) Is this Mandelbrot sequence *escaping, periodic,* or *attracted*? Explain.

52. Consider the Mandelbrot sequence with seed $s = -1.75$.

(a) Using a calculator, find s_1 through s_{12}, rounded to five decimal places.

(b) Is this Mandelbrot sequence *escaping, periodic,* or *attracted*? Explain.

53. Suppose $s_N = 6$, and $s_{N+1} = 38$ are two consecutive terms of a Mandelbrot sequence.

  (a) Find the seed s.

  (b) If $s_N = 6$, what is the value of N?

  (**Hint:** *Use the seed you found in (a)*.)

54. Suppose $s_N = -15/16$, and $s_{N+1} = -159/256$ are two consecutive terms of a Mandelbrot sequence.

  (a) Find the seed s.

  (b) If $s_N = -15/16$, what is the value of N?

  (**Hint:** *Use the seed you found in (a)*.)

JOGGING

55. The total number of holes at step N of the construction of the Sierpinski gasket is given by $(3^N - 1)/2$. Explain how this formula can be derived.

  (**Hint:** *You need to use the formula for the sum of the terms in a geometric sequence given in Chapter 10*.)

56. Find a formula that gives the total number of holes at step N of the construction of the Sierpinski carpet.

  (**Hint:** *Do Exercise 28 first. You will need to use the formula for the sum of the terms in a geometric sequence given in Chapter 10*.)

*Exercises 57 and 58 refer to the Menger sponge, a three-dimensional cousin of the Sierpinski carpet. (See Exercises 25–28.) The **Menger sponge** is defined by the following recursive procedure.*

Menger Sponge

  ■ **Start.** *Start with a solid cube.*

  ■ **Step 1.** *Subdivide the cube into 27 equal subcubes and remove the cube in the center and the 6 cubes in the centers of each face. (Call this Procedure MS.)*

  ■ **Step 2, 3, etc.** *Apply Procedure MS to each cube of the "sponge" obtained in the previous step.*

The first two steps of the construction of the Menger sponge are illustrated in the following figure.

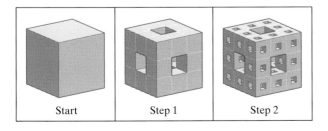

| | Start | Step 1 | Step 2 |

57. Complete the following table.

	Start	Step 1	Step 2	Step 3	...	Step N
Number of cubes removed		0	7			

58. Suppose the starting cube has sides of length 1.

  (a) Complete the following table.

	Start	Step 1	Step 2	Step 3	...	Step N
Volume of the figure		1	$\dfrac{20}{27}$			

  (b) Explain why the Menger sponge has zero volume.

  (c) Calculate the surface area at the start and at step 1 in the construction of the Menger sponge.

  (d) Explain why the surface area at each step of the construction of the Menger sponge increases.

Exercises 59 and 60 refer to reflection and rotation symmetries as discussed in Chapter 11 and thus require a good understanding of the material in that chapter.

59. **(a)** Describe all the reflection symmetries of the Koch snowflake.

  (b) Describe all the rotation symmetries of the Koch snowflake.

  (c) What is the symmetry type of the Koch snowflake?

60. This exercise refers to the Sierpinski carpet discussed in Exercises 25–28.

  (a) Describe all the reflection symmetries of the Sierpinski carpet.

  (b) Describe all the rotation symmetries of the Sierpinski carpet.

  (c) What is the symmetry type of the Sierpinski carpet?

61. Explain why each of the following statements is true. (You will need to use the formula for adding consecutive terms of a geometric sequence given in Chapter 10.)

  (a) $1 + \left(\dfrac{4}{9}\right) + \left(\dfrac{4}{9}\right)^2 + \cdots + \left(\dfrac{4}{9}\right)^{N-1} = \dfrac{9}{5}\left[1 - \left(\dfrac{4}{9}\right)^N\right]$

  (b) $\dfrac{A}{3} + \dfrac{A}{3}\cdot\left(\dfrac{4}{9}\right) + \dfrac{A}{3}\cdot\left(\dfrac{4}{9}\right)^2 + \cdots + \dfrac{A}{3}\cdot\left(\dfrac{4}{9}\right)^{N-1} =$
  $\dfrac{3}{5}A\left[1 - \left(\dfrac{4}{9}\right)^N\right]$

62. Consider the Mandelbrot sequence with seed $s = -0.75$. Show that this Mandelbrot sequence is attracted to the value -0.5.

63. Consider the Mandelbrot sequence with seed $s = 0.2$. Is this Mandelbrot sequence *escaping, periodic,* or *attracted?* If attracted, to what number?

64. Consider the Mandelbrot sequence with seed $s = 0.25$. Is this Mandelbrot sequence *escaping, periodic,* or *attracted?* If attracted, to what number?

65. Consider the Mandelbrot sequence with seed $s = -1.25$. Is this Mandelbrot sequence *escaping, periodic,* or *attracted?* If attracted, to what number?

66. Consider the Mandelbrot sequence with seed $s = \sqrt{2}$. Is this Mandelbrot sequence *escaping, periodic,* or *attracted?* If attracted, to what number?

RUNNING

67. Find the area of the Koch antisnowflake discussed in Exercises 11 through 14. Assume the area of the starting triangle is A.

68. Find the area of the Koch square snowflake discussed in Exercises 7 through 10. Assume the area of the starting square is A.

69. Suppose that we play the chaos game using triangle ABC and that M_1, M_2, and M_3 are the midpoints of the three sides of the triangle. Explain why it is impossible at any time during the game to land inside triangle $M_1 M_2 M_3$.

70. Show that the complex number $s = -0.25 + 0.25i$ is in the Mandelbrot set.

71. Show that the complex number $s = -0.25 - 0.25i$ is in the Mandelbrot set.

72. Show that the Mandelbrot set has a reflection symmetry.

    (***Hint:*** *Compare the Mandelbrot sequences with seeds $a + bi$ and $a - bi$.*)

*Exercises 73 through 75 refer to the concept of **fractal dimension**. The fractal dimension of a geometric fractal consisting of N self-similar copies of itself each reduced by a scaling factor of S is D = log N/log S. (The fractal dimension is described in a little more detail in Project A.)*

73. Compute the fractal dimension of the *Koch curve.*

74. Compute the fractal dimension of the *Sierpinski carpet.* (The Sierpinski carpet is discussed in Exercises 25 through 28.)

75. Compute the fractal dimension of the *Menger sponge.* (The Menger sponge is discussed in Exercises 57 and 58.)

Projects and Papers

A. Fractal Dimension

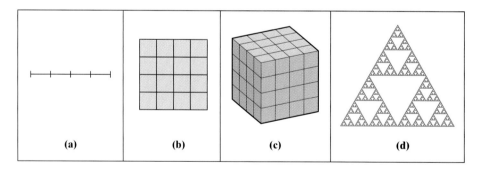

(a)   (b)   (c)   (d)

The dimensions of a line segment, a square, and a cube are, as we all learned in school, 1, 2, and 3, respectively. But what is the dimension of the Sierpinski gasket?

The line segment of size 4 shown in (a) above is made of 4 smaller copies scaled down by a factor of 4; the square shown in (b) above is made of $16 = 4^2$ smaller copies scaled down by a factor of 4; and the cube shown in (c) above is made of 4^3 smaller copies scaled down by a factor of 4. In all of these cases, if N is the number of smaller copies of the object reduced by a scaling factor S, the dimension D is the exponent to which we need to raise S to get N (i.e., $N = S^D$). If we apply the same argument to the Sierpinski gasket shown in (d) above, we see that the Sierpinski gasket is made of $N = 3$ smaller copies of itself, and each copy has been reduced by a scaling factor $S = 2$. If we want to be consistent, the dimension of the Sierpinski gasket should be the exponent D in the equation $3 = 2^D$. To solve for D you have to use *logarithms*. When you do, you get $D = \log 3/\log 2$. Crazy but true: The dimension of the Sierpinski gasket is not a whole number, not even a rational number. It is the irrational number $\log 3/\log 2$ (about 1.585)!

For a geometric fractal with exact self-similarity, we will define its dimension as $D = \dfrac{\log N}{\log S}$, where N is the number of self-similar pieces that the parent fractal is built out of, and S is the scale by which the pieces are reduced (if the pieces are half the size of the parent fractal $S = 2$, if the pieces are one third the size of the parent fractal $S = 3$, and so on).

In this project, you should discuss the meaning and importance of the concept of dimension as it applies to geometric fractals having exact self-similarity.

B. Fractals and Music

The hallmark of a fractal shape is the property of *self-similarity*—there are themes that repeat themselves (either exactly or approximately) at many different scales. This type of repetition also works in music, and the application of fractal concepts to musical composition has produced many intriguing results.

Write a paper discussing the connections between fractals and music.

C. Book Review: *The Fractal Murders*

If you enjoy mystery novels, this project is for you.

The Fractal Murders, by Mark Cohen (Muddy Gap Press, 2002) is a *who-done-it* with a mathematical backdrop. In addition to the standard elements of a classic murder mystery (including brilliant but eccentric detective) this novel has a fractal twist: The victims are all mathematicians doing research in the field of fractal geometry.

Read the novel and write a review of it. Include in your review a critique of both the literary and the mathematical merits of the book. To get some ideas as to how to write a good book review, you should check out the *New York Times Book Review* section, which appears every Sunday in the *New York Times* (*www.nytimes.com*).

D. Fractal Antennas

One of the truly innovative practical uses of fractals is in the design of small but powerful antennas that go inside wireless communication devices such as cell phones, wireless modems, and GPS receivers. The application of fractal geometry to antenna design follows from the discovery in 1999 by radio astronomers Nathan Cohen and Robert Hohlfeld of Boston University that an antenna that has a self-similar shape has the ability to work equally well at many different frequencies of the radio spectrum.

Write a paper discussing the application of the concepts of fractal geometry to the design of antennas.

References and Further Readings

1. Berkowitz, Jeff, *Fractal Cosmos: The Art of Mathematical Design*. Oakland, CA: Amber Lotus, 1998.

2. Briggs, John, *Fractals: The Patterns of Chaos*. New York: Touchstone Books, 1992.

3. Cohen, Mark, *The Fractal Murders*. Boulder, CO: Muddy Gap Press, 2002.

4. Dewdney, A. K., "Computer Recreations: A Computer Microscope Zooms in for a Look at the Most Complex Object in Mathematics," *Scientific American*, 253 (August 1985), 16–24.

5. Dewdney, A. K., "Computer Recreations: A Tour of the Mandelbrot Set Aboard the Mandelbus," *Scientific American*, 260 (February 1989), 108–111.

6. Dewdney, A. K., "Computer Recreations: Beauty and Profundity. The Mandelbrot Set and a Flock of Its Cousins Called Julia," *Scientific American*, 257 (November 1987), 140–145.

7. Flake, Gary W., *The Computational Beauty of Nature: Computer Explorations of Fractals, Chaos, Complex Systems, and Adaptation*. Cambridge, MA: MIT Press, 2000.

8. Gleick, James, *Chaos: Making a New Science*. New York: Viking Penguin, Inc., 1987, Chap. 4.

9. Hastings, Harold, and G. Sugihara, *Fractals: A User's Guide for the Natural Sciences*. New York: Oxford University Press, 1995.

10. Jurgens, H., H. O. Peitgen, and D. Saupe, "The Language of Fractals," *Scientific American*, 263 (August 1990), 60–67.

11. Mandelbrot, Benoit, *The Fractal Geometry of Nature*. New York: W. H. Freeman & Co., 1983.

12. Musser, George, "Practical Fractals," *Scientific American*, 281 (July 1999), 38.

13. Peitgen, H. O., H. Jurgens, and D. Saupe, *Chaos and Fractals: New Frontiers of Science*. New York: Springer-Verlag, Inc., 1992.

14. Peitgen, H. O., H. Jurgens, and D. Saupe, *Fractals for the Classroom*. New York: Springer-Verlag, Inc., 1992.

15. Peitgen, H. O., and P. H. Richter, *The Beauty of Fractals*. New York: Springer-Verlag, Inc., 1986.

16. Peterson, Ivars, *The Mathematical Tourist*. New York: W. H. Freeman & Co., 1988, Chap. 5.

17. Schechter, Bruce, "A New Geometry of Nature," *Discover*, 3 (June 1982), 66–68.

18. Schroeder, Manfred, *Fractals, Chaos, Power Laws: Minutes from an Infinite Paradise*. New York: W. H. Freeman & Co., 1991.

19. Wahl, Bernt, *Exploring Fractals on the Macintosh*. Reading, MA: Addison-Wesley, 1994.

Chapter-Opener Photo Credits

p. 410:

- *Bird of Paradise*. Computer-generated image of a region around the boundary of the Mandelbrot set. Magnification: approximately ×100,000 the original Mandelbrot set. (Image courtesy of Rollo Silver.)

p. 411 (clockwise from the top):

- *Lake Carnegie* (Western Australia). Photograph taken from the Landsat satellite. Scale: 1 inch = 2 miles. (Image courtesy of NASA Landsat Project Science Office and U.S. Geological Survey.)

- *Dasht-e Kevir* (Great Salt Desert, Iran). Photograph taken from the Landsat satellite. Scale: 1 inch = 2.3 miles. (Image courtesy of NASA Landsat Project Science Office and U.S. Geological Survey.)

- *The Brain's Own Transmitters*. Light micrograph image of brain tissue. Magnification: approximately ×200. (Image courtesy of Manfred Kage/Peter Arnold, Inc.)

- *Number 1, 1950 (Lavender Mist)*. Oil on canvas by Jackson Pollock. Size: 87 by 118 inches. (Image courtesy of The National Gallery of Art, Washington D.C.)

part 4

Statistics

13

Collecting Statistical Data

Censuses, Surveys, and Clinical Studies

Statistical reasoning will one day be as necessary for efficient citizenship as the ability to read and write.

H. G. Wells

Information has become the primary currency of the twenty-first century, and, by and large, wherever there is information, statistics are not far behind. Open today's paper and look at the business section—you'll find plenty of statistics there. Not interested in the stock market? Check the health section, or the sports section. All of them are spiked with statistics.

W hat is *statistics*? At its most basic level, statistics is the blending of two fundamental skills we learn separately in school: communicating and manipulating numbers. When we use numbers as a tool to transmit information, we are doing something statistical. If you prefer a more formal description, here it is: *Statistics is the science of dealing with data*. And what is *data*? Data is any type of information packaged in numerical form. [In modern usage, the word *data* is used for the singular form (a single piece of numerical information) as well as the plural form (many pieces of numerical information).]

Behind every statistical statement there is a story, and, like any story, it has a beginning, a middle, an end, and a moral. In this chapter we will

discuss the beginning of the story, which, in statistics, typically means the process of gathering or collecting data. Collecting data seems deceptively simple, but history has repeatedly shown that doing so in an accurate, efficient, and timely manner can be the most difficult part of the statistical story. One of the special features of this chapter is the use of case studies that will help us learn a few of the do's and don'ts of data collection from the past experience of others.

The first step in proper data collection is to identify the population to which the data apply, and the concepts of *population, N-value*, and *census* are discussed in Section 13.1. In Sections 13.2, 13.3, and 13.4 the basic concepts of sampling are introduced (*sampling frame, selection bias, nonresponse bias, sampling error, sampling variability, chance error*) as well as various sampling methods (*quota sampling, simple random sampling, stratified sampling*). A method for estimating the size of a population using sampling (the *capture-recapture method*) is discussed in Section 13.5. When the goal of data collection is to study the relationship between a cause and an effect (e.g., does taking a math class improve your chances of success in life?), a special type of data collection process called a *clinical study* (or just *study*) is required. Some of the key concepts behind clinical studies (*control group, treatment group, randomization, placebo effect, blind and double-blind*) are discussed in Section 13.6.

"Data! Data! Data!" he cried impatiently. "I can't make bricks without clay."

Sherlock Holmes, *The Adventure of the Copper Beeches*

13.1 The Population

Every statistical statement refers, directly or indirectly, to some group of individuals or objects. In statistical terminology, this collection of individuals or objects is called the **population**. The first question we should ask ourselves when trying to make sense of a statistical statement is, "What is the population to which the statement applies?"

In an ideal world, the specific population to which a statistical statement applies is clearly identified within the story itself. In the real world, this rarely happens because the details are skipped (mostly to keep the story moving along but sometimes with an intent to confuse or deceive) or, alternatively, because two (or more) related populations are involved in the story.

> **EXAMPLE 13.1** The Return of the Bald Eagle

Bald Eagles Come Back from the Brink

BY JOHN L. ELIOT,

They ruled the skies on seven-foot (two-meter) wingspans when 17th-century Europeans arrived in North America. Throughout the continent, half a million bald eagles may have soared. But settlers blamed them for killing livestock, so shooting began—and the proud birds' numbers began to plunge....

The Bald Eagle Protection Act of 1940 prohibited shooting or otherwise harming the birds in the [lower 48 states] but didn't cover the pesticides that within a decade began to destroy eagles' eggs. By the 1960s only about 400 breeding pairs of bald eagles remained in the lower 48 [states]....
The banning of DDT in 1972 and other measures launched an amazing comeback by the eagles, whose status changed from endangered to threatened in 1995. Today, with more than 6,000 breeding pairs, bald eagles may soon be taken off the endangered species list entirely, their survival as an icon secured—for now.

National Geographic, Magazine, July 2002

Because of its iconic importance—the bald eagle is the national bird and the emblem of the United States—much has been said and written about bald eagle populations in North America, from their near extinction to their miraculous recovery. The preceding is a representative excerpt from a 2002 *National Geographic* article—just one of many bald eagle stories.

In the context of populations, this is, in spirit, a story about the bald eagle population in the United States—more specifically the lower 48 states. (The bald eagle population of Alaska is not really a part of the story, as bald eagles never reached endangered levels there, and by the way, there are no bald eagles in Hawaii.) Unlike many other animals and birds, bald eagles stay within the confines of a reasonably small geographical area, so it is possible to discuss bald eagle populations in a regional context such as the lower 48 states, or even populations within a specific state.

Interestingly enough, the case about the comeback of the bald eagle population is often made through the use of a *proxy*—the number of bald eagle *breeding pairs* (400 in the 1960s; over 6000 in 2000). Thus, upon closer scrutiny the story ties together two different but closely related populations—the overall bald eagle population within the lower 48 states (including chicks, adolescent birds, etc.) and the population of breeding pairs. The former is the population of interest, but the latter is the population of convenience—breeding pairs are much easier to identify, track, and count.

The *N*-Value

Given a specific population, an obviously relevant question is, "How many individuals or objects are there in that population?" This number is called the **N-value** of the population. (It is common practice in statistics to use capital N to denote population sizes.) It is important to keep in mind the distinction between the N-value—a number specifying the size of the population—and the population itself.

As we learned in Chapter 10, populations change, and thus N-values must often be discussed within the context of time. Example 13.1 illustrates this point well. The N-value of the bald eagle population of North America in the seventeenth century is very different from the N-value in the 1960s, which in turn is very different from the current N-value.

> **EXAMPLE 13.2** *N*-Values of Bald Eagle
> Breeding Pair Populations (1963–2000)

Over the last 45 years, the United States Fish and Wildlife Service has been able to keep a remarkably accurate tally of the number of bald eagle breeding pairs in the lower 48 states. (As we discussed in Example 13.1, breeding pairs are used as a useful proxy for the health of the overall population.) A tremendous amount of effort has gone into collecting and verifying these N-values, which, for a wildlife population, are of remarkable accuracy. Figure 13-1 summarizes the population numbers over the period 1963–2000. (No tallies were conducted in 1964–1973; 1975–1980; 1983; 1985.)

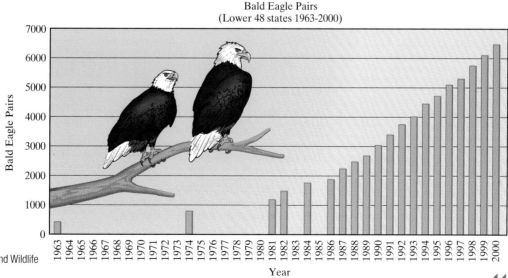

FIGURE 13-1
Source: United States Fish and Wildlife Service.

Our next example illustrates a basic truism: We cannot determine an N-value if we do not first identify the population.

> ### EXAMPLE 13.3 *N* Is in the Eye of the Beholder

Andy has a coin jar full of quarters. He is hoping that there is enough money in the jar to pay for a new baseball glove. Dad says to go count them, and if there isn't enough, he will lend Andy the difference. Andy dumps the quarters out of the jar, makes a careful tally and comes up with a count of 116 quarters.

So what is the *N*-value here? The answer depends on how we define the population—are we counting coins or money? To dad, who will end up stuck with all the quarters, the total number of coins might be the most relevant issue. Thus, to dad $N = 116$. Andy, on the other hand, is concerned with how much money is in the jar. If he were to articulate his point of view in statistical language, he would say that $N = 29$ (dollars). ◀◀

The process of collecting data by going through every member of the population is called a **census**. (When Andy dumped all the quarters in the jar and counted them, he was essentially conducting a small census.) The idea behind a census is simple enough, but in practice a census requires a great deal of "cooperation" from the population. (The quarters in Andy's jar were nice enough to stay put and let themselves be counted.) For larger, more dynamic populations (wildlife, humans, etc.), accurate tallies are inherently difficult if not impossible, and in these cases the best we can hope for is a good estimate of the *N*-value.

> ### EXAMPLE 13.4 2000 Census Undercounts

The most notoriously difficult *N*-value question around is, "What is the *N*-value of the national population of the United States?" This is a question the United States Census tries to answer every 10 years—with very little success.

The 2000 U.S. Census was the largest single peacetime undertaking of the federal government—it employed over 850,000 people and cost about 6.5 billion dollars—and yet, it missed counting between 3 and 4 million people.

Political Fight Brews Over Census Correction

By Haya El Nasser

The 2000 Census did a better job counting people than the last Census, especially minorities and children, but it still missed about 2.7 million to 4 million people. . . .

Preliminary estimates show that the net number of people missed falls between 0.96% and 1.4%. In 1990, the undercount was 1.6%, or 4 million people. There was a significant drop in the undercount of blacks, Hispanics, American Indians and children, population groups that were disproportionately missed in 1990. . . . The estimates are bound to heat up political infighting. The Census Bureau must decide whether the numbers should be adjusted to compensate for the undercount. . . .

Census numbers are used to redraw political districts. An adjusted count would include more minorities, which could reshape key political districts. Republicans worry an adjusted count would help Democrats. . . . The Census Bureau estimates the number of people missed through the same method that would be used to adjust the numbers. It surveys 314,000 sample households and checks to see whether those households filled out Census forms.

Source: *New York Times, February 15, 2001*

Given the critical importance of the U.S. Census and given the tremendous resources put behind the effort by the federal government, why is the head count so far off? How can the best intentions and tremendous resources of our government fail so miserably in an activity that on a smaller scale can be carried out by a child trying to buy a baseball glove? ◀◀

Our first case study gives a brief overview of the ins and outs of the U.S. Census and illustrates the difficulties faced by censuses in general.

Case Study 1: The U.S. Census

Article 1, Section 2, of the Constitution of the United States mandates that a national census be conducted every 10 years. The original intent of the census was to "count heads" for a twofold purpose: taxes and political representation. Like everything else in the Constitution, Article 1, Section 2, was a compromise of many competing interests: The count was to exclude "Indians not taxed" and to count slaves as "three-fifths of a free Person." Since then, the scope and purpose of the U.S. Census has been modified and expanded by the 14th Amendment and the courts in many ways:

- Besides counting heads, the U.S. Census Bureau now collects additional information about the population: sex, age, race, ethnicity, marital status, housing, income, and employment data. Some of this information is updated on a regular basis, not just every 10 years.

- Census data are now used for many important purposes beyond its original ones of *taxation* and *representation*: the allocation of billions of federal dollars to states, counties, cities, and municipalities; the collection of other important government statistics such as the Consumer Price Index and the Current Population Survey; the redrawing of legislative districts within each state; and the strategic planning of production and services by business and industry.

- For the purposes of the Census, the United States population is defined as consisting of "all persons *physically present* and *permanently residing* in the United States." Citizens, legal resident aliens, and even illegal aliens are meant to be included.

Nowadays, the notion that if we put enough money and effort into it, all individuals living in the United States can be counted like coins in a jar is unrealistic. In 1790, when the first U.S. Census was carried out, the population was smaller and relatively homogeneous, as people tended to stay in one place, and, by and large, they felt comfortable in their dealings with the government. Under these conditions it might have been possible for census takers to count heads accurately. Today's conditions are completely different. People are constantly on the move. Many distrust the government. In large urban areas many people are homeless or don't want to be counted. And then there is the apathy of many people who think of a census form as another piece of junk mail.

If the Census undercount was consistent among all segments of the population, the undercount problem could be solved easily. Unfortunately, the modern U.S. Census is plagued by what is known as a *differential undercount*. Ethnic minorities, migrant workers, and the urban poor populations have significantly larger undercount rates than the population at large, and the undercount rates vary significantly within these groups. Using modern statistical techniques, it is possible to make adjustments to the raw Census figures that correct some of the inaccuracy caused by the differential undercount, but in 1999 the Supreme Court ruled that only the raw numbers, and not statistically adjusted numbers, can be used for the purposes of apportionment of Congressional seats among the states. (*Department of Commerce et al. v. United States House of Representatives et al.*)

13.2 Sampling

The practical alternative to a census is to collect data only from some members of the population and use that data to draw conclusions and make inferences about the entire population. Statisticians call this approach a **survey** (or a **poll** when the data collection is done by *asking questions*). The subgroup chosen to provide the data is called the **sample**, and the act of selecting a sample is called **sampling**.

Ideally, every member of the population should have an opportunity to be chosen as part of the sample, but this is possible only if we have a mechanism to identify each and every member of the population. In many situations this is impossible. Say we want to conduct a public opinion poll before an election. The population for the poll consists of every voter in the upcoming election, but how can we identify who is and is not going to vote ahead of time? We know who the eligible voters are (citizens over 18, no felons, etc.) but among this group there are still many nonvoters.

The first important step in a survey is to distinguish the population for which the survey applies (the **target population**) and the actual subset of the population from which the sample will be drawn, called the **sampling frame**. The ideal scenario is when the sampling frame is the same as the target population—that would mean that every member of the target population is a candidate for the sample. When this is impossible (or impractical), then an appropriate sampling frame must be chosen. In a pre-election poll, for example, the sampling frame could be the list of all registered voters or the telephone directory (or directories) for the area in which the election is held. In either case, the sampling frame will miss some voters and include some nonvoters.

The basic philosophy behind sampling is simple and well understood—if we have a sample that is "representative" of the entire population, then whatever we want to know about a population can be found out by getting the information from the sample. In practice, however, things are not always as simple. If we are to draw reliable data from a sample, we must (a) find a sample that is representative of the population, and (b) determine how big the sample should be. These two issues go hand in hand, and we will discuss them next.

Sometimes a very small sample can be used to get reliable information about a population, no matter how large the population is. This is the case when the population is highly homogeneous. For example, with the exception of identical twins, every person's DNA is different. Yet the DNA sampled from just one human cell is sufficient to characterize and identify all the DNA of that individual. Similarly, a person's blood is essentially the same everywhere in the body, which explains why a small blood sample drawn from an arm can be used to draw reliable data about the patient's blood sugar levels, cholesterol levels, and so on.

The more heterogeneous a population gets, the more difficult it is to find a representative sample. The difficulties can be well illustrated by taking a look at the history of *public opinion polls*.

Public Opinion Polls

You are probably familiar with public opinion polls, such as the Gallup poll, the Harris poll, and many others. A public opinion poll is a survey in which the members of the sample provide information by answering specific questions from an "interviewer." The question-answer exchange can be done through a questionnaire, a personal telephone interview, or a direct face-to-face interview.

Nowadays, public opinion polls are used regularly to measure "the pulse of the nation." They give us statistical information ranging from voters' preferences before an election to opinions on issues such as the environment, abortion, and the economy.

Given their widespread use and the influence they exert, it is important to ask how much we can trust the information that we get from public opinion polls. This is a complex question that goes to the very heart of mathematical statistics. We'll start our exploration of it with some historical examples.

Case Study 2: The 1936 *Literary Digest* Poll

The U.S. presidential election of 1936 pitted Alfred Landon, the Republican governor of Kansas, against the incumbent Democratic President, Franklin D. Roosevelt. At the time of the election, the nation had not yet emerged from the Great Depression, and economic issues such as unemployment and government spending were the dominant themes of the campaign.

The *Literary Digest*, one of the most respected magazines of the time, conducted a poll a couple weeks before the election. The magazine had used polls to accurately predict the results of every presidential election since 1916, and their 1936 poll was the largest and most ambitious poll ever. The *sampling frame* for the *Literary Digest* poll consisted of an enormous list of names that included (i) every person listed in a telephone directory anywhere in the United States, (ii) every person on a magazine subscription list, and (iii) every person listed on the roster of a club or professional association. From this sampling frame a list of about 10 million names was created, and every name on this list was mailed a mock ballot and asked to mark it and return it to the magazine.

One cannot help but be impressed by the sheer scope and ambition of the 1936 *Literary Digest* poll, as well as the magazine's unbounded confidence in its accuracy. In its issue of August 22, 1936, the *Literary Digest* crowed:

> *Once again, [we are] asking more than ten million voters—one out of four, representing every county in the United States—to settle November's election in October.*
>
> *Next week, the first answers from these ten million will begin the incoming tide of marked ballots, to be triple-checked, verified, five-times cross-classified and totaled. When the last figure has been totted and checked, if past experience is a criterion, the country will know to within a fraction of 1 percent the actual popular vote of forty million [voters].*

Based on the poll results, the *Literary Digest* predicted a landslide victory for Landon with 57% of the vote, against Roosevelt's 43%. Amazingly, the election turned out to be a landslide victory for Roosevelt with 62% of the vote, against 38% for Landon. The difference between the poll's prediction and the actual election results was a whopping 19%, the largest error ever in a major public opinion poll. The results damaged the credibility of the magazine so much so that soon after the election its sales dried up and it went out of business—the victim of a major statistical blunder.

For the same election, a young pollster named George Gallup (for more on Gallup, see the biographical profile at the end of the chapter) was able to predict accurately a victory for Roosevelt using a sample of "only" 50,000 people. In fact, Gallup *also* publicly predicted, to within 1%, the incorrect results that the *Literary Digest* would get using a sample of just 3000 people taken from the same sampling frame the magazine was using. What went wrong with the *Literary Digest* poll and why was Gallup able to do so much better?

The first thing seriously wrong with the *Literary Digest* poll was the sampling frame, consisting of names taken from telephone directories, lists of magazine subscribers, rosters of club members, and so on. Telephones in 1936 were something of a luxury. At a time when 9 million people were unemployed, magazine subscriptions and club memberships were even more so. When it came to economic status the *Literary Digest* sample was far from being a representative cross section of the voters. This was a critical problem, because voters often vote on economic issues, and given the economic conditions of the time, this was especially true in 1936.

When the choice of the sample has a built-in tendency (whether intentional or not) to exclude a particular group or characteristic within the population, we say that a survey suffers from **selection bias**. It is obvious that selection bias must be avoided, but it is not always easy to detect it ahead of time. Even the most scrupulous attempts to eliminate selection bias can fall short (as will become apparent in our next case study).

The second serious problem with the *Literary Digest* poll was the issue of *nonresponse bias*. In a typical survey it is understood that not every individual is willing to respond to the survey request (and in a democracy we cannot force them to do so). Those individuals who do not respond to the survey request are called *nonrespondents*, while those who do are called *respondents*. The percentage of respondents out of the total sample is called the **response rate**. For the *Literary Digest* poll out of a sample of 10 million people that were mailed a mock ballot only about 2.4 million mailed a ballot back, resulting in a 24% response rate. When the response rate to a survey is low, the survey is said to suffer from **nonresponse bias**. (Exactly at what point the response rate is to be considered low depends on the circumstances and nature of the survey, but a response rate of 24% is generally considered very low.)

Nonresponse bias can be viewed as a special type of selection bias—it excludes from the sample reluctant and uninterested people. Since reluctant and uninterested people can represent a significant slice of the population, we don't want them excluded from the sample. But getting reluctant, uninterested, and apathetic slugs to participate in a survey is a conundrum—in a free country we cannot force people to participate, and bribing them with money or chocolate chip cookies is not always a practical solution.

One of the significant problems with the *Literary Digest* poll was that the poll was conducted by mail. This approach is the most likely to magnify nonresponse bias, because people often consider a mailed questionnaire just another form of junk mail. Of course, given the size of their sample, the *Literary Digest* hardly had a choice. This illustrates another important point: Bigger is not better, and a big sample can be more of a liability than an asset.

The *Literary Digest* story has two morals: (1) *You'll do better with a well-chosen small sample than with a badly chosen large one*, and (2) *watch out for selection bias and nonresponse bias*. ◀◀

Convenience Sampling

There is always a cost (effort, time, money) associated with collecting data, and it is a truism that this cost is proportional to the quality of the data collected—the better the data the more effort required to collect it. It follows that the temptation to take shortcuts when collecting data is always there, and that data collected "on the cheap" should always be scrutinized carefully. One commonly used shortcut in sampling is known as **convenience sampling**. In convenience sampling the selection of which individuals are in the sample is dictated by what is easiest for the data collector, never mind trying to get a representative sample.

A classic example of convenience sampling is when interviewers set up at a fixed location such as a mall or outside a supermarket and ask passersby to be part of a public opinion poll. A different type of convenience sampling occurs when the sample is based on *self-selection*—the sample consists of those individuals who proactively seek the opportunity to be in it. Self-selection is the reason why many Area Code 800 polls ("Call 1-800-YOU-NUTS to express your opinion on the new tax proposal . . . ") are not to be trusted. Even worse are the Area Code 900 polls, where an individual has to actually pay (sometimes as much as $2) to be part in the sample and express his or her opinion.

Convenience sampling is not always bad—there are times when there is no other choice or the alternatives are so expensive that they have to be ruled out. We should keep in mind, however, that data collected through convenience sampling are naturally tainted and should always be scrutinized (that's why we always want to get to the details of *how* the data were collected). More often than not, convenience sampling gives us data that are too unreliable to be of any scientific value. With data, as with so many other things, *you get what you pay for*.

Quota Sampling

Quota sampling is a systematic effort to force the sample to be representative of a given population through the use of quotas—the sample should have so many women, so many men, so many blacks, so many whites, so many people living in urban areas, so many people living in rural areas, and so on. The proportions in each category in the sample should be the same as those in the population. If we can assume that every important characteristic of the population is taken into account when the quotas are set up, it is reasonable to expect that the sample will be representative of the population and produce reliable data.

Our next case study illustrates some of the difficulties with the assumptions behind quota sampling.

Case Study 3: The 1948 Presidential Election

George Gallup had introduced quota sampling as early as 1935 and had used it successfully to predict the winner of the 1936, 1940, and 1944 presidential elections. Quota sampling thus acquired the reputation of being a "scientifically reliable" sampling method, and by the 1948 presidential election all three major national polls—the Gallup poll, the Roper poll, and the Crossley poll—used quota sampling to make their predictions.

For the 1948 election between Thomas Dewey and Harry Truman, Gallup conducted a poll with a sample of approximately 3250 people. Each individual in the sample was interviewed in person by a professional interviewer to minimize nonresponse bias, and each interviewer was given a very detailed set of quotas to meet—for example, 7 white males under 40 living in a rural area, 5 black males over 40 living in a rural area, 6 white females under 40 living in an urban area, and so on. By the time all the interviewers met their quotas, the entire sample was expected to accurately represent the entire population in every respect: gender, race, age, and so on.

Based on his sample, Gallup predicted that Dewey, the Republican candidate, would win the election with 49.5% of the vote to Truman's 44.5% (with third-party candidates Strom Thurmond and Henry Wallace accounting for the remaining 6%). The Roper and Crossley polls also predicted an easy victory for Dewey. (In fact, after an early September poll showed Truman trailing Dewey by 13 percentage points, Roper announced that he would discontinue polling since the outcome was already so obvious.) The actual results of the election turned out to be almost the exact reverse of Gallup's prediction: Truman got 49.9% and Dewey 44.5% of the national vote.

"Ain't the way I heard it." Truman gloats while holding an early edition of the *Chicago Daily Tribune* in which the headline erroneously claimed a Dewey victory based on the predictions of all the polls.

Truman's victory was a great surprise to the nation as a whole. So convinced was the *Chicago Daily Tribune* of Dewey's victory that it went to press on its early edition for November 4, 1948, with the headline "Dewey defeats Truman." The picture of Truman holding aloft a copy of the *Tribune* and his famous retort "Ain't the way I heard it" have become part of our national folklore.

To pollsters and statisticians, the erroneous predictions of the 1948 election had two lessons: (1) *Poll until election day*, and (2) *quota sampling is intrinsically flawed*.

What's wrong with quota sampling? After all, the basic idea behind it appears to be a good one: Force the sample to be a representative cross section of the population by having each important characteristic of the population proportionally represented in the sample. Since income is an important factor in determining how people vote, the sample should have all income groups represented in the same proportion as the population at large. The same should be true for gender, race, age, and so on. Right away, we can see a potential problem: Where do we stop? No matter how careful we might be, we might miss some criterion that would affect the way people vote, and the sample could be deficient in this regard.

An even more serious flaw in quota sampling is that, other than meeting the quotas, the interviewers are free to choose whom they interview. This opens the door to selection bias. Looking back over the history of quota sampling, we can see a clear tendency to overestimate the Republican vote. In 1936, using quota sampling, Gallup predicted that the Republican candidate would get 44% of the vote, but the actual number was 38%. In 1940, the prediction was 48%, and the actual vote was 45%; in 1944, the prediction was 48%, and the actual vote was 46%. Gallup was able to predict the winner correctly in each of these elections, mostly because the spread between the candidates was large enough to cover the error. In 1948, Gallup (and all the other pollsters) simply ran out of luck. It was time to ditch quota sampling.

The failure of quota sampling as a method for getting representative samples has a simple moral: *Even with the most carefully laid plans, human intervention in choosing the sample can result in selection bias.*

◀◀

13.3 Random Sampling

The best alternative to human selection is to let the *laws of chance* determine the selection of a sample. Sampling methods that use randomness as part of their design are known as **random sampling** methods, and any sample obtained through random sampling is called a **random sample** (or a *probability sample*).

The idea behind random sampling is that the decision as to which individuals should or should not be in the sample is best left to chance because the laws of chance are better than human design in coming up with a representative sample. At first, this idea seems somewhat counterintuitive. How can random chance guarantee an absence of bias? Isn't it possible to get by sheer bad luck a sample that is very biased (say, for example, mostly male, or all college students, etc.)? In theory, such an outcome is possible, but in practice, when the sample is large enough, the odds of it happening are so low that we can pretty much rule it out.

Most present-day methods of quality control in industry, corporate audits in business, and public opinion polling are based on random sampling. The reliability of data collected by random sampling methods is supported by both practical experience and mathematical theory. (We will discuss some of the details of this theory in Chapter 16.)

Simple Random Sampling

The most basic form of random sampling is called **simple random sampling**. It is based on the same principle a lottery is. Any set of numbers of a given size has an equal chance of being chosen as any other set of numbers of that size. Thus, if a lottery ticket consists of six winning numbers, a fair lottery is one in which any combination of six numbers has the same chance of winning as any other combination of six numbers. In sampling, this means that any group of members of the population should have the same chance of being the sample as any other group of the same size.

In theory, simple random sampling is easy to implement. We put the name of each individual in the population in "a hat," mix the names well, and then draw as many names as we need for our sample. Of course "a hat" is just a metaphor. If our population is 100 million voters and we want to choose a simple random sample of 2000, we will not be putting all the names in a real hat and then drawing 2000 names one by one. The modern way to do any serious simple random sampling is by computer. Make a list of members of the population, enter it into the computer, and then let the computer randomly select the names. This is a fine idea for small, compact populations, but a hopeless one when it comes to national surveys and public opinion polls.

Implementing simple random sampling in national public opinion polls raises problems of expediency and cost. Interviewing hundreds of individuals chosen by simple random sampling means chasing people all over the country, a task that requires an inordinate amount of time and money. For most public opinion polls—especially those done on a regular basis—the time and money needed to do this are simply not available.

Stratified Sampling

The alternative to simple random sampling used nowadays for national surveys and public opinion polls is a sampling method known as **stratified sampling**. The basic idea of stratified sampling is to break the sampling frame into categories, called **strata**, and then (unlike quota sampling) *randomly* choose a sample from these strata. The chosen strata are then further divided into categories, called substrata, and a random sample is taken from these substrata. The selected substrata are further subdivided, a random sample is taken from them, and so on. The process goes on for a predetermined number of steps (usually four or five).

Our next case study illustrates how stratified sampling works in the case of a national public opinion poll. Basic variations of the same idea can be used at the state, city or local level. The specific details, of course, will be different.

Case Study 4: National Public Opinion Polls

In national public opinion polls the *strata* and *substrata* are defined by a combination of geographic and demographic criteria. For example, the nation is first divided into "size of community" *strata* (big cities, medium cities, small cities, villages, rural areas, etc.). The strata are then subdivided by geographical region (New England, Middle Atlantic, East Central, etc.). This is the first layer of substrata. Within each geographical region and within each size of community stratum, some communities (called *sampling locations*) are selected by simple random sampling. The selected sampling locations are the only places where interviews will be conducted. Next, each of the selected sampling locations is further subdivided into geographical units called *wards*. This is the second layer of substrata. Within each sampling location some of the wards are selected using simple random sampling. The selected wards are then divided into smaller units, called *precincts* (third layer), and within each ward some of its precincts are selected by simple random sampling. At the last stage, *households* (fourth layer) are selected from within each precinct by simple random sampling. The interviewers are then given specific instructions as to which households in their assigned area they must conduct interviews in and the order that they must follow.

The Gallup Poll

THE GALLUP REPORT. PRINCETON, NJ: AMERICAN INSTITUTE OF PUBLIC OPINION

The design of the sample used by the Gallup Poll for its standard surveys of public opinion is that of a replicated area probability *stratified sample* ... After stratifying the nation geographically and by size of community in order to insure conformity of the sample with the 2000 Census distribution of the population, over 360 different sampling locations or areas are selected on a mathematically random basis from within cities, towns, and countries which have in turn, been selected on a mathematically random basis. The interviewers have no choice whatsoever concerning the part of the city, town, or country in which they conduct their interviews.

Approximately five interviews are conducted in each randomly selected sampling point. Interviewers are given maps of the area to which they are assigned and are required to follow a specific travel pattern on contacting households. At each occupied dwelling unit, interviewers are instructed to select respondents by following a prescribed systematic method. This procedure is followed until the assigned number of interviews with male and female adults have been completed....

The efficiency of stratified sampling compared to simple random sampling in terms of cost and time is clear. The members of the sample are clustered in well-defined and easily manageable areas, significantly reducing the cost of conducting interviews as well as the response time needed to collect the data. For a large, heterogeneous nation like the United States, stratified sampling has generally proved to be a reliable way to collect national data.

What about the size of the sample? Surprisingly, it does not have to be very large. Typically, a Gallup poll is based on samples consisting of approximately 1500 individuals, and roughly the same size sample can be used to poll the population of a small city as the population of the United States. *The size of the sample does not have to be proportional to the size of the population.*

13.4 Sampling: Terminology and Key Concepts

As we now know, except for a *census*, the common way to collect statistical information about a population is by means of a *survey*. (When the survey consists of asking people their opinion on some issue, we refer to it as a *public opinion poll*.) In a survey, we use a subset of the population, called a *sample*, as the source of our information, and from this sample, we try to generalize and draw conclusions about the entire population. Statisticians use the term **statistic** to describe any kind of numerical information drawn from a sample. A statistic is always an estimate for some unknown measure, called a **parameter**, of the population. Let's put it this way: A *parameter* is the numerical information we would like to have—the pot of gold at the end of the statistical rainbow, so to speak. Calculating a parameter is difficult and often impossible, since the only way to get the exact value for a parameter is to use a census. If we use a sample, then we can get only an estimate for the parameter, and this estimate is called a *statistic*.

We will use the term **sampling error** to describe the difference between a parameter and a statistic used to estimate that parameter. In other words, the sampling error measures how much the data from a survey differs from the data that would have been obtained if a census had been used. Of course, the very point of sampling is to avoid using a census, so sampling errors can only be estimated. Usually the estimates are given in terms of a margin of error, as in "The margin of sampling error for the poll was plus or minus 3%." We will discuss in greater detail the exact meaning of such statements in Chapter 16.

Sampling error can be attributed to two factors: *chance error* and *sampling bias*. **Chance error** is the result of the basic fact that a sample, being just a sample, can only give us approximate information about the population. In fact, different samples are likely to produce different statistics for the same population, even when the samples are chosen in exactly the same way—a phenomenon known as **sampling variability**. While sampling variability, and thus chance error, are unavoidable, with careful selection of the sample and the right choice of sample size they can be kept to a minimum.

Sample bias is the result of choosing a bad sample and is a much more serious problem than chance error. Even with the best intentions, getting a sample that is representative of the entire population can be very difficult and can be affected by many subtle factors. Sample bias is the result. As opposed to chance error, sample bias can be eliminated by using proper methods of sample selection.

Lastly, we shall make a few comments about the size of the sample, typically denoted by the letter n (to contrast with N, the size of the population). The ratio

See Exercises 21–24

n/N is called the **sampling proportion.** A sampling proportion of $x\%$ tells us that the size of the sample is intended to be $x\%$ of the population. Some sampling methods are conducive for choosing the sample so that a given sampling proportion is obtained, but in many sampling situations it is very difficult to predetermine what the exact sampling proportion is going to be (we would have to know the exact values of both N and n). In any case, it is not the sampling proportion that matters but rather the absolute sample size and quality. Typically, modern public opinion polls use samples of n between 1000 and 1500 to get statistics that have a margin of error of less than 5%, be it for the population of a city, a region, or the entire country. (We will see more details about this idea in Chapter 16.)

13.5 The Capture-Recapture Method

We have already observed that finding the exact N-value of a large and elusive population can be extremely difficult and sometimes impossible. In many cases, a good estimate is all we really need, and such estimates are possible through sampling methods. The simplest sampling method for estimating the N-value of a population is called the **capture-recapture method.** The method consists of two steps, which we will describe in the jargon of the field in which it is most frequently used: population biology.

- **Step 1. Capture (sample):** Capture (choose) a sample of size n_1, *tag* (mark, identify) the animals (objects, people), and release them back into the general population.
- **Step 2. Recapture (resample):** After a certain period of time, capture a new sample of size n_2, and take an exact head count of the *tagged* individuals (i.e., those that were also in the first sample). Let's call this number k.

If we can assume that the recaptured sample is representative of the entire population, then the proportion of tagged individuals in it is approximately equal to the proportion of the tagged individuals in the population. In other words, the ratio k/n_2 is approximately equal to the ratio n_1/N. From this we can solve for N and get $N \approx n_1 \cdot n_2/k$.

▶ EXAMPLE 13.5 Small Fish in a Big Pond

A large pond is stocked with catfish. As part of a research project we need to estimate the number of catfish in the pond. An actual headcount is out of the question (short of draining the pond), so our best bet is the capture-recapture method.

Step 1. For our first sample we capture a predetermined number n_1 of catfish, say $n_1 = 200$. The fish are tagged and released unharmed back in the pond.

Step 2. After giving enough time for the released fish to mingle and disperse throughout the pond, we capture a second sample of n_2 catfish. While n_2 does not have to equal n_1, it is a good idea for the two samples to be of approximately the same order of magnitude. Let's say that $n_2 = 250$. Of the 250 catfish in the second sample, 35 have tags (were part of the original sample).

Assuming the second sample is representative of the catfish population in the pond, the ratio of tagged fish in the second sample (35/250) is approximately the same as the ratio of tagged fish in the pond $(200/N)$. This gives the approximate proportion

$$35/250 \approx 200/N$$

which in turn gives

$$N \approx 200 \times 250/35 \approx 1428.57$$

Obviously, the foregoing value of N cannot be taken literally, since N must be a whole number. Besides, even in the best of cases, the computation is only an estimate. A sensible conclusion is that there are approximately $N = 1400$ catfish in the pond.   ◀◀

13.6 Clinical Studies

A survey typically deals with issues and questions that have direct and measurable answers. *If the election were held today, would you vote for candidate X or candidate Y? How many people live in your household? How many catfish have tags?* In these situations data collection involves some combination of *observing*, *measuring*, and *recording* but no active involvement or interference with the phenomenon being observed.

A different type of data collection process is needed when we are trying to establish connections between a cause and an effect. *Does taking a math class increase your chances of getting a good paying job? Does repeated exposure to second-hand smoke significantly increase your risk for developing lung cancer? Does a daily dose of aspirin reduce your chances of a heart attack? Do the benefits of hormone replacement therapy for women over 50 outweigh the risks?* These kinds of cause-effect questions cannot be answered by means of an immediate measurement and require observation over an extended period of time. Moreover, in these situations the data collection process requires the active involvement of the experimenter—in addition to *observation*, *measurement*, and *recording*, there is also *treatment*.

When we want to know if a certain cause X produces a certain effect Y, we set up a *study* in which cause X is produced and its effects are observed. If the effect Y is observed, then it is possible that X was indeed the cause of Y. We have established an *association* between the cause X and the effect Y. The problem, however, is the nagging possibility that some other cause Z different from X produced the effect Y and that X had nothing to do with it. Just because we established an association, we have not established a cause-effect relation between the variables. Statisticians like to explain this by a simple saying: *Association is not causation.*

Let's illustrate with a fictitious example. Suppose we want to find out if eating too much candy increases the chances of developing diabetes in adults. Here the cause X is eating candy, and the effect Y is developing diabetes. We set up an experiment in which 1000 adult laboratory rats are given, in addition to their normal diet, 8 ounces of candy a day for a period of six months. At the end of the six-month period, 150 of the 1000 rats have developed diabetes. Since in the general rat population only 3% of rats are diabetic, we have definitely established an association between the cause (candy) and the effect (increased incidence of diabetes). It is very tempting to conclude that the diabetes in the rats was indeed caused by the candy, but if we did that, we would be doing really bad science. How can we be so sure that the candy was really the cause of the increased diabetes? Could there be another hidden cause that we haven't noticed? What about their regular food? Lack of exercise? Cage conditions?

If you think that our fictitious candy story is too far-fetched to be realistic, consider the next case study.

Case Study 5: The Alar Scare

Alar is a chemical used by apple growers to regulate the rate at which apples ripen. Until 1989, practically all apples sold in grocery stores were sprayed with Alar. But in 1989 Alar became bad news, denounced in newspapers and on TV as a potent cancer-causing agent and a primary cause of cancer in children. As a result of these reports, people stopped buying apples, schools all over the country removed apple juice from their lunch menus, and the Washington state apple industry lost an estimated $375 million.

The case against Alar was based on a single 1973 study in which laboratory mice were exposed to the active chemicals in Alar. The dosage used in the study was eight times greater than the maximum tolerated dosage—a concentration at which even harmless substances can produce tissue damage. In fact, a child would have to eat about 200,000 apples a day to be exposed to an equivalent dosage of the chemical. Subsequent studies conducted by the National Cancer Institute and the Environmental Protection Agency failed to show any cause-and-effect relationship between Alar and cancer in children.

While it is generally accepted now that Alar does not cause cancer, because of potential legal liability, it is no longer used. The Alar scare turned out to be a false alarm based on a poor understanding of the statistical evidence. Unfortunately, it left in its wake a long list of casualties, among them the apple industry, the product's manufacturer, the media, and the public's confidence in the system.

For most cause-and-effect situations, especially those complicated by the involvement of human beings, a single effect can have many possible and actual causes. What causes cancer? Unfortunately, there is no single cause—diet, lifestyle, the environment, stress, and heredity are all known to be contributory causes. The extent to which each of these causes contributes individually and the extent to which they interact with each other are extremely difficult questions that can be answered only by means of carefully designed statistical studies. ◀◀

For the remainder of this chapter we will illustrate an important type of study called a **clinical study** or **clinical trial**. Generally, clinical studies are concerned with determining whether a single variable or treatment (usually a vaccine, a drug, therapy, etc.) can cause a certain effect (a disease, a symptom, a cure, etc.). The importance of such clinical studies is self-evident: Every new vaccine, drug, or treatment must prove itself by means of a clinical study before it is officially approved for public use. Likewise, almost everything that is bad for us (cigarettes, caffeine, cholesterol, etc.) gets its official certification of badness by means of a clinical study.

Properly designing a clinical study can be both difficult and controversial, and as a result, we are often bombarded with conflicting information produced by different studies examining the same cause-and-effect question. The basic principles guiding a clinical study, however, are pretty much established by statistical practice.

The first and most important issue in any clinical study is to isolate the cause (treatment, drug, vaccine, therapy, etc.) that is under investigation from all other possible contributing causes (called **confounding variables**) that could produce the same effect. Generally, this is best done by *controlling* the study.

In a **controlled study**, the subjects are divided into two different groups: the *treatment group* and the *control group*. The **treatment group** consists of those subjects receiving the actual treatment; the **control group** consists of subjects that are not receiving any treatment—they are there for comparison purposes only (that's why the control group is sometimes also called the *comparison* group). If a real cause-and-effect relationship exists between the treatment and the effect being studied, then the treatment group should show the effects of the treatment and the control group should not.

To eliminate the many potential confounding variables that can bias its results, a well-designed controlled study should have control and treatment groups that are similar in every characteristic other than the fact that one group is being treated and the other one is not. (It would be a very bad idea, for example, to have a treatment group that is all female and a control group that is all male.) The most reliable way to get equally representative treatment and control groups is to use a *randomized controlled study*. In a **randomized controlled study**, the subjects are assigned to the treatment group or the control group randomly (typically by a computer program).

When the randomization part of a randomized controlled study is properly done, treatment and control groups can be assumed to be statistically similar. But there is still one major difference between the two groups that can significantly affect the validity of the study—a critical confounding variable known as the *placebo effect*. The **placebo effect** follows from the generally accepted principle that *just the idea that one is getting a treatment, can produce positive results*. Thus, when subjects in a study are getting a pill or a vaccine or some other kind of treatment, how can the researchers separate positive results that are consequences of the treatment itself from those that might be caused by the placebo effect? When possible, the standard way to handle this problem is to give the control group a *placebo*. A **placebo** is a *make-believe* form of treatment—a harmless pill, an injection of saline solution, or any other fake type of treatment intended to look like the real treatment. A controlled study in which the subjects in the control group are given a placebo is called a **controlled placebo study**.

By giving all subjects a seemingly equal treatment (the treatment group gets the real treatment and the control group gets a placebo which looks like the real treatment), we do not eliminate the placebo effect but rather control it—whatever its effect might be, it impacts all subjects equally. It goes without saying that the use of placebos is pointless if the subject knows he or she is getting a placebo. Thus, a second key element of a good controlled placebo study is that all subjects be kept in the dark as to whether they are being treated with a real treatment or a placebo. A study in which neither the members of the treatment group nor the members of the control group know to which of the two groups they belong is called a **blind** study.

Blindness is a key requirement of a controlled placebo study, but not the only one. To keep the interpretation of the results (which can often be ambiguous) totally objective, it is important that the scientists conducting the study and collecting the data also be in the dark when it comes to who got the treatment and who got the placebo. A controlled placebo study in which neither the subjects nor the scientists conducting the experiment know which subjects are in the treatment group and which are in the control group is called a **double-blind study**.

Our next case study illustrates one of the most famous and important double-blind studies in the annals of clinical research.

Case Study 6: The 1954 Salk Polio Vaccine Field Trials

Polio (infantile paralysis) has been practically eradicated in the Western world. In the first half of the twentieth century, however, it was a major public health problem. Over one-half million cases of polio were reported between 1930 and 1950, and the actual number may have been considerably higher.

Because polio attacks mostly children and because its effects can be so serious (paralysis or death), eradication of the disease became a top public health priority in the United States. By the late 1940s, it was known that polio is a virus and, as such, can best be treated by a vaccine which is itself made up of a virus. The vaccine virus can be a closely related virus that does not have the same harmful effects, or it can be the actual virus that produces the disease but which has been killed by a special treatment. The former is known as a *live-virus vaccine*, the latter as a *killed-virus vaccine*. In response to either vaccine, the body is known to produce *antibodies* that remain in the system and give the individual immunity against an attack by the real virus.

Both the live-virus and the killed-virus approaches have their advantages and disadvantages. The live-virus approach produces a stronger reaction and better immunity, but at the same time, it is also more likely to cause a harmful reaction and, in some cases, even to produce the very disease it is supposed to prevent. The killed-virus approach is safer in terms of the likelihood of producing a harmful reaction, but it is also less effective in providing the desired level of immunity.

These facts are important because they help us understand the extraordinary amount of caution that went into the design of the study that tested the effectiveness of the polio vaccine. By 1953, several potential vaccines had been developed, one of the more promising of which was a killed-virus vaccine developed by Jonas Salk at the University of Pittsburgh. The killed-virus approach was chosen because there was a great potential risk in testing a live-virus vaccine in a large-scale study. (A large-scale study was needed to collect enough information on polio, which, in the 1950s, had a rate of incidence among children of about 1 in 2000.)

The testing of any new vaccine or drug creates many ethical dilemmas that have to be taken into account in the design of the study. With a killed-virus vaccine the risk of harmful consequences produced by the vaccine itself is small. So one possible approach would have been to distribute the vaccine widely among the population and then follow up on whether there was a decline in the national incidence of polio in subsequent years. This approach, which was not possible at the time because supplies were limited, is called the *vital statistics* approach and is the simplest way to test a vaccine. This is essentially the way the smallpox vaccine was determined to be effective. The problem with such an approach for polio is that polio is an epidemic type of disease, which means that there is a great variation in the incidence of the disease from one year to the next. In 1952, there were close to 60,000 reported cases of polio in the United States, but in 1953, the number of reported cases had dropped to almost half that (about 35,000). Since no vaccine or treatment was used, the cause of the drop was the natural variability typical of epidemic diseases. But if an ineffective polio vaccine had been tested in 1952 without a control group, the observed effect of a large drop in the incidence of polio in 1953 could have been incorrectly interpreted as statistical evidence that the vaccine worked.

The final decision on how best to test the effectiveness of the Salk vaccine was left to an advisory committee of doctors, public officials, and statisticians convened by the National Foundation for Infantile Paralysis and the Public Health Service. It was a highly controversial decision, but at the end, a large-scale, randomized, double-blind, controlled placebo study was chosen. Approximately 750,000 children were randomly selected to participate in the study. Of these, about 340,000 declined to participate, and another 8500 dropped out in the middle of the experiment. The remaining children were randomly divided into two groups—a treatment group and a control group—with approximately 200,000 children in each group. Neither the families of the children nor the researchers collecting the data knew if a particular child was getting the actual vaccine or a shot of harmless solution. The latter was critical because polio is not an easy disease to diagnose—it comes in many different forms and degrees. Sometimes it can be a borderline call, and if the doctor collecting the data had prior knowledge of whether the subject had received the real vaccine or the placebo, the diagnosis could have been subjectively tipped one way or the other.

A summary of the results of the Salk vaccine field trials is shown in Table 13-1. These data were taken as conclusive evidence that the Salk vaccine was an effective treatment for polio and on the basis of this study, a massive inoculation campaign was put into effect. Today, all children are routinely inoculated against polio, and the disease has essentially been eradicated in the United States. Statistics played a key role in this important public health breakthrough.

TABLE 13-1 Results of the Salk Vaccine Field Trials

	Number of children	Number of reported cases of polio	Number of paralytic cases of polio	Number of fatal cases of polio
Treatment group	200,745	82	33	0
Control group	201,229	162	115	4
Declined to participate in the study	338,778	182*	121*	0*
Dropped out in the middle	8,484	2*	1*	0*
Total	749,236	428	270	4

*These figures are not a reliable indicator of the actual number of cases—they are only self-reported cases.
Source: Adapted from Thomas Francis, Jr., et al., "An Evaluation of the 1954 Poliomyelitis Vaccine Trials—Summary Report." *American Journal of Public Health*, 45 (1955), 25.

Conclusion

In this chapter we have discussed different methods for collecting data. In principle, the most accurate method is a *census*, a method that relies on collecting data from each member of the population. In most cases, because of considerations of cost and time, a census is an unrealistic strategy. When data are collected from only a subset of the population (called a *sample*), the data collection method is called a *survey*. The most important rule in designing good surveys is to eliminate or minimize *sample bias*. Today, almost all strategies for collecting data are based on surveys in which the laws of chance are used to determine how the sample is selected, and these methods for collecting data are called *random sampling* methods. Random sampling is the best way known to minimize or eliminate sample bias. Two of the most common random sampling methods are *simple random sampling* and *stratified sampling*. In some special situations, other more complicated types of random sampling can be used.

Sometimes identifying the sample is not enough. In cases in which cause-and-effect questions are involved, the data may come to the surface only after an extensive study has been carried out. In these cases, isolating the cause variable under consideration from other possible causes (called *confounding variables*) is an essential prerequisite for getting reliable data. The standard strategy for doing this is a *controlled study* in which the sample is broken up into a *treatment group* and a *control group*. Controlled studies are now used (and sometimes abused) to settle issues affecting every aspect of our lives. We can thank this area of statistics for many breakthroughs in social science, medicine, and public health, as well as for the constant and dire warnings about our health, our diet, and practically anything that is fun.

Profile   George Gallup (1901–1984)

George Gallup was a man of many talents—journalist, sociologist, political analyst, businessman, and statistician. It was the combination of all these talents, together with a strong entrepreneurial spirit and an unlimited amount of self-confidence, that made him a unique fixture of twentieth-century American life—the man trusted by generations to best measure the nation's pulse. Gallup did not invent the modern-day public opinion poll, but he certainly set the standard for how to do it right. Under his tenure, the poll that bears his name became the most widely read and credible public opinion poll in the world.

George Horace Gallup, Jr., was born in Jefferson, Iowa, a small farm town in America's heartland. He studied journalism at the University of Iowa, paying his way through school partly by working as the editor of the student newspaper. He earned a bachelor's degree in journalism in 1923, a master's degree in psychology in 1925, and a Ph.D. in journalism in 1928. In his doctoral thesis, entitled *About Unbiased Methodology of Exploring Readers' Interest in the Content of Newspapers*, Gallup showed that public opinion could be scientifically collected from a very small sample. His doctoral work was a precursor of the sampling methods he would later develop to conduct public opinion polls.

After working as a professor of journalism at Drake University and at Northwestern University, Gallup was recruited by New York advertising agency Young and Rubicam to become the head of its newly developed market-research department. But Gallup was a journalist at heart, and by 1935 he had enough of the advertising business, notwithstanding the fact that he had made millions in it. He moved to Princeton, New Jersey, where he founded the American Institute of Public Opinion and started a weekly syndicated newspaper column immodestly entitled "America Speaks," which featured the results of public opinion polls he designed and conducted. From these humble beginnings, the Gallup poll was born. The organization he founded (now called the Gallup Organization and run by his two sons, George III and Alec) has grown to become the largest and most respected private polling organization in the world.

Today, public opinion polling is used extensively throughout the world, but the increased use of polls in modern life has generated many criticisms about their impact and influence on the political process. As the father of polling, Gallup spent his later years trying to defend polls as an important and useful instrument of democracy. "When a president, or any other leader, pays attention to poll results, he is, in effect, paying attention to the views of the people," he once said.

George Gallup died in 1984 in Switzerland, where he had spent a good part of his last years, partly for its beauty and partly because, as he put it, the country was "virtually run by polls." Once, as a young editor of the University of Iowa student newspaper, Gallup wrote, "Doubt everything, Question everything. Be a radical!" Even as he became an American institution, George Gallup remained always true to this credo.

Key Concepts

Exercises

WALKING

A. Surveys and Polls

Exercises 1 through 4 refer to the following story: As part of a sixth-grade statistics project, the teacher brings to class a candy jar full of gumballs of two different colors: red and green. The assignment is to estimate the proportion of red gumballs in the jar. To do this, the jar is shaken well, and one of the students draws 25 gumballs from the jar. Of these, 8 are red and 17 are green.

1. (a) Describe the population for this survey.

(b) Describe the sample for this survey.

(c) Give the sample statistic for the *proportion* of red gumballs in the jar.

(d) Name the sampling method used for this survey.

2. Given that the total number of gumballs in the jar is 200,

(a) give the sampling proportion for this survey.

(b) give the sample statistic for the *number* of red gumballs in the jar.

3. Given that the total number of gumballs in the jar is 200 and the number of red gumballs is 50,

(a) give the parameter for the proportion of red gumballs in the jar.

(b) give the sampling error, expressed as a percent.

(c) Is the sampling error found in (b) a result of sampling variability or sampling bias? Explain.

4. (a) Compare and contrast the target population and sampling frame for this survey.

(b) What data collection method could be used to find the exact value of the parameter?

Exercises 5 through 8 refer to the following story: The city of Cleansburg has 8325 registered voters. There is an election for mayor of Cleansburg, and there are three candidates for the position: Smith, Jones, and Brown. The day before the election, a telephone poll of 680 randomly chosen registered voters produced the following results: 306 people surveyed indicated that they would vote for Smith, 272 indicated that they would vote for Jones, and 102 indicated that they would vote for Brown.

5. (a) Give the sampling proportion for this survey.

(b) Give the sample statistic estimating the *percentage* of the vote going to Smith.

6. (a) Describe the population for this survey.

(b) Describe the sample for this survey.

(c) Name the sampling method used for this survey.

7. Given that in the actual election Smith received 42% of the vote, Jones 43% of the vote, and Brown 15% of the vote, find the sampling errors in the survey expressed as percentages.

8. Do you think that the sampling error in this example is due primarily to chance error or to sample bias? Explain your answer.

Exercises 9 through 12 refer to the following story: The 1250 students at Eureka High School are having an election for Homecoming King. The candidates are Tomlinson (captain of the football team), Garcia (class president), and Marsalis (member of the marching band). At the football game a week before the election, a pre-election poll is taken of students as they enter the gates. Of the students that attended the game, 203 planned to vote for Tomlinson, 42 planned to vote for Garcia, and 105 planned to vote for Marsalis.

9. (a) Describe the sample for this survey.

(b) Give the sampling proportion for this survey.

10. Name the sampling method used for this survey.

11. **(a)** Compare and contrast the population and the sampling frame for this survey.

    (b) Is the sampling error a result of sampling variability or sampling bias? Explain.

12. **(a)** Give the sample statistics estimating the *percentage* of the vote going to each candidate.

    (b) A week after this survey, Garcia was elected Homecoming King with 51% of the vote; Marsalis got 30% of the vote, and Tomlinson came in last with 19% of the vote. Find the sampling errors in the survey expressed as percentages.

Exercises 13 through 16 refer to the following true story: In 1988, "Dear Abby" asked her readers to let her know whether they had cheated on their spouses or not. The readers' responses are summarized in the following table.

Status	Women	Men
Faithful	127,318	44,807
Unfaithful	22,468	15,743
Total	149,786	60,550

Based on the results of this survey, "Dear Abby" concluded that the amount of cheating among married couples is much less than people believe. (In her words, "The results were astonishing. There are far more faithfully wed couples than I had surmised.")

13. **(a)** Describe as specifically as you can the sampling frame for this survey.

    (b) Compare and contrast the target population and the sampling frame for this survey.

    (c) How was the sample chosen?

    (d) Eighty-five percent of the women who responded to this survey claimed to be faithful. Is 85% a parameter? A statistic? Neither? Explain your answer.

14. **(a)** Explain why this survey was subject to selection bias.

    (b) Explain why this survey was subject to nonresponse bias.

15. **(a)** Based on the "Dear Abby" data, estimate the percentage of married men who are faithful to their spouses.

    (b) Based on the "Dear Abby" data, estimate the percentage of married people who are faithful to their spouses.

    (c) How accurate do you think these estimates are? Explain.

16. If money were no object, could you devise a survey that might give more reliable results than the "Dear Abby" survey? Describe briefly what you would do.

Exercises 17 through 20 refer to the following story: The Cleansburg Planning Department is trying to determine what percent of the people in the city want to spend public funds to revitalize the downtown mall. To do so, it decides to conduct a survey. Five professional interviewers are hired. Each interviewer is asked to pick a street corner of his or her choice within the city limits and every day between 4:00 and 6:00 P.M. the interviewers are supposed to ask each passerby if he or she wishes to respond to a survey sponsored by Cleansburg City Hall. If the response is yes, the follow-up question is, "Are you in favor of spending public funds to revitalize the downtown mall?" The interviewers are asked to return to the same street corner as many days as are necessary until each one has conducted a total of 100 interviews. The results of the survey are shown in the following table.

Interviewer	Yes[a]	No[b]	Nonrespondents[c]
A	35	65	321
B	21	79	208
C	58	42	103
D	78	22	87
E[d]	12	63	594

[a] In favor of spending public funds to revitalize the downtown mall.
[b] Opposed to spending public funds to revitalize the downtown mall.
[c] Declined to be interviewed or had no opinion.
[d] Got frustrated and quit.

17. **(a)** Describe as specifically as you can the target population for this survey.

    (b) Compare and contrast the target population and the sampling frame for this survey.

18. **(a)** What is the size of the sample?

    (b) Calculate the response rate in this survey. Was this survey subject to nonresponse bias?

19. **(a)** Can you explain the big difference in the data from interviewer to interviewer?

    (b) One of the interviewers conducted the interviews at a street corner downtown. Which interviewer? Explain.

    (c) Do you think the survey was subject to selection bias? Explain.

    (d) Was the sampling method used in this survey the same as quota sampling? Explain.

20. Do you think this was a good survey? If you were a consultant to the Cleansburg Planning Department, could you suggest some improvements? Be specific.

Systematic Sampling. *Exercises 21 through 24 refer to the notion of systematic sampling, as illustrated by the following story: The dean of students at Tasmania State University (TSU) wants to determine the percent of undergraduates who tried but could not enroll in Math 101 this semester because of insufficient space. There are 15,000 undergraduates at TSU, so it is decided that the cost of checking with each and every one would be prohibitive. The following method (called systematic sampling) is proposed to choose a representative sample of undergraduates to interview. Start with the registrar's alphabetical listing containing the names of all undergraduates. Randomly pick a number between 1 and 100, and count that far down the list. Take that name and every 100th name after it. (For example, if the random number chosen is 73, then pick the 73rd, 173rd, 273rd, etc., names on the list.) Assume that the survey has a response rate of 0.90.*

21. (a) Compare and contrast the sampling frame and the target population for this survey.

    (b) Give the exact *N*-value of the population.

22. (a) Find the size *n* of the sample.

    (b) Find the sampling proportion.

23. (a) Explain why the method used for choosing the sample is not simple random sampling.

    (b) If all those responding claimed they could not enroll in Math 101, is it more likely the result of sampling variability or sampling bias? Explain.

24. (a) If eight of the students who responded said that they were unable to enroll in Math 101, give a reasonable estimate for the total number of students at the university that were unable to enroll in Math 101.

    (b) Do you think the results of this survey will be reliable? Explain.

Exercises 25 and 26 refer to the following story: An orange grower wishes to compute the average yield from his orchard. The orchard contains three varieties of trees—50% of his trees are of variety A, 25% of variety B, and 25% of variety C.

25. (a) Suppose the grower samples randomly from 300 trees of variety A, 150 trees of variety B, and 150 trees of variety C. What type of sampling is being used?

    (b) Suppose the grower selects for his sample a 10-by-30 rectangular block of 300 trees of variety A, a 10-by-15 rectangular block of 150 trees of variety B, and a 10-by-15 rectangular block of 150 trees of variety C. What type of sampling is being used?

26. (a) Suppose that in his survey, the grower found that each tree of variety A averages 100 oranges, each tree of variety B averages 50 oranges, and each tree of variety C averages 70 oranges. Estimate the average yield per tree of his orchard.

    (b) Is the yield you found in (a) a parameter or a statistic? Explain.

27. Name the sampling method that best describes each situation. Choose your answer from the following list: simple random sampling, convenience sampling, quota sampling, stratified sampling, census.

(a) George wants to know how the rest of the class did on the last quiz. He peeks at the scores of a few students sitting right next to him. Based on what he sees, he concludes that nobody did very well.

(b) On a day when every student is present, the students are given a Teacher Evaluation Questionnaire. Every student fills out the Questionnaire.

(c) Eureka High School has 400 freshmen, 300 sophomores, 300 juniors, and 200 seniors. The student newspaper conducts a poll of student opinion regarding the rumor that the football coach is about to be fired. In the poll 20 freshmen, 15 sophomores, 15 juniors, and 10 seniors are randomly selected to be interviewed.

(d) For the last football game of the season the coach chooses the three captains by putting the names of all the players in a hat and drawing three names. (Maybe that's why they are trying to fire him!)

(e) For the last football game of the season the coach chooses the three captains by randomly selecting a senior offensive player, a senior defensive player, and a senior special teams player.

28. Name the sampling method that best describes each situation. Choose your answer from the following list: simple random sampling, convenience sampling, quota sampling, stratified sampling, census.

(a) The shipper inspects every orange that comes from the suppliers before packaging and shipping.

(b) A few randomly chosen crates of oranges are opened and the oranges on the top layer are inspected.

(c) Of the 250 crates of oranges received by the shipper, 75 crates came from supplier A, 75 from supplier B, and 100 from supplier C. Six crates are randomly chosen for inspection from A's shipment, six crates are randomly chosen from B's shipment, and eight crates are randomly chosen from C's shipment.

B. The Capture-Recapture Method

29. You want to estimate how many fish there are in a small pond. Let's suppose that you capture $n_1 = 500$ fish, tag them, and throw them back in the pond. After a couple of days, you go back to the pond and capture $n_2 = 120$ fish, of which $k = 30$ are tagged. Give an estimate of the *N*-value of the fish population in the pond.

30. To estimate the population in a rookery, 4965 fur seal pups were captured and tagged in early August. In late August, 900 fur seal pups were captured. Of these, 218 had been tagged. Based on these figures, estimate the population of fur seal pups in the rookery to the nearest hundred.

    [*Source: Chapman and Johnson, "Estimation of Fur Seal Pup Populations by Randomized Sampling," Transactions of the American Fisheries Society, 97 (July 1968), 264–270]*

Exercises 31 through 34 refer to the following story: You have a very large coin jar full of nickels, dimes, and quarters. You want to have an approximate idea of how much money you have, but you don't want to go through the trouble of counting all the coins, so you decide to use the capture-recapture method. For the first sample, you shake the jar well and randomly draw 50 coins. You get 12 quarters, 15 nickels, and 23 dimes. Using a black marker, you mark the 50 coins with a black dot and put them back in the jar. For the second sample, you shake the jar well and randomly draw another set of 100 coins. You get 28 quarters, 4 of which have black dots; 29 nickels, 5 of which have black dots; and 43 dimes, 8 of which have black dots.

31. Estimate the total number of quarters in the jar.

32. Estimate the total number of nickels in the jar.

33. Estimate the total number of dimes in the jar.

34. Do you think the capture-recapture method is a reliable way to estimate the number of coins in the jar? Explain your answer. Discuss some of the potential pitfalls and issues one should be concerned about.

35. Starting in 2004, a collaborative study between the Canadian Ministry of Natural Resources, Minnesota's Department of Natural Resources and the Rainy River First Nations investigated the populations of lake sturgeon on the Rainy River and the Lake of the Woods on the U.S.-Canadian border. During the capture phase of the project 1700 lake sturgeon were caught, tagged, and released. During the recapture phase of the project, 660 lake sturgeon were caught and released. Seven of the 660 were part of the original capture group that had been tagged. Based on these figures, estimate the population of lake sturgeon in the Lake of the Woods to the nearest thousand.

[**Source:** *Gauthier, Dan, "Lake of the Woods Sturgeon Population Recovering," Daily Miner and News (Kenora, Ontario), June 11, 2005, p. 31*]

36. Efforts at Utah Lake in 2004 show the carp population dominates fish life in the lake. In 15 days, workers captured, tagged, and released 24,000 carp. Of the 10,300 carp that were later recaptured, 208 had tags. Give an estimate for the *N*-value of the carp population in Utah Lake in 2004.

[**Source:** *Prettyman, Brett, "With Carp Cooking Utah Lake, It's Time to Eat," Salt Lake Tribune (Salt Lake City, Utah), July 15, 2004, p. D3*]

Exercises 37 and 38 refer to the following story: In 1991, a study using the capture-recapture method was used to estimate the number of HIV-infected drug users in Bangkok, Thailand. The "capture" consisted of a list of 4064 names of individuals treated at 18 methadone treatment facilities in Bangkok between April 17 and May 17. The "recapture" consisted of a list of 1540 persons held at 72 Bangkok police stations between June 3 and September 30 that tested positive for methadone. There were 171 persons included on both lists.

[**Source:** *Mastro et al, "Estimating the Number of HIV-Infected Injection Drug Users in Bangkok: A Capture-Recapture Method," American Journal of Public Health, (July 1994), 84–87*]

37. (a) Estimate the number of drug users in Bangkok during the time of the study.

(b) In 1991 it was estimated that the 89% of drug users in Bangkok injected drugs, and that one-third of these were infected with HIV. Using this information estimate the number of HIV-infected drug users in Bangkok in 1991.

38. Discuss the potential pitfalls of this study. Is the general approach used a reasonable approach for estimating populations of drug users?

C. Clinical Studies

Exercises 39 through 42 refer to the following story: The manufacturer of a new vitamin (vitamin X) decides to sponsor a study to determine its effectiveness in curing the common cold. Five hundred college students in the San Diego area who are suffering from colds are paid to participate as subjects in this study. They are all given two tablets of vitamin X a day. Based on information provided by the subjects themselves, 457 out of the 500 subjects are cured of their colds within three days. The average number of days a cold lasts is 4.87 days. As a result of this study, the manufacturer launches an advertising campaign, claiming that "vitamin X is more than 90% effective in curing the common cold."

39. (a) Describe as specifically as you can the target population for this study.

(b) Compare and contrast the target population and the sampling frame for this study.

(c) Is selection bias present in the sample?

40. (a) Was this study controlled?

(b) List three possible causes other than the effectiveness of vitamin *X* itself that could have confounded the results of this study.

41. List four different problems with this study that indicate poor design.

42. Make some suggestions for improving the study.

Exercises 43 through 46 refer to the following story: A team of researchers and surgeons at the Houston VA Medical Center randomly divided 180 potential knee surgery patients into three groups. (There were 324 participants who met inclusion criteria for the study, but 144 declined to participate.) The first group received arthroscopic debridement. A second group received arthroscopic lavage. Patients in the third group received skin incisions and underwent a simulated procedure ("sham" surgery) without actual insertion of the arthroscope. The patients in the study did not know which group they were being divided into and therefore did not know if they were receiving the real or simulated surgery. All the patients who participated in the study were evaluated for two years after the procedure. In the two-year follow-up, all three groups said they had slightly less pain and better knee movement. However, the sham-surgery group often reported the best results.

[***Source:****New England Journal of Medicine, 347, 2 (July 11, 2002), 81–88.]*

43. Describe as specifically as you can the target population for this study.

44. (a) Describe the sample.

   (b) What was the size n of the sample?

   (c) Was the sample chosen by random sampling? Explain.

45. (a) Was this study a controlled placebo experiment? Explain.

   (b) Describe the treatment group(s) in this study.

   (c) Could this study be considered a randomized controlled experiment? Explain.

   (d) Was this experiment blind, double blind, or neither?

46. (a) Discuss any ethical dilemmas in this type of study involving sham surgery.

   (b) Carefully state what a legitimate conclusion from this study might be.

Exercises 47 through 50 refer to the following story: A college professor has a theory that a dose of about 500 milligrams of caffeine a day can actually improve students' performance in their college courses. To test his theory, he chooses the 13 students in his Psychology 101 class who failed the first midterm and asks them to come to his office three times a week for individual tutoring. When the students come to his office, he engages them in friendly conversation while at the same time pouring them several cups of strong coffee (a total of 500 milligrams of caffeine per student). After a month of doing this, he observes that of the 13 students, 8 show significant improvement in their second midterm scores, 3 show some improvement, and 2 show no improvement at all. Based on this, he concludes that his theory about caffeine is correct.

47. Which of the following terms best describes the professor's study: (i) randomized controlled experiment, (ii) double-blind experiment, (iii) controlled placebo experiment, or (iv) clinical study? Explain your choice and why you ruled out the other choices.

48. (a) Describe the target population and the sample of this study.

   (b) What was the value of n?

   (c) Which of the following percentages best describes the sampling proportion for this study: (i) 10%, (ii) 1%, (iii) 0.1%, (iv) 0.01%, or (v) less than 0.01%? Explain.

49. (a) Was the study blind, double blind, or neither? Explain.

   (b) List at least three possible causes other than caffeine that could have confounded the results of this study.

50. Make some suggestions to the poor professor as to how he might improve the study.

Exercises 51 through 54 refer to the following true story: On September 30, 2004 the pharmaceutical giant Merck announced it was pulling the arthritis and acute pain medication Vioxx off the market. The decision was based on the results of a clinical trial named APPROVe designed to test whether Vioxx was effective in preventing the recurrence of colorectal polyps in patients with a history of colorectal adenomas. The 2586 participants in the clinical trial (all of whom had a history of colorectal adenomas) were randomly divided into two groups: 1287 patients were given 25 daily milligrams of Vioxx for the duration of the clinical trial (originally intended to last three years), and 1299 patients were given a placebo. Neither the participants nor the doctors involved in the clinical trial knew who was in which group. During the trial 72 of the participants had cardiovascular events (mostly heart attacks or strokes). Later it was found that 46 of these people were from the group taking the Vioxx and 26 from the group taking the placebo. Based on these results, Merck pulled Vioxx off the market.

51. Describe as specifically as you can the target population for APPROVe.

52. (a) Describe the sample for APPROVe.

   (b) Give the size n of the sample.

53. (a) Describe the control and treatment groups in APPROVe.

   (b) APPROVe can be described as a *double-blind randomized controlled placebo experiment.* Explain why each of these terms applies.

54. Carefully state what a legitimate conclusion from this study might be for a person suffering from arthritis.

D. Miscellaneous

55. Super Size Me. In 2003, Morgan Spurlock set out to study the effect McDonald's has on the average American, a study he documented in the award-winning film *Super Size Me.* Spurlock ate three meals at McDonald's every day for 30 days, taking the "super-size" option whenever it was offered. He also curtailed his physical activity to better match the exercise habits of the average American. In the end, his health declined dramatically—he gained 25 pounds, suffered severe liver dysfunction, and developed symptoms of depression.

   (a) Was Spurlock's study a survey or a clinical trial? Explain.

   (b) Describe as specifically as you can the target population for the study.

   (c) Describe the sample for the study.

   (d) List three problems with this study that indicate poor design.

56. Super Size Me II. Thinking that Spurlock's documentary *Super Size Me* (see Exercise 55) unfairly targeted McDonald's, Merab Morgan set out to do a fast-food study of her own. Morgan ate only at McDonald's for 90 days, sticking to meal plans of no more than 1400 calories a day and choosing to eat mostly burgers and salads (no french fries!). By the end of her study Morgan had dropped 37 pounds and felt well.

(a) Describe the treatment group in Morgan's study.

(b) List some of the possible confounding variables in Morgan's study.

(c) Carefully state what a legitimate conclusion from this study might be.

57. A study by the Center for Academic Transformation showed that students using electronic learning tools have seen marked progress in their test scores. At the University of Alabama, the passing rate for an intermediate algebra class doubled from 40% to 80% when the class was redesigned to rely heavily on supplemental materials, online practice exercises, and interactive tutorials.

(a) Was this study a survey or a clinical trial? Explain.

(b) List some of the possible confounding variables in this study.

58. There are 50 students in the Math 101 class at Tasmania State University—30 females and 20 males. The professor chooses a sample of ten students from the class as follows: Six students are randomly chosen from among the females and four students are randomly chosen from among the males.

(a) Does every student in the class have an equal likelihood of being selected for the sample? Explain.

(b) Does every set of 10 students in the class have an equal likelihood of being selected as the sample? Explain.

59. Determine in each case if the data is a population parameter or a sample statistic.

(a) In 2001, 25% of students taking the SAT math test scored above 590.

(b) In crash testing of a new automobile model, 20% of the crashes would have caused severe neck injury.

(c) Mr. Johnson's blood tested positive for Type II diabetes.

(d) A recent Gallup poll shows that 6 in 10 Americans have attempted to lose weight.

60. Choose the most appropriate concept for each statement from the following list: (i) association is not causation; (ii) sampling variability; (iii) selection bias; (iv) placebo effect.

(a) "In the early part of the twentieth century it was discovered that when viewed over time, the number of crimes increased with membership in the Church of England."

(b) "Half of the subjects in the study were given sugar pills. These subjects responded with fewer days ill than those receiving the treatment."

(c) "Jack chose a simple random sample of 100 widgets, and 11 of the widgets in his sample were defective. Jill chose a simple random sample of 100 widgets, and 19 of the widgets in her sample were defective."

(d) "Jay didn't understand how Jones could have been elected mayor when almost every person he knows voted for Brown."

JOGGING

61. Informal surveys. In everyday life, we are constantly involved in activities that can be described as *informal surveys*, often without even realizing it. Here are some examples.

(i) Al gets up in the morning and wants to know what kind of day it is going to be, so he peeks out the window. He doesn't see any dark clouds, so he figures it's not going to rain.

(ii) Betty takes a sip from a cup of coffee and burns her lips. She concludes the coffee is too hot and decides to add a tad of cold water to it.

(iii) Carla got her first Math 101 exam back with a C grade on it. The students sitting on each side of her also received C grades. She concludes that the entire Math 101 class received a C on the first exam.

For each of the preceding examples,

(a) describe the population.

(b) discuss whether the sample is random or not.

(c) discuss the validity of the conclusions drawn. (There is no right or wrong answer to this question, but you should be able to make a reasonable case for your position.)

62. Read the examples of informal surveys given in Exercise 61. Give three new examples of your own. Make them as different as possible from the ones given in Exercise 61 [changing coffee to soup in (ii) is not a new example].

63. Leading-question bias. The way the questions in many surveys are phrased can itself be a source of bias. When a question is worded in such a way as to predispose the respondent to provide a particular response, the results of the survey are tainted by a special type of bias called *leading-question bias*. The following is an extreme hypothetical situation intended to drive the point home.

In an effort to find out how the American tax-payer feels about a tax increase, the institute conducts a "scientific" one-question poll.

Are you in favor of paying higher taxes to bail the federal government out of its disastrous economic policies and its mismanagement of the federal budget?
Yes _____. *No* _____.

Ninety-five percent of the respondents answered no.

(a) Explain why the results of this survey might be invalid.

(b) Rephrase the question in a neutral way. Pay particular attention to highly charged words.

(c) Make up your own (more subtle) example of leading-question bias. Analyze the critical words that are the cause of bias.

64. Consider the following hypothetical survey designed to find out what percentage of people cheat on their income taxes.

Fifteen hundred taxpayers are randomly selected from the Internal Revenue Service (IRS) rolls. These individuals are then interviewed in person by representatives of the IRS and read the following statement.

This survey is for information purposes only. Your answer will be held in strict confidence. Have you ever cheated on your income taxes?
Yes _____. *No* _____.

Twelve percent of the respondents answered yes.

(a) Explain why the 12% statistic might be unreliable.

(b) Can you think of ways in which a survey of this type might be designed so that more reliable information could be obtained? In particular, discuss who should be sponsoring the survey and how the interviews should be carried out.

65. Listing bias. Today, most consumer marketing surveys are conducted by telephone. In selecting a sample of households that are representative of all the households in a given geographical area, the two basic techniques used are (i) randomly selecting telephone numbers to call from the local telephone directory or directories, and (ii) using a computer to randomly generate seven-digit numbers to try that are compatible with the local phone numbers.

(a) Briefly discuss the advantages and disadvantages of each technique. In your opinion, which of the two will produce the more reliable data? Explain.

(b) Suppose that you are trying to market burglar alarms in New York City. Which of the two techniques for selecting the sample would you use? Explain your reasons.

66. The following two surveys were conducted in January 1991 in order to assess how the American public viewed media coverage of the Persian Gulf war. Survey 1 was an Area Code 900 telephone poll survey conducted by *ABC News*. Viewers were asked to call a certain 900 number if they felt that the media was doing a good job of covering the war and a different 900 number if they felt that the media was not doing a good job in covering the war. Each call cost 50 cents. Of the 60,000 respondents, 83% felt that the media was not doing a good job. Survey 2 was a telephone poll of 1500 randomly selected households across the United States conducted by the *Times-Mirror* survey organization. In this poll, 80% of the respondents indicated that they approved of the press coverage of the war.

(a) Briefly discuss survey 1, indicating any possible types of bias.

(b) Briefly discuss survey 2, indicating any possible types of bias.

(c) Can you explain the discrepancy between the results of the two surveys?

(d) In your opinion, which of the two surveys gives the more reliable data?

67. An article in the *Providence Journal* about automobile accident fatalities includes the following observation: "Forty-two percent of all fatalities occurred on Friday, Saturday, and Sunday, apparently because of increased drinking on the weekends."

(a) Give a possible argument as to why the conclusion drawn may not be justified by the data.

(b) Give a different possible argument as to why the conclusion drawn may be justified by the data after all.

68. (a) For the capture-recapture method to give a reasonable estimate of N, what assumptions about the two samples must be true?

(b) Give reasons why in many situations, the assumptions in (a) may not hold true.

Projects and Papers

A. What's the Latest?

In this project you are to do an in-depth report on a recent study. Find a recent article from a newspaper or news magazine reporting the results of a major study and write an analysis of the study. Discuss the extent to which the article gives the reader enough information to assess the validity of the study's conclusions. If in your opinion there is information missing in the article, generate a list of questions that would help you further assess the validity of the study's conclusions. Pay particular attention to the ideas and concepts discussed in this chapter, including the target population, sample size, sampling bias, randomness, controls, and so on.

Note: The best reporting on clinical studies can be found in major newspapers such as the *New York Times, Washington Post*, and *Los Angeles Times* or weekly news magazines such as *Time* or *Newsweek*. All of these have Web sites where their most recent articles can be downloaded for free.

B. The Governing "Body"

One of the biggest political upsets in recent years was the election of former wrestler Jesse "The Body" Ventura to governor of Minnesota in 1998. Ventura, the Reform Party candidate, won the three-way race against Republican Norm Coleman and Democrat Hubert "Skip" Humphrey III by a vote of 37 to 34 to 29%, respectively. But the day before the election, the *Star-Tribune*/KMSP-TV Minnesota poll showed Coleman leading at 36% and Ventura tied with Humphrey at 29%. A few days earlier, the *St. Paul Pioneer Press* showed 34% support for Humphrey, 33% for Coleman, and only 23% for Ventura. In fact, not one major poll before the election showed that Ventura had much of a chance.

Write an analysis paper on the 1998 Minnesota gubernatorial election. Discuss the methodology used by pollsters leading up to this election and hypothesize about why these polls failed to predict a Ventura victory. Then develop a list of lessons learned by polling organizations from this election.

C. The U.S. Census: A Continuing Political Battle

In January 1999, the United States Supreme Court ruled 5 to 4 (*Department of Commerce et al. v. U.S. House of Representatives et al.*) that the state population figures used for the apportionment of seats in the House of Representatives (see Chapter 4) cannot be determined by means of statistical sampling methods and that an actual census of the population is required by the Census Act. The ruling was based on arguments presented in the earlier case *Glavin v. Clinton*. In an effort to gain a seat in the House of Representatives, the state of Utah has recently unsuccessfully raised similar arguments with the Court ([*Utah et al. v. Evans, Secretary of Commerce, No. 01-283, 2002*] and [*Utah et al. v. Evans, Secretary of Commerce, No. 01-714, 2002*]).

In this project, you are to use the Court decisions cited as a guide in writing a summary of arguments both for and against the Census Bureau's current methodologies. Such arguments may mirror those made by Republicans and Democrats in this continuing political battle.

D. The Placebo Effect: Myth or Reality?

There is no consensus among researchers conducting clinical studies as to the true impact of the placebo effect. According to some researchers, as many as 30% of patients in a clinical trial can be impacted by it; according to others the placebo effect is a myth.

Write a paper on the *placebo effect*. Discuss the history of this idea, the most recent controversies regarding whether it truly exists or not and why is it important to determine its true impact.

E. The Pepsi Challenge

In 1975, Pepsi introduced a marketing campaign called the Pepsi Challenge. During blind taste tests, participants sipped both Coca-Cola and Pepsi-Cola. Even when Coca-Cola tried running its own such blind tests, a majority of participants chose Pepsi over Coke. Despite market research that indicated a change in the taste of Coca-Cola would hurt sales, these taste tests and the associated campaign led Coca-Cola to introduce "New Coke" in 1985. Of course, "New Coke" turned out to be one of the biggest marketing disasters in corporate history.

In this project, you are to discuss the scientific accuracy of the cola taste tests, reasons why the taste test results may have been skewed, and reasons why "New Coke" failed. Then assume you are a statistical consultant to Coca-Cola and develop a strategy using surveys and clinical studies designed to improve the company's market share.

F. Ethical Issues in Clinical Studies

Until the world was exposed to the atrocities committed by Nazi physicians during World War II, there was little consensus regarding the ethics of medical experiments involving human subjects. Following the trial of the Nazi physicians in 1946, the Nuremberg Code of Ethics was developed to deal with the ethics of clinical studies on human subjects. Sadly, unethical experimentation continued. In the Tuskegee syphilis study starting in 1932, 600 low-income African American males (400 infected with syphilis) were monitored. Even though a proven cure (penicillin) became available in the 1950s, the study continued until 1972 with participants denied treatment. As many as 100 subjects died from the disease.

In this project you are to write a paper summarizing current ethical standards for the treatment of humans in experimental studies. Sources for your summary may include the Belmont Report (*ohsr.od.nih.gov/guidelines/belmont.html*), the Declaration of Helsinki (*www.cirp.org/library/ethics/helsinki*), and the Nuremberg Code (*ohsr.od.nih.gov/guidelines/nuremberg.html*).

References and Further Readings

1. Anderson, M. J., and S. E., Fienberg, *Who Counts? The Politics of Census-Taking in Contemporary America*. New York: The Russell Sage Foundation, 1999.

2. Day, Simon, *Dictionary for Clinical Trials*. New York: John Wiley and Sons, Inc., 1999.

3. Francis, Thomas, Jr., et al., "An Evaluation of the 1954 Poliomyelitis Vaccine Trials—Summary Report, *American Journal of Public Health*, 45 (1955), 1–63.

4. Freedman, D., R. Pisani, R. Purves, and A. Adhikari, *Statistics*, 2nd ed. New York: W. W. Norton, Inc., 1991, chaps. 19 and 20.

5. Gallup, George, *The Sophisticated Poll Watcher's Guide*. Princeton, NJ: Princeton Public Opinion Press, 1972.

6. Gleick, James, "The Census: Why We Can't Count," *New York Times Magazine* (July 15, 1990), 22–26, 54.

7. Kalton, Graham, *Introduction to Survey Sampling*. Newbury Park, CA: SAGE Publications, 1983.

8. Kish, Leslie, *Survey Sampling*. New York: Wiley Interscience, 1995.

9. Matthews, J. N. S., *An Introduction to Randomized Controlled Clinical Trials*. London, England: Edward Arnold, 2000.

10. Meier, Paul, "The Biggest Public Health Experiment Ever: The 1954 Field Trial of the Salk Poliomyelitis Vaccine," in *Statistics: A Guide to the Unknown*, 3rd ed., eds. Judith M. Tanur et al. Belmont, CA: Wadsworth, Inc., 1989, 3–14.

11. Mosteller, F., et al., *The Pre-election Polls of 1948*. New York: Social Science Research Council, 1949.

12. Paul, John, *A History of Poliomyelitis*. New Haven, CT: Yale University Press, 1971.

13. Scheaffer, R. L., W. Mendenhall, and L. Ott, *Elementary Survey Sampling*. Boston: PWS-Kent, 1990.

14. Utts, Jessica M., *Seeing Through Statistics*. Belmont, CA: Wadsworth, Inc., 1996.

15. Yates, Frank, *Sampling Methods for Censuses and Surveys*. New York: Macmillan Publishing Co., Inc., 1981.

16. Zivin, Justin A., "Understanding Clinical Trials," *Scientific American*, 282 (April 2000), 69–75.

14

Descriptive Statistics

Graphing and Summarizing Data

> It is a proof of high culture to say the greatest matters in the simplest way.
>
> Ralph Waldo Emerson

The primary purpose of collecting data is to give meaning to a statistical story, to uncover some new fact about our world, and—last but certainly not least—to make a point, no matter how outlandish. But what do we do when we have too much data? One important purpose of statistics is to describe large amounts of data in a way that is understandable, useful, and, if need be, convincing. This is called *descriptive statistics* and is the subject of this chapter.

Imagine that our data consist of the test scores of a group of students in a standardized exam. If we are dealing with a small group of students—say a class—then it is reasonable to look at the collection of test scores of the group and get the "big picture" (how the group performed compared to other groups, how many are at grade level, etc.). On the other hand, if we are dealing with a large group (hundreds, thousands, or even millions), trying to get to the big picture by looking at the individual scores of the students is hopeless. The amount of data to deal with becomes overwhelming—a huge babble of numbers.

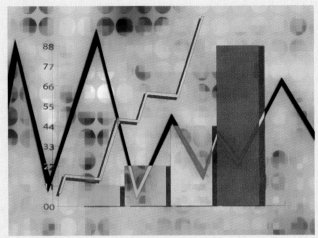

There are two strategies for describing large data sets. One is to present the data in the form of pictures or graphs; the other is to use numerical summaries that serve as "snapshots" of the data set. Graphical descriptions of data (*bar graphs*, *pictograms*, and *pie charts*) are introduced in Section 14.1. Section 14.2 is a brief detour into the types of variables that need to be considered when graphing data—*categorical*, *numerical*, *discrete*, and *continuous*. In Sections 14.3 and 14.4 we discuss the numerical summaries of a data set. *Means*, *medians*, *quartiles*, and *percentiles* tell us something about the numerical value of the data (they are called *measures of location*) and are discussed in Section 14.3. *Ranges*, *interquartile ranges*, and *standard deviations* provide information about the spread within the data (they are known as *measures of spread*) and are discussed in Section 14.4.

14.1 Graphical Descriptions of Data

Data Sets

A **data set** is a collection of data values. Statisticians often refer to the individual data values in a data set as **data points**. For the sake of simplicity, we will work with data sets in which each data point consists of a single number, but in more complicated settings, a single data point can consist of many numbers.

As usual, we will use the letter N to represent the size of the data set. In real-life applications, data sets can range in size from reasonably small (a dozen or so data points) to very large (hundreds of millions of data points), and the larger the data set is, the more we need a good way to describe and summarize it.

To illustrate many of the ideas of this chapter we will need a reasonable data set—big enough to be realistic, but not so big that it will bog us down. Example 14.1, which we will revisit several times in the chapter, provides such a data set. This is a fictitious data set from a hypothetical statistics class, but except for the details, it describes a situation that is familiar to every college student.

> **EXAMPLE 14.1** Stat 101 Test Scores: Part 1

As usual, the day after the midterm exam in his Stat 101 class, Professor Black-beard has posted the results in the hallway outside his office (Table 14-1). The data set consists of $N = 75$ data points (the number of students that took the test). Each data point (listed in the second column) is a raw score on the midterm between 0 and 25 (Professor Blackbeard gives no partial credit). Note that the student IDs in Table 14-1 are numbers but not data—they are used as a substitute for names to protect the students' rights of privacy.

TABLE 14-1 Stat 101/Prof. Blackbeard

Midterm Scores (25 Points Possible)

ID	Score	ID	Score	ID	Score	ID	Score	ID	Score
1257	12	2651	10	4355	8	6336	11	8007	13
1297	16	2658	11	4396	7	6510	13	8041	9
1348	11	2794	9	4445	11	6622	11	8129	11
1379	24	2795	13	4787	11	6754	8	8366	13
1450	9	2833	10	4855	14	6798	9	8493	8
1506	10	2905	10	4944	6	6873	9	8522	8
1731	14	3269	13	5298	11	6931	12	8664	10
1753	8	3284	15	5434	13	7041	13	8767	7
1818	12	3310	11	5604	10	7196	13	9128	10
2030	12	3596	9	5644	9	7292	12	9380	9
2058	11	3906	14	5689	11	7362	10	9424	10
2462	10	4042	10	5736	10	7503	10	9541	8
2489	11	4124	12	5852	9	7616	14	9928	15
2542	10	4204	12	5877	9	7629	14	9953	11
2619	1	4224	10	5906	12	7961	12	9973	10

Like students everywhere, the students in the Stat 101 class have one question foremost on their mind when they look at Table 14-1: How did I do? Each student can answer this question directly from the table. It's the next question that is statistically much more interesting. How did the class as a whole do? To answer this last question, we will have to find a way to package the information in Table 14-1 into a compact, organized, and intelligible whole. ◀◀

Bar Graphs and Variations Thereof

> ### EXAMPLE 14.2 Stat 101 Test Scores: Part 2

The first step in summarizing the information in Table 14-1 is to organize the scores in a **frequency table** such as Table 14-2. In this table, the number below each score gives the **frequency** of the score—that is, the number of students getting that particular score. We can readily see from Table 14-2 that there was one student with a score of 1, one with a score of 6, two with a score of 7, six with a score of 8, and so on. Note that the scores with a frequency of zero are not listed in the table.

TABLE 14-2 Frequency Table for the Stat 101 Data Set

Score	1	6	7	8	9	10	11	12	13	14	15	16	24
Frequency	1	1	2	6	10	16	13	9	8	5	2	1	1

While Table 14-2 is a considerable improvement over Table 14-1, we can do even better. Figure 14-1 shows the same information in a much more visual way called a **bar graph**, with the test scores listed in increasing order on a horizontal axis and the frequency of each test score displayed by the *height* of the column above that test score. Notice that in the bar graph, even the test scores with a frequency of zero show up—there simply is no column above these scores

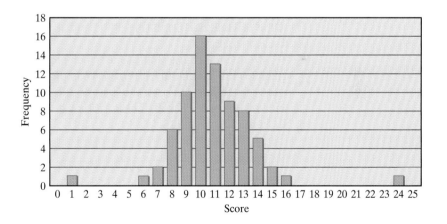

FIGURE 14-1 Bar graph for the Stat 101 data set.

Bar graphs are easy to read, and they are a nice way to present a good general picture of the data. With a bar graph, for example, it is easy to detect **outliers**—extreme data points that do not fit into the overall pattern of the data. In this example there are two obvious outliers—the score of 24 (head and shoulders above the rest of the class) and the score of 1 (lagging way behind the pack).

Sometimes it is more convenient to express the bar graph in term of *relative frequencies*—that is, the frequencies given in terms of percentages of the total population. Figure 14-2 shows a *relative frequency bar graph* for the Stat 101 data set. Note that we indicated on the graph that we are dealing with percentages rather than total counts and that the size of the data set is $N = 75$. This allows anyone who wishes to do so to compute the actual frequencies. For example,

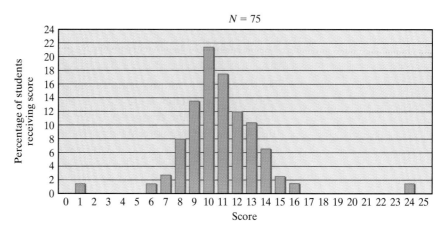

FIGURE 14-2 Relative frequency bar graph for the Stat 101 data set.

Fig. 14-2 indicates that 12% of the 75 students scored a 12 on the exam, so the actual frequency is given by $75 \times 0.12 = 9$ students. The change from actual frequencies to percentages (or viceversa) does not change the shape of the graph—it is basically a change of scale. ⦉⦉

While the term *bar graph* is most commonly used for graphs like the ones in Figs. 14-1 and 14-2, devices other than bars can be used to add a little extra flair or to subtly influence the content of the information given by the raw data. Professor Blackbeard, for example, might have chosen to display the midterm data using a graph like the one shown in Fig. 14-3, which conveys all the information of the original bar graph (Fig. 14-1) and also includes a subtle individual message to each student.

Frequency charts that use icons or pictures instead of bars to show the frequencies are commonly referred to as **pictograms**. The point of a pictogram is that a graph is often used not only to inform but also to impress and persuade, and, in such cases, a well-chosen icon or picture can be a more effective tool than just a bar.

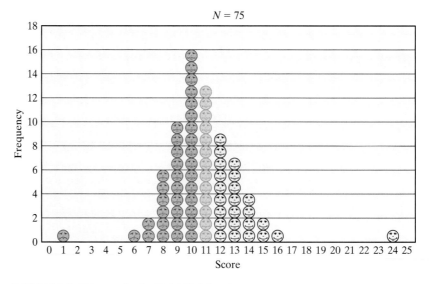

FIGURE 14-3 Pictogram for the Stat 101 data set.

> **EXAMPLE 14.3** Selling the XYZ Corporation

Figure 14-4 is a pictogram showing the growth in yearly sales of the XYZ Corporation between 2001 and 2006. It's a good picture to show at a shareholders meeting, but the picture is actually quite misleading. Figure 14-5 shows a pictogram for exactly the same data with a much more accurate and sobering picture of how well the XYZ Corporation had been doing.

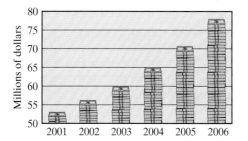

FIGURE 14-4 XYZ Corp. annual sales (in millions).

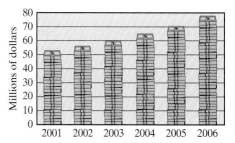

FIGURE 14-5 XYZ Corp. annual sales (in millions).

The difference between the two pictograms can be attributed to a couple of standard tricks of the trade: (i) stretching the scale of the vertical axis, and (ii) "cheating" on the choice of starting value on the vertical axis. As an educated consumer, you should always be on the lookout for these tricks. In graphical descriptions of data, a fine line separates objectivity from propaganda. (For more on this topic, see Project A.) ◀◀

14.2 Variables

Before we continue with our discussion of graphs, we need to discuss briefly the concept of a **variable**. In statistical usage, a variable is any characteristic that varies with the members of a population. The students in Professor Blackbeard's Stat 101 course (the population) did not all perform equally on the exam. Thus, the *test score* is a variable, which in this particular case is a whole number between 0 and 25. In some instances, such as when the instructor gives partial credit or when there is subjective grading, a test score may take on a fractional value, such as 18.5 or 18.25. Even in these cases, however, the possible increments for the values of the variable are given by some minimum amount—a quarter-point, a half-point, whatever. In contrast to this situation, consider a different variable: the *amount of time* each student studied for the exam. In this case the variable can take on values that differ by any amount: an hour, a minute, a second, a tenth of a second, and so on.

A variable that represents a measurable quantity is called a **numerical** (or *quantitative*) variable. When the difference between the values of a numerical variable can be arbitrarily small, we call the variable **continuous**; when possible values of the numerical variable change by minimum increments, the variable is called **discrete**. Examples of *discrete* variables are a person's IQ, an SAT score, a person's shoe size, and the number of points scored in a basketball game. Examples of *continuous* variables are a person's height, weight, foot size (as opposed to shoe size), and the time it takes him or her to run a mile.

Sometimes in the real world the distinction between continuous and discrete variables is blurred. Height, weight, and age are all continuous variables in theory, but in practice they are frequently rounded off to the nearest inch, ounce, and year (or month in the case of babies), at which point they become discrete variables. On the other hand, money, which is in theory a discrete variable (because the difference between two values cannot be less than a penny), is almost always thought of as continuous, because in most real-life situations a penny can be thought of as an infinitesimally small amount of money.

Variables can also describe characteristics that cannot be measured numerically: nationality, gender, hair color, and so on. Variables of this type are called **categorical** (or *qualitative*) variables.

In some ways, categorical variables must be treated differently from numerical variables—they cannot, for example, be added, multiplied, or averaged. In other ways, categorical variables can be treated much like discrete numerical variables, particularly when it comes to graphical descriptions, such as bar graphs and pictograms.

TABLE 14-3

School	Enrollment
Agriculture	2400
Business	1250
Education	2840
Humanities	3350
Science	4870
Other	290

> **EXAMPLE 14.4** Enrollment (by School) at Tasmania State University

Table 14-3 shows undergraduate enrollments in each of the five schools at Tasmania State University. A sixth category ("Other") includes undeclared students, interdisciplinary majors, and so on.

Figure 14-6 shows two equivalent bar graphs describing the information in Table 14-3. The only difference between the two graphs is that in Fig. 14-6(b) the values of the categorical variable are on the vertical axis. There is no prohibition against doing this, and, in fact, many people prefer this approach when dealing with categorical variables.

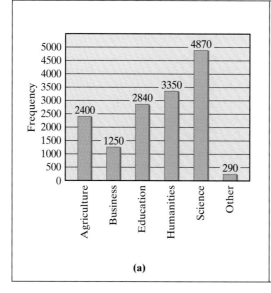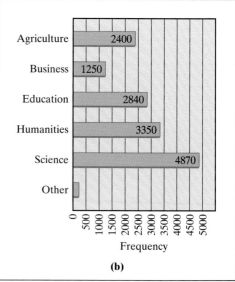

FIGURE 14-6

(a) (b)

When the number of categories is small, as is the case here, another commonly used way to describe the relative frequencies of the categories is by using

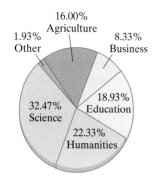

FIGURE 14-7

a **pie chart**. In a pie chart the "pie" represents the entire population (100%), and the "slices" represent the categories, with the size (area) of each slice being proportional to the relative frequency of the corresponding category.

Some relative frequencies, such as 50% and 25%, are very easy to sketch, but how do we accurately draw the slice corresponding to a more complicated frequency, say, 32.47%? Here, a little elementary geometry comes in handy. Since 100% equals 360°, 1% corresponds to an angle of 360°/100 = 3.6°. It follows that the frequency 32.47% is given by 32.47 × 3.6° = 117° (rounded to the nearest degree, which is generally good enough for most practical purposes). Figure 14-7 shows an accurate pie chart for the school-enrollment data given in Table 14-3. ◀◀

Bar graphs and pie charts are excellent ways to graphically display categorical data, but graphs and charts can be deceiving, and we should be very careful about the conclusions we draw when the data are used to compare different populations. Our next example illustrates this point.

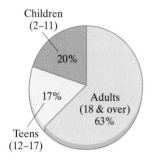

FIGURE 14-8 Audience composition for prime-time TV viewership by age group.

▶ EXAMPLE 14.5 Who's Watching the Boob Tube Tonight?

According to Nielsen Media Research data, the percentages of the TV audience watching TV during prime time (8 P.M. to 11 P.M.), broken up by age group, are as follows: adults (18 years and over), 63%; teenagers (12–17 years), 17%; children (2–11 years), 20%. (The exact figures vary from year to year. These figures are averaged over several years.)

The pie chart in Fig. 14-8 shows this breakdown of audience composition by age group. A pie chart such as this one might be used to make the point that children and teenagers really do not watch as much TV as it is generally believed. The problem with this conclusion is that children make up only 15% of the population at large and teens only 8%. In relative terms, a higher percentage of teenagers (taken out of the total teenage population) watch prime-time TV than any other group, with children second and adults last. ◀◀

The moral of Example 14.5 is that using absolute percentages, as we did in Fig. 14-8, can be quite misleading. When comparing characteristics of a population that is broken up into categories, it is essential to take into account the relative sizes of the various categories.

Class Intervals

While the distinction between qualitative and quantitative data is important in many aspects of statistics, when it comes to deciding how best to display graphically the frequencies of a population, a critical issue is the number of categories into which the data can fall. When the number of categories is too big (say, in the dozens), a bar graph or pictogram can become muddled and ineffective. This happens more often than not with quantitative data—numerical variables can take on infinitely many values, and even when they don't, the number of values can be too large for any reasonable graph. Our next example illustrates how to deal with this situation.

> ### EXAMPLE 14.6 2004 SAT Math Scores: Part 1

The college dreams and aspirations of millions of high school seniors often ride on their SAT scores. Up until 2004 the SAT consisted of a verbal section and a math section, with the scores for each section ranging from a minimum of 200 to a maximum of 800 and going up in increments of 10 points. (The SAT changed in 2005—the "new" SAT has an additional essay section and a range of 600–2400.)

In 2004, 1,419,007 college-bound seniors took the SAT. How do we describe the math section results for this group of students? In one way, this is the same problem as the one illustrated in Example 14.1 (Stat 101 midterm scores)—just a different test and a much larger number of test-takers. We could set up a frequency table (or a bar graph) with the number of students scoring each of the possible scores—200, 210, 220, . . . , 790, 800. The problem is that there are 61 different possible scores between 200 and 800, and this number is too large for an effective bar graph.

In situations such as this one it is customary to present a more compact picture of the data by grouping together, or *aggregating*, sets of scores into categories called **class intervals**. The decision as to how the class intervals are defined and how many there are will depend on how much or how little detail is desired, but as a general rule of thumb, the number of class intervals should be somewhere between 5 and 20.

SAT math scores are usually aggregated into 12 class intervals of essentially the same size: 200–240, 250–290, 300–340, . . . , 700–740, 750–800 (note that the class interval 750–800 has one more test score than the others). Using these class intervals, the distribution of scores in the math section of the 2004 SAT is given in Table 14-4, and the associated bar graph is shown in Fig. 14-9.

TABLE 14-4 2004 SAT Math Scores					
Score	200–240	250–290	300–340	350–390	400–440	450–490
Frequency	12,192	21,380	57,249	106,748	184,665	226,745
Score	500–540	550–590	600–640	650–690	700–740	750–800
Frequency	237,262	210,766	155,423	111,592	63,669	31,316

Source: The College Board.

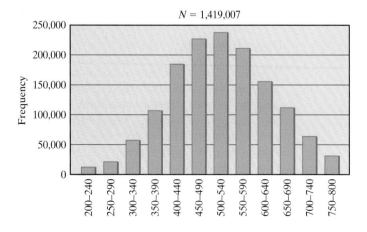

FIGURE 14-9 2004 SAT Math: Distribution of scores. (*Source:* The College Board)

Our next example deals with a topic every college student can relate to—*grades!*

▶ EXAMPLE 14.7 Stat 101 Test Scores: Part 3

The process of converting test scores (a *numerical* variable) into grades (a *categorical* variable) requires setting up *class intervals* for the various letter grades. Typically, the professor has the latitude to decide how to do this. One standard approach is to use an *absolute* grading scale, usually with class intervals of (almost) equal length for all grades except F (for example, A = 90–100, B = 80–89, C = 70–79, D = 60–69, F = 0–59). Another frequently used approach is to use a *relative* grading scale. Here the professor fits the class intervals for the grades to the performance of the class in the test, often using class intervals of different lengths. Some people call this "grading on the curve," although this terminology is somewhat misused.

To illustrate relative grading in action, let's revisit the Stat 101 midterm scores discussed in Example 14.1. After looking at the overall class performance, Professor Blackbeard chooses to "curve" the test scores using class intervals of his own creation. The class intervals and corresponding results (obtained using Table 14-2) are shown in Table 14-5.

TABLE 14-5

Grade	A	B	C	D	F
Test score	18–25	14–17	11–13	9–10	8 or less
Frequency	1	8	30	26	10
Percent	1.33%	10.67%	40%	34.67%	13.33%

The grade distribution in the Stat 101 midterm can now be best seen by means of the bar graph shown in Fig. 14-10. The picture speaks for itself—this was a very tough exam!

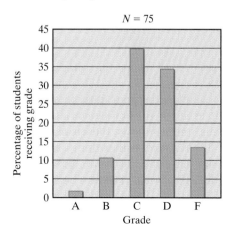

FIGURE 14-10

Histograms

When a numerical variable is continuous, its possible values can vary by infinitesimally small increments. As a consequence, there are no gaps between the class intervals, and our old way of doing things (using separated columns or stacks) will no longer work. In this case we use a variation of a bar graph called a **histogram**. We illustrate the concept of a histogram in the next example.

> ### EXAMPLE 14.8 Starting Salaries of TSU Graduates

Suppose we want to use a graph to display the distribution of starting salaries for last year's graduating class at Tasmania State University.

The starting salaries of the $N = 3258$ graduates range from a low of $40,350 to a high of $74,800. Based on this range and the amount of detail we want to show, we must decide on the length of the class intervals. A reasonable choice would be to use class intervals defined in increments of $5000. Table 14-6 is a frequency table for the data based on these class intervals. We chose a starting value of $40,000 for convenience. (The third column in the table shows the data as a percentage of the population.)

TABLE 14-6 Starting Salaries of First-Year TSU Graduates

Salary	Number of students	Percentage
40,000$^+$–45,000	228	7%
45,000$^+$–50,000	456	14%
50,000$^+$–55,000	1043	32%
55,000$^+$–60,000	912	28%
60,000$^+$–65,000	391	12%
65,000$^+$–70,000	163	5%
70,000$^+$–75,000	65	2%
Total	3258	100%

The histogram showing the relative frequency of each class interval is shown in Fig. 14-11. As we can see, a histogram is very similar to a bar graph. Several important distinctions must be made, however. To begin with, because a histogram is used for continuous variables, there can be no gaps between the class intervals, and it follows, therefore, that the columns of a histogram must touch each other. Among other things, this forces us to make an arbitrary decision as to what happens to a value that falls exactly on the boundary between two class intervals. Should it always belong to the class interval to the left or to the one to the right? This is called the *endpoint convention*. The superscript "plus" marks in Table 14-6 indicate how we chose to deal with the endpoint convention in Fig. 14-11. A starting salary of exactly $50,000, for example, would be listed under the 45,000$^+$–50,000 class interval rather than the 50,000$^+$–55,000 class interval.

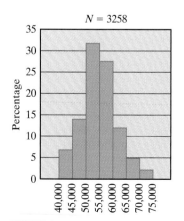

FIGURE 14-11 Histogram for starting salaries of first-year graduates of TSU (with class intervals of $5000).

For more details on histograms with class intervals of unequal lengths, see Exercises 73 and 74.

When creating histograms, we should try, as much as possible, to define class intervals of equal length. When the class intervals are of unequal length, the rules for creating a histogram are considerably more complicated, since it is no longer appropriate to use the heights of the columns to indicate the frequencies of the class intervals.

14.3 Numerical Summaries of Data

As we have seen, a picture can be an excellent tool for summarizing large data sets. Unfortunately, circumstances do not always lend themselves equally well to the use of pictures, and bar graphs and pie charts cannot be readily used in everyday conversation. A different and very important approach is to use a few well-chosen numbers to summarize an entire data set.

In the next couple of sections we will discuss two types of *numerical summaries* of a data set: **measures of location** and **measures of spread**. Measures of location such as the *mean* (or *average*), the *median*, and the *quartiles*, are numbers that provide information about the values of the data. Measures of spread such as the *range*, the *interquartile range*, and the *standard deviation* are numbers that provide information about the spread within the data set. In this section we will focus on measures of location. In Section 14.4 we will discuss measures of spread.

The Mean (Average)

The best known of all numerical summaries of data is the *average*, also called the *mean*. (There is no universal agreement as to which of these names is a better choice—in some settings *mean* is a better choice than *average*, in other settings it's the other way around. In this chapter we will use whichever seems the better choice at the moment.)

The **average** of a set of N numbers is found by adding the numbers and dividing the total by N. In other words, the average of the numbers $d_1, d_2, d_3, \ldots, d_N$ is $A = (d_1 + d_2 + \cdots + d_N)/N$.

> ## EXAMPLE 14.9 Stat 101 Test Scores: Part 4

In this example we will find the average test score in the Stat 101 exam first introduced in Example 14.1. To find this average we need to add all the test scores and divide by 75. The addition of the 75 test scores can be simplified considerably if we use a frequency table. (Table 14-7 is the same as Table 14-2, shown again for the reader's convenience.)

TABLE 14-7 Frequency Table for the Stat 101 Data Set													
Exam score	1	6	7	8	9	10	11	12	13	14	15	16	24
Frequency	1	1	2	6	10	16	13	9	8	5	2	1	1

From the frequency table we can find the sum of all the test scores as follows: multiply each test score by its corresponding frequency and then add these products. Thus, the sum of all the test scores is

$$(1 \times 1) + (6 \times 1) + (7 \times 2) + (8 \times 6) + \cdots + (16 \times 1) + (24 \times 1) = 814$$

If we divide this sum by 75 we get the average test score (rounded to two decimal places), which we will denote by A:

$$A = 814/75 \approx 10.85 \text{ points.}$$

In general, to find the average A of a data set given by a frequency table such as Table 14-8 we do the following steps:

TABLE 14-8

Data value	Frequency
d_1	f_1
d_2	f_2
$\vdots$	$\vdots$
d_k	f_k

- **Step 1.** Find the sum: $Sum = d_1 \cdot f_1 + d_2 \cdot f_2 + \cdots + d_k \cdot f_k$.
- **Step 2.** Find N: $N = f_1 + f_2 + \cdots + f_k$.
- **Step 3.** Find A: $A = Sum/N$.

When dealing with data sets that have outliers, averages can be quite misleading. As our next example illustrates, even a single outlier can have a big effect on the average.

EXAMPLE 14.10 Starting Salaries of Philosophy Majors

Imagine you just read in the paper the following remarkable tidbit: *The average starting salary of philosophy majors who recently graduated from Tasmania State University is $76,400 a year!* This is quite an impressive number, but before we all rush out to change majors, let's point out that one of the graduating philosophy majors happens to be basketball star "Hoops" Tallman, who is doing his thing in the NBA for a starting salary of $3.5 million a year.

If we were to take this one outlier out of the population of 75 philosophy majors, we would have a more realistic picture of what philosophy majors are making. Here is how we can do it:

- The total of all 75 salaries is 75 times the average salary:

$$75 \times \$76,400 = \$5,730,000.$$

- The total of the other 74 salaries (excluding Hoops's cool 3.5 mill) is

$$\$5,730,000 - \$3,500,000 = \$2,230,000$$

- The average of the remaining 74 salaries is

$$\$2,230,000/74 \approx \$30,135.$$

So far, all our examples have involved data values that are positive, but negative data values are also possible, and when both negative and positive data values are averaged, the results can be misleading.

> ## EXAMPLE 14.11 Living Beyond Your Means

Table 14-9 shows the monthly balance (monthly income minus monthly spending) in Billy's budget over the past year. A negative amount indicates that Billy spent more than what he had coming in (adding to his credit card debt).

TABLE 14-9	
Month	**Balance**
Jan.	−732 ← Christmas bills come in!
Feb.	−158
Mar.	−71
Apr.	−238
May	1839 ← $2000 lottery winnings
Jun.	−103
Jul.	−148
Aug.	−162
Sep.	−85
Oct.	−147
Nov.	−183
Dec.	500 ← Christmas present from mom

In spite of his consistent overspending, Billy's average monthly balance for the year is $26 (check it out!). This average hides the true picture of what is going on. Billy is living well beyond his means but was bailed out by a lucky break and a generous mom. ≪

Percentiles

While a single numerical summary—such as the average—can be useful, it is rarely sufficient to give a meaningful description of a data set. A better picture of the data set can be presented by using a well-organized cadre of numerical summaries. The most common way to do this is by means of *percentiles*.

The pth **percentile** of a data set is a value such that p percent of the numbers fall *at or below* this value and the rest fall *at or above* it. It essentially splits a data set into two parts: the lower p% of the data values and the upper $(100 - p)$% of the data values.

Many college students are familiar with percentiles, if for no other reason than the way they pop up in SAT reports. In all SAT reports, a given score—say a score of 610 in the math section—is identified with a percentile, say the 81st percentile. This can be interpreted to mean that 81% of those taking the test scored 610 or less, or, looking up instead of down, that 19% of those taking the test scored 610 or more.

There are several different ways to compute percentiles that will satisfy the definition, and different statistics books describe different methods. We will illustrate one such method below.

The first step in finding the *p*th *percentile* of a data set of *N* numbers is to *sort the numbers by size*. In other words, we *must* rewrite the numbers in the data set in *increasing* order from smallest to largest.

To simplify the explanation, we will now introduce a bit of notation. Let's denote the sorted data set by $\{d_1, d_2, d_3, \ldots, d_N\}$. In this notation d_1 represents the first number in the sorted data set (the smallest number), d_2 the second number, d_{10} the tenth number, and so on. Sometimes we will also need to talk about the average of two consecutive numbers in the sorted list, so we will use more exotic subscripts such as $d_{3.5}$ to represent the average of the data values d_3 and d_4; $d_{7.5}$ to represent the average of the data values d_7 and d_8, and so on.

The next, and most important, step is to identify which *d* represents the *p*th percentile of the data set. To do this, we compute the *p*th *percent of N*, which we will call the **locator** and denote by the letter *L*. [In other words, $L = (p/100) \cdot N$.] If *L* happens to be a whole number, then the *p*th *percentile* will be $d_{L.5}$ (the average of d_L and d_{L+1}). If *L* is not a whole number, then the *p*th *percentile* will be d_{L^+}, where L^+ represents the value of *L* rounded up.

The procedure for finding the *p*th *percentile* of a data set is summarized as follows:

> **Finding the *p*th Percentile of a Data Set**
>
> - **Step 0.** Sort the data set. Let $\{d_1, d_2, d_3, \ldots, d_N\}$ represent the sorted data set.
> - **Step 1.** Find the locator: $L = (p/100) \cdot N$.
> - **Step 2.** Find the *p*th percentile: (i) if *L* is a whole number, the *p*th *percentile* is given by $d_{L.5}$; (ii) if *L* is not a whole number, the *p*th *percentile* is given by d_{L^+} (L^+ is *L* rounded up).

The following example illustrates the procedure for finding percentiles of a data set.

▶ EXAMPLE 14.12 Scholarships by Percentile

To reward good academic performance from its athletes, Tasmania State University has a program where athletes with GPAs in the top 20th percentile of their team's GPAs get a $5000 scholarship, and athletes with GPAs in the top forty-fifth percentile of their team's GPAs that did not get the $5000 scholarship get a $2000 scholarship.

The women's soccer team has $N = 15$ players. A list of their GPAs is as follows:

3.42, 3.91, 3.33, 3.65, 3.57, 3.45, 4.0, 3.71, 3.35, 3.82, 3.67, 3.88, 3.76, 3.41, 3.62

When we sort these GPAs we get the list

3.33, 3.35, 3.41, 3.42, 3.45, 3.57, 3.62, 3.65, 3.67, 3.71, 3.76, 3.82, 3.88, 3.91, 4.0

Since this list goes from lowest to highest GPA, we are looking for the 80th percentile and above (top 20th percentile) for the $5000 scholarships and the 55th percentile and above (top 45th percentile) for the $2000 scholarships.

$5000 Scholarships: The locator for the 80th percentile is $(0.8) \times 15 = 12$. Here the locator is a whole number, so the 80th percentile is given by $d_{12.5} = 3.85$ (the average between $d_{12} = 3.82$ and $d_{13} = 3.88$). Thus, three students (the ones with GPAs of 3.88, 3.91 and 4.0) get $5000 scholarships.

$2000 Scholarships: The locator for the 55th percentile is $(0.55) \times 15 = 8.25$. This locator is not a whole number, so we round it up to 9, and the 55th percentile is given by $d_9 = 3.67$. Thus, the students with GPAs of 3.67, 3.71, 3.76, and 3.82 get $2000 scholarships.   ≪

The Median and the Quartiles

The 50th percentile of a data set is known as the **median** and denoted by M. The median splits a data set into two halves—half of the data is at or below the median and half of the data is at or above the median.

We can find the median by simply applying the definition of percentile with $p = 50$, but the bottom line comes down to this: (i) when N is *odd* the median is the data value in position $(N + 1)/2$ of the sorted data set; (ii) when N is *even* the median is the average of the data values in position $N/2$ and $(N/2) + 1$ of the sorted data set. [All of the preceding follows from the fact that the locator for the median is $L = (0.5)N$. When N is even, L is an integer; when N is odd, L is not an integer.]

> **Finding the Median of a Data Set**
>
> ▪ Sort the data set. Let $\{d_1, d_2, d_3, \ldots, d_N\}$ represent the sorted data set.
> ▪ If N is odd, the median is $d_{(N+1)/2}$. If N is even, the median is the average of $d_{N/2}$ and $d_{(N/2)+1}$.

After the median, the next most commonly used set of percentiles are the *first* and *third quartiles*. The **first quartile** (denoted by Q_1) is the 25th percentile, and the **third quartile** (denoted by Q_3) is the 75th percentile.

▶ EXAMPLE 14.13 Home Prices in Green Hills

During the last year, 11 homes sold in the Green Hills subdivision. The selling prices, in chronological order, were $167,000, $152,000, $128,000, $134,000, $192,000, $163,000, $121,000, $145,000, $170,000, $138,000, and $155,000. We are going to find the *median* and the *quartiles* of the $N = 11$ home prices.

Sorting the home prices from smallest to largest (and dropping the 000's) gives the sorted list

$$121, 128, 134, 138, 145, 152, 155, 163, 167, 170, 192.$$

The locator for the median is $(0.5) \times 11 = 5.5$, the locator for the first quartile is $(0.25) \times 11 = 2.75$, and the locator for the third quartile is $(0.75) \times 11 = 8.25$. None of the locators are whole numbers, so they are all rounded up. This means that the median home price is given by $d_6 = 152$

(i.e., $M = \$152,000$). Likewise, the first quartile is given by $d_3 = 134$ (i.e., $Q_1 = \$134,000$), and the third quartile is given by $d_9 = 167$ (i.e., $Q_3 = \$167,000$).

> **EXAMPLE 14.13** (continued) Another Home Sells in Green Hills

---

Oops! Just this morning a home sold in Green Hills for $164,000. We need to re-calculate the median and quartiles for what are now $N = 12$ home prices.

We can use the sorted data set that we already had—all we have to do is insert the new home price (164) in the right spot (remember, we drop the 000's!). This gives

$$121, 128, 134, 138, 145, 152, 155, 163, \textbf{164,} 167, 170, 192$$

Now $N = 12$ and in this case the median is the average of $d_6 = 152$ and $d_7 = 155$. It follows that the median home price is $M = \$153,500$. The locator for the first quartile is $0.25 \times 12 = 3$ Since the locator is a whole number, the first quartile is the average of $d_3 = 134$ and $d_4 = 138$ (i.e., $Q_1 = \$136,000$). Similarly, the third quartile is $Q_3 = 165,500$ (the average of $d_9 = 164$ and $d_{10} = 167$). ◀◀

> **EXAMPLE 14.14** Stat 101 Test Scores: Part 5

We will now find the median and quartile scores for the Stat 101 data set (shown again in Table 14-10).

TABLE 14-10 Frequency Table for the Stat 101 Data Set

Exam score	1	6	7	8	9	10	11	12	13	14	15	16	24
Frequency	1	1	2	6	10	16	13	9	8	5	2	1	1

Having the frequency table available eliminates the need for sorting the scores—the frequency table has, in fact, done this for us. Here $N = 75$ (odd), so the median is the thirty-eighth score (counting from the left) in the frequency table. To find the thirty-eighth number in Table 14-10, we tally frequencies as we move from left to right: $1 + 1 = 2$; $1 + 1 + 2 = 4$; $1 + 1 + 2 + 6 = 10$; $1 + 1 + 2 + 6 + 10 = 20$; $1 + 1 + 2 + 6 + 10 + 16 = 36$. At this point, we know that the 36th test score on the list is a 10 (the last of the 10's) and the next 13 scores are all 11's. We can conclude that the 38th test score is 11. Thus, $M = 11$.

The locator for the first quartile is $L = (0.25) \times 75 = 18.75$. Thus, $Q_1 = d_{19}$. To find the nineteenth score in the frequency table, we tally frequencies from left to right: $1 + 1 = 2$; $1 + 1 + 2 = 4$; $1 + 1 + 2 + 6 = 10$; $1 + 1 + 2 + 6 + 10 = 20$. At this point we realize that $d_{10} = 8$ (the last of the 8's) and that d_{11} through d_{20} all equal 9. Hence, the first quartile of the Stat 101 midterm scores is $Q_1 = d_{19} = 9$.

Since the first and first quartiles are at an equal "distance" from the two ends of the sorted data set, a quick way to locate the third quartile now is to look for the nineteenth score in the frequency table when we count frequencies *from right to left*. We leave it to the reader to verify that the third quartile of the Stat 101 data set is $Q_3 = 12$. ◀◀

> ## EXAMPLE 14.15 2004 SAT Math Scores: Part 2

In this example we continue the discussion of the 2004 SAT math scores introduced in Example 14.6. Recall that the number of college-bound high school seniors taking the test was $N = 1,419,007$. As reported by the College Board, the median score in the test was $M = 510$, the first quartile score was $Q_1 = 440$, and the third quartile was $Q_3 = 600$. What can we make of this information?

Let's start with the median. From $N = 1,419,007$ (an odd number), we can conclude that the median (510 points) is the 709,504th score in the sorted list of test scores. This means that there were *at least* 709,504 students that scored 510 or less in the math section of the 2004 SAT. Why did we use "at least" in the preceding sentence? Could there have been more than that number who scored 510 or less? Yes, almost surely. Since the number of students who scored 510 is in the thousands, it is very unlikely that the 709,504th score is the last of the 510s.

In a similar vein, we can conclude that there were at least 354,752 scores of $Q_1 = 400$ or less [the locator for the first quartile is $(0.25) \times 1,419,007 = 354,751.25$], and at least 1,064,256 scores of $Q_3 = 600$ or less. «

The details are left to the reader—see Exercise 39.

*A **note of warning:** Medians, quartiles, and general percentiles are often computed using statistical calculators or statistical software packages, which is all well and fine since the whole process can be a bit tedious. The problem is that there is no universally agreed upon procedure for computing percentiles, so different types of calculators and different statistical packages may give different answers from each other and from those given in this book for quartiles and other percentiles (everyone agrees on the median). *Keep this in mind when doing the exercises*—the answer given by your calculator may be slightly different from the one you would get from the procedure we use in the book.

The Five-Number Summary

A common way to summarize a large data set is by means of its *five-number summary*. The **five-number summary** is given by (i) the smallest value in the data set (called the *Min*), (ii) the *first quartile* (Q_1), (iii) the *median* (M), (iv) the *third quartile* (Q_3), and (v) the largest value in the data set (called the *Max*). These five numbers together often tells us a great deal about the data.

> ## EXAMPLE 14.16 Stat 101 Test Scores: Part 6

For the Stat 101 data set, the five-number summary is $Min = 1, Q_1 = 9, M = 11, Q_3 = 12, Max = 24$ (see Example 14.14). What useful information can we get out of this?

Right away we can see that the $N = 75$ test scores were not evenly spread out over the range of possible scores. For example, from $M = 11$ and $Q_3 = 12$ we can conclude that at least 25% of the class (that means at least 19 students) scored either 11 or 12 on the test. At the same time, from $Q_3 = 12$ and $Max = 24$ we can conclude that less than one-fourth of the class (i.e., at most 18 students) had scores in the 13–24 point range. Using similar arguments, we can conclude that at least 19 students had scores between $Q_1 = 9$ and $M = 11$ points and no more than 18 students scored in the 1–8 point range.

The "big picture" we get from the five-number summary of the Stat 101 test scores is that there was a lot of bunching up in a narrow band of scores (at least half of the students in the class scored in the range 9–12 points), and the rest of the class was all over the place. In general, this type of "lumpy" distribution of test scores is indicative of a test with an uneven level of difficulty—a bunch of easy questions and a bunch of really hard questions with little in between. (Having seen the data, we know that the *Min* and *Max* scores were both outliers and that if we disregard these two outliers the test results don't look quite so bad. Of course, there is no way to pick this up from just the five-number summary.) ◀◀

Box Plots

Invented in 1977 by statistician John Tukey, a *box plot* (also known as a *box-and-whisker* plot) is a picture of the five-number summary of a data set. The **box plot** consists of a rectangular box that sits above a scale and extends from the first quartile Q_1 to the third quartile Q_3 on that scale. A vertical line crosses the box, indicating the position of the median M. On both sides of the box are "whiskers" extending to the smallest value, *Min*, and largest value, *Max*, of the data.

Figure 14-12 shows a generic box plot for a data set. Figure 14-13(a) shows a box plot for the Stat 101 data set (see Example 14.14). The long whiskers in this box plot are largely due to the outliers 1 and 24. Figure 14-13(b) shows a variation of the same box plot, but with the two outliers, marked with two crosses, segregated from the rest of the data. (When there are outliers it is useful to segregate them from the rest of the data set—we think of outliers as "anomalies" within the data set.)

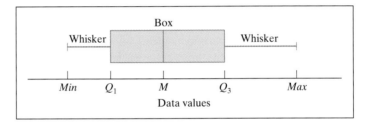

FIGURE 14-12

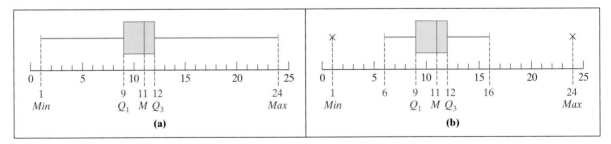

FIGURE 14-13 (a) Box plot for the Stat 101 data set. (b) Same box plot with the outliers separated from the rest of the data.

Box plots are particularly useful when comparing similar data for two or more populations. This is illustrated in the next example.

▶ EXAMPLE 14.17 Comparing Agriculture and Engineering Salaries

Figure 14-14 shows box plots for the starting salaries of two different populations: first-year agriculture and engineering graduates of Tasmania State University.

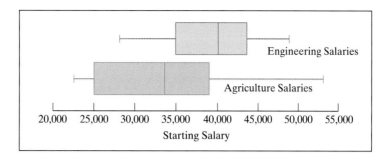

FIGURE 14-14 Comparison of starting salaries of first-year graduates in agriculture and engineering.

Superimposing the two box plots on the same scale allows us to make some useful comparisons. It is clear, for instance, that engineering graduates are doing better overall than agriculture graduates, even though at the very top levels agriculture graduates are better paid. Another interesting point is that the median salary of agriculture graduates is less than the first quartile of the salaries of engineering graduates. The very short whisker on the left side of the agriculture box plot tells us that the bottom 25% of agriculture salaries are concentrated in a very narrow salary range. We can also see that agriculture salaries are much more spread out than engineering salaries, even though most of the spread occurs at the higher end of the salary scale. ◀◀

14.4 Measures of Spread

There are several different ways to describe the spread of a data set; in this section we will describe the three most commonly used ones.

The Range

An obvious approach to describing the spread of a data set is to take the difference between the highest and lowest values of the data. This difference is called the **range** of the data set and usually denoted by R. Thus, $R = Max - Min$.

The range of a data set is a useful piece of information when there are no outliers in the data. In the presence of outliers the range tells a distorted story. For example, the range of the test scores in the Stat 101 exam is $24 - 1 = 23$ points, an indication of a big spread within the scores (i.e., a very heterogeneous group of students). True enough, but if we discount the two outliers, the remaining 73 test scores would have a much smaller range of $16 - 6 = 10$ points.

The Interquartile Range

To eliminate the possible distortion caused by outliers, a common practice when measuring the spread of a data set is to use the **interquartile range**, denoted by the acronym IQR. The interquartile range is the difference between the third quartile and the first quartile ($IQR = Q_3 - Q_1$), and it tells us how spread out the middle 50% of the data values are. For many types of real-world data, the interquartile range is a useful measure of spread.

> ▶ **EXAMPLE 14.18** 2004 SAT Math Scores: Part 3

The five-number summary for the 2004 SAT math scores was $Min = 200$ (yes, there were a few jokers that missed every question!), $Q_1 = 440$, $M = 510$, $Q_3 = 600$, $Max = 800$ (there are still a few geniuses around!). It follows that the

2004 SAT math scores had a range of 600 points ($R = 800 - 200 = 600$) and an interquartile range of 160 points ($IQR = 600 - 440 = 160$). ◀◀

The Standard Deviation

The most important and most commonly used measure of spread for a data set is the *standard deviation*. The key concept for understanding the standard deviation is the concept of *deviation from the mean*. If A is the average of the data set and x is an arbitrary data value, the difference $x - A$ is x's **deviations from the mean**. The deviations from the mean tell us how "far" the data values are from the average value of the data. The idea is to use this information to figure out how spread out the data is. There are, unfortunately, several steps before we can get there.

The deviations from the mean are themselves a data set, which we would like to summarize. One way would be to average them, but if we do that, the negative deviations and the positive deviations will always cancel each other out so that we end up with an average of 0. This, of course, makes the average useless in this case. The cancellation of positive and negative deviations can be avoided by squaring each of the deviations. The squared deviations are never negative, and if we average them out, we get an important measure of spread called the **variance**, denoted by V. Finally, we take the square root of the variance and get the **standard deviation**, denoted by the Greek letter σ (and sometimes by the acronym SD).

The following is an outline of the definition of the standard deviation of a data set.

See Exercise 76.

The Standard Deviation of a Data Set

- Let A denote the mean of the data set. For each number x in the data set, compute its *deviation from the mean* $(x - A)$, and *square* each of these numbers. These are called the *squared deviations*.
- Find the average of the squared deviations. This number is called the *variance* V.
- The *standard deviation* is the square root of the variance $\left(\sigma = \sqrt{V} \right)$.

Another Note of Warning: Sometimes the variance is computed by dividing the sum of the squared deviations by $N - 1$ (instead of by N, as one would in an ordinary average). There are reasons that this definition is appropriate in some circumstances, but a full explanation would take us beyond the purpose and scope of this chapter. In any case, except for small values of N, the difference between the two definitions tends to be very small.

Standard deviations of large data sets are not fun to calculate by hand, and they are rarely found that way. The standard procedure for calculating standard deviations is to use a computer or a good scientific or business calculator, which often are preprogrammed to do all the steps automatically. Be that as it may, it is still important to understand what's behind the computation of a standard deviation, even when the actual grunt work is going to be performed by a machine. One way to accomplish this is to pay some dues.

See Exercises 55–58.

▶ EXAMPLE 14.19 Paying Some Dues

Over the course of the semester, Angela turned in all of her homework assignments. Her grades in the 10 assignments (sorted from lowest to highest) were 85, 86, 87, 88, 89, 91, 92, 93, 94, and 95. Our goal in this example is to compute the standard deviation of this data set the old-fashioned way (i.e., doing our own grunt work).

TABLE 14-11

x	(x − 90)	(x − 90)²
85	−5	25
86	−4	16
87	−3	9
88	−2	4
89	−1	1
91	1	1
92	2	4
93	3	9
94	4	16
95	5	25

See Exercise 57(d).

The first step is to find the mean A of the data set. It's not hard to see that $A = 90$. We are lucky—this is a nice round number! The second step is to calculate the *deviations from the mean* and then the *squared deviations*. The details are shown in the second and third columns of Table 14-11. When we average the squared deviations, we get $(25 + 16 + 9 + 4 + 1 + 1 + 4 + 9 + 16 + 25)/10 = 11$. This means that the variance is $V = 11$ and thus the standard deviation is (rounded to one decimal place) is $\sigma = \sqrt{11} \approx 3.3$ points. ◀◀

Standard deviations are measured in the same units as the original data, so in Example 14.19 the standard deviation of Angela's homework scores was roughly 3.3 points. What should we make of this fact? It is clear from just a casual look at Angela's homework scores that she was pretty consistent in her homework, never straying too much above or below her average score of 90 points. The standard deviation is, in effect, a way to measure this degree of consistency (or lack thereof). A small standard deviation tells us that the data are consistent and the spread of the data is small, as is the case with Angela's homework scores.

The ultimate in consistency within a data set is when all the data values are the same (like Angela's friend Chloe, who got a 20 in every homework assignment). When this happens the standard deviation is 0. On the other hand, when there is a lot of inconsistency within the data set, we are going to get a large standard deviation. This is illustrated by Angela's other friend, Tiki, whose homework scores were 5, 15, 25, 35, 45, 55, 65, 75, 85, and 95. We would expect the standard deviation of this data set to be quite large—in fact, it is almost 29 points.

The standard deviation is arguably the most important and frequently used measure of data spread. Yet it is not a particularly intuitive concept. Here are a few basic guidelines that recap our preceding discussion:

- The standard deviation of a data set is measured in the same units as the original data. For example, if the data are points on a test, then the standard deviation is also given in points. Conversely, if the standard deviation is given in dollars, we can conclude that the original data must have been money—home prices, salaries, or something like that. For sure, the data couldn't have been test scores on an exam.

- It is pointless to compare standard deviations of data sets that are given in different units. Even for data sets that are given in the same units, say, for example, test scores, the underlying scale should be the same. We should not try to compare standard deviations for SAT scores measured on a scale of 200–800 points with standard deviations of a set of homework assignments measured on a scale of 0–100 points.

- For data sets that are based on the same underlying scale, a comparison of standard deviations can tell us something about the spread of the data. If the standard deviation is small, we can conclude that the data points are all bunched together—there is very little spread. As the standard deviation increases, we can conclude that the data points are beginning to spread out. The more spread out they are, the larger the standard deviation becomes. If the standard deviation is 0, it means that all data values are the same.

As a measure of spread, the standard deviation is particularly useful for analyzing real-life data. We will come to appreciate its importance in this context in Chapter 16.

Conclusion

Whether we like to or not, as we navigate through life in the information age, we are awash in a sea of data. Today, data are the common currency of scientific, social, and economic discourse. Powerful satellites constantly scan our planet, collecting prodigious amounts of weather, geological, and geographical data. Government agencies, such as the Bureau of the Census and the Bureau of Labor Statistics, collect millions of numbers a year about our living, working, spending, and dying habits. Even in our less serious pursuits, such as sports, we are flooded with data, not all of it great.

Faced with the common problem of data overload, statisticians and scientists have devised many ingenious ways to organize, display, and summarize large amounts of data. In this chapter we discussed some of the basic concepts in this area of statistics.

Graphical summaries of data can be produced by bar graphs, pictograms, pie charts, histograms, and so on. (There are many other types of graphical descriptions that we did not discuss in the chapter.) The kind of graph that is the most appropriate for a situation depends on many factors, and creating a good "picture" of a data set is as much an art as a science.

Numerical summaries of data, when properly used, help us understand the overall pattern of a data set without getting bogged down in the details. They fall into two categories: (i) measures of location, such as the *average*, the *median*, and the *quartiles*, and (ii) measures of spread, such as the *range*, the *interquartile range*, and the *standard deviation*. Sometimes we even combine numerical summaries and graphical displays, as in the case of the *box plot*. We touched upon all of these in this chapter, but the subject is a big one, and by necessity we only scratched the surface.

In this day and age, we are all consumers of data, and at one time or another, we are likely to be providers of data as well. Thus, understanding the basics of how data are organized and summarized has become an essential requirement for personal success and good citizenship.

Profile W. Edwards Deming (1900–1993)

W. E. Deming was a pioneer in the application of statistics to industry. He is best known for his theories of quality control in manufacturing, which have been widely adopted, first by Japanese and later by American companies such as Xerox and Ford. Like all great ideas, Deming's ideas about quality control were built on a simple observation: All industrial processes are subject to some level of statistical variation, and this variation negatively impacts quality. From this Deming developed the principle that to improve manufacturing quality one has to reduce the causes of statistical variation.

William Edwards Deming was born in Sioux City, Iowa, in 1900, into a family of very modest means. As a young man, much of his life revolved around study and work. As an undergraduate at the University of Wyoming, he supported himself by doing all sorts of odd jobs, from janitor's aide to cleaning boilers at an oil refinery. He earned a bachelor's of science in electrical engineering from Wyoming in 1921, a master's degree in mathematics and physics from the University of Colorado in 1925, and a Ph.D. in mathematical physics from Yale University in 1928. Deming became famous as a statistician and management guru, but he was trained in classical mathematics and physics and was an accomplished musician who played several instruments and composed religious music.

With a doctorate from Yale in hand and a family to support, Deming joined the Department of Agriculture in Washington, D.C., where he performed laboratory research on fertilizers and developed statistical methods to boost productivity. In 1939, Deming joined the Census Bureau, where, as head mathematician, he helped develop many currently used statistical sampling methods. In 1946 he retired from the Census Bureau and became a statistical consultant and Professor at New York University.

After World War II, the methods of *statistical process control* (SPC) developed by Deming were quickly implemented by Japanese industry. Deming's methods focused on reducing variation in assembly line production, thereby minimizing defects and increasing productivity. Though a folk hero in Japan, Deming was virtually ignored by the American business community until 1980, when NBC aired the documentary "If Japan Can, Why Can't We?" detailing the superior quality and growing popularity of Japanese products. Though already 80 years old at the time, many American industrial giants belatedly adopted Deming's techniques.

Deming remained active up until his death at the age of 93, and his lectures and management seminars were attended by the thousands. He was the author of many successful books and over 170 papers and was the recipient of many honors and awards. The Deming Prize, established in his honor by the Japanese Union of Scientists and Engineers, is today the highest award a company can achieve for its excellence in management.

Key Concepts

average (mean), **487**
bar graph, **479**
box plot (box-and-whisker plot), **494**
categorical (qualitative) variable, **482**
category (class), **483**
class interval, **484**
continuous variable, **481**
data set, **477**
data values (data points), **477**

deviations from the mean, **496**
discrete variable, **481**
five-number summary, **493**
frequency, **479**
frequency table, **479**
histogram, **486**
interquartile range, **495**
locator, **490**
mean (average), **487**
measure of location, **487**
measure of spread, **487**

median, **491**
numerical (quantitative) variable, **481**
outlier, **479**
percentile, **489**
pictogram, **480**
pie chart, **483**
quartiles, **491**
range, **495**
standard deviation, **496**
variable, **481**
variance, **496**

Exercises

WALKING

A. Frequency Tables, Bar Graphs, and Pie Charts

Exercises 1 through 4 refer to the scores in a Chem 103 final exam consisting of 10 questions worth 10 points each. The scores on the exam are given in the following table.

Chem 103 Final Exam Scores

Student ID	Score	Student ID	Score	Student ID	Score	Student ID	Score
1362	50	2877	80	4315	70	6921	50
1486	70	2964	60	4719	70	8317	70
1721	80	3217	70	4951	60	8854	100
1932	60	3588	80	5321	60	8964	80
2489	70	3780	80	5872	100	9158	60
2766	10	3921	60	6433	50	9347	60

1. Make a frequency table for the Chem 103 final exam scores.

2. Make a bar graph showing the actual frequencies of the scores on the exam.

3. Using the scale A: 80–100. B: 70–79, C: 60–69, D: 50–59, and F: 0–49,

   (a) find the grade distribution for the final exam.

   (b) make a bar graph showing the grade distribution for the final exam.

4. Using the scale A: 80–100. B: 70–79, C: 60–69, D: 50–59, and F: 0–49,

   (a) determine the percent of students that earned a grade of D on the final exam.

   (b) find the size of the central angle (in degrees) of the "D's wedge" in a pie chart of the grade distribution for the final exam.

   (c) make a pie chart showing the grade distribution for the final exam.

Exercises 5 and 6 refer to the following situation. Every year, the first-grade students at Cleansburg Elementary are given a musical aptitude test. Based on the results of the test, the children are scored from 0 (no musical aptitude) to 5 (extremely talented). This year's results are as follows:

Aptitude score	0	1	2	3	4	5
Frequency	24	16	20	12	5	3

5. (a) How many children were given the aptitude test?

   (b) What percent of the students tested showed no musical aptitude?

   (c) Make a relative frequency bar graph showing the results of the musical aptitude test.

6. Make a pie chart showing the results of the musical aptitude test.

Exercises 7 through 10 refer to the following table, which gives the distance from home to school (measured to the closest half-mile) for each kindergarten student at Cleansburg Elementary School.

ID	d	ID	d	ID	d	ID	d
1362	1.5	2877	1.0	4355	1.0	6573	0.5
1486	2.0	2964	0.5	4454	1.5	8436	3.0
1587	1.0	3491	0.0	4561	1.5	8592	0.0
1877	0.0	3588	0.5	5482	2.5	8964	2.0
1932	1.5	3711	1.5	5533	1.0		
1946	0.0	3780	2.0	5717	8.5		
2103	2.5	3921	5.0	6307	1.5		

d = Distance from home to school (miles).

7. Make a frequency table for the data set.

8. Make a bar graph for the data set.

9. Suppose that class intervals for the distances from home to school for the kindergarteners at Cleansburg Elementary School are defined as follows.

   Very close: Less than 1 mile

   Close: 1 mile up to and including 1.5 miles

   Nearby: 2 miles up to and including 2.5 miles

   Not too far: 3 miles up to and including 4.5 miles

   Far: 5 miles or more

   (a) Make a frequency table for the class intervals.

   (b) Make a pie chart for the percentage of students in each class interval.

10. Make a bar graph for the class intervals defined in Exercise 9.

Exercises 11 and 12 refer to the scores on a math quiz, the results of which are shown in the following bar graph.

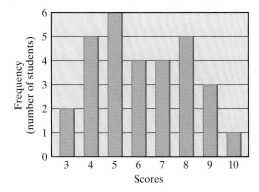

11. (a) How many students took the math quiz?

    (b) What percentage of the students scored 2 points?

    (c) If a grade of 6 or more was needed to pass the quiz, what percentage of the students passed?

12. Make a pie chart showing the results of the quiz.

Exercises 13 and 14 refer to the following pie chart showing the cause of death for 20-to 24-year-olds in the United States in 2002.

(**Source:** *Centers for Disease Control and Prevention, www.cdc.gov*)

Cause of Death in U.S. Among 20–24 Year Olds (2002)
N = 19,234

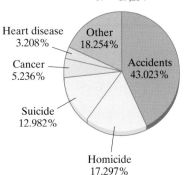

13. Calculate the size of the angle (to the nearest degree) for each of the slices shown in the pie chart.

14. **(a)** Give a frequency table showing the actual frequencies for each category.

    (b) Draw the bar graph corresponding to the frequency table in (a).

Exercises 15 and 16 refer to the following pie chart showing how the 2003 federal government budget of $2.1 trillion was spent.

(***Source:*** *Office of Management and Budget, www.gpo.gov/usbudget/index.html*)

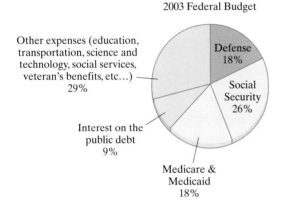

2003 Federal Budget

15. **(a)** Approximately how much money did the federal government spend on interest on the public debt in 2003?

    (b) Approximately how much money did the federal government spend on entitlement programs (Social Security, Medicare, and Medicaid) in 2003?

16. Calculate the size of the central angle for each of the slices shown in the pie chart. Round your answer to the nearest degree.

17. The following table shows the median percentage of wives' contribution to total family income for the years 1995–2003.

    (***Source:*** *Bureau of Labor Statistics, www.bls.gov.*)

Year	1995	1996	1997	1998	1999
Percent	31.9	32.6	32.7	32.8	32.8
Year	2000	2001	2002	2003	
Percent	33.5	34.4	34.8	35.2	

Using the ideas of Example 14.3, make two different-looking pictograms showing the growth in the percentage contribution of wives' income to total family income. In the first pictogram, you are trying to convince your audience that wives' percentage contributions are growing very fast. The second pictogram should give a more conservative picture.

18. The following table gives the percentage of recorded music sales made at record stores from 1995 to 2004.

    (***Source:*** *Recording Industry Association of America, www.riaa.com.*)

Year	1995	1996	1997	1998	1999
Percent	52.0	49.9	51.8	50.8	44.5
Year	2000	2001	2002	2003	2004
Percent	42.4	42.5	36.8	33.2	32.5

Using the ideas of Example 14.2, make two different-looking pictograms showing the decline in the percentage of recorded music sold in record stores from 1995 to 2004. In the first pictogram, you are trying to convince your audience that sales in record stores are declining very fast. The second pictogram should give a more realistic picture.

B. Histograms

Exercises 19 and 20 refer to the data in the following table, which shows the birth weights (in ounces) of the 625 babies born in the city of Cleansburg in 2005.

Weight (in ounces)		
More than	**Up to**	**Frequency**
48	60	15
60	72	24
72	84	41
84	96	67
96	108	119
108	120	184
120	132	142
132	144	26
144	156	5
156	168	2

19. **(a)** Give the length of each class interval (in ounces).

    (b) Suppose a baby weighs exactly 5 pounds 4 ounces. What class interval does she belong to? Describe the endpoint convention.

    (c) Draw the histogram describing the 2005 birth weights in Cleansburg using the class intervals given in the table.

20. **(a)** Write a new frequency table for the birth weights in Cleansburg using class intervals of length equal to 24 ounces.

(b) Draw the histogram corresponding to the data given in the table found in (a).

Exercises 21 and 22 refer to the following histogram showing the payrolls of the 30 teams in Major League Baseball.

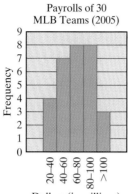

Payrolls of 30
MLB Teams (2005)

21. **(a)** How many teams in Major League Baseball had a payroll of 80 million dollars or more?

(b) Give the length of each class interval (in dollars).

(c) Describe the endpoint convention used in the histogram.

22. Write a frequency table corresponding to the data shown in the histogram using the same class intervals.

C. Means and Medians

23. Consider the data set $\{3, -5, 7, 4, 8, 2, 8, -3, -6\}$.

(a) Find the average.

(b) Find the median.

(c) Consider the data set $\{3, -5, 7, 4, 8, 2, 8, -3, -6, 2\}$ obtained by adding one more data point to the original data set. Find the average and median of this data set.

24. Consider the data set $\{-3.8, -7.3, -4.5, 8.3, 8.3, -9.1, -3.8, 13.2\}$.

(a) Find the average.

(b) Find the median.

(c) Consider the data set $\{-3.8, -7.3, -4.5, 8.3, 8.3, -9.1, -3.8\}$ having one less data point than the original set. Find the average and the median of this data set.

25. For each data set, find the average and the median.

(a) $\{0, 1, 2, 3, 4, 5, 6, 7, 8, 9\}$

(b) $\{1, 2, 3, 4, 5, 6, 7, 8, 9\}$

(c) $\{1, 2, 3, 4, 5, 6, 7, 8, 9, 10\}$

26. For each data set, find the average and the median.

(a) $\{1, 2, 1, 2, 1, 2, 1, 2, 1, 2\}$

(b) $\{1, 2, 3, 4, 1, 2, 3, 4, 1, 2, 3, 4, 1, 2, 3, 4\}$

(c) $\{1, 2, 3, 4, 5, 5, 4, 3, 2, 1\}$

27. For the data set $\{1, 2, 3, 4, 5, \ldots, 98, 99\}$, find

(a) the average.

(***Hint:*** $1 + 2 + 3 + \cdots + N = N \cdot (N + 1)/2$.)

(b) the median.

28. For the data set $\{1, 2, 3, 4, 5, \ldots, 997, 998, 999, 1000\}$, find

(a) the average.

(***Hint:*** $1 + 2 + 3 + \cdots + N = N \cdot (N + 1)/2$.)

(b) the median.

29. This exercise refers to the musical aptitude test discussed in Exercises 5 and 6. The results of the test are given by the following frequency table.

Aptitude score	0	1	2	3	4	5
Frequency	24	16	20	12	5	3

(a) Find the average aptitude score.

(b) Find the median aptitude score.

30. The ages of the firefighters in the City of Cleansburg Fire Department are given in the following frequency table.

Age	25	27	28	29	30	31	32	33	37	39
Frequency	2	7	6	9	15	12	9	9	6	4

(a) Find the average age rounded to two decimal places.

(b) Find the median age.

31. The results of a 10-point philosophy quiz are shown in the following pie chart.

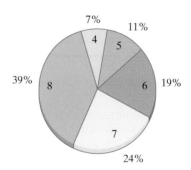

(a) Find the average quiz score.

(b) Find the median quiz score.

32. The results of a 10-point math quiz are shown in the following bar graph.

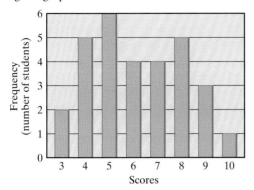

(a) Find the average quiz score.

(b) Find the median quiz score.

D. Percentiles and Quartiles

33. Consider the data set $\{3, -5, 7, 4, 8, 2, 8, -3, -6\}$.

(a) Find the first quartile.

(b) Find the third quartile.

(c) Consider the data set $\{3, -5, 7, 4, 8, 2, 8, -3, -6, 2\}$ obtained by adding one more data point to the original data set. Find the first and third quartiles of this data set.

34. Consider the data set $\{-3.8, -7.3, -4.5, 8.3, 8.3, -9.1, -3.8, 13.2\}$.

(a) Find the first quartile.

(b) Find the third quartile.

(c) Consider the data set $\{-3.8, -7.3, -4.5, 8.3, 8.3, -9.1, -3.8\}$ obtained by deleting one data point from the original data set. Find the first and third quartiles of this data set.

35. For each data set, find the 75th and the 90th percentiles.

(a) $\{1, 2, 3, 4, \ldots, 98, 99, 100\}$

(b) $\{0, 1, 2, 3, 4, \ldots, 98, 99, 100\}$

(c) $\{1, 2, 3, 4, \ldots, 98, 99\}$

(d) $\{1, 2, 3, 4, \ldots, 98\}$

36. For each data set, find the 10th and the 25th percentiles.

(a) $\{1, 2, 3, \ldots, 49, 50, 50, 49, \ldots, 3, 2, 1\}$

(b) $\{1, 2, 3, \ldots, 49, 50, 49, \ldots, 3, 2, 1\}$

(c) $\{1, 2, 3, \ldots, 49, 49, \ldots, 3, 2, 1\}$

37. This exercise refers to the data given in Exercise 30.

Age	25	27	28	29	30	31	32	33	37	39
Frequency	2	7	6	9	15	12	9	9	6	4

(a) Find the first quartile.

(b) Find the third quartile.

(c) Find the 90th percentile.

38. This exercise refers to the math quiz discussed in Exercise 32. The results of the quiz are shown in the following bar graph.

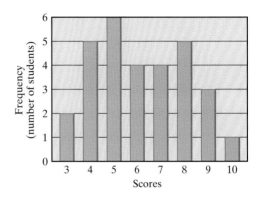

(a) Find the first quartile.

(b) Find the third quartile.

(c) Find the 70th percentile.

39. In 2004, a total of $N = 1{,}419{,}007$ college-bound seniors took the SAT test. Assume that the test scores are sorted from lowest to highest and that the sorted data set is $\{d_1, d_2, \ldots, d_{1{,}419{,}007}\}$.

(a) Determine the position of the median.

(b) Determine the position of the first quartile.

(c) Determine the position of the third quartile.

40. In 2001, a total of $N = 1{,}276{,}320$ college-bound seniors took the SAT test. Assume that the test scores are sorted from lowest to highest and that the sorted data set is $\{d_1, d_2, \ldots, d_{1{,}276{,}320}\}$.

(a) Determine the position of the median.

(b) Determine the position of the first quartile.

(c) Determine the position of the third quartile.

E. Box Plots and Five-Number Summaries

41. For the data set $\{3, -5, 7, 4, 8, 2, 8, -3, -6,\}$,

(a) find the five-number summary. [Use the results of Exercises 23(b) and 33.]

(b) draw a box plot.

42. For the data set
$\{-3.8, -7.3, -4.5, 8.3, 8.3, -9.1, -3.8, 13.2\}$,

(a) find the five-number summary. [See Exercises 24(b) and 34.]

(b) draw a box plot.

43. This exercise refers to the ages of the firefighters discussed in Exercises 30 and 37.

(a) Find the five-number summary.

(b) Draw a box plot.

44. This exercise refers to the math quiz discussed in Exercises 32 and 38. The results of the quiz are shown in the following bar graph.

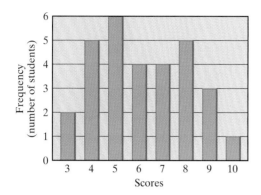

(a) Find the five-number summary for the quiz scores.

(b) Draw a box plot for the quiz scores.

Exercises 45 and 46 refer to the following figure showing the box plots for the starting salaries of Tasmania State University first-year graduates in agriculture and engineering. (These are the two box plots discussed in Example 14.17.)

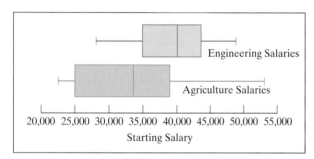

45. (a) Approximately how much is the median salary for agriculture majors?

(b) Approximately how much is the median salary for engineering majors?

(c) Explain how we can tell that the median salary for engineering majors is more than the third quartile of the salaries for agriculture majors.

46. (a) Fill in the blank: Of the 612 engineering graduates, at most _____ had a starting salary greater than $35,000.

(b) Fill in the blank: If there were 240 agriculture graduates with starting salaries of $25,000 or less, the total number of agriculture graduates is approximately _____.

F. Ranges and Interquartile Ranges

47. For the data set $\{3, -5, 7, 4, 8, 2, 8, -3, -6\}$, find

(a) the range.

(b) the interquartile range (see Exercise 33).

48. For the data set $\{-3.8, -7.3, -4.5, 8.3, 8.3, -9.1, -3.8, 13.2\}$, find

(a) the range.

(b) the interquartile range (see Exercise 34).

49. A realty company has sold $N = 341$ homes in the last year. The five-number summary for the sale prices is $Min = \$97,000$, $Q_1 = \$115,000$, $M = \$143,000$, $Q_3 = \$156,000$, and $Max = \$249,000$.

(a) Find the interquartile range of the home sale prices.

(b) How many homes sold for a price between $115,000 and $156,000 (inclusive)?

(*Note:* *If you don't believe you have enough information to give an exact answer, you should give the answer in the form of "at least _____" or "at most _____.")*

50. This exercise refers to the starting salaries of Tasmania State University first-year graduates in agriculture and engineering discussed in Exercises 45 and 46.

(a) Estimate the range for the starting salaries of agriculture majors.

(b) Estimate the interquartile range for the starting salaries of engineering majors.

(c) There were 612 engineering majors. Determine how many starting salaries of engineering majors were between $Q_1 = \$35,000$ and $Q_3 = \$43,500$ (inclusive).

(*Note:* *If you don't believe you have enough information to give an exact answer, you should give the answer in the form of "at least _____" or "at most _____.")*

Exercises 51 through 54 refer to the following definition of an **outlier**: *An outlier is any data value that is above the third quartile by more than 1.5 times the IQR* $[OUT > Q_3 + 1.5(IQR)]$ *or below the first quartile by more than 1.5 times the IQR* $[OUT < Q_1 - 1.5(IQR)]$.

(*Note:* *There is no one universally agreed upon definition of an outlier, and this is but one of several definitions used by statisticians.*)

51. Suppose the preceding definition of outlier is applied to the Stat 101 data set discussed in Example 14.16.

(a) Fill in the blank: Any score bigger or equal to _____ is an outlier.

(b) Fill in the blank: Any score smaller or equal to _____ is an outlier.

(c) Find the outliers (if there are any) in the Stat 101 data set.

52. Using the preceding definition, find the outliers (if there are any) in the City of Cleansburg Fire Department data set discussed in Exercises 30 and 37.

(*Hint: Do Exercise 37 first.*)

53. The five-number summary of 2001 SAT math scores was $Min = 200$, $Q_1 = 440$, $M = 510$, $Q_3 = 590$, and $Max = 800$. Using the preceding definition, determine which SAT scores (if any) correspond to outliers.

54. The five-number summary of 2004 SAT math scores was $Min = 200, Q_1 = 440, M = 510, Q_3 = 600$, and $Max = 800$. Using the preceding definition, determine which SAT scores (if any) correspond to outliers.

G. Standard Deviations

The purpose of Exercises 55 through 58 is to practice computing standard deviations by using the definition. Granted, computing standard deviations this way is not the way it is generally done in practice—a good calculator (or a computer package) will do it much faster and more accurately. The point is that computing a few standard deviations the old-fashioned way should help you understand the concept a little better. If you use a calculator or a computer to answer these exercises, you are defeating their purpose.

55. Find the standard deviation of each of the following data sets.

(a) $\{5, 5, 5, 5\}$

(b) $\{0, 5, 5, 10\}$

(c) $\{-5, 0, 0, 25\}$

56. Find the standard deviation of each of the following data sets.

(a) $\{10, 10, 10, 10\}$

(b) $\{1, 6, 13, 20\}$

(c) $\{1, 1, 18, 20\}$

57. Find the standard deviation of each of the following data sets.

(a) $\{0, 1, 2, 3, 4, 5, 6, 7, 8, 9\}$

(b) $\{1, 2, 3, 4, 5, 6, 7, 8, 9, 10\}$

[**Hint:** *Try comparing what happens here with what happens in (a).*]

(c) $\{6, 7, 8, 9, 10, 11, 12, 13, 14, 15\}$

[**Hint:** *Try comparing what happens here with what happens in (a).*]

(d) $\{5, 15, 25, 35, 45, 55, 65, 75, 85, 95\}$

58. Find the standard deviation of each of the following data sets.

(a) $\{-4, -3, -2, -1, 1, 2, 3, 4\}$

(b) $\{-3, -2, -1, 0, 2, 3, 4, 5\}$

[**Hint:** *Try comparing what happens here with what happens in (a).*]

(c) $\{1, 2, 3, 4, 6, 7, 8, 9\}$

[**Hint:** *Try comparing what happens here with what happens in (a).*]

H. Miscellaneous

The Mode

The mode of a data set is the data point that occurs with the highest frequency. In a frequency table, we look for the largest number in the frequency row; the corresponding data point is the mode. In a bar graph, we look for the tallest bar; the corresponding data point (or category) is the mode. In a pie chart, we look for the largest slice; the corresponding category is the mode.

When there are several data points (or categories) tied for the most frequent, each of them is a mode, but if all data points have the same frequency, rather than every data point being a mode, it is customary to say that there is no mode.

In Exercises 59 through 64, you should find the mode or modes of the given data sets. If there is no mode, your answer should indicate that.

59. The Stat 101 data set given by the following frequency table.

Exam score	1	6	7	8	9	10	
Frequency	1	1	2	6	10	16	
Exam score	11	12	13	14	15	16	24
Frequency	13	9	8	5	2	1	1

60. The history midterm exam scores given by the following table.

Student ID	Score	Student ID	Score
1075	74%	4713	83%
1367	83%	4822	55%
1587	70%	5102	78%
1877	55%	5381	13%
1946	76%	5717	74%
1998	75%	6234	77%
2103	59%	6573	55%
2169	92%	7109	51%
2381	56%	7986	70%
2741	50%	8436	57%
3491	57%		
3711	70%		
3827	52%		
4355	74%		
4531	77%		

61. The Math 100 quiz scores given by the following bar graph.

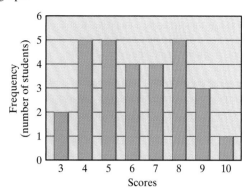

62. The data set given by the following bar graph.

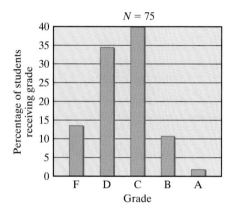

63. The data set given by the following pie chart.

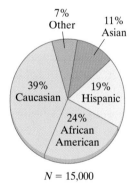

64. The data set given by the following pie chart.

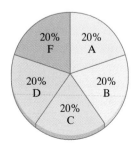

JOGGING

65. Mike's average on the first five exams in Econ 1A is 88. What must he earn on the next exam in order to raise his overall average to 90?

66. Sarah's overall average in Physics 101 was 93%. Her average was based on four exams each worth 100 points and a final worth 200 points. What is the lowest possible score she could have made on the first exam?

67. Josh and Ramon each have an 80% average on the five exams given in Psychology 4. Ramon, however, did better than Josh on all of the exams except one. Give an example that illustrates this situation.

68. Kelly and Karen each have an average of 75 on the six exams given in Botany 1. Kelly's scores have a small standard deviation, and Karen's scores have a large standard deviation. Give an example that illustrates this situation.

69. (a) Give an example of 10 numbers with an average less than the median.

   (b) Give an example of 10 numbers with a median less than the average.

   (c) Give an example of 10 numbers with an average less than the first quartile.

   (d) Give an example of 10 numbers with an average more than the third quartile.

70. Suppose that the average of 10 numbers is 7.5 and that the smallest of them is $Min = 3$.

   (a) What is the smallest possible value of Max?

   (b) What is the largest possible value of Max?

71. What happens to the five-number summary of the Stat 101 data set (see Example 14.16) if

   (a) two points are added to each score?

   (b) ten percent is added to each score?

72. A data set is called **constant** if every value in the data set is the same. A constant data set can be described by $\{a, a, a, \ldots, a.\}$

   (a) Show that the standard deviation of a constant data set is 0.

   (b) Show that if the standard deviation of a data set is 0, it must be a constant data set.

Exercises 73 and 74 refer to histograms with unequal class intervals. When sketching such histograms, the columns must be drawn so that the frequencies or percentages are proportional to the area of the column. The accompanying figure illustrates this. If the column over class interval 1 represents 10% of the population, then the column over class interval 2, also representing 10% of the population, must be one-third as high because the class interval is three times as large.

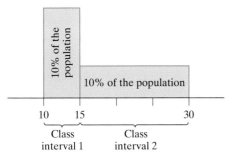

73. If the height of the column over the class interval 20–30 is 1 unit and the column represents 25% of the population, then

(a) how high should the column over the interval 30–35 be if 50% of the population falls into this class interval?

(b) how high should the column over the interval 35–45 be if 10% of the population falls into this class interval?

(c) how high should the column over the interval 45–60 be if 15% of the population falls into this class interval?

74. Two hundred senior citizens are tested for fitness and rated on their times on a 1-mile walk. These ratings and associated frequencies are given in the following table.

Time	Rating	Frequency
6^+-10 minutes	Fast	10
10^+-16 minutes	Fit	90
16^+-24 minutes	Average	80
24^+-40 minutes	Slow	20

Draw a histogram for these data based on the categories given by the ratings in the table.

75. News media have accused Tasmania State University of discriminating against women in the admission policies in its schools of architecture and engineering. The *Tasmania Gazette* states that "68% of all male applicants to the schools of architecture or engineering are admitted, while only 51% of the female applicants to these same schools are admitted." The actual data are given in the following table.

	Architecture		Engineering	
	Applied	Admitted	Applied	Admitted
Male	200	20	1000	800
Female	500	100	400	360

(a) What percent of the male applicants to the School of Architecture were admitted? What percent of the female applicants to this same school were admitted?

(b) What percent of the male applicants to the School of Engineering were admitted? What percent of the female applicants to this same school were admitted?

(c) How did the *Tasmania Gazette* come up with its figures?

(d) Explain how it is possible for the results in (a) and (b) and the *Tasmania Gazette* statement all to be true.

76. Let A denote the mean of the data set $\{x_1, x_2, x_3, \ldots, x_N\}$.

(a) Find the mean of the data set $\{x_1 + c, x_2 + c, x_3 + c, \ldots, x_N + c\}$.

(b) Use the results of (a) to explain why the mean of $\{x_1 - A, x_2 - A, x_3 - A, \ldots, x_N - A\}$ is 0.

(***Note:*** *The data set* $\{x_1 - A, x_2 - A, x_3 - A, \ldots, x_N - A\}$ *represents the set of deviations from the mean*).

(c) Find the median of $\{x_1 + c, x_2 + c, x_3 + c, \ldots, x_N + c\}$.

77. Let M denote the median of the data set $\{x_1, x_2, x_3, \ldots, x_N\}$. Find the median of the data set $\{x_1 + c, x_2 + c, x_3 + c, \ldots, x_N + c\}$. Explain your answer.

78. Suppose that the data set $\{x_1, x_2, x_3, \ldots, x_N\}$ has range R and standard deviation σ.

(a) Explain why the range of $\{x_1 + c, x_2 + c, x_3 + c, \ldots, x_N + c\}$ is R.

(b) Explain why the standard deviation of $\{x_1 + c, x_2 + c, x_3 + c, \ldots, x_N + c\}$ is σ.

(***Hint:*** *Try Exercises 57 and/or 58 first.*)

79. For this exercise you will have to do a tiny bit of research. Let x_1 denote the number of exercises in Chapter 1 of this book, x_2 the number of exercises in Chapter 2, and so on. For the data set $\{x_1, x_2, \ldots, x_{16}\}$,

(a) find the mean.

(b) find the median.

(c) find the five-number summary.

(d) find the standard deviation.

(***Note:*** *You may want to use a calculator here.*)

RUNNING

80. Consider a data set of 10 numbers with $Min = 2$, $Max = 12$, and mean $A = 7$. Let σ denote the standard deviation of this data set.

(a) What is the smallest possible value of σ?

(b) What is the largest possible value of σ?

81. Show that the standard deviation of any set of numbers is always less than or equal to the range of the set of numbers.

82. (a) Show that if $\{x_1, x_2, x_3, \ldots, x_N\}$ is a data set with mean A and standard deviation σ, then $\sigma\sqrt{N} \geq |x_i - A|$ for every data value x_i.

(b) Use (a) to show that every data value is bigger than or equal to $A - \sigma\sqrt{N}$ and smaller than or equal to $A + \sigma\sqrt{N}$ (i.e., for every data value x, $A - \sigma\sqrt{N} \leq x \leq A + \sigma\sqrt{N}$).

83. (a) Given two numbers A and $\sigma > 0$, find two other numbers (expressed in terms of A and σ) whose mean is A and whose standard deviation is σ.

(b) Find three equally spaced numbers whose mean is A and whose standard deviation is σ.

(c) Generalize the preceding by finding N equally spaced numbers whose mean is A and whose standard deviation is σ.

(**Hint:** *Consider N even and N odd separately.*)

84. Show that if A is the mean and M is the median of the data set $\{1, 2, 3, \ldots, N\}$, then for all values of N, $A = M$.

85. Suppose that the mean of the numbers $x_1, x_2, x_3, \ldots, x_N$ is A and that the variance of these same numbers is V. Suppose also that the mean of the numbers $x_1^2, x_2^2, x_3^2, \ldots, x_N^2$ is B. Show that $V = B - A^2$. (In other words, for any data set, if we take the mean of the squared data values and subtract the square of the mean of the data values, we get the variance.)

86. Suppose that the standard deviation of the data set $\{x_1, x_2, x_3, \ldots, x_N\}$ is σ. Explain why the standard deviation of the data set $\{a \cdot x_1, a \cdot x_2, a \cdot x_3, \ldots, a \cdot x_N\}$ (where a is a positive number) is $a \cdot \sigma$.

87. Using the formula
$$1^2 + 2^2 + 3^2 + \cdots + N^2 = N(N + 1)(2N + 1)/6,$$

(a) find the standard deviation of the data set $\{1, 2, 3, \ldots, 98, 99\}$.

(**Hint:** *Use Exercise 85.*)

(b) find the standard deviation of the data set $\{1, 2, 3, \ldots, N\}$.

88. (a) Find the standard deviation of the data set $\{315, 316, \ldots, 412, 413\}$.

(**Hint:** *Use Exercises 78(b) and 87.*)

(b) Find the standard deviation of the data set $\{k + 1, k + 2, \ldots, k + N\}$.

89. Chebyshev's Theorem. The Russian mathematician P. L. Chebyshev (1821–1894) showed that for any data set and any constant k greater than 1, at least $1 - \dfrac{1}{k^2}$ of the data must lie within k standard deviations on either side of the mean A. For example, when $k = 2$, this says that $1 - \dfrac{1}{4} = \dfrac{3}{4}$ (i.e., 75%) of the data must lie within two standard deviations of A (i.e., somewhere between $A - 2\sigma$ and $A + 2\sigma$).

(a) Using Chebyshev's theorem, what percentage of a data set must lie within three standard deviations of the mean?

(b) How many standard deviations on each side of the mean must we take to be assured of including 99% of the data?

(c) Suppose the average of a data set is A. Explain why there is no number k of standard deviations for which Chebyshev's Theorem assures inclusion of 100% of the data.

Projects and Papers

A. Lies, Damn Lies, and Statistics

Statistics are often used to exaggerate, distort, and misinform, and this is most commonly done by the misuse of graphs and charts. In this project you are to discuss the different graphical "tricks" that can be used to mislead or slant the information presented in a picture. Attempt to include items from recent newspapers, magazines, and other media. ***Note:*** References 2, 9, and 14 are good sources for this project.

B. Skewness

Measures of location and *measures of spread* are not the only types of numerical summaries of a data set. It is also possible to measure the amount of asymmetry in a data set by means of a statistical measure called the **skewness** of the data set. Consider the three data sets shown to the right. The top one has a positive skew (skews right), the middle one has no skew, and the bottom one has a negative skew (skews left).

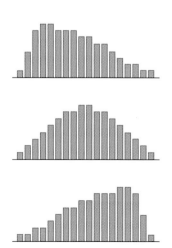

In this project, you are to discuss the concept of *skewness* in a data set. You should include a precise mathematical definition of skewness, and give examples of five different data sets (each containing at least 10 data points) and having skewness of -1, -0.5, 0, 0.5, and 1, respectively.

Note: You may want to use a spreadsheet program such as Excel™ to experiment with the data sets, compute their skewness, and generate graphs. You can also do this with a graphing calculator.

C. Data in Your Daily Life

Which month is the best one to invest in the stock market? During which day of the week are you most likely to get into an automobile accident? Which airline is the safest to travel with? Who is the best place-kicker in the National Football League?

In this project, you are to formulate a question from everyday life (similar to one of the aforementioned questions) that is amenable to a statistical analysis. Then you will need to find relevant data that attempt to answer this question and summarize these data using the methods discussed in this chapter. Present your final conclusions and defend them using appropriate charts and graphs.

D. Book Review: *Curve Ball*

Baseball is the ultimate statistical sport, and statistics are as much a part of baseball as peanuts and Cracker Jack. The book *Curve Ball: Baseball, Statistics, and the Role of Chance in the Game* (reference 1), written by statisticians and baseball lovers Jim Albert and Jay Bennett, is a compilation of all sorts of fascinating statistical baseball issues, from how to better measure a player's offensive performance to what is the true value of home field advantage.

In this project you are to pick one of the many topics discussed in *Curve Ball* and write an analysis paper on the topic.

References and Further Readings

1. Albert, Jim, and Jay Bennett, *Curve Ball: Baseball, Statistics, and the Role of Chance in the Game*. New York: Springer-Verlag, 2001, chap. 2.
2. Cleveland, W. S., *The Elements of Graphing Data*, rev. ed. New York: Van Nostrand Reinhold Co., 1994.
3. Cleveland, W. S., *Visualizing Data*. Summit, NJ: Hobart Press, 1993.
4. Gabor, Andrea, *The Man Who Discovered Quality*. New York: Penguin Books, 1992.
5. Harris, Robert, *Information Graphics: A Comprehensive Illustrated Reference*. New York: Oxford University Press, 2000.
6. Mosteller, F., and W. Kruskal, et al., *Statistics by Example: Exploring Data*. Reading, MA: Addison-Wesley, 1973.
7. Tanner, Martin, *Investigations for a Course in Statistics*. New York: Macmillan Publishing Co., Inc., 1990.
8. Tufte, Edward, *Envisioning Information*. Cheshire, CT: Graphics Press, 1990.
9. Tufte, Edward, *The Visual Display of Quantitative Information*. Cheshire, CT: Graphics Press, 1983.
10. Tufte, Edward, *Visual Explanations: Images and Quantities, Evidence and Narrative*. Cheshire, CT: Graphics Press, 1997.
11. Tukey, John W., *Exploratory Data Analysis*. Reading, MA: Addison-Wesley, 1977.
12. Utts, Jessica, *Seeing Through Statistics*. Belmont, CA: Wadsworth Publishing Co., 1996.
13. Wainer, H., "How to Display Data Badly," *The American Statistician*, 38 (1984), 137–147.
14. Wainer, H., *Visual Revelations*. New York: Springer-Verlag, 1997.
15. Wildbur, Peter, *Information Graphics*. New York: Van Nostrand Reinhold Co., 1989.

15

Chances, Probabilities, and Odds

Measuring Uncertainty

Nothing in life is certain except death and taxes.

Ben Franklin

With the possible exception of death and taxes, pretty much everything else in our lives is shrouded with some degree of uncertainty. Thus, we go through life unwittingly "rolling the dice"—when we apply to college, when we buy a used car, when we shoot a free throw, and so on.

While we are all familiar with the idea of uncertainty, we don't always have a good grasp on how to measure it. Sometimes the degree of uncertainty is small (Steve Nash is shooting the free throw), and sometimes there is a lot of uncertainty (Shaquille O'Neal is shooting the free throw), and sometimes the uncertainty is somewhere in between. To quantify the amount of uncertainty in the many uncertain events that affect our lives, we use the concept of *probability*.

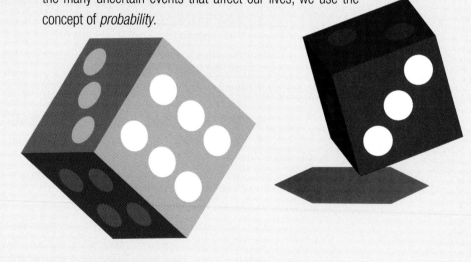

*C*hance, probability, odds—we use these words carelessly in casual conversation, and most of the time we can get away with it. Technically speaking, however, each of these words represents a slightly different way to express the likelihood of an event. When we speak of the *chance* of some event happening, we state the answer in terms of percentages, such as when the weatherperson reports, "the chance of rain tomorrow is 40%." When we speak of the *probability* of some event happening, we should state the answer in terms of a fraction (such as $\frac{2}{5}$) or a decimal (0.4). Finally, when we speak of the *odds in favor* of an

event, we use a pair of numbers, as in "the odds of the Red Sox winning the World Series are 2 to 5."

In this chapter we will learn how to interpret and work with probabilities, chances, and odds in a formal mathematical context. This will be our very brief introduction to the mathematical theory of probability, a relatively young branch of mathematics that has become critically important to many aspects of modern life. Insurance, public health, science, sports, gambling, the stock market—wherever there is uncertainty to be tamed—the mathematical theory of probability plays a significant role.

Our discussion in this chapter is divided into two parts. In the first part we introduce the basic theoretical framework needed for a meaningful discussion of probabilities: the concepts of *random experiment* and *sample space* (Section 15.1), the basic rules of *counting* (Section 15.2), and the dual concepts of *permutation* and *combination* (Section 15.3). In the second part of the chapter we discuss general *probability spaces* (Section 15.4) and probabilities in spaces where all outcomes are equally likely (Section 15.5). The chapter concludes with a brief discussion of *odds* and their relationship to probabilities (Section 15.6).

15.1 Random Experiments and Sample Spaces

In broad terms, probability is the *quantification of uncertainty*. To understand what that means, we may start by formalizing the notion of uncertainty.

We will use the term **random experiment** to describe an activity or process *whose outcome cannot be predicted ahead of time*. Typical examples of random experiments are tossing a coin, rolling a pair of dice, drawing cards out of a deck of cards, predicting the result of a game (football, basketball, etc.), or forecasting the path of a hurricane.

Associated with every random experiment is the *set* of all of its possible outcomes, called the **sample space** of the experiment. For the sake of simplicity, we will concentrate on experiments for which there is only a finite set of outcomes, although experiments with infinitely many outcomes are both possible and important.

We illustrate the importance of the sample space by means of several examples. Since the sample space of any experiment is a set of outcomes, we will use set notation to describe it. We will consistently use the letter S to denote a sample space, and N to denote its size (i.e., the number of outcomes in S).

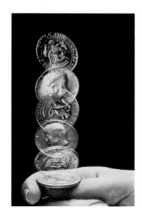

> ## EXAMPLE 15.1 Coin Tossing: Part 1

One simple random experiment is to *toss a quarter* and *observe whether it lands heads or tails*. The sample space can be described by $S = \{H, T\}$, where H stands for *Heads* and T for *Tails*. Here $N = 2$. ◀◀

A couple of comments about coins are in order here. First, the fact that the coin in Example 15.1 was a quarter is essentially irrelevant. Practically all coins have an obvious "heads" side (and thus a "tails" side), and even when they don't—as in a "buffalo nickel"—we can agree ahead of time which side is which. Second, there are fake coins out there on which both sides are "heads." Tossing such a coin does not fit our definition of a random experiment, so from now on, we will assume that all coins used in our experiments have two different sides, which we will call H and T.

> ## EXAMPLE 15.2 Coin Tossing: Part 2

Suppose we toss a coin *twice* and *record* the outcome of each toss (H or T) in the order it happens. The sample space now is $S = \{HH, HT, TH, TT\}$, where HT means that the first toss came up H and the second toss came up T, which is a different outcome from TH (first toss T and second toss H). In this sample space $N = 4$.

Suppose now we *toss two distinguishable coins* (say, a nickel and a quarter) *at the same time* (tricky but definitely possible). This random experiment appears different from the one where we toss one coin twice, but the sample space is still $S = \{HH, HT, TH, TT\}$. (Here we must agree what the order of the symbols is—for example, the first symbol describes the quarter and the second the nickel.)

Since they have the same sample space, we will consider the two random experiments just described as the same random experiment.

Now let's consider a different random experiment. We are still tossing a coin twice, but we only care now about the *number of heads* that come up. Here there are only three possible outcomes (no heads, one head, or both heads) and symbolically we might describe this sample space as $S = \{0, 1, 2\}$. ◀◀

The important point made in Example 15.2 is that a random experiment is defined by two things: the action (such as tossing coins) and the observation that follows that action (what it is that we are really looking for).

> ### EXAMPLE 15.3 Shooting Free Throws: Part 1

Here is a familiar scenario: Your team is down by 1, clock running out, and one of your players is fouled and goes to the line to shoot free throws, with the game riding on the outcome. Not a good time to think of sample spaces, but let's do it anyway.

Clearly this is a random experiment, but what is the sample space? As in Example 15.2, the answer depends on a few subtleties.

In one scenario (the *penalty situation*) your player is going to shoot two free throws no matter what. In this case one could argue that what really matters is how many free throws he or she makes (make both and win the game, miss one and tie and go to overtime, miss both and lose the game). When we look at it this way the sample space is $S = \{0, 1, 2\}$.

A somewhat more stressful scenario is when your player is shooting a *one-and-one*. This means that the player gets to shoot the second free throw only if he or she makes the first one. In this case there are also three possible outcomes, but the circumstances are different because the order of events is relevant (miss the first free throw and lose the game, make the first free throw but miss the second one and tie the game, make both and win the game). We can describe this sample space as $S = \{f, sf, ss\}$, where we use f to indicate *failure* (missed the free throw) and s to indicate *success*. ◀◀

We will now discuss a few examples of random experiments involving dice. A die is a cube, usually made of plastic, whose six faces are marked with dots (from 1 to 6) called "pips." Random experiments using dice have a long-standing tradition in our culture and are a part of both gambling and recreational games (such as Monopoly, Yahtzee, etc.).

> ### EXAMPLE 15.4 Rolling the Dice: Part 1

One of the most common things we do with dice is to *roll a pair of dice* (at the same time) and consider just the *total* of the two die. In this situation we don't really care how a particular total comes about—two "fours" are the same as a "five" and a "three." Here there are 11 possible outcomes: from a low of 2 (when we roll two "aces") to a maximum of 12 (when we roll two "sixes"), and the sample space can be described by $S = \{2, 3, 4, 5, 6, 7, 8, 9, 10, 11, 12\}$.

A more general scenario is when we do care what number each individual die turns up (in certain bets in craps, for example, two "fours" is a winner but a "five" and a "three" is not). Here we have a sample space with 36 different outcomes, shown in Fig. 15-1. When looking at Fig. 15-1 you will notice that we are treating the dice as *distinguishable* objects (as if one were white and the other red), so that ⚀⚅ and ⚄⚅ are considered different outcomes.

FIGURE 15-1

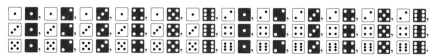

> **EXAMPLE 15.5** Ranking the Candidates in an Election: Part 1

Five candidates (*A*, *B*, *C*, *D*, and *E*) are running in an election. The top three finishers are chosen President, Vice President, and Secretary, in that order. The election itself can be considered a random experiment with sample space $S = \{ABC, ACB, BAC, BCA, CAB, CBA, ABD, ADB, \ldots, CDE\}$, where the outcome *ABC* signifies that candidate *A* is elected President, *B* is elected Vice President, and *C* is elected Secretary. The "..." in the description of the sample space is another way of saying "and so on—you get the point." ◀◀

Example 15.5 illustrates the point that sample spaces can have a lot of outcomes and that we are often justified in not writing each and every one of them down. This is where the "..." comes in handy. The key thing is to understand what the sample space looks like without necessarily writing all the outcomes down. Our real goal is to find *N*, the size of the sample space. If we can do it without having to list all the outcomes then so much the better. We will discuss how this is done in the next two sections.

15.2 Counting Sample Spaces

> **EXAMPLE 15.6** Coin Tossing: Part 3

If we *toss a coin three times* and separately record the outcome of each toss, the sample space is given by $S = \{HHH, HHT, HTH, HTT, THH, THT, TTH, TTT\}$. Here we can just count the outcomes and get $N = 8$.

Now let's look at an expanded version of the same idea and *toss a coin 10 times* (and record the outcome of each toss). In this case the sample space *S* is too big to write down but we can "count" the number of outcomes in *S* without having to tally them one by one. Here is how we do it:

Each outcome in *S* can be described by a string of 10 consecutive letters, where each letter is either *H* or *T*. (This is a natural extension of the ideas introduced in Example 15.2.) For example, *THHTHTHHTT* represents a *single* outcome in our sample space—the one in which the first toss came up *T*, the second toss came up *H*, the third toss came up *H*, and so on. To count all the outcomes, we will argue as follows: There are two choices for the first letter (*H* or *T*), two choices for the second letter, two choices for the third letter, and so on down to the tenth letter. The total number of possible strings is found by *multiplying* the number of choices for each letter. Thus $N = 2^{10} = 1024$. ◀◀

The rule we used in Example 15.6 is an old friend from Chapter 2—the **multiplication rule**, and for a formal definition of the multiplication rule the reader is referred to Section 2.4. Informally, the multiplication rule simply says that *when something is done in stages, the number of ways it can be done is found by multiplying the number of ways each of the stages can be done*. The easiest way to understand this simple but powerful idea is by looking at some examples.

EXAMPLE 15.7 The Making of a Wardrobe: Part 1

Dolores is a young saleswoman planning her next business trip. She is thinking about packing three different pairs of shoes, four skirts, six blouses, and two jackets. If all the items are color coordinated, how many different *outfits* will she be able to create by combining these items?

To answer this question, we must first define what we mean by an "outfit." Let's assume that an outfit consists of one pair of shoes, one skirt, one blouse, and one jacket. Then to make an outfit Dolores must choose a pair of shoes (three choices), a skirt (four choices), a blouse (six choices), and a jacket (two choices). By the multiplication rule the total number of possible outfits is $3 \times 4 \times 6 \times 2 = 144$.

Think about it—Dolores can be on the road for over four months and never have to wear the same outfit twice! And it all fits in a small suitcase.

EXAMPLE 15.8 The Making of a Wardrobe: Part 2

Once again, Dolores is packing for a business trip. This time, she packs three pairs of shoes, four skirts, three pairs of slacks, six blouses, three turtlenecks, and two jackets. As before, we can assume that she coordinates the colors so that everything goes with everything else. This time, we will define an outfit as consisting of a pair of shoes, a choice of "lower wear" (either a skirt *or* a pair of slacks), a choice of "upper wear" (it could be a blouse *or* a turtleneck *or both*), and, finally, she may or may not choose to wear a jacket. How many different such outfits are possible?

This is a more sophisticated variation of Example 15.7. Our strategy will be to think of an outfit as being put together in stages and to draw a box for each of the stages. We then separately count the number of choices at each stage and enter that number in the corresponding box. (Some of these calculations can themselves be mini-counting problems.) The last step is to multiply the numbers in each box. The details are illustrated in Fig. 15-2. The final count for the number of different outfits is $N = 3 \times 7 \times 27 \times 3 = 1701$.

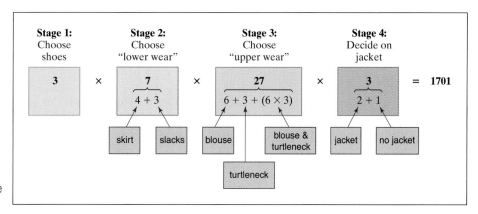

FIGURE 15-2 Counting all possible outfits using a box model.

The method of drawing boxes representing the successive stages in a process and putting the number of choices for each stage inside the box is a convenient device that often helps clarify one's thinking. Silly as it may seem, we strongly recommend it. For ease of reference, we will call it the *box model* for counting.

> **EXAMPLE 15.9** Ranking the Candidates in an Election: Part 2

This example is a follow-up to Example 15.5. Five candidates are running in an election, with the top three getting elected (in order) as President, Vice President, and Secretary. We want to know how big the sample space is. Using a box model, we see that this becomes a reasonably easy counting problem, as illustrated in Fig. 15-3.

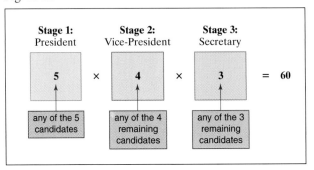

FIGURE 15-3

15.3 Permutations and Combinations

Many counting problems can be reduced to a question of counting the number of ways in which we can choose groups of objects from a larger group of objects. Often these problems require somewhat more sophisticated counting methods than the plain vanilla multiplication rule. In this section we will discuss the dual concepts of *permutation* (a group of objects where the ordering of the objects within the group *makes a difference*) and *combination* (a group of objects in which the ordering of the objects is *irrelevant*).

> **EXAMPLE 15.10** The Pleasures of Ice Cream: Part 1

Baskin-Robbins offers 31 different flavors of ice cream. A "true double" is the name we will use for two scoops of ice cream of two *different* flavors. Say you want a true double in a bowl—how many different choices do you have?

The natural impulse is to count the number of choices using the multiplication rule (and a box model) as shown in Fig. 15-4. This would give an answer of $31 \times 30 = 930$. Unfortunately, this answer is *double counting* each of the true doubles. Why?

When we use the multiplication rule, *there is a well-defined order to things*, and a scoop of strawberry followed by a scoop of chocolate is count-

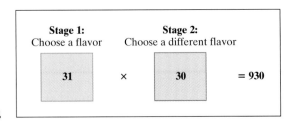

FIGURE 15-4

ed separately from a scoop of chocolate followed by a scoop of strawberry. But by all reasonable standards, a bowl of chocolate-strawberry is the same as a bowl of strawberry-chocolate. (This is not necessarily true when the ice cream is served in a cone. Fussy people can be very picky about the order of the flavors when the scoops are stacked up.)

The good news is that now that we understand why the count of 930 is wrong we can fix it. All we have to do to get the correct answer is to divide the original count by 2. It follows that the number of true double choices at Baskin-Robbins is $(31 \times 30)/2 = 465$.   ◀◀

Example 15.10 is an important one. It warns us that we have to be careful about how we use the multiplication rule and box models in counting problems where the order in which we choose the objects (ice cream flavors) does not affect the answer. Let's take this idea to the next level.

▶ EXAMPLE 15.11 The Pleasures of Ice Cream: Part 2

Some days you have a real craving for ice cream, and on such days you like to go to Baskin-Robbins and order a *true triple* (in a bowl). How many different choices do you have? (As you might have guessed, a "true triple" consists of three scoops of ice cream each of a different flavor.)

Starting with the multiplication rule, we have 31 choices for the "first" flavor, 30 choices for the "second" flavor, and 29 choices for the "third" flavor, for an apparent grand total of $31 \times 30 \times 29 = 26,970$. But once again this answer counts each true triple more than once, in fact, it does so *six times*! (More on that shortly.) If we accept this, the correct answer must be $26,970/6 = 4495$.

Why is it that the count of 26,970 counts each true triple *six* times? The answer comes down to this: Any three flavors (call them *X*, *Y*, and *Z*) can be listed in $3 \times 2 \times 1 = 6$ different ways (*XYZ, XZY, YXZ, YZX, ZXY*, and *ZYX*). The multiplication rule counts each of these separately, but when you think in terms of ice cream scoops in a bowl, they are the same true triple regardless of the order.

The bottom line is that there are 4495 different possibilities for a true triple at Baskin-Robbins. We can better understand where this number comes from by looking at it in its raw, uncalculated form $[(31 \times 30 \times 29)/(3 \times 2 \times 1)]$. The numerator $(31 \times 30 \times 29)$ comes from counting *ordered* triples using the multiplication rule; the denominator $(3 \times 2 \times 1)$ comes from counting the number of ways in which three things (in this case, the three flavors in a triple) can be reordered. The denominator $3 \times 2 \times 1$ is already familiar to us—it is the *factorial* of 3. (We discussed the factorial in Chapters 2 and 6, so we won't dwell on it here.)   ◀◀

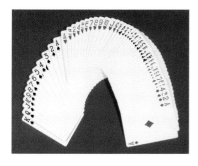

Our next example will deal with the game of poker. Despite the great deal of exposure poker gets on television these days, there are a lot of misconceptions about the mathematics behind poker hands. For readers not familiar with the game, poker is a betting game played with a standard deck of cards [a standard deck of cards has 52 cards divided into 4 *suits* (Clubs, Diamonds, Hearts, and Spades) and with 13 *values* in each suit $(2, 3, \ldots, 10, J, Q, K, A)$]. There are many variations of poker depending on how many cards are dealt and whether all the cards are *down* cards (only the player receiving the card can see it) or some of the cards are *up* cards (dealt face up so that they can be seen by all the players).

> **EXAMPLE 15.12** Five-Card Poker Hands: Part 1

In this example we will compare two types of games: five-card *stud poker* and five-card *draw poker*. In both of these games a player ends up with five cards, but there is an important difference when analyzing the mathematics behind the games: In five-card *draw* the order in which the cards come up is irrelevant; in five-card *stud* the order in which the cards come up is extremely relevant. The reason for this is that in five-card draw all cards are dealt down, but in five-card stud only the first card is dealt down—the remaining four cards are dealt up, one at a time. This means that players can assess the relative strengths of the other players' hands as the game progresses and play their hands accordingly.

Left: Five-card *draw* poker hand: The order in which the cards are dealt is irrelevant. *Center and Right:* Five-card *stud* poker hands. The order of the up cards is important. The hand on the right is better than the hand in the center.

Counting the number of five-card *stud* poker hands is a direct application of the multiplication rule: 52 possibilities for the first card, 51 for the second card, 50 for the third card, 49 for the fourth card, and 48 for the fifth card, for an awesome total of $52 \times 51 \times 50 \times 49 \times 48 = 311{,}875{,}200$ possible hands.

Counting the number of five-card *draw* poker hands requires a little more finesse. Here a player gets five down cards and the hand is the same regardless of the order in which the cards are dealt. There are $5! = 120$ different ways in which the same set of five cards can be ordered, so that one draw hand corresponds to 120 different stud hands. Thus, the stud hands count is exactly 120 times bigger than the draw hands count. Great! All we have to do then is divide the 311,875,200 (number of stud hands) by 120 and get our answer: There are 2,598,960 possible five-card draw hands. [As before, it's more telling to look at this answer in the uncalculated form $(52 \times 51 \times 50 \times 49 \times 48)/5!$] ◄◄

We are now ready to generalize the ideas developed in Examples 15.11 and 15.12. Suppose we have a set of *n* distinct objects and that we want to select *r* different objects from this set. The number of ways that this can be done depends on whether the selections are ordered or unordered. Ordered selections are the generalization of stud poker hands—change the order of selection of the same objects and you get something different. Unordered selections are the generalization of draw poker hands—change the order of selection of the same objects and you get nothing new. To distinguish between these two scenarios, we use the term **permutation** to describe an ordered selection and **combination** to describe an unordered selection. (One way to remember which is which is to remember that there are many more permutations than there are combinations.)

For a given number of objects *n* and a given selection size *r* (where $0 \leq r \leq n$), we can talk about the "number of permutations of *n* objects taken *r* at a time" and the "number of combinations of *n* objects taken *r* at a time," and these two

You should take a look at your calculator and figure out the correct sequence of keystrokes to compute these numbers, and then try Exercises 29 and 30.

extremely important families of numbers are denoted $_nP_r$ and $_nC_r$, respectively. (Some calculators use variations of this notation, such as $P_{n,r}$ and $C_{n,r}$, respectively.)

A summary of the essential facts about the numbers $_nP_r$ and $_nC_r$ is given in Table 15-1.

TABLE 15-1		
Notation	$_nP_r$	$_nC_r$
Formula	$n \cdot (n-1) \cdot \cdots \cdot (n-r+1)$	$n \cdot (n-1) \cdot \cdots \cdot (n-r+1)/r!$
Formula (using factorials)	$\dfrac{n!}{(n-r)!}$	$\dfrac{n!}{(n-r)!r!}$
Applications	Stud poker hands; rankings; committees with assignments.	Draw poker hands; lottery tickets; coalitions; subsets

> **EXAMPLE 15.13** The Florida Lotto: Part 1

Like many other state lotteries, the Florida Lotto is a game in which for a small investment of just one dollar a player has a chance of winning tens of millions of dollars. Enormous upside, hardly any downside—that's why people love playing the lottery and, like they say, "everybody has to have a dream." But, in general, lotteries are a very bad investment, even if it's only a dollar, and the dreams can turn to nightmares. Why so?

In a Florida Lotto ticket, one gets to select *six* numbers from 1 through 53. To win the jackpot (there are other lesser prizes we won't discuss here), those six numbers have to match the winning numbers drawn by the lottery in any order. Since a lottery draw is just an unordered selection of six objects (the winning numbers) out of 53 objects (the numbers 1 through 53), the number of possible draws is $_{53}C_6$ (see Table 15-1). Doesn't sound too bad until we do the computation (or use a calculator) and realize that $_{53}C_6 = 22,957,480$. ⟪

15.4 Probability Spaces

What Is a Probability?

If we toss a coin in the air, *what is the probability that it will land heads up*? This one is not a very profound question, and almost everybody agrees on the answer, although not necessarily for the same reason. The standard answer given is 1/2 (or the equivalent 0.5, or 50%, or 1 out of 2). But why is the answer 1/2, and what does such an answer mean?

One common explanation given for the answer of 1/2 is that when we toss a coin, there are two possible outcomes (H and T), and since H represents one of the two possibilities, the probability of the outcome H must be 1 out of 2 or 1/2. This logic, while correct in the case of an honest coin, has a lot of holes in it.

Consider how the same argument would sound in a different scenario.

> ### EXAMPLE 15.14 Shooting Free Throws: Part 2

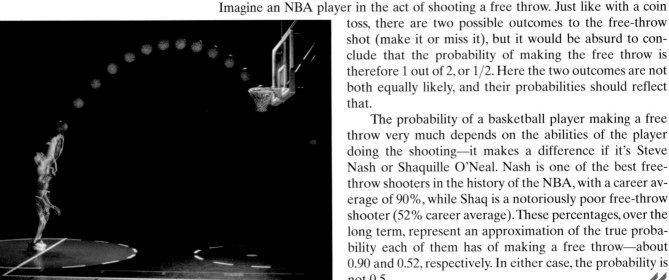

Imagine an NBA player in the act of shooting a free throw. Just like with a coin toss, there are two possible outcomes to the free-throw shot (make it or miss it), but it would be absurd to conclude that the probability of making the free throw is therefore 1 out of 2, or 1/2. Here the two outcomes are not both equally likely, and their probabilities should reflect that.

The probability of a basketball player making a free throw very much depends on the abilities of the player doing the shooting—it makes a difference if it's Steve Nash or Shaquille O'Neal. Nash is one of the best free-throw shooters in the history of the NBA, with a career average of 90%, while Shaq is a notoriously poor free-throw shooter (52% career average). These percentages, over the long term, represent an approximation of the true probability each of them has of making a free throw—about 0.90 and 0.52, respectively. In either case, the probability is not 0.5.

Example 15.14 leads us to what is known as the *empirical* interpretation of the concept of probability. Under this interpretation when tossing an honest coin the probability of *Heads* is 1/2 not because *Heads* is one out of two possible outcomes but because, if we were to toss the coin over and over—hundreds, possibly thousands of times—in the long run about half of the tosses will turn out to be heads, a fact that has been confirmed by experiment many times.

The argument as to exactly how to interpret the statement "the probability of *X* is such and such" goes back to the late 1600s, and it wasn't until the 1930s that a formal theory for dealing with probabilities was developed by the Russian mathematician A. N. Kolmogorov (1903–1987). This theory has made probability one of the most useful and important concepts in modern mathematics. In the remainder of this chapter we will discuss some of the basic concepts of probability theory.

Events

An **event** is any subset of the sample space. That is, an event is any set of individual outcomes. (This definition includes the possibility of an "event" that has no outcomes as well as events consisting of a single outcome.) By definition, events are sets (subsets of the sample space), and we will deal with events using set notation as well as the basic rules of set theory.

A convenient way to think of an event is as a package in which outcomes with some common characteristic are bundled together. Say you are rolling a pair of dice and hoping that the next roll is a 7—that would make you temporarily rich. The best way to describe what really matters (to you) is by packaging together all the different ways to "roll a 7" as a single event *E*:

$$E = \{ \boxdot \; \boxplus, \; \boxdot \; \boxplus, \; \boxdot \; \boxplus, \; \boxplus \; \boxdot, \; \boxplus \; \boxplus, \; \boxplus \; \boxdot \}$$

Sometimes an event consists of just one outcome. We will call such an event a **simple event**. In some sense simple events are the building blocks for all other events (more on that later). There is also the special case of the empty set { }, corresponding to an event with no outcomes. Such an event can never happen, and thus, we call it the **impossible event**.

EXAMPLE 15.15 Coin Tossing: Part 4

In Example 15.6 we considered the random experiment of tossing a coin three times. The sample space for this random experiment is $S = \{HHH, HHT, HTH, HTT, THH, THT, TTH, TTT\}$, and this set has $2^8 = 256$ subsets, each of which represents a different event. Table 15-2 shows just a few of these events.

TABLE 15-2

Event (in words)	Event (as a set)
E_1: Toss 2 or more heads.	$E_1 = \{HHT, HTH, THH, HHH\}$
E_2: Toss more than 2 heads.	$E_2 = \{HHH\}$
E_3: Toss 2 heads or less.	$E_3 = \{TTT, TTH, THT, HTT, THH, HTH, HHT\}$
E_4: Toss no tails.	$E_4 = \{HHH\}$
E_5: Toss exactly 1 tail.	$E_5 = \{HHT, HTH, THH\}$
E_6: Toss exactly 1 head.	$E_6 = \{HTT, THT, TTH\}$
E_7: First toss is heads.	$E_7 = \{HHH, HHT, HTH, HTT\}$
E_8: Toss same number of heads as tails.	$E_8 = \{ \}$ [*Note:* This is the **impossible** event.]
E_9: Toss 3 heads or less.	$E_9 = S$ [*Note:* This event is called the **certain** event.]

Probability Assignments

Let's return to free throw shooting, as it is a useful metaphor for many probability questions.

EXAMPLE 15.16 Shooting Free Throws: Part 3

A player is going to shoot a free throw. We know nothing about his or her abilities—for all we know, the player could be Steve Nash, or Shaquille O'Neal, or Joe Schmoe.

How can we describe the probability that he or she will make the free throw? It seems that there is no way to answer this question, since we know nothing about the ability of the shooter. We could argue that the probability could be just about any number between 0 and 1. No problem—we make our unknown probability a variable, say p.

What can we say about the probability that our shooter misses the free throw? A lot. Since there are only two possible outcomes in the sample space $S = \{s, f\}$, the probability of success (s) and the probability of failure (f) must complement each other—in other words, must add up to 1. This means that the probability of missing the free throw must be $1 - p$.

Table 15-3 is a summary of the line on a generic free-throw shooter. Humble as it may seem, Table 15-3 gives a complete model of free throw shooting. It works when the free-throw shooter is Steve Nash (make it $p = 0.90$), Shaquille

TABLE 15-3

Event	Probability
$\{ \}$	0
$\{s\}$	p
$\{f\}$	$1 - p$
$\{s, f\}$	1

O' Neal (make it $p = 0.52$), or the author of this book (make it $p = 0.30$). Each one of the choices results in a different assignment of numbers to the outcomes in the sample space.   ◀◀

Example 15.16 illustrates the concept of a *probability assignment*. A **probability assignment** is a function that assigns to each event E a number between 0 and 1, which represents the probability of the event E and which we denote by $\Pr(E)$. A probability assignment always assigns probability 0 to the *impossible event* $[\Pr(\{\ \}) = 0]$, and probability 1 to the whole sample space $[\Pr(S) = 1]$. [In the case of a simple event $\{a\}$ we cheat a little and for the sake of simplicity we speak of "the probability of the outcome a" when we really should say "the probability of the event $\{a\}$." Accordingly, we will write $\Pr(a)$ in lieu of the technically correct but awkward $\Pr(\{a\})$.]

With finite sample spaces a probability assignment is defined by assigning probabilities to just the simple events in the sample space. Once we do this, we can find the probability of any event by simply *adding the probabilities of the individual outcomes that make up that event.* There are only two requirements for a valid probability assignment: (i) *All probabilities are numbers between 0 and 1*, and (ii) *the sum of the probabilities of the simple events equals 1.*

When bookies or professional odds-makers handicap a sporting event, they do so by essentially defining a probability assignment for the sample space of all possible outcomes of that event. The next example is a simple illustration of how this might be done.

▶ EXAMPLE 15.17 Handicapping a Tennis Tournament: Part 1

There are six players playing in a tennis tournament: *Ana* (Russian, female), *Ivan* (Croatian, male), *Lleyton* (Aussie, male), *Roger* (Swiss, male), *Serena* (American, female), and *Venus* (American, female).

To handicap the winner of the tournament we need a probability assignment on the sample space $S = \{Ana, Ivan, Lleyton, Roger, Serena, Venus\}$. With sporting events the probability assignment is subjective (it reflects an opinion), but professional odds-makers are usually very good at getting close to the right probabilities. For example, imagine that a professional odds-maker comes up with the following probability assignment: $\Pr(Ana) = 0.08$, $\Pr(Ivan) = 0.16$, $\Pr(Lleyton) = 0.20$, $\Pr(Roger) = 0.25$, $\Pr(Venus) = 0.16$ [We are missing $\Pr(Serena)$ from the list, but since the probabilities of the simple events must add up to 1, we can do the arithmetic: $\Pr(Serena) = 0.15$.]

Once we have the probabilities of the simple events, the probabilities of all other events follow by addition. For example, the probability that an American will win the tournament is given by $\Pr(Serena) + \Pr(Venus) = 0.15 + 0.16 = 0.31$. Likewise, the probability that a male will win the tournament is given by $\Pr(Ivan) + \Pr(Lleyton) + \Pr(Roger) = 0.16 + 0.20 + 0.25 = 0.61$. The probability that an American male will win the tournament is $\Pr(\{\ \}) = 0$, since this one is an impossible event—there are no American males in the tournament!   ◀◀

The probability assignment discussed in Example 15.17 reflects the opinion of one specific observer. A different odds-maker might have a slightly different perspective and come up with a different probability assignment. This underscores the fact that, sometimes, there is no one single "correct" probability assignment on a sample space.

Once a specific probability assignment is made on a sample space, the combination of the sample space and the probability assignment is called a **probability space**. (In this book the term *probability space* will always refer to a *finite* probability space. Infinite probability spaces are important and interesting but require a much higher level of mathematics and we will not discuss them here.) The following is a summary of the key facts related to probability spaces.

Elements of a Probability Space

- **Sample space:** $S = \{o_1, o_2, \ldots, o_N\}$.

- **Probability assignment:** $\Pr(o_1), \Pr(o_2), \ldots, \Pr(o_N)$.
  [Each of these is a number between 0 and 1 satisfying $\Pr(o_1) + \Pr(o_2) + \cdots + \Pr(o_N) = 1$.]

- **Events:** These are all the subsets of S, including $\{\ \}$ and S itself. The probability of an event is given by the sum of the probabilities of the individual outcomes that make up the event. [In particular, $\Pr(\{\ \}) = 0$ and $\Pr(S) = 1$.]

15.5 Equiprobable Spaces

One of the most common uses of randomness in the real world is as a mechanism to guarantee fairness. That's why it is universally accepted that tossing a coin, rolling a die, or drawing cards is a fair way of choosing among equally deserving choices. This is true as long as the coin, die, or deck of cards are all "honest."

What does *honesty* mean when applied to coins, dice, or decks of cards? It essentially means that all individual outcomes in the sample space are equally probable—*Heads* has the same probability as *Tails*, rolling a 6 has the same probability as rolling a 1 (or a 2, or a 3, …), and so on. A probability space where each simple event has an equal probability is called an **equiprobable** space. (Informally, you can think of an equiprobable space as an "equal opportunity" probability space.)

In equiprobable spaces, calculating probabilities of events becomes simply a matter of counting. First, we need to find N, the size of the sample space. Each individual outcome in the sample space will have probability equal to $1/N$ (they all have the same probability and the sum of the probabilities must equal 1). To find the probability of the event E we then find k, the number of outcomes in E. Since each of these outcomes has probability $1/N$, the probability of E is then k/N.

Probabilities in Equiprobable Spaces

$\Pr(E) = k/N$ (where k denotes the size of the event E and N denotes the size of the sample space S).

▶ **EXAMPLE 15.18** Coin Tossing: Part 5

This example is a follow-up of Example 15.15. Suppose that a coin is tossed three times, and we have been assured that the coin is an honest coin. If this is true, then each of the eight possible outcomes in the sample space has probability 1/8.

The probabilities of each of the events E_1 through E_9 discussed in Example 15.15 (the reader is referred to Table 15-2) can be found by just tallying the outcomes in each: $\Pr(E_1) = 4/8 = 1/2$, $\Pr(E_2) = 1/8$, $\Pr(E_3) = 7/8$, and so on. Note that $\Pr(E_8) = 0$, and $\Pr(E_9) = 1$. ◀◀

▶ **EXAMPLE 15.19** Rolling the Dice: Part 2

This example is a follow-up of Example 15.4. Suppose that we roll a pair of honest dice. The sample space has $N = 36$ individual outcomes, each with probability 1/36.

See Exercise 49.

We will use the notation $T_2, T_3, \ldots, T_{12}$ to describe the events "roll a total of 2," "roll a total of 3," $\ldots$, "roll a total of 12," respectively. We will show how to find $\Pr(T_7)$ and $\Pr(T_{11})$ here, and leave the others as an exercise.

$$T_{11} = \{ \boxdot\ \boxdot, \boxdot\ \boxdot \}. \text{ Thus,}$$

$$\Pr(T_{11}) = 2/36 \approx 0.056.$$

$$T_7 = \{ \boxdot\ \boxdot, \boxdot\ \boxdot, \boxdot\ \boxdot, \boxdot\ \boxdot, \boxdot\ \boxdot, \boxdot\ \boxdot \}. \text{ Thus,}$$

$$\Pr(T_7) = 6/36 = 1/6 \approx 0.167.$$ ◀◀

▶ **EXAMPLE 15.20** Rolling the Dice: Part 3

Once again, we are rolling a pair of honest dice. We are now going to find the probability of the event E: "at least one of the die comes up an Ace." (In dice jargon, a $\boxdot$ is called an Ace.)

This is a slightly more sophisticated counting question than the ones in Example 15.19. We will show three different ways to answer the question.

- **Tallying.** We can just write down all the individual outcomes in the event E and tally their number. This approach gives

$$E = \{ \boxdot\ \boxdot, \boxdot\ \boxdot, \boxdot\ \boxdot, \boxdot\ \boxdot, \boxdot\ \boxdot, \boxdot\ \boxdot, \boxdot\ \boxdot, \boxdot\ \boxdot, \boxdot\ \boxdot, \boxdot\ \boxdot, \boxdot\ \boxdot \}$$

  and $\Pr(E) = 11/36$. There is not much to it, but in a larger sample space it will take a lot of time and effort to list every individual outcome in the event.

- **Complementary Event.** Behind this approach is the germ of a really important idea. Imagine that you are playing a game, and you win if at least one of the two numbers comes up an Ace (that's event E). Otherwise you lose (call that event F). The two events E and F are called **complementary** events, and E is called the **complement** of F (and vice versa). The key idea is that *the probabilities of complementary events add up to* 1. Thus, when you want $\Pr(E)$ but it's actually easier to find $\Pr(F)$, then you first do that. Once you have $\Pr(F)$, you are in business: $\Pr(E) = 1 - \Pr(F)$.

Here we can find $\Pr(F)$ by a direct application of the multiplication principle. There are 5 possibilities for the first die (it can't be an Ace but it can be anything else) and likewise there are five possibilities for the second die. This means that there are 25 different outcomes in the event F, and thus $\Pr(F) = 25/36$. It follows that $\Pr(E) = 1 - (25/36) = 11/36$.

■ **Independent Events.** In this approach, we look at each die separately and call them die #1 and die #2. We will let F_1 denote the event "die #1 does *not* come up an Ace," and F_2 denote the event "die #2 does *not* come up an Ace." Clearly, $\Pr(F_1) = 5/6$ and $\Pr(F_2) = 5/6$. Now comes a critical observation: The probability that both events F_1 and F_2 happen can be found by *multiplying* their respective probabilities. This means that $\Pr(F) = (5/6) \times (5/6) = 25/36$. We can now find $\Pr(E)$ exactly as before: $\Pr(E) = 1 - \Pr(F) = 11/36$. ◀◀

Of the three approaches used in Example 15.20, the last approach appears to be the most convoluted, but, in fact, it is the one with the most promise. It is based on the concept of *independence* of events. Two events are said to be **independent events** if the occurrence of one event does not affect the probability of the occurrence of the other. When events E and F are independent, the probability that both occur is the product of their respective probabilities; in other words, $\Pr(E \text{ and } F) = \Pr(E) \cdot \Pr(F)$. This is called **the multiplication principle for independent events**.

The multiplication principle for independent events in an important and useful rule, but be forewarned—it works *only* with independent events! For events that are *not* independent, multiplying their respective probabilities gives us a bunch of nonsense.

Our next example illustrates the real power of the multiplication principle for independent events. As we just mentioned, this principle can only be applied when the events in question are independent, and in many circumstances this is not the case. Fortunately, in the examples we are going to consider the independence of the events in question is intuitively obvious. For example, if we roll an honest die several times, what happens on each roll has no impact whatsoever on what is going to happen on subsequent rolls (i.e., the rolls are independent events).

▶ EXAMPLE 15.21 Rolling the Dice: Part 4

Imagine a game where you roll an honest die four times. If at least one of your rolls comes up an Ace, you are a winner. Let E denote the event "you win," and F denote the event "you lose." We will find $\Pr(E)$ by first finding $\Pr(F)$, using the same ideas we developed in Example 15.20.

We will let F_1, F_2, F_3, and F_4 denote the events "first roll is not an Ace," "second roll is not an Ace," "third roll is not an Ace," and "fourth roll is not an Ace," respectively. Then

$$\Pr(F_1) = 5/6, \Pr(F_2) = 5/6, \Pr(F_3) = 5/6, \text{ and } \Pr(F_4) = 5/6.$$

Now we use the multiplication principle for independent events:

$$\Pr(F) = (5/6) \times (5/6) \times (5/6) \times (5/6) = (5/6)^4 \approx 0.482.$$

Finally, we find $\Pr(E)$: $\Pr(E) = 1 - \Pr(F) \approx 0.518$.

EXAMPLE 15.22 Five-Card Poker Hands: Part 2

One of the best hands in five-card draw poker is a *four-of-a-kind* (four cards of the same value plus a fifth card called a *kicker*). Among all four-of-a-kind hands, the very best is the one consisting of four Aces plus a kicker. We will let *F* denote the event of drawing a hand with four Aces plus a kicker in five-card draw poker. (The *F* stands for "fabulously lucky.") Our goal in this example is to find $\Pr(F)$.

The size of our sample space is $N = 2{,}598{,}960$ (see Example 15.12). Of these roughly 2.6 million possible hands, there are only 48 hands in *F*: four of the five cards are the Aces; the kicker can be any one of the other 48 cards in the deck. Thus, $\Pr(F) = 48/2{,}598{,}960 \approx 0.0000185$, roughly 1 in 50,000. ◄◄

EXAMPLE 15.23 Coin Tossing: Part 6

Imagine that a friend offers you the following bet: Toss an honest coin 10 times. If the tosses come out in an even split (5 *Heads* and 5 *Tails*), you win and your friend buys the pizza; otherwise you lose (and you buy the pizza). Sounds like a reasonable offer—let's check it out.

We are going to let *E* denote the event "5 *Heads* and 5 *Tails* are tossed," and we will compute $\Pr(E)$. We already found (Example 15.6) that there are $N = 1024$ equally likely outcomes when a coin is tossed 10 times and that each of these outcomes can be described by a string of 10 *H*s and *T*s.

We now need to figure out how many of the 1024 strings of *H*s and *T*s have 5 *H*s and 5 *T*s. There are of course, quite a few: *HHHHHTTTTT, HTHTHTHTHT, TTHHHTTTHH,* and so on. Trying to make a list of all of them is not a very practical idea. The trick is to think about these strings as 10 slots each taken up by an *H* or a *T*, but once you determine which slots have *H*s you automatically know which slots have *T*s. Thus, each string of 5 *H*s and 5 *T*s is determined by the choice of the 5 slots for the *H*s. These are unordered choices of 5 out of 10 slots, and the number of such choices is $_{10}C_5 = 252$.

Now we are in business: $\Pr(E) = 252/1024 \approx 0.246$.

The moral of this example is that your friend is a shark. When you toss an honest coin 10 times, about 25% of the time you are going to end up with an even split, the other 75% of the time the split is uneven. ◄◄

You may want to think about this probability and take a guess before you read on—you might be surprised when you see the answer.

15.6 Odds

Dealing with probabilities as numbers that are always between 0 and 1 is the mathematician's way of having a consistent terminology. To the everyday user, consistency is not that much of a concern, and we know that people often express the likelihood of an event in terms of *odds*. Probabilities and odds are not identical concepts but are closely related.

EXAMPLE 15.24 Shooting Free Throws: Part 4

In Example 15.14 we discussed the fact that Steve Nash (one of the most accurate free-throw shooters in NBA history) shoots free throws with a probability of

$p = 0.90$. We can interpret this to mean that on the average, out of every 100 free throws, Nash is going to make 90 and miss about 10, for a hit/miss ratio of 9 to 1. This ratio of hits to misses gives what is known as the *odds of* the event (in this case Nash making the free throw). We can also express the odds in a negative context and say that the *odds against* Nash making the free throw are 1 to 9.

In contrast, the odds of Shaquille O'Neal making a free throw can be described as being roughly 1 to 1. To be more precise, O'Neal shoots free throws with a probability of $p = 0.52$ (see Example 15.14), which represents a hit to miss ratio of 13 to 12 (52 to 48 simplified). Thus, the exact odds of Shaq making a free throw are 13 to 12. «

When *F* and *U* have a common factor, it is customary (and desirable) to simplify the numbers before giving the odds. One rarely hears odds of 15 to 5—they are much better expressed as odds of 3 to 1.

Odds

Let E be an arbitrary event. If F denotes the number of ways that event E can occur (the *favorable* outcomes or *hits*), and U denotes the number of ways that event E does not occur (the *unfavorable* outcomes, or *misses*), then the **odds of** (also called the **odds in favor of**) the event E are given by the ratio F to U, and **the odds against** the event E are given by the ratio U to F.

> ### EXAMPLE 15.25 Rolling the Dice: Part 5

Suppose you are playing a game in which you roll a pair of honest dice. If you roll a total of 7 or a total of 11, you win the game. What are your odds of winning in this game?

If we let E denote the event "roll a total of 7 or 11," we can check that out of 36 possible outcomes 8 are favorable (the roll totals a 7 or an 11) and the other 28 are unfavorable. It follows that the odds of you winning are 2 to 7 (originally 8 to 28 and then simplified to 2 to 7). «

We leave the details to the reader—see Exercise 43(c).

It is easy to convert odds into probabilities: If the odds of the event E are F to U, then $\Pr(E) = F/(F + U)$.

Converting probabilities into odds is also easy when the probability is given in the form of a fraction: If $\Pr(E) = A/B$, the odds of E are A to $B - A$. (When the probability is given in decimal form, the best thing to do is to first convert the decimal form into fractional form.)

> ### EXAMPLE 15.26 Handicapping a Tennis Tournament: Part 2

This example is a follow-up to Example 15.17. Recall that the probability assignment for the tennis tournament was as follows: $\Pr(Ana) = 0.08$, $\Pr(Ivan) = 0.16$, $\Pr(Lleyton) = 0.20$, $\Pr(Roger) = 0.25$, $\Pr(Venus) = 0.16$, and $\Pr(Serena) = 0.15$. We will now express each of these probabilities as odds. (Notice that to do so we first convert the decimals into fractions in reduced form.)

- $\Pr(Ana) = 0.08 = 8/100 = 2/25$. Thus, the odds of Ana winning the tournament are 2 to 23.
- $\Pr(Ivan) = 0.16 = 16/100 = 4/25$. The odds of Ivan winning the tournament are 4 to 21.

- Pr(*Lleyton*) = 0.20 = 20/100 = 1/5. The odds of Lleyton winning the tournament are 1 to 4.

- Pr(*Roger*) = 0.25 = 25/100 = 1/4. The odds of Roger winning the tournament are 1 to 3.

- Pr(*Venus*) = Pr(*Ivan*) = 4/25. The odds of Venus winning the tournament are the same as Ivan's (4 to 21).

- Pr(*Serena*) = 0.15 = 15/100 = 3/20. The odds of Serena winning the tournament are 1 to 17. ◀◀

A final word of caution: There is a difference between odds as discussed in this section and the *payoff odds* posted by casinos or bookmakers in sports gambling situations. Suppose we read in the newspaper, for example, that the Las Vegas bookies have established that "the odds that the Los Angeles Lakers will win the NBA championship are 5 to 2." What this means is that if you want to bet in favor of the Lakers, for every $2 that you bet, you can win $5 if the Lakers win. This ratio may be taken as some indication of the actual odds in favor of the Lakers winning, but several other factors affect payoff odds, and the connection between payoff odds and actual odds is tenuous at best.

Conclusion

While the average citizen thinks of probabilities, chances, and odds as vague, informal concepts that are useful primarily when discussing the weather or playing the lottery, scientists and mathematicians think of probability as a formal framework within which the laws that govern chance events can be understood. The basic elements of this framework are a *sample space* (which represents a precise mathematical description of all the possible outcomes of a *random experiment*), *events* (collections of these outcomes), and a *probability assignment* (which associates a numerical value to each of these events).

Of the many ways in which probabilities can be assigned, a particularly important case is the one in which all individual outcomes have the same probability (*equiprobable spaces*). When this happens, the critical steps in calculating probabilities revolve around two basic (but not necessarily easy) questions: (i) What is the size of the sample space? and (2) what is the size of the event in question? To answer these kinds of questions, knowing the basic principles of "counting" is critical.

When we stop to think how much of our lives is ruled by fate and chance, the importance of probability theory in almost every walk of life is hardly surprising. Understanding the basic mathematical principles behind this theory can help us better judge when taking a chance is a smart move and when it is not. In the long run, this will make us not just better card players, but also better and more successful citizens.

His Sacred Majesty, Chance, decides everything.

Voltaire

Profile Persi Diaconis (1945–)

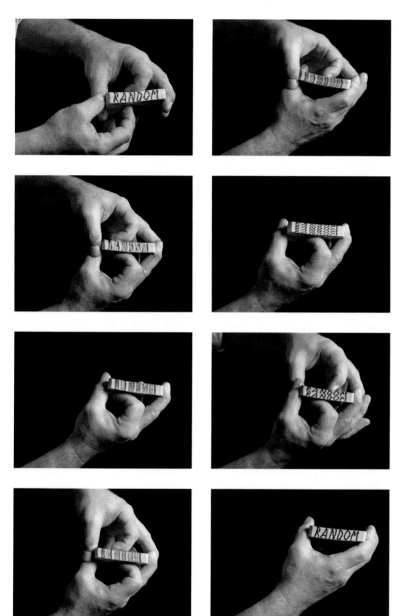

After one more shuffle, the original letters materialize at the original size. Diaconis turns the cards over and spreads them out with a magician's flourish, and there they are in their exact original sequence, from the ace of spades to the king of diamonds.

Diaconis has just performed eight perfect shuffles in a row. There's no hocus-pocus, just skill perfected in his youth: Diaconis ran away from home at 14 to become a magician's assistant and later became a professional magician and blackjack player. Even now, at 57, he is one of a couple of dozen people on the planet who can do eight perfect shuffles in less than a minute.

Diaconis's work these days involves much more than nimbleness of hand. He is a professor of mathematics and statistics at Stanford University. But he is also the world's leading expert on shuffling. He knows that what seems to be random often isn't, and he has devoted much of his career to exploring the difference. His work has applications to filing systems for computers and the reshuffling of the genome during evolution. And it has led him back to Las Vegas, where instead of trying to beat the casinos, he now works for them.

A card counter in blackjack memorizes the cards that have already been played to get better odds by making bets based on his knowledge of what has yet to turn up. If the deck has a lot of face cards and 10's left in it, for instance, and he needs a 10 for a good hand, he will bet more because he's more likely to get it. A good card counter, Diaconis estimates, has a 1 to 2 percent advantage over the casino. On a bad day, a good card counter can still lose $10,000 in a hurry. And on a good day, he may get a tap on the shoulder by a large person who will say, "You can call it a day now." By his mid-twenties, Diaconis had figured out that doing mathematics was an easier way to make a living.

Persi Diaconis picks up an ordinary deck of cards, fresh from the box, and writes a word in Magic Marker on one side: RANDOM. He shuffles the deck once. The letters have reformed themselves into six bizarre runes that still look vaguely like the letters R, A, and so on. Diaconis shuffles again, and the markings on the side become indecipherable. After two more shuffles, you can't even tell that there used to be six letters. The side of the pack looks just like the static on a television set. It didn't look random before, but it sure looks random now.

Keep watching. After three more shuffles, the word RANDOM miraculously reappears on the side of the deck—only it is written twice, in letters half the original size.

Two years ago, Diaconis himself got a tap on the shoulder. A letter arrived from a manufacturer of casino equipment, asking him to figure out whether its card-shuffling machines produced random shuffles. To Diaconis's surprise, the company gave him and his Stanford colleague, Susan Holmes, carte blanche to study the inner workings of the machine. It was like taking a Russian spy on a tour of the CIA and asking him to find the leaks.

When shuffling machines first came out, Diaconis says, they were transparent, so gamblers could actually see the cutting and riffling inside. But gamblers stopped caring after a while, and the shuffling machines turned into closed boxes. They also stopped shuffling cards the way humans do. In the machine that Diaconis and Holmes looked at, each card gets randomly directed, one at a time, to one of 10 shelves. The shuffling machine can put each new card either on the top of the cards already on the shelf or on the bottom, but not between them.

"Already I could see there was something wrong," says Holmes. If you start out with all the red cards at the top of the deck and all the black cards at the bottom, after one pass through the shuffling machine you will find that each shelf contains a red-black sandwich. The red cards, which got placed on the shelves first, form the middle of each sandwich. The black cards, which came later, form the outside. Since there are only 10 shelves, there are at most 20 places where a red card is followed by a black one or vice versa—fewer than the average number of color changes (26) that one would expect from a random shuffle.

The nonrandomness can be seen more vividly if the cards are numbered 1 to 52. After they have passed through the shuffling machine, the numbers on the cards form a zigzag pattern. The top card on the top shelf is usually a high number. Then the numbers decrease until they hit the middle of the first red-black sandwich; then they increase and decrease again, and so on, at most 10 times.

Diaconis and Holmes figured out the exact probability that any given card would end up in any given location after one pass through the machine. But that didn't indicate whether a gambler could use this information to beat the house.

So Holmes worked out a demonstration. It was based on a simple game. You take cards from a deck one by one and each time try to predict what you've selected before you look at it. If you keep track of all the cards, you'll always get the last one right. You'll guess the second-to-last card right half the time, the third-to-last a third of the time, and so on. On average, you will guess about 4.5 cards correctly out of 52.

By exploiting the zigzag pattern in the cards that pass through the shuffling machine, Holmes found a way to double the success rate. She started by predicting that the highest possible card (52) would be on top. If it turned out to be 49, then she predicted 48—the next highest number—for the

second card. She kept going this way until her prediction was too low—predicting, say, 15 when the card was actually 18. That meant the shuffling machine had reached the bottom of a zigzag and the numbers would start climbing again. So she would predict 19 for the next card. Over the long run, Holmes (or, more precisely, her computer) could guess nine out of every 52 cards correctly.

To a gambler, the implications are staggering. Imagine playing blackjack and knowing one-sixth of the cards before they are turned over! In reality, a blackjack player would not have such a big advantage, because some cards are hidden and six full decks are used. Still, Diaconis says, "I'm sure it would double or triple the advantage of the ordinary card counter."

Diaconis and Holmes offered the equipment manufacturer some advice: Feed the cards through the machine twice. The alternative would be more expensive: Build a 52-shelf machine.

A small victory for shuffling theory, one might say. But randomization applies to more than just cards. Evolution randomizes the order of genes on a chromosome in several ways. One of the most common mutations is called a "chromosome inversion," in which the arm of a chromosome gets cut in two random places, flipped over end-to-end, and reattached, with the genes in reverse order. In fruit flies, inversions happen at a rate of roughly one per every million years. This is very similar to a shuffling method called transposition that Diaconis studied 20 years ago. Using his methods, mathematical biologists have estimated how many inversions it takes to get from one species of fruit fly to another, or to a completely random genome. That, Diaconis suggests, is the real magic he ran away from home to find. "I find it amazing," he says, "that mathematics developed for purely aesthetic reasons would mesh perfectly with what engineers or chromosomes do when they want to make a mess."

Source: *Dana Mackenzie, "The Mathematics of Shuffling," Discover (October 2002), pp. 22–23. Reprinted by permission.*

Key Concepts

Exercises

WALKING

A. Random Experiments and Sample Spaces

1. Write out the sample space for each of the following random experiments.

   (a) A coin is tossed three times in a row and we observe on each toss whether it lands heads or tails.

   (b) A coin is tossed three times in a row and we observe the number of times it lands tails.

   (c) A person shoots three consecutive free throws and we observe the number of missed free throws.

2. Write out the sample space for each of the following random experiments.

   (a) A coin is tossed four times in a row and we observe on each toss whether it lands heads or tails.

   (b) A student randomly guesses the answers to a four-question true–false quiz and we observe the student's answers.

3. Four names (A, B, C, and D) are written each on a separate slip of paper, put in a hat, and mixed well. The slips are randomly taken out of the hat, one at a time, and the names recorded.

   (a) Write out the sample space for this random experiment.

   (b) Find N (the size of the sample space).

4. A gumball machine has gumballs of four different flavors: apple (A), blackberry (B), cherry (C), and Doublemint (D). The gumballs are well mixed and when you drop a quarter in the machine you get two random gumballs. Write out the sample space for this random experiment.

   (***Hint:*** $N = 10$.)

In Exercises 5 through 8, the sample spaces are too big to write in full. In these exercises, you should describe the sample space either by describing a generic outcome or by listing some outcomes and then using the … notation. In the latter case, you should write down enough outcomes to make the description reasonably clear.

5. A coin is tossed 10 times in a row and we observe on each toss whether it lands heads or tails. Describe the sample space.

6. A student randomly guesses the answers to a 10-question true–false quiz and we observe the student's answers. Describe the sample space.

7. A die is rolled four times in a row and we observe the number that comes up on each roll. Describe the sample space.

8. A student randomly guesses the answers to a 10-question multiple-choice test (the possible choices to each question are A, B, C, D, or E). Describe the sample space.

B. The Multiplication Rule

9. A California license plate starts with a digit other than 0, followed by three capital letters followed by three more digits (0 through 9).

   (a) How many different California license plates are possible?

   (b) How many different California license plates start with a 5 and end with a 9?

   (c) How many different California license plates have no repeated symbols (all the digits are different and all the letters are different)?

10. A computer password consists of four letters (A through Z) followed by a single digit (0 through 9). Assume that the passwords are not case sensitive (i.e., that an uppercase letter is the same as a lowercase letter).

    (a) How many different passwords are possible?

    (b) How many different passwords end in 1?

    (c) How many different passwords do not start with Z?

    (d) How many different passwords have no Zs in them?

11. A computer password consists of four letters (A through Z) followed by a single digit (0 through 9). Assume that the passwords are case sensitive (i.e., uppercase letters are considered different from lowercase letters).

    (a) How many different passwords are possible?

    (b) How many different passwords start with Z?

    (c) How many different passwords do not start with either z or Z?

    (d) How many different passwords have no Zs in them (uppercase or lowercase)?

12. A French restaurant offers a menu consisting of three different appetizers, two different soups, four different salads, nine different main courses, and five different desserts.

    (a) A fixed-price lunch meal consists of a choice of appetizer, salad, and main course. How many different lunch fixed-price meals are possible?

    (b) A fixed-price dinner meal consists of a choice of appetizer, a choice of soup or salad, a main course, and a dessert. How many different dinner fixed-price meals are possible?

    (c) A dinner special consists of a choice of soup, or salad, or both, and a main course. How many dinner specials are possible?

13. A set of reference books consists of eight volumes numbered 1 through 8.

    (a) In how many ways can the eight books be arranged on a shelf?

    (b) In how many ways can the eight books be arranged on a shelf so that at least one book is out of order?

14. Four men and four women line up at a checkout stand in a grocery store.

    (a) In how many ways can they line up?

    (b) In how many ways can they line up if the first person in line must be a woman?

    (c) In how many ways can they line up if they must alternate woman, man, woman, man, and so on and if a woman must always be first in line?

15. The ski club at Tasmania State University has 35 members (15 females and 20 males). A committee of three members—a President, a Vice President, and a Treasurer—must be chosen.

    (a) How many different three-member committees can be chosen?

    (b) How many different three-member committees can be chosen if the President must be a female?

    (c) How many different three-member committees can be chosen if the committee cannot have all females or all males?

16. The ski club at Tasmania State University has 35 members (15 females and 20 males). A committee of four members—a President, a Vice President, a Treasurer, and a Secretary—must be chosen.

    (a) How many different four-member committees can be chosen?

    (b) How many different four-member committees can be chosen if the President and Treasurer must be females?

    (c) How many different four-member committees can be chosen if the President and Treasurer must both be female and the Vice President and Secretary must both be male?

    (d) How many different four-member committees can be chosen if the committee must have two females and two males?

17. How many seven-digit numbers (i.e., numbers between 1,000,000 and 9,999,999)

    (a) are even?

    (b) are divisible by 5?

    (c) are divisible by 25?

18. How many 10-digit numbers (i.e., numbers between 1,000,000,000 and 9,999,999,999)

    (a) have no repeated digits?

    (b) are palindromes? (A palindrome is a number such as 37473 that reads the same whether read from left to right or from right to left.)

C. Permutations and Combinations

Exercises 19 through 28 should be done without using a calculator.

19. (a) $_{10}P_2$

    (b) $_{10}C_2$

    (c) $_{10}P_3$

    (d) $_{10}C_3$

20. (a) $_{11}P_2$

    (b) $_{11}C_2$

    (c) $_{20}P_2$

    (d) $_{20}C_2$

21. (a) $_{10}C_9$

    (b) $_{10}C_8$

    (c) $_{100}C_{99}$

    (d) $_{100}C_{98}$

22. (a) $_{12}P_2$

    (b) $_{12}P_3$

    (c) $_{12}P_4$

    (d) $_{12}P_5$

23. (a) $_{20}C_2$

    (b) $_{20}C_{18}$

    (c) $_{20}C_3$

    (d) $_{20}C_{17}$

24. (a) $_{10}C_3 + {}_{10}C_4$

    (b) $_{11}C_4$

    (c) $_9C_6 + {}_9C_7$

    (d) $_{10}C_7$

25. (a) $_3C_0 + {}_3C_1 + {}_3C_2 + {}_3C_3$

    (b) $_4C_0 + {}_4C_1 + {}_4C_2 + {}_4C_3 + {}_4C_4$

    (c) $_5C_0 + {}_5C_1 + {}_5C_2 + {}_5C_3 + {}_5C_4 + {}_5C_5$

    (d) $_{10}C_0 + {}_{10}C_1 + {}_{10}C_2 + \cdots + {}_{10}C_{10}$

    [***Hint:*** *Look for the pattern in (a) through (c).*]

26. (a) $_9C_4 - {}_9C_5$

    (b) $_{10}C_3 - {}_{10}C_7$

    (c) $_{100}C_{10} - {}_{100}C_{90}$

    (d) $_{1000}C_{498} - {}_{1000}C_{592}$

27. Given that $_{150}P_{50} \approx 6.12 \times 10^{104}$, find an approximate value for $_{150}P_{51}$.

28. Given that $_{150}C_{50} \approx 2.01 \times 10^{40}$, find an approximate value for $_{150}C_{51}$.

For Exercises 29 and 30 you will need a scientific or a business calculator with built-in $_nP_r$ and $_nC_r$ keys.

29. Use a calculator to compute each of the following.

(a) $_{20}P_{10}$

(b) $_{52}C_{20}$

(c) $_{52}C_{32}$

(d) $_{53}C_{20}$

30. Use a calculator to compute each of the following.

(a) $_{18}P_{10}$

(b) $_{51}C_{21}$

(c) $_{51}C_{30}$

(d) $_{51}P_{30}$

In Exercises 31 through 34, give your answer in symbolic form, using the notation $_nP_r$ or $_nC_r$.

31. The board of directors of the XYZ Corporation has 15 members. In how many ways can one choose

(a) a committee of four members (President, Vice President, Treasurer, and Secretary)?

(b) a delegation of four members where all members have equal standing?

32. There are 10 horses entered in a race. In how many ways can one pick

(a) the top three finishers regardless of order?

(b) the first-, second-, and third-place finishers in the race?

33. There are 20 singers auditioning for a musical. In how many different ways can the director choose

(a) a duet?

(b) a lead singer and a backup?

(c) a quintet?

34. There are 119 Division 1A college football teams.

(a) How many Top 25 rankings are possible?

(b) How many ways are there to choose eight teams and seed them (1 through 8) for a playoff?

(c) How many ways are there to choose eight teams (unseeded) for a playoff?

D. General Probability Spaces

35. Consider the sample space $S = \{o_1, o_2, o_3, o_4, o_5\}$. Suppose you are given $\Pr(o_1) = 0.22$ and $\Pr(o_2) = 0.24$.

(a) If o_3, o_4, and o_5 all have the same probability, find $\Pr(o_3)$.

(b) If o_3 has the same probability as o_4 and o_5 combined, find $\Pr(o_3)$.

(c) If o_3 has the same probability as o_4 and o_5 combined and if $\Pr(o_5) = 0.1$, give the probability assignment for this probability space.

36. Consider the sample space $S = \{o_1, o_2, o_3, o_4\}$. Suppose you are given $\Pr(o_1) + \Pr(o_2) = \Pr(o_3) + \Pr(o_4)$.

(a) If $\Pr(o_1) = 0.15$, find $\Pr(o_2)$.

(b) If $\Pr(o_1) = 0.15$ and $\Pr(o_3) = 0.22$, give the probability assignment for this probability space.

37. Seven players are entered in a tennis tournament. According to one expert handicapper, $P_2, P_3, \ldots, P_7$ all have the same probability of winning, and P_1 is twice as likely to win as one of the other players. Write down the sample space, and find the probability assignment for the probability space defined by this handicapper's opinion.

38. Six players are entered in a chess tournament. According to an expert handicapper, P_1 has a probability of 0.25 of winning the tournament, P_2 a probability of 0.15, P_3 a probability of 0.09, and P_4, P_5, and P_6 all have an equal probability of winning. Write down the sample space, and find the probability assignment for the probability space defined by this handicapper's opinion.

39. The circular spinner shown in the figure has a red sector with a central angle of 108°, blue and white sectors both with central angles of 72°, and green and yellow sectors both with central angles of 54°. Assume that the needle is spun so that it randomly stops at one of these sectors (if it stops exactly on the line between sectors then the needle is spun again).

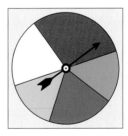

(a) Describe a sample space for this game.

(b) Give the probability assignment for this probability space.

40. The circular spinner shown in the figure is divided into five sectors. The white sector has central angle of 36°, the yellow sector has central angle of 54°, and the blue sector has central angle of 144°. Assume that the needle is spun so that it randomly stops at one of the four sectors (if it stops exactly on the line between sectors then the needle is spun again).

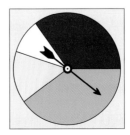

(a) Describe a sample space for this game.

(b) Give the probability assignment for this probability space.

E. Events

41. Consider the random experiment of tossing a coin three times in a row. [See Exercise 1(a).] Write out the event described by each of the following statements as a set.

(a) E_1: "toss exactly 2 heads."

(b) E_2: "all tosses come out the same."

(c) E_3: "half of the tosses are heads, and half are tails."

(d) E_4: "first two tosses are tails."

42. Consider the random experiment from Exercise 2(b), where a student takes a four-question true–false quiz. Write out the event described by each of the following statements as a set.

(a) E_1: "exactly 2 of the answers given are Ts."
 (**Note:** $T =$ True, $F =$ False)

(b) E_2: "at least 2 of the answers given are Ts."

(c) E_3: "at most 2 of the answers given are Ts."

(d) E_4: "the first 2 answers given are Ts."

43. Consider the random experiment from Example 15.4 of rolling a pair of dice. Write out the event described by each of the following statements as a set.

(a) E_1: "roll two of a kind." (That is, both numbers are equal.)

(b) E_2: "roll a total of 3 or less."

(c) E_3: "roll a total of 7 or 11."

44. Consider the random experiment of drawing 1 card out of an ordinary deck of 52 cards. Write out the event described by each of the following statements as a set.

(a) E_1: "the card drawn is the queen of hearts."

(b) E_2: "the card drawn is a queen."

(c) E_3: "the card drawn is a heart."

(d) E_4: "the card drawn is a face card." (A face card is a jack, queen, or king.)

45. Consider the random experiment from Exercise 5 of tossing a coin 10 times in a row. Write out the event described by each of the following statements as a set.

(a) E_1: "toss no tails."

(b) E_2: "toss exactly 1 tail."

(c) E_3: "toss exactly twice as many heads as tails."

46. Consider the sample space $S = \{A, B, C, D\}$. Make a list of all the possible events for this sample space. (Remember that an event is any subset of S including $\{ \ \}$ and S itself.)

F. Equiprobable Spaces

47. Consider the random experiment of tossing an honest coin 3 times in a row. Find the probability of each of the following events.

(**Hint:** See Exercise 41.)

(a) E_1: "toss exactly 2 heads."

(b) E_2: "all tosses come out the same."

(c) E_3: "half of the tosses are heads, and half are tails."

(d) E_4: "first two tosses are tails."

48. Consider the random experiment where a student takes a four-question true–false quiz. Assume now that the student randomly guesses the answer for each question. Find the probability of each of the following events.

(**Hint:** See Exercise 42.)

(a) E_1: "exactly 2 of the answers given are Ts."
 (**Note:** $T =$ True, $F =$ False)

(b) E_2: "at least 2 of the answers given are Ts."

(c) E_3: "at most 2 of the answers given are Ts."

(d) E_4: "the first 2 answers given are Ts."

49. Consider the random experiment of rolling a pair of honest dice. Let $T_2, T_3, \ldots, T_{12}$ represent the events "roll a total of 2," "roll a total of 3," $\ldots$, "roll a total of 12," respectively.

(a) Find $\Pr(T_6)$ and $\Pr(T_8)$.

(b) Find $\Pr(T_5)$ and $\Pr(T_9)$.

(c) Let E_1 denote the event "roll two of a kind" (i.e., both numbers are equal). Find $\Pr(E_1)$.

(d) Let E_2 denote the event "roll a total of 3 or less." Find $\Pr(E_2)$.

(e) Let E_3 denote the event "roll a total of 7 or 11." Find $\Pr(E_3)$.

50. Consider the random experiment of drawing 1 card out of a well-shuffled, honest deck of 52 cards. Find the probability of each of the following events.

(**Hint:** See Exercise 44.)

(a) E_1: "the card drawn is the queen of hearts."

(b) E_2: "the card drawn is a queen."

(c) E_3: "the card drawn is a heart."

(d) E_4: "the card drawn is a face card." (A face card is a jack, queen, or king.)

51. Consider the random experiment of tossing an honest coin 10 times in a row. Find the probability of each of the following events.

(**Hint:** See Exercises 5 and 45.)

(a) E_1: "toss no tails."

(b) E_2: "toss exactly 1 tail."

(c) E_3: "toss exactly twice as many heads as tails."

52. Consider the random experiment of drawing two cards out of a well-shuffled, honest deck of 52 cards. (Here the order of the cards does not matter.) Find the probability of each of the following events.

(a) E_1: "draw a pair of queens."

(b) E_2: "draw a pair." (A pair is two cards of the same value—two 7's, two jacks, etc.)

53. If a pair of honest dice are rolled once, find the probability of

(a) rolling a total of 8.

(b) not rolling a total of 8.

(c) rolling a total of 8 or 9.

(d) rolling a total of 8 or more.

54. A gumball machine has gumballs of four different flavors: Apple (A), Berry (B), Cherry (C), and Doublemint (D). When a 50-cent piece is put into the machine, five random gumballs come out. Find the probability that

(a) each gumball is a different flavor.

(b) at least two of the gumballs are the same flavor.

55. A student takes a 10-question true–false quiz and randomly guesses the answer to each question. Suppose a correct answer is worth 1 point, an incorrect answer is worth -0.5 points. Find the probability that the student

(a) gets 10 points.

(b) gets -5 points.

(c) gets 8.5 points.

(d) gets 8 or more points.

(e) gets 5 points.

(f) gets 7 or more points.

56. Ten names ($A, B, C, D, E, F, G, H, I, J$) are written, each on a separate slip of paper, put in a hat, and mixed well. Four names are randomly taken out of the hat, one at a time. Assume that the order in which the names are drawn matters. Find the probability that

(a) A is the first name chosen.

(b) A is one of the four names chosen.

(c) A is not one of the four names chosen.

(d) The four names chosen are $A, B, C,$ and D in that order.

57. A club has 15 members. A delegation of four members must be chosen to represent the club at a convention. All delegates are equal, so the order in which they are chosen doesn't matter. Assume that the delegation is chosen randomly by drawing the names out of a hat. Find the probability that

(a) Alice (one of the members of the club) is selected.

(b) Alice is not selected.

(c) club members Alice, Bert, Cathy, and Dale are selected.

58. Suppose that the probability of giving birth to a boy and the probability of giving birth to a girl are both 0.5. In a family of four children, what is the probability that

(a) all four children are girls.

(b) there are two girls and 2 boys.

(c) the youngest child is a girl.

G. Odds

59. Find the odds of each of the following events.

(a) An event E with $\Pr(E) = 4/7$

(b) An event E with $\Pr(E) = 0.6$

60. Find the odds of each of the following events.

(a) An event E with $\Pr(E) = 3/11$

(b) An event E with $\Pr(E) = 0.375$

61. In each case, find the probability of an event E having the given odds.

(a) The odds of E are 3 to 5.

(b) The odds against E are 8 to 15.

(c) The odds of E are 1 to 1.

62. In each case, find the probability of an event E having the given odds.

(a) The odds of E are 4 to 3.

(b) The odds against E are 12 to 5.

(c) The odds of E are the same as the odds against E.

JOGGING

63. The 35-member ski club at Tasmania State University needs to select a planning committee for an upcoming ski trip. The committee will have a chair, a secretary, and three at-large members. How many different planning committees can be chosen?

64. Andre and Roger are playing in the final match of the U.S. Open Tennis Championships. (A tennis match is a best-of-five games contest—the first player to win three games wins the match, and there are no ties.) We can describe the outcome of the tennis match by a string of letters (A or R) that indicate the winner of each game. For example, the string *RARR* represents an outcome where Roger wins games 1, 3 and 4, at which point the match is over (game 5 is not played).

(a) Describe the event "Roger wins the match in game 5."

(b) Describe the event "Roger wins the match."

(c) Describe the event "the match goes 5 games."

65. Two teams (call them X and Y) play against each other in the World Series. The World Series is a best-of-seven series—the first team to win four games wins the series. (Games cannot end in a tie.) We can describe an outcome for the World Series by writing a string of letters that indicate (in order) the winner of each game. For example, the string $XYXXYX$ represents the outcome: X wins game 1, Y wins game 2, X wins game 3, and so on.

(a) Describe the event "X wins in 5 games."

(b) Describe the event "the series is over in game 5."

(c) Describe the event "the series is over in game 6."

66. A pizza parlor offers six toppings—pepperoni, Canadian bacon, sausage, mushroom, anchovies, and olives—that can be put on their basic cheese pizza. How many different pizzas can be made? (A pizza can have anywhere from no toppings to all six toppings.)

67. (a) In how many different ways can 10 people form a line?

(b) In how many different ways can 10 people be seated around a circular table?

(**Hint:** *The answer to (b) is much less than the answer to (a). There are many different ways in which the same circle of 10 people can be broken up to form a line. How many?*)

(c) In how many different ways can five boys and five girls get in line so that boys and girls alternate (boy, girl, boy, ..., or girl, boy, girl, ...)?

(d) In how many different ways can five boys and five girls sit around a circular table so that boys and girls alternate?

68. Eight points are taken on a circle.

(a) How many chords can be drawn by joining all possible pairs of the points?

(b) How many triangles can be made using these points as vertices?

69. A study group of 15 students is to be split into 3 groups of 5 students each. In how many ways can this be done?

70. An urn contains seven red balls and three blue balls.

(a) If three balls are selected all at once, what is the probability that two are blue and one is red?

(b) If three balls are selected by pulling out a ball, noting its color, and putting it back in the urn before the next selection, what is the probability that only the first and third balls drawn are blue?

(c) If three balls are selected one at a time without putting them back in the urn, what is the probability that only the first and third balls drawn are blue?

71. If we toss an honest coin 20 times, what is the probability of

(a) getting 10 Hs and 10 Ts?

(b) getting 3 Hs and 17 Ts?

(c) getting 3 or more Hs?

Exercises 72 through 76 refer to five-card draw poker hands. (See Example 15.22.)

72. What is the probability of getting "4 of a kind" (four cards of the same value)?

73. What is the probability of getting all five cards of the same color?

74. What is the probability of getting a "flush" (all five cards of the same suit)?

75. What is the probability of getting an "ace-high straight" (10, J, Q, K, A of any suit but not all of the same suit)?

76. What is the probability of getting a "full house" (three cards of equal value and two other cards of equal value)?

77. Consider the following game. We roll a pair of honest dice five times. If we roll a total of seven at least once, we win; otherwise, we lose. What is the probability that we will win?

78. Which makes the better bet, *Bet 1* or *Bet 2*? *Bet 1*: In four consecutive rolls of an honest die, a 6 will come up at least once. *Bet 2*: In 24 consecutive rolls of a pair of dice, two 6's (called *boxcars*) will come up at least once.

Historical note: *A French gambler by the name of Chevalier de Méré (1607–1684) wanted to know the answer to this question because he had large sums of money riding on these bets. He asked his friend, the famous mathematician Blaise Pascal, for the answer. Pascal's work on this question is considered by many historians to have laid the groundwork for the formal mathematical theory of probability* (see Project B).

79. A factory assembles car stereos. From random testing at the factory, it is known that, on the average, 1 out of every 50 car stereos will be defective (which means that the probability that a car stereo randomly chosen from the assembly line will be defective is 0.02). After manufacture, car stereos are packaged in boxes of 12 for delivery to the stores.

(a) What is the probability that in a box of 12, there are no defective car stereos? What assumptions are you making?

(b) What is the probability that in a box of 12, there is at most 1 defective car stereo?

80. There are 500 tickets sold in a raffle. If you have three of these tickets and five prizes are to be given, what is the probability that you will win at least one prize? (Give your answer in symbolic notation.)

RUNNING

81. If an honest coin is tossed N times, what is the probability of getting the same number of Hs as Ts?

(**Hint:** *Consider two cases: N even and N odd.*)

82. How many different "words" (they don't have to mean anything) can be formed using all the letters in

(a) the word PARSLEY.

(**Note:** *This one is easy!*)

(b) the word PEPPER.

[**Note:** *This one is much harder! Think about the difference between (a) and (b).*]

83. Shooting craps. In the game of craps, the player's first roll of a pair of dice is very important. If the first roll is 7 or 11, the player wins. If the first roll is 2, 3, or 12, the player loses. If the first roll is any other number (4, 5, 6, 8, 9, 10), this number is called the player's "point." The player then continues to roll until either the point reappears, in which case the player wins, or until a 7 shows up before the point, in which case the player loses. What is the probability that the player will win? (Assume that the dice are honest.)

84. The birthday problem. There are 30 people in a room. What is the probability that at least two of these people have the same birthday—that is, have their birthdays on the same day and month?

85. Yahtzee. Yahtzee is a dice game in which five standard dice are rolled at one time.

(a) What is the probability of scoring "Yahtzee" with one roll of the dice? (You score Yahtzee when all five dice match.)

(b) What is the probability of a *four of a kind* with one roll of the dice? (A *four of a kind* is rolled when four of the five dice match.)

(c) What is the probability of rolling a *large straight* in one roll of the dice? (A *large straight* consists of five numbers in succession on the dice.)

(d) What is the probability of rolling *trips* with one roll of the dice? (*Trips* are rolled when three of the five dice match, and the other two dice do not match the trips or each other.)

86. Heartless poker. Complete the following table for five-card draw poker hands from a deck in which all the hearts have been removed.

Hand	Probability
Four of a kind	
Flush (five cards of same suit)	
Full house (three cards of equal value and two others of equal value)	
Three of a kind	
Two pair	

Projects and Papers

A. A History of Gambling

Games of chance, be they for religious purposes or for profit, can be traced all the way back to some of the earliest civilizations—the Babylonians, Assyrians, and ancient Egyptians were all known to play primitive dice games as well as board games of various kinds.

In this project you are to write a paper discussing the history of gambling and the historical role of games of chance in the development of probability theory. Good references for this project are references 1 and 2.

B. The Letters Between Pascal and Fermat

In the summer of 1654, Pierre de Fermat and Blaise Pascal—two of the most famous mathematicians of their time—exchanged a series of famous letters discussing certain mathematical problems related to gambling. The problems were originally proposed to Pascal by his friend, a certain Chevalier de Méré—a nobleman and notorious gambler with a flair for raising interesting mathematical questions. The exchange of letters between Pascal and Fermat is of particular interest because it laid the foundation for the future development of the mathematical theory of probability. While not all of the correspondence between Fermat and Pascal has survived, enough of the letters exist to see how both mathematicians approached, each in his own way, de Méré's questions.

In this project, you are to write a paper discussing the famous exchange of letters between Pascal and Fermat. Describe the questions that were discussed and the approach that each mathematician took to solve them. Good references for this project are references 4 and 16.

C. The Largest Number Game

Suppose you are offered a chance to play the following game: A box contains 100 tickets. Each ticket has a different number written on it. The numbers can be any 100 numbers, large or small, integer or rational—anything goes as long as they are all different. Among all the tickets in the box, the one with the largest number is the *winning ticket*—if you can turn it in, you win $100. For any other ticket, you get nothing. The ground rules for this game are as follows: You can draw a ticket out of the box. If you don't like it, you can draw another ticket from the box, but first you must tear up the other ticket. You can continue drawing from the box until you decide to stop or there are no more tickets in the box, but you must always tear up the previous ticket before you

can draw again. In other words, the only ticket you can use is the one you drew last.

In this project you are to discuss and explain a strategy for playing this game that will give you a 25% or better chance of winning the $100. You should also discuss what is a reasonable sum that you are willing to pay for the right to play the game.

Note: This game and variations of it are sometimes discussed in the literature under the name *The Secretary Problem.*

D. Book Review: *Conned Again, Watson! Cautionary Tales of Logic, Math, and Probability*

As its title suggests, *Conned Again, Watson! Cautionary Tales of Logic, Math, and Probability*, by Colin Bruce (Perseus Publishing, 2002), is a collection of mathematical tales set in the form of Sherlock Holmes adventures. Each of the stories has a mathematical connection and several touch on issues related to the theme of this chapter (particularly relevant are "The Case of the Gambling Nobleman" and "The Case of the Surprise Heir," but several others deal with probability topics as well).

Choose one of the stories in the book with a probabilistic connection and write a review of the story. Pay special attention to the mathematical issues raised in the story.

References and Further Readings

1. Bennett, Deborah, *Randomness*. Cambridge, MA: Harvard University Press, 1998.

2. Bernstein, Peter, *Against the Gods: The Remarkable Story of Risk*. New York: John Wiley & Sons, 1996.

3. Bruce, Colin, *Conned Again, Watson! Cautionary Tales of Logic, Math, and Probability*. Cambridge, MA: Perseus Publishing, 2002.

4. David, Florence N., *Games, Gods, and Gambling*. New York: Dover Publications, 1998.

5. de Finetti, B., *Theory of Probability*. New York: John Wiley & Sons, 1970.

6. Everitt, Brian S., *Chance Rules: An Informal Guide to Probability, Risk, and Statistics*. New York: Springer-Verlag, 1999.

7. Gigerenzer, Gerd, *Calculated Risks*. New York: Simon & Schuster, 2002.

8. Gnedenko, B. V., and A. Y. Khinchin, *An Elementary Introduction to the Theory of Probability*. New York: Dover Publications, 1962.

9. Haigh, John, *Taking Chances*. New York: Oxford University Press, 1999.

10. Keynes, John M., *A Treatise on Probability*. New York: Harper and Row, 1962.

11. Krantz, Les, *What the Odds Are*. New York: HarperCollins Publishers, Inc., 1992.

12. Levinson, Horace, *Chance, Luck and Statistics: The Science of Chance*. New York: Dover Publications, 1963.

13. McGervey, John D., *Probabilities in Everyday Life*. New York: Ivy Books, 1992.

14. Mosteller, Frederick, *Fifty Challenging Problems in Probability with Solutions*. New York: Dover Publications, 1965.

15. Packel, Edward, *The Mathematics of Games and Gambling*. Washington, DC: Mathematical Association of America, 1981.

16. Renyi, Alfred, *Letters on Probability*. Detroit, MI: Wayne State University Press, 1972.

17. Weaver, Warren, *Lady Luck: The Theory of Probability*. New York: Dover Publications, Inc., 1963.

16

Normal Distributions

Everything Is Back to Normal (Almost)

> The normal is
> what you find
> but rarely. The
> normal is an
> ideal.
>
> W. Somerset
> Maugham

Some 60 years ago, the South African mathematician John Kerrich spent several years as a German prisoner of war in Jutland. To pass the time, Kerrich tossed a coin—over and over, ... and over, keeping meticulous records of the results. By the time he was done he had tossed the same coin 10,000 times. Kerrich broke up his 10,000 tosses into 100 sets of 100 tosses and recorded the number of Heads in each set of 100. The results of Kerrich's coin-tossing experiment are shown in the top bar graph on the opposite page. (*Source:* John Kerrich, *An Experimental Introduction to the Theory of Probability,* 1964.)

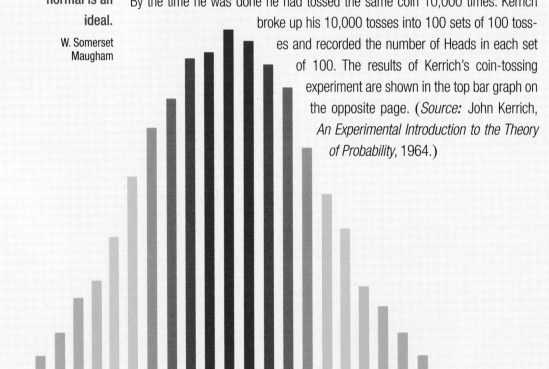

Admittedly, the bar graph shown at the top of this page does not seem in particularly interesting, so why does Kerrich's coin-tossing experiment deserve our attention? Essentially, the only problem with Kerrich's data is that he didn't toss the coin enough times! What would have happened if Kerrich had continued the coin-tossing experiment for a few more years? The much more interesting picture shown at the bottom of this page shows the distribution of the number of *Heads* in 10,000 sets of 100 tosses each (that's *1 million* coin tosses!) These results were obtained by the author using a computer simulation rather than an actual coin toss marathon but are just as valid as if they were done

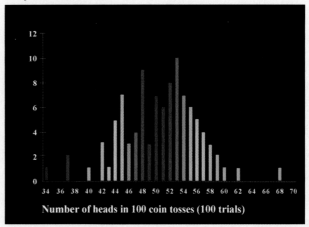

Number of heads in 100 coin tosses (100 trials)

for real. In fact, had Kerrich tossed a coin 1 million times, his data would have produced a bell-shaped bar graph looking very much like the bottom graph on this page. We can say this with full confidence even though it never happened. Like many other phenomena, the long-term behavior of a tossed coin is guaranteed (when tossed enough times) to produce a bell-shaped bar graph—it is a fundamental law of mathematics.

The bell-shaped bar graph at the bottom of this page is but one example of real-world data that approximates what is known as a *normal distribution*. More than any other type of regular pattern, normal distributions rule the statistical world, and the purpose of this chapter is to gain an understanding of the meaning of "normal" in the context of statistical data. Section 16.1 is an introductory section illustrating the ubiquity of *approximately normal* distributions among real-world data sets. Sections 16.2 through 16.4 discuss the basic mathematical properties of *normal curves*. Section 16.5 revisits the connection between *normal curves and real-world data* sets introduced in Section 16.1. Section 16.6 deals with the connection between *random events and normal distributions*. The chapter concludes with a brief introduction to the important topic of *statistical inference* (Section 16.7).

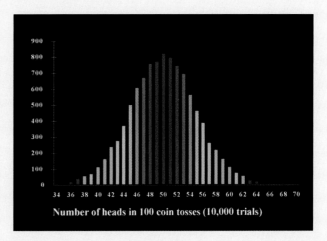

Number of heads in 100 coin tosses (10,000 trials)

16.1 Approximately Normal Distributions of Data

> **EXAMPLE 16.1** Distribution of Heights of NBA Players

Table 16-1 is a frequency table giving the heights of the 430 National Basketball Association players listed on team rosters at the start of the 2005–2006 season. (*Source:* National Basketball Association, *www.nba.com.*) The bar graph for this data set is shown in Fig. 16-1. We can see that the bar graph fits roughly the pattern of a somewhat skewed (off-center) bell-shaped curve (the orange curve). An idealized bell-shaped curve for this data (the red curve) is shown for comparison purposes.

The data would be even more bell-shaped if it weren't for the abundance of 6'7" to 6'11" players in the NBA. This is not a quirk of nature but rather a reflection of modern playing styles.

TABLE 16-1 Heights of NBA Players (2005–2006); *N* = 430

Height	Frequency	Height	Frequency	Height	Frequency	Height	Frequency
5'5"	1	6'2"	19	6'8"	45	7'2"	5
5'9"	1	6'3"	23	6'9"	48	7'3"	3
5'10"	4	6'4"	24	6'10"	44	7'5"	1
5'11"	4	6'5"	28	6'11"	42	7'6"	1
6'0"	9	6'6"	33	7'0"	26		
6'1"	18	6'7"	44	7'1"	7		

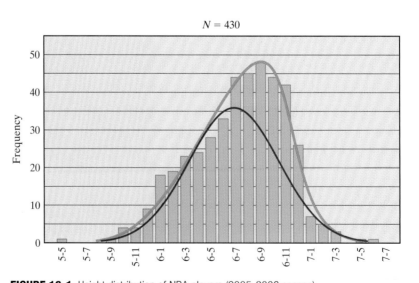

FIGURE 16-1 Height distribution of NBA players (2005-2006 season).

▶ **EXAMPLE 16.2** 2005 SAT Scores: Part 1

Table 16-2 shows the scores of $N = 1,475,623$ college-bound seniors on the verbal section of the 2005 SAT. (Scores range from 200 to 800 and are grouped in class intervals of 50 points.) The table shows the relative frequency (percentage of test takers) in each class interval. A bar graph for the data is shown in Fig. 16-2. The orange bell-shaped curve traces the pattern of the data in the bar graph. If the data followed a perfect bell curve, it would follow the red curve shown in the figure. Unlike the first example, the orange and red curves are very close.

TABLE 16-2 2005 Verbal SAT Scores ($N = 1,475,623$)		
Score ranges	**Frequency**	**Percentage**
750–800	30,479	2.1%
700–740	47,546	3.2%
650–690	95,510	6.5%
600–640	158,136	10.7%
550–590	209,333	14.2%
500–540	252,057	17.1%
450–490	255,028	17.3%
400–440	198,912	13.5%
350–390	124,871	8.5%
300–340	61,160	4.1%
250–290	26,553	1.8%
200–240	16,038	1.1%

Source: *The College Board National Report*, 2005

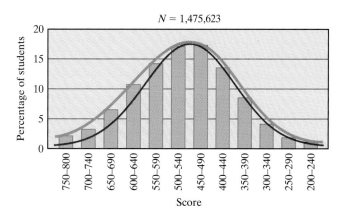

FIGURE 16-2

The two very different data sets discussed in Examples 16.1 and 16.2 have one thing in common—both can be described as having bar graphs that roughly fit a bell-shaped pattern. In Example 16.1, the fit is crude; in Example 16.2, it is very good. In either case, we say that the data set has an **approximately normal distribution**. The word *normal* in this context is to be interpreted as meaning that the data fits into a special type of bell-shaped curve; the word *approximately* is a reflection of the fact that with real-world data we should not expect an absolutely perfect fit. A distribution of data that has a perfect bell shape is called a **normal distribution**.

Perfect bell-shaped curves are called **normal curves**. Every approximately normal data set can be idealized mathematically by a corresponding normal curve (the red curves in Examples 16.1 and 16.2). This is important because we can then use the mathematical properties of the normal curve to analyze and draw conclusions about the data. The tighter the fit between the approximately normal distribution and the normal curve, the better our analysis and conclusions are going to be. Thus, to understand real-world data sets that have an approximately normal distribution, we first need to understand some of the mathematical properties of normal curves.

16.2 Normal Curves and Normal Distributions

Carl Friedrich Gauss (1777–1855). A biographical profile of Gauss can be found at the end of this chapter.

The study of normal curves can be traced back to the work of the great German mathematician Carl Friedrich Gauss, and for this reason, normal curves are sometimes known as *Gaussian curves*. Normal curves all share the same basic shape—that of a bell—but otherwise they can differ widely in their appearance. Some bells are tall and skinny, others are short and squat, and others fall somewhere in between (Fig. 16-3). Mathematically speaking, however, they all have the same underlying structure. In fact, whether a normal curve is skinny and tall or short and squat depends on the choice of units on the axes, and any two normal curves can be made to look the same by just fiddling with the scales of the axes.

What follows is a summary of some of the essential facts about normal curves and their associated normal distributions. These facts are going to help us greatly later on in the chapter.

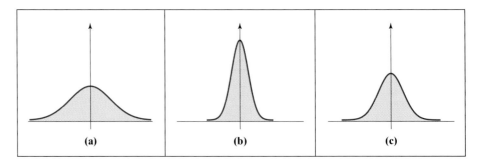

FIGURE 16-3 Normal curves may look different, but they are all cut out of the same cloth.

(a)   (b)   (c)

■ **Symmetry.** Every normal curve has a vertical axis of symmetry, splitting the bell-shaped region outlined by the curve into two identical halves. This is the only line of symmetry of a normal curve, so we can refer to it without ambiguity as *the line of symmetry*.

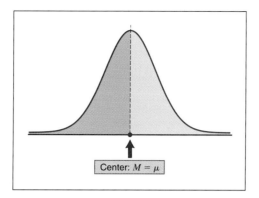

Center: $M = \mu$

FIGURE 16-4

■ **Median/mean.** We will call the point of intersection of the horizontal axis and the line of symmetry of the curve the **center** of the distribution (Fig. 16-4). The center is both the median and the mean (average) of the data. We use the Greek letter μ (mu) to denote this value. Thus, for data fitting a normal distribution, the median and the mean of the data set are indeed the same number. The fact that the median equals the mean implies that 50% of the data are less than or equal to the mean and 50% of the data are greater than or equal to the mean. For data fitting an approximately normal distribution, the median and the mean should be close to each other but we should not expect them to be equal.

Median and Mean of a Normal Distribution

In a normal distribution, $M = \mu$. (If the distribution is approximately normal, then $M \approx \mu$.)

■ **Standard deviation.** The standard deviation—traditionally denoted by the Greek letter σ (sigma)—is an important measure of spread, and it is particularly useful when dealing with normal (or approximately normal) distributions, as we will see shortly. The easiest way to describe the standard deviation of a normal distribution is to look at the normal curve. If you were to bend a piece of wire into a bell-shaped normal curve at the very top, you would be bending the wire downward [Fig. 16-5(a)], but at the bottom you would be bending the wire upward [Fig. 16-5(b)]. As you move your hands down the wire, the curvature gradually changes, and there is one point on each side of the curve where the transition from being bent downward to being bent upward takes place. Such a point [P in Fig. 16-5(c)] is called a **point of inflection**

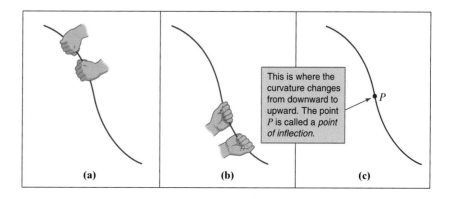

This is where the curvature changes from downward to upward. The point P is called a *point of inflection.*

P

(a)　　　　(b)　　　　(c)

FIGURE 16-5

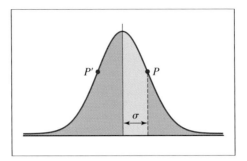

FIGURE 16-6

of the curve. The standard deviation of a normal distribution is the horizontal distance between the line of symmetry of the curve and one of the two points of inflection (*P* or *P'* in Fig. 16-6).

Standard Deviation of a Normal Distribution

In a *normal distribution*, the standard deviation σ equals the distance between a point of inflection and the line of symmetry of the curve.

■ **Quartiles.** We learned in Chapter 14 how to find the quartiles of a data set. When the data set has a normal distribution, the first and third quartiles can be approximated using the mean μ and the standard deviation σ. The magic number to memorize is 0.675. Multiplying the standard deviation by 0.675 tells us how far to go to the right or left of the mean to locate the quartiles.

Quartiles of a Normal Distribution

In a *normal distribution*, $Q_3 \approx \mu + (0.675)\sigma$ and $Q_1 \approx \mu - (0.675)\sigma$.

> **EXAMPLE 16.3** A Mystery Normal Distribution

Imagine you are told that a data set of $N = 1{,}475{,}623$ data points has a normal distribution with mean $\mu = 508$ and standard deviation $\sigma = 113$. For now, let's not worry about the source of this data—we'll discuss this soon.

Just knowing the mean and standard deviation (and that it is normally distributed) will allow us to draw a few useful conclusions already (and even more later) about this data set.

■ Since the median equals the mean, we have $M = 508$. This implies that half of the data (737,812 data points) are smaller than or equal to 508 and 737,812 data points are greater than or equal to 508.

■ The first quartile is given by $Q_1 \approx 508 - 0.675 \times 113 \approx 432$. This implies that 25% of the data (36,906 data points) are smaller than or equal to 432.

■ The third quartile is given by $Q_3 \approx 508 + 0.675 \times 113 \approx 584$. This implies that 25% of the data (36,906 data points) are bigger than or equal to 584. ≪

16.3 Standardizing Normal Data

We have seen that normal curves don't all look alike, but this is only a matter of perception. In fact, all normal distributions tell the same underlying story but use slightly different dialects to do it. One way to understand the story of any given normal distribution is to rephrase it in a simple common language—a language that uses the mean μ and the standard deviation σ as its only vocabulary. This process is called **standardizing** the data.

To standardize a data value x, *we measure how far x has strayed from the mean μ using the standard deviation σ as the unit of measurement.* A standardized data value is often referred to as a ***z*-value**.

The best way to illustrate the process of standardizing normal data is by means of a few examples.

▶ EXAMPLE 16.4 From *x* to *z*: Part 1

Let's consider a normally distributed data set with mean $\mu = 45$ ft and standard deviation $\sigma = 10$ ft. We will standardize several data values, starting with a couple of easy cases.

■ $x_1 = 55$ ft is a data point located 10 ft *above* the mean $\mu = 45$ ft. Coincidentally, 10 ft happens to be exactly *one* standard deviation. The fact that $x_1 = 55$ ft is located one standard deviation above the mean (point A in Fig. 16-7) can be rephrased by saying that the *standardized value* of $x_1 = 55$ is $z_1 = 1$.

■ $x_2 = 35$ ft is a data point that is 10 ft (i.e., one standard deviation) *below* the mean (point B in Fig. 16-7). This means that the standardized value of $x_2 = 35$ is $z_2 = -1$.

■ $x_3 = 50$ ft is a data point that is 5 ft (i.e., half a standard deviation) above the mean (point C in Fig 16-7). This means that the standardized value of $x_3 = 50$ ft is $z_3 = 0.5$. (We could have also said that the standardized value is $z_3 = 1/2$ but it is customary to use decimals to describe standardized values.)

■ $x_4 = 21.58$ is ... uh, this is a slightly more complicated case. How do we handle this one? First, we find the signed distance between the data value and the mean by taking their difference $(x_4 - \mu)$. In this case we get 21.58 ft − 45 ft = −23.42 ft. (Notice that for data values smaller than the mean this difference will be negative.) If we now divide this difference by $\sigma = 10$ ft, we get the standardized value $z_4 = -2.342$. This tells us that the data point x_4 is −2.342 standard deviations from the mean $\mu = 45$ ft (point D in Fig. 16-7).

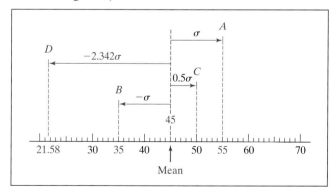

FIGURE 16-7

In Example 16.4 we were somewhat fortunate in that the standard deviation was $\sigma = 10$, an especially easy number to work with. It helped us get our feet wet. What do we do in more realistic situations, where the mean and standard deviation may not be such nice round numbers? Other than the fact that we may need a calculator to do the arithmetic, the basic idea we used in Example 16.4 remains the same:

> **Standardizing Rule**
>
> In a normal distribution with mean μ and standard deviation σ, the standardized value of a data point x is $z = (x - \mu)/\sigma$.

▶ EXAMPLE 16.5 From x to z: Part 2

This time we will consider a normally distributed data set with mean $\mu = 63.18$ lb and standard deviation $\sigma = 13.27$ lb. What is the standardized value of $x = 91.54$ lb?

This looks nasty, but with a calculator, it's a piece of cake:

$$z = (91.54 - 63.18)/13.27 = 28.36/13.27 = 2.13715\ldots \approx 2.14$$

One important point to note is that while the original data is given in pounds, there are no units given for the z-value. (The units for the z-value are standard deviations, and this is implicit in the very fact that it is a z-value.) ◀◀

The process of standardizing data can also be reversed, and given a z-value we can go back and find the corresponding x-value. All we have to do is take the relation $z = (x - \mu)/\sigma$ and solve for x in terms of z. When we do this we get the relationship $x = \mu + \sigma \cdot z$. Given μ, σ, and a value for z, this relationship allows us to "unstandardize" z and find the original data value x.

▶ EXAMPLE 16.6 From z to x

Consider a normal distribution with mean $\mu = 235.7$ m and standard deviation $\sigma = 41.58$ m. What is the data value x that corresponds to the standardized value of $z = -3.45$?

We first compute the value of -3.45 standard deviations: $-3.45\sigma = -3.45 \times 41.58$ m $= -143.451$ m. The negative value indicates that the data point is located -143.451 m below the mean $\mu = 235.7$ m. Thus, $x = 235.7$ m $- 143.451$ m $= 92.249$ m. ◀◀

16.4 The 68–95–99.7 Rule

When we look at a typical bell-shaped distribution, we can see that most of the data are concentrated near the center. As we move away from the center the heights of the columns drop rather fast, and if we mover far enough away from the center there are essentially no data to be found. These are all rather informal observations, but there is a more formal way to phrase this called the **68–95–99.7 rule**. This useful rule is obtained by using one, two, and three standard deviations above and below the mean as special landmarks. In effect, the 68–95–99.7 rule is three separate rules in one.

1. In every normal distribution, about 68% of all the data values fall within one standard deviation above and below the mean. In other words, 68% of all the data have standardized values between $z = -1$ and $z = 1$. The remaining 32% of the data are divided equally between data with standardized values $z \leq -1$ and data with standardized values $z \geq 1$ [Fig. 16-8(a)].

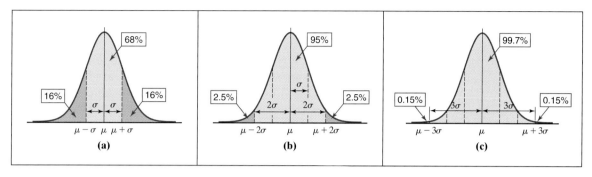

FIGURE 16-8 The 68–95–99.7 rule.

2. In every normal distribution, about 95% of all the data values fall within two standard deviations above and below the mean. In other words, 95% of all the data have standardized values between $z = -2$ and $z = 2$. The remaining 5% of the data are divided equally between data with standardized values $z \leq -2$ and data with standardized values $z \geq 2$ [Fig. 16-8(b)].

3. In every normal distribution, about 99.7% (i.e., practically 100%) of all the data values fall within three standard deviations above and below the mean. In other words, 99.7% of all the data have standardized values between $z = -3$ and $z = 3$. There is a minuscule amount of data with standardized values outside this range. [Fig. 16-8(c)].

For approximately normal distributions, it is often convenient to round off the 99.7% to 100% and work under the assumption that essentially all of the data fall within three standard deviations above and below the mean. This means that if there are no outliers in the data, we can figure that there are approximately six standard deviations separating the smallest (*Min*) and the largest (*Max*) values of the data. In Chapter 14 we defined the range R of a data set ($R = Max - Min$), and, in the case of an approximately normal distribution, we can conclude that the range is about six standard deviations. Remember that this is true as long as we can assume there are no outliers.

16.5 Normal Curves as Models of Real-Life Data Sets

The reason we like to idealize a real-life, approximately normal data set by means of a normal distribution is that we can use many of the properties we just learned about normal distributions to draw useful conclusions about our data. For example, the 68–95–99.7 rule for normal curves can be reinterpreted in the context of an approximately normal data set as follows:

1. About 68% of the data values fall within (plus or minus) one standard deviation of the mean.

2. About 95% of the data values fall within (plus or minus) two standard deviations of the mean.

3. About 99.7%, or practically 100%, of the data values fall within (plus or minus) three standard deviations of the mean.

> ### EXAMPLE 16.7 2005 SAT Scores: Part 2

This example is a follow-up of Example 16.2. It is also related to the discussion in Example 16.3. We are now going to use what we learned so far to analyze the 2005 SAT (verbal) scores. As you may recall, there were $N = 1,475,623$ scores, distributed in a very nice, bell-shaped distribution (the distribution of the scores in class intervals of 50 points was given in Example 16.2). The two new pieces of information that we are going to use in this example are the mean ($\mu = 508$ points) and the standard deviation ($\sigma = 113$ points) of the scores, provided courtesy of the College Board. (*Source:* The College Board, National Report, 2005.)

Just knowing the mean and the standard deviation (and that the distribution of test scores is approximately normal) allows us to draw a lot of useful conclusions:

■ **Median.** In an approximately normal distribution the mean and the median should be about the same. Since we are told that the mean was $\mu = 508$, we can assume the median score was close to 508. Moreover, the median has to be an actual test score when N is odd (which it is in this example), and SAT scores come in multiples of 10, so a reasonable guess for the median would be 510.

■ **First quartile.** Recall that the first quartile is located 0.675 standard deviations below the mean. This means that in this example the first quartile should be close to $508 - 0.675 \times 113 \approx 432$ points (see Example 16.3). But again, the first quartile has to be an actual test score (N is not divisible by 4), so $Q_1 = 430$ points is our best guess.

■ **Third quartile.** We know that the third quartile is located 0.675 standard deviations above the mean. In this case this gives $508 + 0.675 \times 113 \approx 584$. The most reasonable guess is that the actual third quartile was $Q_3 = 580$.

Here is a remarkable thing: In all three cases, our guess was right on the money—as reported by the College Board, the 2005 SAT verbal scores had median $M = 510$, first quartile $Q_1 = 430$, and third quartile $Q_3 = 580$!

We are now going to go analyze the 2005 SAT scores in a little more depth using the 68–95–99.7 rule.

■ **The middle 68.** Approximately 68% of the scores should have fallen within plus or minus *one* standard deviation from the mean. In this case, this range of scores goes from $508 - 113 = 395$ to $508 + 113 = 621$ points. Since SAT scores can only come in multiples of 10, we can estimate that a little over two-thirds of students had scores between 400 and 620 points. The remaining third were equally divided between those scoring 620 points or more (about 16% of test takers) and those scoring 400 points or less (the other 16%).

■ **The middle 95.** Approximately 95% of the scores should have fallen within plus or minus *two* standard deviations from the mean. In this case, that means scores between $508 - 226 = 282$ and $508 + 226 = 734$ points, which really means SAT scores between 280 and 730 points. The remaining 5% of the scores were 730 points or above (about 2.5%) and 280 points or below (the other 2.5%).

■ **Everyone.** The 99.7 part of the 68–95–99.7 rule is not much help in this example. Essentially, it says that all test scores fell between $508 - 339 = 169$ points and $508 + 339 = 847$ points. Duh! SAT (verbal) scores range from 200 to 800 points, so what else is new?

16.6 Distributions of Random Events

We are now ready to take up another important aspect of normal curves—their connection with random events and, through that, their critical role in margins of error of public opinion polls. Our starting point is the following important example.

> **EXAMPLE 16.8** Coin-Tossing Experiments: Part 1

In the opening of this chapter, we discussed the coin-tossing experiment performed by John Kerrich while he was a prisoner of war during World War II. Kerrich tossed a coin 10,000 times and kept records of the number of *Heads* in groups of 100 tosses.

With modern technology, we can repeat Kerrich's experiment and take it much further. Practically any computer can imitate the tossing of a coin by means of a random-number generator. If we use this technique we can "toss coins" in mind-boggling numbers—millions of times if we so choose.

We will start modestly. We will toss our make-believe coin 100 times and count the number of *Heads*, which we will denote by X. Before we do that, let's say a few words about X. Since we cannot predict ahead of time its exact value—we are tempted to think that it should be 50, but, in principle, it could be anything from 0 to 100—we call X a **random variable**. The possible values of the random variable X are governed by the laws of probability: Some values of X are extremely unlikely ($X = 0$, $X = 100$); others are much more likely ($X = 50$), although the likelihood of $X = 50$ is not as great as one would think. It also seems reasonable that (assuming that the coin is fair and *Heads* and *Tails* are equally likely) the likelihood of $X = 49$ should be the same as the likelihood $X = 51$, the likelihood of $X = 48$, should be the same as the likelihood of $X = 52$, and so on.

While all of the preceding statements are true, we still don't have a clue as to what is going to happen when we toss the coin 100 times. One way to get a sense of the probabilities of the different values of X is to repeat the experiment many times and check the frequencies of the various outcomes. Finally, we are ready to do some experimenting!

Our first trial results in 46 *Heads* out of 100 tosses ($X = 46$). The first 10 trials give, in order, $X = 46, 49, 51, 53, 49, 52, 47, 46, 53, 49$. Figure 16-9(a) shows a bar graph for these data.

Continuing this way, we collect data for the values of X in 100, 500, 1000, 5000, and 10,000 trials. The bar graphs are shown in Figs. 16-9(b)–(f), respectively.

Figure 16-9 paints a pretty clear picture of what happens: As the number of trials increases, the distribution of the data becomes more and more bell shaped. At the end, we have data from 10,000 trials, and the bar graph gives an almost perfect normal distribution!

What would happen if someone else decided to repeat what we did—toss an honest coin (be it by hand or by computer) 100 times, count the number of heads, and repeat this experiment a few times? The first 10 trials are likely to produce results very different from ours, but as the number of trials increases, their results and our results will begin to look more and more alike. After 10,000 trials, their bar graph will be almost identical to the bar graph shown in Fig. 16-9(f). In a sense, this says that doing the experiments a second time is a total waste of time—in fact, it was even a waste the first time! *Everything that happened at the end could have been predicted without ever tossing a coin!*

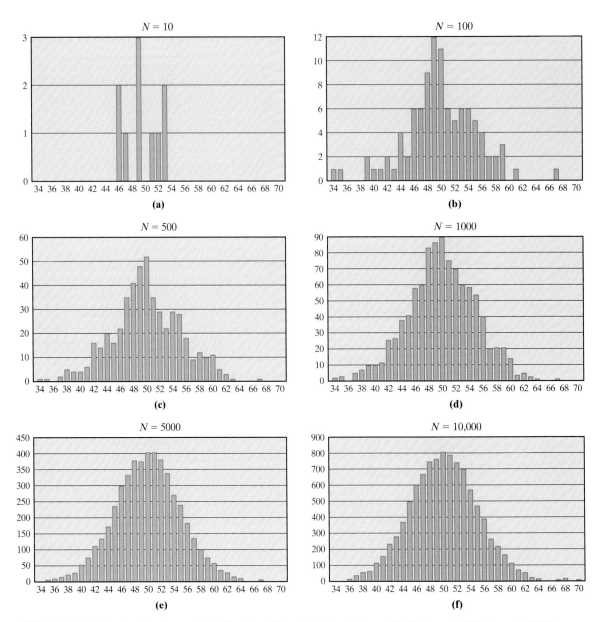

FIGURE 16-9 Distribution of random variable *X* (number of *Heads* in 100 coin tosses) (a) 10 times, (b) 100 times, (c) 500 times, (d) 1000 times, (e) 5000 times, and (f) 10,000 times.

Knowing that the random variable *X* has an approximately normal distribution is, as we have seen, quite useful. The clincher would be to find out the values of the mean μ and the standard deviation σ of this distribution. Looking at Fig. 16-9(f), we can pretty much see where the mean is—right at 50. This is not surprising, since the axis of symmetry of the distribution has to pass through 50 as a simple consequence of the fact that the coin is honest. The value of the standard deviation is less obvious. For now, let's accept the fact that it is $\sigma = 5$. We will explain how we got this value shortly.

Let's summarize what we now know. An honest coin is tossed 100 times. The number of *Heads* in the 100 tosses is a random variable, which we call *X*. If we repeat this experiment a large number of times (say *N*), the random variable *X* will have an approximately normal distribution with mean $\mu = 50$ and standard deviation $\sigma = 5$, and the larger the value of N is, the better this approximation will be.

The real significance of these facts is that they are true not because we took the trouble to toss a coin a million times. Even if we did not toss a coin at all, all of these statements would still be true. *For a sufficiently large number of repetitions of the experiment of tossing an honest coin 100 times, the number of Heads X is a random variable that has an approximately normal distribution with center $\mu = 50$ Heads and standard deviation $\sigma = 5$ Heads.* This is a mathematical, rather than an experimental, fact.

$\blacktriangleleft\blacktriangleleft$

16.7 Statistical Inference

Next, we are going to take our first tentative leap into statistical inference. Suppose that we have an honest coin and intend to toss it 100 times. We are going to do this just once, and we will let X denote the resulting number of *Heads*. Been there, done that! What's new now is that we a have a solid understanding of the statistical behavior of the random variable X—it has an approximately normal distribution with mean $\mu = 50$ and standard deviation $\sigma = 5$—and this allows us to make some very reasonable predictions about the possible values of X.

For starters, we can predict the chance that X will fall somewhere between 45 and 55 (one standard deviation below and above the mean)—it is 68%. Likewise, we know that the chance that X will fall somewhere between 40 and 60 is 95%, and between 35 and 65 is a whopping 99.7%.

What if, instead of tossing the coin 100 times, we were to toss it n times? (After all, there is nothing unique about tossing 100 times.) Not surprisingly, the bell-shaped distribution we saw in Example 16.8 would still be there—only the values of μ and σ would change. Specifically, for n sufficiently large (typically $n \geq 30$), the number of *Heads* in n tosses would be a random variable with an approximately normal distribution with mean $\mu = n/2$ *Heads* and standard deviation $\sigma = (\sqrt{n})/2$ *Heads*. This is an important fact for which we have coined the name the **honest-coin principle**.

> **The Honest-Coin Principle**
>
> Suppose an honest coin is tossed n times ($n \geq 30$), and let X denote the number of *Heads* that come up. The random variable X has an approximately normal distribution with mean $\mu = n/2$ *Heads* and standard deviation $\sigma = (\sqrt{n})/2$ heads.

Note that when we apply the honest-coin principle to $n = 100$ tosses, the distribution of the number of *Heads* has mean $\mu = 100/2 = 50$ *Heads* and standard deviation $\sigma = \sqrt{100}/2 = 10/2 = 5$ heads, confirming what we already knew.

▶ **EXAMPLE 16.9** Coin-Tossing Experiments: Part 2

An honest coin is going to be tossed 256 times. Before this is done, we have the opportunity to make some bets. Let's say that we can make a bet (with even odds) that if the number of *Heads* tossed falls somewhere between 120 and 136, we will win; otherwise we will lose. Should we make such a bet?

Let X denote the number of *Heads* in 256 tosses of an honest coin. By the honest-coin principle, X is a random variable having a distribution that is approximately normal with mean $\mu = 256/2 = 128$ *Heads* and standard deviation $\sigma = \sqrt{256}/2 = 8$ heads. The values 120 to 136 are exactly one standard deviation below and above the mean of 128, which means that *there is a 68% chance that the number of Heads will fall somewhere between 120 and 136.* We should indeed make this bet! A similar calculation tells us that *there is a 95% chance that the number of Heads will fall somewhere between 112 and 144*, and *the chance that the number of Heads will fall somewhere between 104 and 152 is 99.7%.* ◀◀

What happens when the coin being tossed is not an honest coin? Surprisingly, the distribution of the number of *Heads* X in n tosses of such a coin is still approximately normal, as long as the number n is not too small (a good rule of thumb is $n \geq 30$). All we need now is a **dishonest-coin principle** to tell us how to find the mean and the standard deviation.

> **The Dishonest-Coin Principle**
>
> Suppose an arbitrary coin is tossed n times ($n \geq 30$), and let X denote the number of *Heads* that come up. Suppose also that p is the probability of the coin landing heads, and $(1 - p)$ is the probability of the coin landing tails. Then the random variable X has an approximately normal distribution with mean $\mu = n \cdot p$ *Heads* and standard deviation $\sigma = \sqrt{n \cdot p \cdot (1 - p)}$ *Heads.*

▶ EXAMPLE 16.10 Coin-Tossing Experiments: Part 3

A coin is rigged so that it comes up *Heads* only 20% of the time (i.e., $p = 0.20$). The coin is tossed 100 times ($n = 100$), and X is the number of *Heads* in the 100 tosses. What can we say about X?

According to the dishonest-coin principle, the distribution of the random variable X is approximately normal with mean $\mu = 100 \times 0.20 = 20$ *Heads* and standard deviation $\sigma = \sqrt{100 \times 0.20 \times 0.80} = 4$ heads.

Applying the 68–95–99.7 rule with $\mu = 20$ and $\sigma = 4$ gives the following facts:

- There is about a 68% chance that X will be somewhere between 16 and 24 ($\mu - \sigma \leq X \leq \mu + \sigma$).

- There is about a 95% chance that that X will be somewhere between 12 and 28 ($\mu - 2\sigma \leq X \leq \mu + 2\sigma$).

- The number of *Heads* is almost guaranteed (about 99.7% chance) to fall somewhere between 8 and 32 ($\mu - 3\sigma \leq X \leq \mu + 3\sigma$).

Note that in this example, *Heads* and *Tails* are no longer interchangeable concepts—*Heads* is an outcome with probability $p = 0.2$, while *Tails* is an outcome with much higher probability (0.8). We can, however, apply the principle equally well to describe the distribution of the number of *Tails* in 100 coin tosses of the same dishonest coin: The distribution for the number of *Tails* is approximately normal with mean $\mu = 100 \times 0.80 = 80$ and standard deviation $\sigma = \sqrt{100 \times 0.80 \times 0.20} = 4$. Note that σ is still the same. ◀◀

See Exercise 77.

The dishonest-coin principle can be applied to any coin, even one that is fair (i.e., with $p = 1/2$). In the case $p = 1/2$, the honest- and dishonest-coin principles say the same thing.

The dishonest-coin principle is a special version of one of the most important laws in statistics, a law generally known as the *central limit theorem*. We will now briefly illustrate why the importance of the dishonest-coin principle goes beyond the tossing of coins.

> ## EXAMPLE 16.11 Sampling for Defective Light Bulbs

An assembly line produces 100,000 light bulbs a day, 20% of which generally turn out to be defective. Suppose we draw a random sample of $n = 100$ light bulbs. Let X represent the *number of defective light bulbs* in the sample. What can we say about X?

A moment's reflection will show that, in a sense, this example is completely parallel to Example 16.10—think of selecting defective light bulbs as analogous to tossing *Heads* with a dishonest coin. We can use the dishonest-coin principle to infer that the number of defective light bulbs in the sample is a random variable having an approximately normal distribution with a mean of 20 light bulbs and standard deviation of 4 light bulbs. Thus,

- There is a 68% chance that the number of defective light bulbs in the sample will fall somewhere between 16 and 24.

- There is a 95% chance that the number of defective light bulbs in the sample will fall somewhere between 12 and 28.

- The number of defective light bulbs in the sample is practically guaranteed (a 99.7% chance) to fall somewhere between 8 and 32.

Probably the most important point here is that each of the preceding facts can be rephrased in terms of sampling errors, a concept we first discussed in Chapter 13. For example, say we had 24 defective light bulbs in the sample; in other words, 24% of the sample (24 out of 100) are defective light bulbs. If we use this statistic to estimate the percent of defective light bulbs overall, then the sampling error would be 4% (because the estimate is 24% and the value of the parameter is 20%). By the same token, if we had 16 defective light bulbs in the sample, the sampling error would be −4%. Coincidentally, the standard deviation is $\sigma = 4$ light bulbs, or 4% of the sample. (We computed it in Example 16.10.) Thus, we can rephrase our previous assertions about sampling errors as follows:

- When estimating the proportion of defective light bulbs coming out of the assembly line by using a sample of 100 light bulbs, there is a 68% chance that the sampling error will fall somewhere between −4 and 4%.

- When estimating the proportion of defective light bulbs coming out of the assembly line by using a sample of 100 light bulbs, there is a 95% chance that the sampling error will fall somewhere between −8 and 8%.

- When estimating the proportion of defective light bulbs coming out of the assembly line by using a sample of 100 light bulbs, there is a 99.7% chance that the sampling error will fall somewhere between −12 and 12%.   ◀◀

▶ **EXAMPLE 16.12** Sampling with Larger Samples

Suppose we have the same assembly line as in Example 16.11, but this time we are going to take a really big sample of $n = 1600$ light bulbs. Before we even count the number of defective light bulbs in the sample, let's see how much mileage we can get out of the dishonest-coin principle. The standard deviation for the distribution of defective light bulbs in the sample is $\sqrt{1600 \times 0.2 \times 0.8} = 16$, which just happens to be exactly 1% of the sample ($16/1600 = 1\%$). This means that when we estimate the proportion of defective light bulbs coming out of the assembly line using this sample, we can have some sort of a handle on the sampling error.

- We can say with some confidence (68%) that the sampling error will fall somewhere between -1 and 1%.
- We can say with a lot of confidence (95%) that the sampling error will fall somewhere between -2 and 2%.
- We can say with tremendous confidence (99.7%) that the sampling error will fall somewhere between -3 and 3%. ◀◀

The next and last example shows how the dishonest-coin principle can be used to estimate the margin of error in a public opinion poll, an issue of considerable importance in modern statistics.

▶ **EXAMPLE 16.13** Measuring the Margin of Error of a Poll

In California, school bond measures require a 66.67% vote for approval. Suppose that an important school bond measure is on the ballot in the upcoming election. In the most recent poll of 1200 randomly chosen voters, 744 of the 1200 voters sampled, or 62%, indicated that they would vote for the school bond measure. Let's assume that the poll was properly conducted and that the 1200 voters sampled represent an unbiased sample of the entire population. What are the chances that the 62% statistic is the result of sampling variability and that the actual vote for the bond measure will be 66.67% or more?

Here, we will use a variation of the dishonest-coin principle, with each voter being likened to a coin toss. Voting for the bond measure is like the coin coming up *Heads*; against is *Tails*. The probability (p) of "heads" for this "coin" will turn out to be the proportion of voters in the population that support the bond measure: If p turns out to be 0.6667 or more, the bond measure will pass. Our problem is that we don't know p, so how can we use the dishonest-coin principle to estimate the mean and standard deviation of the sampling distribution?

The idea here is to use the 62% (0.62) statistic from the sample as an estimate for the actual value of p in the formula for the standard deviation given by the dishonest-coin principle. (Even though we know that this is only a rough estimate for p, this generally turns out to give us a good estimate for the standard deviation.) In our example, the approximate standard deviation for the number of "heads" in the sample turns out to be $\sqrt{1200 \times 0.62 \times 0.38} \approx 16.8$ voters. When we convert this number to percentages, we get a standard deviation that is approximately 1.4% of the sample ($16.8/1200 = 0.014$).

The standard deviation for the sampling distribution of the proportion of voters in favor of the measure expressed as a percentage of the entire sample is called the **standard error**. (For our example, we have found above that the standard error is approximately 1.4%.) In sampling and public opinion polls, it is customary to express the information about the population in terms of **confidence**

intervals, which are themselves based on standard errors: A 95% confidence interval is given by two standard errors below and above the statistic obtained from the sample; a 99.7% confidence interval is given by going three standard errors below and above the sample statistic.

In our example, we have a 95% confidence interval of 62% plus or minus 2.8%, which means that we can say with 95% confidence (we would be right approximately 95 out of 100 times) that the actual vote for the bond measure will fall somewhere between 59.2% (62 − 2.8) and 64.8% (62 + 2.8), and thus that the bond measure will lose. Want even more certainty? Take a 99.7% confidence interval of 62% plus or minus 4.2%—it is almost certain that the actual vote will turn out somewhere in that range. Even in the most optimistic scenario, the vote will not reach the 66.67% needed to pass the bond measure. ≪

Conclusion

From coin tosses to test scores, many real-life data sets follow the call of the bell. In this chapter we studied bell-shaped (normal) curves, some of their mathematical properties, and how these properties can be used to analyze real-life bell-shaped data sets. It was a brief introduction to what is undoubtedly one of the most widely used and sophisticated tools of modern mathematical statistics.

In this chapter we also got a brief glimpse of the concept of statistical inference. The process of drawing conclusions and inferences based on limited data is an essential part of statistics. It gives us a way not only to analyze what has already taken place, but also to make reasonably accurate large-scale predictions of what will happen in certain random situations. Casinos know, without any shadow of a doubt, that in the long run, they will make a profit—it is a mathematical law! A similar law gives us the confidence to trust the results of surveys and public opinion polls (up to a point!), the quality of the products we buy, and even the statistical data our government uses to make many of its decisions. In all of these cases, bell-shaped distributions of data and the laws of probability come together to give us insight into what was, is, and most likely will be.

Profile Carl Friedrich Gauss (1777–1855)

In spite of the obvious connection between mathematics and money, mathematicians don't usually make the cut when it comes to getting their picture featured in a bank note. The notable exception is Carl Friedrich Gauss, the man often referred to as "the prince of mathematicians." The German ten-mark note shown in the figure celebrates both Gauss and one of his most famous "discoveries"—the normal curve (shown above the 10).

Carl Friedrich Gauss was born in the Duchy of Brunswick (now Germany) into a poor, working-class family of little education. From a very early age, young Gauss showed a prodigious talent for numbers and abstract mathematical thinking. One of the most famous stories about Gauss was his solution to the addition problem $1 + 2 + 3 + \cdots + 99 + 100 = ?$. The problem was assigned one day in class by a teacher hoping to keep the children busy for a while. No sooner had the teacher finished assigning the problem, Gauss had the solution. His insight: The sum consists of 50 pairs $(1 + 100, 2 + 99, \ldots, 50 + 51)$, each of which adds up to 101 and thus equals $50 \times 101 = 5050$. Gauss was seven years old at the time.

Recognizing Gauss's mathematical genius, the Duke of Brunswick agreed to be the young man's sponsor and financially support his education. Gauss first attended the Collegium Carolinum in Brunswick, but eventually transferred to the more prestigious Göttingen University. By the time he was 21, Gauss had produced several major mathematical discoveries, including the complete characterization of all regular polygons that can be constructed with straightedge and compass, a classic problem that could be traced all the way back to the ancient Greeks. In 1799 Gauss received his doctorate in mathematics from the University of Helmstedt for his work on what is now known as the *fundamental theorem of algebra*. Two years later he published his first and greatest book, *Disquisitiones Arithmeticae*, a pioneering treatise on the theory of numbers. With the publication of *Disquisitiones Arithmeticae*, Gauss came to be recognized as one of the greatest mathematicians of his age. He was 24 years old at the time.

Gauss's scientific genius went well beyond theoretical mathematics, and he made major contributions to empirical sciences such as astronomy and surveying, where precise observation and measurement are critical. He greatly added to his growing scientific reputation when he was able to predict the exact orbit of the asteroid Ceres, which had been discovered in 1801 by the Italian astronomer Giuseppe Piazzi. Piazzi had observed and tracked what he thought was a new planet for only a few weeks before it disappeared behind the

Sun. Using Piazzi's limited data and a method of his own invention, Gauss predicted where and when Ceres would be seen again. Just about a year after its disappearance Ceres reappeared, almost in the exact position Gauss had predicted. On the strength of this achievement, Gauss was appointed Director of the Göttingen Observatory and Professor of Astronomy at Göttingen University.

As part of his astronomical calculations, Gauss became interested in the study of measurement errors and their probability distributions. This led him to the discovery of the bell-shaped normal distribution, which is also called the *Gaussian distribution* in his honor. Always a practical man, toward the later part of his life Gauss became increasingly interested in the applications of mathematics to physics and made important discoveries in the theory of magnetism, developed Kirchhoff's laws in electricity, and constructed an early version of the telegraph. He also became interested in the analysis and prediction of financial markets and made a tremendous amount of money speculating in stocks.

By the time he died in 1855 at the age of 78, Gauss had achieved a remarkable measure of fame and fortune—he was the preeminent mathematician of his time as well as an extremely wealthy man.

Key Concepts

Exercises

WALKING

A. Normal Curves

1. For the normal distribution described by the curve shown in the figure,

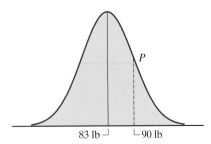

(a)   find μ.

(b)   find M (the median).

(c)   estimate the value of σ.

   (***Note:*** *P is a point of inflection of the curve.*)

2. For the normal distribution described by the curve shown in the figure,

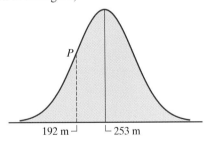

(a)   find μ.

(b)   find M (the median).

(c)   estimate the value of σ.

   (***Note:*** *P is a point of inflection of the curve.*)

3. For a normal distribution with $\mu = 81.2$ lb and $\sigma = 12.4$ lb,

(a)   estimate the value of Q_1 (rounded to the nearest tenth of a pound).

(b) estimate the value of Q_3 (rounded to the nearest tenth of a pound).

4. For a normal distribution with $\mu = 2354$ points and $\sigma = 468$ points,

(a) estimate the value of Q_1 (rounded to the nearest point).

(b) estimate the value of Q_3 (rounded to the nearest point).

5. For a normal distribution with $Q_1 = 72.8$ lb and $\sigma = 12.4$ lb, estimate the value of Q_3 (rounded to the nearest tenth of a pound).

6. For a normal distribution with $Q_3 = 2670$ points and $\sigma = 468$ points, estimate the value of Q_1 (rounded to the nearest point).

7. For a normal distribution with $\mu = 81.2$ in and $Q_3 = 94.7$ in, estimate the value of σ (rounded to the nearest tenth of an inch).

8. For a normal distribution with $\mu = \$18,565$ and $Q_1 = \$15,514$, estimate the value of σ (rounded to the nearest dollar).

9. For the normal distribution described by the curve shown in the figure,

(**Note:** P and P' are the inflection points of the curve.)

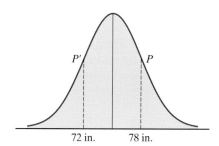

72 in.   78 in.

(a) find the center.

(b) estimate the value of σ (rounded to the nearest inch).

(c) estimate the values of Q_1 and Q_3 (rounded to the nearest inch).

10. For the normal distribution described by the curve shown in the figure,

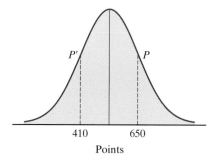

410   650

Points

(a) find the center.

(b) estimate the value of σ (rounded to the nearest point).

(**Note:** P and P' are the inflection points of the curve.)

(c) estimate the values of Q_1 and Q_3 (rounded to the nearest point).

In Exercises 11 through 14, you are given some information about a distribution. Explain why the distribution is not normal. (**Note:** M denotes the median.)

11. A distribution with median $M = 82$, mean $\mu = 71$, and standard deviation $\sigma = 11$

12. A distribution with median $M = 210$, mean $\mu = 195$, and standard deviation $\sigma = 15$

13. A distribution with median $M = 453$, mean $\mu = 453$, first quartile $Q_1 = 343$, and third quartile $Q_3 = 553$

14. A distribution with center $M = \mu = 47$, first quartile $Q_1 = 35$, and standard deviation $\sigma = 10$

B. Standardizing Data

15. A normal distribution has $\mu = 30$ kg and $\sigma = 15$ kg. Find the standardized value of

(a) 45 kg.

(b) 54 kg.

(c) 0 kg.

(d) 3 kg.

16. A normal distribution has $\mu = 110$ points and $\sigma = 12$ points. Find the standardized value of

(a) 128 points.

(b) 100 points.

(c) 110 points.

(d) 71 points.

17. A normal distribution has $\mu = 253.45$ ft and $Q_3 = 278.58$ ft. Find the standardized value of

(a) 261.71 ft

(b) 185.79 ft

(c) 253.45 ft

18. A normal distribution has $\mu = 49.5$ lb and $Q_1 = 44.1$ lb. Find the standardized value of

(a) 41.5 lb

(b) 61.5 lb

(c) 35.1 lb

19. Find the standardized value of Q_1 in any normal distribution.

20. Find the standardized value of Q_3 in any normal distribution.

21. In a normal distribution with $\mu = 183.5$ ft and $\sigma = 31.2$ ft, find the data value corresponding to each of the following standardized values.

(a) -1

(b) 0.5

(c) -2.3

(d) 0

22. In a normal distribution with $\mu = 83.2$ gal and $\sigma = 4.6$ gal, find the data value corresponding to each of the following standardized values.

(a) 2

(b) -1.5

(c) -0.43

(d) 0

23. In a normal distribution with $\mu = 50$ lb, a weight of 84 lb has a standardized value of 2. Find the standard deviation σ.

24. In a normal distribution with mean $\mu = 30$, the data value -60 has a standardized value of -3. Find the standard deviation σ.

25. In a normal distribution with standard deviation $\sigma = 15$, the data value 50 has a standardized value of 3. Find the mean μ.

26. In a normal distribution with standard deviation $\sigma = 20$, the data value 10 has a standardized value of -2. Find the mean μ.

27. In a normal distribution the data value 20 has a standardized value of -2, and the data value 100 has a standardized value of 3. Find the mean μ and the standard deviation σ.

28. In a normal distribution the data value -10 has a standardized value of 0, and the data value 50 has a standardized value of 2. Find the mean μ and the standard deviation σ.

C. The 68–95–99.7 Rule

29. Find the mean μ and standard deviation σ for the normal distribution described by the curve shown in the figure.

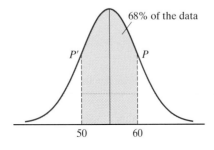

30. Find the mean μ and standard deviation σ for the normal distribution described by the curve shown in the figure.

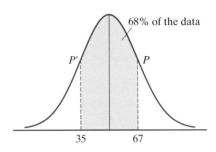

31. Find the mean μ and standard deviation σ for the normal distribution described by the curve shown in the figure.

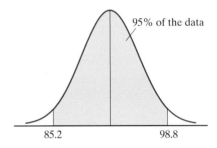

32. Find the mean μ and standard deviation σ for the normal distribution described by the curve shown in the figure.

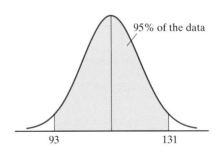

33. Find the mean μ and standard deviation σ for the normal distribution described by the curve shown in the figure.

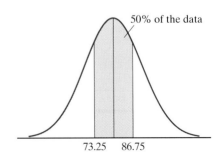

34. Find the mean μ and standard deviation σ for the normal distribution described by the curve shown in the figure.

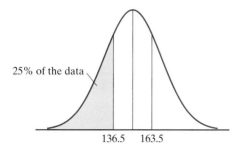

25% of the data

136.5   163.5

35. In a normal distribution, what percent of the data fall

(a) below the point two standard deviations above the mean?

(b) between one and two standard deviations above the mean?

36. In a normal distribution, what percent of the data fall

(a) above the point three standard deviations below the mean?

(b) between two and three standard deviations below the mean?

37. A normal distribution has standard deviation $\sigma = 6.1$ cm, and 84% of the data fall above 50.2 cm. Find the mean μ.

38. A normal distribution has mean $\mu = 56.3$ cm, and 84% of the data fall above 50.2 cm. Find the standard deviation σ.

39. A normal distribution has mean $\mu = 12.6$ and standard deviation $\sigma = 4.0$. Approximately what percent of the data fall between 9.9 and 16.6?

40. A normal distribution has mean $\mu = 500$ and standard deviation $\sigma = 35$. Approximately what percent of the data fall between 465 and 605?

D. Approximately Normal Data Sets

Exercises 41 through 44 refer to the following: 2500 students take a college entrance exam. The scores on the exam have an approximately normal distribution with mean $\mu = 52$ points and standard deviation $\sigma = 11$ points.

41. (a) Estimate the average score on the exam.

(b) Estimate what percent of the students scored 52 points or more.

(c) Estimate what percent of the students scored between 41 and 63 points.

(d) Estimate what percent of the students scored 63 points or more.

42. (a) Estimate how many students scored between 30 and 74 points.

(b) Estimate how many students scored 30 points or less.

(c) Estimate how many students scored 19 points or less.

43. (a) Estimate the first-quartile score for this exam.

(b) Estimate the third-quartile score for this exam.

(c) Estimate the interquartile range for this exam.
   (***Note:*** $IQR = Q_3 - Q_1$.)

44. For each of the following scores, estimate in what percentile of the students taking the exam the score would place you.

(a) 52

(b) 63

(c) 60

(d) 85

Exercises 45 through 48 refer to the following: As part of a research project, the blood pressures of 2000 patients in a hospital are recorded. The systolic blood pressures (given in millimeters) have an approximately normal distribution with mean $\mu = 125$ and standard deviation $\sigma = 13$.

45. (a) Estimate the number of patients whose blood pressure was between 99 and 151 millimeters.

(b) Estimate the number of patients whose blood pressure was between 112 and 151 millimeters.

46. (a) Estimate the number of patients whose blood pressure was 99 millimeters or higher.

(b) Estimate the number of patients whose blood pressure was between 99 and 138 millimeters.

47. For each of the following blood pressures, estimate the percentile of the patient population to which they correspond.

(a) 100 millimeters

(b) 112 millimeters

(c) 138 millimeters

(d) 164 millimeters

48. (a) Assuming that there were no outliers, estimate the value of the lowest (*Min*) and the highest (*Max*) blood pressures.

(b) Assuming that there were no outliers, give an estimate of the five-number summary (*Min*, Q_1, μ, Q_3, *Max*) of the distribution of blood pressures.

Exercises 49 through 52 refer to the following: Packaged foods sold at supermarkets are not always the weight indicated on the package. Variability always crops up in the manufacturing and packaging process. Suppose that the exact weight of a "12-ounce" bag of potato chips is a random variable that has an approximately normal distribution with mean $\mu = 12$ ounces and standard deviation $\sigma = 0.5$ ounce.

49. If a "12-ounce" bag of potato chips is chosen at random, what are the chances that

(a) it weighs somewhere between 11 and 13 ounces?

(b) it weighs somewhere between 12 and 13 ounces?

(c) it weighs more than 11 ounces?

50. If a "12-ounce" bag of potato chips is chosen at random, what are the chances that

(a) it weighs somewhere between 11.5 and 12.5 ounces?

(b) it weighs somewhere between 12 and 12.5 ounces?

(c) it weighs more than 12.5 ounces?

51. Suppose that 500 "12-ounce" bags of potato chips are chosen at random. Estimate the number of bags with weight

(a) 11 ounces or less.

(b) 11.5 ounces or less.

(c) 12 ounces or less.

(d) 12.5 ounces or less.

(e) 13 ounces or less.

(f) 13.5 ounces or less.

52. Suppose that 1500 "12-ounce" bags of potato chips are chosen at random. Estimate the number of bags of potato chips with weight

(a) between 11 and 11.5 ounces.

(b) between 11.5 and 12 ounces.

(c) between 12 and 12.5 ounces.

(d) between 12.5 and 13 ounces.

(e) between 13 and 13.5 ounces.

Exercises 53 through 56 refer to the following: The distribution of weights for children of a given age and sex is approximately normal. This fact allows a doctor or nurse to find from a child's weight the weight percentile of the population (all children of the same age and sex) to which the child belongs. Typically, this is done using special charts provided to the doctor or nurse, but these percentiles can also be computed using facts about approximately normal distributions, such as the ones we learned in this chapter.

(***Note:*** *The numbers in these examples are 1977 figures taken from charts produced by the National Center for Health Statistics, U.S. Department of Health and Human Services.*)

53. The distribution of weights for 6-month-old baby boys is approximately normal with mean $\mu = 17.25$ pounds and standard deviation $\sigma = 2$ pounds.

(a) Suppose that a 6-month-old boy weighs 15.25 pounds. Approximately what weight percentile is he in?

(b) Suppose that a 6-month-old boy weighs 21.25 pounds. Approximately what weight percentile is he in?

(c) Suppose that a 6-month-old boy is in the 75th percentile in weight. Estimate his weight.

54. The distribution of weights for 12-month-old baby girls is approximately normal with mean $\mu = 21$ pounds and standard deviation $\sigma = 2.2$ pounds.

(a) Suppose that a 12-month-old girl weighs 16.6 pounds. Approximately what weight percentile is she in?

(b) Suppose that a 12-month-old girl weighs 18.8 pounds. Approximately what weight percentile is she in?

(c) Suppose that a 12-month-old girl is in the 75th percentile in weight. Estimate her weight.

55. The distribution of weights for 1-month-old baby girls is approximately normal with mean $\mu = 8.75$ pounds and standard deviation $\sigma = 1.1$ pounds.

(a) Suppose that a 1-month-old girl weighs 11 pounds. Approximately what weight percentile is she in?

(b) Suppose that a 1-month-old girl weighs 12 pounds. Approximately what weight percentile is she in?

(c) Suppose that a 1-month-old girl is in the 25th percentile in weight. Estimate her weight.

56. The distribution of weights for 12-month-old baby boys is approximately normal with mean $\mu = 22.5$ pounds and standard deviation $\sigma = 2.2$ pounds.

(a) Suppose that a 12-month-old boy weighs 24 pounds. Approximately what weight percentile is he in?

(b) Suppose that a 12-month-old boy weighs 21 pounds. Approximately what weight percentile is he in?

(c) Suppose that a 12-month-old boy is in the 84th percentile in weight. Estimate his weight.

E. The Honest- and Dishonest-Coin Principles

57. An honest coin is tossed $n = 3600$ times. Let the random variable Y denote the number of tails tossed.

(a) Find the mean and the standard deviation of the distribution of the random variable Y.

(b) What are the chances that Y will fall somewhere between 1770 and 1830?

(c) What are the chances that Y will fall somewhere between 1800 and 1830?

(d) What are the chances that Y will fall somewhere between 1830 and 1860?

58. An honest coin is tossed $n = 6400$ times. Let the random variable X denote the number of *Heads* tossed.

(a) Find the mean and the standard deviation of the distribution of the random variable X.

(b) What are the chances that X will fall somewhere between 3120 and 3280?

(c) What are the chances that X will fall somewhere between 3080 and 3200?

(d) What are the chances that X will fall somewhere between 3240 and 3280?

59. Suppose a random sample of $n = 7056$ adults is to be chosen for a survey. Assume that the gender of each adult in the sample is equally likely to be male as it is female. Find the probability that the number of females in the sample is

(a) between 3486 and 3570.

(b) less than 3486.

(c) less than 3570.

60. An honest die is rolled. If the roll comes out even (2, 4, or 6), you will win $1; if the roll comes out odd (1, 3, or 5), you will lose $1. Suppose that in one evening you play this game $n = 2500$ times in a row.

(a) What is the probability that by the end of the evening you will not have lost any money?

(b) What is the probability that the number of even rolls will fall between 1250 and 1300?

(c) What is the probability that you will win $100 or more?

(d) What is the probability that you will win exactly $101?

61. A dishonest coin with probability of *Heads* $p = 0.4$ is tossed $n = 600$ times. Let the random variable X represent the number of times the coin comes up heads.

(a) Find the mean and standard deviation for the distribution of X.

(b) Find the first and third quartiles for the distribution of X.

(c) Find the probability that the number of *Heads* will fall somewhere between 216 and 264.

62. A dishonest coin with probability of *Heads* $p = 0.75$ is tossed $n = 1200$ times. Let the random variable X represent the number of times the coin comes up *Heads*.

(a) Find the mean and standard deviation for the distribution of X.

(b) Find the first and third quartiles for the distribution of X.

(c) Find the probability that the number of *Heads* will fall somewhere between 900 and 945.

63. Suppose that an honest die is rolled $n = 180$ times. Let the random variable X represent the number of times the number 6 is rolled.

(a) Find the mean and standard deviation for the distribution of X.

(b) Find the probability that a 6 will be rolled more than 40 times.

(c) Find the probability that a 6 will be rolled somewhere between 30 and 35 times.

64. Suppose that one out of every ten cereal boxes has a prize. Out of a shipment of $n = 400$ cereal boxes, find the probability that there are

(a) somewhere between 34 and 40 prizes.

(b) somewhere between 40 and 52 prizes.

(c) more than 52 prizes.

65. Each day a machine produces 90,000 widgets. Quality control data shows that the probability that one of the widgets produced by the machine is defective is $p = 0.10$. Shop rules state that those days when the machine produces more than 9180 defective widgets it must be recalibrated. Estimate the percentage of the time that this machine must be recalibrated.

66. At Tasmania State University, the probability that an entering freshman will graduate in four years is 0.89. What are the chances that of a class of 2000 freshmen, 1750 or more will graduate in four years?

JOGGING

Percentiles.

*The **pth percentile** of a sorted data set is a number x_p such that p % of the data fall at or below x_p and $(100 - p)\%$ of the data fall at or above x_p. (For details, see Chapter 14, Section 14.3.) For normally distributed data sets, there are detailed statistical tables that give the location of the pth percentile for every possible p between 1 and 99. The following table is an abbreviated version giving the approximate location of some of the more frequently used percentiles in a normal distribution with mean μ and standard deviation σ. For approximately normal distributions, the table can be used to estimate these percentiles.*

Percentile	Approximate location	Percentile location	Approximate
99th	$\mu + 2.33\sigma$	1st	$\mu - 2.33\sigma$
95th	$\mu + 1.65\sigma$	5th	$\mu - 1.65\sigma$
90th	$\mu + 1.28\sigma$	10th	$\mu - 1.28\sigma$
80th	$\mu + 0.84\sigma$	20th	$\mu - 0.84\sigma$
75th	$\mu + 0.675\sigma$	25th	$\mu - 0.675\sigma$
70th	$\mu + 0.52\sigma$	30th	$\mu - 0.52\sigma$
60th	$\mu + 0.25\sigma$	40th	$\mu - 0.25\sigma$
50th	μ		

In Exercises 67 through 73, you should use the table to make your estimates.

67. The distribution of weights for 6-month-old baby boys is approximately normal with mean $\mu = 17.25$ pounds and standard deviation $\sigma = 2$ pounds.

(a) Suppose that a 6-month-old baby boy weighs in the 95th percentile of his age group. Estimate his weight in pounds approximated to two decimal places.

(b) Suppose that a 6-month-old baby boy weighs in the 40th percentile of his age group. Estimate his weight in pounds approximated to two decimal places.

68. Several thousand students took a college entrance exam. The scores on the exam have an approximately normal distribution with mean $\mu = 55$ points and standard deviation $\sigma = 12$ points.

   (a) For a student that scored in the 99th percentile, estimate the student's score on the exam.

   (b) For a student that scored in the 30th percentile, estimate the student's score on the exam.

69. Consider again the distribution of weights of 6-month-old baby boys discussed in Exercise 67.

   (a) Jimmy is a 6-month-old-baby who weighs 17.75 lb. Estimate the percentile corresponding to Jimmy's weight.

   (b) David is a 6-month-old baby who weighs 16.2 lb. Estimate the percentile corresponding to David's weight.

70. Consider again the college entrance exam discussed in Exercise 68.

   (a) Mary scored 83 points on the exam. Estimate the percentile in which this score places her.

   (b) Adam scored 45 points on the exam. Estimate the percentile in which this score places him.

   (c) If there were 250 students that scored 35 points or less on the exam, estimate the total number of students that took the exam.

71. In 2005, 1,475,623 college-bound seniors took the SAT exam. The distribution of scores in the math section of the SAT was approximately normal with mean $\mu = 520$ and standard deviation $\sigma = 115$.

   (*Source*: www.collegeboard.org)

   (a) Estimate the 75th percentile score on the exam.

   (b) Find the percentile corresponding to a test score of 750.

72. Consider a normal distribution with mean $\mu = 0$ and standard deviation $\sigma = 1$.

   (a) Find the 90th percentile (rounded to two decimal places).

   (b) Find the 10th percentile (rounded to two decimal places).

   (c) Find the 80th percentile (rounded to two decimal places).

   (d) Find the 20th percentile (rounded to two decimal places).

   (e) Suppose that you are given that the 85th percentile is approximately 1.04. Find the 15th percentile.

73. The grade breakdown in Professor Blackbeard's Stat 101 class is 10% As, 20% Bs, 40% Cs, 25% Ds, and 5% Fs. The numerical class scores had an approximately normal distribution with mean $\mu = 65.2$ and standard deviation $\sigma = 10$.

   (a) What is the minimum numerical score needed to get an A?

   (b) What is the minimum numerical score needed to get a B?

   (c) What is the minimum numerical score needed to get a C?

   (d) What is the minimum numerical score needed to get a D?

74. Estimate how many standard deviations describe the interquartile range of an approximately normal distribution.

75. An honest coin is tossed n times. Let the random variable X denote the number of *Heads* tossed. Find the value of n so that there is a 95% chance that X will be between $n/2 - 15$ and $n/2 + 15$.

76. An honest coin is tossed n times. Let the random variable Y denote the number of tails tossed. Find the value of n so that there is a 16% chance that Y will be at least $n/2 + 10$.

77. Explain why when the dishonest-coin principle is applied to an honest coin ($p = 0.5$), we get the honest-coin principle.

RUNNING

78. A dishonest coin with probability of *Heads* $p = 0.1$ is tossed n times. Let the random variable X denote the number of *Heads* tossed. Find the value of n so that there is a 95% chance that X will be between $n/10 - 30$ and $n/10 + 30$.

79. An honest pair of dice is rolled n times. Let the random variable Y denote the number of times a total of 7 is rolled. Find the value of n so that there is a 95% chance that Y will be between $n/6 - 20$ and $n/6 + 20$.

80. On an American roulette wheel, there are 18 red numbers, 18 black numbers, plus 2 green numbers (0 and 00). Thus, the probability of a red number coming up on a spin of the wheel is $p = 18/38 \approx 0.47$. Suppose that we go on a binge and bet \$1 on red 10,000 times in a row. (A \$1 bet on red wins \$1 if red comes up; otherwise, we lose the \$1.)

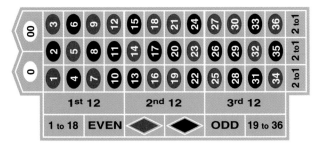

   (a) Let Y represent the number of times we lose (i.e., the number of times that red does not come up). Use the dishonest-coin principle to describe the distribution of the random variable Y.

(b) Approximately what are the chances that we will lose 5300 times or more?

(c) Approximately what are the chances that we will lose somewhere between 5150 and 5450 times?

(d) Explain why the chances that we will break even or win in this situation are essentially zero.

81. After polling a random sample of 800 voters during the most recent gubernatorial race, the *Tasmania Gazette* reports: *As the race for governor of Tasmania heads into its final days, our most recent poll shows Mrs. Butterworth ahead of the incumbent Mrs. Cubbison by 6 percentage points—53% to 47%. The results of the poll indicate with near certainty that if the election had been held at the time the poll was taken, Mrs. Butterworth would be the next governor of Tasmania.*

(a) Estimate the standard error for this poll.

(b) Compute a 95% confidence interval for this poll.

(c) Compute a 99.7% confidence interval for this poll.

Projects and Papers

A. Bernouilli Trials and Binomial Distributions

A random experiment with only two outcomes (which for convenience we call "success" and "failure") is called a **Bernouilli trial**. Typical examples of Bernouilli trials are tossing a coin (be it honest or dishonest) and shooting a free throw. One of the important questions discussed in this chapter is what happens when we independently repeat the same Bernouilli trial many times (see Examples 16.8, 16.10, and 16.11). The number of *successes* when we independently repeat a given Bernouilli trial n times is given by an important distribution called the **binomial distribution**. The rule that defines the binomial distribution is as follows:

Binomial Distribution

In n independent Bernouilli trials each with probability of success p, the probability of having exactly r successes is given by $_nC_r p^r (1 - p)^{n-r}$.

In this project you are to explore binomial distributions and their connection with normal curves. In the first part of the project you should explain how binomial distributions and normal curves are related. In the second part of the project you are to illustrate how one can use some of the facts about normal curves we learned in this chapter to compute approximate values for the probability of any number of successes in n Bernouilli trials, regardless of how big n is. For the purposes of illustration choose a real-life example that involves a large number of repeated Bernouilli trials in a field that is of special interest to you (sports, the economy, politics, public health, etc.).

B. Confidence Intervals

The concept of a *confidence interval* was introduced in Example 16.13 in this chapter. The two most frequently used levels of confidence intervals are 95% confidence intervals (sometimes described as intervals at a 95% *confidence level*) and 90% confidence intervals (intervals at a 90% *confidence level*).

In this project, you are to describe the process of constructing confidence intervals in general. Given a target confidence level 95% (or 90%, or x%), how do you construct the corresponding confidence interval? Conversely, given a specified interval, how do you find the confidence level that best fits that interval? Illustrate the relationship between confidence level and the size of the confidence interval using a real-life poll. Conclude with a discussion of some common misconceptions about confidence intervals. When no confidence interval is mentioned, how should we properly interpret the results of the poll?

C. Book Review: *The Bell Curve* and Rebuttals

In 1994, the late Richard J. Herrnstein of Harvard and Charles Murray of the Massachusetts Institute of Technology (MIT) wrote a book entitled *The Bell Curve: Intelligence and Class Structure in American Life*. The book used social statistics (in particular, bell-shaped distributions of data on IQs and other tests of human intelligence) to justify the conclusion that intelligence levels vary across racial, ethnic, and economic groups. The book became one of the most discussed books of its time and raised a firestorm of controversy and many rebuttals.

In this project you should (1) summarize the main arguments presented by Herrnstein and Murray in their book and the justification given for these arguments, and (2) summarize some of the main rebuttal arguments to the premises of the book. Most of the information you will need for this project you will find in references 3, 6, and 8.

References and Further Readings

1. Clemons, T., and M. Pagano, "Are Babies Normal?" *The American Statistician*, 53:4 (1999), 298–302.

2. Converse, P. E., and M. W. Traugott, "Assessing the Accuracy of Polls and Surveys." *Science*, 234 (1986), 1094–1098.

3. Devlin, B., S. E. Fienberg, D. P. Resnick, and K. Roeder, "Wringing the Bell Curve," *Chance*, 8, 1995, 27–36.

4. Frankel, Max, "Margins of Error," *New York Times Magazine*, December 15, 1996, 34.

5. Freedman, D., R. Pisani, R. Purves, and A. Adhikari, *Statistics*, 2d ed. New York: W. W. Norton, 1991, chaps. 16 and 18.

6. Gould, Stephen Jay, *The Mismeasure of Man*. New York: W. W. Norton, 1996.

7. Herrnstein, R. J., and C. Murray, *The Bell Curve: Intelligence and Class Structure in American Life*. New York: Free Press, 1994.

8. Jacoby, Russell, and Naomi Glauberman (eds.), *The Bell Curve Debate: History, Documents, Opinions*. New York: Times Books, 1995.

9. Kerrich, John, *An Experimental Introduction to the Theory of Probability*. Witwatersrand, South Africa: University of Witwatersrand Press, 1964.

10. Larsen, J., and D. F. Stroup, *Statistics in the Real World*. New York: Macmillan Publishing Co., Inc., 1976.

11. Mosteller, F., W. Kruskal, et al., *Statistics by Example: Detecting Patterns*. Reading, MA: Addison-Wesley Publishing Co., Inc., 1973.

12. Mosteller, F., R. Rourke, and G. Thomas, *Probability and Statistics*. Reading, MA: Addison-Wesley Publishing Co., Inc., 1961.

13. Tanner, Martin, *Investigations for a Course in Statistics*. New York: Macmillan Publishing Co., Inc., 1990.

Answers to Selected Problems

Chapter 1

Walking

A. Ballots and Preference Schedules

1. (a) 7

(b) The Country Cookery: plurality.

(c)

Number of voters	5	3	1	3
1st choice	A	C	B	C
2nd choice	B	B	D	B
3rd choice	C	A	C	D
4th choice	D	D	A	A

3. (a) 21 **(b)** 11 **(c)** A **(d)** E

5. (a) B and E

(b)

Number of voters	8	3	5	5
1st choice	A	A	C	D
2nd choice	C	D	D	C
3rd choice	D	C	A	A

(c) A

7.

Number of voters	255	480	765
1st choice	L	C	M
2nd choice	M	M	L
3rd choice	C	L	C

9.

Number of voters	47	36	24	13	5
1st choice	B	A	B	E	C
2nd choice	E	B	A	B	E
3rd choice	A	D	D	C	A
4th choice	C	C	E	A	D
5th choice	D	E	C	D	B

B. Plurality Method

11. (a) It's a tie between B and D.

(b) D

13. (a) Twenty-three votes will guarantee A at least a tie for first; 24 votes guarantee that A is the only winner. (With 23 of the remaining 30 votes, A has 49 votes. The only candidate with a chance to have that many votes is C. Even if C gets the other 7 remaining votes, C would not have enough votes to beat A.)

(b) Eleven votes will guarantee C at least a tie for first; 12 votes guarantee C is the only winner. (With 11 of the remaining votes, C has 53 votes. The only candidate with a chance to have that many votes is D. Even if D gets the other 19 remaining votes, D would not have enough votes to beat C.)

15. (a) 361

(b) 145

(c) 73

C. Borda Count Method

17. (a) Professor Chavez.

(b)

Number of voters	5	3	5	5	3
1st choice	A	A	C	D	B
2nd choice	B	D	D	C	A
3rd choice	C	B	A	B	C
4th choice	D	C	B	A	D

The winner is now Professor Argand.

(c) Professor Chavez was the winner of an election and when an irrelevant alternative (Epstein) was disqualified and the ballots recounted, Professor Argand was the winner of the reelection.

19. (a) Borrelli.

(b) Dante has a majority (13) of the first-place votes but does not win the election.

(c) Dante, having a majority of the first-place votes, is a Condorcet candidate but does not win the election.

21. (a) 80, 50, 40, 30; Winner: B.

(b) 8N, 5N, 4N, 3N; Winner: B.

(c) No. The number of points for each candidate is just multiplied by N.

23. (a) 200 points **(b)** 50 points

25. (a) 10 points **(b)** 1100 points **(c)** 310 points

D. Plurality-with-Elimination Method

27. (a) Professor Argand

(b)

Number of voters	5	3	5	3	2	3
1st choice	A	A	E	D	D	B
2nd choice	B	D	D	B	B	E
3rd choice	D	B	A	E	A	A
4th choice	E	E	B	A	E	D

Professor Epstein is now the winner.

(c) The independence-of-irrelevant-alternatives criterion.

29. (a) Dante is the winner.

(b) Dante has a majority of first-place votes (13 first-place votes out of a total of 24 votes).

(c) If there is a choice that has a majority of the first-place votes, then that candidate will be the winner under the plurality-with-elimination method.

568 Answers to Selected Problems

31. B

33. (a) Año Nuevo, California

(b) Cloudbreak, Fiji is the Condorcet candidate.

(c) Cloudbreak, Fiji, which is the Condorcet candidate, fails to win the election under the plurality-with-elimination method.

E. Pairwise Comparisons Method

35. (a) Chad

(b)

Number of voters	8	6	5	5	2
1st choice	C	E	E	D	D
2nd choice	B	D	C	C	E
3rd choice	D	B	D	E	B
4th choice	E	C	B	B	C

Dora is now the winner.

(c) The independence-of-irrelevant-alternatives criterion.

37. Professor Argand

39. E

F. Ranking Methods

41. (a) Winner A. Second place: C. Third place: D. Last place: B.

(b) Winner: A. Second place: C. Third place: D. Last place: B.

(c) Winner: C. Second place: A. Third place: D. Last place: B.

(d) Winner: D. Second place: C. Third place: A. Last place: B.

43. (a) Winner A. Second place: B. Third place: C. Last place: D.

(b) Winner: B. Second place: A. Third place: D. Last place: C.

(c) Winner: B. Second place: A. Third place: C. Last place: D.

(d) Winner: B. Second place: A. Third place: D. Last place: C.

45. Winner: B. Second place: C. Third place: D. Last place: A.

47. (a) Winner: A. Second place: C. Third place: D. Last place: B.

(b) Winner: A. Second place: C. Third place: D. Last place: B.

(c) Winner: C. Second place: D. Third place: A. Last place: B.

(d) Winner: D. Second place: C. Third place: A. Last place: B.

49. (a) Winner: A. Second place: D. Third place: B. Last place: C.

(b) Winner: B. Second place: A. Third place: D. Last place: C.

(c) Winner: B. Second place: A. Third place: D. Last place: C.

(d) Winner: B. Second place: A. Third place: D. Last place: C.

G. Miscellaneous

51. 125,250

53. 5,060,560

55. (a) 105

(b) 1 hour and 45 minutes

57. 6

59. (a) A

(b) B

(c) A

(d) Condorcet criterion; majority criterion; independence-of-irrelevant-alternatives criterion

Jogging

61. Suppose the two candidates are A and B and that A gets a first-place votes and B gets b first-place votes and suppose that $a > b$. Then A has a majority of the votes and the preference schedule is

Number of voters	a	b
1st choice	A	B
2nd choice	B	A

It is clear that candidate A wins the election under the plurality method, the plurality-with-elimination method, and the method of pairwise comparisons. Under the Borda count method, A gets $2a + b$ points while B gets $2b + a$ points. Since $a > b$, $2a + b > 2b + a$ and so again A wins the election.

63. If X is the winner of an election using the plurality method and, in a reelection, the only changes in the ballots are changes that only favor X, then no candidate other than X can increase his/her first-place votes and so X is still the winner of the election.

65. If X is the winner of an election using the method of pairwise comparisons and, in a reelection, the only changes in the ballots are changes that favor X and only favor X, then candidate X will still win every pairwise comparison that he/she won in the original election and possibly even some new ones—while no other candidate will win any new pairwise comparisons. So X will gain more points than any other candidate will gain and hence will remain the winner of the election.

67. (a) Suppose a candidate, C, gets v_1 first-place votes, v_2 second-place votes, v_3 third-place votes, …, v_N Nth-place votes. For this candidate $p = v_1N + v_2(N - 1) + v_3(N - 2) + \cdots + v_{N-1} \cdot 2 + v_N \cdot 1$, and $r = v_1 \cdot 1 + v_2 \cdot 2 + v_3 \cdot 3 + \cdots + v_{N-1}(N - 1) + v_N N$.
So $p + r = v_1(N + 1) + v_2(N + 1) + v_3(N + 1) + \cdots + v_{N-1}(N + 1) + v_N(N + 1)$
$= (v_1 + v_2 + \cdots + v_N)(N + 1)$
$= k(N + 1)$

(b) Suppose candidates C_1 and C_2 receive p_1 and p_2 points, respectively, using the Borda count as originally described in the chapter and r_1 and r_2 points under the variation described in this exercise. Then if $p_1 < p_2$, we have $-p_1 > -p_2$ and so $k(N + 1) - p_1 > k(N + 1) - p_2$, which implies [using part (a)], $r_1 > r_2$. Consequently the relative ranking of the candidates is not changed.

69. (a) This follows since $1610 + 1540 + 1530 = 65 \times (25 + 24 + 23)$. That is, these three teams combined received as many first, second, and third-place votes as possible.

(b) USC: 11 second-place votes; 2 third-place votes

Oklahoma: 31 second-place votes; 27 third-place votes

Auburn: 23 second-place votes; 36 third-place votes

71. 5 points for each first-place vote; 3 points for each second-place vote; 1 point for each third-place vote

73. (a) C

(b) A is a Condorcet candidate but is eliminated in the first round

Number of voters	10	6	6	3	3
1st choice	B	A	A	D	C
2nd choice	C	B	C	A	A
3rd choice	D	D	B	C	B
4th choice	A	C	D	B	D

(c) B wins under the Coombs method. However, if 8 voters move B from their 3rd choice to their 2nd choice, then C wins

Number of voters	10	8	7	4
1st choice	B	C	C	A
2nd choice	A	A	B	B
3rd choice	C	B	A	C

Chapter 2

Walking

A. Weighted Voting Systems

1. (a) 6   **(b)** 20   **(c)** 4   **(d)** 65%

3. (a) 14   **(b)** 27   **(c)** 18   **(d)** 19

5. (a) $[10:8,4,2,1]$

   (b) $[11:8,4,2,1]$

   (c) $[12:8,4,2,1]$

   (d) $[13:8,4,2,1]$

7. (a) There is no dictator; P_1 and P_2 have veto power; P_3 is a dummy.

(b) P_1 is a dictator; P_2 and P_3 are dummies.

(c) There is no dictator, no one has veto power, and no one is a dummy.

9. (a) P_1 and P_2 have veto power; P_5 is a dummy.

(b) P_1 is a dictator; P_2, P_3, and P_4 are dummies.

(c) P_1 and P_2 have veto power; P_3 and P_4 are dummies.

(d) All 4 players have veto power.

B. Banzhaf Power

11. (a) 10

(b) $\{P_1, P_2\}, \{P_1, P_3\}, \{P_1, P_2, P_3\}, \{P_1, P_2, P_4\},$
$\{P_1, P_3, P_4\}, \{P_2, P_3, P_4\}, \{P_1, P_2, P_3, P_4\}$

(c) P_1 only

(d) $P_1: \dfrac{5}{12}; P_2: \dfrac{1}{4}; P_3: \dfrac{1}{4}; P_4: \dfrac{1}{12}$

13. (a) $P_1: \dfrac{3}{5}; P_2: \dfrac{1}{5}; P_3: \dfrac{1}{5}$ **(b)** $P_1: \dfrac{3}{5}; P_2: \dfrac{1}{5}; P_3: \dfrac{1}{5}$

15. (a) $P_1: \dfrac{1}{3}; P_2: \dfrac{1}{4}; P_3: \dfrac{1}{6}; P_4: \dfrac{1}{6}; P_5: \dfrac{1}{12}$

(b) $P_1: \dfrac{7}{19}; P_2: \dfrac{5}{19}; P_3: \dfrac{3}{19}; P_4: \dfrac{3}{19}; P_5: \dfrac{1}{19}$

17. (a) $P_1: 1; P_2: 0; P_3: 0; P_4: 0$

(b) $P_1: \dfrac{7}{10}; P_2: \dfrac{1}{10}; P_3: \dfrac{1}{10}; P_4: \dfrac{1}{10}$

(c) $P_1: \dfrac{3}{5}; P_2: \dfrac{1}{5}; P_3: \dfrac{1}{5}; P_4 = 0$

(d) $P_1: \dfrac{1}{2}; P_2: \dfrac{1}{2}; P_3: 0; P_4: 0$

(e) $P_1: \dfrac{1}{3}; P_2: \dfrac{1}{3}; P_3: \dfrac{1}{3}; P_4: 0$

19. $A: \dfrac{1}{3}; B: \dfrac{1}{3}; C: \dfrac{1}{3}; D: 0$

21. (a) $\{P_1, P_2\}, \{P_1, P_3\}, \{P_2, P_3\}, \{P_1, P_2, P_3\}$

(b) $\{P_1, P_2, P_4\}, \{P_1, P_3, P_4\}, \{P_2, P_3, P_4\}, \{P_1, P_2, P_3, P_4\}$
$\{P_1, P_2, P_4, P_5\}, \{P_1, P_2, P_4, P_6\}, \{P_1, P_3, P_4, P_5\},$
$\{P_1, P_3, P_4, P_6\}, \{P_2, P_3, P_4, P_5\}, \{P_2, P_3, P_4, P_6\},$
$\{P_1, P_2, P_3, P_4, P_5\}, \{P_1, P_2, P_3, P_4, P_6\}, \{P_1, P_2, P_3, P_4, P_5, P_6\}$

(c) P_4 is never a critical player since every time it is part of a winning coalition, that coalition is a winning coalition without P_4 as well.

(d) $P_1: \dfrac{1}{3}; P_2: \dfrac{1}{3}; P_3: \dfrac{1}{3}; P_4: 0; P_5: 0; P_6: 0$

C. Shapley-Shubik Power

23. (a) $\langle P_1, \underline{P_2}, P_3 \rangle, \langle P_1, \underline{P_3}, P_2 \rangle, \langle P_2, \underline{P_1}, P_3 \rangle, \langle P_2, P_3, \underline{P_1} \rangle,$
$\langle P_3, \underline{P_1}, P_2 \rangle, \langle P_3, \underline{P_2}, \underline{P_1} \rangle$

(b) $P_1: \dfrac{4}{6}; P_2: \dfrac{1}{6}; P_3: \dfrac{1}{6}$

25. $P_1: \dfrac{7}{12}; P_2: \dfrac{3}{12}; P_3: \dfrac{1}{12}; P_4: \dfrac{1}{12}$

27. (a) $P_1: 1; P_2: 0; P_3: 0$

(b) $P_1: \dfrac{4}{6}; P_2: \dfrac{1}{6}; P_3: \dfrac{1}{6}$ **(c)** $P_1: \dfrac{4}{6}; P_2: \dfrac{1}{6}; P_3: \dfrac{1}{6}$

(d) $P_1: \dfrac{1}{2}; P_2: \dfrac{1}{2}; P_3: 0$ **(e)** $P_1: \dfrac{1}{3}; P_2: \dfrac{1}{3}; P_3: \dfrac{1}{3}$

29. (a) $P_1: 1; P_2: 0; P_3: 0$

(b) $P_1: \dfrac{4}{6}; P_2: \dfrac{1}{6}; P_3: \dfrac{1}{6}$ **(c)** $P_1: \dfrac{1}{2}; P_2: \dfrac{1}{2}; P_3: 0$

(d) $P_1: \dfrac{1}{2}; P_2: \dfrac{1}{2}; P_3: 0$ **(e)** $P_1: \dfrac{1}{3}; P_2: \dfrac{1}{3}; P_3: \dfrac{1}{3}$

31. (a) $P_1: \dfrac{5}{12}; P_2: \dfrac{3}{12}; P_3: \dfrac{3}{12}; P_4: \dfrac{1}{12}$

(b) $P_1: \dfrac{5}{12}; P_2: \dfrac{3}{12}; P_3: \dfrac{3}{12}; P_4: \dfrac{1}{12}$

(c) $P_1: \dfrac{5}{12}; P_2: \dfrac{3}{12}; P_3: \dfrac{3}{12}; P_4: \dfrac{1}{12}$

33. $A: \dfrac{1}{3}; B: \dfrac{1}{3}; C: \dfrac{1}{3}; D: 0$

D. Miscellaneous

35. (a) 6,227,020,800

(b) $6,402,373,705,728,000 \approx 6.402374 \times 10^{15}$

(c) $15,511,210,043,330,985,984,000,000 \approx 1.551121 \times 10^{25}$

(d) Roughly 500 billion years

37. (a) 362,880 **(b)** 11 **(c)** 110 **(d)** 504 **(e)** 10,100

39. (a) 11.1 **(b)** 101.01

41. (a) 63 **(b)** 31 **(c)** 15 **(d)** 48

43. (a) 720 **(b)** 120 **(c)** 600

45. (a) $13 \leq q \leq 18$

(b) 19

(c) $23 \leq q \leq 24$

(d) $23 \leq q \leq 24$

(e) The answers to (c) and (d) are the same. The only winning coalition is the grand coalition if and only if every player has veto power.

Jogging

47. $P_1: \frac{3}{8}; P_2: \frac{3}{8}; P_3: \frac{1}{8}; P_4: \frac{1}{8}; P_5: 0$

49. (a) 720 **(b)** 120 **(c)** $\frac{1}{6}$ **(d)** $\frac{1}{6}$

(e) If the quota equals the sum of all the weights, then the only way a player can be pivotal is for the player to be the last player in the sequential coalition. Since every player will be the last player in the same number of sequential coalitions, each of the N players must have the same Shapley-Shubik power index.

51. (a) $[4:2,1,1,1]$ and $[9:5,2,2,2]$ are among the possible answers.

(b) $H: \frac{1}{2}; A_1: \frac{1}{6}; A_2: \frac{1}{6}; A_3: \frac{1}{6}$

53. (a) Suppose that a winning coalition that contains P is not a winning coalition without P. Then, by definition, P would be a critical player in that coalition.

(b) Suppose that P is a member of every winning coalition. Then removing P from any winning coalition cannot result in another winning coalition. So P is critical in every winning coalition. Since there is at least one winning coalition (the grand coalition), P is not a dummy.

(c) Suppose that P is a pivotal member of some sequential coalition. Let's say that P is the kth player listed. Then, consider the k-player coalition consisting of the previous $k - 1$ players listed in that sequential coalition and P. P would be a critical player in that coalition and is therefore not a dummy.

(d) Suppose that P is never a pivotal member in a sequential coalition. If P were ever critical in some coalition S, then a sequential coalition that lists the members of S other than P, then P, and then the voters not in S would have P as its pivotal member.

55. (a) $7 \leq q \leq 13$

(b) For $q = 7$ or $q = 8$, P_1 is a dictator.

(c) For $q = 9$, only P_1 has veto power since P_2 and P_3 together have just 5 votes.

(d) For $10 \leq q \leq 12$, both P_1 and P_2 have veto power since no motion can pass without both of their votes. For $q = 13$, all three players have veto power.

(e) For $q = 7$ or $q = 8$, both P_2 and P_3 are dummies. For $10 \leq q \leq 12$, P_3 is a dummy since all winning coalitions contain $\{P_1, P_2\}$, which is itself a winning coalition.

57. (a) Both have Banzhaf power distribution

$P_1: \frac{2}{5}; P_2: \frac{1}{5}; P_3: \frac{1}{5}; P_4: \frac{1}{5}.$

(b) In the weighted voting system $[q: w_1, w_2, \ldots, w_N]$, if P_k is critical in a coalition then the sum of the weights of all the players in that coalition (including P_k) is at least q, but the sum of the weights of all the players in the coalition except P_k is less than q. Consequently, if the weights of all the players in that coalition are multiplied by $c > 0$ ($c \leq 0$ would make no sense), then the sum of the weights of all the players in the coalition (including P_k) is at least cq but the sum of the weights of all the players in the coalition except P_k is less than cq. Therefore, P_k is critical in the same coalition in the weighted voting system $[cq: cw_1, cw_2, \ldots, cw_N]$. Since the critical players are the same in both weighted voting systems, the Banzhaf power distributions will be the same.

59. (a) 120 **(b)** 120 **(c)** $\frac{1}{3}$

(d) $P_1: \frac{1}{3}; P_2: \frac{2}{15}; P_3: \frac{2}{15}; P_4: \frac{2}{15}; P_5: \frac{2}{15}; P_6: \frac{2}{15}$

61. You should buy your vote from P_1. The following table explains why.

Buying a vote from	Resulting weighted voting system	Resulting Banzhaf power distribution	Your power
P_1	$[6: 3, 2, 2, 2, 2]$	$P_1: \frac{1}{5}; P_2: \frac{1}{5}; P_3: \frac{1}{5}; P_4: \frac{1}{5}; P_5: \frac{1}{5}$	$\frac{1}{5}$
P_2	$[6: 4, 1, 2, 2, 2]$	$P_1: \frac{1}{2}; P_2: 0; P_3: \frac{1}{6}; P_4: \frac{1}{6}; P_5: \frac{1}{6}$	$\frac{1}{6}$
P_3	$[6: 4, 2, 1, 2, 2]$	$P_1: \frac{1}{2}; P_2: \frac{1}{6}; P_3: 0; P_4: \frac{1}{6}; P_5: \frac{1}{6}$	$\frac{1}{6}$
P_4	$[6: 4, 2, 2, 1, 2]$	$P_1: \frac{1}{2}; P_2: \frac{1}{6}; P_3: \frac{1}{6}; P_4: 0; P_5: \frac{1}{6}$	$\frac{1}{6}$

63. (a) You should buy your vote from P_2.

(b) You should buy two votes from P_2.

(c) Buying a single vote from P_2 raises your power from $\frac{1}{25} = 4\%$ to $\frac{3}{25} = 12\%$. Buying a second vote from P_2 raises your power to $\frac{2}{13} \approx 15.4\%$. The increase in power is less with the second vote, but if you value power over money, it might still be worth it to you to buy that second vote.

65. (a) The losing coalitions are $\{P_1\}, \{P_2\},$ and $\{P_3\}$. The complements of these coalitions are $\{P_2, P_3\}, \{P_1, P_3\},$ and $\{P_1, P_2\}$, respectively, all of which are winning coalitions.

(b) The losing coalitions are $\{P_1\}, \{P_2\}, \{P_3\}, \{P_4\}, \{P_2, P_3\}, \{P_2, P_4\},$ and $\{P_3, P_4\}$. The complements of these coalitions are $\{P_2, P_3, P_4\}, \{P_1, P_3, P_4\}, \{P_1, P_2, P_4\}, \{P_1, P_2, P_3\}, \{P_1, P_4\}, \{P_1, P_3\},$ and $\{P_1, P_2\}$, respectively, all of which are winning coalitions.

(c) If P is a dictator, the losing coalitions are all the coalitions without P; the winning coalitions are all the coalitions that include P. The complement of any coalition without P (losing) is a coalition with P (winning).

(d) 2^{N-1}

67. (a) In each nine-member winning coalition, every member is critical. In each coalition having 10 or more members, only the five permanent members are critical.

(b) At least nine members are needed to form a winning coalition. So, there are $210 + 638 = 848$ winning coalitions. Since every member is critical in each nine-member coalition, the nine-member coalitions yield a total of $210 \times 9 = 1890$ critical players. Since only the permanent members are critical in coalitions having 10 or more members, there are $638 \times 5 = 3190$ critical players in these coalitions. Thus, the total number of critical players in all winning coalitions is 5080.

(c) Each permanent member is critical in each of the 848 winning coalitions. Thus, the Banzhaf power index of a permanent member is 848/5080.

(d) The five permanent members together have $5 \times 848/5080 = 4240/5080$ of the power. The remaining 840/5080 of the power is shared equally among the 10 nonpermanent members, giving each a Banzhaf power index of 84/5080.

(e) In the given weighted voting system, the quota is 39, each permanent member has seven votes, and each nonpermanent member has one vote. The total number of votes is 45 so if any one of the permanent members does not vote for a measure, there would be at most $45 - 7 = 38$ votes and the measure would not pass. Thus all permanent members have veto power. On the other hand, all five permanent members' votes only add up to 35 so at least four nonpermanent members' votes are needed for a measure to pass.

Chapter 3

Walking
A. Fair Division Concepts

1. (a) $9.00 **(b)** $3.00 **(c)** $3.00

3. (a) $6.00 **(b)** $4.00 **(c)** $2.00

(d) Piece 1: $1.00; piece 2: $1.50; piece 3: $2.00; piece 4: $2.50; piece 5: $3.00; piece 6: $2.00

5. (a) Ana: s_2, s_3

(b) Ben: s_3

(c) Cara: s_1, s_2, s_3

(d) Ana: s_2; Ben s_3; Cara: s_1

7. (a) Adams: s_1, s_4

(b) Benson: s_1, s_2

(c) Cagle: s_1, s_3

(d) Duncan: s_4

(e) Adams: s_1; Benson: s_2; Cagle: s_3; Duncan: s_4

9. (a) Abe: s_2, s_3

(b) Betty: s_1, s_2, s_3, s_4

(c) Cory: s_3

(d) Dana: s_1

(e) Abe: s_2; Betty: s_4; Cory: s_3; Dana: s_1

B. The Divider–Chooser Method

11. (a) The left piece would be all vegetarian and 1/3 of the meatball part.

(b) One piece would be half of the vegetarian part and the other would be half of the vegetarian part and the entire meatball part.

(c) Jared: $4.00
Karla: $8.00

(d) Jared: $7.00
Karla: $4.00

13. (a) (iii) only

(b) either piece

15. (a) Answers may vary. For example,

(I)	(I)	(I)
(II)	(II)	(II)
(i)	**(ii)**	**(iii)**

(b) (i) either piece; **(ii)** II; **(iii)** I

17. (a) 50%

(b) the left piece: 40%; the right piece: 60%

(c) Jamie takes the right piece; Mo gets the left piece.

C. The Lone-Divider Method

19. (a) Divine: s_1; Chase: s_2; Chandra: s_3

(b) Divine: s_2; Chase: s_3; Chandra: s_1

(c) Divine: s_3; Chase: s_2; Chandra: s_1

21. (a) DiPalma: s_1; Childs: s_2; Choate: s_3; Chou: s_4

(b) DiPalma: s_2; Childs: s_3; Choate: s_4; Chou: s_1

(c) DiPalma: s_3; Childs: s_2; Choate: s_4; Chou: s_1

(d) DiPalma: s_4; Childs: s_2; Choate: s_3; Chou: s_1

23. (a) Desi: s_4; Cher: s_2; Cheech: s_3; Chong: s_1

(b) Desi: s_4; Cher: s_3; Cheech: s_1; Chong: s_2

(c) None of the choosers chose s_4, which can only be given to the divider.

25. (a) D: s_1; C_1: s_2; C_2: s_4; C_3: s_3; C_4: s_5

(b) D: s_1; C_1: s_4; C_2: s_2; C_3: s_3; C_4: s_5

(c) There are only two possible cases: C_1 receives s_2 or C_1 receives s_4 (since those were the slices in C_1's bid list). The only fair divisions in each case appear in (a) and (b).

27. (a) D: s_4; C_1: s_5; C_2: s_1; C_3: s_6; C_4: s_2; C_5: s_3

(b) C_5 must get s_3, which forces C_4 to get s_2. This leaves only s_5 for C_1, which in turn leaves only s_6 for C_3. Consequently, only s_1 is left for C_2 and the divider D must get s_4.

29. (a) Gong; he is the only player that could possibly value each piece equally.

(b) Egan: $\{s_3, s_4\}$; Fine: $\{s_1, s_3, s_4\}$; Hart: $\{s_3\}$

(c) Egan: s_4; Fine: s_1; Gong: s_2; Hart: s_3

(d) Hart must receive s_3, so that Egan in turn must receive s_4. It follows that Fine must receive s_1 so that Gong winds up with s_2.

D. The Lone-Chooser Method

31. (a) One possible second division by Angela is

(b) One possible second division by Boris is

(c)

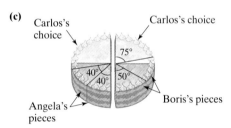

(d) Angela: $12.00; Boris: $10.00; Carlos: $22.67

33. (a) One possible second division by Boris is

(b) The second division by Angela is

(c) One possible fair division is

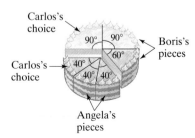

(d) Angela: $12.00; Boris: $12.00; Carlos: $14.67

35. (a) One possible second division by Brian is

(b) One possible second division by Arthur is

(c) One possible fair division is

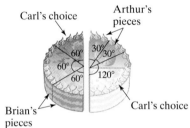

(d) Arthur: $33\frac{1}{3}\%$; Brian: $66\frac{2}{3}\%$; Carl: $83\frac{1}{3}\%$

37. (a) One possible second division by Carl is

Carl's piece

(b) One possible second division by Brian is

Brian's piece

(c) One possible fair division is

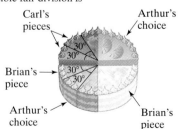

(d) Arthur: $66\frac{2}{3}\%$; Brian: $33\frac{1}{3}\%$; Carl: $33\frac{1}{3}\%$

39. (a) Karla would divide the sandwich into one piece consisting of half of the vegetarian part and another piece consisting of half of the vegetarian part and the entire meatball part. Jared would choose the larger slice containing the meatball part and half of the vegetarian part.

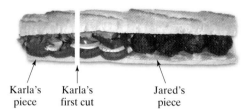

Karla's piece   Karla's first cut   Jared's piece

(b) Jared would divide his piece in three equally sized pieces – two pieces would be all meat and the other would be all vegetarian.

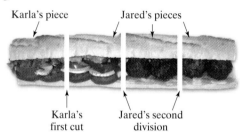

Karla's piece   Jared's pieces

Karla's first cut   Jared's second division

(c) Karla would divide her piece in three equally sized pieces.

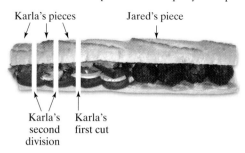

Karla's pieces   Jared's piece

Karla's second division   Karla's first cut

(d) Lori will select one of Jared's meatball parts and one of Karla's vegetarian parts.

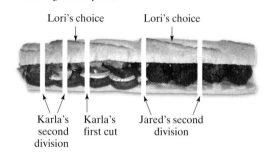

Lori's choice   Lori's choice

Karla's second division   Karla's first cut   Jared's second division

Jared: 50%; Karla: $33\frac{1}{3}$%; Lori: $38\frac{8}{9}$%

E. The Last-Diminisher Method

41. (a) P_4 **(b)** $6.00   **(c)** P_1

43. (a) P_5 **(b)** $7.00   **(c)** P_1

45. (a) P_9 **(b)** P_1 **(c)** P_5

     (d) P_1 **(e)** P_2

47. (a) 45°

     (b) No players are diminishers in round 1.

(c) Arthur receives his original claim at the end of round 1.

(d) Brian   **(e)** Carl

49. (a) 4/9   **(b)** Karla

     (c) 1/3 of the vegetarian part

     (d) 1/2 of the vegetarian part

     (e) the entire meatball part and 1/6 of the vegetarian part

F. The Method of Sealed Bids

51. (a) In the first settlement, Ana gets the desk and receives $120 in cash; Belle gets the Dresser; Chloe gets the vanity and the tapestry and pays $360. The surplus is $240.

     (b) Ana gets the desk and receives $200 in cash; Belle gets the dresser and receives $80; Chloe gets the vanity and the tapestry and pays $280.

53. Bob gets the business and pays $155,000. Jane gets $80,000 and Ann gets $75,000.

55. (a) In the first settlement, A gets items 4 and 5 and pays $765, B receives $582; C gets items 1 and 3 and pays $287, D receives $606, and E gets items 2 and 6 and pays $266. The surplus is $130.

     (b) A receives items 4 and 5 and pays $739; B receives $608; C receives items 1 and 3 and pays $261; D receives $632; E receives items 2 and 6 and pays $240.

57. Angelina bid $1800 on the laptop.

G. The Method of Markers

59. (a) A gets items 10, 11, 12, 13; B gets items 1, 2, 3; C gets items 5, 6, 7.

     (b) Items 4, 8, and 9 are left over.

61. (a) A gets items 1, 2; B gets items 10, 11, 12; C gets items 4, 5, 6, 7.

     (b) Items 3, 8, and 9 are left over.

63. (a) A gets items 19, 20; B gets items 15, 16, 17; C gets items 1, 2, 3; D gets items 11, 12, 13; E gets items 5, 6, 7, 8.

(b) Items 4, 9, 10, 14, and 18 are left over.

65. (a)

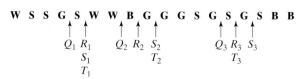

W   S   S   G   S   W   W   B   G   G   G   S   G   S   G   S   B   B

Q_1 R_1 Q_2 R_2 S_2 Q_3 R_3 S_3
     S_1 T_2 T_3
     T_1

(b)

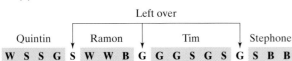

Left over

Quintin	Ramon	Tim	Stephone
W S S G	S W W B	G G G S G S	G S B B

(c) A coin toss would be needed to decide if Quentin or Stephone should receive S. If Stephone were to receive S, then Stephone would receive S and both G's and would pay $13.32. In that case, Quentin and Ramon would receive $4.69 and Tim would receive $3.94.

67. (a)

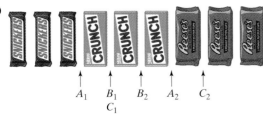

A_1 B_1 B_2 A_2 C_2
C_1

(b)

Ana	Belle	Chloe

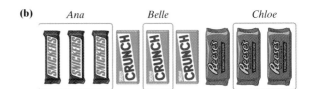

Two Nestle Crunch bars and one Reese's Peanut Butter Cup are left over.

(c) Belle would select one of the Nestle Crunch bars, Chloe would then select the Reese's Peanut Butter Cup, and then Ana would be left with a Nestle Crunch bar.

Jogging

69. (a) Two divisions are possible.

Abe	Babe	Cassie
P, S	Q	R
P, S	R	Q

(b) Four divisions are possible.

Abe	Babe	Cassie
R, S	Q	P
P, R	Q	S
Q, S	R	P
P, Q	R	S

(c) Out of P, Q, and S, Abe gets two of the three. Combine the left over one with R and divide it fairly between Babe and Cassie using the divider-chooser method.

71. (a) The total area is 30,000 m^2 and the area of C is only 8000 m^2. Since P_2 and P_3 value the land uniformly, each thinks that a fair share must have an area of at least 10,000 m^2.

(b) Any cut that divides the remaining property in parts of 11,000 m^2 will work. For example,

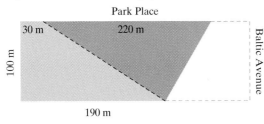

(c) The cut parallel to Park Place which divides the parcel in half is approximately

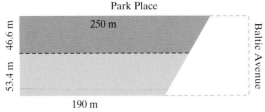

(d) The cut parallel to Baltic Avenue which divides the parcel in half is

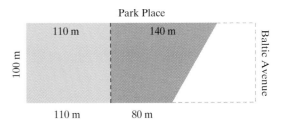

73. (a) If P_2 and P_3 each value the land (except for the 20 m by 20 m plot in the upper left corner) at v dollars per square meter, then the total value of the land to each of them is $(300)(100)v - 2(20)(20)v = 29{,}200v$. Thus a fair share to each of them would have value at least $9{,}733.33v$. But the value of piece C to each of them is $100\left(\dfrac{70 + 90}{2}\right)v = 8000v$ and so they would both pass and P_1 would end up with piece C.

(b) The cut parallel to Baltic Avenue which divides the parcel in half is

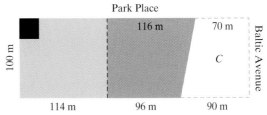

75. (a) No. The chooser is not assured a fair share in her own value system.

(b) The divider. The divider has the advantage that he can assure himself a fair share of the cake if he divides it appropriately. The chooser is essentially at the mercy of the laws of chance.

77. (a) A: Allen, Bryant; B: Evans, Francis; C: Carter, Duncan

(b) A: Allen, Duncan; B: Evans, Bryant; C: Carter, Francis

(c) No. In (a), A and B do as well as they can. In (b), C does as well as they can and cannot be improved. This means that B cannot be improved since Francis must go to C.

Chapter 4

Walking

A. Standard Divisors and Quotas

1. (a) 50,000

(b) Apure: 66.2; Barinas: 53.4; Carabobo: 26.6; Dolores: 13.8

(c)

State	Apure	Barinas	Carabobo	Dolores
Upper Quota	67	54	27	14
Lower Quota	66	53	26	13

3. (a) The "states" are the 6 bus routes and the "seats" are the 130 buses.

(b) 1000; The standard divisor represents the average number of passengers per bus per day.

(c) A: 45.30; B: 31.07; C: 20.49; D: 14.16; E: 10.26; F: 8.72

5. (a) 119

(b) 200,000

(c) A: 8,100,000; B: 5,940,000; C: 4,730,000; D: 2,920,000; E: 2,110,000

7. 32.32

9. (a) 0.5%

(b) A: 22.74; B: 16.14; C: 77.24; D:29.96; E: 20.84; F: 33.08

B. Hamilton's Method

11. A: 66; B: 53; C: 27; D: 14

13. A: 45; B: 31; C: 21; D: 14; E: 10; F: 9

15. A: 40; B: 30; C: 24; D: 15; E: 10

17. A: 23; B: 16; C: 77; D: 30; E: 21; F: 33

19. (a) Bob: 0; Peter: 3; Ron: 8

(b) Bob: 1; Peter: 2; Ron: 8

(c) The population paradox. For studying an extra 2 minutes (an increase of 3.70%), Bob gets a piece of candy. However, Peter, who studies an extra 12 minutes (an increase of 4.94%), has to give up a piece.

21. (a) Bob: 0; Peter: 3; Ron: 8

(b) Bob: 1; Peter: 3; Ron: 7; Jim: 6

(c) The new-states paradox. Ron loses a piece of candy to Bob when Jim enters the discussion and is given his fair share (six pieces) of candy.

C. Jefferson's Method

23. A: 67; B: 54; C: 26; D: 13

25. A: 46; B: 31; C: 21; D: 14; E: 10; F: 8

27. A: 41; B: 30; C: 24; D: 14; E: 10

29. A: 22; B: 16; C: 78; D: 30; E: 21; F: 33

31. From Exercise 7, the standard quota of Texas is 32.32. Jefferson's method only allows for violations of the upper quota. This means that Texas could not receive fewer than 32 seats.

D. Adams's Method

33. A: 66; B: 53; C: 27; D: 14

35. A: 45; B: 31; C: 20; D: 14; E: 11; F: 9

37. A: 40; B: 29; C: 24; D: 15; E: 11

39. A: 23; B: 16; C: 77; D: 30; E: 21; F: 33

41. This fact illustrates that Adams's method can produce lower-quota violations.

E. Webster's Method

43. A: 66; B: 53; C: 27; D: 14

45. A: 45; B: 31; C: 21; D: 14; E: 10; F: 9

47. A: 40; B: 30; C: 24; D: 15; E: 10

49. A: 23; B: 16; C: 77; D: 30; E: 21; F: 33

Jogging

51. (a) the total number of seats available

(b) the standard divisor

(c) the percentage of the total population in state N

53. (a) Take, for example, $q_1 = 3.9$ and $q_2 = 10.1$ (with $M = 14$). Under both Hamilton's method and Lowndes's method, A gets 4 seats and B gets 10 seats.

(b) Take, for example, $q_1 = 3.4$ and $q_2 = 10.6$ (with $M = 14$). Under Hamilton's method, A gets 3 seats and B gets 11 seats. Under Lowndes's method, A gets 4 seats and B gets 10 seats.

(c) Assuming that $f_1 > f_2$, the methods produce different apportionments if $\dfrac{f_2}{q_2 - f_2} > \dfrac{f_1}{q_1 - f_1}$ or $\dfrac{q_1}{q_2} > \dfrac{f_1}{f_2}$

55. (a) In Jefferson's method the modified quotas are larger than the standard quotas and so rounding downward will give each state at least the integer part of the standard quota for that state.

(b) In Adams's method the modified quotas are smaller than the standard quota and so rounding upward will give each state at most one more than the integer part of the standard quota for that state.

(c) If there are only two states, an upper quota violation for one state results in a lower quota violation for the other state (and vice versa). Since neither Jefferson's nor Adams's method can have both upper and lower violations of the quota rule, neither can violate the quota rule when there are only two states.

57. (a) 49,374,462

(b) 1,591,833

59. (a) A: 5; B: 10; C: 15; D: 21

(b) For $D = 100$, the modified quotas are A: 5, B: 10, C: 15, D: 20. For $D < 100$, each of the modified quotas will increase, so rounding upward will give at least A: 6, B: 11, C: 16, D: 21 for a total of at least 54 seats. But for $D > 100$, each of the modified quotas will decrease, and so rounding upward will give at most A: 5, B: 10, C: 15, D: 20 for a total of 50 at most.

(c) From part (b), we see that there is no divisor such that the sum of the modified upper quotas will be 51.

61. Apportionment 1: Webster's method; Apportionment 2: Adams's method; Apportionment 3: Jefferson's method. Adams's method favors small states and Jefferson's method favors large states.

Chapter 5

Walking

A. Graphs: Basic Concepts

1. (a) Vertex set: $\mathcal{V} = \{A, B, C, X, Y, Z\}$;

Edge set: $\mathcal{E} = \{AX, AY, AZ, BX, BY, BZ, CX, CY, CZ\}$;

$\deg(A) = 3, \deg(B) = 3, \deg(C) = 3, \deg(X) = 3,$
$\deg(Y) = 3, \deg(Z) = 3.$

(b) Vertex set: $\mathcal{V} = \{A, B, C\}$;

Edge set: $\mathcal{E} = \{\ \}$;

$\deg(A) = 0, \deg(B) = 0, \deg(C) = 0.$

(c) Vertex set: $\mathcal{V} = \{V, W, X, Y, Z\}$;

Edge set: $\mathcal{E} = \{XX, XY, XZ, XV, XW, WY, YZ\}$;

$\deg(V) = 1, \deg(W) = 2, \deg(X) = 6,$
$\deg(Y) = 3, \deg(Z) = 2.$

3. (a)

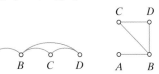

(b)

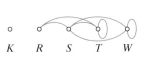

5. (a) Both graphs have the same vertex set
$\mathcal{V} = \{A, B, C, D\}$ and the same edge set
$\mathcal{E} = \{AB, AC, AD, BD\}$.

(b)

7. (a)   **(b)** **(c)**

9. (a)   **(b)**

(c) **(d)**

11. (a) C, B, A, H, F **(b)** C, B, D, A, H, F

(c) C, B, A, H, F **(d)** C, D, B, A, H, G, G, F

(e) 4 **(f)** 3 **(g)** 12

13. E, E;

B, C, B;

A, B, C, A;

A, B, C, A;

A, B, E, D, A;

A, B, E, E, D, A;

A, C, B, E, D, A;

A, C, B, E, D, A;

A, C, B, E, E, D, A;

A, C, B, E, E, D, A;

A, B, C, B, E, D, A;

A, B, C, B, E, E, D, A

B. Graph Models

15.

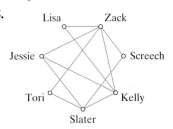

17.

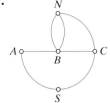

19.

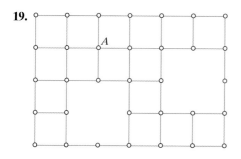

C. Euler's Theorems

21. (a) has an Euler circuit because all vertices have even degree

  (b) has neither an Euler circuit nor an Euler path because there are four vertices of odd degree

23. (a) has neither an Euler circuit nor an Euler path because there are more than two vertices of odd degree

  (b) has no Euler circuit, but has an Euler path because there are exactly two vertices of odd degree

(c) has neither an Euler circuit nor an Euler path because the graph is not connected

25. (a) has neither an Euler circuit nor an Euler path because there are eight vertices of odd degree

  (b) has no Euler circuit, but has an Euler path because there are exactly two vertices of odd degree

  (c) has no Euler circuit, but has an Euler path because there are exactly two vertices of odd degree

D. Finding Euler Circuits and Euler Paths

27.

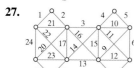

29.

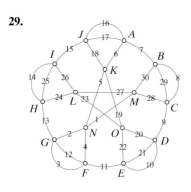

31.

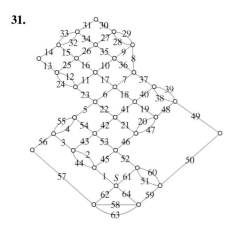

33.

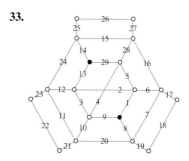

35. (a)

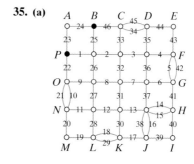

(b)

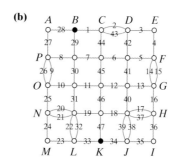

E. Unicursal Tracings

37. (a) The drawing has neither because there are more than two odd vertices.

(b) Open unicursal tracing. For example,

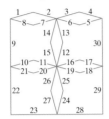

(c) Open unicursal tracing. For example,

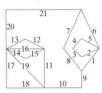

39. (a) Open unicursal tracing. For example,

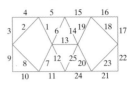

(b) Open unicursal tracing. For example,

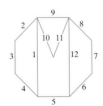

(c) Has neither because there are more than two odd vertices.

F. Eulerizations and Semi-Eulerizations

41. (a)

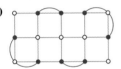

(b)

43. (a)

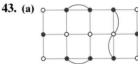

(b)

G. Miscellaneous

45. (a)

(b)

(c)

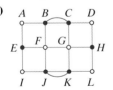

47. (a)

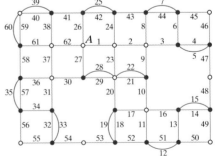

(b) *BC* and *JK*

49. Four times. There are 10 vertices of odd degree. Two can be used as the starting and ending vertices. The remaining eight odd degree vertices can be paired so that each pair forces one lifting of the pencil.

51.

Jogging

53. (a) An edge *XY* contributes 2 to the sum of the degrees of all vertices (1 to the degree of *X*, and 1 to the degree of *Y*).

(b) If there were an odd number of vertices, the sum of the degrees of all the vertices would be odd.

55. (a) If each vertex were of odd degree, then the graph has an odd number of odd vertices. This is not possible! So, it must be that each vertex is in fact of even degree. By Euler's circuit theorem, the graph would have an Euler circuit.

(b) The following graph is regular (all vertices are of degree 2) and it has an Euler circuit.

The graph below is also regular (all vertices have degree 3) but it does not have an Euler circuit.

57. A connected graph having all even vertices must have an Euler circuit. But if a graph has an Euler circuit, it cannot have a bridge (because then it would be impossible to get from one component of the graph to another and back).

59. (a) 12

(b)

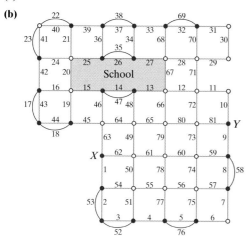

61. (a) Connect all but one of the vertices with a complete set of edges and to leave a single vertex isolated (incident to no edges).

(b) $\dfrac{(N-2)(N-1)}{2}$

63. (a) One optimal route: $R, L, R, A, C, L, C, D, L, D, B, R, B$. Listing the bridges in the order they are crossed: Washington, Jefferson, Truman, Wilson, Grant, Roosevelt, Monroe, Adams, Adams, Kennedy, Hoover, Lincoln.

(b) One optimal route: $B, R, B, R, L, R, A, C, L, C, D, B, D, L$. Listing the bridges in the order they are crossed: Hoover, Lincoln, Lincoln, Washington, Jefferson, Truman, Wilson, Grant, Roosevelt, Monroe, Kennedy, Kennedy, Adams.

Chapter 6

Walking

A. Hamilton Circuits and Hamilton Paths

1. (a) 1. A, B, D, C, E, F, G, A;
2. A, D, C, E, B, G, F, A;
3. A, D, B, E, C, F, G, A

(b) A, G, F, E, C, D, B

(c) D, A, G, B, C, E, F

3. 1. A, B, C, D, E, F, G, A
2. A, B, E, D, C, F, G, A
3. A, F, C, D, E, B, G, A
4. A, F, E, D, C, B, G, A
Mirror-image circuits:
5. A, G, F, E, D, C, B, A
6. A, G, F, C, D, E, B, A
7. A, G, B, E, D, C, F, A
8. A, G, B, C, D, E, F, A

5. (a) A, F, B, C, G, D, E

(b) A, F, B, C, G, D, E, A

(c) A, F, B, E, D, G, C

(d) F, A, B, E, D, C, G

7. (a) 1. A, B, C, D, E, F, A
2. A, B, E, D, C, F, A
Mirror-image circuits:
3. A, F, E, D, C, B, A
4. A, F, C, D, E, B, A

(b) 1. D, E, F, A, B, C, D
2. D, C, F, A, B, E, D
Mirror-image circuits:
3. D, C, B, A, F, E, D
4. D, E, B, A, F, C, D

(c) The circuits in (b) are the same as the circuits in (a), just rewritten with a different starting vertex.

9. Any Hamilton path passing through vertex A must contain edge AB. Any path passing through vertex E must contain edge BE. And, any path passing through vertex C must contain edge BC. Consequently, any path passing through vertices A, C, and E must contain at least three edges meeting at B and hence would pass through vertex B more than once.

11. (a) $A, I, J, H, B, C, F, E, G, D$

(b) $G, D, E, F, C, B, A, I, J, H$

(c) If such a path were to start by heading left, it would not contain C, D, E, F, or G since it would need to pass

through B again in order to do so. On the other hand, if a path were to start by heading right, it could not contain A, I H, or J.

(d) Any circuit would need to cross the bridge BC twice (in order to return to where it started). But then B and C would be included twice.

13. (a) 6

(b) B, D, A, E, C, B (weight 27)

(c) the mirror image B, C, E, A, D, B (weight 27)

15. (a) A, D, F, E, B, C (weight 29)

(b) A, B, E, D, F, C (weight 30)

(c) A, D, F, E, B, C (weight 29)

B. Factorials and Complete Graphs

17. (a) 3,628,800

(b) 2,432,902,008,176,640,000 $\approx 2.43 \times 10^{18}$

(c) 670,442,572,800

19. (a) 2,432,902,008,176,640,000 $\approx 2.43 \times 10^{18}$

(b) 40! $\approx 8.16 \times 10^{47}$

(c) 39! $\approx 2.04 \times 10^{46}$

21. (a) 362,880 **(b)** 11 **(c)** 110 **(d)** 10,100

23. (a) 11.1 **(b)** 101.01

25. (a) 190 **(b)** 210 **(c)** 50

27. (a) 6 **(b)** 10 **(c)** 201

C. Brute-Force and Nearest Neighbor-Algorithms

29. (a) A, C, B, D, A (weight 62)

(b) A, D, C, B, A (weight 80)

(c) B, D, C, A, B (weight 74)

(d) C, D, B, A, C (weight 74)

31. (a) B, C, A, E, D, B (cost $722)

(b) C, A, E, D, B, C (cost $722)

(c) D, B, C, A, E, D (cost $722)

(d) E, C, A, D, B, E (cost $741)

33. (a) A, D, E, C, B, A (cost $11,656)

(b) A, D, B, C, E, A (cost $9,760)

(c) A, B, C, E, D, A (cost $11,656)

35. (a) Atlanta, Columbus, Kansas City, Tulsa, Minneapolis, Pierre, Atlanta (cost $2915.25)

(b) Atlanta, Kansas City, Tulsa, Minneapolis, Pierre, Columbus, Atlanta (cost $2804.25)

D. Repetitive Nearest-Neighbor Algorithm

37. B, C, D, E, A, B (weight 9.8)

39. A, D, B, C, E, A (cost $9,760)

41. Atlanta, Columbus, Minneapolis, Pierre, Kansas City, Tulsa, Atlanta (3252 miles)

E. Cheapest-Link Algorithm

43. B, E, D, C, A, B (weight 9.9)

45. B, E, C, A, D, B (cost $10,000)

47. Atlanta, Columbus, Pierre, Minneapolis, Kansas City, Tulsa, Atlanta (3465 miles)

F. Miscellaneous

49. (a) A, B, C, D, E, A (weight 104)

(b) A, B, C, D, E, A (weight 104)

(c) A, C, D, E, B, A (weight 7)

51. (a)

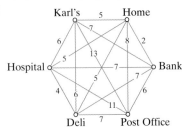

(b) Home, Bank, Post Office, Deli, Hospital, Karl's , Home (30 miles)

53. (a)

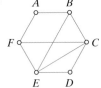

(b)
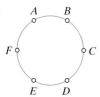

(c)

55. If we draw the graph describing the friendships among the guests (see figure), we can see that the graph does not have a Hamilton circuit, which means it is impossible to seat everyone around the table with friends on both sides.

Jogging

57.
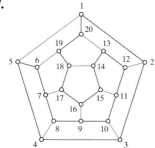

59. $A, B, C, D, J, I, F, G, E, H$

61. The 2-by-2 grid graph cannot have a Hamilton circuit because each of the four corner vertices as well as the interior vertex I must be preceded and followed by a boundary vertex. But there are only four boundary vertices—not enough to go around.

63. (a)
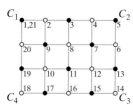

(b)
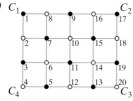

(c) Think of the vertices of the graph as being colored like a checker board with C_1 being a red vertex. Then each time we move from one vertex to the next we must move from a red vertex to a black vertex or from a black vertex to a red vertex. Since there are 10 red vertices and 10 black vertices and we are starting with a red vertex, we must end at a

black vertex. But C_2 is a red vertex. Therefore, no such Hamilton path is possible.

65.
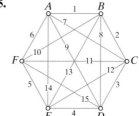

67. (a) Any circuit would need to cross the bridge twice (in order to return to where it started). But then each vertex at the end of the bridge would be included twice.

(b) The following graph has a Hamilton path from B to C.

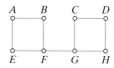

69. Dallas, Houston, Memphis, Louisville, Columbus, Chicago, Kansas City, Denver, Atlanta, Buffalo, Boston, Dallas.

The process starts by finding the smallest number in the Dallas row. This is 243 miles to Houston. We cross out the Dallas column and proceed by finding the smallest number in the Houston row (other than that representing Dallas which has been crossed out). This is 561 miles to Memphis. We now cross out the Houston column and continue by finding the smallest number in the Memphis row (other than those representing Dallas and Houston which have been crossed out). The process continues in this fashion until all cities have been reached.

71. (a) Detroit-Flint-Lansing-Grand Rapids-Cheboygan-Sault Ste. Marie-Marquette-Escanaba-Menominee-Sault-Ste. Marie-Cheboygan-Detroit (cost $540.86)

(b) Detroit–Flint-Lansing-Grand Rapids–Cheboygan-Sault Ste. Marie-Marquette–Escanaba-Menominee (cost $389.11)

Chapter 7

Walking

A. Trees

1. (a) Is a tree
(b) Is not a tree because it is not connected
(c) Is not a tree because it has a circuit
(d) Is a tree

3. (a) (II) A tree with 8 vertices must have 7 edges.
(b) (III)

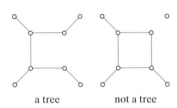

a tree   not a tree

(c) (I) If every edge of a graph is a bridge, then the graph must be a tree.

5. (a) (I) If there is exactly one path joining any two vertices of a graph, the graph must be a tree.
(b) (II) A tree with 8 vertices must have 7 edges and every edge must be a bridge.
(c) (I) If every edge is a bridge, then the graph has no circuits. Since the graph is also connected, it must be a tree.

7. (a) (III)

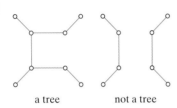

a tree   not a tree

(b) (II) A tree has no circuits.

(c) (I) A graph with 8 vertices, 7 edges, and no circuits, must also be connected and hence must be a tree.

9. (a) (II) Since the degree of each vertex is even, it must be at least 2. Thus, the sum of the degrees of all 8 vertices must be at least 16. But a tree with 8 vertices must have 7 edges, and the sum of the degrees of all the vertices would have to be 14.

(b) (III)

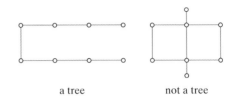

a tree   not a tree

(c) (III)

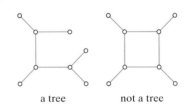

a tree   not a tree

B. Spanning Trees

11. (a)

(b)

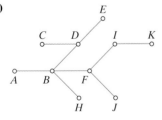

(c)

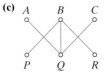

13. (a)

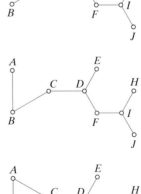

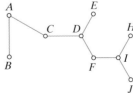

(b)

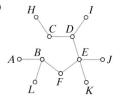

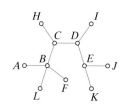

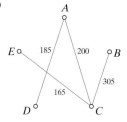

15. (a) 12

(b) 15

17. (a) 36

(b) 45

C. Minimum Spanning Trees and Kruskal's Algorithm

19. (a)

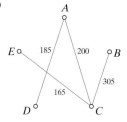

(b) 855

21. (a)

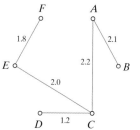

(b) 9.3

23. Kansas City–Tulsa, Pierre–Minneapolis, Minneapolis–Kansas City, Atlanta–Columbus, Columbus–Kansas City.

25. $380,000; A spanning tree for the 20 vertices will have exactly 19 edges.

D. Steiner Points and Shortest Networks

27. (a) 580 miles

(b) 385 miles

29. (a) 366 km.

(b) 334 km.

31. 340 miles; Since the measure of angle CAB is 120°, there is no Steiner point. Therefore, the shortest network is the MST, which consists of the edges AB and AC. In a 30-60-90 triangle, the hypotenuse is twice as long as the shortest leg (see Exercise 45(a)). Since the shortest leg of both of the 30-60-90 triangles in the figure is 85 miles, $AB = 170$ miles and $AC = 170$ miles.

33. Z; The sum of the distances from Z to A, B, and C is 232 miles, the sum of the distances from X to A, B, and C is 240 miles, and the sum of the distances from Y to A, B, and C is 243 miles.

35. (a) $CE + ED + EB$ is larger since $CD + DB$ is the shortest network connecting the cities C, D, and B.

(b) $CD + DB$ is the shortest network connecting the cities C, D, and B since angle CDB is 120° and so the shortest network is the same as the MST.

(c) $CE + EB$ is the shortest network connecting the cities C, E, and B since angle CEB is more than 120° and so the shortest network is the same as the MST.

E. Miscellaneous

37. (a) $k = 0, 1, 2,$ and 5 are all possible.

(b) $0 \leq k \leq 120$ and $k = 123$ are all possible.

39. (a) 0

(b) 1   **(c)** 118

41. (a)

(b)

(c)

43. 15°

45. (a) $h = 2s$ **(b)** $l = \sqrt{3}s$

(c) $h = 42$ cm

$l \approx 36.3$ cm

(d) $s = 15.2$ cm

$l \approx 26.3$ cm

47. In a 30°-60°-90° triangle, the side opposite the 30° angle is one half the length of the hypotenuse and the side opposite the 60° angle is $\dfrac{\sqrt{3}}{2}$ times the length of the hypotenuse (see Exercise 45). Therefore, the distance from C to J is $\dfrac{\sqrt{3}}{2} \times 500$ miles ≈ 433 miles and the total length of the T-network (rounded to the nearest mile) is 433 miles + 500 miles = 933 miles.

Jogging

49. (a) 99

(b) 171

51. (a)

(b) Suppose that a tree is regular. Since two vertices of a tree must have degree 1 (see Exercise 50(b)), every vertex must have degree 1. So the sum of the degrees is N. However, the sum of the degrees is also twice the number of edges. So there are $N/2$ edges. In order to be a tree, it must be that $N - 1 = N/2$ (the number of edges must be one fewer than the number of vertices). Solving this equation gives $N = 2$ as the only solution. So, if $N \geq 3$, the graph cannot be regular.

53. (a) They are trees.

(b) If $R = 1$, then $M = N$. So the network is not a tree. From the definition of a tree, there must be at least one circuit. Suppose that the graph had two (or more) circuits. Then, there would be an edge of one of the circuits that is not an edge of another circuit. Removing such an edge would leave us with a connected graph with the number of vertices being one more than the number of edges (i.e., a tree). But this tree would have a circuit. This is impossible, so there cannot be more than one circuit.

(c) The maximum redundancy occurs when the degree of each vertex is as large as possible (i.e., $N - 1$). In that case, the number of edges is $(N - 1) + (N - 2) + \cdots + 3 + 2 + 1$. So, $R = (N - 1) + (N - 2) + \cdots + 3 + 2 + 1 - (N - 1) = (N - 2) + \cdots + 3 + 2 + 1 = (N^2 - 3N + 2)/2$.

55. (a) According to Cayley's theorem, there are $3^{3-2} = 3$ spanning trees for a complete graph with 3 vertices.

Likewise, for the completed graph with 4 vertices, Cayley's theorem predicts $4^{4-2} = 16$ spanning trees.

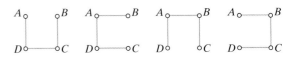
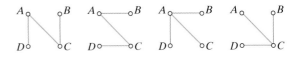
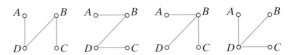
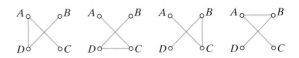

(b) The number of spanning trees is larger than the number of Hamilton circuits. The complete graph with N vertices has $(N - 1)!$ Hamilton circuits and N^{N-2} spanning trees. Since $(N - 1)! = 2 \times 3 \times 4 \times \cdots \times (N - 1)$ and $N^{N-2} = N \times N \times N \times \cdots \times N$, and both expressions have the same number of factors, with each factor in $(N - 1)!$ smaller than the corresponding factor in N^{N-2}, we know that for $N = 3$, $(N - 1)! < N^{N-2}$.

57. The minimum cost network connecting the 4 cities has a 3-way junction point at A and has a total cost of 205 million dollars.

59. (a)

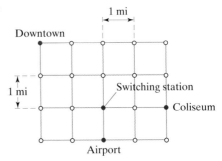

(b) \$7.5 million

61. (a) $m(\angle BFA) = m(\angle AEC) = m(\angle CGB) = 120°$

(b) In $\triangle ABF, m(\angle BFA) = 120°$ and so $m(\angle BAF) + m(\angle ABF) = 60°$. So, $m(\angle BAF) < 60°$ and $m(\angle ABF) < 60°$. Similarly, $m(\angle ACE) < 60°, m(\angle CAE) < 60°, m(\angle BCG) < 60°$, and $m(\angle CBG) < 60°$. Consequently, $m(\angle A) = m(\angle BAF) + m(\angle CAE) < 60° + 60° = 120°$. Likewise, $m(\angle B) < 120°$ and $m(\angle C) < 120°$. So, triangle ABC has a Steiner point S.

(c) Any point X inside or on $\triangle ABF$ (except vertex F) will have $m(\angle AXB) > 120°$. Any point X inside or on $\triangle ACE$ (except vertex E) will have $m(\angle AXC) > 120°$. Any point X inside or on $\triangle BCG$ (except vertex G) will have $m(\angle BXC) > 120°$. If S is the Steiner point, $m(\angle ASB) = m(\angle ASC) = m(\angle BSC) = 120°$, and so S cannot be inside or on $\triangle ABF$ or $\triangle ACE$ or $\triangle BCG$. It follows that the Steiner point S must lie inside $\triangle EFG$.

63. The length of the network is $4x + (500 - x) = 3x + 500$, where $250^2 + \left(\dfrac{x}{2}\right)^2 = x^2$ (see figure). Solving this gives

$x = \dfrac{500\sqrt{3}}{3}$, and so the length of the network is $3\dfrac{500\sqrt{3}}{3} + 500 = 500\sqrt{3} + 500 \approx 1366$ miles.

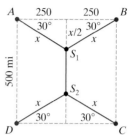

65. (a) **(b)**

Chapter 8

Walking

A. Directed Graphs

1. (a) Vertex set: $\mathcal{V} = \{A, B, C, D, E\}$;
Arc set: $\mathcal{A} = \{AB, AC, BD, CA, CD, CE, EA, ED\}$

(b) $\text{indeg}(A) = 2, \text{indeg}(B) = 1, \text{indeg}(C) = 1, \text{indeg}(D) = 3, \text{indeg}(E) = 1$.

(c) $\text{outdeg}(A) = 2, \text{outdeg}(B) = 1, \text{outdeg}(C) = 3, \text{outdeg}(D) = 0, \text{outdeg}(E) = 2$.

3. (a) C and E

(b) B and C

(c) $B, C,$ and E

(d) No vertices are incident from D.

(e) CD, CE and CA

(f) No arcs are adjacent to CD.

5. (a) **(b)**

7. (a) **(b)**

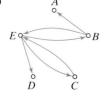
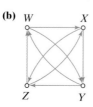

9. (a) 2 **(b)** 1 **(c)** 1 **(d)** 0

11. (a) A, B, D, E, F is one possible path.

(b) A, B, D, E, C, F

(c) B, D, E, B

(d) The outdegree of vertex F is 0, so it cannot be part of a cycle.

(e) The indegree of vertex A is 0, so it cannot be part of a cycle.

(f) B, D, E, B is the only cycle

13.

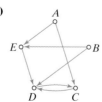

15. (a) B; B is the only person that everyone respects.

(b) A; A is the only person that no one respects.

B. Project Digraphs

17.

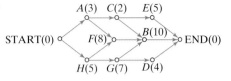
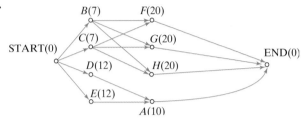

21.

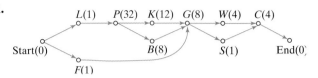

19.

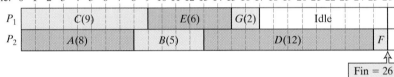
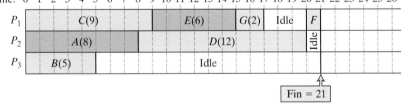

C. Schedules and Priority Lists

23. (a) 18 hours

   (b) There is a total of 75 hours of work to be done. Three processors working without any idle time would take $\frac{75}{3} = 25$ hours to complete the project.

25. There is a total of 75 hours of work to be done. Dividing the work equally between the six processors would require each processor to do $\frac{75}{6} = 12.5$ hours of work. Since there are no $\frac{1}{2}$-hour jobs, the completion time could not be less than 13 hours.

27. Time: 0 1 2 3 4 5 6 7 8 9 10 11 12 13 14 15 16 17 18 19 20 21 22 23 24 25 26

| P_1 | C(9) | E(6) | G(2) | Idle |
| P_2 | A(8) | B(5) | D(12) | F |

Fin = 26

29. Time: 0 1 2 3 4 5 6 7 8 9 10 11 12 13 14 15 16 17 18 19 20 21 22 23 24 25 26

P_1	C(9)	E(6)	G(2)	Idle	F
P_2	A(8)	D(12)	Idle
P_3	B(5)	Idle

Fin = 21

31. B, C, A, E, D, G, F; B, C, A, E, D, F, G; B, C, A, E, G, F, D; B, C, A, E, G, D, F; B, C, A, D, E, G, F

33. According to the precedence relations, G cannot be started until K is completed.

35. Time: 0 1 2 3 4 5 6 7 8 9 10 11 12 13 14 15 16 17 18 19 20 21 22 23 24 25 26 27 28 29 30 31 32 33

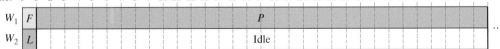

33 34 35 36 37 38 39 40 41 42 43 44 45 46 47 48 49 50 51 52 53 54 55 56 57 58 59 60 61

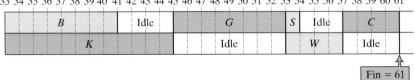

Fin = 61

D. The Decreasing-Time Algorithm

37. Time: 0 1 2 3 4 5 6 7 8 9 10 11 12 13 14 15 16 17 18 19 20 21 22 23 24 25 26

P_1	C(9) — E(6) — G(2) — Idle
P_2	A(8) — B(5) — D(12) — F

Fin = 26

39. Time: 0 1 2 3 4 5 6 7 8 9 10 11 12 13 14 15 16 17 18 19 20 21 22 23 24 25 26 27 28 29 30 31 32 33

W_1	F — P
W_2	L — Idle
W_3	Idle

...

33 34 35 36 37 38 39 40 41 42 43 44 45 46 47 48 49 50 51 52 53 54 55 56 57 58 59 60 61

| K — G — W — C |
| B — Idle — Idle — S — Idle |
| Idle |

Fin = 61

41. (a) Time: 0 2 4 6 8 10 12 14 16 18 20 22 24 26 28 30 32 34 36 38 40 42 44 46 48 50 52

P_1	A(12) — D(6) — F(5) — H(5) — J(4) — L(3) — Idle
P_2	B(7) — C(7) — E(6) — G(5) — I(5) — K(4) — M(3)

Fin = 37

(b) Time: 0 2 4 6 8 10 12 14 16 18 20 22 24 26 28 30 32 34 36 38 40 42 44 46 48 50 52

P_1	A(12) — D(6) — F(5) — H(5) — J(4) — K(4)
P_2	B(7) — C(7) — E(6) — G(5) — I(5) — L(3) — M(3)

Opt = 36

43. (a) Time: 0 1 2 3 4 5 6 7 8 9 10 11 12 13 14 15 16 17 18 19 20 21 22 23 24

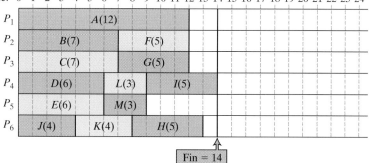

P_1	A(12)
P_2	B(7) — F(5)
P_3	C(7) — G(5)
P_4	D(6) — L(3) — I(5)
P_5	E(6) — M(3)
P_6	J(4) — K(4) — H(5)

Fin = 14

(b) Time: 0 1 2 3 4 5 6 7 8 9 10 11 12 13 14 15 16 17 18 19 20 21 22 23 24

P_1	M(3)	K(4)	I(5)
P_2	L(3)	J(4)	H(5)
P_3	C(7)		G(5)
P_4	B(7)		F(5)
P_5	E(6)		D(6)
P_6	A(12)		

Opt = 12

(c) For six copiers, the optimal completion time is 12 hours and one of the tasks takes 12 hours, thus the job cannot be completed in less than 12 hours.

E. Critical Paths and Critical-Path Algorithm

45. (a)

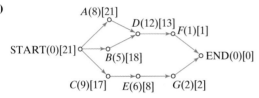

$A(8)[21]$
$D(12)[13]$ $F(1)[1]$
START(0)[21]
$B(5)[18]$
END(0)[0]
$C(9)[17]$ $E(6)[8]$ $G(2)[2]$

(b) $START, A, D, F, END$

(c) Time: 0 1 2 3 4 5 6 7 8 9 10 11 12 13 14 15 16 17 18 19 20 21 22 23 24 25 26

P_1	A(8)	D(12)	G(2)	
P_2	B(5)	C(9)	E(6)	F Idle

Fin = 22

(d) There are a total of 43 work units, so the shortest time the project can be completed by two workers is $\frac{43}{2} = 21.5$ time units. Since there are no tasks with less than one time unit, the shortest time the project can actually be completed is 22 hours.

47. Time: 0 2 4 6 8 10 12 14 16 18 20 22 24 26 28 30 32 34 36 38 40 42 44 46 48 50 52

P_1	B	E	F	I		K
P_2	A	C	Idle	G	Idle	
P_3	D		H	J	Idle	

Fin = 49

49. Time: 0 1 2 3 4 5 6 7 8 9 10 11 12 13 14 15 16 17 18 19 20 21 22 23 24 25 26

P_1	C	A	E	
P_2	G	D	F	Idle
P_3	Idle	H	Idle B	Idle

Fin = 25

F. Scheduling with Independent Tasks

51. (a) Time: 0 1 2 3 4 5 6 7 8 9 10 11 12 13 14 15 16 17 18 19 20 21 22 23 24 25 26

P_1	8 · 5 · 4 · 1
P_2	7 · 6 · 3 · 2

Fin = 18

(b) $Opt = 18$

(c) 0%

53. (a) Time: 0 1 2 3 4 5 6 7 8 9 10

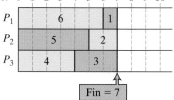

Fin = 7

(b) $Opt = 7$

55. (a) Time: 0 1 2 3 4 5 6 7 8 9 10 11 12 13 14 15 16 17 18 19 20

P_1	E · A · F
P_2	I · B · Idle
P_3	D · C · Idle
P_4	H · G · Idle

Fin = 15

(b) Time: 0 1 2 3 4 5 6 7 8 9 10 11 12 13 14 15 16 17 18 19 20

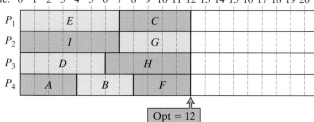

Opt = 12

(c) 25%

57. (a) Time: 0 2 4 6 8 10 12 14 16 18 20 22 24 26 28 30 32 34 36 38 40 42 44 46 48 50 52

P_1	34 · 8 · 2
P_2	21 · 13 · 5 · 3 · 1 1

Opt = 44

(b) Time: 0 3 6 9 12 15 18 21 24 27 30 33 36 39 42 45 48 51 54 57 60 63 66 69 72 75 78

Opt = 72

G. Miscellaneous

59. Each arc of the graph contributes 1 to the indegree sum and 1 to the outdegree sum.

61. (a)

N	ε
7	$\dfrac{6}{21} \approx 28.57\%$
8	$\dfrac{7}{24} \approx 29.17\%$
9	$\dfrac{8}{27} \approx 29.63\%$
10	$\dfrac{9}{30} \approx 30\%$

(b) Since $M - 1 < M$, we have for every M that

$$\frac{M-1}{3M} \leq \frac{M}{3M} = \frac{1}{3}.$$

63. See Figure 8-14.

65. See Figure 8-23.

Jogging

67. (a)

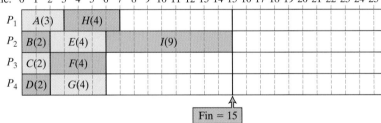

(b)

(c) An extra processor was used and yet the finishing time of the project increased.

69. (a) Same as 67(a).

(b)

(c) The project had fewer restrictions on the order of the assignments to the processors and yet the finishing time of the project increased.

71. (a) The finishing time of a project is always more than or equal to the number of hours of work to be done divided by the number of processors doing the work.

(b) The schedule is optimal with no idle time.

(c) The total idle time in the schedule.

Chapter 9

Walking

A. Fibonacci Numbers

1. (a) 55   (b) 57   (c) 144   (d) 27.5   (e) 5

3. (a) 12   (b) 610   (c) 6   (d) 144   (e) 2

5. (a) One more than three times the Nth Fibonacci number.
   (b) Three times the $(N + 1)$st Fibonacci number.
   (c) One more than the $(3N)$th Fibonacci number.
   (d) The $(3N + 1)$st Fibonacci number.

7. (a) 39,088,169   (b) 9,227,465

9. I. This is an equivalent way to express the fact that each term of the Fibonacci sequence is equal to the sum of the two preceding terms.

11. (a) $47 = 34 + 13$
   (b) $48 = 34 + 13 + 1$
   (c) $207 = 144 + 55 + 8$
   (d) $210 = 144 + 55 + 8 + 3$

13. (a) $1 + 2 + 5 + 13 + 34 + 89 = 144$
   (b) 22   (c) $N + 1$

15. (a) Answers will vary. One example is
$$F_7 + F_4 = 13 + 3 = 16 = 2 \times 8 = 2F_6$$
   (b) $F_{N+3} + F_N = 2F_{N+2}$

17. (a) $2F_6 - F_7 = 2 \times 8 - 13 = 3 = F_4$
   (b) $2F_{N+2} - F_{N+3} = F_N$

B. The Golden Ratio

19. (a) 46.97871   (b) 46.97871   (c) 21

21. (a) 21   (b) 34   (c) 13

23. (a) $a = 34, b = 21$ (b) $a = 17, b = 38$

25. 1.394×10^{104}

27. (a) 169   (b) 2.41429
   (c) 2.41421   (d) 2.41421

C. Fibonacci Numbers and Quadratic Equations

29. (a) $x = 1 + \sqrt{2} \approx 2.41421, x = 1 - \sqrt{2} \approx -0.41421$
   (b) $x = -1, x = \dfrac{8}{3} \approx 2.66667$

31. (a) $x = 1$
   (b) $x = -\dfrac{21}{55}$

33. (a) Putting $x = 1$ in the equation gives $F_N = F_{N-1} + F_{N-2}$ which is a defining equation for the Fibonacci numbers.
   (b) The sum of the roots of the equation is $-\dfrac{-F_{N-1}}{F_N} = \dfrac{F_{N-1}}{F_N}$ and so the other root is $x = \left(\dfrac{F_{N-1}}{F_N}\right) - 1$.

D. Similarity

35. (a) 124.5 in.
   (b) 945 in.^{2}

37. (a) 156 m   (b) 2880 m^2

39. 3, 5

E. Gnomons

41. $c = 24$

43. $x = 4$

45. 20 by 30

47. (a) $m(\angle ACD) = 72°; m(\angle CAD) = 72°; m(\angle CDA) = 36°$
   (b) $AC = 1, CD = \phi, AD = \phi$

49. $x = 12, y = 10$

Jogging

51. $A_N = 5F_N$

53. (a) $T_1 = 11, T_2 = 18, T_3 = 29, T_4 = 47, T_5 = 76, T_6 = 123, T_7 = 199, T_8 = 322$
   (b) $T_N = 7F_{N+1} + 4F_N$
$$= 7(F_N + F_{N-1}) + 4(F_N + F_{N-2})$$
$$= (7F_N + 4F_{N-1}) + (7F_{N-1} + 4F_{N-2})$$
$$= T_{N-1} + T_{N-2}$$
   (c) $T_1 = 11, T_2 = 18, T_N = T_{N-1} + T_{N-2}$

55. $\dfrac{1 + \sqrt{5}}{2} + \dfrac{1 - \sqrt{5}}{2} = 1$

Since the first term is positive and the second term is negative and the sum is a whole number, the decimal parts are the same.

57. $\dfrac{l}{s} = \dfrac{144\phi + 89}{89\phi + 55} = \dfrac{\phi^{12}}{\phi^{11}} = \phi$

59. $\sqrt{2}$

61. $x = 6, y = 12, z = 10$

63. $x = 3, y = 5$

65. **(a)** Since we are given that $AB = BC = 1$, we know that $m(\angle BAC) = 72°$ and so $m(\angle BAD) = 180° - 72° = 108°$. This makes $m(\angle ABD) = 180° - 108° - 36° = 36°$ and so ΔABD is isosceles with $AD = AB = 1$. Therefore, $AC = x - 1$. Using these facts and the similarity of triangle ABC and triangle BCD, we have $\dfrac{x}{1} = \dfrac{1}{x - 1}$ or, $x^2 = x + 1$ so that $x = \phi$.

(b) $36°, 36°, 108°$

(c) $\dfrac{\text{longer side}}{\text{shorter side}} = \dfrac{x}{1} = \phi$

67. **(a)** $\dfrac{10}{\phi}$

(b) $\dfrac{10r}{\phi}$

Chapter 10

Walking

A. Linear Growth and Arithmetic Sequences

1. **(a)** $P_1 = 205, P_2 = 330, P_3 = 455$
(b) $P_{100} = 12{,}580$
(c) $P_N = 80 + 125N$

3. **(a)** $P_{30} = 225$ **(b)** 185 **(c)** 186

5. **(a)** $d = 3$ **(b)** $P_{50} = 158$ **(c)** $P_N = 8 + 3N$

7. **(a)** $A_3 = -19$
(b) $A_0 = 26$
(c) None. The sequence is decreasing and starts at 26.

9. **(a)** $P_N = P_{N-1} + 5; P_0 = 3$
(b) $P_N = 3 + 5N$
(c) $P_{300} = 1503$

11. **(a)** 497 **(b)** 24,950

13. **(a)** 100th term **(b)** 16,050

15. **(a)** 3,519,500 **(b)** 3,482,550

17. **(a)** 213 **(b)** $137 + 2N$
(c) \$7,124 **(d)** \$2,652

B. Exponential Growth and Geometric Sequences

19. **(a)** 13.75
(b) $11 \times (1.25)^9 \approx 81.956$
(c) $P_N = 11 \times (1.25)^N$

21. **(a)** $P_1 = 20, P_2 = 80, P_3 = 320$
(b) $P_N = 5 \times 4^N$
(c) 9

23. **(a)** $P_N = 1.50P_{N-1}, P_0 = 200$

(b) $P_N = 200 \times 1.5^N$
(c) approximately 11,500

25. **(a)** 3×2^{100}
(b) $P_N = 3 \times 2^N$
(c) $3 \times (2^{101} - 1)$
(d) $3 \times 2^{101} - 3 \times 2^{50} = 3 \times 2^{50} \times (2^{51} - 1)$

C. Financial Applications

27. **(a)** \$40.50 **(b)** 40.5% **(c)** 40.5%

29. \$4,587.64

31. \$5,184.71

33. \$1,874.53

35. **(a)** 0.01 (i.e., 1%)
(b) \$9,083.48
(c) approximately 12.6825%

37. Great Bulldog Bank: 6%;
First Northern Bank: approximately 5.904%;
Bank of Wonderland: approximately 5.654%

39. **(a)** \$10,834.71 **(b)** \$11,338.09 **(c)** \$10,736.64

41. \$1133.56

43. **(a)** \$6,209.21 **(b)** \$6,102.71 **(c)** \$6,077.89

D. Logistic Growth Model

45. **(a)** $p_1 = 0.168$
(b) $p_2 \approx 0.1118$
(c) approximately 7.945%

47. **(a)** $p_1 = 0.1680, p_2 \approx 0.1118, p_3 \approx 0.0795,$
$p_4 \approx 0.0585, p_5 \approx 0.0441, p_6 \approx 0.0337,$
$p_7 \approx 0.0261, p_8 \approx 0.0203, p_9 \approx 0.0159,$
$p_{10} \approx 0.0125$
(b) extinction

49. (a) $p_1 = 0.4320$, $p_2 \approx 0.4417$, $p_3 \approx 0.4439$,
$p_4 \approx 0.4443$, $p_5 \approx 0.4444$, $p_6 \approx 0.4444$,
$p_7 \approx 0.4444$, $p_8 \approx 0.4444$, $p_9 \approx 0.4444$,
$p_{10} \approx 0.4444$

(b) The population becomes stable at $\frac{4}{9} \approx 44.44\%$ of the habitat's carrying capacity.

51. (a) $p_1 = 0.3570$, $p_2 \approx 0.6427$, $p_3 \approx 0.6429$,
$p_4 \approx 0.6428$, $p_5 \approx 0.6429$, $p_6 \approx 0.6428$,
$p_7 \approx 0.6429$, $p_8 \approx 0.6428$, $p_9 \approx 0.6429$,
$p_{10} \approx 0.6428$

(b) The population becomes stable at $\frac{9}{14} \approx 64.29\%$ of the habitat's carrying capacity.

53. (a) $p_1 = 0.5200$, $p_2 = 0.8112$, $p_3 \approx 0.4978$,
$p_4 \approx 0.8125$, $p_5 \approx 0.4952$, $p_6 \approx 0.8124$,
$p_7 \approx 0.4953$, $p_8 \approx 0.8124$, $p_9 \approx 0.4953$,
$p_{10} \approx 0.8124$

(b) The population settles into a two-period cycle alternating between a high-population period at approximately 81.24% and a low-population period at approximately 49.53% of the habitat's carrying capacity.

E. Miscellaneous

55. (a) exponential   **(b)** linear   **(c)** logistic

(d) exponential   **(e)** logistic   **(f)** linear

(g) linear, exponential, or logistic (they all apply!)

57. (a) $P_2 = 22$, $P_3 = 42$

(b) The first two numbers in this sequence are even. Thereafter, the sum and product of two even numbers is also even.

Jogging

59. 6%

61. (a) a 6, 8, 10 triangle   **(b)** a 1, $\sqrt{\Phi}$, Φ triangle

63. 39.15%

65. \$10,737,418.23

67. $r = \dfrac{1 \pm \sqrt{5}}{2}$

69. $(a + ar + ar^2 + \cdots + ar^{N-1})(r - 1)$
$= ar + ar^2 + ar^3 + \cdots + ar^{N-1} + ar^N$
$\quad -a - ar - ar^2 - ar^3 - \cdots - ar^{N-1}$
$= -a + ar^N$
$= a(r^N - 1)$

After dividing both sides by $(r - 1)$,

$$a + ar + ar^2 + \cdots + ar^{N-1} = \frac{a(r^N - 1)}{r - 1}.$$

71. One example is given by $P_N = 8 \times \left(\dfrac{1}{2}\right)^N$. Here $r = \dfrac{1}{2}$ and

$$P_0 = 8, P_1 = 4, P_2 = 2, P_3 = 1, P_4 = \frac{1}{2}, P_5 = \frac{1}{4} \cdots$$

73. No. This would require $p_0 = p_1 = 0.8(1 - p_0)p_0$ and so $p_0 = 0$ or $p_0 = -0.25$, neither of which are possible.

75. approximately 14,619 snails

Chapter 11

Walking

A. Reflections

1. (a) C **(b)** F **(c)** E **(d)** B

3. (a)

(b)

(c)

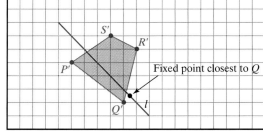

Fixed point closest to Q

5. (a)

(b)

(c)

(d)

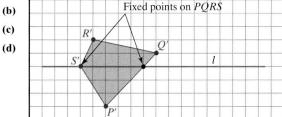

Fixed points on $PQRS$

7.

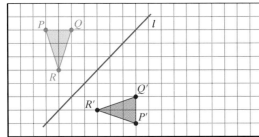

9.

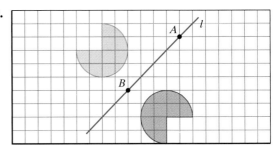

B. Rotations

11. (a) *I*   (b) *E*   (c) *G*   (d) *A*

    (e) *F*   (f) *C*   (g) *E*   (h) *D*

13. (a) $110°$ (b) $350°$ (c) $10°$ (d) $81°$

15. (a)

    (b)

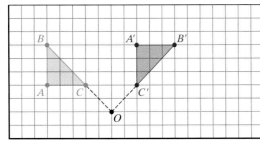

17. (a)

    (b)

    (c)

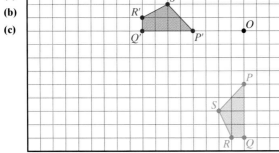

This is a 90° clockwise rotation.

19. (a)

    (b)

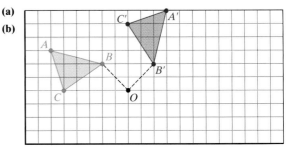

C. Translations

21. (a) *C*   (b) *C*   (c) *A*   (d) *D*

23. (a)

    (b)

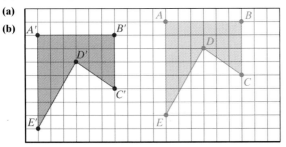

25. One possible answer: One translation is right (East) 4 miles. A second translation is up (North) 3 miles (see figure).

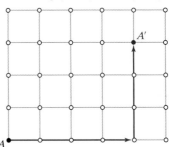

D. Glide Reflections

27.

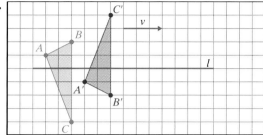

31. (a)
(b)

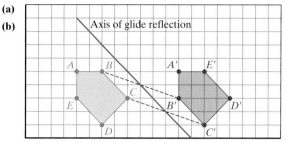

29. (a)
(b)

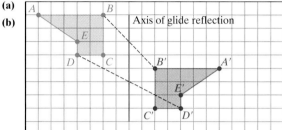

33. (a)
(b)

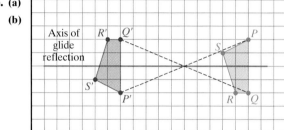

E. Symmetries of Finite Shapes

35. (a) Reflection with axis going through the midpoints of AB and DC; reflection with axis going through the midpoints of AD and BC; rotations of $180°$ and $360°$ with rotocenter the center of the rectangle.

(b) No reflections. Rotations of $180°$ and $360°$ with rotocenter the center of the parallelogram.

(c) Reflection with axis going through the midpoints of AB and DC; rotation of $360°$ with rotocenter the center of the trapezoid.

37. (a) Reflections (three of them) with axis going through pairs of opposite vertices; reflections (three of them) with axis going through the midpoints of opposite sides of the hexagon; rotations of $60°, 120°, 180°, 240°, 300°, 360°$ with rotocenter the center of the hexagon.

(b) Reflections with axis AD, GJ, BE, HK, CF, IL; rotations of $60°, 120°, 180°, 240°, 300°, 360°$ with rotocenter the center of the star.

39. (a) D_2 **(b)** Z_2 **(c)** D_1

41. (a) D_6 **(b)** D_6

43. (a) D_1 **(b)** D_1 **(c)** Z_1 **(d)** Z_2 **(e)** D_1 **(f)** D_2

45. (a) J **(b)** T **(c)** Z **(d)** I

47. (a) Symmetry type D_5 is common among many types of flowers (daisies, geraniums, etc.). The only requirements are that the flower have five equal, evenly spaced petals and that the petals have a reflection symmetry along their long axis. In the animal world, symmetry type D_5 is less common, but it can be found among certain types of starfish, sand dollars, and in some single-celled organisms called diatoms.

(b) The Chrysler Corporation logo is a classic example of a shape with symmetry D_5. Symmetry type D_5 is also common in automobile wheels and hubcaps. One of the largest and most unusual buildings in Washington, D.C. has symmetry of type D_5.

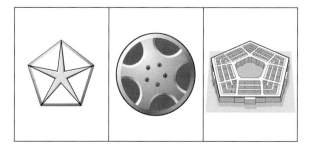

(c) Objects with symmetry type Z_1 are those whose only symmetry is the identity. Thus, any "irregular" shape fits the bill. Tree leaves, seashells, plants, and rocks more often than not have symmetry type Z_1.

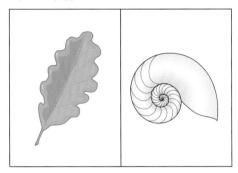

(d) Examples of manmade objects with symmetry of type Z_1 abound.

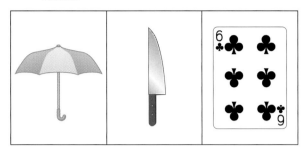

F. Symmetries of Border Patterns

49. (a) $m1$ **(b)** $1m$ **(c)** 12   **(d)** 11

51. (a) $m1$ **(b)** 12   **(c)** $1g$ **(d)** mg

53. 12

G. Miscellaneous

55. Since every proper rigid motion is equivalent to either a rotation or a translation, and a translation has no fixed points, the specified rigid motion must be equivalent to a rotation.

57. (a) D **(b)** D **(c)** B **(d)** E **(e)** G

59. (a) improper   **(b)** proper   **(c)** improper   **(d)** proper

61. The combination of two improper rigid motions is a proper rigid motion. Since C is a fixed point, the rigid motion must be a rotation with rotocenter C.

Jogging

63. (a) A clockwise rotation with center C and angle of rotation 2α.

(b) A counterclockwise rotation with center C and angle of rotation 2α.

65. (a)

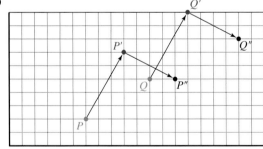

(b)

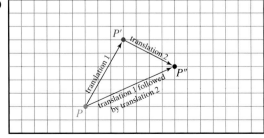

67. (a) By definition, a border has translation symmetries in exactly one direction (let's assume the horizontal direction). If the pattern had a reflection symmetry along an axis forming 45° with the horizontal direction, there would have to be a second direction of translation symmetry (vertical).

(b) If a pattern had a reflection symmetry along an axis forming an angle of $\alpha°$ with the horizontal direction, it would have to have translation symmetry in a direction that forms an angle of $2\alpha°$ with the horizontal. This could only happen for $\alpha = 90°$ or $\alpha = 180°$ (since the only allowable direction for translation symmetries is the horizontal).

69. (a)

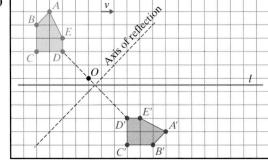

(b) Rotations and translations are proper rigid motions, and hence preserve clockwise-counterclockwise orientations. The given motion is an improper rigid motion (it reverses the clockwise-counterclockwise orientation). Since performing the same rotation and glide reflection again results in the original figure, this is a reflection (rather than a glide reflection) whose axis is shown in (a).

71. Rotations and translations are proper rigid motions, and hence preserve clockwise-counterclockwise orientations. The given motion is an improper rigid motion (it reverses the clockwise-counterclockwise orientation). If the rigid motion was a reflection, then PP', RR', and QQ' would all be perpendicular to the axis of reflection and hence would all be parallel. It must be a glide reflection (the only rigid motion left).

Chapter 12

Walking

A. The Koch Snowflake and Variations

1.

	Start	Step 1	Step 2	Step 3	Step 4	. . .	Step 30	Step N
Number of sides	3	12	48	192	768		3×4^{30}	3×4^N
Length of each side	1 in.	1/3 in.	1/9 in.	1/27 in.	1/81 in.		$\frac{1}{3^{30}}$ in.	$\frac{1}{3^N}$ in.
Length of the boundary	3 in.	4 in.	48/9 in.	192/27 in.	768/81 in.		≈ 0.265 mi	$3 \times \left(\frac{4}{3}\right)^N$ in.

3.

	Start	Step 1	Step 2	Step 3	Step 4
Area	24 in.^{2}	32 in.^{2}	$\frac{320}{9}$ in.^{2}	$\frac{3,008}{81}$ in.^{2}	$\frac{27,584}{729}$ in.^{2}

5. 38.4 in.^{2}

7.

	Start	Step 1	Step 2	Step 3	Step 4	. . .	Step 50	Step N
Number of sides	4	20	100	500	2500		4×5^{50}	4×5^N
Length of each side	1 in.	1/3 in.	1/9 in.	1/27 in.	1/81 in.		$\frac{1}{3^{50}}$ in.	$\frac{1}{3^N}$ in.
Length of the boundary	4 in.	20/3 in.	100/9 in.	500/27 in.	2500/81 in.		$\approx 7,810,562$ mi	$4 \times \left(\frac{5}{3}\right)^N$

9.

	Start	Step 1	Step 2	Step 3	Step 4
Area	1 in.^{2}	$\frac{13}{9}$ in.^{2}	$\frac{137}{81}$ in.^{2}	$\frac{1333}{729}$ in.^{2}	$\frac{12,497}{6561}$ in.^{2}

11.

	Start	Step 1	Step 2	Step 3	Step 4	. . .	Step 40	Step N
Number of sides	3	12	48	192	768		3×4^{40}	3×4^N
Length of each side	a	$a/3$	$a/9$	$a/27$	$a/81$		$\frac{a}{3^{40}}$	$\frac{a}{3^N}$
Length of the boundary	$3a$	$4a$	$\frac{48}{9}a$	$\frac{192}{27}a$	$\frac{768}{81}a$		$\frac{3 \times 4^{40}}{3^{40}}a$	$3a \times \left(\frac{4}{3}\right)^N$

13.

	Start	Step 1	Step 2	Step 3	Step 4	Step 5
Area	81 in.^{2}	54 in.^{2}	42 in.^{2}	$\frac{110}{3}$ in.^{2}	$\frac{926}{27}$ in.^{2}	$\frac{8078}{243}$ in.^{2}

15.

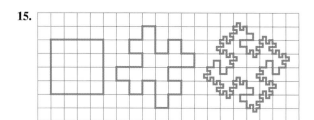

Start　　　Step 1　　　Step 2

17. (a)

	Start	Step 1	Step 2	Step 3	Step 4
Area	A	A	A	A	A

(b) If the area of the starting square is A, then the area of the resulting quadratic Koch island is also A.

B. The Sierpinski Gasket and Variations

19.

	Start	Step 1	Step 2	Step 3	Step 4	. . .	Step N
Area	1	$\dfrac{3}{4}$	$\left(\dfrac{3}{4}\right)^2$	$\left(\dfrac{3}{4}\right)^3$	$\left(\dfrac{3}{4}\right)^4$		$\left(\dfrac{3}{4}\right)^N$

21. The area of the Sierpinski gasket is smaller than the area of the gasket formed during any step of construction. That is, if the area of the original triangle is 1, then the area of the Sierpinski gasket is less than $\left(\dfrac{3}{4}\right)^N$ for every positive value of N. Since $0 < 3/4 < 1$, the value of $\left(\dfrac{3}{4}\right)^N$ can be made smaller than any positive quantity for a large enough choice of N. It follows that the area of the Sierpinski gasket is smaller than any positive quantity.

23.

	Start	Step 1	Step 2	Step 3	Step 4	. . .	Step N
Number of triangles	1	3	9	27	81		3^N
Perimeter of each triangle	24 in.	12 in.	6 in.	3 in.	1.5 in.		$24 \times \left(\dfrac{1}{2}\right)^N$ in.
Length of the boundary	24 in.	36 in.	54 in.	81 in.	121.5 in.		$24 \times \left(\dfrac{3}{2}\right)^N$ in.

25.

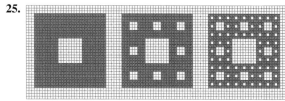

Step 1　　　Step 2　　　Step 3

27. (a)

	Start	Step 1	Step 2	Step 3	Step 4
Perimeter	4	16/3	80/9	496/27	3536/81

(b) $L + 8^N \cdot \dfrac{4}{3^{N+1}}$

29.

	Start	Step 1	Step 2	Step 3	Step 4	. . .	Step N
Number of triangles	1	6	6^2	6^3	6^4		6^N
Area of each triangle	A	$\dfrac{1}{9}A$	$\dfrac{1}{9^2}A$	$\dfrac{1}{9^3}A$	$\dfrac{1}{9^4}A$		$\dfrac{1}{9^N}A$
Area of gasket	A	$\dfrac{2}{3}A$	$\left(\dfrac{2}{3}\right)^2 A$	$\left(\dfrac{2}{3}\right)^3 A$	$\left(\dfrac{2}{3}\right)^4 A$		$\left(\dfrac{2}{3}\right)^N A$

31. The area of the Sierpinski ternary gasket is smaller than the area of the gasket formed during any step of construction. That is, if the area of the original triangle is A, then the area of the Sierpinski ternary gasket is less than $\left(\dfrac{2}{3}\right)^N A$ for every positive value of N. Since $0 < 2/3 < 1$, the value of $\left(\dfrac{2}{3}\right)^N A$ can be made smaller than any positive quantity for a large enough choice of N. It follows that the Sierpinski ternary gasket can be made smaller than any positive quantity.

C. The Chaos Game and Variations

33.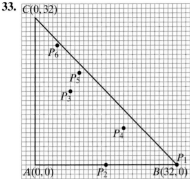

35.

Number rolled	3	1	2	3	5	5
Point	P_1	P_2	P_3	P_4	P_5	P_6
Coordinates	(32, 0)	(16, 0)	(8, 0)	(20, 0)	(10, 16)	(5, 24)

37. **(a)**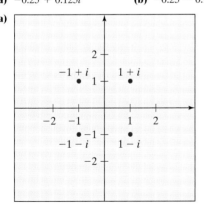

(b)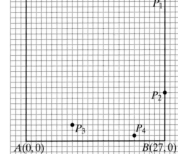

(c)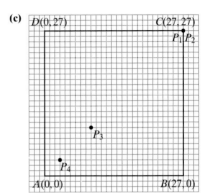

39. **(a)** P_1: (0, 27); P_2: (18, 9); P_3: (6, 3); P_4: (20, 1)

(b) P_1: (27, 27); P_2: (9, 9); P_3: (3, 3); P_4: (19, 19)

(c) P_1: (0, 0); P_2: (18, 18); P_3: (6, 24); P_4: (20, 8)

D. Operations with Complex Numbers

41. **(a)** $1 + 3i$ **(b)** $1 - 3i$ **(c)** $-1 - i$

43. **(a)** $-0.25 + 0.125i$ **(b)** $-0.25 - 0.125i$

45. **(a)**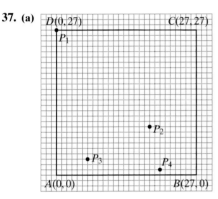

(b)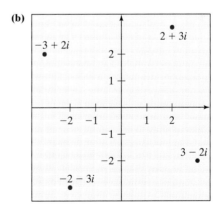

(c) It is a 90-degree counterclockwise rotation.

E. Mandelbrot Sequences

47. **(a)** $2, 2, 2, 2$

(b) 2

(c) Attracted to 2. Each number in the sequence is 2.

49. **(a)** $s_1 = -0.25$; $s_2 = -0.4375$; $s_3 \approx -0.3086$; $s_4 \approx -0.4048$; $s_5 \approx -0.3362$

(b) -0.3660

(c) Attracted to -0.3660. From parts (a) and (b), the sequence gets close to -0.3660 and then $s_{N+1} = s_N$ (approximately).

51. (a) $s_1 = \dfrac{3}{4}; s_2 = \dfrac{17}{16}; s_3 = \dfrac{417}{256}$

(b) If $s_N > 1$, then $(s_N)^2 > s_N$, and so
$$s_{N+1} = (s_N)^2 + \frac{1}{2} > s_N + \frac{1}{2} > s_N.$$

(c) Escaping. Notice from part (a) that $s_2 > 1$. Part (b) then implies that the sequence is escaping.

53. (a) 2   **(b)** 1

Jogging

55. At each step, one new hole is introduced for each solid triangle and the number of solid triangles is tripled.

Step	Holes	Solid triangles
1	1	3
2	$1 + 3$	3^2
3	$1 + 3 + 3^2$	3^3
4	$1 + 3 + 3^2 + 3^3$	3^4

Using the formula for the sum of the terms in a geometric sequence (with common ratio $r = 3$), we see that at the Nth step there will be
$$1 + 3 + 3^2 + 3^3 + \cdots + 3^{N-1} = \frac{1 - 3^N}{1 - 3} = \frac{3^N - 1}{2} \text{ holes.}$$

57.

	Start	Step 1	Step 2	Step 3	...	Step N
Number of cubes removed	0	7	20×7	$20^2 \times 7$		$20^{N-1} \times 7$

59. (a) Reflection with axis a vertical line passing through the center of the snowflake; reflections with axes lines making $30°$, $60°$, $90°$, $120°$, $150°$ angles with the vertical axis of the snowflake and passing through the center of the snowflake.

(b) Rotations of $60°, 120°, 180°, 240°, 300°, 360°$ with rotocenter the center of the snowflake.

(c) D_6

61. (a) Using the formula for adding the terms in a geometric sequence (with common ratio $r = 4/9$),
$$1 + \left(\tfrac{4}{9}\right) + \left(\tfrac{4}{9}\right)^2 + \cdots + \left(\tfrac{4}{9}\right)^{N-1} = \frac{1 - \left(\tfrac{4}{9}\right)^N}{\left(\tfrac{5}{9}\right)} = \tfrac{9}{5}\left[1 - \left(\tfrac{4}{9}\right)^N\right].$$

(b) Using the result in (a), we have
$$\left(\frac{A}{3}\right) + \left(\frac{A}{3}\right)\left(\frac{4}{9}\right) + \left(\frac{A}{3}\right)\left(\frac{4}{9}\right)^2 + \cdots + \left(\frac{A}{3}\right)\left(\frac{4}{9}\right)^{N-1} =$$
$$\left(\frac{A}{3}\right)\left[1 + \left(\frac{4}{9}\right) + \left(\frac{4}{9}\right)^2 + \cdots + \left(\frac{4}{9}\right)^{N-1}\right] =$$
$$\left(\frac{A}{3}\right)\left(\frac{9}{5}\right)\left[1 - \left(\frac{4}{9}\right)^N\right] = \left(\frac{3}{5}\right)A\left[1 - \left(\frac{4}{9}\right)^N\right].$$

63. It is attracted to $\dfrac{1 - \sqrt{0.2}}{2} \approx 0.276393$.

65. periodic

Chapter 13

Walking

A. Surveys and Public Opinion Polls

1. (a) the gumballs in the jar

(b) the gumballs drawn out of the jar

(c) 32%

(d) simple random sampling

3. (a) 25%

(b) 7%

(c) Sampling variability; simple random sampling, which does not suffer from selection bias, was used.

5. (a) $\dfrac{680}{8325}$ **(b)** 45%

7. Smith: 3%; Jones: 3%; Brown: 0%

9. (a) the 350 students attending the Eureka High football game the week before the election

(b) $\dfrac{350}{1250}$

11. (a) The population consists of all 1250 students at Eureka High School while the sampling frame consists only of those 350 students that attended the football game the week prior to the election.

(b) Mainly sampling bias; the sampling frame is not representative of the population.

13. (a) all married people who read Dear Abby's column

(b) Abby's target population appears to be all married people. However, she is sampling from a (nonrepresentative) subset of the population—a sampling frame that consists of those married couples that read her column.

(c) self-selection

(d) A statistic. It is based on data taken from a sample.

15. (a) 74%

(b) 81.8%

(c) Not very accurate. The sample was not representative of the target population.

17. (a) the citizens of Cleansburg

(b) The sampling frame is limited to that part of the target population that passes by a city street corner between 4:00 P.M. and 6:00 P.M.

19. (a) The choice of street corner could make a great deal of difference in the responses collected.

(b) *D*; We are assuming that people who live or work downtown are much more likely to answer yes than people in other parts of town.

(c) Yes, for two main reasons: (i) People out on the street between 4 P.M. and 6 P.M. are not representative of the population at large. For example, office and white-collar workers are much more likely to be in the sample than homemakers and school teachers. (ii) The five street corners were chosen by the interviewers and the passersby are unlikely to represent a cross section of the city.

(d) No. No attempt was made to use quotas to get a representative cross section of the population.

21. (a) The target population and the sampling frame both consist of all TSU undergraduates.

(b) 15,000

23. (a) In simple random sampling, any two members of the population have as much chance of both being in the sample as any other two. But in this sample, two people with the same last name—say Len Euler and Linda Euler—have no chance of being in the sample together.

(b) Sampling variability. The students sampled appear to be a representative cross section of all TSU undergraduates that would attempt to enroll in Math 101.

25. (a) stratified sampling

(b) quota sampling

27. (a) convenience sampling

(b) census

(c) stratified sampling

(d) simple random sampling

(e) quota sampling

B. The Capture-Recapture Method

29. 2000

31. 84

33. 124

35. 160,000

37. (a) 36,600

(b) 10,860

C. Clinical Studies

39. (a) anyone who could have a cold and would consider buying vitamin *X* (i.e., pretty much all adults)

(b) The sampling frame is only a small portion of the target population consisting only of college students in the San Diego area that are suffering from colds.

(c) Yes. This sample would likely under represent older adults and those living in colder climates.

41. (i) Using college students. (College students are not a representative cross section of the population in terms of age and therefore in terms of how they would respond to the treatment.)

(ii) using subjects only from the San Diego area

(iii) offering money as an incentive to participate

(iv) allowing self-reporting (the subjects themselves determine when their colds are over)

43. all potential knee surgery patients

45. (a) Yes. There was a control group receiving the sham-surgery (placebo) and two treatment groups.

(b) The first treatment group consisted of those patients receiving arthroscopic debridement. The second treatment group consisted of those patients receiving arthroscopic lavage.

(c) Yes. The 180 patients in the study were assigned to a treatment group at random.

(d) blind

47. (iv) clinical study; The professor was trying to establish the connection between a cause (10 milligrams of caffeine a day) and

an effect (improved performance in college courses). Other than that, the experiment had little going for it: it was not controlled (no control group); not randomized (the subjects were chosen because of their poor grades); no placebo was used and consequently the study was not double-blind.

49. (a) It is likely that the study was blind but not double-blind since the professor knew who was in the study.

(b) (i) A regular visit to the professor's office could in itself be a boost to a student's self-confidence and help improve his or her grades.

(ii) The "individualized tutoring" that took place during the office meetings could also be the reason for improved performance.

(iii) The students selected for the study all got F's on their first midterm, making them likely candidates to show some improvement.

51. all people having a history of colorectal adenomas

53. (a) The treatment group consisted of the 1287 patients that were given 25 daily milligrams of Vioxx. The control group consisted of the 1299 patients that were given a placebo.

(b) This is an experiment since members of the population received a treatment. It is a controlled placebo experiment since there was a control group that did not receive the treatment, but instead received a placebo. It is a randomized controlled experiment since the 2586 participants were randomly divided into the treatment and control groups. The study was double blind since neither the participants nor the doctors involved in the clinical trial knew who was in each group.

D. Miscellaneous

55. (a) A clinical trial. A treatment was imposed (eating three meals at McDonald's every day for 30 days) on a sample of the population.

(b) The target population is the set of "average Americans."

(c) The sample consisted of one person (Mr. Spurlock).

(d) (i) a sample that is not representative of the population
(ii) small sample size
(iii) tThe lack of a control group

57. (a) This study was a clinical trial since a treatment was imposed (the heavy reliance on supplemental materials, online practice exercises, and interactive tutorials) on a sample of the population.

(b) (i) The instructors used in the treatment group may be more excited about this new curricular approach or they may be better teachers.

(ii) If students in this particular intermediate algebra class were self-selected, they may not be representative of the target population.

(iii) Students in the treatment group may have benefited simply by being forced to put more time studying into the course. To eliminate this possible confounding variable, the control groups should be asked to spend the same amount of time studying out of class.

59. (a) parameter

(b) statistic

(c) statistic

(d) statistic

Jogging

61. (a) (i) the entire sky; (ii) all the coffee in the cup; (iii) the entire Math 101 class

(b) In none of the three examples is the sample random.

(c) (i) In some situations one can have a good idea as to whether it will rain or not by seeing only a small section of the sky, but in many other situations rain clouds can be patchy and one might draw the wrong conclusions by just peeking out the window. (ii) If the coffee is burning hot on top, it is likely to be pretty hot throughout, so Betty's conclusion is likely to be valid. (iii) Since Carla used convenience sampling and those students sitting next to her are not likely to be a representative sample, her conclusion is likely to be invalid.

63. (a) The question was worded in a way that made it almost impossible to answer yes.

(b) "Will you support some form of tax increase if it can be proven to you that such a tax increase is justified?" is better, but still not neutral. "Do you support or oppose some form of tax increase?" is bland but probably as neutral as one can get.

(c) Many such examples exist. A very real example is a question such as "Would you describe yourself as pro-life or not?" or "Would you describe yourself as pro-choice or not?"

65. (a) Under method 1, people whose phone numbers are unlisted are automatically ruled out from the sample. At the same time, method 1 is cheaper and easier to implement than method 2. Method 2 will typically produce more reliable data since the sample selected would better represent the population.

(b) Method 2. The two main reasons are (i) people with unlisted phone numbers are very likely to be the same kind of people that would seriously consider buying a burglar alarm, and (ii) the listing bias is more likely to be significant in a place like New York City.

67. (a) Fridays, Saturdays, and Sundays make up 3/7 or about 43% of the week. It follows that proportionately, there are fewer fatalities due to automobile accidents on Friday, Saturday, and Sunday (42%) than there are on Monday through Thursday.

(b) On Saturday and Sunday there are fewer people commuting to work. The number of cars on the road and miles driven is significantly less on weekends. The number of accidents due to fatalities should be proportionally much less. The fact that it is 42% indicates that there are other factors involved and increased drinking is one possible explanation.

Chapter 14

Walking

A. Frequency Tables, Bar Graphs, and Pie Charts

1.

Score	10	50	60	70	80	100
Frequency	1	3	7	6	5	2

3. (a)

Grade	A	B	C	D	F
Frequency	7	6	7	3	1

(b)

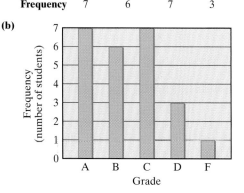

5. (a) 80

(b) 30%

(c)

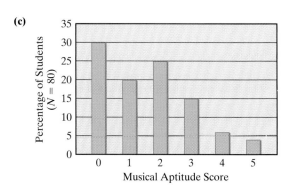

7.

Distance (miles) to school	0.0	0.5	1.0	1.5	2.0	2.5	3.0	5.0	8.5
Frequency	4	3	4	6	3	2	1	1	1

9. (a)

Class interval	Very close	Close	Nearby	Not too far	Far
Frequency	7	10	5	1	2

(b)

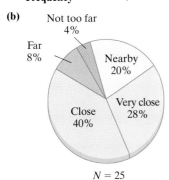

N = 25

11. (a) 30

(b) 0%

(c) approximately 56.7%

13. Accidents: 155°; Homicide: 62°; Suicide: 47°; Cancer: 19°; Heart Disease: 12°; Other: 66°

15. (a) $189,000,000,000 ($189 billion)

(b) $924,000,000,000 ($924 billion)

17.

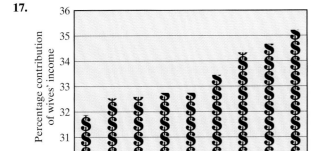

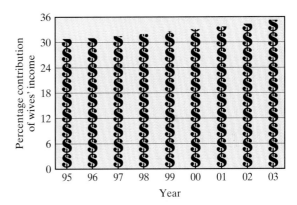

B. Histograms

19. (a) 12 ounces

(b) The third class interval: "more than 72 ounces and less than or equal to 84 ounces." Values that fall exactly on the boundary between two class intervals belong to the class interval to the left.

(c)

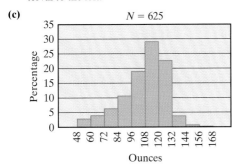

N = 625

21. (a) 11

(b) $20,000,000 ($20 million)

(c) Payrolls of $20 million, $40 million, $60 million, $80 million, or $100 million belong to the class interval on the right.

C. Means and Medians

23. (a) 2

(b) 3

(c) average = 2; median = 2.5

25. (a) average = 4.5; median = 4.5

(b) average = 5; median = 5

(c) average = 5.5; median = 5.5

27. (a) 50

(b) 50

29. (a) 1.5875

(b) 1.5

31. (a) 6.77

(b) 7

D. Percentiles and Quartiles

33. (a) $Q_1 = -3$

(b) $Q_3 = 7$

(c) $Q_1 = -3; Q_3 = 7$

35. (a) 75th percentile = 75.5; 90th percentile = 90.5

(b) 75th percentile = 75; 90th percentile = 90

(c) 75th percentile = 75; 90th percentile = 90

(d) 75th percentile = 74; 90th percentile = 89

37. (a) $Q_1 = 29$

(b) $Q_3 = 32$

(c) 37

39. (a) the 709,504th score, $d_{709,504}$

(b) the 354,752nd score, $d_{354,752}$

(c) the 1,064,256th score, $d_{1,064,256}$

E. Box Plots and Five-Number Summaries

41. (a) $Min = -6, Q_1 = -3, M = 3, Q_3 = 7, Max = 8$

(b)

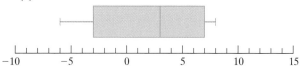

43. (a) $Min = 25, Q_1 = 29, M = 31, Q_3 = 32, Max = 39$

(b)

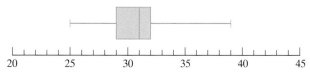

45. (a) between $33,000 and $34,000

(b) $40,000

(c) The vertical line indicating the median salary in the engineering box plot is to the right of the box in the agriculture box plot.

F. Ranges and Interquartile Ranges

47. (a) 14   **(b)** 10

49. (a) $41,000

(b) At least 171 homes

51. (a) 16.5   **(b)** 4.5

(c) Outliers are 1 and 24.

53. 200 and 210

G. Standard Deviations

55. (a) 0

(b) $\sqrt{\dfrac{50}{4}} = \dfrac{5\sqrt{2}}{2} \approx 3.5$

(c) $\sqrt{\dfrac{550}{4}} \approx 11.7$

57. (a) $\sqrt{\dfrac{82.5}{10}} \approx 2.87$

(b) $\sqrt{\dfrac{82.5}{10}} \approx 2.87$

(c) $\sqrt{\dfrac{82.5}{10}} \approx 2.87$

(d) $\sqrt{\dfrac{8250}{10}} \approx 28.7$

H. Miscellaneous

59. 10

61. 4, 5, and 8

63. Caucasian

Jogging

65. 100

67. Ramon gets 85 out of 100 on each of the first four exams and 60 out of 100 on the fifth exam. Josh gets 80 out of 100 on all 5 of the exams.

69. **(a)** $\{1, 1, 1, 1, 6, 6, 6, 6, 6, 6\}$ (Average $= 4$; Median $= 6$)

    (b) $\{1, 1, 1, 1, 1, 1, 6, 6, 6, 6\}$ (Average $= 3$; Median $= 1$)

    (c) $\{1, 1, 6, 6, 6, 6, 6, 6, 6, 6\}$ (Average $= 5$; $Q_1 = 6$)

    (d) $\{1, 1, 1, 1, 1, 1, 1, 1, 6, 6\}$ (Average $= 2$; $Q_3 = 1$)

71. **(a)** Two points will be added to each number in the five-number summary.

    (b) Each number in the five-number summary will be multiplied by 1.1.

73. **(a)** 4   **(b)** 0.4   **(c)** 0.4

75. **(a)** Male: 10%, Female: 20%

    (b) Male: 80%, Female: 90%

    (c) The figures for both schools were combined. A total of 820 males were admitted out of a total of 1200 that applied—an admission rate for males of approximately 68.3%. Similarly, a total of 460 females were admitted out of a total of 900 that applied—an admission rate for females of approximately 51.1%.

(d) In this example, females have a higher percentage $\left(\frac{100}{500} = 20\% \right)$ than males $\left(\frac{20}{200} = 10\% \right)$ for admissions to the School of Architecture and also a higher percentage $\left(\frac{360}{400} = 90\% \right)$ than males $\left(\frac{800}{1000} = 80\% \right)$ for the School of Engineering. When the numbers are combined, however, females have a lower percentage $\left(\frac{100 + 360}{500 + 400} \approx 51.1\% \right)$ than males $\left(\frac{20 + 800}{200 + 1000} \approx 68.3\% \right)$ in total admissions. The reason that this apparent paradox can occur is purely a matter of arithmetic: Just because $\frac{a_1}{a_2} > \frac{b_1}{b_2}$ and $\frac{c_1}{c_2} > \frac{d_1}{d_2}$ it does not necessarily follow that $\frac{a_1 + c_1}{a_2 + c_2} > \frac{b_1 + d_1}{b_2 + d_2}$.

77. $M + c$; the relative sizes of the numbers are not changed by adding a constant c to every number.

79. **(a)** $\frac{1245}{16} = 77.8125$ exercises   **(b)** 78

    (c) $Min = 67, Q_1 = 75, M = 78, Q_3 = 80.5, Max = 89$

    (d) $\sqrt{\dfrac{518.4375}{16}} \approx 5.7$ exercises

Chapter 15

Walking

A. Random Experiments and Sample Spaces

1. **(a)** $\{HHH, HHT, HTH, THH, TTH, THT, HTT, TTT\}$

    (b) $\{0, 1, 2, 3\}$ **(c)** $\{0, 1, 2, 3\}$

3. **(a)** $\{ABCD, ABDC, ACBD, ACDB, ADBC, ADCB,$
$BACD, BADC, BCAD, BCDA, BDAC, BDCA, CABD,$
$CADB, CBAD, CBDA, CDAB, CDBA, DABC, DACB,$
$DBAC, DBCA, DCAB, DCBA\}$

    (b) $N = 24$

5. Answers will vary. A typical outcome is a string of 10 - letters each of which can be either an H or a T. An answer like $\{HHHHHHHHHH, \ldots, TTTTTTTTTT\}$ is not sufficiently descriptive. An answer like

$\{\ldots HTTHHHTHTH, \ldots, TTHTHHTTHT, \ldots,$
$HHHTHTTHHT, \ldots\}$ is better. An answer like $\{(X_1 X_2 X_3 X_4 X_5 X_6 X_7 X_8 X_9 X_{10}): \text{ each } X_i \text{ is either } H \text{ or } T\}$ is best.

7. Answers will vary. An outcome is an ordered sequence of four numbers, each of which is an integer between 1 and 6 inclusive. The best answer would be something like $\{(n_1, n_2, n_3, n_4): \text{each } n_i \text{ is } 1, 2, 3, 4, 5, \text{ or } 6\}$. An answer such as $\{(1, 1, 1, 1), \ldots, (1, 1, 1, 6), \ldots, (1, 2, 3, 4), \ldots, (3, 2, 6, 2), \ldots, (6, 6, 6, 6)\}$ showing a few typical outcomes is possible, but not as good. An answer like $\{(1, 1, 1, 1), \ldots, (2, 2, 2, 2), \ldots, (6, 6, 6, 6)\}$ is not descriptive enough.

B. The Multiplication Rule

9. **(a)** 159,184,000

    (b) 1,757,600

    (c) 70,761,600

11. **(a)** 73,116,160

    (b) 1,406,080

    (c) 70,304,000

    (d) 62,500,000

13. **(a)** 40,320

    (b) 40,319

15. **(a)** 39,270

    (b) 16,830

    (c) 29,700

17. **(a)** 4,500,000

    (b) 1,800,000

    (c) 360,000

C. Permutations and Combinations

19. (a) 90 **(b)** 45 **(c)** 720 **(d)** 120

21. (a) 10 **(b)** 45 **(c)** 100 **(d)** 4,950

23. (a) 190 **(b)** 190 **(c)** 1,140 **(d)** 1,140

25. (a) 8 **(b)** 16 **(c)** 32 **(d)** 1,024

27. 6.12×10^{106}

29. (a) $670,442,572,800 \approx 6.70 \times 10^{11}$

(b) $125,994,627,894,135 \approx 1.26 \times 10^{14}$

(c) $125,994,627,894,135 \approx 1.26 \times 10^{14}$

(d) $202,355,008,436,035 \approx 2.02 \times 10^{14}$

31. (a) $_{15}P_4$ **(b)** $_{15}C_4$

33. (a) $_{20}C_2$ **(b)** $_{20}P_2$ **(c)** $_{20}C_5$

D. General Probability Spaces

35. (a) 0.18 **(b)** 0.27

(c) $\Pr(o_1) = 0.22, \Pr(o_2) = 0.24, \Pr(o_3) = 0.27,$
$\Pr(o_4) = 0.17, \Pr(o_5) = 0.1$

37. $S = \{P_1, P_2, P_3, P_4, P_5, P_6, P_7\}; \Pr(P_1) = \dfrac{2}{8},$

$\Pr(P_2) = \Pr(P_3) = \Pr(P_4) = \Pr(P_5) = \Pr(P_6) = \Pr(P_7) = \dfrac{1}{8}$

39. (a) $S = \{\text{red, blue, white, green, yellow}\}$

(b) $\Pr(\text{red}) = 0.3, \Pr(\text{blue}) = \Pr(\text{white}) = 0.2,$
$\Pr(\text{green}) = \Pr(\text{yellow}) = 0.15$

E. Events

41. (a) $E_1 = \{HHT, HTH, THH\}$

(b) $E_2 = \{HHH, TTT\}$

(c) $E_3 = \{\ \}$

(d) $E_4 = \{TTH, TTT\}$

43. (a) $E_1 = \{(1,1), (2,2), (3,3), (4,4), (5,5), (6,6)\}$

(b) $E_2 = \{(1,1), (1,2), (2,1)\}$

(c) $E_3 = \{(1,6), (2,5), (3,4), (4,3), (5,2), (6,1), (5,6), (6,5)\}$

45. (a) $E_1 = \{HHHHHHHHHH\}$

(b) $E_2 = \{HHHHHHHHHT, HHHHHHHHTH,$
$HHHHHHHTHH, HHHHHHTHHH,$
$HHHHHTHHHH, HHHHTHHHHH,$
$HHHTHHHHHH, HHTHHHHHHH,$
$HTHHHHHHHH, THHHHHHHHH\}$

(c) $E_3 = \{\ \}$

F. Equiprobable Spaces

47. (a) $\dfrac{3}{8} = 0.375$ **(b)** $\dfrac{2}{8} = 0.25$

(c) 0 **(d)** $\dfrac{2}{8} = 0.25$

49. (a) $\Pr(T_6) = \dfrac{5}{36}; \Pr(T_8) = \dfrac{5}{36}$

(b) $\Pr(T_5) = \dfrac{1}{9}; \Pr(T_9) = \dfrac{1}{9}$

(c) $\dfrac{1}{6}$ **(d)** $\dfrac{1}{12}$ **(e)** $\dfrac{2}{9}$

51. (a) $\dfrac{1}{1024}$ **(b)** $\dfrac{5}{512}$ **(c)** 0

53. (a) $\dfrac{5}{36}$ **(b)** $\dfrac{31}{36}$ **(c)** $\dfrac{1}{4} = 0.25$ **(d)** $\dfrac{5}{12}$

55. (a) $\dfrac{1}{1024}$ **(b)** $\dfrac{1}{1024}$ **(c)** $\dfrac{5}{512}$ **(d)** $\dfrac{11}{1024}$

(e) 0 **(f)** $\dfrac{7}{128}$

57. (a) $\dfrac{4}{15}$ **(b)** $\dfrac{11}{15}$ **(c)** $\dfrac{1}{1365}$

G. Odds

59. (a) 4 to 3 **(b)** 3 to 2

61. (a) $\dfrac{3}{8}$ **(b)** $\dfrac{15}{23}$ **(c)** $\dfrac{1}{2}$

Jogging

63. 6,492,640

65. (a) $\{YXXXX, XYXXX, XXYXX, XXXYX\}$

(b) $\{YXXXX, XYXXX, XXYXX, XXXYX,$
$XYYYY, YXYYY, YYXYY, YYYXY\}$

(c) $\{YYXXXX, YXYXXX, YXXYXX, YXXXYX,$
$XYYXXX, XYXYXX, XYXXYX, XXYYXX,$
$XXYXYX, XXXYYX, XXXYYY, XYXYYY,$
$XYYXYY, XYYYXY, YXXXYY, YXYXYY,$
$YXYYXY, YYXXYY, YYXYXY, YYYXXY\}$

67. (a) 3,628,800 **(b)** 362,880

(c) 28,800 **(d)** 2,880

69. 126,126

71. (a) $\dfrac{46,189}{262,144} \approx 0.176$ **(b)** $\dfrac{285}{262,144} \approx 0.001$

(c) $\dfrac{1,048,365}{1,048,576} \approx 0.9998$

73. $\dfrac{253}{4,998} \approx 0.05$

75. $\dfrac{1}{2548} \approx 0.00039$

77. $\dfrac{4,651}{7,776} \approx 0.6$

79. (a) $(0.98)^{12} \approx 0.78$; We are assuming that the events are independent.

    (b) $(0.98)^{12} + 12(0.98)^{11}(0.02) \approx 0.98$

Chapter 16

Walking

A. Normal Curves

1. (a) 83 lb   **(b)** 83 lb   **(c)** 7 lb

3. (a) 72.8 lb   **(b)** 89.6 lb

5. 89.5 lb

7. 20 in.

9. (a) 75 in.   **(b)** 3 in.   **(c)** $Q_1 \approx 73$ in.; $Q_3 \approx 77$ in.

11. $\mu \neq M$

13. $\mu - Q_1 = 453 - 343 = 110$;

    $Q_3 - \mu = 553 - 453 = 100$

B. Standardizing Data

15. (a) 1   **(b)** 1.6   **(c)** -2 **(d)** -1.8

17. (a) approximately 0.22   **(b)** approximately -1.82

    (c) 0

19. approximately -0.675

21. (a) 152.3 ft   **(b)** 199.1 ft   **(c)** 111.74 ft   **(d)** 183.5 ft

23. 17 lb

25. 5

27. $\mu = 52$; $\sigma = 16$

C. The 68–95–99.7 Rule

29. $\mu = 55$; $\sigma = 5$

31. $\mu = 92$; $\sigma = 3.4$

33. $\mu = 80$; $\sigma \approx 10$

35. (a) 97.5%   **(b)** 13.5%

37. $\mu = 56.3$ cm

39. 59%

D. Approximately Normal Data Sets

41. (a) 52 points   **(b)** 50%   **(c)** 68%   **(d)** 16%

43. (a) $Q_1 \approx 44.6$ points

    (b) $Q_3 \approx 59.4$ points

    (c) $IQR \approx 14.8$ points

45. (a) 1900   **(b)** 1630

47. (a) approximately the 3rd percentile

    (b) approximately the 16th percentile

    (c) approximately the 84th percentile

    (d) approximately the 99.85th percentile

49. (a) 95%   **(b)** 47.5%   **(c)** 97.5%

51. (a) 13   **(b)** 80   **(c)** 250   **(d)** 420

    (e) 488   **(f)** 499

53. (a) 16th percentile

    (b) 97.5th percentile

    (c) 18.6 lb

55. (a) 97.5th percentile

    (b) 99.85th percentile

    (c) 8 lb

E. The Honest and Dishonest Coin Principles

57. (a) $\mu = 1800$, $\sigma = 30$ **(b)** about 68%

    (c) about 34%   **(d)** about 13.5%

59. (a) about 68%   **(b)** about 16%

    (c) about 84%

61. (a) $\mu = 240$, $\sigma = 12$ **(b)** $Q_1 \approx 232$; $Q_3 \approx 248$

    (c) about 0.95

63. (a) $\mu = 30$, $\sigma = 5$

    (b) approximately 0.025

    (c) approximately 0.34

65. 2.5%

Jogging

67. (a) 20.55 lb   **(b)** 16.75 lb

69. (a) 60th percentile   **(b)** 30th percentile

71. (a) 600 points   **(b)** 97.5th percentile

73. (a) 78 points   **(b)** 70.4 points

    (c) 60 points   **(d)** 48.7 points

75. 225

77. For $p = \dfrac{1}{2}$, $\mu = np = \dfrac{n}{2}$, and

$$\sigma = \sqrt{np(1-p)} = \sqrt{n \cdot \frac{1}{2} \cdot \frac{1}{2}} = \sqrt{\frac{n}{4}} = \frac{\sqrt{n}}{2}.$$

Index

Photo Credits